PROGRESS IN UNDERWATER ACOUSTICS

PROGRESS IN UNDERWATER ACOUSTICS

Edited by
HAROLD M. MERKLINGER
Defence Research Establishment Atlantic
Dartmouth, Nova Scotia, Canada

PLENUM PRESS • NEW YORK AND LONDON

Library of Congress Cataloging in Publication Data

Associated Symposium on Underwater Acoustics (1986: Halifax, N.S.)
Progress in underwater acoustics.

"Proceedings of the Twelfth International Congress on Acoustics Associated Symposium on Underwater Acoustics, held July 16–18, 1986, in Halifax, Nova Scotia, Canada"—T.p. verso.
"Sponsored by the Canadian Acoustical Association"—Pref.
Includes bibliographical references and indexes.
1. Underwater acoustics—Congresses. I. Merklinger, Harold, M. II. Associated Symposium on Underwater Acoustics (1986: Halifax, N.S.) III. Canadian Acoustical Association. IV. Title.
QC242.A87 1986 621.389′5 87-11020
ISBN 0-306-42552-1

Proceedings of the Twelth International Congress on Acoustics
Associated Symposium on Underwater Acoustics, held July 16–18, 1986,
in Halifax, Nova Scotia, Canada

Printed in the United States of America

FOREWORD

IMAGE TRACKS AT HALIFAX

by L.B. Felsen

All living kind much effort spend
To cope with their environment.
Some use their eyes, some use their nose
To sense where other things repose.
For one group, nothing's more profound
Than to explore the world with sound.
These audio diagnosticians
Go by the name of acousticians.

They regularly meet to check
Whether their sonogram's on track.
With images stored in their packs,
This year they came to Halifax.
There they combined with ocean types
And each could hear the other's gripes.

A meeting naturally does start
Reviewing present state of art.
What we found out is where it's at:
We cannot hope to match the bat.

Computer printouts by the reams
Document new inversion schemes.
Each wiggle gets processed with care
To image what is actually there.
The ill-posed problem gives us grief,
It's science laced with strong belief.
The lowly bat has no such doubt:
Ill-posed or not, it sorts things out.

After two days of imagery,
The sonic thrusters went to sea.
The ocean bottom, smooth or rough,
Makes tracking sonic signal tough.

Some model modes, some model rays,
Some feel that spectra all portrays.
Then there are those who with despatch,
Take refuge in the ocean wedge.

If things get messy, randomize.
What's partly smooth, determinize.
You ponder, is it this or that?
And wish you were a lowly bat.

The meeting's hosts did treat us well.
They let the climate cast its spell.
No weath'ry hope was placed in vain.
We were exposed to wind and rain,
We glimpsed blue sky through clouds dispersed.
But rainy sequence was reversed:
The ocean types would like it wet
Yet they got stuck with sun instead.

Each conf'rence has the same refrain:
It has been fun to meet again.
May Nova Scotia long survive
And let the ocean addicts thrive.
To our hosts who labored hard,
I offer thanks, ere we depart.

Halifax, Nova Scotia, 14-18 July 1986.

PREFACE

This book contains the proceedings of the 12th ICA Associated Symposium on Underwater Acoustics held at the Chateau Halifax hotel in Halifax, Nova Scotia Canada, 16-18 July 1986. The proceedings is not quite complete, 98 papers were presented, 94 of these are contained in this book.

The 12th International Congress on Acoustics (ICA) itself was held in Toronto during the period 24-31 July 1986. It and the several associated symposia were sponsored by the Canadian Acoustical Association. About three years ago, in 1983, the community of acousticians in Halifax agreed to organize two of these associated symposia under a single organizing committee. The 15th in an annual series of symposia on Acoustical Imaging was one of these, and Prof. Hugh Jones of the Technical University of Nova Scotia took on the management of the technical program. Prof. Leif Bjørnø of Denmark pressed us to include more emphasis on underwater acoustics and, in the end, I agreed to look after the technical program for a separate symposium for Underwater Acoustics.

I was not alone in this task of organizing the symposium. Assisting immeasurably were Academician L.M. Brekhovskikh (USSR), Prof. Leif Bjørnø (Denmark), Dr. R.P. Chapman (Canada), and Dr. W.A. Kuperman (USA), the members of our Technical Committee. Local arrangements were the responsibility of a Local Organizing Committee chaired by Prof. Jones and co-chaired by myself. Other members were Mr. H.W. Kwan, Dr. L.J. Leggat, Mr. David M.F. Chapman, Dr. D.D. Ellis, Dr. N.A. Cochrane and Ms. S. Robertson, with Ms. L. Buckly, Ms. S. Forbes and Mr. R. Raven helping with respect to the hotel.

The original call-for-papers proposed the following technical areas for special attention:

- Underwater Acoustic Imaging
- Acoustic Propagation (Deep water, Shallow water, Slope water)
- Scattering of Rough Ocean Boundaries
- Acoustic Signal Processing and Beamforming
- Inverse Methods; and
- Acoustics in the Offshore Industry: Petroleum and Fishing.

During the interval between the call-for-papers and the symposium, the falling price of oil on the world market resulted in a noticeable decline of activity in the offshore petroleum industry. This change in emphasis, I believe, carried over to our symposium. The greatest emphasis - in terms of papers submitted - was on sound propagation, with a healthy respect paid to the interaction of sound with the ocean bottom.

In the event, 106 papers appeared in the program and, of these, 98 were presented. The authors are to be commended on the quality of their presentations. And the Chairmen did a fine job managing the discussion and keeping to the published schedule. Over two hundred delegates representing 22 countries attended the symposium. The papers held the attention of delegates to the very end; even the two final afternoon sessions were very well attended.

The organization of this proceedings does not conform precisely with that of the Symposium itself. The constraints imposed by scheduling two parallel sessions around mealtimes and social functions forced certain groupings that were not based entirely on subject matter. A proceedings is not so constrained, and so the papers are grouped under the following general classifications:

- Sound Scattering from Ocean Boundaries
- Sound Scattering from Biological and Other Objects
- Acoustic Characterization of the Ocean and Ocean Floor
- Sound Propagation in the Ocean
    - a) General
    - b) Shallow Water
    - c) Slope Water
- Transducers, Radiation and Instrumentation
- Signal Processing and Beamforming.

It may be noted that there is no Underwater Acoustic Imaging classification. Most of the papers presented under the imaging classification will appear in the proceedings of the 15th Symposium on Acoustical Imaging, a companion to this volume. The three imaging papers slated for inclusion in the underwater acoustics proceedings fit equally well under one of the above classifications. Within each classification, the order of the papers is essentially the same as for the oral presentations.

Many people assisted in the process of editing this proceedings. Mary Jo Delaney did most of the checking while Linda Kenney did most of the corrections. Assisting with the editing were David Chapman, Dale Ellis, Joe Farrell, Garry Heard, Phil Staal, John Stockhausen and Jim Theriault. Thank you all.

Harold M. Merklinger

Halifax, Nova Scotia
Canada
October 1986

ACKNOWLEDGEMENT

The preparation of this book was supported in part by the Nova Scotia Department of Development.

CONTENTS

OPENING

SOUND SCATTERING FROM OCEAN BOUNDARIES

SCATTERING BY BIOLOGICAL AND OTHER BODIES

ACOUSTIC CHARACTERIZATION OF THE OCEAN AND THE OCEAN FLOOR

b) Shallow Water

SIGNAL PROCESSING AND BEAMFORMING

OPENING ADDRESS

D. Schofield

Chief, Research and Development
National Defence Headquarters, 101 Colonel By Drive
Ottawa, Ontario, Canada K1A 0K2

ABSTRACT

This opening address reviews a few highlights in the progress of underwater acoustics. Reference is made to Canadian contributions as well as to the author's personal recollections. - Ed.

INTRODUCTION

First let me say that it is a pleasure to be here, to open this Underwater Acoustics Symposium and the joint session on Underwater Acoustic Imaging.

I would also like to welcome our visitors to Canada, to Nova Scotia, and to Halifax, particularly those of you who have just arrived for the Underwater Acoustics Symposium. I spent many years here in Halifax and I can assure you that Halifax has a long, established tradition of welcoming visitors from throughout the world. You will find a visit to Halifax in the summer to be a cheerful, and more usually, a festive occasion. However, Halifax weather is notorious for being highly changeable. At this time of year the warm Gulf Stream and the cold Labrador Current do battle off our coast. Frequently the result is a standoff known as fog.

Some of you will be planning to move on to the main 12th International Congress on Acoustics next week. It does not start until a week from tomorrow; I would encourage you to see more of Halifax and its surroundings before you go.

A LITTLE HISTORY

If you take me up on my invitation to visit a while in Nova Scotia, you may happen upon a very famous name in acoustics: that of Alexander Graham Bell. The famous first words spoken over a telephone, "Mr. Watson, come here, I want you" were spoken in 1876 over Canadian wires in Brantford, Ontario - a small city west of Toronto. The following year Bell established a two-way link between "the deeps" and the surface at a coal mine in Glace Bay, Nova Scotia. Thereafter, Alexander Graham Bell established a summer home and laboratory at Baddeck on Cape Breton Island. In later years in Nova Scotia he devoted much of his time to manned flight and to hydrofoil craft. Today there is a museum in Baddeck devoted to Bell's Nova Scotia accomplishments.

In my words to you, printed in the program for this meeting, I made reference to Reginald Fessenden. R. A. Fessenden is a Canadian native who has many inventions to his credit. He was the first person to broadcast voice modulated radio waves. He was also the first to utilize an electrically operated transmitter/receiver for the purpose of underwater echo ranging. His motivation was the tragic sinking of the TITANIC in the year 1912. Fessenden had been working on using underwater sound for telegraphy. His "Fessenden Oscillator" was then and for many years after the most powerful (2 kilowatts) and efficient (40-50%) device available for underwater signalling. He reasoned that one might be able to locate an iceberg by transmitting a pulse and listening for the echo. His experiments proved to be successful and under good conditions, he was able to detect icebergs up to two miles away - as well as numerous false targets. With the commencement of war in 1914, Fessenden was summoned by the British admiralty to assist in the development of a sonar to detect submarines. Given his interest in radio broadcasting, it is no surprise that he was also probably the first man to demonstrate the underwater telephone. As many of you are aware, we are still trying to accomplish Fessenden's main goals - especially under other-than-ideal conditions.

We Canadians also claim to be the inventors of the Variable Depth Sonar for naval applications. There are probably one or two scientists here today - myself being one of them - who took part in that development.

I have given only three examples of Canadian contributions to acoustics but sufficient I think to justify the claim that Canada has a long and distinguished tradition in acoustics. I believe it is entirely appropriate for these meetings associated with the 12th International Congress on Acoustics to be held in Canada. And I am personally proud to have been part both of acoustics in Canada and of this present meeting of the ICA in Canada.

MY WORDS TODAY

In my few minutes this morning I would like to review some of the changes I have witnessed in acoustics - especially underwater acoustics - over the past thirty years or so.

When I first began to work in underwater acoustics in 1952, the thought that one could actually form the equivalent of visual images using sound, seemed perhaps plausible in principle, but rather remote in practice. We have come a long way. The concept of acoustic imaging has moved from an idea to reality, in the underwater field as well as in other media. Advances in ultrasonic transducers and in signal processing techniques, have led to the ultrasonic microscope, the medical ultrasonic scanner, acoustic holography and acoustic tomography. I will not dwell on the history of acoustical imaging; I'm sure Glenn Wade covered this quite adaquately when he opened the 15th Symposium on Acoustical Imaging last Monday morning. I would like to spend most of my time on underwater acoustics.

UNDERWATER ACOUSTICS

Our understanding of the behaviour of sound in the ocean environment has improved tremendously. Thirty years ago ray tracing was a major effort, as was trying to find the right information on oceanographic parameters such as temperature and salinity. Of necessity, our methods required many simplifying assumptions. And, of course, analytic solutions to simple problems were much preferred over numerical methods. Analytic normal mode methods were just beginning to provide new insight on sound propagation for some areas such as shallow water. Advances in the numerical modeling of wave propagation in a realistic ocean have allowed us to include diffraction and scattering effects as well as the very complex interaction of sound with the ocean bottom. However, we still have problems in providing the detailed oceanographic information needed by these models. Nevertheless, the science of predicting underwater sound transmission characteristics has made very significant progress. I note that some of you are turning the problem about - using observed acoustic effects to chart the oceanographic and geological conditions.

We have learned about sound absorption at low frequencies. We used to worry about using explosive sound sources, the pulses from which clearly exhibit nonlinear propagation effects, in such studies. I note that there is a paper in this program which addresses the importance of nonlinear effects in long range ocean sound propogation. In the last 30 years we have even learned how to turn nonlinear propagation effects to our advantage.

We have also learned a lot about noise in the ocean, its sources and characteristics. This is especially true for the frequencies below a few hundred hertz where uncontaminated data were, for many years, so elusive. In my early years, the simplifying assumption was generally made that ambient sea noise was omnidirectional, and its origins were only vaguely understood. Our source of data were the classic Knudsen curves. While it would be foolish to say we have all the answers: the mechanics of noise generation by the sea surface is still not completely clear, but our scientific knowledge of noise has increased by orders of magnitude. We have reached a stage where we can begin to interpret the noise characteristics in terms of wind, precipitation, shipping, and perhaps ice conditions.

When I recall the state of knowledge in that other key acoustic background parameter, reverberation, it was little better than a rule of thumb. We knew that when pulses were transmitted scattering in the ocean generated reverberation and we had some feel for the level of reverberation to be expected. We were also aware that contributions to reverberation could be received from the surface, the bottom and the volume of the ocean. During the late 50's and the 60's, extensive work was carried out on all types of reverberation at least sufficient to develop good empirical descriptions, of scattering in the ocean from the total volume and from scattering layers and from the boundaries of the oceans.

Basically the elements of underwater sonar have remained unchanged for a long time. I think that the first recorded utilization of underwater sound is attributed to Leonardo da Vinci in the 16th century when with the aid of a long tube between the ear and the water, ships could be heard at great distances. Sometime later underwater bells were used for navigation purposes. A stethescope applied to the hull of the boat acted as the receiver. Active sonars still consist of the same elements as they did 30 or 40 years ago - some sort of transmitter or projector and some sort of receiver plus a detector - sometimes it is still the human ear.

Technology has changed the form and capability of every component. Transmitting transducers have become lighter in weight though more efficient and of higher power. The "art" of transducer design is being supplemented by very rational - if computer intensive - methods. We have learned how better to direct the sound energy where we want it to go. We can place the transducer where we want it to be (and get it back); we can transmit greater power at lower frequencies - and hence hope for longer propagation ranges. We can generate signals as complex as those used by whales and seals and sort through the returned echos. We can have receiving arrays consisting of hundreds of sensors for high spatial resolution. Perhaps most surprising of all, we have the electronic tools to process those hundreds of received signals - without sinking or unduly heating the ship. I recall during some discussions some time ago, a colleague observed that to process the signals from a proposed sonar one would need to have a second ship in tow, just to carry the extra electronics. That has changed. We are no longer forced to make the approximations and compromises forced upon us by the signal processing technology of thirty years ago.

APPLICATIONS

That these advances have taken place should not be a surprise. There has been a great deal of research and development in many countries over the past thirty years. What I do find surprising is the broad range of applications of acoustics that one finds today. It is no surprise that naval vessels should possess sophisticated sonars. But today fish finding sonars are part of fisheries research and commercial fishing operations. Even the sports fisherman in his small boat can have a fish finder. My neighbor, knowing about my background in sonar, recently admitted to me that when he bought his new fishing boat he had not equipped it with the latest in fish finding sonar. The salesman had told him that next year he would be able to buy a fish finder which would identify the species of the fish beneath his boat.

We all see acoustics being used routinely for non destructive testing, for crack detection in metals and so on.

We see active sonar being used to measure ship's speed - in the thwartship direction as well as fore and aft.

Ocean noise is being analysed to provide information on wind and precipitation at remote ocean locations. The purpose is of course to provide better weather and climate forcasting.

We see sidescan sonar being used to chart bottom conditions, to find and identify wrecks, and to find other objects of commercial interest on the sea floor.

Acoustic telemetry is being used to control free-roving unmanned submersibles and to allow divers to communicate with one another.

The acoustic microscope is being used to examine specimens which fail to reveal their secrets to the optical or electron microscopes. We even see a cross between the electron and acoustic microscopes.

The two areas where we find really extensive use of acoustics are the diverse fields of medicine and off-shore exploration.

In medicine, the ultrasound examination has become a standard alternative to the x-ray. It even has certain advantages such as detecting motion within the human body and avoiding radiation damage. We can measure blood flow and test divers for the onset of bends.

In off-shore resource operations, sound is used to look at the geology of the ocean bottom, to guide ships and instruments, to provide precision navigation facilities, to relocate well heads and pipelines, and even to conduct precision measurements for the cutting of pipe and so on.

As an acoustician it is very gratifying to see my field of endeavour being utilized on such a broad scale. Perhaps I am biased, but I believe that it has been largely our efforts in underwater acoustics which have provided the groundwork for many of these developments. I believe that it is 60 or so years of transducer and signal processing development for sonar that has made medical ultrasound possible. The growth in oceanography - the understanding of our ocean environment - received significant support from our need to understand the sonar environment. Offshore mineral exploration and exploitation would be almost non existant without acoustic methods. I, for one, am pleased to have been part of these scientific and technological developments.

CONCLUDING REMARKS

I hope that my remarks this morning will serve to emphasize to you and to others - the growing importance of underwater and imaging acoustics in our world.

In closing let me wish you a most productive interchange. Symposia such as these are very important in clarifying old thoughts and in stimulating new ones.

ACOUSTICAL VISUALISATION OF THE OCEAN BOTTOM

L.M. Brekhovskikh, V.V. Krasnoborodko and V.Ch. Kiriakov

P.P. Shirshov Institute of Oceanology
the USSR Academy of Sciences
Moscow, V-218, USSR

ABSTRACT

Measurements of an acoustical field of signals returned back from the bottom insonified by vertically emitted acoustic pulses were carried out. The acoustical images of a deep ocean bottom ground were obtained. Comparisons with the results of geological explorations give a good correspondence between the acoustical images and geomorphological structure of the bottom.

INTRODUCTION

Practically all the knowledge about the relief of the ocean's bottom we have at present was obtained by the simple echo sounder. During the last decade more efficient and sophisticated systems were developed such as the multibeam echo sounder and the side-scan sonar. The latter delivers some information about the scattering properties of the bottom also. This appears to be very useful in particular for developing methods for the express exploration for manganese nodules. The importance of ocean mineral resources of this kind has greatly increased in recent years.

We have tried to approach the problem of the acoustical imaging of the bottom from another side. It was found some time ago that the sound signals returned by the deep ocean's bottom display very strong spatial inhomogeneities of various scales. When a research ship is moving together with its sound receiving system, these irregularities become the cause of temporal signal fluctuations (Gazey, 1963; Volovov and Lysanov, 1969). The received signal may change considerably when the receiver's position is displaced only a fraction of the sound wave length. The question arises what information about the bottom one can obtain by measuring the sound field simultaneously over some area by a system consisting of many hydrophones. This is the idea motivating the experiments described below.

EXPERIMENTS

Experiments were carried out from a drifting ship. The bottom was insonified at normal incidence by the regular narrow-beam echo sounder (NBS) VM-13 type. The transducer of the latter was mounted on a gyrostabilized platform. Sound pulses of 40 ms duration and at a carrier frequency of 12 kHz were sent down to the bottom each 2.67 s. The half-width of the echo-sounder directional pattern (at the level of -3 dB) was $\delta = 6.75°$.

Plane equidistant rectangular hydrophone gratings were used as the receiving systems. One of these antennas had 64 hydrophones (grating dimensions 2.4×2.4 m, distance between neighbouring hydrophones 30 cm), the other had 256 hydrophones (dimensions 2.7×2.7 m, distance between hydrophones 17 cm). Gratings were towed at a depth of 50–300 m not far from the research ship, while the latter was drifting. The signal reflected from the ocean surface could be separated and excluded at these depths. A special flexible suspension for the gratings together with their centering was used to keep the normal to the grating plane in the vertical position. A fixed orientation of the smaller antenna in the horizontal plane was ensured by the use of a rudder. In reality the rudder was a thin plate with an area of 1 m^2. The larger antenna maintained a fixed orientation with respect to the drift direction automatically due to some of its constructional peculiarities.

The mode of operation of the receiving system with the smaller grating was the following. The inner generator of the antenna began to work at the moment the front of the sound pulse returned by the bottom arrived at the grating. This generator secured, one by one, the connection of each hydrophone with the amplifier whose output in turn was connected by cable to the data processing system on board the ship. The time required for the "interrogation" of all hydrophones was 32 ms. After that, the system waited for the next pulse's arrival when the new cycle of interrogation began. The duration of the interrogation cycle was the same for the larger grating also, but this time after the completion of one cycle a special signal was transmitted and the next cycle of interrogation began immediately.

The signals received on board the ship were digitized and handled by the computer. The result was that one could record the amplitude and the phase of the acoustic field at each hydrophone in this manner. The distribution of the signal amplitude and phase over the gratings was then converted into an "image" of the ocean bottom below the ship.

DATA ANALYSIS AND SOME RESULTS

Let us introduce rectangular coordinate systems in the plane of the receiving grating and in the horizontal bottom (object) plane (when averaged through the small irregularities). Let $\mathbf{x}$ and $\mathbf{X}$ be two-dimensional vectors in these planes respectively. Let $P(\mathbf{X})$ be the sound field of the pulse returned by the bottom immediately at the bottom (at the object plane). This field is formed by the interaction of the incident pulse with the upper, near surface, part of the bottom. According to Vidmar (1980) the sound absorption coefficient in the ocean bottom sediments varies from 0.01 dB/m/kHz for clay sediments up to 0.2 dB/m/kHz for sand and silt. We observed the latter case in our experiments, so that sound waves reflected from a sublayer at the depth say 2 m inside the bottom was attenuated due to absorption by 10 dB and can be neglected.

Considering $p(\mathbf{X})$ as a field of secondary sources, the field in the grating plane will be, according to Huygen's principle,

$$p(\mathbf{x}) = A \int P(\mathbf{X}) \exp(ikr)/r \, d\mathbf{X}, \tag{1}$$

where k is the wave number of the sound wave. A is a constant.

The depth of the ocean R is large compared with the radius of the insonified zone of the bottom $r_0 \sim R\delta$ ($R \sim 4000$ m, $r_0 \sim 500$ m in our case). Hence r in the denominator in Eq. (1) can be replaced by R and included in the constant. The exponent, in turn, can be transformed as

$$\exp[ikR(1 + |\mathbf{X} - \mathbf{x}|^2/R^2)^{1/2}] \approx \exp[ik(R + |\mathbf{X} - \mathbf{x}|^2/2R - |\mathbf{X} - \mathbf{x}|^4/8R^3)]. \tag{2}$$

The first term here gives a constant factor which can also be included in the constant. The second term can be written as $(|\mathbf{X}|^2 - 2\mathbf{x}\mathbf{X})/(2R)$ since $kD^2/(2R) \ll 1$, where D is the linear dimension of the grating. From the third term we retain only $|\mathbf{X}|^4/8R^3$ since $3kr_0^3D/8R^3 \ll 1$. Now, Eq. (1) can be rewritten as

$$p(\mathbf{x}) = A \int P(\mathbf{X}) \exp[(ik/2R)|\mathbf{X}|^2 - (ik/R)\mathbf{x}\mathbf{X} - (ik/8R^3)|\mathbf{x}|^4]\, d\mathbf{X}. \tag{3}$$

Consider now the expression (for the bottom image):

$$\begin{aligned} I(\mathbf{x}) = \ & (1/4\pi^2 A) \ \exp[(ik/8R^3)|\mathbf{X}|^4 - (ik/2R)|\mathbf{X}|^2] \\ & \times \int p(\mathbf{X}) \exp[(ik/R)\mathbf{x}\mathbf{X}]\, d\mathbf{x}, \end{aligned} \tag{4}$$

where the integration is over the grating aperture. Being obtained by the inverse Fourier transform of the Eq. (3), $I(\mathbf{X})$ can be considered an estimate of $P(\mathbf{X})$. Only the amplitude of the acoustical image $I(\mathbf{X})$ will be of interest in what follows. Then the exponent before the integral in Eq. (4) may be omitted and obtaining the bottom's image reduces to finding the Fourier transform of the field over the grating aperture. According to the Rayleigh criterion the linear resolution of two point objects in the bottom plane will be $l = 2\pi R/(kD)$.

We replace the integration in Eq. (4) by a summation. Let the directions of the coordinate axes in the grating plane be along its axes. Then we obtain from Eq. (4):

$$I(\mathbf{x}) = (d^2/4\pi^2 A) \sum_{n=0}^{N-1} \sum_{m=0}^{N-1} p(n,m) \exp[(ikd/R)(nX_1 + mX_2)], \tag{5}$$

where $\mathbf{X} = (X_1, X_2)$, d is the distance between neighbouring hydrophones (the same along both axes), $p(n,m)$ is the sound field observed at the point $\mathbf{x} = (nd, md)$ in the grating, and N is the number of hydrophones along each axis (the same in both directions).

In our experiment, the inequality $kd > \pi$ applies. Under this condition the substitution of integration for the summation gives, according to Eq. (5), the relation $|I(\mathbf{X})| = |I(\mathbf{X} + \Delta\mathbf{X})|$, where $\Delta\mathbf{X} = (Lp, Lq)$, $L = 2\pi R/(kd)$, and where p, q are integer numbers. In other words, the grating can be considered, under this condition, as an antenna with many main lobes accompanied by their side lobes. The area of the bottom, enveloped by the main (zero order) lobe together with its side lobes, had in our case a linear dimension L of about 1500 m. It exceeded the linear dimension r_0 of the zone insonified by the echo sounder, and hence the antenna was operating within the limits of its main lobe.

Fig. 1 shows the bottom's image obtained in such a manner by the use of the 64-element grating in the ocean (depth about 4000 m). More precisely, this is the distribution of the quantity $S = 20 \log(|I(\mathbf{X})|/\max|I(\mathbf{X})|)$, which is the amplitude of the returned sound field over the bottom expressed in a logarithmic scale relative to its maximum value in this area. An asterisk locates the point where $S = 0$ dB. Lines in the figure are contours of equal S. Peripheral parts of this figure correspond to poorly insonified parts of the bottom. It would be desirable certainly to correct the results taking into account the degree of insonification throughout the area under consideration. This has not been done yet, but for the area inside the curve $S = -10$ dB, this correction does not exceed 3 dB.

Let us see now what these results mean as compared with results of an ordinary geological survey of the bottom carried out at the same place. First, undersea photography showed that the bottom was covered by silt with inclusions of manganese nodules. The concentration of the latter at many places in the area was determined as a next step. All these acoustical and geological investigations, which lasted several days, needed rather exact knowledge of the location of the ship and of the different

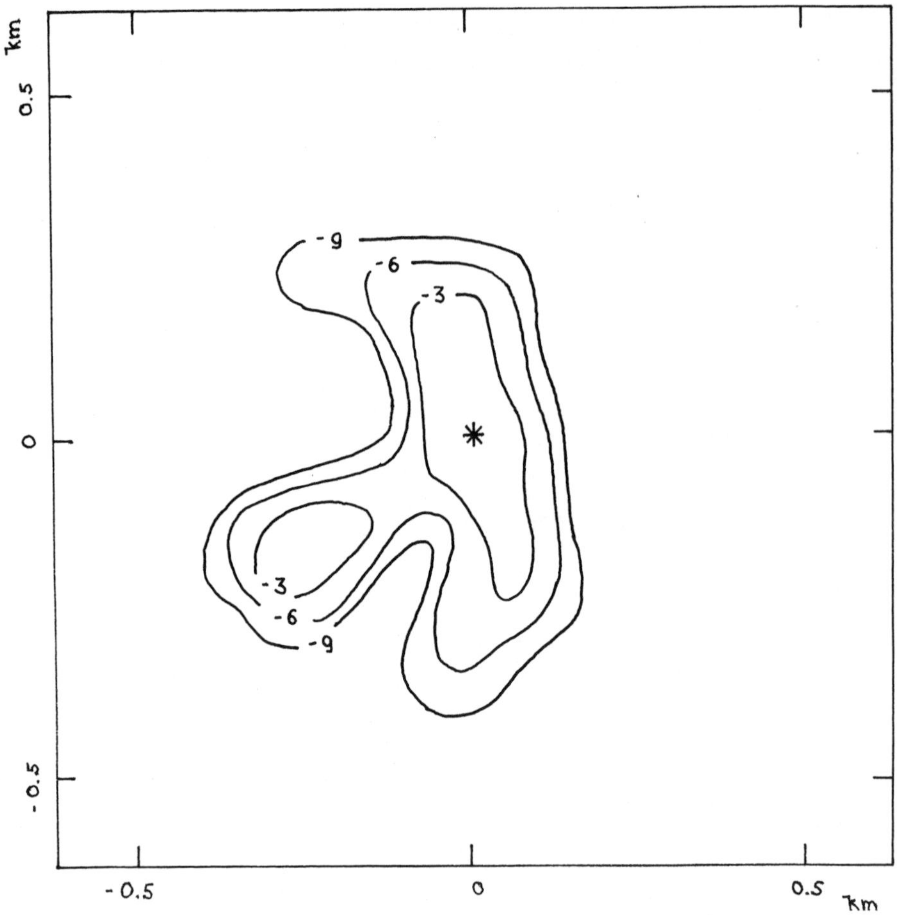

Fig. 1. Acoustical image of the bottom.

devices used during the experiment. This was achieved by using various methods. In particular, an anchored buoy was installed at the beginning of the experiment and its position was controlled throughout the experiment.

Fig. 2 shows the same area as for Fig. 1. The full lines are the isobaths. We see that the central part of the area has an almost constant depth of 4100 m. The points are the places where the concentration of nodules was determined. Concentration is indicated in the units kg/m^2. The broken line in Fig. 2 surrounds the region of the high intensity for the returned acoustical signal. It is, in fact, the curve $S = -10$ dB from Fig. 1. Not considering other details of the figure, which will be explained below, we see that the regions of the bottom with high acoustical intensity are also those with high concentrations of the nodules.

The acoustical image in Fig. 1 is of not very high resolution ($l \sim 200$ m). Higher resolution can be achieved by using a synthesized receiving aperture. When the ship is drifting, the grating moves and is in a somewhat different position for the reception of two successive pulses. At the same time, there exists some common part for the two positions. Combining two or more successive grating positions, one can obtain an effective antenna aperture which is greater than the grating dimension itself. Dash-and-dot lines in Fig. 2 envelope the bottom regions having a high amplitude of returned signal ($S \gg -10$ dB) obtained by combining three successive pulses. In this way, however, we obtain the higher resolution only in one direction, namely in the drift (east-west) direction. This is probably the reason why these regions in Fig. 2 have lesser dimensions in this direction.

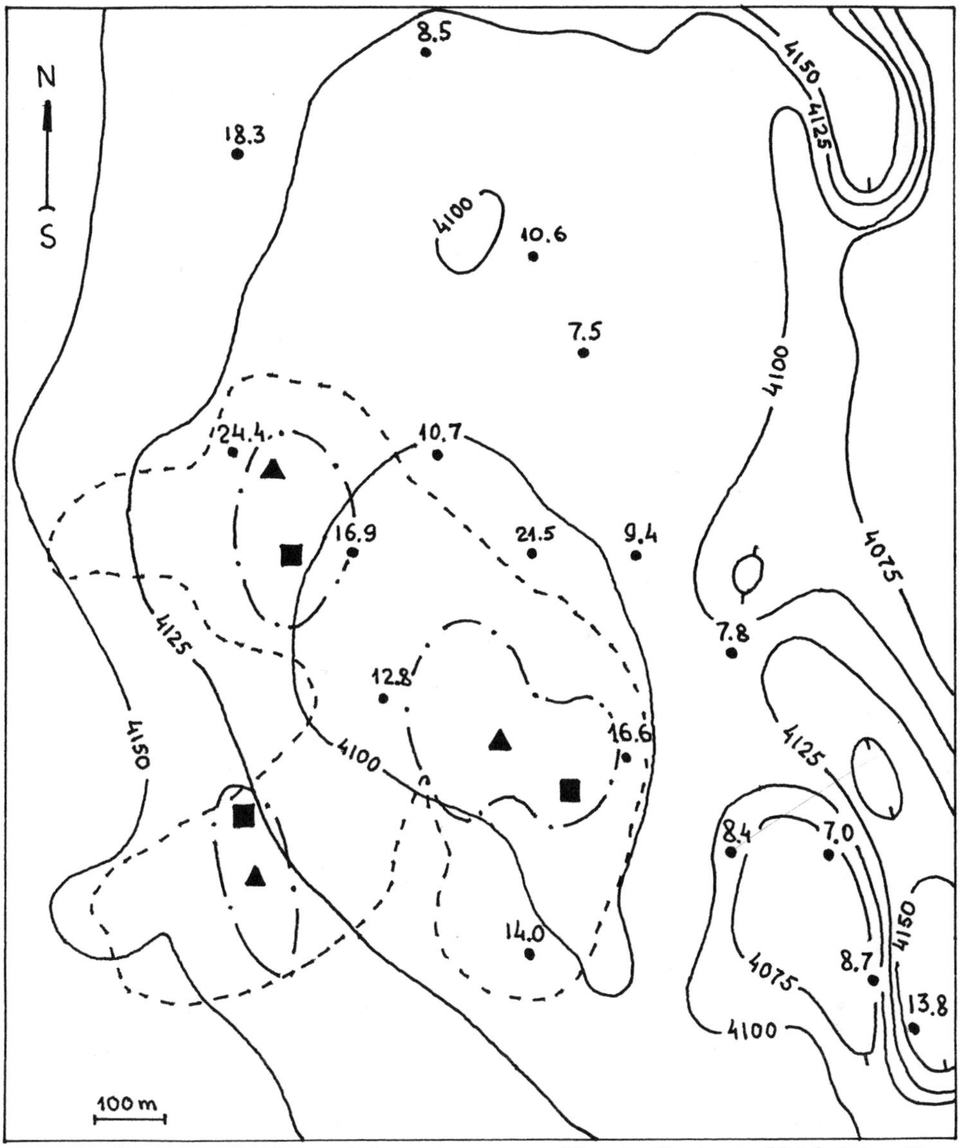

Fig. 2. Bathymetric chart of the same area as Fig. 1. Regions of high intensity of the returned acoustical signal as detected by various methods are indicated.

There exist also other methods for the more precise location of those regions associated with a high intensity of acoustical signal at the bottom. We have used adaptive methods of spectrum estimation (McClellan, 1982) for this purpose.

It is convenient to change $\mathbf{X}$ in (4) to a new variable $\alpha = (\alpha_1, \alpha_2)$ where $\alpha_1 = X_1/R$ and $\alpha_2 = X_2/R$ are the angles in two perpendicular planes. Then Eq. (4) reduces to

$$I(\alpha R) = I'(\alpha) \sim \int p(\mathrm{x}) \exp(-ik\alpha\mathrm{x})\, d\mathrm{x}. \tag{6}$$

The last expression is the two-dimensional Fourier transform of the field registered by the hydrophones of the grating. Hence, obtaining the acoustical image (distribution of amplitude of the return signal) of the bottom is the same problem as estimating the spatial spectrum of acoustical signals at the grating.

Formula (5) gives the traditional spectrum estimation. Using it one can express the field at the object (bottom) plane in terms of the polynomials in $\exp(-ikd\alpha_1)$ and $\exp(-ikd\alpha_2)$, thus securing comparatively smooth variation of the field at this plane. If the image of the object is high in contrast, adaptive methods of spectrum estimation are more adequate. Using such methods one can obtain a higher resolution of the "bright" spots. For traditional methods of spectrum estimation side lobes appear due to the assumption that the field is zero outside the antenna's aperture. Adaptive methods do not have such a handicap. The distinguishing feature of these methods is that the estimation of the intensity spectrum is looked for in the form:

$$|I'(\alpha)|^2 \sim 1/\Phi(\alpha), \tag{7}$$

where now $\Phi(\alpha)$ is expressed in terms of the polynomials noted above. This expression may have narrow maxima at the points where $\Phi(\alpha)$ is small, hence high resolution can be achieved. Among adaptive methods, the likelihood method, autoregressive method and maximum entropy method are well-known. We have used the last two of these.

The autoregressive method is based on the assumption that the real field $p(n_1, n_2)$ at the point (n_1, n_2) of the grating may be replaced by

$$\hat{p}(n_1, n_2) = - \sum_{\substack{m_1=0 \\ m_1 = m_2 \neq 0 \text{ simultaneously}}}^{M_1} \sum_{m_2=0}^{M_2} a(m_1, m_2) p(n_1 - m_1, n_2 - m_2), \tag{8}$$

which is the model of linear prediction of the $M_1 \times M_2$ type. The error of prediction (8) is $\varepsilon(n_1, n_2) = p(n_1, n_2) - \hat{p}(n_1, n_2)$. The coefficients $a(m_1, m_2)$ are found under the condition that the sum of squared modulus of the errors,

$$E = \sum_{m_1=M_1+1}^{N-1} \sum_{m_2=M_2+1}^{N-1} |\varepsilon(m_1, m_2)|^2,$$

is a minimum. This gives the equations

$$\delta E/\delta a(m_1, M_2) = 0; \qquad \begin{aligned} m_1 &= 0, \ldots M_1; \\ m_2 &= 0, \ldots M_2; \\ m_1 &= m_2 \neq 0 \text{ simultaneously.} \end{aligned}$$

The succession of sound field values $p(n_1, n_2)$ is an autoregressive one if the succession of the errors $\varepsilon(n_1, n_2)$ is white noise. Under this condition the power spectrum of the field is

$$|I'(\alpha)|^2 \sim \sum_{n_1} \sum_{n_2} |a(n_1, n_2) \exp[-ikd(\alpha_1 n_1 + \alpha_2 n_2)]|^{-2}. \tag{9}$$

If this is not exactly the case, the procedure is merely the adjusting of the original succession to the autoregressive one and Eq. (9) is only the estimate of the spectrum.

The maximum entropy method proceeds from the correlation function of the sound field at the grating plane:

$$R(n_1, n_2) = \sum_{m_1} \sum_{m_2} p(m_1 + n_1, m_2 + n_2) p^*(m_1, m_2),$$

which relates to the power spectrum $|I'(\alpha)|^2$ as

$$\int |I'(\alpha)|^2 \exp[ik(n_1\alpha_1 + n_2\alpha_2)d] k \, d\alpha = R(n_1, n_2). \tag{10}$$

Under this condition the maximum number of the field realizations $p(n_1, n_2)$ are possible for the same correlation function. Hence, as compared with other methods this one does not impose additional restrictions on the field $p(n_1, n_2)$. It can be shown that the solution of the problem (9) under the condition (10) gives the spectrum estimation of the type (7).

Adaptive methods do not necessarily give exact spectra. The positions of the spectrum's maxima and their values could be incorrect. Besides, these methods are more sensitive to noise. Therefore it is reasonable to combine adaptive methods with the traditional method. In Fig. 2 the positions of peak image intensities determined with the autoregressive and maximum entropy methods are marked by black triangles and squares respectively. We see that all methods supplement each other very well.

We have not discussed the results obtained by the using the 256-hydrophone grating. They are in the process of analysis yet. Due to the continuous interrogation of hydrophones realized in this system, we could particularly observe the variation of backscattering ability of the bottom when the sound pulse is spreading along it.

The sound imaging method using hydrophone gratings could be further developed in other directions. The source and grating may be installed in different places. The use of the source sending the pulse in different directions (not only in vertical one) would allow us to obtain acoustical pictures from many different parts of the bottom. Additional information could be obtained using a source with variable frequency.

It is worthwhile mentioning in conclusion that the experiments described above are only in the initial stage and very far from practical applications. Our hope is that by developing further this method combined with other ones, we can obtain additional interesting information about the ocean bottom.

REFERENCES

Gazey, B. K., 1963, Sea-bed echo amplitude fluctuations arising from ship motion, Rad. Electron. Engr., 26:125.

Volovov, V. I. and Lysanov, Yu. P., 1969, Correlation of the fluctuations of sound signal reflected from the ocean bottom, Sov. Phys. Acoust., 15:179.

Vidmar, P.J., 1980, The dependence of bottom reflection loss on the geoacoustic parameters of deep sea (solid) sediments, J. Acoust. Soc. Am., 68:1442.

McClellan, J. H., 1982, Multidimensional spectral estimation, Proc. IEEE, 70:1029.

THE SEA BOTTOM BACKSCATTERING OF SOUND

(THE HISTORY AND MODERN STATE)

Yu. Yu. Zhitkovskii

P.P. Shirshov Oceanology Institute
the USSR Academy of Sciences
Moscow, V-218, USSR

ABSTRACT

In the 25 years from 1960 to 1985, vast studies of the processes of sound scattering by ocean bottom were carried out. A connection between the angular and frequency dependences of the sound scattering strength by the ocean bottom and its geomorphological characteristics is discussed.

INTRODUCTION

The first published data on sound scatter by the shallow sea bottom were, to all appearances, obtained in the USA during World War II ("Physics . . .", 1946). Later a paper on sound scattering by a harbour bottom appeared (Urick, 1954). In these papers the main features of angular and frequency dependences of the scattered signals were pointed out as well as the effect of the bottom material type (rock, sand and so on) on them. In 1961, results of the first deep-water measurements at frequencies of 530 and 1030 Hz were published (MacKenzie, 1961).

We started a systematic examination of sound scattering by the bottom of the deep ocean in 1960 during the eighth trip of the scientific-research ship (SRS) "Mikhail Lomonosov". Later these studies were continued during the first and the second trips of the SRS "Petr Lebedev" in 1961 and 1962.

In our experiments we have used near-surface explosions as a powerful wide-band sound source. The receiver was situated close to the point of the explosion. This technique enabled us to obtain quickly (during half an hour) information about angular and frequency dependences of sea bottom sound backscattering strength for angles of acoustic wave incidence from 10° to 60° and frequencies from 0.1 to 20 kHz (Fig. 1).

By conventional definition the scattering strength M is a scattering coefficient m expressed in a logarithmic scale:

$$M = 10 \log m, \qquad (1)$$

where $m = W/SJ$, W is the power backscattered by a portion of the bottom of area S per unit solid angle and J is the intensity of the incident sound wave at the bottom.

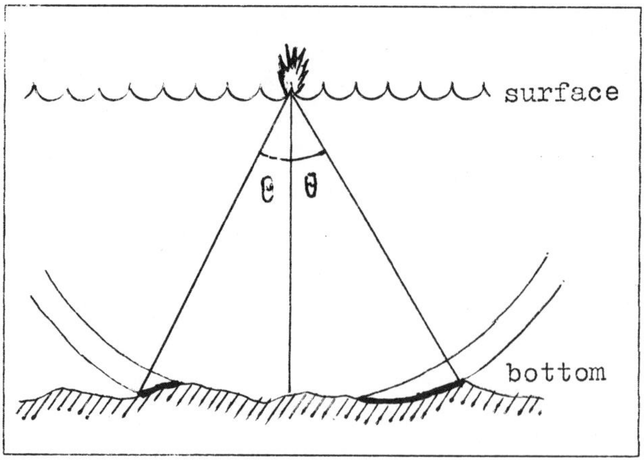

Figure 1: Geometry of the experiment.

Using the method described above, we, during the 3 years 1960-62, examined several hundred geomorphological regions of the Atlantic Ocean. Angular and frequency dependences obtained by us differed greatly from those obtained by other researchers in shallow water.

To our regret we had no opportunity to compare our results with any others as there were at that time no published papers relating to the deep ocean. This disturbed us to some extent, as a pulse (or explosion) method is a so-called absolute method and is liable to errors.

During the second trip of SRS "Petr Lebedev", when we were in Halifax in May of 1962, Professor Ernest Guptill kindly invited us to visit Dalhousie University. In the library we saw a Urick and Sailing paper in the latest JASA issue. In the region of the Hatteras abyssal plain they also had used an explosion method to study backscatter from the ocean bottom sound (Urick and Sailing, 1962). Two of their working points appeared to be only a few dozen kilometers from the region of our studies in 1960. A comparison of the results showed good agreement, and we lost our doubts as to validity of own results. Soon a paper by Burstein and Keane (1964) and some other papers appeared.

Characteristics of the scattering strength obtained in different geomorphological regions of the World's Oceans allowed us to draw some general conclusions about the relationship of the backscattering characteristics to geomorphological parameters of the ocean bottom. Our concept was reported at the 5th International Congress on Acoustics held in 1965 in Liege (Zhitkovskii and Volovova, 1965).

Briefly it may be summarized as follows. In the deep ocean, it is the shape of ocean bottom surface that determines, as a rule, the characteristics of ocean bottom backscattering of sound. In regions with a smoothed bottom, for small angles of incidence, scattering is due to large-scale irregularities (Fig. 2). Theoretically, such scattering was described in the Kirchhoff approximation by Isakovich (1952) and somewhat later independently by Eckart (1953). In this case the scattering strength doesn't depend on frequency and decreases rapidly as the angle of incidence of the wave increases (Fig. 3). In this connection, at

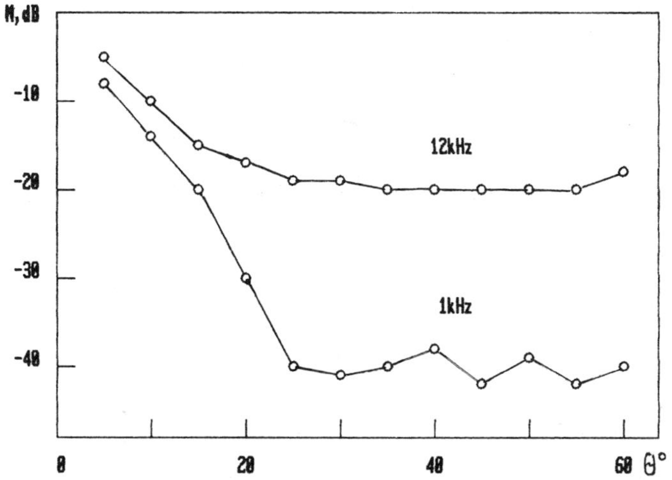

Figure 2: Angular dependences of the scattering strength, obtained in a region with a smoothed bottom.

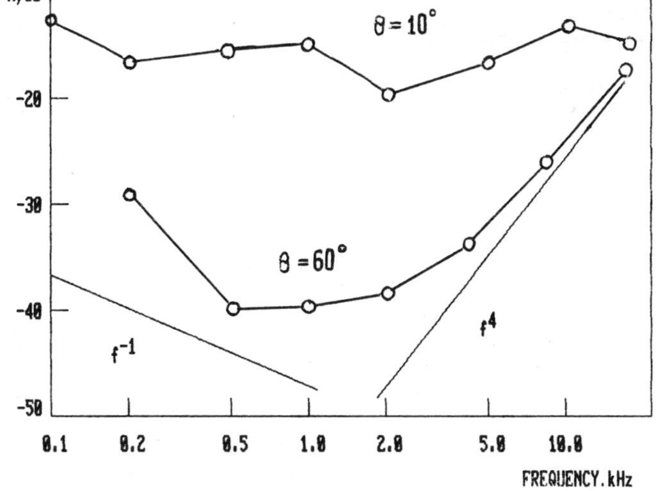

Figure 3: Frequency dependences of the scattering strength, obtained in a region with a smoothed bottom.

grazing incidence, scattering from irregularities which are small compared to the wavelength plays the main role. For small angles of incidence it is masked by strong scattering from large irregularities. Scattering from small irregularities has a considerably weaker angular dependence than scattering from large ones and a pronounced frequency dependence, which sometimes mounts to the fourth power of the frequency (Fig. 3).

In regions of very rough relief (crests of mountain ridges, breaks and so on), where the inclination angles of the irregularities can have arbitrary values and the irregularities themselves are large compared to the wavelength of sound, with frequencies of the order of 1 kHz, scattering follows the Lambert law ($m \sim \cos^2\theta$, θ is angle of incidence) and doesn't depend on frequency (Fig. 4).

In regions of moderately structured relief (abyssal hills, foothills and so on) the scattering strength features are in general similar to those obtained in regions with smoothed relief but peculiarities in dependences are not so emphasized: the scattering strength decreases more slowly for small incidence angles, the frequency dependence of the scattering coefficient doesn't exceed the second power of frequency and so on. It is worth noting that in regions of smooth and moderately rough relief the peculiarities indicated above are observed at frequencies higher than 2 or 4 kHz. At lower frequencies, due to the diminishing rate of sound absorption in the bottom material, somewhat different laws are observed. At frequencies from 0.1 to 1 kHz the frequency dependence of the scattering coefficient is close to f^{-1}(Fig. 3). This fact is possibly connected with the frequency dependence of absorption in the bottom material if sound is due to volume inhomogeneities of the bottom and irregularities of its internal boundaries.

An estimate of scattering coefficient values, made for regions of highly rough bottom and using a scattering model following Lambert's law, showed that for data to coincide quantitatively, it was necessary to assume that some portion of the scattered signals was scattered multiply by large relief irregularities.

In the collection edited by V.M. Albers a paper by Chapman appeared (1967), where he reported that in the Northern Atlantic numerous measurements of the ocean bottom sound scattering strength were made by his group also. Comparison of this data with ours, carried out by Chapman, showed a very good agreement of the results.

Unfortunately a method using shallow explosions, enables one to obtain angular dependences of the scattering strength only for incidence angles up to 60° because the signal scattered at large angles is overlapped by the second bottom reflection (bottom-surface-bottom). That is why in 1966-67 we developed equipment and methods to use a submerged explosion and receiver. Electronic equipment to receive the sound and a controller were housed in a submerged container. The controller enabled, after receiving an operator's signal from onboard the ship, the firing, in turn, of twelve 200 g tolite blasting charges suspended under the container like the rungs of a rope-ladder. It was possible to submerge the system by three-strand cable up to 6000 m. We obtained our first data in the Atlantic in 1967. Then the system was used in 1969 in the Indian Ocean. It allowed us to obtain angular dependences for incidence angles up to 85°. Independently, and at about the same time an analogous system was used by H.M. Merklinger, who reported his results in JASA (Merklinger, 1968). Roughly at the same time studies using submerged explosions were conducted by J.P. Buckley and R.J. Urick (1968).

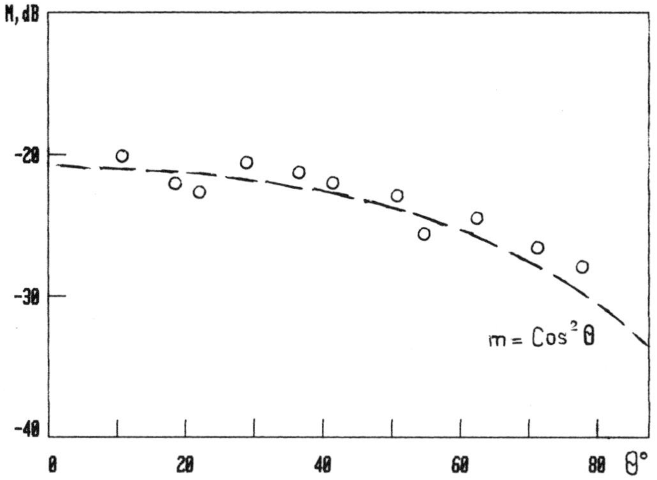

Figure 4: The angular dependence of the scattering strength, obtained in a region of the Arabian-Indian mountain ridge crest.

It should be noted that research on sound scattering by the ocean bottom developed in different countries not only in the same way but even almost simultaneously, even though the researchers themselves learned of it a posteriori due to the time delay between running the study and reporting its results.

Since 1969 we began systematic studies of sound scatter by the ocean bottom in shallow water regions and have obtained data significantly different from those for deep water. A bottom in shelf areas is smooth as a rule. Nevertheless, the angular dependence was closer to results obtained in deep-water regions with highly rough relief, than to those obtained in abyssal plains. That is, the scattering strength decreased even slower as the angle of incidence increased than according to the Lambert Law. In our paper (Zhitkovskii, 1968) it was shown that such angular dependence can take place if scattering is due to an absorbing inhomogeneous layer. In this case the angle dependence follows the Lommel-Zeeliger law $m \sim \cos\theta$.

But there was another question that needed clarification. Nearly all data in the deep ocean were obtained using explosions (except for the works of MacKenzie and Patterson) but absolutely all studies in shallow water were carried out using tone-signal sources. That is why it was advisable to carry out direct experiments to compare these two methods. We made such experiments in 1972 in the Barents Sea at a point with a depth of 300 m (Zhitkovskii, 1973). As is obvious from Fig. 5 both methods gave practically the same result.

Hence, the difference in the angular dependences was caused by the nature of sound scatter by deep ocean bottoms and shallow regions itself. We have supposed that in shallow regions, contrary to deep-water ones, it is bottom structure not relief irregularities which play the main role in sound scatter; that is, scattering due to the structure prevails over scattering due to irregularities. At first, we failed to

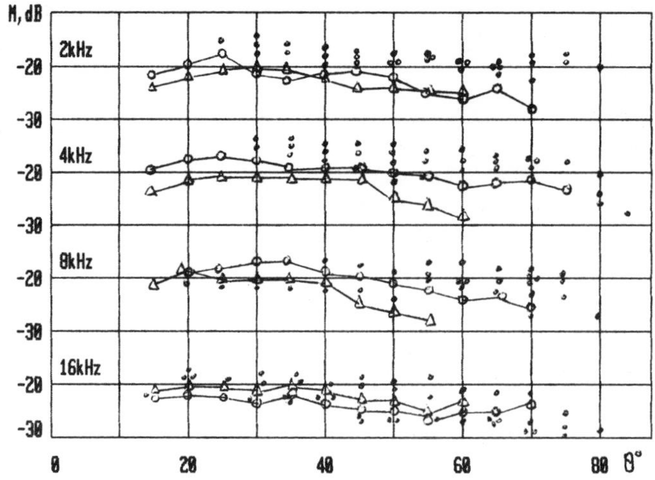

Figure 5: Angular dependences of the scattering strength, obtained in the Barents Sea by two methods. For clarity the angular dependences, obtained at different frequencies, are 20 dB shifted. Numbers in the second column from the left correspond to the radiated frequency for tone-signal sources (dots) or middle filter frequencies for explosive sources (circles and triangles).

explain the lack of the sound scattering strength frequency dependence in a wide frequency range in shallow regions (Bunchuk and Zhitkovskii, 1980). But later in Yu. P. Lysanov and others works this phenomenon found its interpretation (Lysanov, 1980; Ivakin and Lysanov, 1981).

It turned out that it is possible, for instance, if the vertical scale of inhomogeneities is smaller, but the horizontal scale is greater than the wavelength; that is, the inhomogeneities are thin lenses. This conclusion agrees with the geologists' concepts about the bottom structure but it is worthwhile to confirm the conclusion directly, for example, by model experiment.

At the end of the 70's we began to study local variability of sound scatter by the deep ocean bottom as well as sound scatter in regions where the bottom is covered by ferro-manganese nodules (FMN). Before that time we had studied in some detail numerous regions of the Atlantic Ocean and the north-west half of the Indian Ocean; that is, we had not worked in the Pacific Ocean and in the south-east half of the Indian Ocean just where main FMN fields are situated.

For these studies we developed a deep-water system including a container with a thiristor generator of several kW in power, a receiving amplifier and a controller (Zotov et al., 1984). Outside the container, a piezoceramic receiver and transducer were situated. The system was capable of working at 6000 m. It was lowered by cable and operated with discrete frequencies in the range from 2 to 16 kHz. Using this system in 1980-86 we carried out experiments in the Atlantic, the Pacific and the Indian oceans. It turned out that local variability of the bottom scattering strength is very high. So, in the Atlantic in a region of abyssal hills the scattering strength sometimes differs several times at points only 200 or 300 m apart (Fig. 6) (Zhitkovskii, 1982).

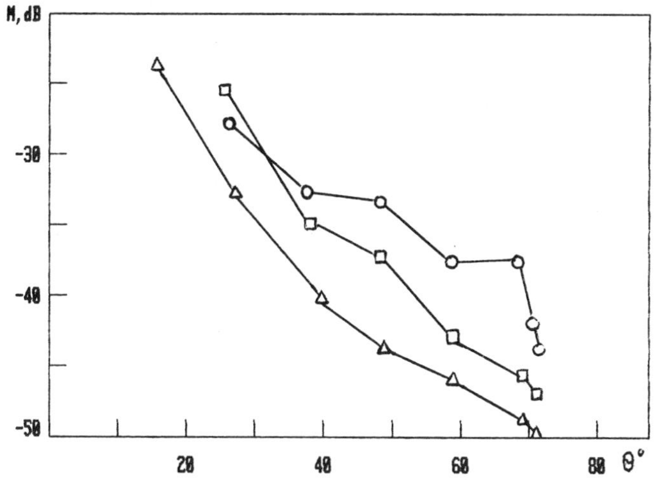

Figure 6: Angular dependences of the scattering strength, obtained at frequency 5 kHz in a region of abyssal hills (different curves correspond to areas situated from 200 to 300 m apart).

Rather interesting results have been obtained in regions covered by FMN. It has turned out, that areas covered by FMN can have comparatively small sizes, for example, of several hundred meters extent. In areas covered by FMN the scattering strength doesn't depend on angle and is from 10 to 15 dB greater than in areas without FMN (Fig. 7). In this case the scattering coefficient is proportional to the fourth power of frequency (Fig. 8). This is not surprising, as the nodule dimensions never were greater than 10 cm and as a rule were from 3 to 4 cm, and the wavelength for the highest frequency was 10 cm (Zotov and Fokin , 1985; Brekhovskikh et al., 1985). One may suppose that sound scatter,in areas where there were no FMN on the bottom surface, was caused by a small number of FMN buried in sediments. This conclusion follows from the fact that the sediment material had very high porosity (about 95%), its sound velocity and density were very close to that of water. Hence, the scattering from the sediment material itself should be very small. Angular and frequency dependences of the scattering strength, obtained in areas without FMN, also provide evidence in favour of the hypothesis of buried nodules. Angular dependences correspond to volume scatter, and frequency ones correspond to volume scatter due to small (compared to wavelength) inhomogeneities located in an absorbing medium. It is known that in sediments absorption increases as for frequency to the first power. In agreement with Fig. 8, it causes a decrease from four to three in the power of frequency to which the scattering coefficient is proportional.

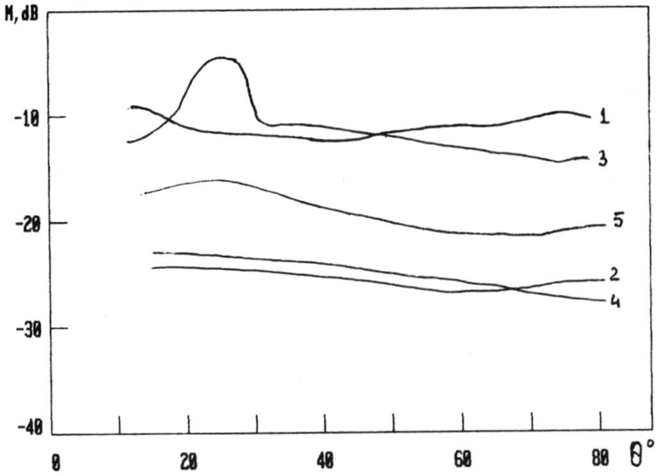

Figure 7: Angular dependences of the scattering strength, obtained at a frequency of 16 kHz in a region possessing FMN. The curves 1 and 3 correspond to areas where there were many FMN at the surface. The curves 2 and 4 correspond to areas where there were no FMN, and the curve 5 corresponds to an area with moderate content of FMN.

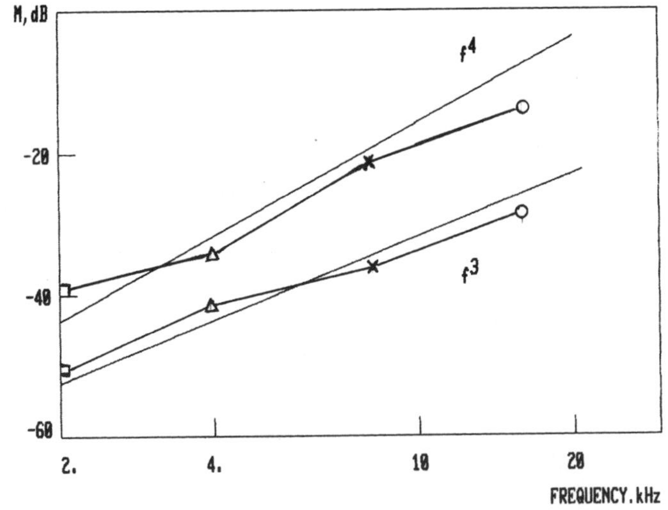

Figure 8: Frequency dependences of the scattering strength at an angle of incidence of 40° in a region containing FMN. The upper curve corresponds to a bottom area covered by FMN, the lower curve corresponds to an area where FMN are absent.

REFERENCES

Brekhovskikh L.M., Zhitkovskii Yu. Yu., Zakhlestin A.Yu. and Savel'ev V.V., 1985, Sound scattering by ferro - manganese nodules, Sov. Phys. Acoust., 31:N4.

Buckley J.P. and Urick R.J., 1968, Backscattering from the deep-sea bed at small grazing angles, J. Acoust. Soc. Am., 44:648.

Bunchuk A.V. and Zhitkovskii Yu. Yu., 1980, Sound scattering by ocean bottom in shallow water regions, Sov. Phys. Acoust., 26:N5.

Burstein A.W. and Keane J.J., 1964, Backscattering of explosive sound from ocean bottoms, J. Acoust. Soc. Am., 36:1956.

Chapman, R.P., 1967, Sound Scattering in the Ocean, Chapter 9 in "Underwater Acoustics Vol 2", V.M. Albers Ed., Plenum, New York.

Eckart C., 1953, The scattering of sound from the sea surface, J. Acoust. Soc. Am., 25:566.

Isakovich M.A., 1952, Wave scattering from a statistically rough surface, ZhETF, 23:305 (In Russian).

Ivakin A.N. and Lysanov Yu. P. 1981, To the theory of sound scattering by stochastic inhomogeneities of underwater soil, Sov. Phys. Acoust., 27:110.

Lysanov Yu. P. 1980, On an geoacoustical model of sediments upper layer in shallow sea, Dokl. AN SSSR, 251:714 (In Russian).

MacKenzie K.V., 1961, Bottom reverberation for 530 and 1030 cps sound in deep water, J. Acoust. Soc. Am., 33:1498.

Merklinger H.M., 1968, Bottom reverberation measured with explosive charges fired deep in the ocean, J. Acoust. Soc. Am. 44:508.

"Physics of Sound in the Sea", 1946, Summary Tech. Rept., Washington, D.C.

"Underwater Acoustics", N2, 1970, Mir, Moscow (In Russian).

Urick R.J., 1954, The backscattering of sound from a harbour bottom, J. Acoust. Soc Am. 26:231

Urick R.J., 1954, Backscattering of explosive sound from the deep-sea bed, J. Acoust. Soc. Am., 34:1721.

Zhitkovskii Yu. Yu., 1968, Sound scattering by of inhomogeneities of ocean bottom soil, Izv. Atm. Ocean. Sci., 4:N5.

Zhitkovskii Yu. Yu., 1973, Comparison of pulse (explosive) and tone-signal methods of measuring the ocean bottom sound scattering strength, in "Proc. VIII All-Union Acoustical Conference", Moscow (In Russian).

Zhitkovskii Yu. Yu., 1982, Sound scattering by ocean bottom, in: "Ocean Acoustics. State of art", Nauka, Moscow (In Russian).

Zhitkovskii Yu. Yu. and Volovova L.A., 1965, Sound scattering from the ocean bottom, in: "Rapports du 5-e Congress International d'Acoustique", Liege.

Zotov A.I. and Fokin A.V., 1985, On the investigation of local scattering by the bottom of the deep-sea, Oceanol., 25:219 (In Russian).

Zotov A.I., Kuznetzov V.N. and Savel'ev V.V., 1984, Deep-water system for hydroacoustical studies, Oceanol., 24:175. (In Russian).

A UNIFIED DESCRIPTION OF WAVE SCATTERING AT BOUNDARIES WITH LARGE AND SMALL SCALE ROUGHNESS

A.G. Voronovich

P.P. Shirshov Institute of Oceanology
the USSR Academy of Sciences
Moscow, V-218, USSR

ABSTRACT

The small slope approximation for investigation of wave scattering by rough surfaces is described and compared with other approaches to the problem. The validity condition for the method consists generally only in the smallness of the elevation slopes without any restrictions to the magnitude of wavelength. A comparison with the exact numerical solution for a saw-type roughness is made. The approximate inverse problem of determining the parameters of roughness from scattering data is solved on the basis of the results obtained.

INTRODUCTION

The process of wave scattering at rough surfaces is encountered in many branches of physics. Particularly in underwater acoustics the problem of sound scattering by a rough pressure-release surface is of considerable importance. There is a great bulk of literature on this problem. The few existing theoretical approaches which examine it most frequently use different versions of one of two methods (Brekhovskikh and Lysanov, 1982; Bass and Fuks, 1978): the method of small perturbations (MSP) and the tangent plane approximation (TPA) (or Kirchhoff, or quasiclassical approximation).

We are, however, often faced with situations which cannot be considered in the framework of one of these methods alone. Such is the problem of scattering of sound or of electromagnetic waves by a rough sea surface, which is of great practical importance. In such cases the so called two-scale surface model is used (Brekhovskikh and Lysanov, 1982; Bass and Fuks, 1978; Kur'yanov, 1962), and the scattering is examined with the use of both methods: the TPA for scattering by large-scale smooth components and the MSP for scattering by small-scale components.

When introducing two classes of roughness we inevitably insert into the theory at least one parameter which is to some degree arbitrary. This is sometimes inconvenient and precludes considering the corresponding inverse problem - determination of the roughness spectrum through the characteristics of a scattered field.

It is, however, possible to develop an approach which enables one to consider the problem of wave scattering by rough surfaces and which gets over these difficulties in a radical way. This approach assumes smallness of a single geometrical parameter, namely a small slope of undulations, without any restriction as to the wavelength. Hence the Rayleigh parameter may be arbitrarily large. As a result, we can devise a universal method for obtaining an expression for the scattering amplitude under the small slope approximation (SSA), using an appropriate formula arising in the usual MSP.

Another method for the unified consideration of scattering of electromagnetic waves at a rough boundary was put forward by Bahar (1972a, b).

SCATTERING AMPLITUDE

The rough surface is supposed to be plane, on average (z=0), and is specified by the equation $z=h(\mathbf{r})$, where $\mathbf{r}=(x,y)$ is the horizontal component of the radius-vector $\mathbf{R}=(\mathbf{r},z)$. We assume that the z-axis is directed upward and that the medium at $z<h(\mathbf{r})$ is homogeneous and possesses a wave propagation velocity c. Let a plane monochromatic wave of frequency f be incident on the boundary from below (i.e. from $z=-\infty$). Then the total field may be represented in the form

$$P=P_{in}+P_{sc}$$
$$=q_0^{-1/2}\exp(i\mathbf{k}_0\cdot\mathbf{r}+iq_0z)+\int S(\mathbf{k},\mathbf{k}_0)q^{-1/2}\exp(i\mathbf{k}\cdot\mathbf{r}-iqz)d\mathbf{k}\,, \quad (1)$$

where $(\mathbf{k}_0,q_0)$ and $(\mathbf{k},-q)$ are the wave vectors of the incident and scattered waves, $q=q(k)=q_k=[K^2-k^2]^{1/2}$, (Im q >0), $q_0=q(k_0)$, and $K=2\pi f/c$. The integral in Eq.(1) exists if $z<h(\mathbf{r})$. If $z>\min(h(\mathbf{r}))$ and the Rayleigh hypothesis doesn't hold, then in the general case, one should use analytic continuation of the solution from the region $z<\min(h(\mathbf{r}))$. In some cases analytic continuation may be accomplished by some regularization of the integral.

Equation (1) is a general solution of the Helmholtz Equation which satisfies the radiation condition. To obtain the scattering amplitude $S(\mathbf{k},\mathbf{k}_0)$ the boundary condition at z=h should be applied. The method of obtaining S is quite analagous to that used in quantum mechanics. In particular, the reciprocity theorem and unitary condition should be fulfilled. Using an expansion of the free space Green's function, $G_0(\mathbf{R})$, as in a superposition of plane waves we easily find, in terms of S, the Green's function for the boundary problem under consideration:

$$G(\mathbf{R},\mathbf{R}_0)=G_0(\mathbf{R}-\mathbf{R}_0)-(i/8\pi^2)\int \{q^{-1/2}\exp(i\mathbf{k}\cdot\mathbf{r}-iqz)S(\mathbf{k},\mathbf{k}_0)$$
$$\times\, q_0^{-1/2}\exp(-i\mathbf{k}_0\cdot\mathbf{r}_0-iq_0z_0)\}d\mathbf{k}d\mathbf{k}_0\,. \quad (2)$$

Thus, to know S means to know the Green's function, but in theoretical investigations of the scattering process, S is much more convenient.

Now consider the statistical case. Taking the ensemble of surfaces to be space-homogeneous, then we generally have

$$\langle S(\mathbf{k},\mathbf{k}_0)\rangle = \bar{V}_k\delta(\mathbf{k}-\mathbf{k}_0)\,, \quad (3)$$

where $\bar{V}$ is the so-called mean reflection coefficient (Brekhovskikh and Lysanov, 1982; Bass and Fuks, 1978). The space-homogeneity condition for the second statistical moment of S may be stated as follows:

$$\langle\Delta S(\mathbf{k}-\mathbf{a}/2,\mathbf{k}_0-\mathbf{a}_0/2)\Delta S^*(\mathbf{k}+\mathbf{a}/2,\mathbf{k}_0+\mathbf{a}_0/2)\rangle= C(\mathbf{k},\mathbf{k}_0;\mathbf{a})\delta(\mathbf{a}-\mathbf{a}_0)\,, \quad (4)$$

where $\Delta S(\mathbf{k}_1,\mathbf{k}_2)=S(\mathbf{k}_1,\mathbf{k}_2)-\langle S(\mathbf{k}_1,\mathbf{k}_2)\rangle$ (the same notation is used in Voronovich, 1983a). The unitary condition (the energy conservation law) yields

$$\int_{|\mathbf{k}'|<K} c(\mathbf{k},\mathbf{k}')d\mathbf{k}' + |\bar{V}_k|^2 = 1\,, \quad (5)$$

where $c(\mathbf{k},\mathbf{k}') = C(\mathbf{k},\mathbf{k}';0)$.

Using Eqs. (1), (2), and (4) it is easy to show that when z and $z_0 \to -\infty$ the following relationship applies

$$<\Delta P_{sc}(\mathbf{r},z)\Delta P^*_{sc}(\mathbf{r}',z')>=(1/16\pi^2)\int\int (1/Kq)\exp(i\mathbf{k}\cdot\Delta\mathbf{r}-iq\Delta z)d\mathbf{k}$$

$$\times (1/Kq_0)c(\mathbf{k},\mathbf{k}_0)K^2\delta(\mathbf{R}-\mathbf{r}_0+\mathbf{k}Z/q+\mathbf{k}_0 z_0/q_0)d\mathbf{k}_0 , \tag{6}$$

where $d\mathbf{k}/Kq=d\mathbf{n}$ is an element of solid angle and $\Delta\mathbf{r}=\mathbf{r}-\mathbf{r}'$, $\Delta z=z-z'$, $\mathbf{R}=(\mathbf{r}+\mathbf{r}')/2$, $Z=(z+z')/2$, and where $(\mathbf{r}_0,z_0)$ are the coordinates of the point source. Equation (6), the correlation function of the scattered field, coincides with that for an ensemble of plane waves δ-correlated in direction. For the mean intensity, formula (6) may be transformed to

$$<|\Delta P_{sc}(\mathbf{r},z)|^2>=\int qq_0 c(\mathbf{k},\mathbf{k}_0)[(\mathbf{r}-\mathbf{r}')^2+z^2]^{-1}J_{in}(\mathbf{r}')d\mathbf{r}' . \tag{7}$$

Here $\mathbf{r}'=\mathbf{r}_0-\mathbf{k}_0 z_0/q_0$ is the horizontal coordinate of the ray emitted from the source at the level z=0,

$$\mathbf{k}=\mathbf{k}(\mathbf{r}')=K(\mathbf{r}-\mathbf{r}')[(\mathbf{r}-\mathbf{r}')^2+Z^2]^{-1/2},$$

$$\mathbf{k}_0=\mathbf{k}_0(\mathbf{r}')=K(\mathbf{r}'-\mathbf{r}_0)[(\mathbf{r}'-\mathbf{r}_0)^2+z_0^2]^{-1/2},$$

and $J_{in}(\mathbf{r}')=[q(\mathbf{k}_0)]^{-1/2}$ is the intensity of the incident wave (Eq.(7) holds for a directional source also). According to the definition of the scattering coefficient m_s (Brekhovskikh and Lysanov, 1982), it follows from Eq.(7) that

$$m_s = qq_0 c(\mathbf{k},\mathbf{k}_0) . \tag{8}$$

So the quantity c, determined by Eqs.(4) and (5), has a practical importance and its computation, along with V from Eq.(3), may be considered as one of the main problems in the theory of wave scattering at statistically rough boundaries.

From the physical point of view, it is quite likely that the asymptotic expansion of S in powers of elevations exists:

$$S(\mathbf{k},\mathbf{k}_0)=S_0\delta(\mathbf{k}-\mathbf{k}_0)+2i(qq_0)^{1/2}B(\mathbf{k},\mathbf{k}_0)h(\mathbf{k}-\mathbf{k}_0)$$

$$+(qq_0)^{1/2}\sum_{n=2}\int\ldots\int B_n(k,k_0;k_1,\ldots k_{n-1})h(\mathbf{k}-\mathbf{k}_1)\ldots h(\mathbf{k}_{n-1}-\mathbf{k}_0)d\mathbf{k}_1\ldots d\mathbf{k}_{n-1} , \tag{9}$$

where

$$h(\mathbf{k}) = \int \exp(-i\mathbf{k}\cdot\mathbf{r})h(\mathbf{r})d\mathbf{r}/(4\pi^2) .$$

(We won't introduce a new designation for the Fourier transform of $h(\mathbf{r})$, because it is clear from the structure of appropriate formulae which quantity is meant.) Equation (9) should be effective if the roughness is small enough. Because the coefficients B_n are independent of $h(\mathbf{r})$, any elevations may be used for their determination. In particular, one may choose very smooth and gentle undulations for which the Rayleigh hypothesis holds and the scattered field at the boundary may be computed immediately from Eq.(1). For example, at the pressure release surface ($P|_{z=h}=0$), S satisfies the equation

$$q_0^{-1/2}\exp[i\mathbf{k}_0\cdot\mathbf{r}+iq_0 h(\mathbf{r})]+\int S(\mathbf{k},\mathbf{k}_0)q_k^{-1/2}\exp[i\mathbf{k}\cdot\mathbf{r}-iq_k h(\mathbf{r})]d\mathbf{k}=0 . \tag{10}$$

Expanding $\exp(\pm iqh)$ in a power series, we easily find

$$S(\mathbf{k},\mathbf{k}_0)=-\delta(\mathbf{k}-\mathbf{k}_0)-2i(qq_0)^{1/2}h(\mathbf{k}-\mathbf{k}_0)+2(qq_0)^{1/2}\int q_{\mathbf{k}'}h(\mathbf{k}-\mathbf{k}')h(\mathbf{k}'-\mathbf{k}_0)d\mathbf{k}'+...$$

In the TPA the scattering amplitude may be expressed in the form (Voronovich, 1983a,b)

$$S^{(K)}(\mathbf{k},\mathbf{k}_0)=g(\mathbf{k},\mathbf{k}_0)B(\mathbf{k},\mathbf{k}_0)2(qq_0)^{1/2}(q+q_0)^{-1}$$

$$\times\int \exp[-i(\mathbf{k}-\mathbf{k}_0)\cdot\mathbf{r}+i(q+q_0)h(\mathbf{r})]d\mathbf{r}/4\pi^2\,, \qquad (11)$$

and for the pressure release surface B=-1 and

$$g(\mathbf{k},\mathbf{k}_0)=1+[(\mathbf{k}-\mathbf{k}_0)^2+(q-q_0)^2]/4qq_0\,. \qquad (12)$$

Considering in the TPA the extreme case $(q+q_0)h \ll 1$, we easily find that due to the factor g even the first order term with respect to h differs from the corresponding term in Eq.(9). Though g is close to unity for the near specular directions, it may differ from unity considerably for large scattering angles. The TPA is not generally correct in this situation.

We now discuss a statistical case. It follows from Eqs.(3), (4), and (9) that for small roughness,

$$\bar{V}_{\mathbf{k}}^{(MSP)}=S_0(\mathbf{k})+q_{\mathbf{k}}\int B_2(\mathbf{k},\mathbf{k}_0;\mathbf{k}')W(\mathbf{k}-\mathbf{k}')d\mathbf{k}'\,, \qquad (13)$$

and that

$$c^{(MSP)}(\mathbf{k},\mathbf{k}_0)=4q_{\mathbf{k}}q_0B^2(\mathbf{k},\mathbf{k}_0)W(\mathbf{k}-\mathbf{k}_0)\,, \qquad (14)$$

where $\langle h(\mathbf{k})h^*(\mathbf{k}')\rangle=W(\mathbf{k}-\mathbf{k}')\delta(\mathbf{k}-\mathbf{k}')$, (W(k) is the spatial spectrum of the roughness). Averaging of Eq.(11) results in the appearance of characteristic functions. To simplify formulae we'll consider further the sufficiently representative case of a Gaussian ensemble. Then we immediately obtain from Eq.(11)

$$\bar{V}_{\mathbf{k}}=S_0\exp[-q^2D(\infty)]\,, \qquad (15)$$

where $D(\mathbf{r})=\langle[h(\mathbf{r}+\mathbf{a})-h(\mathbf{a})]^2\rangle$ is the structure function of elevations (it is supposed that correlation vanishes at $|\mathbf{r}|\to\infty$, thus $D(\infty)=2\langle h^2\rangle$). Also, it is easily shown that

$$c^{(K)}(\mathbf{k},\mathbf{k}_0)=g^2(\mathbf{k},\mathbf{k}_0)B^2(\mathbf{k},\mathbf{k}_0)4qq_0(q+q_0)^{-2}\int \exp[-i(\mathbf{k}-\mathbf{k}_0)\cdot\mathbf{r}]$$

$$\times[\exp(-[q+q_0]^2D(\mathbf{r})/2)-\exp(-[q+q_0]^2D(\infty)/2)]d\mathbf{r}/4\pi^2. \qquad (16)$$

Equations (9) and (11) along with the specific expressions for B and g depending on boundary conditions are, from the principal point of view, the main results of the MSP and the TPA theory of wave scattering at rough surfaces which are, on average, planar. In the statistical case these results consist of Eqs. (13)-(16).

SMALL SLOPE APPROXIMATION

The SSA may be derived in different ways as described by Voronovich (1983a, 1985a, 1986). The condition for validity of the method consists of the requirement that

$$\text{slope of roughness} \ll \min(\sin(\chi), \sin(\chi_0))\,, \qquad (17)$$

where χ and χ_0 are the grazing angles of incident and scattered waves. Thus shadowings are not permitted. Then the expression for S, correct to an accuracy of h^2, may be expressed in the form (Voronovich, 1985a):

$$S^{(SSA)} = S_1^{(SSA)}+S_2^{(SSA)} ,$$

where

$$S_1^{(SSA)}(\mathbf{k},\mathbf{k}_0)=B(\mathbf{k},\mathbf{k}_0)2(qq_0)^{1/2}(q+q_0)^{-1}\int \exp[-i(\mathbf{k}-\mathbf{k}_0)\cdot\mathbf{r}+i(q+q_0)h(\mathbf{r})]d\mathbf{r}/4\pi^2 , \quad (18a)$$

$$S_2^{(SSA)}(\mathbf{k},\mathbf{k}_0)=-i/2(qq_0)^{1/2}(q+q_0)^{-1} \int\int \exp[-i(\mathbf{k}-\mathbf{k}_0-\mathbf{k}')\cdot\mathbf{r}+i(q+q_0)h(\mathbf{r})]d\mathbf{r}/4\pi^2$$

$$\times [B_2(\mathbf{k},\mathbf{k}_0;\mathbf{k}-\mathbf{k}')+B_2(\mathbf{k},\mathbf{k}_0;\mathbf{k}_0+\mathbf{k}')+2(q+q_0)B(\mathbf{k},\mathbf{k}_0)]h(\mathbf{k}')d\mathbf{k}' . \quad (18b)$$

Formula (18a) gives S to an accuracy of h^2 and Eq.(18b) is the correction of the order h^2. Coefficients B, B_2 (which are matrices in the general case of vector fields) are determined by means of Eq.(9) in the usual MSP. In some cases (scattering of scalar waves at pressure release and hard surfaces, or electromagnetic waves at ideally conducting surfaces) B_2 may be expressed with the use of B (Voronovich, 1983a, 1985b)

$$B_2(\mathbf{k},\mathbf{k}_0;\mathbf{k}')=-2q_{k'}B(\mathbf{k},\mathbf{k}')S_0(\mathbf{k}')B(\mathbf{k}',\mathbf{k}_0) . \quad (19)$$

Note that in the general case the following relation applies

$$S_0(\mathbf{k})=B(\mathbf{k},\mathbf{k}) .$$

In the limit $(q+q_0)h \ll 1$, the expansion in Eq.(9), to an accuracy of terms of the order $\sim h^2$, results from Eq.(18). Furthermore it is easy to see that

$$S^{(K)}(\mathbf{k},\mathbf{k}_0)=g(\mathbf{k},\mathbf{k}_0)S_1^{(SSA)}(\mathbf{k},\mathbf{k}_0) . \quad (20)$$

In near specular directions: $\mathbf{k}\sim\mathbf{k}_0$, we have $S^{(K)}\sim S^{(SSA)}$, as it should be. It may be shown (Voronovich, 1985a) that for the pressure release surface in the limit $k\to\infty$, Eq.(9) follows from Eqs.(18a-b).

For the energy characteristics of scattering it is sufficient to take S according to Eq.(18a). Then it follows from relation (20) that the expression for $c(\mathbf{k},\mathbf{k}_0)$ reduces to Eq.(16) in which the factor g^2 should be omitted:

$$c^{(SSA)}(\mathbf{k},\mathbf{k}_0)=c^{(K)}(\mathbf{k},\mathbf{k}_0)/g^2(\mathbf{k},\mathbf{k}_0) . \quad (21)$$

To obtain the mean reflection coefficient both terms in Eq.(18) should be taken into account. The result of averaging may be expressed as follows:

$$\overline{V}^{(SSA)}(\mathbf{k})=\exp[-q_k{}^2D(\infty)][V^{(MSP)}(\mathbf{k})+S_0q_k{}^2D(\infty)] , \quad (22)$$

where $V^{(MSP)}(k)$ should be formally computed according to Eq.(13) (although the Rayleigh parameter may be large). It proves out that Eq.(22) may be represented in the following form too:

$$\ln[S_0^{-1}\overline{V}^{(SSA)}(\mathbf{k})]=S_0^{-1}q_k\int B_2(\mathbf{k},\mathbf{k}_0;\mathbf{k}')W(\mathbf{k}-\mathbf{k}')d\mathbf{k}' . \quad (22a)$$

Note that simple formulae (22) and (22a) arise for the case of Gaussian statistics.

Thus in the SSA, scattering is described by Eq.(18) and in the statistical case by Eqs.(21) and (22).

For examination of the proposed theory some numerical experiments for scattering at periodic pressure release surfaces were carried out by Voronovich (1985c). In this situation Eq.(18) reduces to

$$S(k,k_0)=\sum_{-\infty}^{+\infty} S_N\delta(k-k_N)\ ,$$

$$S_N=-(q_Nq_0)^{1/2}(q_N+q_0)^{-1}\sum_{m=-\infty}^{+\infty}[2\delta_{m0}+i(q_{N-m}+q_m-q_N-q_0)h_m]Q_{N,m}\ , \qquad (23)$$

where

$$k_m=k_0+pm,\ q_m=q(k_m),\ p=2\pi/d,\ h_m=d^{-1}\int_{-d/2}^{d/2} h(x)\exp(-ipmx)dx,$$

$$Q_{N,m}=d^{-1}\int_{-d/2}^{d/2}\exp[-i(N-m)px+i(q_0+q_N)h(x)]dx,$$

and d is the period of the roughness. The case of saw-type (echelette) undulations, $h(x)=a|x|$, $|x|<d/2$, was studied and, employing Eq.(23), the reflection coefficients $R_N=S_N(q_0/q_N)^{1/2}$ were calculated. This geometry is particularly interesting because of the presence of infinite values of the second derivatives, which are not forbidden by Eq.(17). The values of R_N obtained from Eq.(23) were compared to the exact solution of the problem obtained with the aid of the special numerical algorithm of Vaynstein and Sukov (1984). The results of the comparison are shown in Fig. 1(a). The angle of incidence was held constant at 45° and the slope of facets $a=\tan\alpha=0.2$. Numerical computations were made for integral values of d/λ and the corresponding points are connected by heavy solid lines. The same quantities calculated according to the TPA (i.e. according to Eq.(11)), are shown by crosses and are connected by solid lines; those calculated according to Eq.(18a) (SSA to second order) are shown by small circles. Figure 1(a) shows the dependence of R_N on the parameter d/λ for the spectrum of maximum amplitude. The number of appropriate spectra are indicated under the abscissa. Figure 1(b) shows the difference, E, between the exact and approximate values of R_N for N=-5 depending on d/λ and Figs.1(c) and 1(d) - for all homogeneous spectra at $d/\lambda=5$ and $d/\lambda=2$. It is very important that corrections of second order in the slope make the result much more accurate for all values of d/λ and for all spectra. This fact corroborates the asymptotic character of the derived expansion.

There is an opportunity to solve the appropriate inverse problem within the bounds of the SSA in a rather simple way: that is determining the spectrum of the roughness through the scattering coefficient. The essence of the proposed procedure is made clear by considering the case of the pressure release surface. According to Eqs.(8), (11), and (16) we have

$$m_s=4q^2q_0^2(q+q_0)^{-2}\int\exp[-i(\mathbf{k}-\mathbf{k}_0)\cdot\mathbf{r}][\exp(-[q+q_0]^2D(\mathbf{r})/2)$$

$$-\exp(-[q+q_0]^2D(\infty)/2)]d\mathbf{r}/4\pi^2\ . \qquad (24)$$

The quantity $I=(q+q_0)^2m_s(\mathbf{k},\mathbf{k}_0)/4q^2q_0^2$ should depend only on parameters $\mathbf{a}=\mathbf{k}-\mathbf{k}_0$ and $b^2=(q+q_0)^2/4$:

$$I(\mathbf{a},b^2)=\int\exp(-i\mathbf{a}\cdot\mathbf{r})[\exp(-2b^2D(\mathbf{r}))-\exp(-2b^2D(\infty))]d\mathbf{r}/4\pi^2\ .$$

The Fourier transform now gives

$$D(\mathbf{r})=-(2b^2)^{-1}\ln[1+\int I(\mathbf{a},b^2)(\exp[i\mathbf{a}\cdot\mathbf{r}]-1)d\mathbf{a}]\ . \qquad (25)$$

Thus to obtain $D(\mathbf{r})$ it is sufficient, for example, to know the scattering coefficient in the backward direction as a function of frequency and the angle of incidence. In this case

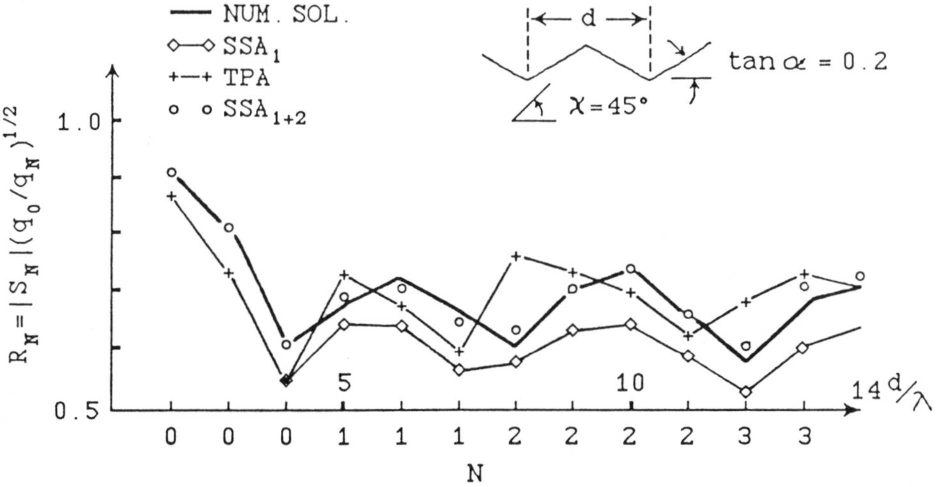

Figure 1(a). Comparison of reflection coefficients R_N obtained with first and first plus second order SSA, and the TPA methods.

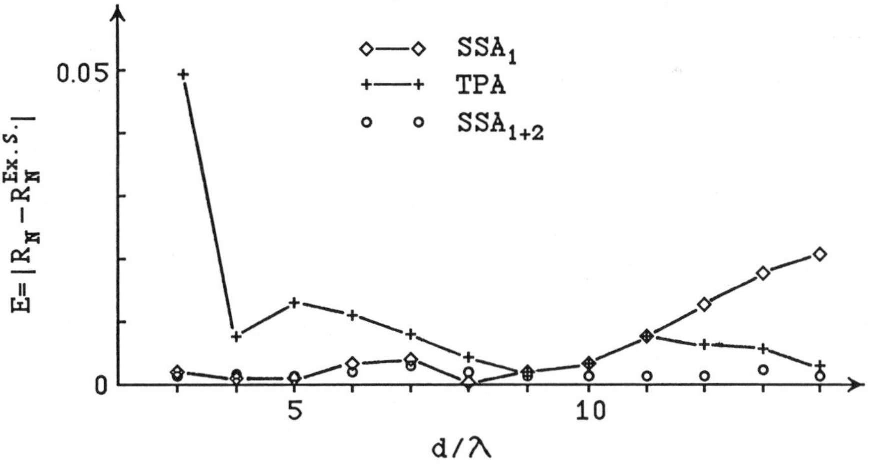

Figure 1(b). Difference between computed values of the reflection coefficient and the exact solution for the reflection coefficient. (N=−5)

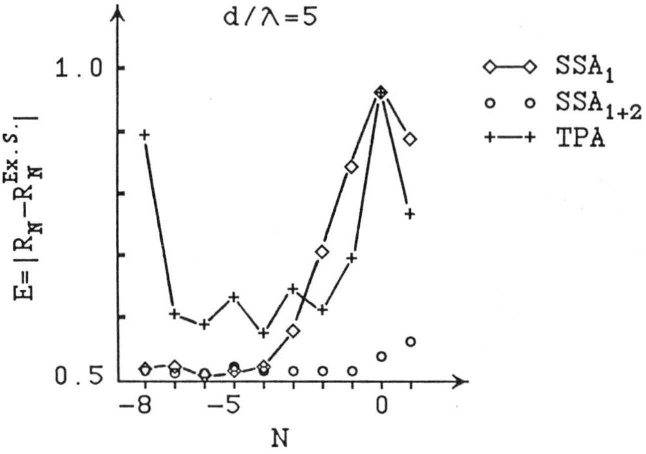

Figure 1(c). Difference between exact and computed solutions for the case of d/λ=5.

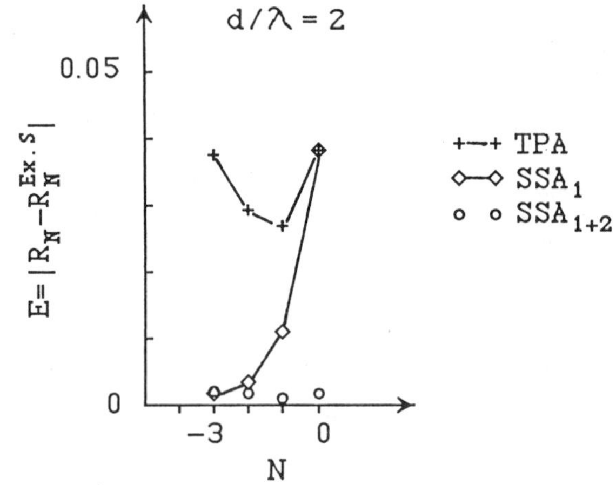

Figure 1(d). Difference between exact and computed solutions for the case of d/λ=2.

$$D(\mathbf{r})=-(2b^2)^{-1}\ln[1+2b^{-2}\int M(\mathbf{k},b^2)(\exp[2i\mathbf{k}\cdot\mathbf{r}]-1)d\mathbf{k}], \qquad (26)$$

where $M(\mathbf{k},b^2)=m_s(-\mathbf{k},\mathbf{k})|_{k^2=K^2-b^2}$. So the vertical component of the wavenumber should be fixed. In MSP, $m_s(\mathbf{k},\mathbf{k}_0) = 4k^2k_0^2W(\mathbf{k}-\mathbf{k}_0)$, $M(\mathbf{k},b^2)=4b^4W(2\mathbf{k})$, and reconstruction of $D(\mathbf{r})$ with the help of Eq.(26) gives the correct result. The right hand sides of Eqs.(25) and (26) shouldn't in fact depend on b^2. This statement may be used for examination of the theory. In more complex cases when the function $B(\mathbf{k},\mathbf{k}_0)$ is non-trivial and contains primarily unknown parameters, this fact may be used for finding these unknowns. In just the same way the inverse problem may be solved in the deterministic case.

CONCLUSIONS

The small slope approximation permits us to study scattering at rough surfaces for arbitrary wavelengths. The condition for validity of the SSA method lies in the smallness of the roughness slopes without any restrictions as to the magnitude of the Rayleigh parameter. If the angles of incidence or of the scattering are near grazing the validity condition is more restrictive - slopes must be less than the minimum of these angles (see Eq.17), so that shadowings are absent. It turns out that the lowest order SSA may be built in a universal manner on the basis of the solution of the scattering problems with the MSP. In this case the expression for the scattering amplitude splits into two factors: the dimensionless kernel $B(\mathbf{k},\mathbf{k}_0)$ (may be a matrix) which depends only on the boundary conditions, and some integral factor which is connected only with the shape of the undulations. Being of pure geometrical origin, the last takes into account distortions of an initially plane wave front which arise due to the roughness. This factorization occurs only in the lowest order of SSA and for the larger terms the shape of the surface and boundary condition enter in a more complex manner.

From this factorization follows one experimental criterion for smallness of surface slopes: the ratio of S for waves of different types (e.g. sound and electromagnetic waves) shouldn't depend on the roughness if the wave vectors of the incident and scattered waves coincide. This ratio should be equal to the quotient of coefficients B for these waves (see Eqs.16 and 21). Note that different polarizations of some vector waves may be considered as waves of different types.

It is possible to give the following qualitative characteristic of waves scattering at rough surfaces with small slopes: large-scale horizontal components of elevations, regardless of the kind of boundary conditions, only destroy the wave front and the "real" scattering occurs at the components of roughness with scales of the order of the wavelength (or shorter).

In the statistical case, the proposed theory enables us to obtain, with the help of the very simple transformation (Eq.22), the mean reflection coefficient in the SSA by the use of its value arising when the MSP is formally applied to the roughness.

For scattering at a pressure release surface, it may be shown (Voronovich, 1984) that if roughness consists of both types of elevation, Eq. (18a) for the scattering amplitude transforms into an appropriate formula arising in the two-scale model.

The SSA, in many respects, solves the problems of a unified description of wave scattering at rough surfaces in the framework of a single method and, in a sense, unifies the MSP and the TPA.

The proposal of smallness of roughness shouldn't be very restrictive from the point of view of unification because, in the general case for an arbitrary surface with slopes of unit order, in the high frequency limit, multiple scattering will arise and the TPA in its usual formulation becomes inadequate.

The numerical experiment supports the main assumption that the SSA is applicable independent of wavelength or scattering angle. The case of the saw-type roughness describes a situation which could not be investigated by means of the two classical approaches in the scattering theory.

Finally, we demonstrated how in the framework of the proposed theory the appropriate inverse problem can be solved. In this case the data for both the angular and frequency dependencies of the scattering coefficient are needed. In principle, we can formulate the problem of determining some characteristics of the roughness by employing Eq.(24) through just one of the aforementioned dependencies, but this problem has yet to be solved.

Of course for the further examination of the proposed theory, comparison of the SSA solutions with exact numerical solutions for other types of boundaries is very desirable.

ACKNOWLEDGEMENTS

The author is very grateful to L.M. Brekhovskikh for his stimulating interest in this work. I am very much obliged to A.I. Sukov for providing the numerical results. The author especially thanks V.V. Vavilova for her help in the English translation of the paper.

REFERENCES

Bahar, E., 1972a, Generalized Fourier transform for stratified media, Can. J. Phys., 50:3123.

Bahar, E., 1972b, Radiowave propagation in stratified media with non-uniform boundaries and varying electromagnetic parameters - full wave analysis, Can. J. Phys., 50:3131.

Bass, F.G., and Fuks, I.M., 1978, "Wave scattering from statistically rough surfaces.", Pergamon, New York.

Brekhovskikh, L, and Lysanov, Yu, 1982, "Fundamentals of ocean acoustics.", Springer-Verlag, New York.

Kur'yanov, B.F., Sov. Phys. Acoust., 8:252.

Vaynstein, L.A., and Sukov, A.I., 1984, Radiotek. Electron., 29:1472.

Voronovich, A.G., 1983a, Dok Akad. Nauk SSSR, 272:1351.

Voronovich, A.G., 1983b, Dok Akad. Nauk SSSR, 273:830.

Voronovich, A.G., 1984, Akust. Zh., 30:747.

Voronovich, A.G., 1985a, Pis'ma Zh. Eksp. Teor. Fiz., 89:116.

Voronovich, A.G., 1985b, Dok Akad. Nauk SSSR, 282:286.

Voronovich, A.G., 1986, Dok Akad. Nauk SSSR, 287:425.

Voronovich, A.G., and Sukov, I.A., 1985, in: "Wolny i Difraktzya 85", Tbilisi, 1:176.

VALIDITY OF THE BORN AND RYTOV APPROXIMATIONS

S. Leeman, P. Chandler°, L.A. Ferrari°, and D.A. Seggie*

King's College School of Medicine and Dentistry, Dept. of Med. Eng. and Phys., Dulwich Hospital, London SE22 8PT U.K.

°University of California Irvine, Dept. of Elec. Eng. Irvine,California 92717, U.S.A.

*University College London, Dept. of Phonetics and Linguistics, 4 Stephenson Way, London NW1 2HE, U.K.

ABSTRACT

The Born and Rytov approximations are well-known for the important roles they play in describing acoustic wave propagation for both the direct and inverse scattering problems. However, considerable dispute still exists as to their relative merits and validity domains. The approximations are re-examined here, and several results are demonstrated without recourse to extensive numerical computations of possibly limited generality.

INTRODUCTION

Most present acoustical imaging procedures are based on the following scenario. An acoustic wave, whose spacetime properties are in principle exactly measurable, is allowed to enter an object or scattering region, whose interaction with that wave are poorly known (if at all). The consequences of the object/wave interaction are recorded, either in the form of external measurements carried out on scattered acoustic waves, or as external measurements made of the modifications suffered by the input wave on its passage through the object. In many cases, the direct display of the recorded data does not result in an intelligible image, and the latter has to be recovered from the measurements by computational techniques. In this sense, much of modern acoustical imaging may be seen as an inverse scattering problem: an image of the object is to be recovered via an algorithm relating the output (measured) and input (known) wave fields. Clearly, "image" denotes a mapping of the wave/object interaction parameters, and it becomes obvious that effective inverse imaging cannot be divorced from physical modelling which aims to accurately specify the pertinent quantities [Leeman and Jones, 1984]. However, a problem of equal importance is the "computational model" whereby, given the framework of a physical model, an inverse imaging algorithm is devised.

In practice, computational models are checked by variations on the following stratagem (see, for example, Slaney et al., 1984). Given a relatively uncomplicated physical model, the *exact* scattered field from a very simple (usually two-dimensional, rotationally symmetric) object is laboriously computed, for the case of a simple input field, such as a continuous plane wave. The emphasis on simplicity is dictated by the extreme difficulty and computational intensity demanded by calculations of this type. This exact, computed, field is then utilised to provide the data set from which the object is recovered via a computational scheme which is, in essence, the inversion procedure under test. The success of the inversion algorithm is assessed by its ability to recover the original, input, object. Virtually all realistically tested computational schemes depend on the validity of either the Born or Rytov approximations (both to be discussed below), and it becomes a matter of some importance to be clear as to the general validity criteria for these two very widely used approximations.

It is, unfortunately, difficult to extrapolate from the results obtained in computer experiments of the type referred to above: these are limited to very simple, two-dimensional objects, and the generality of their results is not clear. However, at present, it does seem that the climate of opinion favours the Rytov over the Born approximation for applications in inverse acoustic imaging [Kaveh et al., 1982]. In the following, some elements of this question are approached via analytic techniques, which enable three-dimensional situations to be analysed, and which clarify the general validity of the conclusions arrived at. Although we have emphasised the inverse problem, it should be observed that a study of the Born and Rytov approximations is important also for the roles they play in describing the direct scattering and acoustic wave propagation problems.

BASIC EQUATIONS

We choose to work within the framework of a relatively simple physical model, viz: the inhomogeneous Helmholtz equation. This wave equation describes acoustic wave propagation in a velocity-inhomogeneous medium. Note that, although it is not explicitly demonstrated here, many of the general conclusions of this communication are not compromised when extensions to more sophisticated models are made.

Consider, therefore, a loss-less medium described by a fluctuating, spatially varying, (acoustic) velocity, $c(\underline{r})$, such that

$$c^2(\underline{r}) = c_0^2\{1 + n(\underline{r})\}^{-1}$$

with c_0 constant, and all velocity variations incorporated into $n(\underline{r})$. The function $n(\underline{r})$ denotes the scattering interaction, and a mapping of n constitutes an "image" of the object. In practice, objects with well-defined boundaries are generally investigated, and it is convenient to assume that $n(\underline{r})$ vanishes identically everywhere outside a region bounded by a surface denoted $B(\underline{r})$. For simplification, it is assumed that the scattering region enclosed within B is embedded in a uniform, loss-less medium with an acoustic velocity equal to c_0, but this is not an essential assumption, and by no means affects the validity of the results derived below. A linear acoustic wave, $\psi(\underline{r})$, of circular frequency ω, will propagate through the above medium according to the dictates of the inhomogeneous Helmholtz equation:

$$\nabla^2\psi(\underline{r}) + k^2\psi(\underline{r}) = -n(\underline{r})k^2\psi(\underline{r})$$

Throughout, $\underline{r}$ denotes the space location vector, and the constant, k, denotes the entity ω/c_0. For more general (linearly) scattering media, the interaction function $k^2n(\underline{r})$ is replaced by some linear operator which more closely embodies the scattering interaction. It is possible to express the Helmholtz equation in its integral formulation:

$$\psi(\underline{r}) = \psi_0(\underline{r}) + k^2\int d\underline{r}'G(\underline{r},\underline{r}')n(\underline{r}')\psi(\underline{r}')$$

where ψ_0 denotes the incident field (i.e. the wave that would exist in the absence of the velocity fluctuations), and G denotes the Green's function appropriate for the scattering problem,

$$G(\underline{r},\underline{r}') = \exp\{ik|\underline{r} - \underline{r}'|\}/4\pi|\underline{r} - \underline{r}'|$$

It is convenient also to define the kernel of the integral equation as

$$K(\underline{r},\underline{r}') \equiv k^2G(\underline{r},\underline{r}')n(\underline{r})$$

In an entirely symbolic way, the integral equation can be succinctly written as

$$\psi = \psi_0 + K\psi$$

with ψ and ψ_0 denoting the appropriate wave functions, and K denoting the kernel *operator*.

The underlying structure of the basic integral equation is now apparent, and it may clearly be solved by iteration to yield the so-called Born-Neumann expansion for the field:

$$\psi = \psi_0 + K\psi_0 + K^2\psi_0 + \ldots\ldots$$

This gives a valid solution, provided that the expansion converges Note the simplified and suggestive operator notation whereby

$$K^m \equiv K.K.K.K.\ldots\ldots..K \qquad (\text{ m factors})$$

To avoid any confusion about this symbolic operator notation, the expression K^2U, for example, is demonstrated in full

$$K^2U \equiv \int d\underline{r}'\int d\underline{r}''K(\underline{r},\underline{r}')K(\underline{r}',\underline{r}'')U(\underline{r}'')$$

THE BORN APPROXIMATION

The full series solution for the field ψ, indicated above, is extremely cumbersome to evaluate in any realistic case. In practice, therefore, the series is terminated, in order to give an approximate solution

$$\psi \simeq \psi_0 + K\psi_0 + \ldots. + K^N\psi_0$$

If the series is terminated after (N+1) terms, the approximation is called the Nth Born approximation. In particular, the first Born approximation is given by

$$\psi_B \equiv \psi_0 + K\psi_0$$

The first Born approximation ("1BA") is of some considerable interest, since the inverse problem can be solved exactly when it is valid [e.g. Leeman,1980]. Clearly, it is important to have some idea of whether the

1BA applies in any given imaging situation, or not. In order to address this problem, we point out that the 1BA cannot be expected to hold unless the full Born-Neumann expansion converges; it is this observation that is developed in the following.

Consider the general equation, symbolically written as

$$y = u + Hy$$

The series solution of this equation is found by iteration

$$y = u + Hu + H^2u + \ldots\ldots.$$

This solution exists provided that the series converges. If the assumption is made that the functions y and u are both square-integrable, i.e.

$$\int d\underline{r} |y(\underline{r})|^2 \equiv \P y \P < \infty$$

and $\P u \P < \infty$

then it is well known [Smithies, 1962] that the series solution for y converges if

$$\P H \P \equiv \int d\underline{r} \int d\underline{r}' |H(\underline{r},\underline{r}')|^2 < 1$$

The notation ¶..¶ denotes the so-called L^2 norm of the indicated function or kernel.

Unfortunately, for the Helmholtz equation considered here, there is no guarantee that the wave functions (formally) have a finite norm (consider, for example, the case that ψ_o is chosen to be a plane wave). Moreover, the form of the Green's function necessitates that ¶K¶ does not exist (the integrals defining the norm are easily seen to diverge). The way around this difficulty is to call upon the powerful symmetrisation technique developed by Scadron, Weinberg, and Wright [1964]. In this approach, the original equation

$$\psi = \psi_o + GV\psi \qquad \text{with } V \equiv k^2 n$$

is rewritten as

$$V^{1/2}\psi = V^{1/2}\psi_o + (V^{1/2}GV^{1/2})V^{1/2}\psi$$

More concisely,

$$\psi^* = \psi_o^* + K^*\psi^*$$

where the starred quantities have obvious meanings. We suggestively refer to this last equation as the Helmholtz* equation. It is readily confirmed that, because n is non-zero only within the finite region bounded by B, the norms $\P\psi^*\P$ and $\P\psi_o^*\P$ exist. Therefore, the series solution to the Helmholtz* equation,

$$\psi^* = \psi_o^* + K^*\psi_o^* + \ldots\ldots$$

will converge provided that $\P K^* \P < 1$. Assume provisionally that this condition is fulfilled.

The final stage in the argument is to note that convergence of the series solution to ψ^* is readily demonstrated to imply the convergence of the series solution to ψ (merely formally premultiply both sides of the equation by $V^{-1/2}$). Thus, it is concluded that the condition

$$\|K^*\| \equiv \|V^{1/2}GV^{1/2}\| < 1$$

is sufficient to ensure that the Born-Neumann expansion for the (scattering) solution to the Helmholtz equation will converge.
By direct substitution, it may be shown that

$$\|K^*\| = k^4\int d\underline{r}\int d\underline{r}'|n(\underline{r})|.|G(\underline{r},\underline{r}')|^2|n(\underline{r}')|$$

Some manipulation of this last expression leads to the following *sufficient* condition for the convergence of the series solution to ψ:

$$(1/4\pi)k^2\sup_r\int d\underline{r}'|n(\underline{r}')|.|\underline{r} - \underline{r}'|^{-1} < 1$$

where $\sup_r$ denotes that the least upper bound of the succeeding function be taken, as the variable $\underline{r}$ ranges over its entire domain.

Consider now a specific example, viz. the scattering of an incident plane wave by a uniform sphere of radius R. Let the acoustic velocity difference between the sphere and its surrounding medium be such that

$$|n(\underline{r})| = \Delta^2 \quad \text{for } |\underline{r}| \leq R$$
$$= 0 \quad \text{otherwise}$$

Computation of the sufficiency conditions derived above, lead to the conclusion that the convergence of the Born series is assured if

$$\tfrac{1}{2}\Delta kR < 1$$

In other words, if this condition is violated, the 1BA is unlikely to be a valid approximation. Under such circumstances, the utilisation of an inversion algorithm which assumes the validity of the 1BA would lead to unacceptably distorted images. This result is to be compared with the remarkably similar validity condition for the 1BA arrived at by Slaney, Kak and Larsen [1984], for a two-dimensional problem (scattering by a uniform disc), after tortuous and numerous computer experiments. Since the treatment presented here is perfectly general, three (or even more!) dimensional, readily applicable to quite complex scattering structures, and even capable of handling more sophisticated physical models - all without the need for recourse to further extensive computing regimes - the decisive advantage of the analytic approach is apparent in this case. For example, the above derivation makes clear that the condition $\tfrac{1}{2}\Delta kR<1$ applies also to the case of scattering from an arbitrarily shaped three-dimensional inhomogeneous region; then, Δ has to be interpreted as the maximum value of $|n(\underline{r})|^{1/2}$, and R has to be understood to be the radius of the smallest sphere that can encompass the inhomogeneity.

THE RYTOV APPROXIMATION

The Rytov approximation derives from a different approach towards solving the Helmholtz equation. First, the possible solution is written as

$$\psi(\underline{r}) = \psi_0(\underline{r})\exp\{\phi(\underline{r})\}$$

where ψ_0 denotes the same function as before, and ϕ is the "complex phase". By direct substitution into the Helmholtz equation, it may be verified that the above decomposition for ψ is permissible provided that the complex phase satisfies

$$\phi(\underline{r}) = \int d\underline{r}' G(\underline{r},\underline{r}')\{k^2 n(\underline{r}') + \nabla'\phi(\underline{r}')\cdot\nabla'\phi(\underline{r}')\}\psi_0(\underline{r}')/\psi_0(\underline{r})$$

This is a complicated, non-linear integral equation, and the Rytov approach obviates the embarrassment of having to find a solution, by the simple expedient of dropping the non-linear terms. Thus, the Rytov approximation ("RA") consists of writing the complex phase as

$$\begin{aligned}\phi(\underline{r}) &\simeq \int d\underline{r}' G(\underline{r},\underline{r}')k^2 n(\underline{r}')\psi_0(\underline{r}')/\psi_0(\underline{r}) \\ &\equiv \phi_R(\underline{r}) \\ &= \psi_B(\underline{r})/\psi_0(\underline{r})\end{aligned}$$

Thus, in the RA, the wave field is given by

$$\psi_R(\underline{r}) = \psi_0(\underline{r})\exp\{\psi_B(\underline{r})/\psi_0(\underline{r})\}$$

This expression forms the basis for the common statement that, since the RA is an exponentiated 1BA, it may be expected that higher order Rytov terms are given by exponentiating the Born series.

Occasionally, the validity condition for the Rytov approximation is essentially expressed as

$$|\nabla\phi(\underline{r})|^2 \ll k^2|n(\underline{r})|$$

It is clear that this condition must hold over the entire range of integration in the exact expression for the complex phase indicated above, i.e. over all space. However, in practice, the object will generally have a scattering distribution, $n(\underline{r})$, which fluctuates through zero, taking on both positive and negative values, inside the object boundary, B. Outside the bounding surface n will everywhere be identically equal to zero. Thus at many possible locations inside B, and everywhere outside B, $n = 0$, and the above condition fails, since it would demand that the positive-definite quantity, $|\nabla\phi|^2$, be very much smaller than zero!

The validity of the Rytov approximation should thus be correctly inferred from the relative magnitudes of the two terms contributing to the complex phase. That is, the RA will be valid provided that

$$\left|\int d\underline{r}' G(\underline{r},\underline{r}')[\nabla'\phi(\underline{r}')]^2\psi_0(\underline{r}')\right| \ll \left|\int d\underline{r}' G(\underline{r},\underline{r}')k^2 n(\underline{r}')\psi_0(\underline{r}')\right|$$

This inequality is difficult to simplify or manipulate. However, an interesting observation may be made. The left-hand-side of the inequality represents the 1BA scattering from an effective "object", $[\nabla\phi]^2$, while the right-hand-side represents the 1BA scattering from the actual object, n. The effective "object" is extended over all space (even though its values may diminish somewhat with distance from the true scattering distribution), but the true object is bounded, in practice. The validity condition for the Rytov approximation may thus be interpreted in terms of the relative magnitudes of the 1BA scattering from the two types of objects. This statement holds whether or not the 1BA is indeed a good approximation to the scattered field.

It is clear that the RA may be expected to be poor when the 1BA to the scattering from the true object is not too strong, while the gradients of the complex phase ($|\nabla\phi|$) take on large local values, and/or have significant values over much of space. One example would be scattering from sharp (on a wavelength scale) corners of the object, where local values of $|\nabla\phi|$ would be expected to become large, even if the overall strength of scattering, as measured by the 1BA, remains small. This prediction is borne out by the (two-dimensional) numerical results of Zapalowski et al. [1986], who compared the accuracy of the 1BA and RA for two dimensional scattering from equal area discs and squares of the same scattering strength. They showed that the RA could be more accurate than the 1BA for scattering from a disc, while the relative merits of the two approximations were reversed for scattering from the equivalent square. This somewhat unexpected result, viz. that object geometry (rather than only size or scattering strength) may also play a role in the applicability of the two approximations, is rendered more plausible in the light of the above discussion.

CONCLUSIONS

It has been demonstrated that sufficient conditions for the convergence of the Born-Neumann expansion for the scattered field may be derived analytically, and that the resultant convergence criterion sheds some interesting light on the validity domain of the 1BA. In this way the limited generality and other drawbacks of laborious computer based numerical experiments may be overcome. In particular, three-dimensional scattering and more complex scattering interactions may be tackled.

Mathematically meaningful validity conditions for the RA may be stated, but are difficult to simplify. However, they do allow for an interesting physical interpretation, which leads to the prediction that geometric considerations (i.e. shape, rather than only size and scattering strength) of the scattering object may be generally involved when comparing the 1BA and RA.

It is certainly clear that the relative merits of the Born and Rytov approximations will continue to be investigated for some time yet.

REFERENCES

Kaveh, M., Soumekh, M., and Mueller, R.K., 1982, Acoustical Imaging 11: (Plenum Press, N.Y.)

Leeman, S., 1980, Acoustical Imaging 9: 513 (Plenum Press, N.Y.)

Leeman, S., and Jones, J.P., 1984, Acoustical Imaging 13: 233 (Plenum Press, N.Y.)

Scadron, M., Weinberg, S., and Wright, J., 1964, Phys. Rev., 135 B: 202

Slaney, M., Kak, A.C., and Larsen, L.E., 1984, IEEE Trans. on Microwave Theory and Techniques, MTT-32, #8: 860

Smithies, F., 1962, "Integral Equations", University Press, Cambridge

Zapalowski, L., Leeman, S., and Fiddy, M.A., 1986, Acoustical Imaging 14: 295 (Plenum Press, N.Y.)

MULTIPLE SCATTERING AT ROUGH OCEAN BOUNDARIES

John A. DeSanto

Center for Wave Phenomena, Mathematics Department
Colorado School of Mines
Golden, CO 80401

INTRODUCTION

This is a brief review paper on several recent theoretical approaches to rough surface scattering with the emphasis on multiple scattering. The limitations of single scattering theories are generally well understood. On this basis, it is our belief that future research on rough surface scattering should concentrate on multiple scattering methods. This includes both formal development of rigorous theories of scattering as well as multiple scattering approximation methods to make the whole development useful. Much of the material we present here can be found further developed in a forthcoming review paper by DeSanto and Brown (1986). Single scattering ideas are also developed in this paper and it is shown how the techniques used in the latter development have a natural extension to multiple scattering theories.

CONNECTED DIAGRAM EXPANSION

This was a method developed by us (Zipfel and DeSanto, 1972; DeSanto, 1973, 1974, 1981a) for random surfaces. We used Fourier transform techniques and Green's theorem to generate exact stochastic equations in transform or $\underset{\sim}{k}$-space. The results generalized the surface diagram results in Bass and Fuks (1979) which were defined using perturbation theory in powers of the surface height. The latter were also presented in coordinate-space. The basic result was that for a stochastic surface $h = h(x,y)$ the scattered part of the Green's function, G, can be written as the solution of a Lippmann-Schwinger integral equation

$$G(\underset{\sim}{k}',\underset{\sim}{k}'') = V(\underset{\sim}{k}',\underset{\sim}{k}'')A(\underset{\sim}{k}'-\underset{\sim}{k}'') + \iiint V(\underset{\sim}{k}',\underset{\sim}{k})A(\underset{\sim}{k}'-\underset{\sim}{k})G^{0}(k)G(\underset{\sim}{k},\underset{\sim}{k}'')d\underset{\sim}{k} \qquad (1)$$

in terms of Fourier transform variables which can be interpreted as incident ($\underset{\sim}{k}''$), final ($\underset{\sim}{k}'$) and intermediate ($\underset{\sim}{k}$) directions of propagation. The latter are integrated over to yield a full wave theory including all orders of multiple scattering. The terms in the equation are the vertex function

$$V(\underset{\sim}{k}',\underset{\sim}{k}) = \frac{-2i}{(2\pi)^3}\left[\frac{k_t'\cdot(k_t'-k_t)}{k_z'-k_z} + K'^2 P\left(\frac{1}{k_z'}\right)\right] , \qquad (2)$$

See page 809 for Abstract.

with the notation $\underset{\sim}{k} = (k_x, k_y, k_z) = (k_t, k_z)$, $K = (k_0^2 - k_t^2)^{1/2}$, k_0 is a reference wavenumber, and P represents the Cauchy Principal value distribution. V can be interpreted as a kinematical factor in the scattering process. We also have the Fourier transform of the free-space Green's function

$$G^0(k) = (k^2 - k_0^2)^{-1} \quad , \tag{3}$$

and the dynamical term containing the surface variability

$$A(\underset{\sim}{k}) = \iint \exp(-i\underset{\sim}{k}\cdot\underset{\sim}{x}_s)\, dxdy \quad , \tag{4}$$

where $\underset{\sim}{x}_s = (x,y,h)$ is a point on the random surface $h(x,y)$. The result is for a perfectly reflecting surface but it can be generalized to include an interface (DeSanto, 1983) or an impedance type boundary (DeSanto, 1985a).

A formal solution of Eq. (1) can be found as a Born expansion in powers of VA. Each term can be associated with a diagram, analogous to the diagrams introduced in random volume scattering theory (Frisch, 1968). The resulting series can be formally averaged and resummed in a particular way ("partial summation"). The resulting diagram interpretation is in terms of "connected" diagrams which are equivalent to the cluster decomposition techniques in statistical mechanics (Huang, 1963). Symbolically we can write (1) as

$$G = VA + LAG \quad , \tag{5}$$

where

$$L = \iiint VG^0 d\underset{\sim}{k} \quad . \tag{6}$$

This has a formal solution given by

$$G = \sum_{n=0}^{\infty} (LA)^n VA \quad . \tag{7}$$

Its ensemble average yields the coherent field. The average of any product of G with itself (for example, the product $G\,G^*$ is related to the incoherent field) can be determined once the averages of products of the A-functions are known. These can all be formally computed. The first moment for homogeneous Gaussian height statistics is

$$E\,A(\underset{\sim}{k}_1) = A_1(\underset{\sim}{k}_1) = (2\pi)^2 \delta(k_{1t})\tilde{p}(k_{1z}) \quad , \tag{8}$$

where $\tilde{p}$ is the Fourier transform of the probability density function (σ = rms height)

$$\tilde{p}(k_z) = \exp(-\sigma^2 k_z^2/2) \quad , \tag{9}$$

and E is the averaging operator. The second moment, A_2, is found via a cluster decomposition

$$E\,A(\underset{\sim}{k}_1)A(\underset{\sim}{k}_2) = A_1(\underset{\sim}{k}_1)A_1(\underset{\sim}{k}_2) + A_2(\underset{\sim}{k}_1,\underset{\sim}{k}_2) \tag{10}$$

where

$$A_2(\underset{\sim}{k}_1,\underset{\sim}{k}_2) = (2\pi)^2\delta(k_{1t} + k_{2t})\tilde{p}(k_{1z})\tilde{p}(k_{2z})R_2(\underset{\sim}{k}_1,k_{2z}) \quad , \tag{11}$$

with

$$R_2(\underset{\sim}{k}_1,k_{2z}) = \iint d\rho_t \exp(-ik_{1t}\cdot\rho_t) \left[\exp\left[-\sigma k_{1z}k_{2z}C(\rho_t) \right]-1\right] \quad . \tag{12}$$

Here $C(\rho_t)$ is the surface correlation function. The result, symbolically, is an integral equation

$$EG = M + \iiint MG^{0}EG\, d\underset{\sim}{k} \quad , \tag{13}$$

where the "mass operator" M is defined by

$$M = \sum_{j=1}^{\infty} M_j = \sum_{j=1}^{\infty} L^{j-1}\, VA_j \quad , \tag{14}$$

as a sum of "connected" or cluster decomposed terms A_j. The integral equation can be shown to be one-dimensional, and solved explicitly when $M \approx M_1$. The result is a larger coherent scatter return than single scatter theories predict (DeSanto, 1981a), and better agreement with data for the reflection coefficient EG. Other approximations are possible by truncating the series for the mass operator, and higher order moments such as EGG^* can be both formally and approximately developed.

SMOOTHING

The method of smoothing was originally developed to treat random equations in coordinate-space (Keller, 1962) but will be applied here in transform space. G is first decomposed into a mean or coherent term plus a fluctuating part

$$G = EG + \delta G \quad , \tag{15}$$

where $E\delta G = 0$. Substituting Eq. (15) into Eq. (5) we get, after some manipulation, an equation on only the coherent part given by

$$EG = M^{s} + LEA \sum_{n=0}^{\infty} (LA-LEA)^{n}\, EG \tag{16}$$

in terms of the smoothing mass operator term

$$M^{s} = VEA + LEA \sum_{n=0}^{\infty} (LA-LEA)^{n} V(A-EA) \tag{17}$$

where the interpretation on E is that it acts on all random quantities occurring to its right. Comparing the first two terms in the smoothing mass operator to those of the connected diagram mass operator we see that using Eq. (14) and Eq. (10)

$$M_1^{s} = VEA = VA_1 = M_1 \quad , \tag{18}$$

and

$$M_2^s = LEAV(A-EA) = M_2 \quad . \tag{19}$$

The two terms are exactly equal. This does not continue however, and we have shown (DeSanto, 1986) that the two methods differ in third and higher orders. As a practical matter, no one considers these higher order terms, so that in transform space the lowest order approximations are the same. The smoothing technique has recently been applied to electromagnetic problems (Brown, 1984) and to scalar problems with several different boundary conditions (Watson and Keller, 1984).

SPECTRAL METHOD

This was a method we developed (DeSanto, 1985b), using analogies with earlier work on periodic surfaces (DeSanto, 1981b). The method works for an interface between media of different densities (ρ_1 and ρ_2 with $\rho = \rho_2/\rho_1$) and different sound speeds or wavenumbers (k_1 and k_2 with $K = k_2/k_1$). We illustrate the method with a one-dimensional surface h(x) which can be stochastic. Above the surface, region 1 ($z > h(x)$), the field ϕ_1 (pressure, velocity potential, etc.) satisfies a Helmholtz equation with wavenumber k_1. Below the surface, region 2 ($z < h$), the field ϕ_2 satisfies a Helmholtz equation with wavenumber k_2. Above the highest surface excursion in region 1 (region A: $z > \max(h)$) we can write an exact spectral representation ϕ_A consisting of incident and scattered terms as

$$\phi_A(x,z) = \phi^{in}(x,z) + \phi^{sc}(x,z) \quad . \tag{20}$$

with

$$\phi^{in}(x,z) = D \exp\left[ik_1(\alpha x-\beta z)\right] \tag{21}$$

and

$$\phi^{sc}(x,z) = \int_{-\infty}^{\infty} A(\mu) \exp\left[ik_1(\mu x + mz)\right]d\mu \tag{22}$$

with α and β the sine and cosine of the incident angle (for this plane wave) and $\mu^2 + m^2 = 1$ with $Re(m), \geq 0$ and $Im(m) \geq 0$. The amplitudes A are unknown.

Below the lowest surface excursion in region 2 (region B, $z \leq \min(h)$) the transmitted field is

$$\phi_B(x,z) = \int_{-\infty}^{\infty} B(p) \exp\left[ik_2(px-qz)\right]dp \quad . \tag{23}$$

with $p^2 + q^2 = 1$ and $Re(q) \geq 0$ and $Im(q) \geq 0$. The B-amplitudes are also unknown. If we now use Green's theorem on ϕ_1 and the auxiliary functions

$$G^{\pm}(x,z) = \exp\left[ik_1(\pm m'z-\mu'x)\right] \tag{24}$$

in region 1, and ϕ_2 with the functions

$$W^{\pm}(x,z) = \exp\left[ik_2(\pm q'z-p'x)\right] \tag{25}$$

in region 2, and use the continuity conditions of pressure and normal velocity at the interface we can derive exact representations for the amplitudes A and B in terms of the boundary values of the field, say F(x), and a term proportional to the normal derivative on the boundary, N(x). For example, for $A(\mu)$ we get

$$A(\mu) = (k_1/4\pi m)\int_{-\infty}^{\infty}\left[(m-\mu h'(x))F(x) + N(x)\right]\cdot$$

$$\cdot \exp\left[-ik_1(mh(x) + \mu x)\right]dx \quad , \tag{26}$$

with an analogous equation for B (DeSanto, 1985b). Two auxiliary equations are used to find the boundary conditions. The resulting equations for A and B are exact representations for these stochastic amplitudes which can be used to find field moments, for example the coherent amplitude $EA(\mu)$ or its second moment $EA(\mu)A^*(\mu')$. Approximate values of F and N can also be found and tested against an energy conservation result also derived using Green's theorem. It is

$$\rho \int dp\,|\beta(p)|^2 \mathrm{Re}(q) + \int d\mu\,|A(\mu)|^2 \mathrm{Re}(m) = \beta \quad . \tag{27}$$

These results are formally exact and approximation methods are presently being explored. The extension to two-dimensional surfaces is also possible.

STOCHASTIC FOURIER TRANSFORM

This is a method developed by Brown (1982). We give only a brief discussion of it here and refer to DeSanto and Brown (1986) for a more thorough review. We contrast it with our previous Fourier transform ideas in the connected diagram approach. The latter was a standard Fourier approach, with the Fourier or $\underset{\sim}{k}$-variables conjugate to the coordinate-space or $\underset{\sim}{x}$-variables. Only the k_z-variable was conjugate to a stochastic quantity h, and the equations we derived from Eq. (1) to Eq. (7) were valid for a deterministic surface as well as a stochastic one.

In the Stochastic Fourier transform approach the $\underset{\sim}{k}$-variables are conjugate to the stochastic quantities of the surface, the height, slopes, curavatures, etc. Since this is so, the method applies only to random surface scattering. In addition the $\underset{\sim}{k}$ can be infinite-dimensional rather than three-dimensional in the standard approach, so in order to be tractable, the important stochastic dependence must be isolated.

As a simple example we assume we have an integral equation of the form

$$\phi(\underset{\sim}{x}_s') = \phi^{in}(\underset{\sim}{x}_s') + \iint K(\underset{\sim}{x}_s',\underset{\sim}{x}_s)\phi(\underset{\sim}{x}_s)dx_t \quad . \tag{28}$$

The function ϕ is a function of all the stochastic variables, h, $\partial_t h$, $\partial_t^2 h$, etc. In order to find its average, $E\phi$, we multiply by the single point joint probability density function

$$p_1(h, \partial_t h, \partial_t^2 h, \ldots)$$

and integrate over all the stochastic variables. To average the term $K\phi$ we must multiply by the two point joint probability density function

$$p_2(h, h', \partial_t h, \partial_t h', \partial_t^2 h, \partial_t^2 h', \ldots)$$

where $h' = h(x', y')$, etc.

Next write each integrated term as the convolution of products of Fourier transforms at zero separation. The Fourier transform variables for the $E\phi$ term are conjugate to the terms h, $\partial_t h$, ∂_t^2, etc. The resulting integral equation is infinite-dimensional and must be truncated using a physical approximation related to the importance of the various stochastic terms. For example if the surface height is the only important variable then there results a one-dimensional equation which is solvable.

SUMMARY

All the methods discussed are exact; that is they model the scattering process completely including all orders of multiple scattering. The major research effort in all of them is to find tractable approximation. This is presently being pursued by several groups.

REFERENCES

Bass, F.G. and Fuks, I.M., 1979, "Scattering of Waves from StatisticallyIrregular Surfaces," Pergamon, New York.

Brown, G.S., 1982, A stochastic Fourier transform approach to scattering from perfectly conducting randomly rough surfaces, **IEEE Trans. AP**, 30:1135.

Brown, G.S., 1984, Application of the integral equation method of smoothing to random surface scattering, **IEEE Trans. AP**, 32:1308.

Brown, G.S., 1985, Simplifications in the stochastic Fourier transform approach to random surface scattering, **IEEE Trans. AP**, 33:48.

DeSanto, J.A.., 1973, Scattering from a random rough surface: diagram methods for elastic media, **J. Math. Phys.**, 14:1566.

DeSanto, J.A., 1974, Green's function for electromagnetic scattering from a random rough surface, **J. Math. Phys.**, 15:283.

DeSanto, J.A., 1981a, Coherent multiple scattering from rough surfaces, in: "Multiple Scattering and Waves In Random Media," P.L. Chow, W. Kohler and G.C. Papanicolaou, eds., North-Holland, Amsterdam.

DeSanto, J.A., 1981b, Scattering from a perfectly reflecting arbitrary periodic surface: An exact theory, **Radio Sci.**, 16:1315.

DeSanto, J.A., 1983, Scattering of scalar waves from a rough interface using a single integral equation, **Wave Motion**, 5:125.

DeSanto, J.A., 1985a, Impedance at a rough waveguide boundary, **Wave Motion**, 7:307.

DeSanto, J.A., 1985b, Exact spectral formalism for rough surface scattering, **J. Opt. Soc. Am.**, A2:2202.

DeSanto, J.A., 1986, Relation between the connected diagram and smoothing methods for rough surface scattering, **J. Math. Phys.**, 27:377.

DeSanto, J.A. and Brown, G.S., 1986, Analytical techniques for multiple scattering from rough surfaces, in: "Progress In Optics, " vol. 23, E. Wolf, ed., North-Holland, Amsterdam.

Frisch, V., 1968, Wave propagation in random media, in: "Probabilistic Methods In Applied Mathematics, I," A.T. Bharucha-Reid, ed., Academic, New York.

Huang, K., 1963, "Statistical Mechanics," Wiley, New York.

Keller, J.B., 1962, Wave propagation in random media, in: Proc. Sym. Appl. Math 13:227, American Mathematical Society, Providence.

Watson, J.G. and Keller, J.B., 1984, Rough surface scattering via the smoothing method, **J. Acoust. Soc. Am.**, 75:1705.

Zipfel, G.G. and DeSanto, J.A., 1972, Scattering of a scalar wave from a random rough surface: a diagrammatic approach, **J. Math. Phys.**, 13:1903.

ROUGH SURFACE SCATTERING AND THE KIRCHHOFF APPROXIMATION

Diana F. McCammon and Suzanne T. McDaniel

Applied Research Laboratory
The Pennsylvania State University
P.O. Box 30, State College
State College, PA 16804

ABSTRACT

Most approximate solutions to surface scattering begin with the Kirchhoff approximation to the unknown surface field. The validity of the Kirchhoff approximation is tested in this paper by the use of exact solutions for scattering from a pressure release sinusoid. Rough surfaces are expressed in a Fourier Series, and series solutions for small Rayleigh parameter are derived. The first order series term of Holford, Uretsky and Rayleigh are shown to be identical. Surface radiation patterns for shallow grazing angles are examined and conclusions are drawn on the region of validity of the Kirchhoff approximation.

INTRODUCTION

The Kirchhoff approximation in pressure release scattering theory consists of assuming that the unknown normal derivative of the pressure on the surface of the scatterer can be approximated by twice the incident field in the illuminated regions and zero in the shadowed regions. In this paper we will derive expressions for the plane wave reflection coefficient from a pressure release periodic surface with a Rayleigh roughness parameter less than unity. By comparison between theories, we will find that the Kirchhoff approximation is valid for reflected orders near specular or for high angles of incidence or low RMS slope surfaces. We will also show that three theoretically exact approaches, those of Holford, Uretsky and Rayleigh, all produce the same first order term for the reflection cofficient for small Rayleigh roughness.

THEORETICAL ANALYSIS

This analysis of scattering from a two-dimensional pressure release surface $\xi(x)$ begins with the Helmholtz integral equation expressed in the form of a second kind Fredholm equation [1]

$$\psi(x) = 2\psi(x)_{inc} - \int_{-\infty}^{\infty} \psi(x')K(x'-x,x)\,dx', \tag{1}$$

where $\psi(x)$ is the particle velocity and $K(x'-x,x')$ is the derivative of free-space Green's function

$$K(\tau,x) = \frac{ik}{2}\ \frac{H_1^{(1)}(k\rho)}{\rho}\ [\xi(x+\tau)-\xi(x)-\tau\xi'(x)],$$

with $\rho = (\tau^2+[\xi(x+\tau)-\xi(x)]^2)^{1/2}$, and $\tau = x'-x$.

The solution to Eq. (1), given by Holford [2] utilizes the method of moments. It assumes the velocity on the periodic surface is $\psi(x) = \sum\psi_n e^{ik\alpha_n x}$, where $\alpha_n = \cos\theta_0+nK_0/k$ is the direction cosine of the Bragg reflected order, $K_0 = 2\pi/\Lambda_0$ is the surface fundamental wave number, $\xi(x) = \sum h_n e^{inK_0 x}$, h_n is the surface waveheight spectrum. Substitution of this series and the application of the operator

$$\frac{1}{\Lambda_o}\int_0^{\Lambda_o} e^{-ik\alpha_m x}dx, \tag{2}$$

results in a system of equations for ψ_n

$$\psi_n = (I+V_{n,m})^{-1}2\psi_{m_{inc}}, \tag{3}$$

$$V_{n,m} = \frac{1}{\Lambda_o}\int_0^{\Lambda_o} e^{-i(n-m)K_0 x}\,dx \int_{-\infty}^{\infty} e^{ik\alpha_m\tau}\,K(\tau,x)d\tau.$$

This can be expanded to $\psi_n = (I-V_{n,m}+\ldots)2\psi_{m_{inc}}$,

if $V_{n,m} < 1$. This equation gives the Fourier coefficients of the unknown particle velocity in terms of a series developed about the known incident velocity coefficients. The first term is twice the incident velocity coefficient (Kirchhoff approximation to the unknown field). The second term gives a correction to Kirchhoff. The plane wave reflection coefficient is obtained from a convolution with the Fourier transform of the surface characteristic.

$$R_q = \frac{1}{2\gamma_q}\sum_n \psi_n\, C_{q-n}(k\gamma_q), \tag{4}$$

$$C_j(t) = \frac{1}{\Lambda_o}\int_0^{\Lambda_o} e^{-it\xi(x)-ijK_0 x}\,dx.$$

For this analysis, we will assume the Rayleigh roughness parameter $k\gamma_q h_{RMS}$ is sufficiently smaller than unity that we can expand $e^{-it\xi(x)}$ as $C_j(k\gamma_q) = \delta_j - ik\gamma_q h_j + \ldots$. The incident velocity coefficients are given by Eq. (2) on $\psi(x)_{inc} = -[\gamma_0 + \alpha_0\xi'(x)]e^{ik[\alpha_0 x-\gamma_0\xi(x)]}$. For small Rayleigh parameter, this becomes to first order

$$\psi_{m_{inc}} = -\gamma_0\delta_m + ih_m(k\gamma_0^2-m\alpha_0 K_0) -\ldots . \tag{5}$$

Similarly, if we assume $\rho\approx\tau$ in the kernel of Eqn (3)

$$V_{n,m} = ih_{n-m}\left[\frac{(n-m)K_0\alpha_m}{\gamma_m} + k(\gamma_n-\gamma_m)\right], \tag{6}$$

Eqs. (5) and (6) give an approximation for the unknown velocity series

$$\psi_n = 2\psi_{n_{inc}} + 2i\gamma_0 h_n\left[\frac{nK_0\alpha_0}{\gamma_0} + k(\gamma_n-\gamma_0)\right]. \tag{7}$$

From these we obtain the reflection coefficients

$$R_q^{KIRC} = -\delta_q + ih_q \left(k\gamma_o\left[1+\frac{\gamma_o}{\gamma_q}\right] - q\,\frac{\alpha_o K_o}{\gamma_q}\right), \quad (8)$$

$$R_q^{EX} = R_{qK}^{KIRC} + ih_q \left(k\gamma_o\left[1-\frac{\gamma_o}{\gamma_q}\right] + q\,\frac{\alpha_o K_o}{\gamma_q}\right), \quad (9)$$

$$= -\delta_q + 2ik\gamma_o h_q. \quad (10)$$

Note that most of the first order correction term in the exact solution cancels the Kirchhoff approximation, leaving a simple expression that depends on the frequency, incident angle and wave height spectrum. We can obtain this same simple expression for plane wave reflection coefficients using theoretical approaches of Rayleigh and Uretsky. In Rayleigh's method[3], we equate the incident pressure to a series of reflected waves

$$e^{-ik\gamma_o \xi(x)} = -\sum A_m\, e^{ik\gamma_m \xi(x)}\, e^{imK_o x}\,. \quad (11)$$

Using perturbation methods, let $A_m = A_m{}^\circ + A_m{}^1 + \ldots$ and expand the exponentials for small Rayleigh parameter

$$1 - ik\gamma_o \xi(x) = -\sum (A_m{}^\circ + A_m{}^1)(1+ik\gamma_m \xi(x))\, e^{imK_o x}.$$

Then $A_m{}^\circ = -\delta_m$ and $-2ik\gamma_o\xi(x) = -\sum A_m{}^1\, e^{imK_o x}$, from which we get $A_m{}^1 = 2ik\gamma_o h_m$, the same as Eq. (10).

The solution of Uretsky [4] uses the method of moments technique on the Helmholtz integral equation expressed as a first kind Fredholm equation. This produces a system of equations for ψ_n as in Eq. (3)

$$\psi_n{}^u = (V^u{}_{n,m})^{-1}\, 2C_m(k\gamma_o), \quad (12)$$

where the kernel for $V^u{}_{n,m}$ is $K^u(\tau,x) = kH_o{}^{(1)}(k\rho)/2$. Also assuming $\rho \approx \tau$, $V^u{}_{m,n} = \delta_{n-m}/\gamma_m$. Thus $\psi^u{}_n = -2\gamma_n C_n(k\gamma_o)$, and $R^u{}_q = -\delta_q + 2ik\gamma_o h_q$.

NUMERICAL COMPARISON

We have shown that three different full wave solutions to pressure release scattering all produce identical first order reflection coefficients, dependent only on the wave height spectrum, frequency and incident angle while the Kirchhoff approximation contains an extra term related to the surface slope. Figure 1 displays scattering strength defined as $10 \log (\pi k\gamma_q R_q{}^2/K_o)$ off specular and $20 \log R_o$ at specular as a function of reflected angle. The Pierson [5] wavenumber spectrum was used with a windspeed of 13 kts, an RMS wave height of 0.26 m, and a fundamental wavenumber of $K_o = 0.01434$. The acoustic frequency was 110 Hz, the grazing angle was 5°, and the Rayleigh roughness parameter was .01. A random phase was included in the surface wave spectral amplitudes. The irregular solid line is a full solution to Eq. (3) using standard matrix inversion techniques with $V_{n,m}$ represented by a 128x128 complex matrix. The smooth solid line is the first order approximation $-\delta_q + 2ik\gamma_o h_q$ where h_q is the waveheight spectral amplitude at wavenumber qK_o. The dashed line is the first order term from the Kirchhoff approximaton. It is apparent that for backscattering greater than 30°, the Kirchhoff approximation is badly in error, with the error being chiefly caused by the slope term $q\alpha_o K_o/\gamma_q$ that grows as the back scattered order q increases.

This derivation is intended to contrast a full wave solution with the popular Kirchhoff approximation to demonstrate the region of validity of

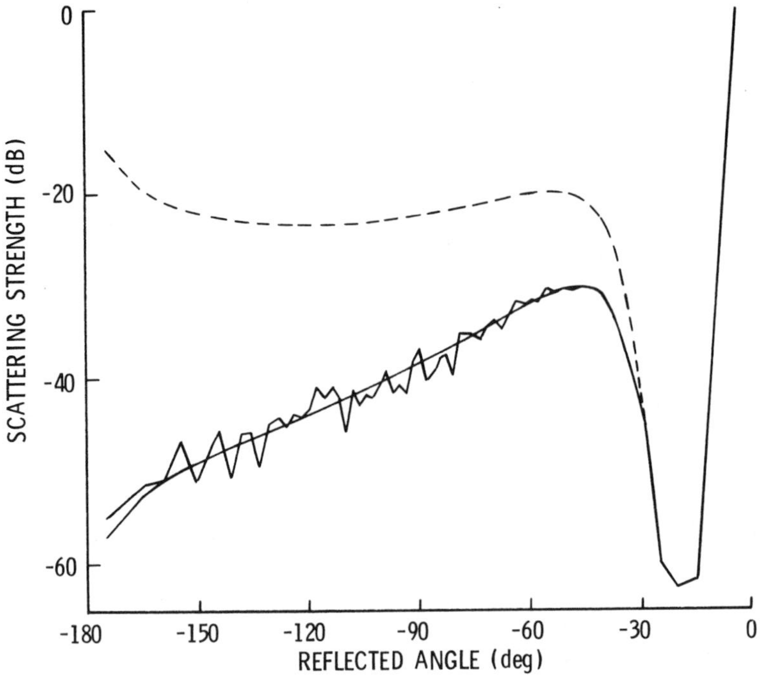

Fig. 1. Scattering strength vs backscattered angle.

the latter. In this formulation it is easy to see that the Kirchhoff term will become dominant when $V_{n,m}$ becomes small with respect to the identity matrix, that is for small slopes ($qK_0 \to 0$) or high angles ($\alpha_0 \to 0$). In addition, for any choice of surfaces, Kirchhoff will always give good agreement near the specular order since here $q \approx 0$ and $R_q^{EX} \to R_q^{KIRC}$.

It is interesting to observe that the Eckart scattering formulation employs the Kirchhoff approximation to the surface velocity and in addition, approximates the normal derivitive by the z derivitive, thereby discarding the slope term. Thus Eckart's formulation is actually improvement over Kirchhoff.

SUMMARY

We have demonstrated that the full wave scattering formulations of Holford and Uretsky that begin with the Helmholtz scattering integral can lead to a simple expression for the reflection coefficient when the Rayleigh roughness parameter is small. This same expression can also be found from the theoretical methods used by Lord Rayleigh. The Kirchhoff approximation, however, only approaches this expression for reflection orders near specular, and in general will differ greatly from the full wave solutions unless the surface slope is small or the grazing angle is near 90°.

ACKNOWLEDGEMENT

This work was sponsored by the EVA support for the Shipboard Sonar Program at Naval Underwater Systems Center, New London, Conn.

REFERENCES

1. W. C. Meecham, "On the Use of the Kirchhoff Approximation for the Solution of Reflection Problems," J. Ration. Mech. Anal. 5, 323-333 (1956).

2. R. L. Holford, "Scattering of Sound Waves at a Periodic Pressure-Release Surface: An Exact Solution," J. Acoust. Soc. Am. 70, 1116-1128 (1981).

3. Lord Rayleigh, Theory of Sound, Dover Publications, New York, 2, 89-96 (1945).

4. J. L. Uretsky, "The Scattering of Plane Waves From Periodic Surfaces," Amn. Phys. 33, 400-427 (1965).

5. W. T. Pierson, "The Theory and applications of ocean wave measuring system at and below the sea surface, on the land, from aircraft and from spacecraft," NASA Contract Rep CR-2646 NASA, Washington, DC, 305-308 (1976).

HIGH FREQUENCY SCATTERING FROM ROUGH BOTTOMS AND THE SECOND BORN APPROXIMATION

Stanley A. Chin-Bing and Michael F. Werby

Naval Ocean Research and Development Activity
Numerical Modeling Division
NSTL, Mississippi 39529-5004, USA

ABSTRACT

High frequency scattering from rough ocean bottoms has been treated with some success using the first Born approximation. For various grazing angles, many rough bottom descriptions require the inclusion of secondary scattering effects which the first Born approximation does not include. However, if only one secondary scattering event is important, the second Born approximation can be used effectively for prediction purposes. Calculations from a representative rough ocean bottom (a sinusoid) for a variety of frequencies are presented and results from the first and second Born approximations are compared.

INTRODUCTION

The scattered field from rough interfaces is described mathematically by the Hemholtz equation with appropriate boundary conditions. In a fluid the equation is

$$(\nabla^2 + k^2) P = 0 \tag{1}$$

where the total field is P and the scattered field, P_s, is the difference between the total field and the incident field P_i, (i.e., $P_s = P - P_i$). The determination of P depends crucially on the type of boundary including topography as well as material constituency. If the bottom is assumed penetrable then one must solve the appropriate equation at each side of the interface and equate relevant bouandary conditions. This could become quite complicated particularly if one side of the interface involves a layered medium, inclusions or an elastic medium. For the high frequency case one can reduce this problem to a simpler form by assuming that the interface is impenetrable (sound soft for an air/water interface and sound hard for a water/material interface where the material is clearly defined and corresponds to an abrupt change) and thus only topographic considerations remain important. Indeed, one could employ rough interface scattering in such a frequency domain as a tool to determine topographic features. There are many considerations that enter into rough interface scattering. That which we seek to address here involves the influence of topographic features in rough interface scattering at suitably high frequencies such that the above simplifying assumptions can be made. In particular, we wish to address the problem of scattering at various grazing angles where sequential scattering becomes possible.

THEORY

One very successful method to describe rough interface scattering at high frequency is the Born approximation (Morse and Ingard, 1968) which is usually done in first order. This corresponds to retaining the first term in the Neumann series solution for a Fredholm integral equation of the second kind (inhomogeneous). Unfortunately, rescattering from one part of the surface to another part is a consideration that the first Born approximation cannot account for. Moreover, shadowing must also be considered as a factor (McCammon and McDaniel, 1986). One, however, can account for secondary and higher order scattering by implementing the second order and higher Born corrections (Chin-Bing, et al., 1986). The object of this study is therefore to include higher order Born corrections in an investigation of various grazing angle phenomena. To do this we first represent Eq. (1) as an integral equation using the Helmholtz integral representation.

The Helmholtz integral representation of the field is

$$P = P_i + \int_\Gamma \left(P_\Gamma \vec{\nabla} G - G\, \vec{\nabla} P_\Gamma \right) \cdot d\vec{S} \tag{2}$$

where $\Gamma = \Gamma(x,y)$ is the surface from which scattering occurs, G is the Greens function defined by the equation

$$(\nabla^2 + k^2)\, G = \delta(\vec{R}) \ . \tag{3}$$

Generally, Eq. (2) can not be solved exactly; thus, approximations are used.

Referring to Fig. 1, we choose

$$P_i = D_o \exp[-(x^2/2\sigma_x^2) - (y^2/2\sigma_y^2)] \exp[i\vec{K}_s \cdot (\vec{R}_s + \vec{r})] \,/\, |\vec{R}_s + \vec{r}| \ , \tag{4}$$

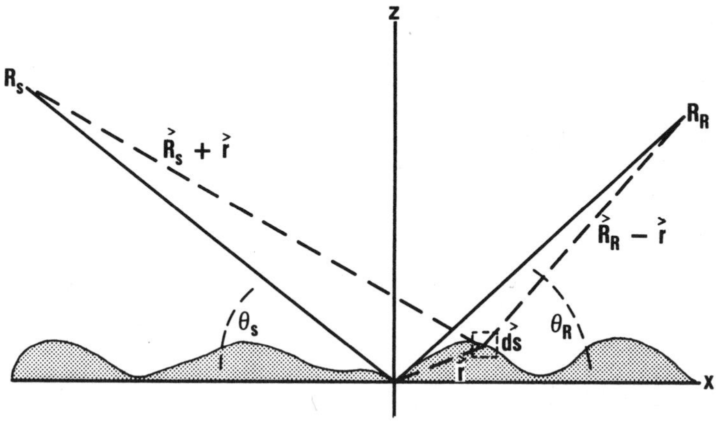

Fig. 1. Schematic showing coordinate system for first Born.

where $\vec{K}_S$ is the propagation vector from the source, σ^2 is the variance, D_o is a constant, and we have allowed for Gaussian spreading on the surface (Tolstoy, 1973).

Choose the outgoing Green's function as

$$G = (1/4\pi)\exp[i\vec{K}_R \cdot (\vec{R}_R - \vec{r})] / |\vec{R}_R - \vec{r}| \tag{5}$$

and the surface element, $d\vec{S} = \hat{n}\, dS$, where $\hat{n}$ is a unit normal vector to the surface of the interface, i.e.,

$$\hat{n} = \vec{\nabla}(\Gamma - z) / |\vec{\nabla}(\Gamma - z)|$$

$$= (\hat{i}\partial\xi/\partial x + \hat{j}\partial\xi/\partial y - \hat{k}) / [(\partial\xi/\partial x)^2 + (\partial\xi/\partial y)^2 + 1]^{1/2}, \tag{6}$$

$$dS = [(\partial\xi/\partial x)^2 + (\partial\xi/\partial y)^2 + 1]\, dx\, dy\ , \tag{7}$$

and $\xi = \xi(x,y)$. This is the first Neumann condition, $P_\Gamma = P_i$, (where we have assumed a reflection coefficient of unity). Thus, the first order is

$$P_S^{(1)} = P^{(1)} - P_i = \int (\, P_i\vec{\nabla}G - G\,\vec{\nabla}P_i)\cdot d\vec{S}\ . \tag{8}$$

This reduces upon calculation to

$$P_S^{(1)} = \int P_i G\, \vec{\Gamma} \cdot \vec{n}\, dx\, dy \tag{9}$$

where

$$\hat{n} = \hat{i}\ \partial\Gamma/\partial x + \hat{j}\ \partial\Gamma/\partial y - \hat{k} \tag{10}$$

and

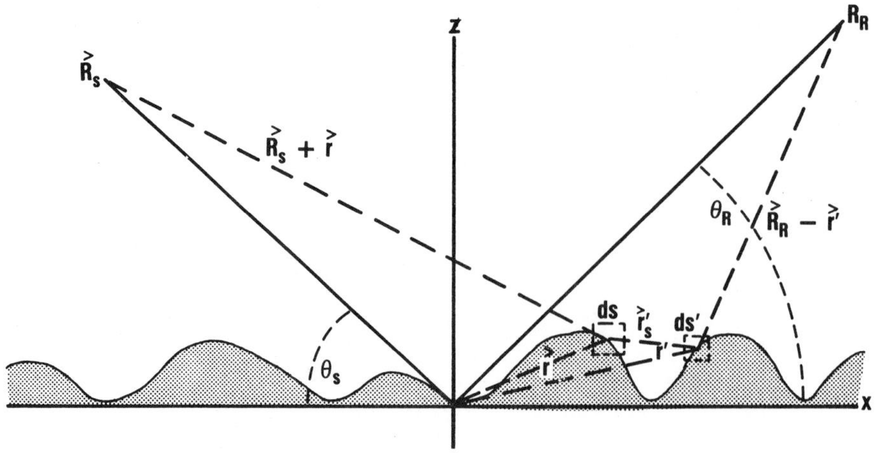

Fig. 2. Schematic showing coordinate system for second Born.

$$\vec{\Gamma} = i(\vec{K}_S - \vec{K}_R) + (\vec{R}_R - \vec{r}) / |\vec{R}_R - \vec{r}|^2$$

$$- (\vec{R}_S + \vec{r}) / |\vec{R}_S + \vec{r}|^2 - \hat{i}x/\sigma_x^2 - \hat{j}y/\sigma_y^2 . \quad (11)$$

Eq. (8) is the first Born approximation. This equation has been frequently applied to high frequency scattering with some success. However, it has the disadvantage of failing for grazing angles where rescattering from one rough surface to an adjacent point can be important. The application of the second Born can however account for a sequential event, i.e., a second scattering.

Referring to Fig. 2, the second Born is given by

$$P_S^{(2)} = \int_\Gamma (P_\Gamma^{(1)} \vec{\nabla}_{r'} G - G \vec{\nabla}_{r'} P_\Gamma^{(1)}) \cdot d\vec{S} , \quad (12)$$

where

$$P_\Gamma^{(1)} = \int_\Gamma (P_\Gamma^{(0)} \vec{\nabla}_r \bar{G} - \bar{G} \vec{\nabla}_r P_\Gamma^{(0)}) \cdot d\vec{S} , \quad (13)$$

$$\bar{G} = (1/4\pi) \exp[i\vec{K}' \cdot (\vec{r}' - \vec{r})] / |\vec{r}' - \vec{r}| , \quad (14)$$

$$G = (1/4\pi) \exp[i\vec{K}_R \cdot (\vec{R}_R - \vec{r}')] / |\vec{R}_R - \vec{r}'| , \quad (15)$$

and $P_\Gamma^{(0)} = P_\Gamma^{(0)}(\vec{R}_S, \vec{r})$ where now $\vec{r}$ is on the surface.

Note that $\vec{K}'$ is a reference wave vector that allows the scattered signal from dS to propagate into dS'. This is done by simple ray trace considerations.

The final results for second Born are

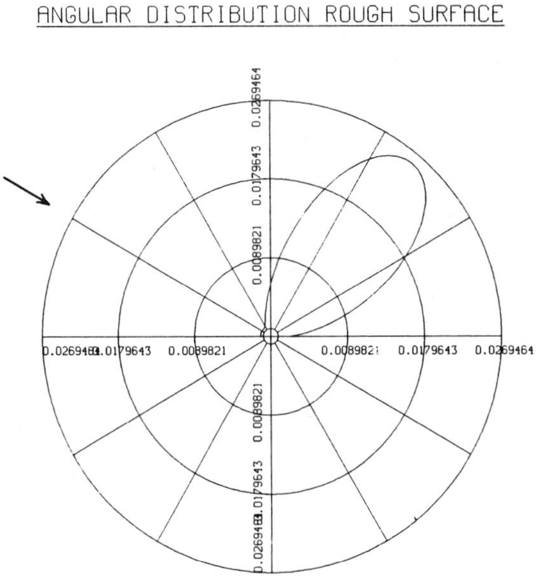

Fig. 3. Polar plot showing the results of the first Born calculation for a Gaussian beam scattered from a rigid sinusoidally corrugated surface.

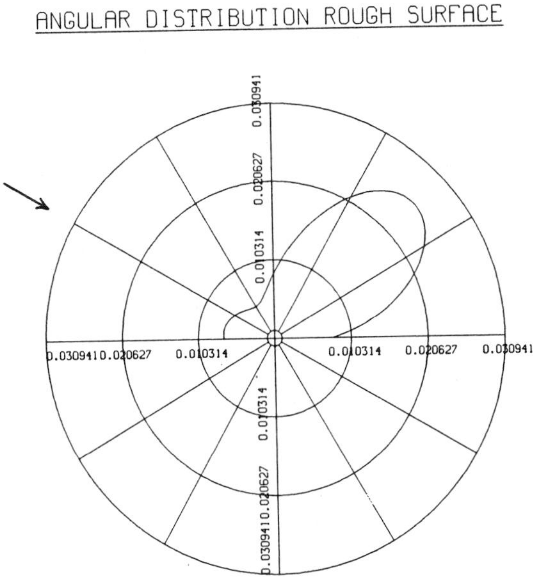

Fig. 4. Polar plot showing the results of the second Born calculation for a Gaussian beam scattered from a rigid sinusoidally corrugated surface.

$$P_S^{(2)} = \int_\Gamma P_i^{(0)} G \bar{G} \gamma' \vec{\Omega} \cdot \vec{n}' \, dx' \, dy' \, dx \, dy \tag{16}$$

where

$$\gamma' = \vec{\Gamma}' \cdot \vec{n}' \quad , \tag{17}$$

$$\vec{\Gamma}' = i(\vec{K}' - \vec{K}_S) + (\vec{r}' - \vec{r}) / |\vec{r}' - \vec{r}|^2$$

$$- (\vec{R}_S + \vec{r}) / |\vec{R}_S + \vec{r}|^2 - \hat{i}x/\sigma_x^2 - \hat{j}y/\sigma_y^2 \quad , \tag{18}$$

and

$$\vec{\Omega} = -i(\vec{K}_R - \vec{K}') + (\vec{R}_R - \vec{r}') / |\vec{R}_R - \vec{r}'|^2$$

$$+ \; n' / |\vec{r}' - \vec{r}|^2 + (\vec{r}' - \vec{r}) / |\vec{r}' - \vec{r}|^2 \quad . \tag{19}$$

In general,

$$P_S = P_S^{(1)} + P_S^{(2)} + \cdots + P_S^{(N)} + \cdots \tag{20}$$

and

$$P_S^{(N)} = \int \left(P_S^{(N-1)} \vec{\nabla} G - G \vec{\nabla} P_S^{(N-1)} \right) \cdot d\vec{S} \quad . \tag{21}$$

NUMERICAL EXAMPLE

An example was chosen that clearly shows the effect of secondary scattering. A rigid sinusoidally corrugated surface given by the equation $\xi = (A)\sin(2\pi x/\lambda)$, where A = 0.5 meters, and λ = 2 meters, was ensonified by a Gaussian beam of variances $\sigma_x = 2$ and $\sigma_y = 2$, with $D_o = 100$. The frequency was 200 Hz. and R_S , the distance from the source to the center of the ensonified area (refer to Fig.2), was 300 meters. The receiver was at R_R = 300 meters, where R_R is measured from the center of the ensonified area to the receiver. The angle made by R_S and the reference scattering surface was 30 degrees. The source was assumed to be in water with a sound velocity of 1500 meters/second.

Figure 3 shows a polar plot of the first Born results. The scattering (rigid sinusoidally corrugated) surface is along the horizontal and the lower half of the polar plot is not physical. The incident beam is from the left as shown by the arrow. The results show a strong forward scatter symmetric lobe centered about 50 degrees as measured from the horizontal. At the chosen incident angle, very little backscatter is seen.

Figure 4 shows a polar plot of the second Born results for the same incident beam on the same scattering surface. A small but significant backscatter contribution can be easily seen. This is due to the secondary scatter from the surface. A comparison of Fig. 3 and Fig. 4 shows that the forward scattered lobes are not coincident. The forward scattered lobe from the second Born calculation is rotated approximately 7 degrees closer to the horizontal plane (than the lobe from the first Born calculation). This shifting of the forward scattered lobe in Fig. 4 is attributed to the secondary scattering component. Note the difference in the plotting scales between the two figures. Figure 4 includes both the first and the second Born and the forward scattered lobe is actually larger in peak magnitude than in Fig. 3 which contains only the first Born results.

SUMMARY

A formulism has been given that can account for a secondary scattering event from a rough surface. The formulism does not require approximations, but rather numerically integrates the scattering integrals. Thus, it is equally suitable for near-field and far-field calculations. It is seen that in some cases, depending on the shape of the scattering surface, this secondary scattering event can be significant. Failure to include this component, as is done in the first Born calculation, can lead to erroneous results in the calculation of the angle and intensity of the forward scattered beam. A numerical example has been given that illustrates these points.

ACKNOWLEDGEMENTS

This work was performed at the Naval Ocean Research and Development Activity (NORDA) and was funded by NORDA through Dr. Steve Stanic.

REFERENCES

Chin-Bing, S. A., Werby, M. F., and Stanic, S., 1986, The Second Born Approximation and High-Frequency Scattering at Low Grazing Angles, J. Acoust. Soc. Am., Suppl. 1, 79:S68.

McCammon, D. F., and McDaniel, S. T., 1986, Surface Velocity, Shadowing, Multiple Scattering, and Curvature on a Sinusoid, J. Acoust. Soc. Am., 79:1778.

McCammon, D. F., and McDaniel, S. T., 1986, Surface Reflection: On the Convergence of a Series Solution to a Modified Helmholtz Integral Equation and the validity of the Kirchhoff Approximation, J. Acoust. Soc. Am., 79:64

McCammon, D. F., and McDaniel, S. T., 1985, Application of a New Theoretical Treatment to an Old Problem, Sinusoidal Boundary Reflection, J. Acoust. Soc. Am., 78:149.

Morse, P. M., and Ingard, K. U., 1968, "Theoretical Acoustics," McGraw-Hill Book, New York.

Pierce, A. D., 1981, "Acoustics: An Introduction to its Physical Principles and Applications," McGraw Hill, New York.

Tolstoy, I., 1973, "Wave Propagation," McGraw Hill, New York.

STATISTICAL CHARACTERISTICS OF ACOUSTIC FIELDS SCATTERED AT THE SEA SURFACE: A HELMHOLTZ-GULIN MODEL

Jerald W. Caruthers†, Stanley A. Chin-Bing†, and Jorge C. Novarini°

†Naval Ocean Research and Development Activity
NSTL, Mississippi 39529-5004

°Servicio de Hidrografia Naval
Depto. Oceanografia
Montes de Oca 2124,
Buenos Aires, Rep. Argentina

ABSTRACT

Statistical properties of acoustic fields scattered from wind-driven surfaces are studied using a numerical simulation technique. The technique consists of the numerical integration of the Helmholtz equation, as further developed by Gulin (Gulin, 1962), over simulated numerical model surfaces. The statistical quantities are calculated as averages over an ensemble of 30 random surfaces.

The surfaces are numerical realizations of the sea surface having a Neumann-Pierson directional spectrum[1] for 5-m/sec (10-knot) wind speed. The surface is assumed to be insonified by a directional source, and the forward-scattered field is sampled at a rectangular receiver array set perpendicular to the wind direction and consisting of 63 omnidirectional hydrophones centered at the specular direction and extending into nonspecular directions. The field is evaluated for frequencies between and including 0.75 and 3.0 kHz. The statistical quantities studied are total intensity, coherent intensity, degree of coherence, phase and amplitude fluctuations, scattering coefficient, and spatial correlation.

INTRODUCTION

An understanding of the statistics of a field scattered by a randomly rough surface is important for realizing the full potential of a variety of signal processing and beamforming techniques that are of current interest.

---

[1]The original computations for this work were done in the early 1970's; however, the results were never published. The results remain unique and relevant today and even the use of the older Neumann-Pierson spectrum will serve for later comparison with computations using newer spectra such as Pierson-Moskovich. Studies using other spectra, wind speeds, cross-wind directions, and formulations for the physics of the scattering are underway.

Among the shortcomings of the techniques used in the studies of surface scattering reported in the literature is that realistic surfaces and a full set of statistical quantities are not being addressed extensively. Moreover, little has been reported about the statistics and spatial distributions of the field scattered in nonspecular, out-of-plane directions nor in the near-field of the scattering surface.

Despite the recent interest in the randomly rough surface scattering problem, a thorough understanding of the statistics of acoustic fields scattered from the sea surface is lacking. This deficiency is due in part to restrictions placed on analytical solutions by the indeterminacy in the statistics and directionality of the scattering surface, as well as to the numerous approximations required. But performing accurate experiments at sea with a simultaneous measurement of the surface statistics, including directionality, is a formidable task. Experiments performed in tanks have not used surfaces that model the real sea surface accurately.

The traditional analytical approach (Baker and Copson, 1939; Eckart, 1953; Gulin, 1962) begins with the Helmholtz theorem and makes the Kirchhoff approximation. For a problem involving a directional source at (X_s, Y_s, Z_s) insonifying a rectangular area (A) of the surface $z = z(x,y)$ and a receiver at (X_r, Y_r, Z_r), Gulin (Gulin, 1962, Eq. (5)) reduces the Helmholtz expression to

$$P(X_r, Y_r, Z_r) = (ik/4\pi) \iint_A (1/R_1R_2)(Z_1/R_1 + Z_2/R_2) \exp[ik(R_1 + R_2)]\, dx\, dy \quad (1)$$

where $R_1^2 = (X_s - x)^2 + (Y_s - y)^2 + (Z_s - z)^2$; $R_2^2 = (X_r - x)^2 + (Y_r - y)^2 + (Z_r - z)^2$,

$Z_1 = Z_s - z$; $Z_2 = Z_r - z$; and k is the acoustic wavenumber. For this model the surface is assumed to be only moderately rough so that, except for near-grazing incidence angles, no part of the surface is shadowed and the effects of multiple scattering are negligible. We shall call this equation the Helmholtz-Gulin equation. (In a companion paper we develop a model designed to correct this latter deficiency (Chin-Bing and Werby, 1986).

Because the surface elevation function, $z = z(x,y)$, is a random variable, a scattering formulation such a Eq. (1) is usually multiplied by its complex conjugate and ensemble averaged. This process leads to a formulation in terms of the surface correlation function. Unfortunately, it also causes a loss of first-moment information (mean pressure amplitude and phase and their fluctuations). Often the surface correlation function is assumed to be statistically isotropic or unidirectional and simple unrealistic forms are assumed. Also far-field assumptions are often made.

To avoid these assumptions and to obtain a variety of statistical parameters that require a knowledge of first-moment quantities, two numerical techniques were developed. For both of these techniques an ensemble of numerical surface models are required. Such surfaces have been developed and described previously (Caruthers and Novarini, 1971). The first scattering technique was based on a facet or image model (Novarini and Caruthers, 1973; Caruthers and Clark, 1973). While that model produced some useful results with a relatively small expenditure of computer time, it did not consider diffraction effects. To achieve a better insight into the detail of the scattering problem, a second technique was developed, which involved the straightforward numerical integration of Eq. (1) using the ensemble of model surfaces in the integrand to obtain ensemble averages for pressure, including phase. The statistical quantities obtained from the application of the new technique are then averaged over the ensemble.

The model has several other advantages: it does not require the usual far-field approximation; it allows for the use of model surfaces that have realistic sea-surface directional power spectra; it allows for calculating the statistical quantities that require first-moment information; and it allows for easy change of experimental geometry and input conditions. Moreover, this specific implementation of rough surface scattering is directly applicable to rough bottom and underice scattering, and the particular equation describing the physics of scattering is easily changed; and the output format is being standardized for ease in intercomparisons of the different spectra, scattering formulations, etc.

SIMULATION TECHNIQUES

As described elsewhere (Caruthers and Novarini, 1971; Novarini and Caruthers, 1973), the model surfaces with the desired directional spectra were generated by applying appropriate smoothing to an array of uncorrelated Gaussian random numbers. The transfer function of the smoothing filter was proportional to the square root of the spectrum of the desired autocorrelation function. The Neumann-Pierson directional sea-surface spectrum for 5-m/sec wind speed, was chosen for this study. Thirty members of the ensemble of surfaces were obtained by changing the input array of random numbers. A second, independent ensemble of 30 surfaces was also constructed to test the statistical significance of the results.

The surfaces were modeled as discrete elevations (z) on an xy plane and the x-axis along the wind direction. As calculated directly from the simluated surface, the standard deviation of the surface elevations (σ) was 0.125 m, the downwind correlation length was 12.0 m, and the crosswind correlation length was 7.0 m.

The geometry of the scattering experiment is shown in Fig. 1. (One needs to imagine the figure inverted to obtain the physical picture of underwater scattering from a sea surface.) The source was simulated to be on the xz plane at the spherical coordinates (R_{10}, Θ_o, χ_o) = (350 m, 45°, 0°), where the origin was the center of the insonified area, Θ_o is the zenith angle, and χ_o the angle with respect to the wind. The region of the surface that was insonified was by 24.7 m by 32.7 m in the downwind and crosswind directions, respectively. The scattered field was sampled at 63 points simulating a 9 x 7 hydrophone array. This array, with specular element at the center (R_{20}, Θ_s, χ_s) = (350 m, 45°, 180°) was in the vertical plane, perpendicular to the wind. The simulated array was 52.5 m wide and 86.5 m high.

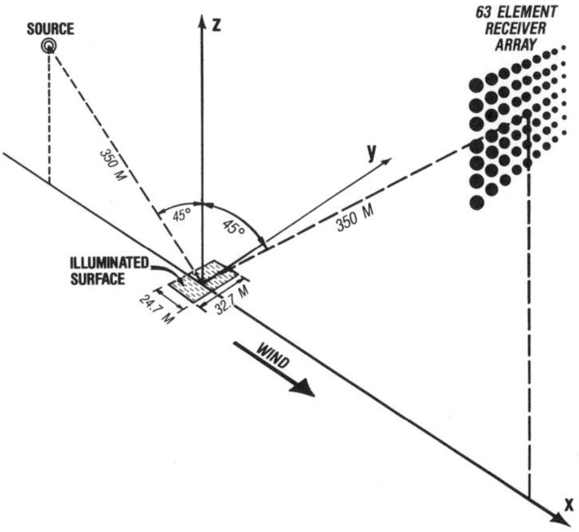

Fig. 1. Geometry for the numerical simulation (coordinate system inverted).

The various statistical quantities that are computed are defined in Table 1. We use the term "spatial correlation function" for the quantity usually referred to as the "coherence function" in recent literature. We use "coherence" to be the square root of the ratio of the coherent and total intensities. The value of this definition is discussed elsewhere (Novarini and Caruthers, 1972). Each of the quantities in Table 1 was computed at each of the hydrophone locations. (Because of space limitations, results for all the quantities are not presented here.)

Table 1. Definition of Statistical Quantities.

Quantity	Symbol	Formula
Mean Pressure	$\langle p\rangle$	$1/M\ \Sigma p_i = a \exp(i\Theta)$
Total Intensity	$\langle pp^*\rangle$	$1/M\ \Sigma p_i p_i^* \doteq A^2$
Intensity of Mirror-Like Surface	$p_o p_o^*$	A_o^2
Coherent Intensity	$\langle p\rangle\langle p^*\rangle$	a^2
Incoherent Intensity	I_s	$\langle(p-\langle p\rangle)^2\rangle = \langle pp^*\rangle - \langle p\rangle\langle p^*\rangle$ $= A^2 - a^2$
Phase Fluctuations	$\delta\Theta$	$\left[1/M\ \Sigma(\Theta - \Theta_i)^2\right]^{1/2}$
Amplitude Fluctuations	$\delta a/a$	$\left[1/M\ \Sigma(\langle p\rangle - p_i)^2\right]^{1/2}/a$ $= \left[A^2 - a^2\right]^{1/2}/a = \left[I_s\right]^{1/2}/a$
Scattering Coefficient	σ_s	$\langle pp^*\rangle/p_o p_o^* = A^2/A_o^2$
Coherence	γ	$\left[\langle p\rangle\langle p^*\rangle/\langle pp^*\rangle\right]^{1/2} = a/A$
Spatial correlation function	$\langle p_a p_b^*\rangle$	$1/M\ \Sigma p_{a,i} p_{b,i}^*$
Roughness (Rayleigh) parameter	R	$k\sigma(\sin\Theta_o + \sin\Theta_s)$.

Notes: i labels the surface realizations, M is the total number of surface realizations in the ensemble, Σ is sum over all realizations.

Calculations were carried out for 0.75, 1.0, 1.5, 2.0, and 3.0 kHz. This frequency range was chosen because it represents a transition between smooth surfaces and rough surfaces. As a measure of roughness we use a roughness parameter defined in Table I. For this work the roughness parameter varies from 0.5 to 2.5.

RESULTS

Fig. 2 shows the spatial distribution of the total and coherent intensities for 1.5 kHz. Note the drop of both types of intensities with displacement away from the specular point. There is, however, a much more rapid drop in coherent intensity as expected. Similar results are obtained for the other frequencies and, as shown in Fig. 3, coherence decreases with frequency. This coherence dependency upon frequency is strong because the range of computations is in the region of transition from smooth to rough surfaces.

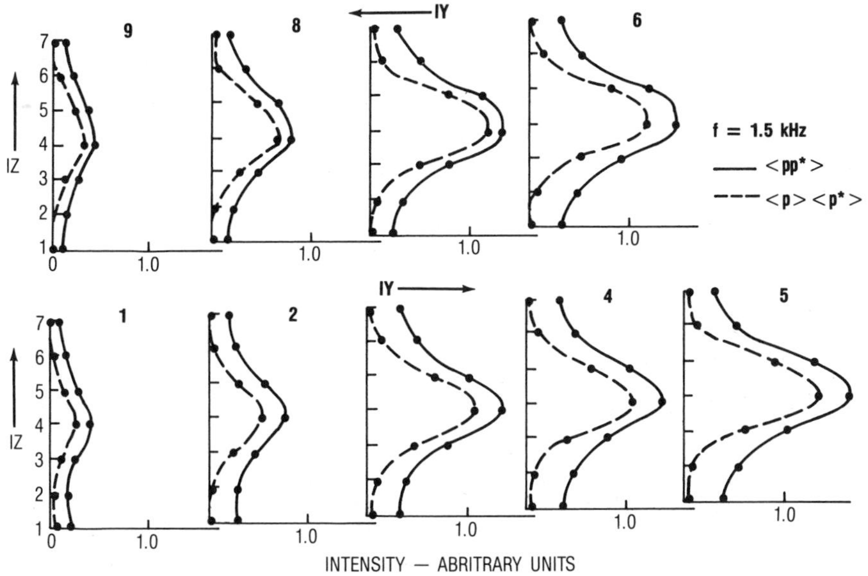

Fig. 2. Profiles of total and coherent intensities for each hydrophone column for 1.50 kHz. (IY labels the column; IZ labels the hydrophone within the column.)

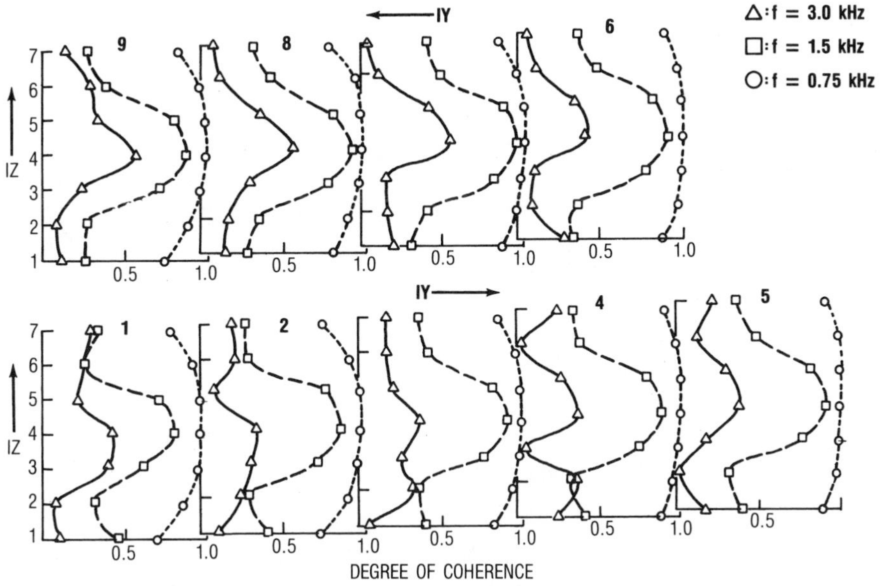

Fig. 3. Profiles of degree of coherence for each hydrophone column for 0.75, 1.50, and 3.00 kHz. (IY labels the column; IZ labels the hydrophone within the column.)

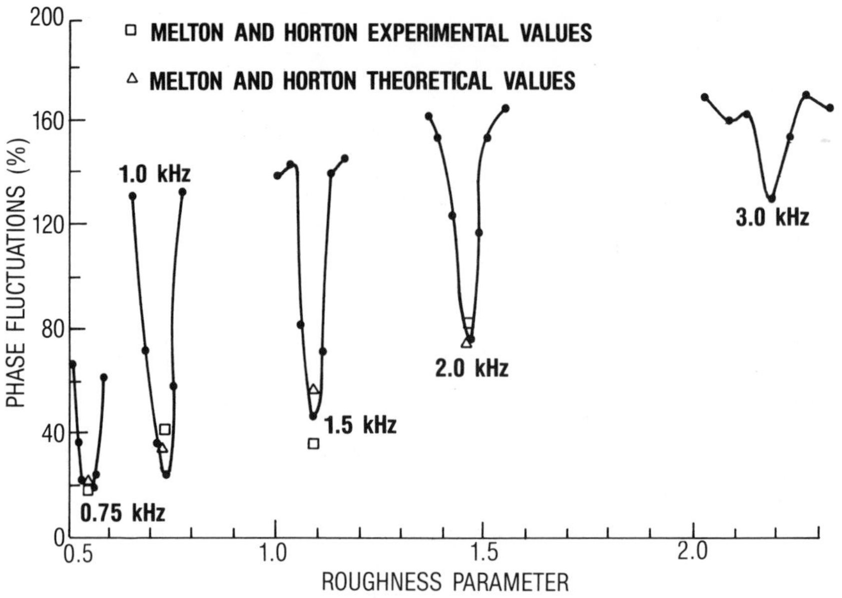

Fig. 4. Phase fluctuations as row averages of standard deviations of the phase in percent of a radian versus roughness paramter.

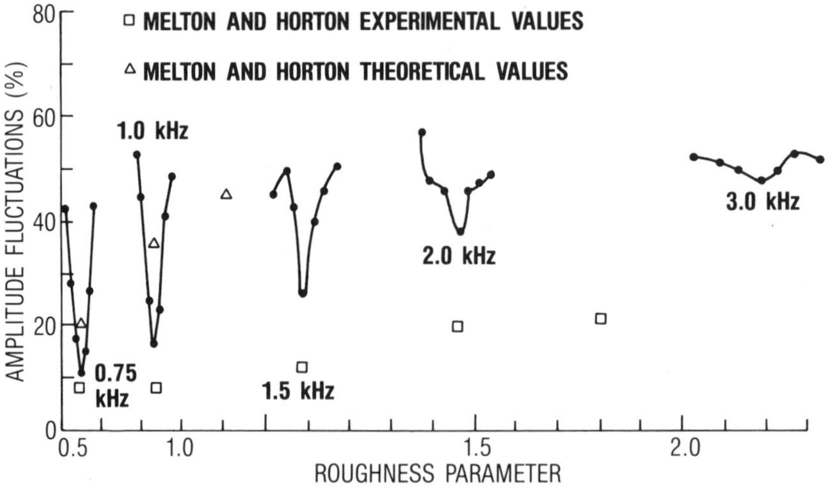

Fig. 5. Amplitude fluctuations as row averages of standard deviations of amplitude in percentage of the average amplitude versus roughness paramter.

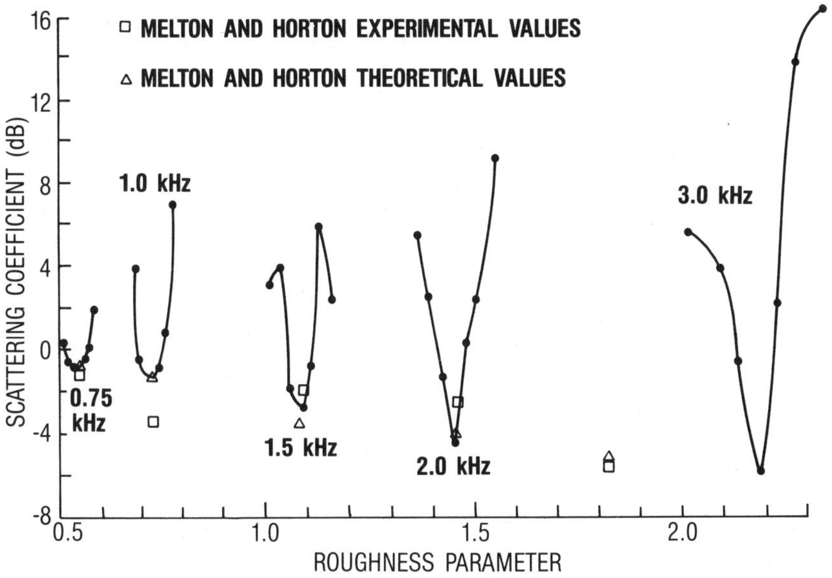

Fig. 6. Scattering coefficient averaged across a row versus roughness parameter. (Each row has a different roughness parameter value at a given frequency. The dip at each frequency is for the specular row.)

The next two figures show the dependence of the phase fluctuations (Fig. 4) and amplitude fluctuations (Fig. 5) upon the roughness parameter. For these graphs the roughness parameter was varied in two ways: by varying the frequency and by varying the scattering angle. To improve the statistical confidence in determining phase and amplitude fluctuations, the values for the seven hydrophones in a row for each roughness parameter were averaged. Plotted for comparison with the specular values obtained here are specular values obtained by Melton and Horton (Melton and Horton, 1970). Note that specular values of phase fluctuations compare very well with the previous studies and that our specular values for amplitude fluctuations are intermediate to experimental and theoretical values given elsewhere (Melton and Horton, 1970).

We found no nonspecular values presented elsewhere with which we could compare our results. Our results in nonspecular directions indicate a rapid deterioration of phase and amplitude stability with displacement from specular. These strong dips in the amplitude and phase fluctuations at the specular direction, at all frequencies, support the conclusion that coherence of a beam scattered at the sea surface deteriorates rapidly in the nonspecular direction. Phase fluctuations appear to approach a saturation value of 1.8 radians. This is to be expected since the standard deviation for uniform phase distribution in 2π cycle, expressed as percentage of a radian, is 1.81 radians.

Theoretical and experimental results (Melton and Horton, 1970; Gulin and Malyshev, 1963) indicate that amplitude fluctuations saturate at about 50% when the roughness parameter exceeds 1.0. Some experimental values of amplitude fluctuations (Melton and Horton, 1970) do not reach their theoretical saturation values even to roughness parameter values of 4.0. Our theoretical results indicate a tendency for amplitude fluctuations to saturate at about 60% for roughness parameter values over 2.0. It is worth noting that the amplitude fluctuation for a Rayleigh distribution is 52%. A plausible explanation for the discrepancies in the various results is that perhaps the roughness parameter is not the proper parameter to be used in such comparisons. Its use has been justified based on the fact that working in the far-field approximation and proceeding along the traditional analytical approach, as discussed in the introduction, amplitude fluctuations reduce to a function of the roughness parameter. The traditional approach might not be valid for large values of the roughness parameter.

As used in this report and others (including Melton and Horton, 1970), the scattering coefficient is the ratio of the total intensity in a given direction to the intensity reflected into that direction from an equivalent mirror-like surface. The row-averaged scattering coefficient is plotted in Fig. 6. The specular values compare favorably with the theoretical and experimental values obtained (Melton and Horton, 1970). The figure shows generally that the energy is increasingly scattered out of the specular direction.

CONCLUSIONS

The numerical integration of the Helmholtz-Gulin equation over an ensemble of realistic representations of sea surfaces is shown to be a powerful technique in the study of the statistical characteristics and spatial distributions of the scattered field. The technique has the advantage over conventional approaches because it provides first-moment information and the only approximation required is the Kirchhoff approximation. The method allows for calculating quantities not usually available to other methods.

The following major findings resulted from this study.

- Coherent and total intensities decrease with increasing frequency in such a way that coherence decreases rapidly as the roughness parameter transition from less than one to greater than one.
- For a given frequency in this transition region, coherence decreases rapidly for increased displacement from the specular direction.
- Phase and amplitude fluctuations increase significantly in nonspecular directions. (No previous study has given their values in nonspecular directions.)
- Phase fluctuations appear to approach a saturation value of 1.8 radian and amplitude fluctuations appear to approach a saturation value of 60% of the average amplitude.

REFERENCES

Baker, B. B., and Copson, E. T., 1938, "The Mathematical Theory of Huygen's Principle," Oxford University Press, London, pp. 23-28.

Caruthers, J. W., and Clark, D. N., 1973, Frequency Dependence of Acoustic Wave Scattering from Randomly Rough Surfaces, J. Acoust. Soc. Am. 54:802.

Caruthers, J. W., and Novarini, J. C., 1971, Numerical Modeling of Randomly Rough Surfaces with Application to Sea Surfaces, Texas A&M Univer., Dept. Oceanography, Tech. Rpt. Ref. 71-13-T.

Chin-Bing, S. A., and Werby, M. F., 1986, High Frequency Scattering from Rough Bottoms and the Second Born Approximation, Proceedings of the 12th International Congress on Acoustics Associated Symposium on Underwater Acoustics, 16-18 July 1986, Halifax, Nova Scotia.

Eckart, C., 1953, The Scattering of Sound from the Sea Surface," J. Acoust. Soc. Am. 25:566.

Gulin, E. P., 1962, Amplitude and Phase Fluctuations of a Sound Wave Reflected from a Statistically Uneven Surface, Sov. Phys.-Acoust. 8:135.

Gulin, E. P., and Malyshev, K. I., 1963, Statistical Characteristics of Sound Signals Reflected from the Undulating Sea Surface, Sov. Phys.-Acoust. 8:228.

Melton, D. R., and Horton, Sr., C. W., 1970, Importance of the Fresnel Correction in Scattering from a Rough Surface. I. Phase and Amplitude Fluctuations, J. Acoust. Soc. Am. 47:290.

Novarini, J. C. and Caruthers, J. W., 1972, The Degree of Coherence of Acoustic Signals Scattered at Randomly Rough Surfaces, J. Acoust. Soc. Am., 51:417.

Novarini, J. C., and Caruthers, J. W., 1973, Numerical Modeling of Acoustic Wave Scattering from Randomly Rough Surfaces: A Facet Model, J. Acoust. Soc. Am. 53:876.

THE IMPORTANCE OF HYBRID RAY PATHS, BOTTOM LOSS, AND FACET REFLECTION ON OCEAN BOTTOM REVERBERATION

Dale D. Ellis and J. B. Franklin

Defence Research Establishment Atlantic
P.O. Box 1012
Dartmouth, Nova Scotia, Canada B2Y 3Z7

ABSTRACT

Mackenzie's deep-water model for bottom reverberation [*J. Acoust. Soc. Am.* **23**, 1498–1504 (1961)] has been extended to handle reverberation arriving later than the second fathometer return. The model uses straight-line ray paths, flat bathymetry, and a short pulse length. Energy is scattered from the bottom in all directions; in particular, the energy can return by a different path from the outgoing one (hence, the term hybrid path). A three-parameter backscattering function has been used which incorporates Lambert's law at low grazing angles augmented by a facet-reflection process at steep angles. Also, three bottom loss curves based on the Rayleigh reflection coefficients for sand, silt and clay have been used for a sensitivity study. The results indicate that: (i) the facet-reflection process broadens the fathometer returns for a short pulse characteristic of an impulsive source; (ii) the hybrid paths are important (adding about 5 dB reverberation) when the bottom loss is low; and (iii) bottom loss is as important as the backscattering in determining the reverberation level. A correction to the isovelocity model to allow for a sound speed gradient indicates that (for omnidirectional sources and receivers) the effect of the sound speed profile is small. Thus, the isovelocity model seems to be a reasonable approximation to longer times than one might at first expect.

INTRODUCTION

In this paper we extend Mackenzie's (1961) model for bottom reverberation to handle reverberation arriving later than the second fathometer return. The model is then used to study the effects of the bottom scattering strength, bottom loss, and the contributions of the various ray paths on the bottom reverberation.

The model assumes a source with a short pulse length in an ocean of constant sound speed and flat bathymetry. The first section of the paper describes the model and its assumptions and gives a derivation of the equations for the reverberation and fathometer returns. Mackenzie's model is derived as a special case. Also described is a correction which handles some of the effects of a sound speed profile.

The second main section of the paper describes the input parameters for the model calculations, in particular the bottom loss and scattering functions used in the sensitivity study. The bottom loss curves are based on the reflection coefficients for a half-space of sand, silt or clay. The bottom scattering function is a three-parameter formula that provides a reasonable fit to backscattering measurements. It incorporates

Lambert's law at low grazing angles with an enhanced facet-like scattering at the steeper angles.

The third section shows the results of the sensitivity study using the simple model. Results are shown for the effects of the bottom loss, the effect of the facet strength and width, the contributions of the various ray paths, and the effect of a sound speed gradient. Among the conclusions are: (i) the facet-reflection process broadens the fathometer returns for a short pulse characteristic of an impulsive source; (ii) the hybrid paths are important (adding about 5 dB reverberation) when the bottom loss is low; (iii) bottom loss is as important as the backscattering in determining the reverberation level; and (iv) the effect of the sound speed profile is small.

In summary, the simple model is computationally very easy to implement, and allows the relative importance of a number of environmental parameters to be easily studied. It seems to provide a reasonable approximation to longer times than one might at first expect.

DESCRIPTION OF THE MODEL

Figure 1 illustrates the main features of the model. The ocean bottom is flat and the water has a constant sound speed, so the ray paths are straight lines. The source and receiver are near the ocean surface. For the calculation of angles and path lengths they are assumed to be *at* the ocean surface; for the calculation of propagation loss they are at sufficient depth to permit incoherent addition of the reverberation arriving by different ray paths (see below). Energy is scattered from the bottom in all directions; in particular, the energy can return by a different path from the outgoing one (hence, the term hybrid path). The figure also qualitatively illustrates that the backscattering is stronger near the steeper angles. The reverberation is assumed to originate from scattering at the ocean–bottom interface, and not from volume reverberation from within the bottom.

There are a few other assumptions in the model. The contributions of all the ray paths are added incoherently; this includes the four nearly identical paths due to the near-surface source and receiver. Multiple scattering effects are not included in the model; that is, scattered energy is not rescattered. This assumption may not be completely valid near normal incidence where the backscattering is relatively strong. The pulse length is of short enough duration that the ray path lengths and bottom grazing angles do not change appreciably; the pulse may, however, be of arbitrary shape. Omnidirectional sources and receivers have been used in the study, although beam patterns can easily be incorporated in the equations.

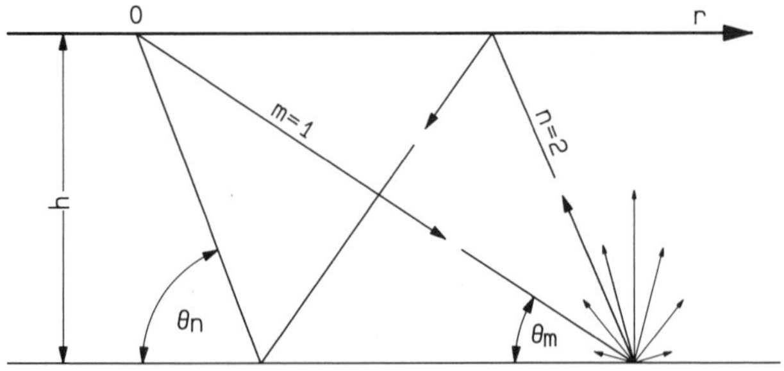

Fig. 1. The hybrid path model.

Derivation of the Equations

Figure 1 shows a source and receiver at the surface of the ocean of depth h and constant sound speed c_0. Consider the (m,n) path in which the ray undergoes $m-1$ bottom reflections on its outgoing path and $n-1$ reflections on its return path. The path lengths are l_m and l_n, and the angles for reflection and scattering are θ_m and θ_n. Using spherical spreading propagation loss, the contribution I_{mn} to the reverberation intensity at time t due to the portion τ of the initial pulse I_0 is

$$dI_{mn}(t,\tau) = I_0(\tau) l_m^{-2} l_n^{-2} R^{m-1}(\theta_m) R^{n-1}(\theta_n) B(\theta_m,\theta_n)\,|dA| \tag{1}$$

where $dA = 2\pi r dr$, r is the horizontal range, $R(\theta)$ is the magnitude of the bottom reflection coefficient, and $B(\theta,\phi)$ is the bottom scattering function. The following relations apply:

$$\begin{aligned} l_k^2 &= (2k-1)^2 h^2 + r^2, & (2)\\ \theta_k &= \arcsin\left[(2k-1)h/l_k\right], & (3)\\ l = l_m + l_n &= c_0(t-\tau), & (4) \end{aligned}$$

where l is the total path length.

Consider the time t as fixed, and the pulse time τ to be the independent variable. From Eqs.(2) and (4) the change in range dr can be obtained in terms of $d\tau$ as

$$-c_0 d\tau = \frac{l_m + l_n}{l_m l_n} r dr; \qquad \text{or equivalently,} \qquad |dA| = 2\pi c_0 l_m l_n l^{-1} d\tau. \tag{5}$$

The range r can be eliminated from the expressions for l_m and l_n by noting from Eq.(2) that $l_n^2 - l_m^2 = 4h^2(n^2 - m^2 - n + m)$ and using Eq.(4) to obtain

$$\begin{aligned} l_m &= (l/2)\left[1 - (2h/l)^2(n^2 - m^2 - n + m)\right], & (6)\\ l_n &= (l/2)\left[1 + (2h/l)^2(n^2 - m^2 - n + m)\right]. & (7) \end{aligned}$$

Substituting Eqs.(4–7) into Eq.(1) and integrating over the duration of the pulse gives the total reverberation due to the (n,m) path

$$I_{mn} = 8\pi c_0 \int_0^{\tau_0} \frac{I_0(\tau) R^{m-1}(\theta_m) R^{n-1}(\theta_n) B(\theta_m,\theta_n)}{l^3\left[1 - (2h/l)^4(n^2 - m^2 - n + m)^2\right]}\, d\tau.$$

To obtain the total reverberation intensity we assume that the contributions from all the ray paths add incoherently; this includes the four nearly-identical ray paths due to the up and down going rays at the near-surface source and receiver. Further, we assume that the pulse length is short so that l, θ_m, and θ_n do not vary appreciably over the duration of the pulse. The total reverberation intensity is then

$$I(t) = \frac{32\pi c_0 E_0}{l^3} \sum_{m,n} \frac{R^{m-1}(\theta_m) R^{n-1}(\theta_n) B(\theta_m,\theta_n)}{\left[1 - (2h/l)^4(n^2 - m^2 - n + m)^2\right]}, \tag{8}$$

where $E_0 = \int_0^{\tau_0} I_0(\tau) d\tau$ and now the path length simplifies to $l = c_0 t$. The limits of the summation are time-dependent. They can be determined by noting from Eq.(2) that $l_m + l_n \geq 2h(n + m - 1)$, or

$$n + m \leq 1 + (c_0 t / 2h). \tag{9}$$

Absorption in the water column is easily handled by multiplying $I(t)$ by the term $\exp(-\alpha c_0 t)$, where α is the loss per unit distance at the frequency of interest. Note that frequency dependence is not included in the model except through the dependence of α, R, and B on frequency.

Mackenzie's Model

A common assumption for backscattering, particularly at low grazing angles is Lambert's law of diffuse scattering: $B(\theta,\theta') = \mu \sin\theta \sin\theta'$. From Eqs.(3),(6), and (7) this becomes

$$B(\theta_m,\theta_n) = \frac{4\mu h^2(2m-1)(2n-1)}{l^2[1-(2h/l)^4(n^2-m^2-n+m)^2]}.$$

If only the $m = n = 1$ path is considered, then Eq.(8) reduces to the t^{-5} time dependence associated with Lambert's law:

$$I(t) = 128\pi E_0 \mu h^2 c_0^{-4} t^{-5}.$$

This is equivalent to Eq.(4) of Mackenzie(1961) when the pulse length is short.

Effect of a Sound Speed Gradient

The approximation of isovelocity water for the deep ocean is obviously of concern when small grazing angles are involved. A simple correction for the sound speed profile can be applied as illustrated in Fig. 2. We assume a sound speed c_1 at the source and receiver depth, a speed $c_2 > c_1$ at the ocean bottom, and a linear gradient between. The ray path is then an arc of a circle with an angle θ_1 at the source–receiver, an angle θ_2 at the bottom, and angle θ for the chord. Using $(\theta_1 + \theta_2) = 2\theta$ and Snell's law $c_2/c_1 = \cos\theta_2/\cos\theta_1$ gives

$$\tan\theta_2 = \frac{c_1/c_2 - \cos 2\theta}{\sin 2\theta}, \tag{10}$$

with a similar expression for θ_1.

If the effects of the curved ray paths on travel time, propagation loss, and scattering area are neglected, then Eq.(8) can still be used except that the angles θ_m and θ_n in the reflection coefficient and scattering function are modified using Eq.(10). The terms in the summation of Eq.(8) are not included if $\theta < \cos^{-1}(c_1/c_2)$. This correction thus alleviates one of the major problems with the isovelocity model by removing those ray paths that do not interact with the bottom.

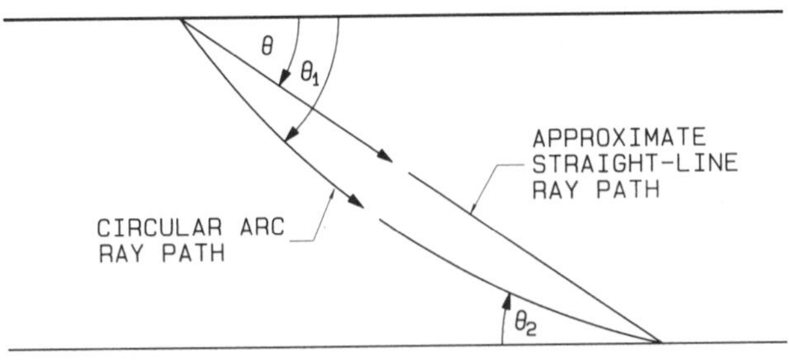

Fig. 2. Sound speed gradient.

Fathometer Returns

In addition to the scattering from the bottom, there are the bottom reflections at normal incidence. For the n-th bottom bounce, the path length is $2nh$ and the received intensity is

$$R_f(t) = \sum_n \frac{I_0(t-t_n)R^n(\pi/2)}{(nh)^2},$$

where $t_n = 2nh/c_0$ and the four similar source/bottom/receiver paths have been included. Note that in this case the received intensity is a series of discrete events, in contrast to the terms of Eq.(8) which are continuous in time except for their onset.

In the reverberation formula of Eq.(8) it was only necessary to assume a total energy of E_0 in the pulse, but here the duration of the pulse is important. For our calculations of the fathometer returns we have used a uniform amplitude over the duration of the return; i.e., $\overline{I} = E_0/\Delta t$:

$$R_f(t) = \begin{cases} \dfrac{E_0}{\Delta t}\dfrac{R^n(\pi/2)}{(nh)^2}, & t_n - \Delta t/2 \le t \le t_n + \Delta t/2 \\ 0 & \text{otherwise.} \end{cases} \tag{11}$$

For calculations of the fathometer returns a Δt of 1 s has been rather arbitrarily assigned to account for source and receiver depths of up to several hundred meters, sub-bottom reflections, and any pulse length of the impulsive source.

ACOUSTIC PROPERTIES

For the calculations to be shown in the next section a water depth of 5000 m and an average sound speed of 1500 m/s have been used. Most of the calculations assume isovelocity water, but results are also shown with positive sound speed gradients that trap rays with source or receiver angles of 10 and 20 degrees with respect to the horizontal. The remainder of this section discusses the various bottom parameters used in the study.

Bottom Reflection Loss

The bottom reflection loss curves for water over a homogeneous half-space of sand, silt, or clay are shown in Fig. 3a. The acoustic parameters are given in Table 1, and are based on those of Hamilton as listed by Eller and Gershfeld(1985). These are intended to be representative of low, medium, and high loss bottoms for the purpose of comparing the effect of bottom loss on reverberation.

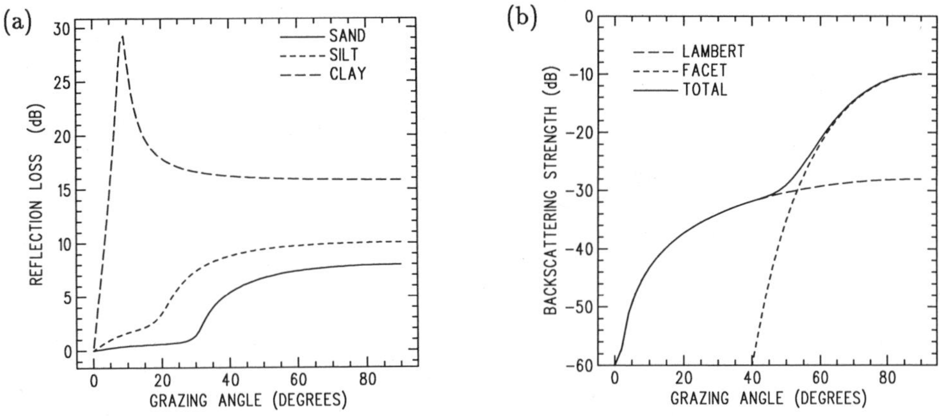

Fig. 3. Bottom properties: (a) bottom loss curves; (b) bottom scattering function.

Table 1. Acoustic parameters used for calculating the reflection coefficients

Material	Specific gravity	Sound speed m/s	Sound speed c/c_0	Attenuation dB/m-kHz	Attenuation dB/λ
water	1.0	1500	1.0	0.	0.
sand	2.0	1750	1.167	0.5	0.9
silt	1.8	1600	1.067	0.7	1.2
clay	1.4	1485	0.99	0.1	0.15

Bottom Backscattering

Bottom backscattering measurements seem to be characterized by a $\sin^q(\theta)$ behaviour at low angles (where $0.5 < q < 2$), with an enhanced scattering at steep angles. Here we refer to this steep-angle scattering as a facet-reflection process, even though the scattering may be due to volume scattering in the bottom (Stockhausen, 1963). The following three-parameter scattering function has been used for the calculations presented here:

$$B(\theta,\phi) = \mu \sin\theta \sin\phi + \nu \exp\left[-\cot\theta \cot\phi/\sigma^2\right]. \tag{12}$$

The first term is Lambert's law which dominates at low grazing angles; the second term provides a reasonable fit to the enhanced backscattering measured near normal incidence. This formulation is a generalization of the monostatic ($\theta = \phi$) backscattering function used by Hoffman(1976), who used the fixed values $\nu = 1$ and $\sigma = 0.1$. Here we allow the facet strength ν and width σ to vary. The bistatic extrapolation ($\theta \neq \phi$) of the facet term is *ad. hoc*, but the $\exp\left[-\cot^2(\theta)/\sigma^2\right]$ dependence appears in various discussions of steep angle ocean surface and bottom backscattering processes (Chapman and Scott, 1964; Ivakin and Lysanov, 1981). A possibly more realistic facet term that depends on the difference between the scattered and specularly reflected rays is given by Brekhovskikh and Lysanov (1982) as $\sin^{-4}(\Delta\theta)\exp\left[-\cot^2(\Delta\theta)/\sigma^2\right]$, where $\Delta\theta = (\theta+\phi)/2$. This produces a slightly narrower facet term when $\theta \neq \phi$, but has a negligible effect on the calculations presented here.

For this study we have used $10\log\mu = -28$ dB which provides an approximate fit to deep ocean backscattering over a wide frequency range (Mackenzie, 1961). A number of arbitrary, but reasonable, values have been used for ν and σ. Figure 3b shows the function for (monostatic) backscattering, including the contributions of the two terms of Eq.(12) with $10\log\nu = -10$ dB and $\sigma = 0.35$. Note that these values give a wider, but weaker, facet term than that of Hoffman, but seem to be a more reasonable fit to measured backscattering (e.g., see Urick, 1983).

REVERBERATION CALCULATIONS

Before discussing the calculations it is worth noting that the water depth is probably the most important factor affecting reverberation. However, it is not necessary to perform any calculations since scaling arguments can be used to determine the effect. In terms of a dimensionless time $\eta = l/2h = c_0t/2h$, the fathometer returns and new terms in the summation of Eq.(8) appear at times $\eta = 1, 2, \ldots$ Furthermore, the angles and other factors inside the summation depend only on the ratio $l/2h$. Thus, $h^3I(\eta)/c_0$ is independent of water depth and sound speed. The h^3 dependence of the reverberation is essentially due to the l^{-4} dependence of the propagation loss and the l^1 dependence of the scattering area.

A number of calculations have been performed to determine the effect of the various quantities on the reverberation. Most of the calculations have been performed using the bottom loss for sand as shown in Fig. 3a and the backscattering strength as shown in Fig. 3b. The results are plotted in dB as reverberation level $RL = 10\log\left[I(t)/E_0\right]$ where $I(t)$ is defined in Eq.(8).

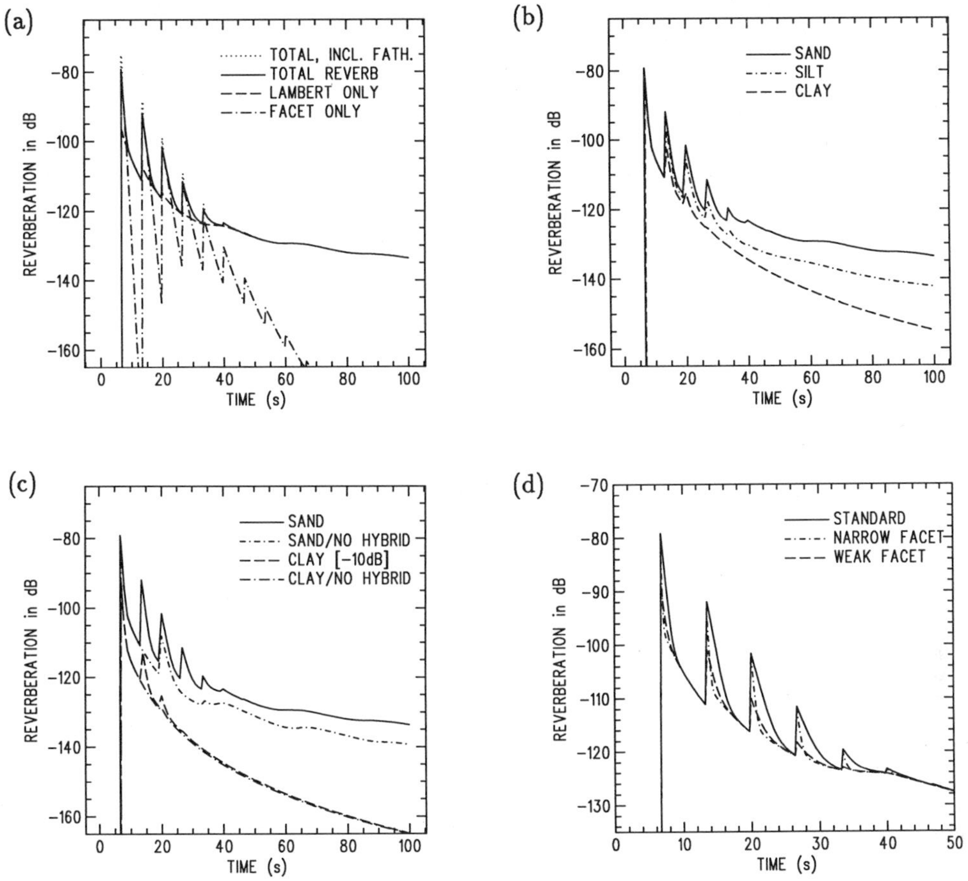

Fig. 4. Reverberation calculations: (a) effect of the various scattering components; (b) effect of bottom loss; (c) effect of the hybrid paths; and (d) effect of the facet strength and width.

Effect of the Various Scattering Components

Using the acoustic properties of sand as given in Fig. 3a and Table 1, and the standard backscattering function as given in Fig. 3b, the contributions of the various components of reverberation are shown in Fig. 4a.

It can be seen that the fathometer returns of width $\Delta t = 1$ s are considerably broadened by the steep angle reverberation from the facet scattering. The low-angle Lambert's scattering dominates the reverberation between the fathometer returns and at times greater than 40 seconds (approximately 5 fathometer returns).

Effect of Bottom Loss

Figure 4b shows that the bottom loss can have a significant influence on the reverberation arriving later than the second fathometer return. Results are shown for the sand, silt, and clay bottom types (see Fig. 3a and Table 1). Note that all curves give the same results out to the first fathometer return since the same scattering function is used in all three cases. The reader is reminded that the curves shown here and in the following figures depict only the reverberation, and do not include the vertical bottom reflections or “fathometer returns”.

Effect of the Hybrid Paths

Figure 4c shows that it is important to include all the scattered rays connecting the receiver and the bottom especially when the bottom loss is low, as in sand. These extra paths contribute about 5 dB to the observed reverberation. For a high loss bottom such as clay these paths do not contribute significantly except at quite early times. Note that when the hybrid paths are not included that the steep angle reverberation is absent at times just after the even "fathometer" returns. The relative contributions of all the paths will be considered later in conjunction with the effects of a sound speed gradient.

Effect of the Facet Strength and Width

The standard facet strength and width used for the previous calculations can be varied to show its effect on the reverberation. Fig. 4d compares the standard facet term ($10 \log \nu = -10$ dB and $\sigma = 0.35$) with (i) a "narrow" facet with the same strength ν but a narrower width $\sigma = 0.175$; and (ii) a "weak" facet with $10 \log \nu = -20$ dB and the standard width $\sigma = 0.35$.

The results are very much what one would expect. The "fathometer-like" returns are reduced in strength or narrowed, but in any case the low angle reverberation, as in Fig. 4a, dominates after the first four or five fathometer returns.

Effect of a Sound Speed Gradient

Figure 5 compares the reverberation in isovelocity water with that from sound speed profiles with linear gradients corresponding to trapped rays of 10° and 20°. The effect of the sound speed gradient is much less than one might expect, being only about 3 dB over the interval shown. The explanation of this can be seen by comparing the relative importance of the various paths discussed in the next section.

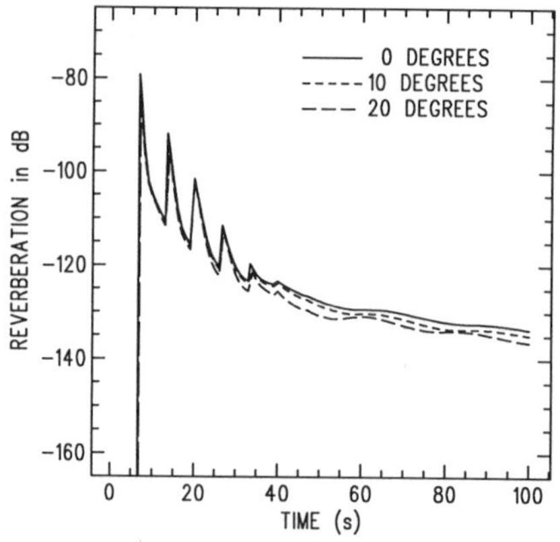

Fig. 5. Effect of sound speed gradients.

Contributions of the Various Paths

The contributions of some of the various paths are shown in Figs. 6a and 6b for isovelocity water and for a sound speed profile that traps rays of 20°. There are 9 curves in each figure but some of them are explicitly labelled in addition to being identified in the legend. In particular we see that when the contribution from the $(1, 1)$ path begins to drop off the other multiple bounce paths $(2, 2)$, $(2, 3)$, etc., remain essentially as strong as before and contribute the major part of the reverberation. Since these higher order paths are travelling at steeper angles, the sound speed gradient is relatively unimportant. This explains why the sound speed profile has such a small effect on the overall reverberation.

(a) (b)

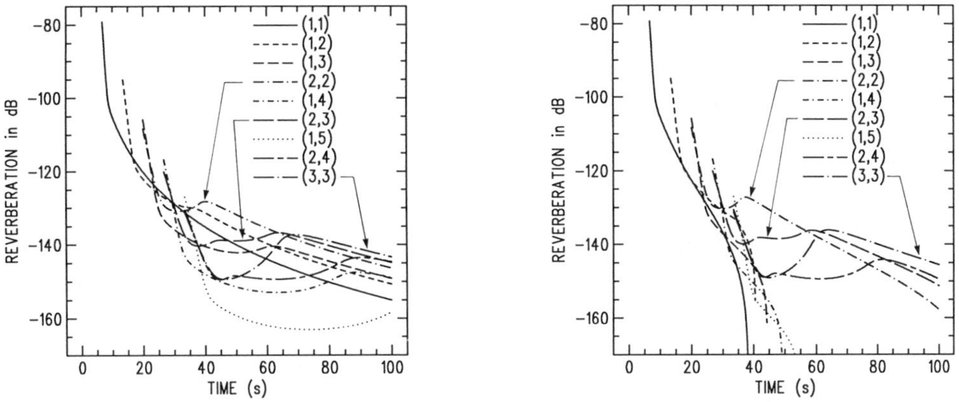

Fig. 6. Contributions of some of the possible paths: (a) for isovelocity water; (b) for a sound speed gradient that traps rays up to 20°.

Note that the hybrid paths ($m \neq n$) occur in pairs (m, n) and (n, m) but only one has been plotted on the graphs. To obtain their combined effect, one can simply add 3 dB to the curves shown. Note also that the small ripples in the reverberation in Fig. 5 at times of approximately 40, 65, and 90 seconds are related to the strong angular dependence of the bottom loss near 30°, (see Fig. 3a).

CONCLUSIONS

A simple model of ocean bottom reverberation has been extended to handle reverberation arriving after the second fathometer return. The model uses an isovelocity water approximation, but a first correction for the sound speed profile using a linear gradient shows that the effect is quite small.

A three-parameter bottom scattering function has been proposed. It provides a reasonable approximation to many measurements of bottom backscattering, and has been extended to the bistatic case.

Calculations have been performed for a deep ocean environment using several realistic examples of the bottom loss and backscattering inputs. The results show that the reverberation depends as much on bottom loss as the scattering parameters. As well, it is surprising that the effect of the sound speed profile is so insignificant when compared with its effect on ray curvature.

In summary, the simple model is computationally very easy to implement, and allows the relative importance of a number of environmental parameters to be easily studied. It seems to provide a reasonable approximation to longer times than one might at first expect.

ACKNOWLEDGEMENTS

The authors appreciate the useful discussions with their colleagues B. A. Trenholm, J. H. Stockhausen and D. V. Crowe from DREA, and S. N. Wolf, exchange scientist from Naval Research Laboratory, Washington, D.C.

REFERENCES

Brekhovskikh, L., and Lysanov, Yu., 1982, "Fundamentals of Ocean Acoustics", Springer-Verlag, Berlin, 1982, Chapter 9.

Chapman, R. P., and Scott, H. D., 1964, Surface backscattering strengths measured over an extended range of frequencies and grazing angles, J. Acoust. Soc. Am., 36:1735–1737.

Eller, A. I., and Gershfeld, D. A., Low-frequency acoustic response of shallow water ducts, J. Acoust. Soc. Am., 78:622–631.

Hoffman, D. W., 1976, "LORA: A model for predicting the performance of long-range active sonar systems," NUC TP 541, Naval Undersea Center, San Diego, CA.

Ivakin, A. N., and Lysanov, Yu. P., 1981, Underwater sound scattering by volume inhomogeneities of a bottom medium bounded by a rough surface, Sov. Phys. Acoust., 27:212–215.

Mackenzie, K. V., 1961, Bottom reverberation for 530- and 1030-cps sound in deep water, J. Acoust. Soc. Am., 33:1498–1504.

Stockhausen, J. H., 1963, "Scattering from the volume of an inhomogeneous half-space," Naval Research Establishment, Dartmouth, N.S., Canada, Report 63/9.

Urick, R. J., 1983, "Principles of Underwater Sound," 3rd edition, McGraw-Hill, New York, Chapter 8.

THE STUDY OF SOUND BACKSCATTERING FROM MICROINHOMOGENEITIES IN SEA WATER

V.A. Akulichev and V.A. Bulanov

Pacific Oceanological Institute
Far East Science Centre
Vladivostok, USSR

ABSTRACT

The authors show that the use of a parametric source to measure backscattering from microinhomogeneities offers several advantages. By analyzing backscattering results obtained at a variety of signal frequencies and pulse lengths, it is possible to obtain the size distribution of both resonant and non resonant scatterers separately. - Ed.

INTRODUCTION

Real liquids contain different phase inclusions in the form of gas and vapour bubbles, solid particles, biological elements and so on. The problems concerned with determining the sizes and concentrations of these phase inclusions in liquids can be solved by means of acoustic methods based on the investigation of scattering or attenuation of acoustic signals with different frequencies f. In this case, the change of frequency allows one to obtain information not only about resonant inclusions, those being gas and vapour bubbles, fishes with swim bladders and others, but also about nonresonant inclusions which are solid particles, zooplankton, plants, emulsion drops, etc. To solve similar problems, it is possible to use acoustical parametric sources characterized by a high directivity in a wide range of difference frequencies f. The study of backscattering or attenuation of acoustic signals with frequency f allows us to determine the sizes and concentration of different phase inclusions in real liquids. In addition to this, in comparison with common acoustical methods of using parametric sound sources, it allows us to realize a distant measurement of backscattering in the pulse mode over a wide frequency range and to determine the spatial distribution of different phase inclusions in liquid.

THEORY

Consider the relationship between the backscattering of pulsed acoustic signals with carrier frequency f and size distribution function of phase inclusions $g(R)$, where R is a typical size of the inclusion. It is assumed here that the length of acoustic pulses τ is large as compared to the time of establishing steady oscillations of the phase inclusions. The intensity of backscattering $\Im_s$ at a distance r from the scattering volume of liquid can be determined in the first approximation by the formula[1]

$$\Im_s = \Im_i V \sigma / r^2 e^{2\alpha r} \qquad (1)$$

where $\Im_i$ is the intensity of the incident wave, α is the coefficient of sound absorption, $\alpha = \alpha(f,g)$, σ is a backscattering cross-section of a unit volume of a liquid medium with phase inclusions expressed in the following integral form

$$\sigma(f) = \int | s(f,R) |^2 g(R)\, dR. \tag{2}$$

The function $s(f,R)$ determines the scattering amplitude of acoustic signals with frequency f on a single scatterer with a characteristic dimension R. The size distribution function $g(R)$ determines the number of bubbles in a unit volume of liquid with the radii being in the interval $(R,R+dR)$. The function $s(f,R)$ can be expressed in the form

$$S(f,R) = \frac{R}{(f_0^2/f^2 - 1) - i\,\delta} \tag{3}$$

where $\delta = \delta(f,R)$ is the attenuation constant. Eq. (3) is true at scatterer dimensions R which are smaller than the acoustic wavelength in the medium. In this case, only monopole sound scattering is expected. When $\delta << 1$, Eq. (3) determines the resonant scattering which is observed, for example, in the case of gas or vapour bubbles. The resonance frequency of a single gas bubble is expressed by the formula

$$f_0 = \frac{1}{2\pi R}\left[\frac{3\gamma P_0}{\rho}(1 + Ah)\right]^{1/2} \tag{4}$$

where P_0 is the pressure in liquid at the sea surface, A is a constant value determining the influence of the depth h, $A = 0.1\text{m}^{-1}$, ρ is the liquid density, and γ is the adiabat constant equal to 1.4 for the air.

It can be shown that the scattering volume of the medium situated at a distance r from the parametric source with a characteristic width of the directivity diagram θ is equal to $V = \pi r^2 c\tau\theta^2/2$, where c is the sound speed in the liquid. So long as the intensity $\Im$ and pressure amplitude P for the scattered and incident waves are connected by the relations $\Im_s \sim P_s^2$ and $\Im_i \sim P_i^2$, from Eq. (1) and Eq. (4) one can obtain an expression for the bubble size distribution function[2]

$$g(R) = \frac{4\,\delta_0(R)}{\pi^2 c\,\tau\,\theta^2 R^3}\left(\frac{P_s}{P_i}\right)^2 \tag{5}$$

where the index "o" indicates that all the values refer to the resonant bubbles. In this case it is supposed that the absorption coefficient α is small and the weakening of the acoustic wave at the expense of absorption at the maximum range of acoustic sounding is insignificant. Eq. (5) allows the determination of the bubble distribution function $g(R)$ on the basis of experimental data on the frequency dependences $P_s(f)$ and $P_i(f)$ and on the basis of attenuation constant $\delta_0(R)$ calculations.

Eq. (5) is obtained on the assumption that the thickness of the scattering layer D exceeds the spatial length of the acoustic pulse, i.e. the condition $D > c\tau$ is valid. Also of some interest is the possibility of determining the bubble size distribution function $g(R)$ under the condition that $D < c\tau$. In this case the value of the scattering volume is independent of the pulse length τ and is equal to $V = \pi r^2\theta^2 D$. Then the expression for the bubble size distribution function can be obtained in the following form

$$g(R) = \frac{2\,\delta_0(R)}{\pi^2 D\ \theta^2 R^3}\left(\frac{P_s}{P_i}\right)^2 . \tag{5'}$$

Thus, the expression for the distribution function $g(R)$ depends on the relationship between the thickness of the D layer and length of the acoustic pulse $c\tau$. Therefore, to obtain the most complete information on the bubble distribution in sea water, measurements should be conducted not only at different frequencies f but also at different values of the pulse length τ, with the quantity τ longer than the characteristic time needed to establish steady-state bubble oscillations.

EXPERIMENTAL RESULTS

We shall present the results of experimental investigations of gas bubble size distribution in sea water. In Fig. 1, a block diagram shows an experimental installation based on a parametric sound source which allows us to form a high radiation directivity in the wide frequency range of 4 to 40 kHz, determined by an average pumping frequency of about 150 kHz.

Fig. 2 presents the distribution function $g(R)$ of gas bubbles in sea water at different depths of 2.5 and 18 m measured under conditions of a shallow gulf at slight sea. It is seen that with the depth the bubble concentration decreases by an order of magnitude. Also evident is a decrease in the fluctuations of $g(R)$ at the greater depth. The distribution function can be approximated by the dependence $g(R) \sim R^{-3.7}$ which is in conformity with the results of other workers[3,4].

Fig. 3 shows the results of our measurements obtained by acoustic methods in the frequency ranges of 4 kHz to 40 kHz and 60 kHz to 150 kHz. Also shown are results by other authors,[3-12] obtained by acoustical and other methods in both tap water and sea water. Gavrilov's results of acoustical measurements in tap water settled for 25 minutes, 3 hours,

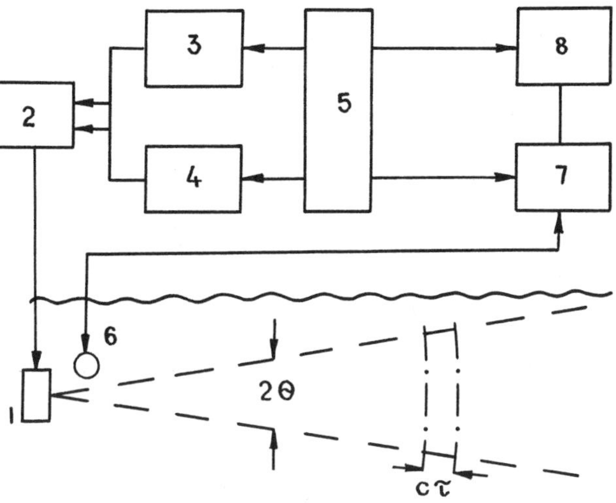

Fig. 1. A schematic of the experimental installation.
1 - *parametric sound source,* **2**- *power amplifier,* **3,4**- *signal generators,* **5**- *computer,* **6**- *hydrophone,* **7**- *band filter,* **8**- *tape recorder.*

and several days, respectively, are presented here too. Medwin's results were obtained by acoustical measurements in sea water in November, February and August. Colobaev's results were obtained by optical measurements in sea water. Johnson and Cook's results were also obtained by means of optical methods. Dalen and Løvik obtained their results by acoustical measurements in sea water and Weitendorf's results were obtained by optical methods and they correspond to measurements in a wake in sea water. Results of nonlinear acoustical measurements in sea water and results of cavitational measurements in tap water are also presented. "a" denotes results of measurements in sea water by optical methods. Results of our measurements are designated by b, c, d, e and f: b corresponds to results of measurements in sea water at the depths of 2.5 m to 18 m in a shallow gulf; c - under the same conditions in a wake behind a launch; d - in an equitorial region of the Pacific at the depths of 2.5 m to 20 m; e - in the north-western part of the Pacific at the depths of 1 m to 16 m; and f - at the depths of 3 m to 5 m in a shallow gulf.

It can be seen from Fig. 3 that with the growth of bubble radii, their concentration in sea water decreases according to the law $g(R) \sim R^{-n}$, where n, the exponent, is within the range of 3.5 to 3.8. The results of the various authors differ mainly due to the different conditions under which the measurements were obtained (different sea state, water gas content, wind speed, etc.).

NONSTATIONARY SOUND BACKSCATTERING

The above mentioned results correspond to the scattering of acoustic signals with a pulse duration of $\tau > \tau_0$, where τ_0 determines the characteristic time of establishing steady-state oscillations of the phase inclusions and is connected with the quality factor by the relation $\tau_0 = Q/\pi f$.

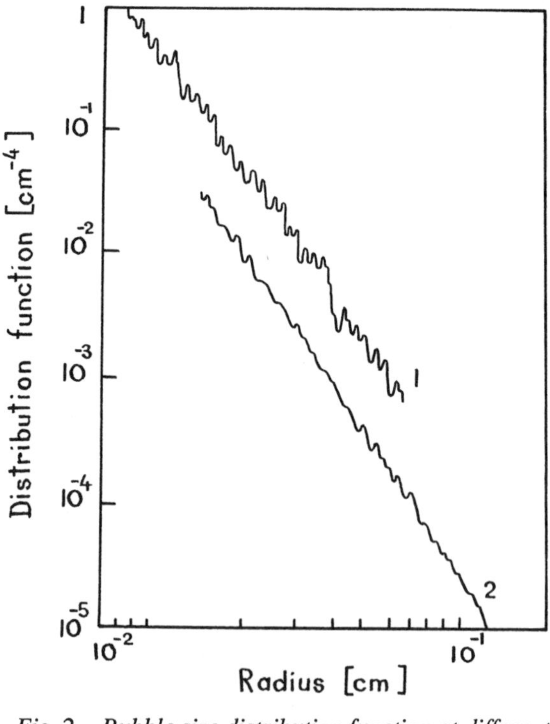

Fig. 2. Bubble size distribution function at different depths in the sea.
1 *- 2.5 m depth,* **2** *- 18 m depth.*

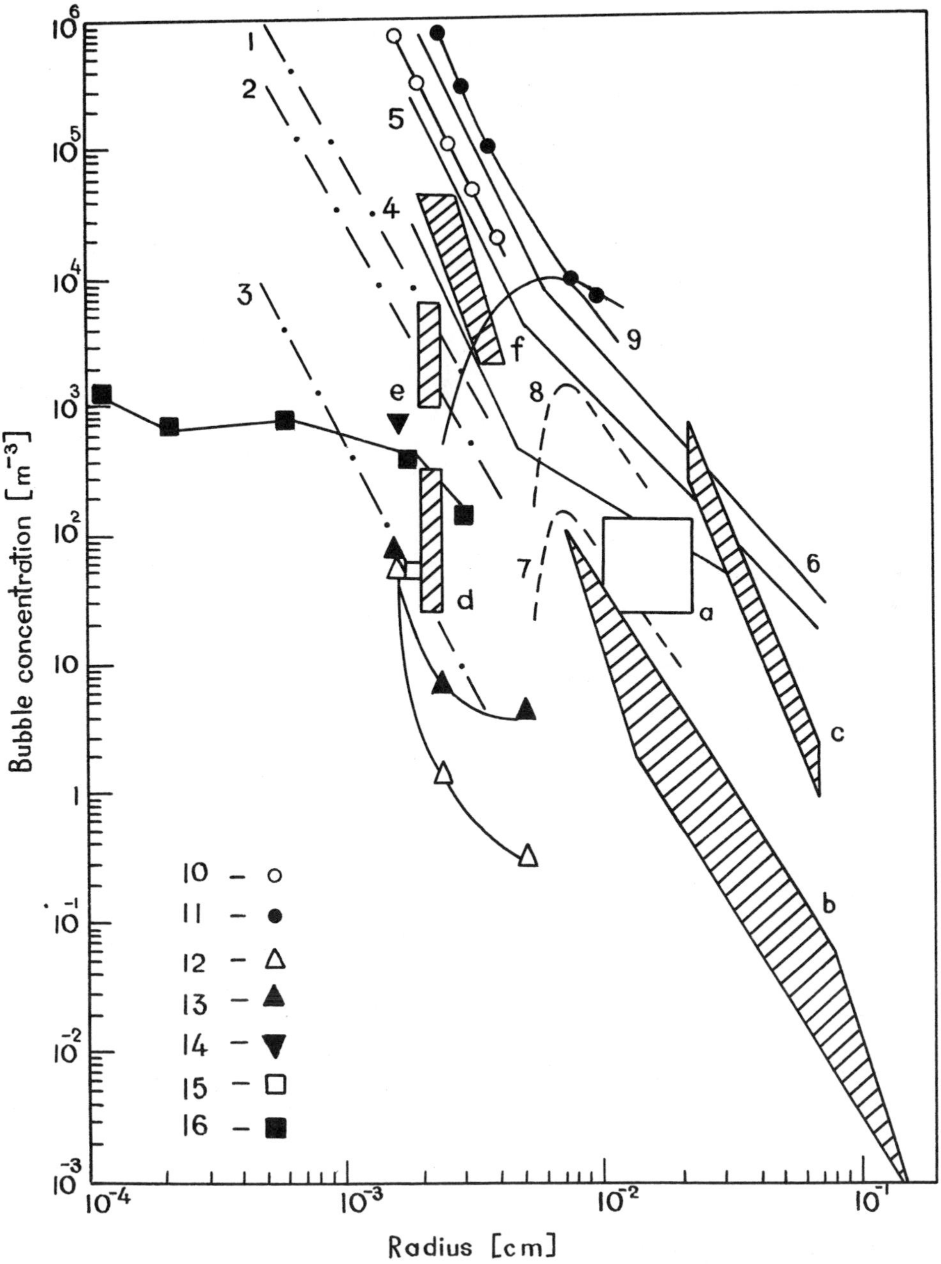

Fig. 3. *Results of the measurement of bubble concentration in water, obtained by different authors.* **1 - 3** - *Gavrilov*[3], **4- 6**- *Medwin*[4], **7** *and* **8**- *Colobaev*[5], **9**- *Johnson and Cook*[6], **10**- *Dalen and Løvik*[7], **11**- *Weitendorf*[8], **12-14**- *Sandler et al.*[9], **15**- *Ostrovsky and Sutin*[10], **16**- *Messino et al.*[11], **a**- *Blanchard and Woodcock*[12], **b,c,d,e,** *and* **f**- *results of our measurements in sea water.*[2]

The combined sound backscattering cross-section of bubbles, σ_1, and solid particles, σ_2, in a single volume of the medium is in general defined by the expression

$$\sigma = \sigma_1 + \sigma_2 = \int [\, |\sigma_1|^2 g_1(R) + |\sigma_2|^2 g_2(R)\,]\, dR. \tag{6}$$

where $g_1(R)$ and $g_2(R)$ are respectively the distribution functions for bubbles and solid particles according to size R. The integration is made from the minimum R_{min} to the maximum R_{max} sizes of the phase inclusions. The scattering section σ proves to be dependent on the pulse length τ. We introduce the function $w(\tau) = (P_s/P_i)/\sqrt{\tau}$. One can show that at different pulse lengths the expression

$$w^2(\tau) = \frac{\pi c\, \theta^2}{2} \left\{ \left[1 - \frac{1 - e^{-\tau/\tau_0}}{\tau/\tau_0} \right] \sigma_1(\infty) + \sigma_2 \right\} \tag{7}$$

is valid, where $\sigma_1(\infty)$ is the cross section of acoustic scattering by bubbles at $\tau > \tau_0$ which has the form

$$\sigma_1(\infty) = \frac{\pi R^3 g(R)}{2\delta} \quad . \tag{8}$$

For long pulses $\tau > \tau_0$ the value $w(\infty)$ is determined by the steady oscillations of the bubbles and the solid particles. For short pulses $\tau < \tau_0$, the value $w(0)$ is determined by unsteady oscillations of the bubbles and steady oscillations of the solid particles. Due to the considerable extent of resonant sound scattering in sea water, the value $w(\infty)$ exceeds $w(0)$ essentially. The bubble size distribution function can be separated from the influence of other background scatterers and can be defined by the formula

$$g(R) = \frac{4\delta_0(R)}{\pi^2 c\, \theta^2 R^3} \left[w^2(\infty) - w^2(0) \right] \quad . \tag{9}$$

Fig. 4 shows the dependences of the function $w(\tau)$ on the τf value which have been obtained experimentally in sea water at a depth of about 3 m for acoustic signals with different frequencies. It is seen that with the reduction of the pulse length τ the value of this function decreases in accordance with the formula

$$w^2(\tau) = w^2(0) + \left[1 - \frac{1 - e^{-\tau/\tau_0}}{\tau/\tau_0} \right] \left[w^2(\infty) - w^2(0) \right] \quad . \tag{10}$$

In Fig. 4 the values of the function $w(\tau)$ are marked with circles which correspond to the length of τ_0. So far as $\tau_0 = Q/\pi f$, one can determine the quality factor of the bubbles.

The values obtained in such a way, Q_e, are compared with the calculated value, Q_c, in the table below.

f [kHz]	5	15	25	35
R [cm]	0.074	0.025	0.015	0.011
Q_e	19.8	17.3	-	16.3
Q_c	33	13.8	11	10

It follows from this table that for the bubbles with resonance frequencies of 5 to 35 kHz, Q reduces from 20 to 16, respectively. Thus, the study of scattering at different pulse lengths allows us to experimentally determine the values of Q for gas bubbles at different frequencies.

Eqs (7) - (10) are obtained for the case of extended bubble layers, where $D > c\tau$. In the case of layers of little extent, when $D < c\tau$, Eqs (7) - (10) do not allow the determination of

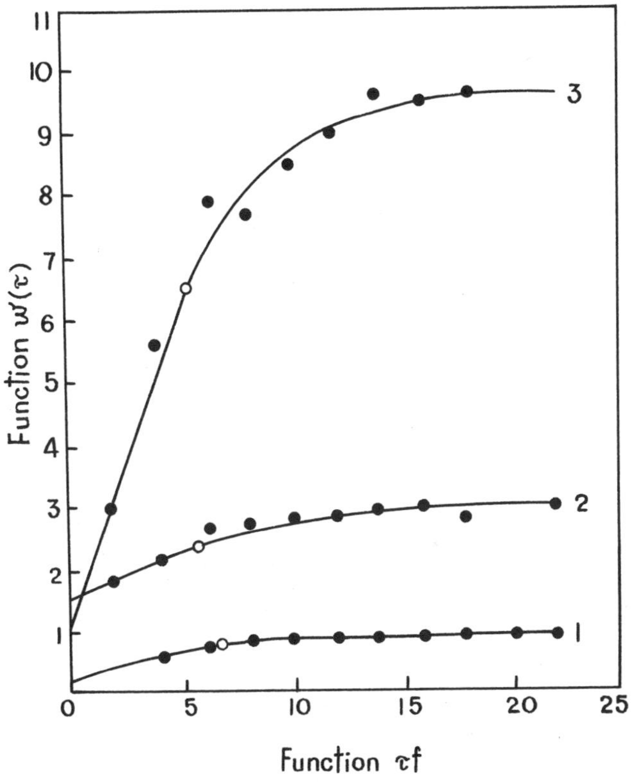

Fig. 4. The function $w(\tau)$ versus pulse length τ at different frequencies f.
1 - f = 5 kHz, 2 - f = 15 kHz, 3 - f = 35 kHz;
black circlets show experimental results;
white circlets show $w(\tau_0)$ values.

the quality factor of the bubbles if the τ value decreases in the above mentioned manner. It should be noted that in sea water under natural conditions, the case of thin bubble layers is very rarely observed when $D < c\tau$.

The sound backscattering cross-section from a unit volume of sea water with nonresonant inclusions can be defined by the formula

$$\sigma_2 = \frac{2w^2(0)}{\pi c \theta^2} \quad . \tag{11}$$

Hence, the possibility of separating scattering from bubbles and solid particles occurs by using pulsed signals with different lengths and different carrier frequencies.

CONCLUSION

Through the application of parametric sound sources one can succeed in obtaining results such as those shown here which are almost completely free from errors caused by pulse length effects due to non-steady-state bubble oscillations. Furthermore, one can succeed in distinguishing between the scattering of sound by gas bubbles and the scattering of sound by solid particles.

REFERENCES

1. A. Ishimary, "Wave Propagation and Scattering in Random Media," Academic Press, New York (1978).
2. V. A. Akulichev, V. A. Bulanov, S. A. Klenin and V. D. Kiselyov, The use of parametric sound sources in the study of sound scattering in sea water, in: "Proc. 10th Intern. Symposium on Nonlinear Acoustics," Kobe (1984).
3. L. R. Gavrilov, On the gas bubble size distribution in sea water, Akust. Zhurnal (Russian), 15:25 (1969).
4. H. Medwin, In situ acoustic measurements of bubble populations in coastal ocean water, J. Geophys. Res., 75:599 (1970).
5. P. A. Colobaev, The study of concentration and statistical size distribution of wind borne bubbles in the near surface ocean layer, Oceanologiya (Russian), 15:1013 (1975).
6. B. D. Johnson and R. C. Cook, Bubble populations and spectra in coastal water: photographic approach, J. Geophys. Res., 84:3761 (1979).
7. J. Dalen and A. Løvik, The influence of wind-induced bubbles on echo integration, J. Acoust. Soc. Am., 69:1653 (1981).
8. E. A. Weitendorf, Complementing discussion contribution to the papers of H. Medwin, P. Schippers and A. Løvik, in: "Cavitation and Inhomogeneities in Underwater Acoustics," W. Lauterborn, ed., Springer-Verlag, Berlin (1980).
9. B. M. Sandler, D. A. Selivanovsky and A. Yu. Sokolov, The measurement of gas bubble concentrations in the near-surface layer of the sea, Doklady AN (Russian), 260:1474 (1981).
10. L. A. Ostrovsky and A. M. Sutin, Nonlinear acoustical methods for the diagnostics of gas bubbles in liquid, in: "Ultrasonic Diagnostics," Inst. Appl. Phys., Gorkii (1983).
11. D. Messino, D. Sette and F. Wanderling, Statistical approach to ultrasonic cavitation, J. Acoust. Soc. Am., 35:1575 (1963).
12. D. C. Blanchard and A. H. Woodcock, Bubble formation and modification in the sea and its meteorological significance, Tellus, 9:145 (1957).

SCARING EFFECTS IN FISH AND HARMFUL EFFECTS ON EGGS, LARVAE AND FRY BY OFFSHORE SEISMIC EXPLORATIONS

John Dalen* and Geir Magne Knutsen

Institute of Marine Research
P.O.Box 1870 N-5011 BERGEN, NORWAY
*Present address: ELAB, O.S. Bragstads pl 6
N-7034 Trondheim-NTH, Norway

ABSTRACT

A co-operating survey between a seismic vessel and a research vessel took place in the North Sea during June, 1984. Changes of the behaviour patterns of the fish along the course lines of the seismic vessel from immediately before to just after airgun shooting proved that the fish were affected. Changes in the overall fish distribution were demonstrated between the situation before shooting started to that after one week of shooting. Eggs, larvae and fry of cod were exposed to a small airgun and fry to a large airgun and a watergun during March-July 1985. Significant damages when exposed, to the small airgun were not observed. The older fry exposed to both airguns at 1 m and to the watergun at 6 m got problems with their balance. The watergun killed 90% of the fry at short distances.

SCARING EFFECTS ON FISH FROM 3-DIMENSIONAL SEISMIC SURVEYS

Introduction

During recent years different configurations of airguns have mostly been applied as offshore seismic sources. The generated sound energy from airguns is distributed at rather low frequencies which are audible to most fish species (Chapman and Hawkins 1969 a, b, Olsen 1969 a, b). The sound pressure which is presented to the fish is a function of the supplied air pressure to the airguns, the volume of the chambers, the number of airguns and the distance between the airguns and the fish.

Until some years ago 2-dimensional exploration methods with inter-course-line distances of 1 to 10 km were mostly in use. By the more recent applied 3-dimensional methods the seismic vessel runs on parallel course lines 50 to 100 m apart. From this we may expect that the fish will be exposed in a more intensive, continuous and systematic manner by 3-dimensional surveys than by 2-dimensional surveys.

Methods

The survey took place in Mid June 1984 with the research equipped stern-trawler/purse seiner MV "Libas" and the seismic vessel MV "Malene

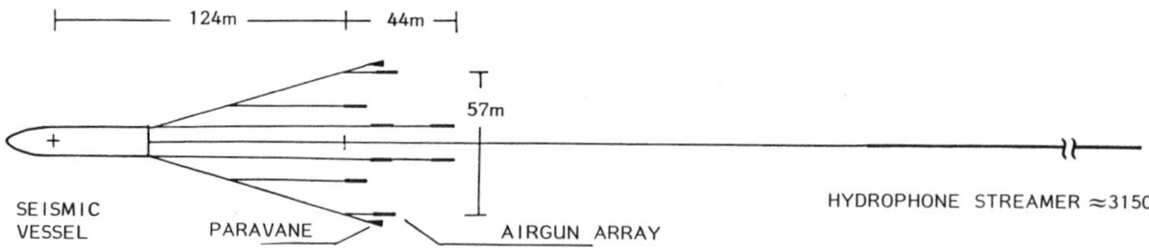

Fig. 1. Seismic vessel, with airgun arrays in a "super wide airgun" (SWAG) configuration and the hydrophone streamer.

Østervold" in the North Sea, block 34/7. Block 34/7 is defined by the latitudes 61°15'N and 61°30'N, and the longitudes 02°00'E and 02°20'E. The seismic explorations took place in the southern part of the block in an area of extension 6·10 n. miles, denoted as "the seismic area" (indicated by dotted lines on Fig. 3).

For acoustic fish observations MV "Libas" was equipped with: The echo-sounder, SIMRAD Ek 38 A, an echo integrator, SIMRAD QM Mk II, the sonars, SIMRAD SM 600 and WESMAR SS 200. For the seismic explorations MV "Malene Østervold" held the following airgun set-up: 40 airguns distributed along 8 arrays with 5 airguns in each array towed at 6 m depth. The airguns used were the Bolt 1500 B, 1500 C, and the 600 B. The total chamber volume was 77932 cm^3 (4752 cu.in.) while the supplied air pressure to the airguns was 138 kg/cm^2 (2000 psi). The firing period was 10 sec corresponding to firing intervals of approximately 25 m at 5 knots speed. The acoustic output of the primary pulse was measured (during calibration) to 3.144 MPa re 1 m which equals 249.9 dB // 1 μPa re 1 m. Fig. 1 shows a horizontal view of the current airgun configuration and the streamer.

The program of the survey comprised the following tasks:

Task 1: The observation vessel should observe the abundance and distribution of fish prior to the airgun shooting along every new course line the seismic vessel was to run. Immediately after the shooting along the line the observation vessel should make a new sweep along the line to observe any changes in the abundance and distribution patterns of the fish.

Task 2: Before the seismic exploration started the fish populations in block 34/7 and adjacent blocks should be acoustically mapped using echo-sounder, echo integrator and sonars to provide information on fish distributions in a practically undisturbed state. A number of trawl stations should be taken to identify the fish by species and size.

Task 3: The acoustic fish mapping should be repeated a number of times during the period to obtain information of any migration or movements of the fish in the seismic area and adjacent areas.

Task 4: The observation vessel should be stationary at specific distances (150 - 300 m) from the course lines of the seismic vessel. While this vessel approaches, passes and moves away, changes of the behaviour of the fish in the area should be observed. This procedure should be repeated under different conditions.

Results and Discussion

Behaviour and distribution of fish along the course lines before and after the airgun shooting. The seismic explorations started at June 16. During this program we worked according to Task 1.

To study the behaviour of each species or groups of species the echo-integrator recordings were divided into groups based on the echograms and the trawl catch compositions:

1 - Demersal fish: Including saithe (Gadus virens L.), cod (Gadus morhua L.), haddock (Melanogrammus aeglefinus L.), whiting (Merlangus merlangus L.), great silversmelt (Argentina silus), ling (Molva molva L.), tusk (Brosme brosme L.), and other rare species.

2 - Blue whiting (Micromesistius poutassou) - a pelagic species.

3 - Small pelagic fish: Including Norway pout (Trisopterus esmarleii L.), lantern fish (Myctophidae sp.), silvery cod (Gadiculus argenteus thori L.), and few and sparse recordings of herring (Clupea harengus L.)

4 - Plankton and spawn. Appearing in the upper 40 metres in parts of the area.

The seismic area was divided into 12 sub-areas of extension 2·2.5 n. miles for statistical analysis and for data scrutinizing. Table 1 shows the mean echo-integrator recordings, i.e. mean echo abundance in each sub- area before, $\overline{I}_b$, and after, $\overline{I}_a$, shooting with standard deviation and number of observations.

Fig. 2 shows a vertical section of the seismic vessel and distributions of pelagic and demersal fish. The hypothesis to be tested is that the pelagic species migrate out to each side of the course line and that the demersal species migrate downwards to the bottom as the vessel runs the course line. If the hypothesis is true, this yields that the echo integrator recordings, will be reduced along each course line after the shooting compared to the values prior to the shooting both for the pelagic species and for the demersal species which cannot be fully observed as some of the fish enter the echo-sounder dead zone close to the bottom.

The depth within the seismic area varied from 100 m to 300 m. Based on the measured resultant source strength, 249.9 dB//1μPa re 1 m, and simply reducing this by spherical spreading out to relevant depths, the estimated sound pressure levels are 210, 204, and 200 dB//1μPa at 100, 200, and 300 m respectively.

Dalen (1973) showed that herring responded to airgun sound pressure levels of 180-186 dB//1μPa by swimming away from the airgun locations. Similar observations were reported by Chapman and Hawkins (1969b) on whiting at airgun sound pressure level of 188 dB//1μPa. The estimated values are at least 12-20 dB above the referred values so positive responces to the current airgun set-up are expected.

When we calculated the averaged echo abundance from Table 1 for the species groups 1-3 for the whole seismic area from the observations along the course lines prior to the shooting (105 observations), and those after the shooting (110 observations) we found: For the demersal species the averaged echo abundance was reduced by 36%, for the blue whiting by 54%, and for the small pelagic species by 13% respectively which confirms the hypothesis.

To indicate possible changes of the compositions of species and of the abundance of fish in the depth-range close to the bottom before and after the airgun shooting, three comparative bottom trawl stations were undertaken June 20. One trawl station was taken before and two immediately after the shooting on one line at nearby-laying positions. The results showed that the number of fish of the demersal species in the trawl catches increased by 34% and 290% from the catch before shooting to the two catches after shooting respectively.

Table 1. Mean relative echo abundance [mm/nm^2] of demersal fish, blue whiting and small pelagic fish before, I_b, and after I_a, the airgun shooting along each course line. S is standard deviation, and N is number of observations in the sub-areas.

	Demersal species							Blue whiting							Small pelagic species							
	Before shooting			After shooting				Before shooting			After shooting				Before shooting			After shooting				
Sub-area	$\bar{I}_b$	S	N	$\bar{I}_a$	S	N	$\bar{I}_b-\bar{I}_a$	$\bar{I}_b$	S	N	$\bar{I}_a$	S	N	$\bar{I}_b-\bar{I}_a$	$\bar{I}_b$	S	N	$\bar{I}_a$	S	N	$\bar{I}_b-\bar{I}_a$	Comments
I	1.5	0.7	2	4.2	3.1	9		4.5	3.5	2	0.1	0.3	9		1.0	1.4	2	3.7	3.8	9		Not used, few recordings before shooting
II	-	-	-	4.2	3.9	10		-	-	-	3.6	5.0	10		-	-	-	1.5	1.7	10		Not used, few recordings before shooting
III	2.8	2.2	9	2.5	1.8	11	0.3	71.7	103.8	9	12.0	17.9	11	59.7	0	-	9	0.4	0.8	11	-0.4	
IV	1.2	0.9	10	1.1	1.1	10	0.1	77.3	101.4	10	34.4	26.5	10	42.9	0	-	10	0	-	10	0	
V	5.0	3.5	17	3.6	2.4	14	1.4	0	-	17	0	-	14	0	2.6	2.8	17	3.0	3.2	14	-0.4	
VI	7.8	8.2	4	1.8	1.2	14		0	-	4	0	-	14		1.3	1.5	4	1.1	0.9	14		Not used, few recordings before shooting
VII	0	-	1	1.9	1.6	15		1.0	-	1	2.0	5.1	15		0	-	1	1.5	1.6	15		Not used, 1 recording before shooting
VIII	1.0	1.4	9	0.9	1.0	15	0.1	29.9	41.5	9	21.7	32.3	15	8.2	0.8	1.4	9	1.9	3.3	15	-1.1	
IX	4.1	4.2	18	2.5	1.9	15	1.6	0	-	18	0	-	15	0	2.1	1.9	18	0.7	0.8	15	1.4	
X	4.7	6.3	16	2.3	1.8	14	2.4	0	-	16	0	-	14	0	1.5	1.8	16	1.3	2.0	14	0.2	
XI	7.3	6.4	12	2.2	2.9	16	5.1	0.2	0.6	12	0	-	16	0.2	2.3	1.9	12	1.4	1.1	16	0.9	
XII	2.9	2.8	14	4.1	6.3	15	-1.2	0.4	0.8	14	1.7	3.2	15	-1.3	0.9	1.1	14	1.3	1.3	15	-0.4	

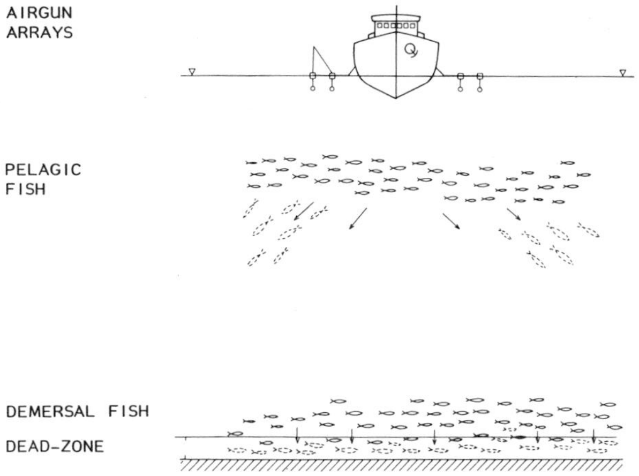

Fig. 2. Sketch of the seismic vessel and distributions of fish.

A binomial test (Zar 1974) and a nonparametric test (Mann-Whitney test) (Lehmann 1975) were performed of the data of Table 1. The hypothesis of the tests were:

$$H_0: \bar{I}_b = \bar{I}_a \tag{1}$$

$$A: \text{(alternative hypothesis)}: \bar{I}_b > \bar{I}_a \tag{2}$$

Sub-areas having less than 5 observations were excluded from the tests and Table 1 thus yields:

Demersal species: $\bar{I}_b > \bar{I}_a$ in 7 of 8 events

Blue whiting: $\bar{I}_b > \bar{I}_a$ in 4 of 5 events

Small pelagic species: $\bar{I}_b > \bar{I}_a$ in 3 of 7 events

The computed probabilities according to the one-sided binomial test are:

$$P(x)_{\text{demersal species}} = 0.0313 \tag{3}$$

$$P(x)_{\text{blue whiting}} = 0.0783 \tag{4}$$

$$P(x)_{\text{small pelagic species}} = 0.2734 \tag{5}$$

On a significance level equal to 0.05 the observed echo abundance of demersal species were higher before shooting compared to that after shooting. For blue whiting and small pelagic species the tests did not confirm the alternative hypothesis at this high significance level.

Nonparametric one-sided test (Mann-Whitney test). This test will show the strength of the hypothesis relative to a t-test since the observations are not normally distributed.

The results showed that the high number of events where H_0 could not be rejected on a significant level of 0.05, meaning that the characteristic changes between the echo abundances before and after the shooting were not that great to verify the alternative hypothesis.

All together we conclude that the statistical tests did not show as clearly as the calculated reduction of the averaged echo abundance that there were consistent changes of the echo abundance from before shooting to

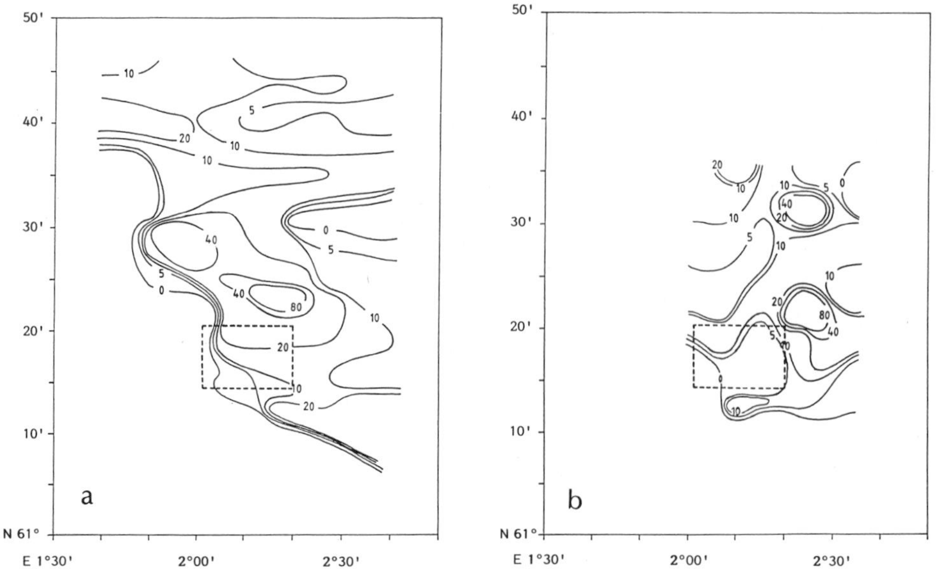

Fig. 3. Echo abundance distributions of blue whiting. a: prior to the airgun shooting, b: after 6 days of airgun shooting, (echo integrator recordings in relative units: mm/sqr. nautical mile).

after shooting.

<u>Geographical distributions of fish in the area.</u> According to Task 2 we carried out an acoustic survey of block 34/7 and the 8 adjacent blocks prior to the seismic activities during June 13-15. The area was covered by zig-zag course lines with intercourse line distances of 5 nautical miles. Based on the fish distributions and the need for sufficient coverage 12 trawl stations were carried out to identify the recordings by species, and each species by size. During the last 18 hrs. of the survey, we made the final acoustic survey of block 34/7 and nearby areas according to Task 3.

Fig. 3 shows as an example the distributions of blue whiting by echo abundance maps. Comparing these maps from the two periods, i.e. prior to the seismic activities and after 6 days of airgun shooting, we can draw the following conclusions:

<u>Demersal species:</u> The distribution had changed to a situation where the fish either were forced to the bottom or the fish had migrated out of the seismic area or both, since the overall echo abundance had been reduced. Results from the comparative trawling indicated evidences for the first mentioned reason.

<u>Blue whiting:</u> The fish had migrated out of the seismic area - to the north and the east since the overall echo abundance had been strongly reduced in the area.

<u>Small pelagic fish:</u> Here we could not demonstrate any particular systematic changes of the fish distributions between the two situations. There were some reduction of the echo abundance after 6 days of airgun shooting although the overall echo abundance was rather low during the whole survey period.

The sub-program expressed in Task 4 could only be undertaken a few number of times due to unfavourable scattered fish distributions, and that

"Libas" could not be as close to the course lines of "Malene Østervold" as these scattered fish distributions required, due to unwanted ship noise picked up by the hydrophone streamer.

Altogether we must express that scaring effects from offshore seismic explorations could have more clearly been demonstrated having higher fish abundances in the topical area - especially schooling fish like herring.

Acknowledgement

We are very grateful to the crew of MV "Libas" for their spirit and will of co-operation. We are also grateful to the crew and the GECO-personnel on board MV "Malene Østervold" for elegant co-operation during the survey. The interest and friendly support of Saga Petroleum A/S during the planning and running of the survey are highly appreciated.

HARMFUL EFFECTS ON EGGS, LARVAE AND FRY BY AIRGUNS AND WATERGUNS

Introduction

Several investigations describe the effects of explosives on fish (Aplin 1947, Hubbs and Rechnitzer 1952, Cocer and Hollis 1952, Jakosky and Jakosky 1956). However, only a few studies report of effects from airguns. Weinhold and Weaver (1972) investigated effects on coho salmon (Onchoryncus kisutch) smolts detecting no harmful effects.

To examine if the most common applied energy sources for seismic surveying might injure fish, we exposed eggs, larvae and fry of cod (Gadus morhua L.) to two different airguns and a watergun.

The intention of this investigation was to detect any harmful effects, and if such effects were observed, to determine at what distances they occurred and what ontogenetic stages which were most vulnerable.

Methods

Compared to an airgun having a chamber charged with high pressure air befor firing, the watergun has a chamber containing seawater of ambient pressure. When firing a watergun the seawater is forced out of the chamber by a shuttle while generating a rather weak acoustic pressure wave. When the shuttle is arrested at the outer end of the chamber, and the driving force thereby removed, inertia of the the fast-moving slugs of water causes cavities to form adjacent to the chamber ports. Subsequent collapse of these cavities results in a strong and discrete acoustic impulse - the primary pulse (Newman 1975, French and Henson 1978). Contrary to the airgun generating a primary pulse of positive pressure value, i.e. an explosion pulse, the watergun generates a primary pulse having a negative pressure value, i.e. an implosion pulse.

Table 2 shows the sound pressure levels of the applied energy sources for relevant depth and supplied high pressure air.

The eggs were collected from the spawning stock of cod at the Aquaculture Station of the Institute of Marine Research at Austevoll near Bergen. The eggs were spawned naturally. The postlarvae were reared in plastic bags in the laboratory on rotatories (Brachiones plicatitis) while the fry were reared in a pond.

The plastic bags containing the eggs, larvae and fry were placed in 6 l vessels in a climatic room having temperature and light controlling

Table 2. Sound pressure levels (nominal values) vs. chamber volume of the airguns and the watergun.

Energy source	Chamber vol. [cm^3]	Sound pressure level [dB//1μPa re 1m]:
Small airgun (Bolt 600B)	640	222.0
Large airgun (Bolt 1500C)	8610	231.0
Watergun (Seismic Systems P400)	8610	229.0

facilities. Prior to the exposure, lids were put on the plastic bags, and the vessels, including those containing the control groups, were brought out on open sea where the exposures took place. Fig. 4 shows a sketch of the set-up during the exposures. The bag and the airgun were lowered to 6 m depth, and the distance between them was varied between 1 and 10 m. Stable temperature was maintained by keeping the vessels in a cooling box while they were away from the climatic room. The largest fry were exposed in small meshed net cages (40x40x40 cm) at 4 m depth. All control groups were treated in the exact same manners as the test groups except for the exposures. Altogether 28 groups were exposed to the guns at the egg and larval stages, 19 at the fry stages - all against 35 control groups at the different stages.

After the exposure the dead specimen were taken away and counted yielding data of direct lethal affects. The fry were dissected after 7 days or after they died yielding potential data of morphological changes. Sublethal effects were detected by observing the behaviour and the feeding success. This was tested on the larvae from the groups exposed at the egg- and larvae stages, by feeding 50 larvae each day from each group with rotatories (500 prey/l) reared in the laboratory. The larvae were in the vessel for 3 hours. The feeding success was defined as the percentage of larvae with food in the stomach (Tilseth and Ellertsen 1984).

Results and discussion

Exposures with the small airgun (Bolt 600B). No significant differences of survival between the test groups and the control groups (nonparametric test (Breslow 1970) and MANTEL-COX test (Mantel 1966)) were observed at any distances at:

1 - the egg stages (2, 5 and 10 days after fertilization)

2 - the larval and postlarval stages (1, 5, 37, 38, 40, and 41 days after hatching)

3 - the age of 56, 69 and 110 days after hatching for the fry.

The feeding success of the larvae which were exposed at the egg and larval stages was not significant from those of the control groups indicating there were no sublethal effects from the small airgun. This was also the case for the behaviour studies of the fry prior to and after the exposures.

The fry at the age of 110 days got problems with their balance after the exposures, but recovered within a few minutes.

Exposures with the large airgun (Bolt 1500C). Only fry of age 110 days were exposed to this airgun. The reason for this was that the large airgun and the watergun were not available during the earlier phases of the experi-

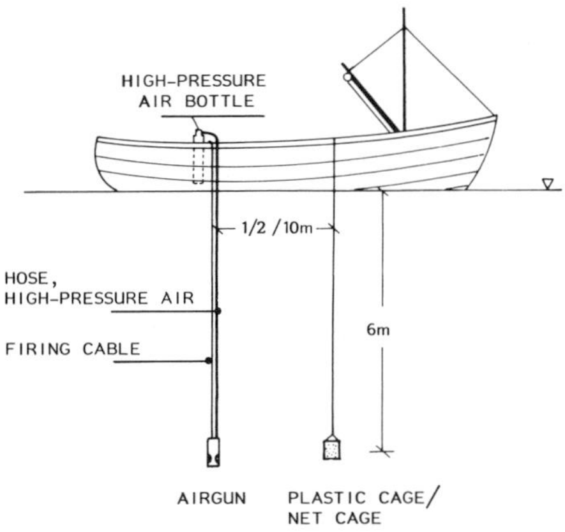

Fig. 4. Sketch of the boat, airgun set-up, and cage containing eggs, larvae or fry.

ments. None of the specimen were killed. They got problems with their balance, but recovered within a few minutes.

Exposures with the watergun (Seismic Systems P400). As mentioned only fry of age 110 days were exposed to this energy source too. At a distance of 2 m 90% of the specimen were killed immediately or died within 2-3 hours after the exposures. When dissected all dead specimen had ruptured swimbladders, with hemorrhaging alongside the swimbladder and on the upper part of the liver. 30% of them had also ruptured bellywalls. At the distance of 6 m all specimen got problems with their balance but recovered except for one that died after 2 days.

When we compare and evaluate the effects from airguns and waterguns on fish, it is important to notice the different polarity of their primary pulses. An airgun generates a positive pressure pulse while a watergun a negative one. The fish and its internal organs will be exposed to a less dangerous compression by the airgun than to the more dangerous expansions by the watergun which can tear the swimbladder and the bellywall.

We believe that the demonstrated effects here are effects at low exposure levels compared to offshore real situations where a great number of guns are fired approximately simultaneously of which the resultant pressure wave will be strongly amplified.

ACKNOWLEDGEMENTS

We are very grateful to the staff of the Aquaculture Station of the Institute of Marine Research, Austevoll, for their kindness and never-ending will of co-operation. The interest and support of Fjord Instruments A/S during the last phase of the experiments are strongly appreciated.

The total project has been financed by the Institute of Marine Research, Bergen, and the Royal Norwegian Department of Oil and Energy.

REFERENCES

Aplin, J.A. 1947. The Effect of Explosives on Marine Life. California Fish and Game, vol. 33, pp 23-30.

Breslow, N. 1970. A generalized Kruskal-Wallis Test for comparing k samples subject to unequal Patterns of Censorship. Biometrika 57, pp 579-594.

Chapman, C.J. and Hawkins, A.D. 1969a. A Field Determination of Hearing Thresholds for the Cod (Gadus morhua L.). 8th I.F. Meeting, Lowestoft. (mimeo.)

1969b. The Importance of Sound in Fish Behaviour in Relation to Capture by Trawls. FAO Fish. Rep., (62) vol. 3, pp 717-729.

Cocer, C.M. and Hollis, E.H. 1950. Fish Mortality caused by a series of heavy Explosions in Chesapeake bay. Jour. Wildlife Management, vol. 14, No. 4, pp 435-444.

Dalen, J. 1973. Stimulering av sildestimer. Forsøk i Hopavågen og Imsterfjorden/Verrafjorden 1973. Rapport for NTNF. (Stimulating Herring Schools. Investigations in Hopavågen and Imsterfjorden/Verrafjorden 1973. Report for the NTNF.) Institute of Technical Cybernetics, The Norwegian Institute of Technology, Trondheim. (In Norwegian).

French, W.S. and Henson, C.G. 1978. Signature Measurements on the Water Gun Marine Seismic Source. Contr. to Offshore Tech. Conf. 1978. Houston, Texas, USA.

Hubbs, C.L. and Rechnitzer, A.B. 1952. Report on Experiments Designed to Determine Effects of Underwater Explosions on Fish Life. California Fish and Game, vol. 38, pp 333-366.

Jakosky, J.J. and Jakosky, J.Jr. 1956. Characteristics of Explosives for Marine Seismic Explorations, Geophysics, Vol. 21, pp 969-991.

Lehmann, E.L. 1975. NONPARAMETRICS: Statistical Methods Based on Ranks. Holden-Day, Inc., McGraw-Hill Int. Book Company, San Francisco, USA.

Mantel, N. 1966. Evaluation of Survival Data and two New Rank Order Statistics arising in its Considerations. Cancer Chemotherapy Reports 50, pp 163-170.

Newman, P. 1978. Watergun fills Marine Seismic Gap. The Oil and Gas Journal, Aug. 7, 1978, pp 138-150.

Olsen, K. 1969a. Directional Hearing in Cod (Gadus morhua L.). 8th I.F. Meeting, Lowestoft, (mimeo).

1969b. Directional Responses in Herring to Sound and Noise Stimuli. ICES, Meet., Gear and Behaviour Comm., Bergen, (mimeo).

Tilseth, S., and Ellertsen, B. 1984. Feeding and Vertical Distribution of Cod Larvae in Relation to Availability of Prey Organisms. Rapp. P.-V. Reun. Cons. int. Explor. Mer, 178, pp 317-319.

Weinhold, R.J., and R.R. Weaver, 1972. Seismic Airguns Effect on Immature Coho Salmon. Contr. to the 42nd Ann. Meet., Society of Exploration Geophysicists, Anaheim, California 1972. Unpubl. 15 p.

Zar, J.H. 1974. Biostatistical Analysis. Prentice-Hall, Inc., Englewood Cliffs, N.J. USA.

FISH STOCK ASSESSMENT BY A STATISTICAL ANALYSIS OF ECHO SOUNDER SIGNALS

P.N. Denbigh and J. Weintroub

Central Acoustics Laboratory
Department of Electrical and Electronic Engineering
University of Cape Town, Rondebosch 7700, R.S.A.

ABSTRACT

In 1984 Wilhelmij and Denbigh[1] described a technique for determining the number density of random scatterers based upon a statistical analysis of acoustic backscattered signals. This paper examines the application of the method to the specific problem of estimating the number density of fish within a shoal using the signals obtained from an echo sounder.

INTRODUCTION

If the return signal from a fish shoal at some given instant is caused by overlapping echoes from a very large number of fish, the central-limit theorem predicts that the amplitude of the echo waveform should have Gaussian statistics. This is equivalent to having an envelope which has Rayleigh statistics or an intensity which has exponential statistics, the intensity being the square of the envelope. If the average number of overlapping echoes contributing to the signal at each instant decreases to about ten or less, the central-limit theorem becomes non-applicable and there is a significant deviation of the statistics from those just mentioned. A suitable measure of this deviation can be used to predict the average number of overlapping echoes, i.e. the average number of fish in the resolution cell of the echo sounder. A division of this number by the volume of the resolution cell gives the number of fish per cubic meter, or the number density.

The original paper by Wilhelmij and Denbigh was based upon the work of Pusey et al[2] for electromagnetic scattering and used the second normalized moment of backscattered intensity as a pertinent measure of the statistical properties of the waveform. The second normalized moment of intensity is the second moment of intensity divided by the square of the first moment, or $\langle I^2\rangle/\langle I\rangle^2$. For the case of identical scatterers having a constant scattering strength and a Poisson volume distribution, Pusey et al showed that if the statistics are stationary this second normalized moment is related to the average number of scatterers by the formula

$$\langle N\rangle = \frac{1}{\langle I^2\rangle/\langle I\rangle^2 - 2} \tag{1}$$

If the number of scatterers is large enough for the central-limit theorem to apply, we would find $\langle I^2\rangle/\langle I\rangle^2 = 2$, giving $\langle N\rangle = \infty$ from the above formula. Table 1 gives a few examples showing the relationship between $\langle N\rangle$ and the second normalized moment based upon a rearrangement of this formula.

Table 1. The second normalized moment of intensity as a function of $\langle N\rangle$.

$\langle N\rangle$	$\langle I^2\rangle/\langle I\rangle^2 = 2 + 1/\langle N\rangle$
64	2.016
32	2.031
16	2.063
8	2.125
4	2.25
2	2.5
1	3
0.5	4

It becomes clear from Table 1 that a very high precision is needed in the value of the second normalized moment if large values of $\langle N\rangle$ are to be deduced from it with any accuracy.

The paper by Wilhelmij and Denbigh served three main purposes. It verified the applicability of the technique using experimental backscattered ultrasonic signals from a random matrix of polystyrene beads in a water tank. It showed how the problem of non-stationary statistics due to beam divergence may be overcome by a suitable averaging procedure. It also predicted theoretically, for different values of $\langle N\rangle$, how many independent measurements of backscattered intensity are needed in order to obtain estimates of $\langle N\rangle$ with given accuracies.

This present paper has four main objectives. Firstly, it presents evidence to suggest that a Rayleigh distribution of echo amplitudes is more appropriate to the scattering from fish in a fish shoal than the assumption of identical echo amplitudes. Secondly, a theoretical derivation is given relating $\langle N\rangle$ to the second normalized moment when the scatterers have a Rayleigh distribution of target strengths. Thirdly, it presents computer simulations to determine the error in estimates as a function of the true value of $\langle N\rangle$ and of the number of independent measurements. Fourthly, it presents results obtained by applying the method to echo sounder signals obtained from real fish shoals at sea and compares them with number density estimates based upon echo integration. Much of the material has been described by the authors at conferences[3,4]. Reference 4 regrettably contains some mistakes in the calculations of the estimates by statistical analysis.

THE PDF OF ECHO ENVELOPES FROM INDIVIDUAL FISH

Considering an individual fish as a flexible line array of many small scattering points,its echo is determined by the sum of contributions from each scattering point. The result of interference between these point scatterers will vary with time and will be different for each fish. If the contributions are of similar amplitude but random phase, the central-limit theorem may be applied once again to suggest that the echo envelope will vary from fish to fish in accordance with a Rayleigh distribution. A recent paper by Clay and Heist gives evidence that the Rician PDF is a more accurate description of acoustic scattering by individual live fish. The physical explanation is that the backscattered signal may be considered to have two components, one which varies in a noise like manner with fish orientation and flexing, and one which does not vary. The noise like component arises from interference between the distributed scattering points along the body of the fish. The constant component arises from the swim bladder when it is small enough in wavelengths to produce constant conditions for interference irrespective of orientation and flexing. The envelope of the backscattered signal can then be thought of as arising from the envelope of two superposed components,a sinewave and noise, thus leading to the classical Rician distribution.

It is proposed that the sonar used for the assessment of fish number density by the statistical technique would have a narrow beam and a good range resolution in order to achieve a small average number of scatterers in the resolution cell. It follows that high operating frequencies are needed, and that the swim bladder is likely therefore to have an appreciable dimension in wavelengths. In these circumstances, and in accordance with the findings of Clay and Heist for fish more than 25λ long, the constant amplitude component will be small and a close approximation to a Rayleigh PDF may be expected. This Rayleigh distribution of echo amplitudes is the case considered in the following analysis.

THEORY

The measures of envelope statistics used in the technique are the first and second moments of intensity, where the intensity is the square of the envelope. Consider initially that there are a fixed number of fish N contributing to the return signal at any instant. As an example let N = 4 so that there are four overlapping echoes a cos $(\omega t + \alpha)$, b cos $(\omega t + \beta)$, c cos $(\omega t + \gamma)$ and d cos $(\omega t + \delta)$ where the amplitudes a, b, c and d may be considered random variables with a Rayleigh PDF, and where the phases α, β, γ and δ may be considered random variables with a uniform distribution between 0 and 2π. The resultant intensity, which is the square of the envelope, will be termed I_4, where the subscript 4 denotes that it results from the sum of 4 echoes. By adding the in-phase components of the four terms, and then the quadrature components, it follows that

$$I_4 = (a \cos\alpha + b \cos\beta + c \cos\gamma + d \cos\delta)^2 + (a \sin\alpha + b \sin\beta + c \sin\gamma + d \sin\delta)^2$$

Making use of the relationships $\cos^2 x + \sin^2 x = 1$, and $\cos x.\cos y + \sin x.\sin y = \cos(x-y)$, this can be expanded to become

$$\begin{aligned} I_4 = & (a^2 + b^2 + c^2 + d^2) \\ & + 2ab\cos(\alpha-\beta) + 2ac\cos(\alpha-\gamma) + 2ad\cos(\alpha-\delta) \\ & + 2bc\cos(\beta-\gamma) + 2bd\cos(\beta-\delta) \\ & + 2cd\cos(\gamma-\delta) \end{aligned}$$

$$\langle I_4 \rangle = \langle a^2 \rangle + \langle b^2 \rangle + \langle c^2 \rangle + \langle d^2 \rangle = 4\langle a^2 \rangle$$

assuming that each amplitude has the same mean value.

Squaring the expression for I_4 and making use of the fact that all cross product terms such as $2ab \cos(\alpha-\beta).2ac \cos(\alpha-\gamma)$ have zero mean, we obtain

$$\begin{aligned} \langle I_4^2 \rangle = & \langle a^4 \rangle + \langle b^4 \rangle + \langle c^4 \rangle + \langle d^4 \rangle \\ & + \langle 4a^2b^2 \rangle + \langle 4a^2c^2 \rangle + \langle 4a^2d^2 \rangle \\ & + \langle 4b^2c^2 \rangle + \langle 4b^2d^2 \rangle \\ & + \langle 4c^2d^2 \rangle \end{aligned}$$

The format of this expression is chosen to show how the terms originate and to enable a generalized prediction of $<I_N^2>$ for values of N other than 4.

In the above example and noting that $<a^2b^2> = <a^2>^2$ if the amplitudes are independent random variables, we obtain

$$<I_4^2> = 4<a^4> + 4(3+2+1)<a^2>^2$$

A similar derivation for N overlapping echoes gives

$$<I_N> = N<a^2>$$

$$<I_N^2> = N<a^4> + 2N(N-1)<a^2>^2$$

which makes use of the simplification that $\{(N-1) + (N-2) + \ldots + 2 + 1\}$ equals $N(N-1)/2$.

For the Rayleigh distribution $p(a) = a/\sigma^2 e^{-a^2/2\sigma^2}$

it is readily shown that $<a^2> = 2\sigma^2$ and $<a^4> = 8\sigma^4$. Arbitrarily putting $\sigma = 1/\sqrt{2}$ it follows that

$$<I_N> = N \qquad (2)$$

$$<I_N^2> = 2N + 2N(N-1) \qquad (3)$$

The next step is to proceed from the situation where the the number of fish in each resolution cell is fixed at N , and to consider instead fish which are Poisson distributed in volume. The Poisson distribution predicts that the probability that there are N fish in a resolution cell and is given by

$$P_{<N>}(N) = \frac{<N>^N e^{-<N>}}{N!} \qquad (4)$$

where $<N>$ is the average number of fish in the resolution cell. We can now take the moments $<I_N>$ and $<I_N^2>$ corresponding to a fixed number of fish N in the resolution cell, and multiply them by the probability $P_{<N>}(N)$ that there are N fish in the resolution cell. We do this for every value of N for which $P_{<N>}(N)$ is significant and then perform summations of these weighted first and second moments. This gives us averaged intensity moments $<I>$ and $<I^2>$ which are appropriate to fish that have a Poisson distribution in volume and have a Rayleigh distribution of amplitudes.

Using the expression for $<I_N>$ given in Eq.2 we obtain

$$<I> = \sum_{N=1}^{\infty} <I_N>P_{<N>}(N) = \sum_{N=1}^{\infty} N\, P_{<N>}(N)$$

$$= <N>\text{ , by the definition of a mean value.}$$

Doing the same operation for the value of $<I_N^2>$ given in Eq.3 we obtain

$$<I^2> = \sum_{N=1}^{\infty} <I_N^2>\, P_{<N>}(N)$$

$$= \sum_{N=1}^{\infty} 2N\, P_{<N>}(N) + \sum_{N=1}^{\infty} 2N(N-1)\, \frac{<N>^N e^{-<N>}}{N!}$$

$$= 2<N> + 2e^{-<N>} \sum_{N=2}^{\infty} \frac{<N>^N}{(N-2)!}$$

$$= 2\langle N\rangle + 2e^{-\langle N\rangle}\left\{\frac{\langle N\rangle^2}{0!} + \frac{\langle N\rangle^3}{1!} + \frac{\langle N\rangle^4}{2!} + \ldots\right\}$$

Making use of the expansion $e^x = 1 + x + \frac{x^2}{2!} + \frac{x^3}{3!} + \ldots$

this simplifies to

$$\langle I^2\rangle = 2\langle N\rangle + 2\langle N\rangle^2$$

$$\langle I^2\rangle/\langle I\rangle^2 = 2/\langle N\rangle + 2$$

$$\text{or} \quad \langle N\rangle = \frac{2}{\langle I^2\rangle/\langle I\rangle^2 - 2} \tag{5}$$

The relationship previously used was given by Eq.1. This assumed scatterers which were Poisson distributed in volume but which had identical backscattering cross-sections. Experimental estimates of the density of a randomized volume of polystyrene spheres were obtained using this formula and have shown good agreement with the true densities . The new equation gives number densities which are exactly twice as great. It is believed that the Rayleigh distribution of echo amplitudes assumed in this new equation is likely to be more appropriate to fish in a shoal. This applies even if the fish do not vary in size. When the fish length drops below 25 wavelengths, it is expected that the amplitude statistics will become non-Rayleigh and that the correct constant in the numerator of Eq.5 will now lie between one and two. An extension to be preceding analysis can determine this constant if the amplitude statistics of the individual fish are known.

COMPUTER SIMULATIONS AND ERRORS IN ESTIMATES

Using either Eq.1 or Eq.5 it it clear that an error in the second normalized moment, arising from the use of too few samples, will cause an error in the estimate of the mean number of scatterers in the resolution cell. The original error analysis by Wilhelmij and Denbigh was valid only for very small errors and should therefore be applied with caution. Although less general than this original analysis a useful insight into errors has been obtained by a computer simulation of a scattering model.

In the simulation it is assumed that there are M independent measurements of intensity, each arising from interference between the scatterers from M different resolution cells. The Poisson distribution of Eq.4 tells us what number Q out of these M measurements arises from cells containing N scatterers, i.e.

$$Q = M\, P_{\langle N\rangle}(N) \tag{6}$$

where $\langle N\rangle$ is the average number of scatterers. For each cell containing N scatterers, an intensity may be computed by performing a vector addition of N complex amplitudes, each having a random phase and a Rayleigh distributed envelope. The M intensities computed in this way may be used to determine a value for the second normalized moment and hence to predict a mean number of scatterers by using Eq.5. One such run must be expected to give an answer which is different from the true value of $\langle N\rangle$, due to the use of too few intensity values in calculating the second normalized moment of intensity. However, very many such simulations may be performed and an analysis of the results produces a sample mean of the estimates and the standard deviation of the estimates. Table 2 is derived from many simulations each based on 500 independent measurements of intensity. The number of simulations was adequate for the results to have converged to be closely equal to the values shown.

Table 2. Simulation results for 500 samples, showing bias and errors in the estimates of <N>.

<N>	mean of the estimate of <N>	standard deviation of the estimate of <N>
0.1	0.11	0.013
0.5	0.52	0.079
1.0	1.1	0.20
2.0	2.2	0.62
4.0	4.8	2.0

An interesting point to note is that there is a bias in the estimate of <N>. This arises because the value of second normalized moment in Eq.5 is always greater than 2 and the subtraction of 2 in the denominator causes a greater error if the error in second normalized moment is too low by some amount than if it is too high by the same amount. Concerning the standard deviation, it is relevant to note that a technique capable of producing an estimate of fish number density to an accuracy of 30% would be considered a useful technique. If 500 independent measurements were available an interpolation of the results of Table 2 shows that, neglecting the bias error, this accuracy is achieved if the number of scatterers in the resolution cell is 2.1 or less.

The corresponding table for 1000 independent measurements is given in Table 3. It is seen that,as is to be expected, the bias and the standard deviation are both reduced. An accuracy of 30% corresponds now to a value of <N> equal to about 4.5.

Table 3. Simulation results for 1000 samples, showing bias and errors in the estimates of <N>.

<N>	mean of the estimate of <N>	standard deviation of the estimate of <N>
0.1	0.1	0.011
0.5	0.52	0.068
1.0	1.04	0.15
2.0	2.11	0.37
4.0	4.23	1.14

SEA TRIALS

In 1985 an Anchovy Recruitment Survey was conducted by the Sea Fisheries Research Institute aboard the R.S. Africana. The main objective of the cruise was to obtain an estimate by acousic survey of the young anchovy biomass on the South and West coasts of South Africa. The method used for the survey was echo integration using the ship's SIMRAD EK-38 and EK-120 scientific echo sounders. One of the authors participated in the first leg of the cruise and the outputs of the echo sounder were recorded and used to calculate fish density estimates using the statistical technique. These were compared with the estimates made by the Sea Fisheries Research Institute using the echo integrator. In the light of the experience gained during the first few days of the cruise a few observations can be made regarding the suitability of the statistical method for acoustic survey of anchovy. During the day,. the fish concentrate into extremely dense shoals which, at a survey speed of 12 knots are insonified typically for 10 pings before the ship has passed over them. The number of independent envelope samples obtained from such a short insonification was found to be inadequate for working out a statistically significant estimate of the normalized moment of intensity. At night, however, the fish spread out into a more diffuse sound scattering layer, which may be several miles in horizontal extent, and several tens of metres deep. Examples of sounder charts of shoals during the day and at night are given in Figure 1.

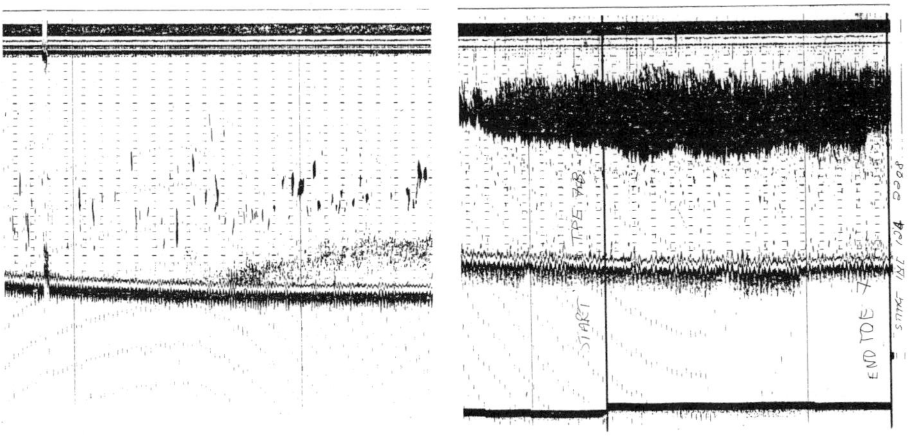

Figure 1. Day and night time echo soundings of anchovy.

Experiments were conducted on suitable night-time aggregations of anchovy. It was arranged that the echo integrator should run concurrently with the statistical method. This achieved a standard against which the estimates using the statistical technique could be compared. It should be noted, however, that echo integration results are generally recognized to be prone to considerable error. A second way of verifying the new technique was to examine the effect of increasing pulse length. An increase in pulse length should not change the estimates of density.

The experiments were conducted when suitable aggregations were present. As many samples as the extent of the shoal would allow were taken. Results are presented in Table 4. The statistical estimates were obtained using the formula

$$\rho = \frac{2}{(\langle I^2\rangle/\langle I\rangle^2{}_{av} - 2)\times(0.125c^3\Omega\tau t_l t_m)} \qquad (7)$$

This is an extension of Eq.5 which takes into account the volume of the resolution cell, including its change with range due to beam divergence. Apart from the modifying factor of 2 already discussed, the derivation of this is as given by Wilhelmij and Denbigh. In this formula c is the velocity of sound; Ω is the beam solid angle taken somewhat arbitrarily to be that corresponding to the 3dB beamwidth of the transducer when used solely as a transmitter; τ is the pulse duration; t_l and t_m are the time limits of the intensity data that are used; $(\langle I^2\rangle/\langle I\rangle^2)_{av}$ is obtained by calculating $\langle I^2\rangle/\langle I\rangle^2$ for each range bin doing the averaging over all pings, and then averaging these second normalized moments over the range bins. In a typical experiment there were 50 pings, and 12 range bins from each ping which gave statistically independent measurements of intensity.

Table 4. Estimates of number density by statistical and echo integration methods.

Experiment number	Pulse duration (ms)	Statistical estimate of ρ (number/m^3)	Integrator estimate of ρ (number/m^3)	Integ. est. / Stat. est.
1	0.14	0.55	2.1	3.8
2	0.31	0.076	0.39	5.1
3	0.68	0.24	0.78	3.2
4	0.80	0.43	1.42	3.3
5	0.80	0.26	0.81	3.1
6	0.68	0.52	-	-

Experiments 1 - 5 were conducted on different night-time anchovy aggregations. Experiment 6 was conducted on a species of small pelagic fish called lightfish, which is present in extremely extensive distributed aggregations on the Atlantic coast. The second column of the table gives the pulse duration. An estimate of number density based on Eq.7 is shown in Column 3. The estimate of number density from the echo integrator is given in Column 4 , and the ratio of echo integrator to statistical method estimates is in Column 5. It will be noted that this ratio is large. No integrator results are available for Expt. 6 due to a lack of target strength data for the lightfish. It may be interesting to note that,for the anchovy, the number of fish in the mid-range resolution cell varied between 0.6 and 1.7 for the five experiments.

Table 5 shows the effect of increased pulse lengths for experiments 2, 3 and 6. It is seen that the density estimates are very similar to those of Table 4.

Table 5. Statistical estimates of number density for increased pulse lengths.

Experiment number	Pulse duration (ms)	Statistical estimate (number/m^3)
2	0.68	0.070
3	0.86	0.28
6	0.86	0.75

DISCUSSION

Estimates of fish number density have been obtained using the echo integrator method and using the new method based upon the statistical properties of the echo waveform. The echo integrator results must be regarded as approximate as they rely on assumptions made of fish target strength. However, errors using echo integrator methods are likely to be insufficient to account for the large discrepancies between the two methods. It appears from column 5 of Table 4 as though the statistical technique is not yet producing useful measures of fish number density. Work is proceeding to explain sources of error in the statistical technique and the most plausible so far,not yet proven, is that they arise because of variations in the density of the fish shoal. There is a non linearity in the method which gives rise to errors. Suppose, for example, that half of the echo sounder measurement correspond to a region where $\langle N \rangle = 2$ and half to where $\langle N \rangle = 4$. The linearity of the echo integration technique is such that the estimate would be the mean value, or $\langle N \rangle = 3$. This is not the case with the statistical technique as can be seen by the following argument. Eq. 5 gives

$$\langle I^2 \rangle / \langle I \rangle^2 = 2 + 2/\langle N \rangle$$

For those measurements corresponding to $\langle N \rangle = 2$, this gives $\langle I^2 \rangle / \langle I \rangle^2 = 3$. Arbitrarily, putting $\langle I \rangle = 1$ this signifies $\langle I^2 \rangle = 3$. For the measurements corresponding to $\langle N \rangle = 4$, we would expect twice the echo intensity or $\langle I \rangle = 2$. Therefore, for this region $\langle I^2 \rangle = 2^2(2+2/4) = 10$. Considering all the measurements together we obtain $\langle I \rangle = 1/2.(1+2) = 1.5$ and $\langle I^2 \rangle = 1/2.(3+10) = 6.5$. It follows that the estimate of $\langle N \rangle$ using all the measurements together would be

$$\langle N \rangle = \frac{2}{6.5/1.5^2 - 2} = 2.25$$

This estimate is not the mean value for the two regions and the dangers of a non-uniform density are clearly revealed. In this example, as always, the effect of a non-uniform density is to produce an estimate which is less than the true average density. This bias is in agreement with the discrepancy between the experimental estimates using the statistical technique and those using the echo integrator. Work is at present under way to overcome this cause of error by dividing the fish shoals into regions of constant density, based upon measurements of returned energy, and to then apply the statistical technique to each of the regions separately. The density estimates for each region will then be combined to obtain an estimate of overall average density.

Another source of error in the estimates is believed to lie in the assumption of a Rayleigh distribution of echo amplitudes. Although almost certainly better than the assumption of constant amplitudes there are several reasons for doubting its strict validity. Firstly the anchovy are typically 10λ long which is less than the 25λ specified by Clay and Heist for the Rician statistics to be approximated by Rayleigh statistics. Secondly the assumption of constant fish size is not strictly accurate in practice. Thirdly there is no account made of the variation of acoustic intensity across the beam of the echo sounder. Further theoretical work needs to be done but it seems likely that these three effects largely compensate one another and that the assumption of a Rayleigh distribution is not grossly in error and not sufficient to account for the major discrepancies described earlier.

A third source of error may lie in the somewhat arbitrary choice of taking the beam solid angle to corresond to the 3 dB beamwidth of the transducer when used as a transmitter. This was done largely to be in keeping with the notation of Ref.1. It is interesting to note that, if the 3 dB beamwidth corresponding to the combined transmit/receive response is used, the beam solid angle is less by a factor of 1.9 and the density estimates are increased by this same large factor to give very much better agreement with the echo integrator estimates. A value of beam solid angle which is perhaps easiest to justify is 4π divided by the transmitter/receiver directivity factor. A calculation on the basis of this increases the estimates by a factor of 1.3. Clearly there is scope for further theoretical work to examine the effects of beam shape.

CONCLUSIONS

It is possible to obtain an estimate of fish number density by a statistical analysis of echo soundings. In contrast to the echo integrator, a knowledge of fish target strength is not needed. Experimental estimates of number density using a tank model of a fish shoal have produced good agreement with the true density. Experimental estimates of the number density of a real fish shoal have, however, shown large errors. The estimates have been considerably and consistently too low. One of causes appears to be a somewhat arbitrary choice of beam solid angle used in the calculation. Another appears to be a non-uniformity in the fish shoal. There is reason to hope that a technique for overcoming both these sources of errors may be found and that useful estimates of fish number density may become feasible.

ACKNOWLEDGEMENTS

The authors would like to thank Mr Ian Hampton of the Sea Fisheries Institute for making possible our participation in the Anchovy Recruitment Survey, for providing the echo integration estimates, and for extensive advice generally.

REFERENCES

1. P. Wilhelmij and P.N. Denbigh, "A Statistical Approach to determining the Number Density of Random Scatterers from Backscattered Pulses", J.Acoust.Soc.Am.76(6),1810 (1984)
2. P.N. Pusey,D.W. Schaefer, and D.E. Koppel, "Single-Interval Statistics of Light Scattered by Identical Independent Scatterers", J.Phys.A Math. Nucl.Gen. 7, 530-540 (1974)
3. P.N. Denbigh, "Scattering from Fish Shoals and the Determination of Fish Density from a Statistical Analysis of Echo Waveforms", Intl. Conf. on Developments in Marine Acoustics, Sydney, Dec 1984
4. J. Weintroub, "Estimation of the Number Density of Random Scatterers with Application to Acoustic Fish Stock Assessment," First South African Congress on Acoustics, Pretoria, October 1985
5. C.S. Clay and B.G. Heist, "Acoustic Scattering by Fish - Acoustic Models and a Two Parameter Fit," J.Acoust.Soc.Am.,75(4),1077-1083 (1984)

PROCESSING OF FISH ECHO-SIGNALS FOR CLASSIFICATION PURPOSE

D. Vray and G. Gimenez

Laboratoire de Traitement du Signal et Ultrasons, INSA 502
20, Avenue Albert Einstein, 69621 Villeurbanne cedex
France

ABSTRACT

Echo-sounding is extensively used during acoustical sea-surveys. Two techniques permit us to process the received echoes in order to estimate fish densities: echo counting and echo integration (evaluation of signal energy). However, it is necessary to know the nature of insonified fishes to interpret the data given by these techniques. We propose to study the characteristics of echoes by means of digital processing without a priori information on the scattering bodies. At first, we digitilize and compact the recorded signals, then pertinent parameters are extracted from the echoes, afterwards, their statistical properties are studied.

INTRODUCTION

Interest given to fishery problems has grown during the last twenty years. Two fields are mainly concerned with improvement of techniques and methods: (i) stock estimates which provide data used for determining fishery quotas (in our country, for example, IFREMER - Institut Français de Recherche pour l'Exploitation de la Mer - conducts each year a survey to estimate the amount of herring in the North Sea); (ii) fishing surveys, which often concern only a few species.

For the above-mentioned problems it is necessary to identify the insonified species. For stock assessments this allows us to obtain a fish density (or an absolute quantity) from the energy of the echo-signals. Because of identification, fishing surveys become more and more selective, the unequal commercial values of the various available species being responsible for this need.

The problems of classification and identification of fish species were recently investigated by our laboratory. This paper begins with a limited review of some previous work followed by the presentation of our approach to these problems.

ACOUSTICAL FISH DETECTION AND SIGNAL CLASSIFICATION

During acoustical surveys, the echo sounder is of common use. It produces an ultrasonic acoustical pulse emitted downward and receives the echo signals backscattered upward by the various targets (fishes, sea-bottom . . .).[1]

After reception, echo signals are processed. Echo counting and above all echo integration are used as above to estimate fish density.[2] For echo integration, an integrator evaluates the area under the envelope of the squared echo signal and the resulting value is assumed to be proportional to the fish density. The coefficient of proportionality includes the average of the target strengths of the various insonified fishes. Fish Target Strength (TS) is the measure of the acoustic size of individual fish. It is a crucial parameter whose value can be estimated by numerous techniques.[3] Note that the major contribution to TS comes from the swimbladder which, due to its variable gas content, helps the animal to regulate its buoyancy.

Usually two methods are used to identify species: (i) an attempt to catch the insonified fish by a simultaneous trawling - this is biased because an unknown number of animals avoid the trawl and escape, and obviously this concerns stock assessment - and (ii) during fishing-surveys, close examination of the echograms by experienced people. This results in a decision to trawl or not to trawl the insonified fish.

To avoid an actual catch of fish or the examination of echograms by experts, some authors suggest classification and identification methods based on extraction of signal characteristics. Some of these studies which involve a statistical approach, are referred to below. Usually, the distribution of each extracted parameter is computed. Then, the question is: "Are these distributions representative of a particular fish?".

To classify echo-signals, FAY[4] uses statistical properties as well as qualitative parameters which exhibit a few modalities like the type of the sea-bottom or the type of the fish-shoal.

In their method DEUSER et al.[5,6] sample the echo envelopes (quadrature sampling) to extract eight parameters and then use them to elaborate a classification algorithm which adapts itself to the environmental state.

GIRYN et al.[7,8,9] propose a vector representation for a given number K of successive transmissions. We define the vector $P\ (P_1, \ldots, P_i, \ldots P_z)$ where the components P_i are given by

$$P_i = \log m_i \text{ with } m_i = \frac{1}{K} \sum_{k=i}^{K} m_i(k) .$$

Here m_i is the average, for the K echo signal envelopes, of the various individual moments of ith order $m_i(k)$. Then each sequence of K successive transmissions, represented by the associated observation vector P is submitted to a classification algorithm.

AZZALI[10] uses the sampled echo envelopes to obtain a binary image (black or white) which is used to compute a set of seven functions of moments. These moments are said to be invariant because they are not affected by depth or vessel speed, nor by rotation and translation of the fish-area image. Then the images are classified according to two different decision rules. This procedure has been applied to data collected during surveys in Adriatic Sea.

Besides the preceeding works, other statistical methods are employed for density estimation. They use the statistical characteristics of the echo signal and permit us to estimate the density of sound scatterers without knowledge of the average target strength. Experimental works,[11] as well as <u>in situ</u> measurements[12,13], are already implemented.

Keeping in mind the preceeding papers, we are trying to extract the greatest number of parameters, some of them characterizing each echo and others characterizing the time occurence of the echoes. Obviously this time repartition is closely related to the spatial repartition of the targets. Now we consider all the transmissions corresponding to the insonification of a single fish-shoal and we compute the mean values of each above-defined parameter. At this stage, information deduced from the observations of echograms and the environment are added and a classical classification algorithm is used.

EXPERIMENTS AND SIGNAL PROCESSING

Until now we have handled three different types of signals recorded during surveys performed in multi-species areas (Cariaco, Venezula; North Sea; or Lake LEMAN, France/Switzerland).

The signals displayed in the present section were obtained with a 120 kHz emitting frequency and a 4 kHz band-pass. An additional frequency shift, from 120 kHz to 12 kHz, was used to record the received analog signal on a conventional audio recorder.

In order to process recorded signals on a computer, we have to digitalize them. However, the considerable amount of resulting data can not be stored on a magnetic memory (disk) without a proper data reduction.

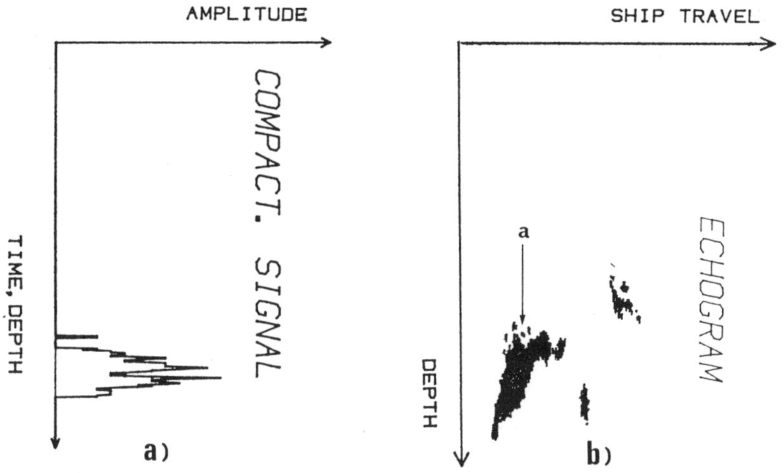

Figure 1. a) Reconstructed signal from compacted data;
b) Echogram elaborated from successive reconstructed signals. Signal of Figure 1.a. produces the vertical line (arrowed "a") on Figure 1.b., the amplitude being coded only with two levels (white or black).

It can be seen from a preliminary observation that echoes backscattered by fish constitute 10% or less of the recorded signal. This is why it is possible to compact our data by means of a comparison between the amplitude of the digitized recorded signal and a given threshold.

This allows us to recognize echoes because, when they occur, the amplitude of the signal exceeds the threshold value. Only in this case, is the signal, i.e. the echo, stored on the disk. This is a method to discriminate between echoes and noise. The time of occurrence of each echo is also stored, the various data being actually separated on the digital record by proper flags. Thus it is possible to restore the original signals by insertion of a zero-amplitude signal of proper length between the echoes.

Figure 1a. shows the echo envelope backscattered by a fish-shoal. This signal has been reconstructed from compacted non-zero amplitude signal samples. This echo belongs to a sequence of echoes backscattered by a single shoal. From this sequence, it's possible to build an echogram (Fig. 1.b). This kind of acoustical image of this shoal is constituted of the contributions brought by the various echoes of each successive transmission. Note that this reconstructed echogram is a filtered version of the one obtained during the original echo-sounding survey.

Figure 2 shows the system which digitizes and compacts data. Signals recorded on a tape recorder (1) are digitized at the sampling rate F_e given by (2). The analog multiplexer permits us to select which channel will be digitized using the sample-and-hold amplifier (4) and the analog to digital converter (5). Data are then transmitted to the central memory (8) of the computer (7) by using a direct memory access device (6). Data compacted are stored on a disk (9) and can be displayed on a plotter (10) to represent either a single transmission or an echogram.

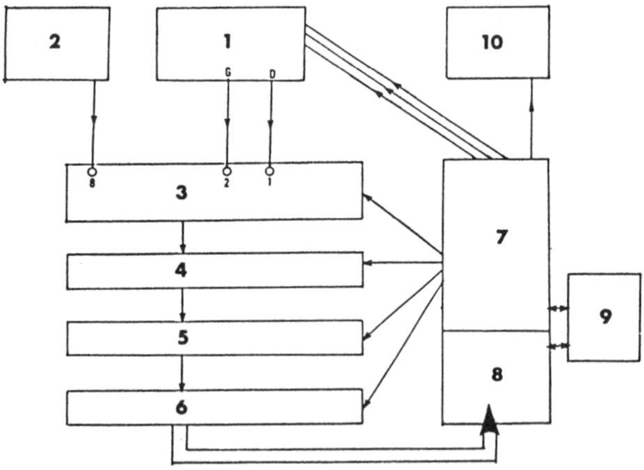

Figure 2: The digitizing system.

Now values of parameters defined in the preceeding section can be calculated either with the data resulting from processing of the analog signal or with its envelope.

Then the mean values are calculated by averaging over the transmissions corresponding to the insonification of a given shoal.

Let us remark that, due to the threshold device, a single echo can be divided into several different ones. These "errors" can be corrected before the computation of the parameter values.

A few parameters extracted from the echo signal are reported hereafter. They belong to the set of parameters which will be submitted to a classification algorithm. N is the number of samples of the considered signal. A_i is the numerical value of the sample i.

- MOY: Average amplitude of echoes: $A = \frac{1}{N} \sum_{i=1}^{N} A_i$

- MCz: Central Moments of the zth order: $\frac{1}{N} \sum_{i=1}^{N} [A_i - A]^z$

- Eci: Peak count i: Number of peaks of amplitude greater than T_i

- HAUT, SURF, FOND: Echo duration and time separating the echo from sea surface or bottom: the number of samples which constitute these durations. This number can be easily related to a distance.

- MAX: Value of the maximum amplitude of the considered echo.

At this stage a second set of parameters is added to the previous one. They concern the whole "object" (shoal) and are deduced from the restored digitized echogram elaborated from the compacted data (see Fig. 1). These parameters are: estimation of the shoal length and height, distance between the shoal and sea-bottom or surface, visual appreciation of shoal form and grey level distribution type. More information is needed to take into account the particular conditions of the experiment, namely: vessel speed, emission angle of the transducer (first lobe), duration of the emission pulse, frequency of pulse repetition, and sampling rate.

PRELIMINARY RESULTS

A good knowledge of the distribution of each parameter extracted from the signal is necessary to handle and to understand results given by data analysis algorithms. To this aim, histograms of each variable (parameter) are computed, as well as elementary statistical characteristics. At this stage, it is not our aim to find a particular probability density function which approximates the actual histogram. Instead of that we try to find some characteristics, which discriminate various situations. Fig. 3, for example, shows an histogram of the maximum amplitude of each echo (this variable is named MAX). Three modes can be seen, which represent three different groups of echoes.

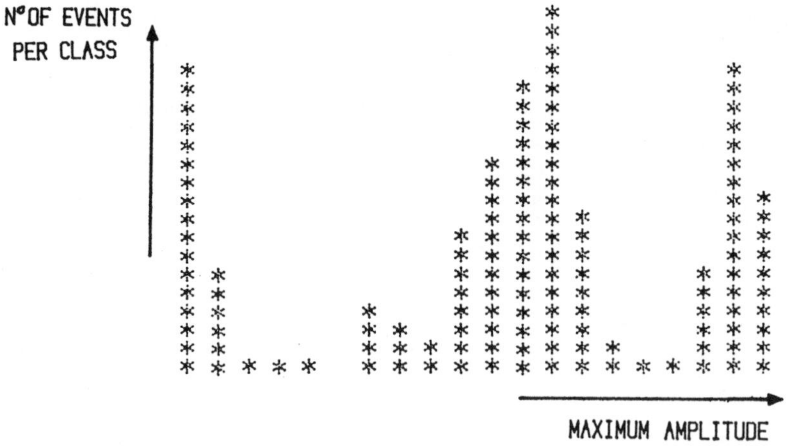

Figure 3: Distribution of the maximum amplitude of the echoes.

Meanwhile, as our final goal is the discrimination of fish species, obviously a single variable cannot perform the task. This is why a dozen variables are evaluated and the correlation matrix is computed in order to determine if one or more variables are strongly correlated. In this case this means that the information brought by a particular variable is already contained in another one. For example, in Figure 4, the coefficient of correlation between "SURF" and "FOND" is 0.98.

	HAUT	SURF	FOND	MOY	MC2	MC3	ECO	EC1	MAX
HAUT	1.00	0.63	-0.77	0.73	0.93	0.64	0.97	0.98	0.77
SURF	0.63	1.00	-0.98	0.14	0.37	0.09	0.53	0.59	0.11
FOND	-0.77	-0.98	1.00	-0.30	-0.54	-0.23	-0.68	-0.73	-0.29
MOY	0.73	0.14	-0.30	1.00	0.88	0.44	0.84	0.78	0.97
MC2	0.93	0.37	-0.54	0.88	1.00	0.66	0.96	0.95	0.91
MC3	0.64	0.09	-0.23	0.44	0.66	1.00	0.54	0.53	0.62
ECO	0.97	0.53	-0.68	0.84	0.96	0.54	1.00	0.99	0.84
EC1	0.98	0.59	-0.73	0.78	0.95	0.53	0.99	1.00	0.78
MAX	0.77	0.11	-0.29	0.97	0.91	0.62	0.84	0.78	1.00
	HAUT	SURF	FOND	MOY	MC2	MC3	ECO	EC1	MAX

Figure 4: Correlation matrix of 9 variables.

Afterwards, data analysis algorithms are useful to understand the relations between the variables. For example, the method of Principal Components Analysis (PCA) is devoted to the computation of factorial axes. Amongst the possible axes, only those associated with the high eigen-values of the correlation matrix are retained. Note that the total information is usually kept constant during the projection of the variables on these axes. Moreover, the best understanding of the relations between variables occurs when variables are projected on the plane made by the first and second factorial axis.

It must be noted that the preceeding methods are only descriptive. They are useful tools to display our information differently and promote a comprehensive view of the data. These methods constitute a preliminary approach before the use of discrimination and classification algorithms.

CONCLUSION AND PERSPECTIVES

We have described several situations where the identification of insonified fish-species is important. To be able to reach the solution of the problem, we are presently developing a method which is based on the extraction of numerous parameters computed from the digitized echoes or from the echogram of the shoal responsible of these echoes. In future work, parameters coming from the frequency-domain will be included specially with data given by echo sounders (currently under development), with larger bandwidth than the present ones. For testing our classification algorithms, data collected in lake LEMAN will be very useful since the number of fish-species of the lake is limited and their behaviour is fairly well-known to researchers incharge of the study of the lake.

ACKNOWLEDGEMENT

This work has been supported by IFREMER (Institut Français de Recherche pour l'Exploitation de la Mer).

REFERENCES

1. K.A. Johannesson, R.B. Mitson, "A Practical Manual for Aquatic Biomass Estimation", FAO Fish. Tech. Pap. No. 240, Rome, 1983.

2. R. Person, E. Marchal, T. Terre, J. Berthe, "Systeme d'echointégration numérique pour l'évaluation des stocks "AGENOR", Paper No. 17, presented at Fisheries Acoustics: a symposium held in Bergen, Norway, 21-24 June 1982.

3. J. E. Ehrenberg, Paper No. 104, "A review of in situ target strength estimation techniques" in Fisheries Acoustics: A symposium held in Bergen, Norway, 21-24 June 1982.

4. D.Q.M. Fay, "Digital Computer Analysis of Echo-Sounder Data for Fish Identification", in Proceedings of the Conference, Acoustics in Fisheries, held at the Hull College of Higher Education, Hull, England, 26-27 Sept. 1978. (Pub. by Univ. of Bath, Bath, UK)

5. L.M. Deuser and D. Middleton, "On the Classification of Underwater Acoustic Signals I: An Environmentally Adaptive Approach", J. Acoust. Soc. Am., 65, 438-443 (1979).

6. L.M. Deuser, D. Middleton, T.D. Plemons, J.K. Vaughan, "On the Classification of Underwater Acoustic Signals II: Experimental Application Involving Fish", J. Acoust. Soc. Am., 65, 444-455 (1979).

7. A. Giryn, M. Rojewski, K. Somla, "About the Possibility of Sea Creature Species Identification on the Basis of Applying Pattern Recognition to Echo-Sounder Signals" in Proceedings of the Meeting on Hydroacoustical Methods for the Estimation of Marine Fish Population, 25-29 June 1979, Vol II, J.B. Suomala Ed. pp 455-466.

8. A. Giryn, K. Somla, M. Rojewski, "The Number Parameters of Echo Sounder Signals with Respect to their Usefulness to Sea Creatures Species Identification" in the Proceedings of the Meeting on Hydroacoustical Methods for the Estimation of Marine Fish Population, 25-29 June 1979, Vol. II, J.B. Suomala Ed., pp 467-489.

9. A. Giryn, "The New Approach to the Biomass Measurement of Hydrocoustic Methods based on Probability Distribution of Creature Weights and the Echo Signals", in the Proceedings of the Meeting on Hydroacoustical Methods for the Estimation of Marine Fish Populations, 25-29 June 1979, Vol. II, J.B. Suomala Ed., pp 491-494.

10. M. Azzali, "Regarding the Possibility of Relating Echo Signal Features to Classes of Marine Organisms: Tests carried out in the north and middle Adriatic Sea", Paper No. 23 presented at Fisheries Acoustics: A symposium held in Bergen, Norway, 21-24 June, 1982.

11. P. Wilhelmij and P. Denbigh, "A Statisical Approach to Determining the Number Density of Random Scatterers from Backscattered Pulses", J. Acoust. Soc. Am., 76, 1810-1818 (1984).

12. M.L. Peterson, C.S. Clay, S.B. Brandt, "Acoustic Estimates of Fish Density and Scattering Function", J. Acoust. Soc. Am., 60, 618-622 (1976).

13. T.K. Stanton, "Density estimates of biological sound scatterers using sonar echo peak PDFs", J. Acoust. Soc. Am., 78, 1868-1873 (1985).

14. M. Jambu and M.O. Lebeaux, Cluster Analysis and Data Analysis, North-Holland, Amsterdam, 1983.

BROADBAND ACOUSTICAL SCATTERING BY INDIVIDUAL FISH

Leif Bjørnø and Niels Kjærgaard

Industrial Acoustics Laboratory
Technical University of Denmark
Building 352, DK-2800 Lyngby, Denmark

ABSTRACT

Acoustic signals scattered at dorsal aspect from anaesthetized roachs of three lengths (14, 20 and 24 cm) are presented. The time-functions of the scattered signals show a strong dependence on the insonified position along the roach. This dependence is assumed to be caused by interaction between the scattered signal, in particular being influenced by the shape of the swimbladder, and signals arising from elastic wave systems produced in various anatomical parts of the fish during insonification.

INTRODUCTION

The use of acoustical methods for the study of biomass in the sea, i.e. the type, size distribution, number, spatial distribution etc. of fish, has increased considerably during recent years. In particular resonance scattering by fish possessing a swimbladder has been in the center of research activities, while multiple-frequency acoustical scattering studies forming the basis of data to be used for validating empirical and in particular conceptual models have been sparse[1]. The application of multiple-frequency acoustical methods in order to estimate a specific target population is, conceptually at least, straightforward, but is significantly more complex than most bioacoustical applications. The accuracy and resolution of multiple-frequency methods also requires more careful design of experiments based on a limited number of free parameters to be varied. These parameters are in particular (a) the frequency, (b) the spatial coverage and resolution of the measurements via the transducer directivity and the positioning of the insonified object and (c) the data processing algorithms.

As the fundamental scattering object is the individual fish, there is still a chronical need for knowledge about the backscattering cross section, or the target strength, of individual fish in order to be able to make an accurate estimation of abundances and size distributions of certain classes of fish.

Apart from measurements on fish at various frequencies[2-4] there has been a substantial interest in measurements on artificial models of fish, frequently using models having simple geometrical shapes[5]. Theoretical computations have for instance also been performed on models based on arrays of point scatterers[6,7] of Huygens wavelets, and recently the relative volume scattering strength of scatterers along the axis of live fish has been measured[8]. These measurements were performed using a focused 220 kHz source moved along the length of yellow perch (12 - 14 cm long) and hog sucker (30 cm long). The resulting dependence of the distribution of volume scatterers on anatomy showed that the swimbladder gave about 80% of the scattered energy for yellow perch and only about 20% of the scattered energy for the hog sucker, while other distributed scatterers were the vertebrae, the head and the flesh. Particular peaks and valleys were found in the volume scattering strength along the fish and were assumed to be caused by constructive and destructive interference of acoustic signals.

Recently, theoretical and experimental studies have been carried out[9] by which a fish has been represented by an ideal pressure-release surface having the exact size and shape as the swimbladder, which was determined morphometrically and was mathematically represented by a finite-element triangulation. The target strengths were computed by means of the Kirchhoff's approximation not taking into consideration the diffraction effects. The scattering model thus developed has proven its applicability at rather high frequencies, i.e. for fish lengths in the nominal range from 8 - 36 acoustical wavelengths. However, as will be shown in the subsequent sections of this paper, also bone, flesh etc. of the fish play an important role as scattering elements, in particular at higher frequencies and at higher angles of incidence.

This paper aims at an illumination of the acoustic target strength of individual fish (roach) of various lengths (14 - 24 cm) measured over a broad frequency range (20 - 200 kHz) along the axis of live fish.

THEORY

For finite signals scattered and by the use of finite-bandwidth receivers the backscattering cross section σ may be written[10] as:

$$\sigma = 4\pi \int_0^\infty |S(\omega)\cdot F(\omega)\cdot H(\omega)|^2 \cdot d\omega \Big/ \int_0^\infty |S(\omega)\cdot H(\omega)|^2 \cdot d\omega \qquad (1)$$

where $S(\omega)$ is the incident signal spectrum, $F(\omega)$ is the monochromatic scattering amplitude of the target and $H(\omega)$ is the receiver frequency response function. ω is the angular frequency related to the velocity of sound of the medium through $\omega \doteq ck$. For narrow-band receivers and for monochromatic signals expression (1) reduces to $\sigma = 4\pi|F(\omega)|^2$.

From (1) the target strength (TS) may be expressed by:

$$TS = 10\cdot\log(\sigma/4\pi) \qquad (2)$$

For narrow-band receivers and for monochromatic signals (2) may be written[11] as:

$$TS = 10\cdot\log(I_r/I_i)_{r=1m} \qquad (3)$$

EXPERIMENTAL SET-UP AND PROCEDURES

The experiments were performed using a parametric acoustic array in order to obtain an improved directivity of the incident signal beam and in order to preserve the directivity over a broad frequency range. The transmitters were based on air-backed PZT elements, 20 mm in diameter and having a natural frequency of 1.2 MHz[12]. Highly directive, low frequency transients were produced by selfdemodulation of the 1.2 MHz primary radiation, pulsed in a Gaussian envelope of duration T = 5 μs. Also incident difference-frequency signals produced by nonlinear interaction in water of a 100% amplitude modulation of a tone-burst of primary frequency 1.2 MHz with a monochromatic primary frequency of half the difference frequency were used for the experimental studies[12]. A block diagram showing the instruments used during the experiments based on the Gaussian envelope primary waves is given in Fig. 1.

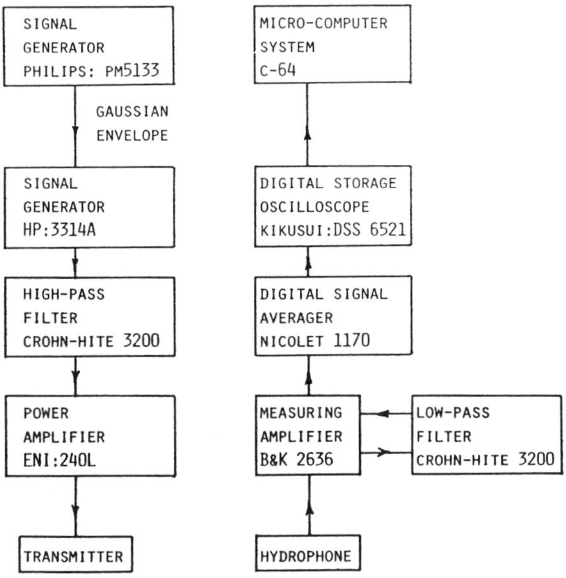

Figure 1. Instruments used during the experiments based on Gaussian envelope primary waves.

The scattered signal was received by means of B&K type 8103 hydrophones positioned outside the incident beam.

Living, but anaesthetized, roachs were positioned vertical to the acoustic axis of the transmitter in a distance of 110 cm from the transmitter. The position of the acoustic axis relative to the axis of the roachs is shown in Fig. 2, where the position numbers refer to series of target strength measurements.

Calibration of the experimental set-up was done using a ping-pong ball (38 mm in diameter) and a solid steel ball (30 mm in diameter), respectively. An excellent agreement was found between the calibration curves for the target strength as a function of frequency for the two primary signal types, the Gaussian and the tone-burst, respectively. The calibration showed that most scattering is of a geometrical nature, which by the ping-pong ball leads to a nearly frequency independent target strength, and by the steel ball leads to target strength values dependent on frequency due to elastic waves produced in the steel ball - liquid system, as also found for short pulses by Dragonette et al[13]. Only dorsal aspect target strengths are reported in this paper.

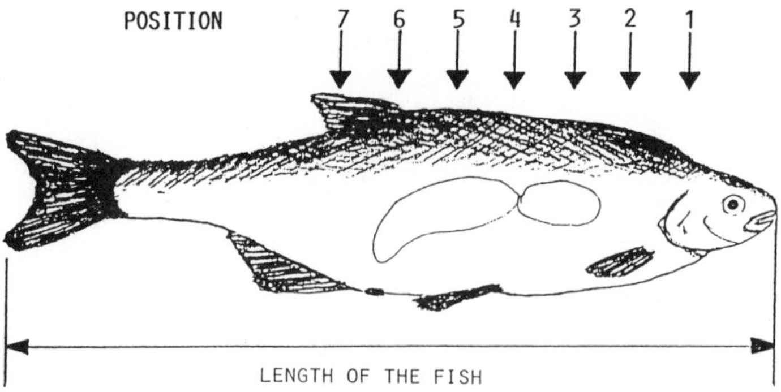

ROACH.

Figure 2. Sketch of a roach showing the measurement positions formed by the intersections between the acoustic axis of the transmitter and the fish axis.

Due to the short time permitted for measurements to be performed during the same period of anaesthesia of the roach the received signals were digitized and stored on a disc for later, in-depth, analysis using a micro computer which during the experiments controlled the sequence of measurements performed.

EXPERIMENTAL RESULTS AND DISCUSSIONS

Figure 3 shows target strength as a function of frequency for a 26 cm long roach measured at dorsal aspect with the acoustic axis at point 1. and 5., see figure 2. The continuous curve originates from the Gaussian pri-

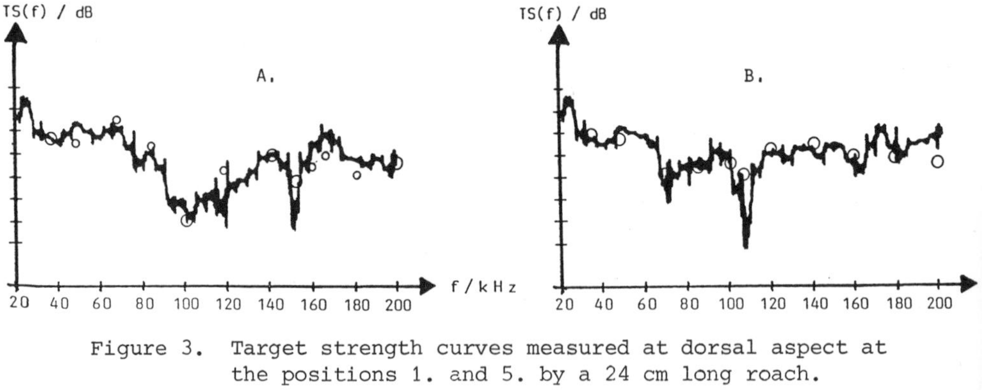

Figure 3. Target strength curves measured at dorsal aspect at the positions 1. and 5. by a 24 cm long roach.

mary signal envelope and o. indicates individual values based on the toneburst primary signal. A good agreement between target strength values obtained using the two primary signal types is observed. As will be seen from figure 3, a considerable difference in the variation of target strengtl with frequency is found for various positions on the same roach and for constant geometrical scattering conditions. As the 6 dB half beamwidth of the difference-frequency signal at the position of the roach is about 6 cm, a considerable insonification position influence on the target strength values is found , probably being caused by scattering contributions from head, gills, vertebrae, flesh and other organs.

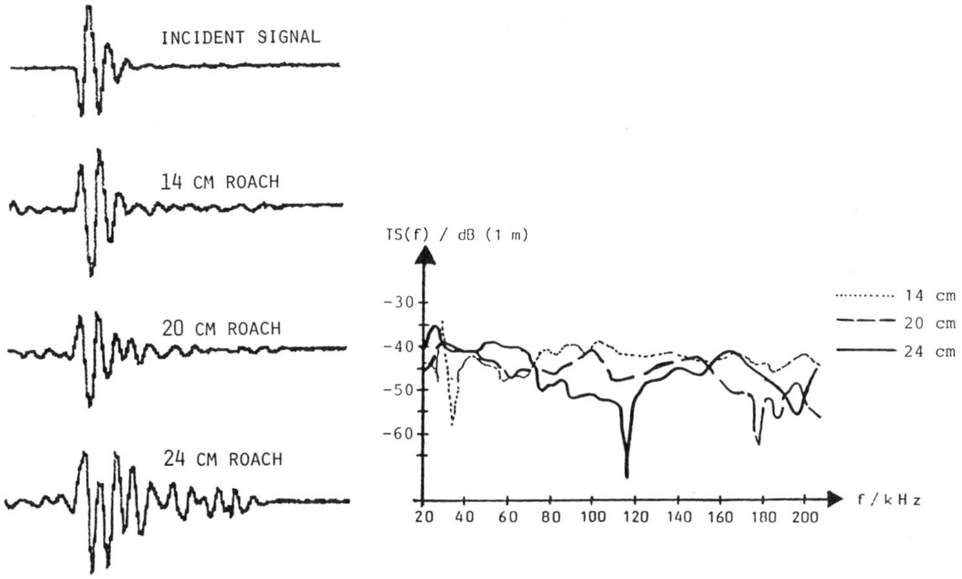

Figure 4. Time-signals and target strength - frequency curves measured at dorsal aspect at the same anatomical position (5) by 3 different roach lengths.

In figure 4 the incident and scattered time-signals from roachs of lengths; 14, 20 and 24 cm for the difference-frequency signal incident at point 5. in all three cases are shown. The target strength as a function of frequency for the 3 roach lengths is moreover shown in figure 4. The target strength courses in figure 4 seem to show, that no simple relation exists between the length (age) of the roach and its target strength spectrum. A peculiar feature in figure 4 is the reduced target strength values at higher frequencies for the roach of greatest length, which in the frequency range 70 - 150 kHz is contradictory to the results found by Love[14].

A systematic study of the scattered time-signals measured for the acoustic axis at various positions 1. to 7. along the 24 cm long roach for the same aspect angle has been performed, and the results are given in figure 5. A considerable variation in the time-signals between more of the measuring positions may be seen in figure 5, and the same variation will be found for various aspect angles keeping the position constant. A comparison of the target strength spectra derived from the signals in figure 5, is given in figure 6, which also shows the considerable variation in target strength spectra with position along the same roach. A particular feature of figure 6 is the strong variation in the target strength spectra for a small variation in frequency. This variation is similar to the spectral variation found by target strength values measured by elastic spheres etc. and discloses the prospective influence of various wave systems set-up in the different parts; swimbladder, flesh etc. of the fish. The interaction between these wave systems and the main scattering effects, probably being most strongly influenced by the swimbladder shape, may form the key to type and size characteristic target strength values in the target strength spectra above swimbladder resonance.

The variation in phase and amplitude found between the signals arising from scattering at various positions along the roach in figure 5 will have a strong influence on the time course, and thus on the target strength values as a function of frequency of a scattered signal arising from the insonification of the whole fish at one time as normally done when standard fish sonars are used. Single frequency scattering will hardly disclose type and size characteristic features of the spectrum above swimbladder resonance and only a systematic study of the continuous target strength spectra may be able to unveil a prospective existence of characteristic spectral features by the target strength, which may be related to fish type, size etc This systematic study is going on for the moment and it includes particular studies of target strength spectra produced by scattering from the swimbladder alone. These studies should give information about the influence on target strength spectra arising from the swimbladder shape and size.

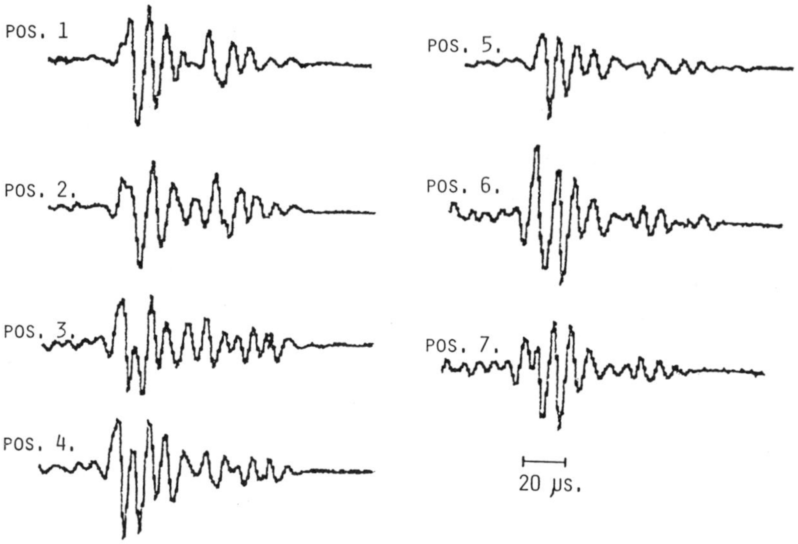

Figure 5. Systematic studies of time-signals by scattering at dorsal aspect from a 24 cm long roach measured at various positions along its axis.

CONCLUSIONS

Several target strength spectra measured at dorsal aspect at various positions along single roachs of various length show a strong influence on the target strength values arising from the position. This variation is probably caused by interaction between the scattered time-signals, influenced in particular by the swimbladder shape, and signals arising from elastic wave systems produced in various anatomical parts of the fish during its insonification. This interaction pattern may be the key to a better understanding of characteristic scattering features by fish and in particular to the disclosure of the existence of prospective target strength - frequency relations connected with type and size of the fish.

ACKNOWLEDGMENTS

The authors gratefully acknowledge the financial support of the project received from the Danish Technical Research Council.
The research reported in this paper has been carried out in a close collaboration with the Danish Institute for Fisheries and Marine Research and the advices and encouragements given by Mr. H. Lassen and Mr. E. Kirkegaard are gratefully acknowledged.

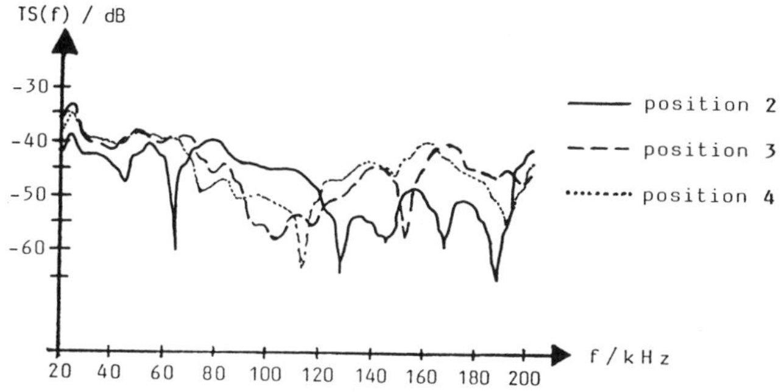

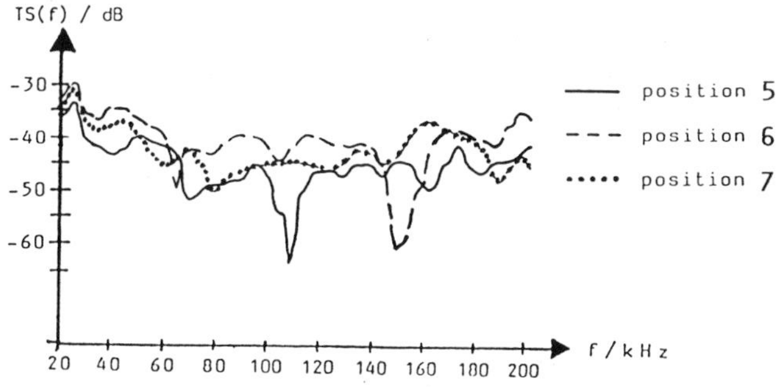

Figure 6. Target strength spectra calculated from the time-signals by scattering at dorsal aspect from a 24 cm long roach measured at various positions along its axis.

REFERENCES

1. Greenlaw, C.F. and Johnson, R.K., 1983, Multiple-frequency acoustical estimation, in "Biological Oceanography", Vol. 2, Crane, Russak & Company, Inc.
2. Haslett, R.W.G., 1965, Acoustic backscattering cross section of fish at three frequencies and their representation on a universal graph. Br. J. Appl. Phys., 16, 1143.
3. Haslett, R.W.G., 1979, The fine structure of sonar echoes from underwater targets, such as fish. Proc. Ultrasonics International 1979, IPC Science & Technology Press Ltd., Guildford, England, 307.
4. Løvik, A. and Hovem, J., 1979, An experimental investigation of swimbladder resonance in fishes. J. Acoust. Soc. Amer., 66, 850.
5. Haslett, R.W.G., 1966, Acoustic backscattering from air-filled cylindrical hole embedded in a sound-translucent cylinder. Br. J. Appl. Phys. 17, 549.
6. Huang, K. and Clay, C.S., 1980, Backscattering cross sections of live fish. PDF and aspect. J. Acoust. Soc. Amer., 67, 795.
7. Clay, C.S. and Heist, B.G., 1984, Acoustic scattering by fish - Acoustic models and a two-parameter fit. J. Acoust. Soc. Amer., 75, 1077.
8. Sun, Y., Nash, R. and Clay, C.S., 1985, Acoustic measurements of the anatomy of fish at 220 kHz. J. Acoust. Soc. Amer., 78, 1772.
9. Foote, K.G., 1985, Rather-high-frequency sound scattering by swimbladder fish. J. Acoust. Soc. Amer., 78, 688.

10. Foote, K.G., 1982, Optimizing copper spheres for precision calibration of hydroacoustic equipment. J. Acoust. Soc. Amer., 71, 742.
11. Urick, R.J., 1975, "Principles of Underwater Sound for Engineers", McGraw-Hill, New York.
12. Bjørnø, L., Christoffersen, B. and Schreiber, M.P., 1976, Some experimental investigations of the parametric acoustic array. Acustica, 35, 99.
13. Dragonette, L.R., Vogt, R.H., Flax, L. and Neubauer, W.G., 1974, Acoustic reflection from elastic spheres and rigid spheres and spheroids. II Transient analysis. J. Acoust. Soc. Amer., 55, 1130.
14. Love, R.H., 1977, Target strength of an individual fish at any aspect. J. Acoust. Soc. Amer., 62, 1377.

MULTICHANNEL FALSE COLOR ECHOGRAMS AS A

BIOLOGICAL INTERPRETATIVE TOOL

N.A. Cochrane+ and D.D. Sameoto*

+Atlantic Oceanographic Laboratory
Department of Fisheries and Oceans
Bedford Institute of Oceanography
P.O. Box 1006, Dartmouth, Nova Scotia
Canada, B2Y 4A2

*Marine Ecology Laboratory
Department of Fisheries and Oceans
Bedford Institute of Oceanography
P.O. Box 1006, Dartmouth, Nova Scotia
Canada, B2Y 4A2

ABSTRACT

Image processing systems permit display of multichannel acoustic backscatter echograms as false color imagery. This technique for rapid visual assessment is illustated with 51 and 200 kHz acoustic data from Emerald Basin located in the central Scotian Shelf. Direct multilevel net sampling shows that strong 200 kHz scattering layers below 100 m depth are compatible with "Rayleigh" scattering from copepods. Layers in the upper 60 m scattering with nearly equal strengths at 51 kHz and 200 kHz are ascribed to comparatively large fish scattering in the "geometric" regime.

INTRODUCTION

Echo measurement techniques are useful for mapping and quantitatively assessing zooplankton and micro-nekton stocks (Greenlaw and Pearcy, 1985; Sameoto, 1982; Holliday and Pieper, 1980; Greenlaw, 1979; and Pieper, 1979). In theoretical fluid sphere modelling of backscattering a critical parameter is ka, the product of the acoustic wavenumber and the target's (organism) effective radius. Theory (Anderson, 1950; Johnson, 1977) indicates that for ka<<1 organisms are "Rayleigh" scatterers with target strengths rising 12 dB/octave. For ka>>1 organisms are "geometric" scatterers with target strengths invariant with frequency. Laboratory studies of live and preserved zooplankton (Richter, 1985; and Greenlaw, 1977) support these models although departures from predicted model responses are frequently observed at large ka values. Because of the frequency dependence of organism target strengths, simultaneous multifrequency observations promise rapid, remote, quantitative, and in-situ assessment. In a single component population, delineation of the Rayleigh-geometric transition frequency (ka=1) roughly defines organism size and target strength which when combined with volume backscattering strength (target strength + 10 log concentration) yields concentration. Size and concentration estimates define biomass density. Since distinctive spectral responses should characterize many classes of marine organisms, false color echograms constructed analogous to the well known LANDSAT multispectral scanner false color imagery should be particularly useful for rapid visual delineation of mixed populations.

INSTRUMENTATION AND PROCESSING

Since 1983 the Atlantic Oceanographic Laboratory and the Marine Ecology Laboratory at the Bedford Institute of Oceanography have routinely collected digital echosounding data on up to 4 frequency channels simultaneously using two modified DATASONICS DFT-210 dual channel sounders aboard survey ship C.S.S. DAWSON. These units incorporate precise, digitally-switched, time variable gain (TVG) to correct for signal spreading and absorption. High frequency transducers are mounted in a hydrodynamic fish towed amidships at 7-10 m depth. Data are digitized from detected outputs or from synchronously demodulated down-converted signals at rates up to 20 kHz/channel. Data are logged on 9 track magnetic tapes for subsequent processing.

Transducer beam patterns, with transducers mounted in a thin walled tow body, were obtained using the calibration facilities of the Defence Research Establishment, Atlantic. The circular 200 kHz transducer pattern closely matched circular piston theory allowing its integrated beam width factor to be derived from formulae (Clay and Medwin, 1977). The 51 kHz rectangular transducer pattern deviated considerably from piston theory necessitating numerical calculation of its factor over a 60 x 60° integration grid. Its factor was equivalent to that of an 18 x 18° rectangular piston. Final total system calibrations were accomplished in a large tank using a self calibration normal surface reflection technique with identical transducer cabling, coupling transformers, and transmit power levels to at-sea deployments.

During post processing, TVG corrections were extended in dynamic (depth) range and absorption corrections refined on detailed analyses of measured temperature and salinity structure. A cancellation technique for system/ambient noise was applied by subtraction of a noise power profile from the signal squared amplitudes on a ping by ping basis and reconversion of absolute differences to echo amplitudes. Processed echograms were output to an EPC model 1600 grey scale recorder in a calibrated discreet grey level mode with predefined spatial block averaging or transfered to a VAX based PERCEPTRON image processing system where grey scale, amplitude modulated pseudocolor, or multichannel false color imagery were displayed in a 512 x 512 pixal format.

EXPERIMENT

The construction and utility of 2 channel false color imagery is illustrated using data collected simultaneously at 51 and 200 kHz from Emerald Basin in October 1984. Figure 1 shows the location of a 6 hour transect steamed at 8.5 knots from which a detailed analysis section spanning 1205-1435 AST local time ($43^{\circ}49.2'$N $62^{\circ}54.5'$W to $43^{\circ}32.6'$N $63^{\circ}12.8'$W) was selected. Relevant acoustic and acquisition parameters appear in Table 1. Figure 2 shows pre-processed 51 and 200 kHz echograms and Figure 3, the same sections after processing. The 200 kHz TVG response was extended beyond 200 m to bottom and noise cancellation

Table 1. Principal operational parameters

Freq.	XDCR Beamwd.	Pulse Leng.	Rep.	TVG Range	Power	Dig. Rate
51 kHz	18.0° rect.	2 ms	10 s	5-500 m	1 KW	10 kHz
200 kHz	5.4° circ.	5 ms	10 s	2-200 m	0.5 KW	10 kHz

applied using a noise power vs depth estimate derived from 300 separate reception cycles (without transmit) immediately following the transect.

To construct false color sections two positive (i.e. intensity increasing with signal level) sections were overlayed in constrasting colors, for example red for 51 kHz and blue for 200 kHz. Chroma and intensity contrasts were reduced to acceptable levels for CRT display and photographic reproduction by overlaying the fourth roots of the signal amplitudes. Note that the large differences in transducer beamwidths (5.4 vs 18°) and transmitted pulse lengths (5 ms vs 2 ms) restricts proper image registration and hence valid false color imagery to the larger diffuse areas of plankton scattering as opposed to point scattering or similar restricted spatial scale features. For Emerald Basin data, as seen in Figure 3, two distinct scattering regimes are present. The first regime corresponds to a reddish zone (evident in our transparencies) above 60 m depth, the second regime corresponds to a bluish zone below 100 m arising from comparatively greater 200 kHz scattering at these depths. Noise cancellation produces noticeable improvement in image contrast above 200 m where the signal-to-noise ratio is set by receive electronics. Below 200 m dominant ambient environmental noise is statistically less stationary making cancellation less effective.

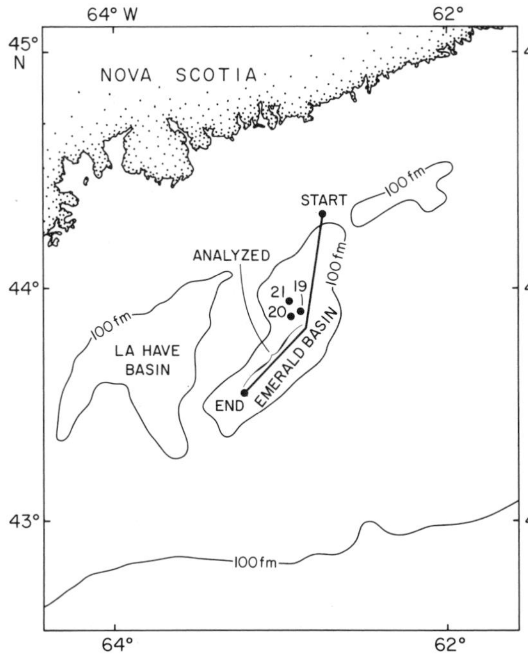

Fig. 1. Location of steamed acoustic profile and BIONESS sampling stations.

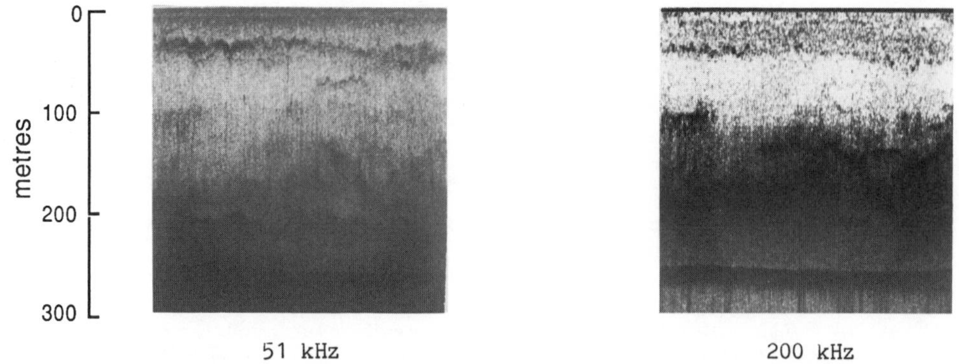

Fig. 2. Preprocessed echograms at 51 and 200 kHz. 512 successive pings are plotted.

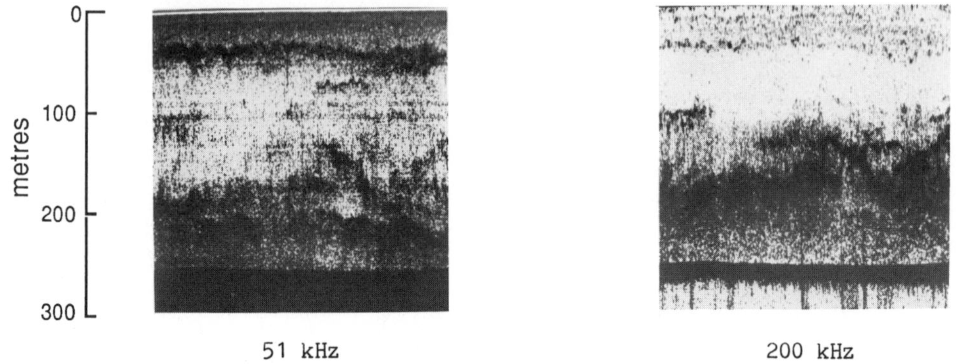

Fig. 3. Post-processed echograms at 51 and 200 kHz.

Figure 4 shows a plot of volume backscattering strength (VBS) vs depth for the first 200 pings (approximately 16 minutes of profile) of our analysis interval. In this plot precise corrections for acoustic absorption were applied using TS data from stations 20 and 21 (Figure 1). Six layer absorption models were employed utilizing the formulations of Clay and Medwin (1977) including pressure terms and the temperature dependencies of shear and bulk viscosities and molecular relaxation frequencies. This revealed that real time absorption corrections were overestimated by 1.9 dB at 250 m range.

INTERPRETATION

Biological sampling is provided by three BIONESS (Sameoto et al., 1980) towed multi-net sampling stations (Figure 1). Station 19 (2330 AST) was sampled to 100 m in 10 m vertical intervals. Station 20 (0100 AST) was sampled between 150 and 250 m at 20-30 m intervals. The sole daylight station, station 21, (1100 AST) was mainly sampled below 180 m at 10 m intervals but also included several large interval samples between 180 m and the surface. Sampling volumes normally ranged from 40 to 150 m^3. Figure 5 shows copepod densities from plotting the mid-points of sampling intervals vs. depth for all three stations while eliminating bins exceeding 40 m. Several features should be noted: (1) Scattering peaks at 10-15 m depth are reverberation from the sea surface and ship's hull. (2) Bottom occurs at about 260 m. (3) Above 60 m backscattering strengths at 51 and 200 kHz are approximately equal. (4) 200 kHz

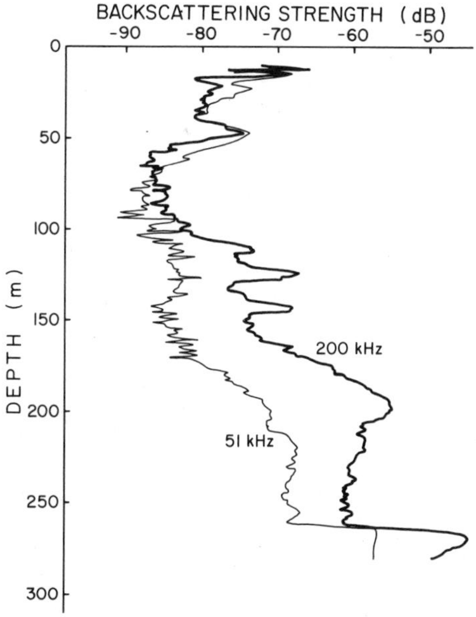

Fig. 4. Volume backscattering strength vs. depth at 51 and 200 kHz averaged over 200 successive pings with noise cancellation.

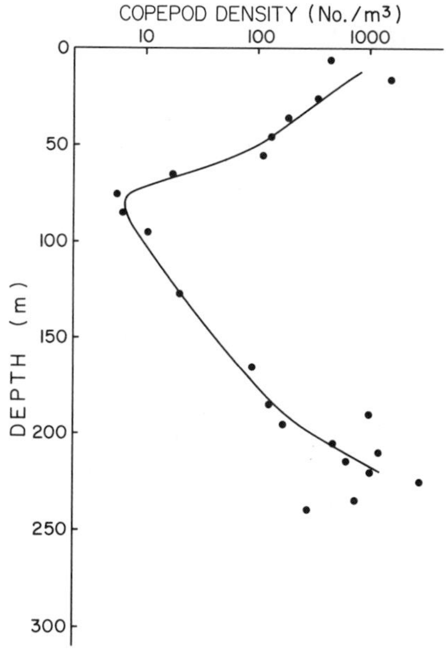

Fig. 5. Copepod densities vs depth for stations 19, 20, and 21 eliminating sampling intervals greater than 40 m in vertical extent.

backscattering levels from 100 to about 210 m generally increase with depth, display localized peaks, and significantly exceed 51 kHz levels. (5) Below 200 m the difference in backscattering strengths declines.

Theoretical VBS's were estimated using the fluid sphere model of Johnson (1977), length to volume zooplankton regression parameters of Greenlaw and Johnson (1982), and the fluid sphere acoustic parameters advanced by Holliday and Pieper (1980) (density ratio 1.12, velocity ratio 1.09) and concentrations from net analyses.

In the top 60 m nighttime scattering populations were dominately euphausiids, mainly adult Meganyctiphanes norvegica (30-35 mm length) at densities of about 1/m^3. However a single daytime sample showed no euphausiids above 100 m consistent with their well known diurnal migration cycle. The abundant upper level nighttime euphausiid population has been observed from submersibles to form a dense layer within about 3 m of the bottom during daylight hours (B. Hargrave, personal communication). Because of the long pulse lengths and high signal gains employed for our observations (Table 1) it is unlikely that such an intense bottom layer, if present, could be distinguished from the true bottom echo. The copepod population from 20-60 m was predominately Metridia lucens with peak densities of about 100/m^3. The dominate scattering components, 60% stage V females (1.5 mm length average) and 13% stage IV females (1.9 mm length average), each contributed about -94 dB or -91 dB total to the VBS at 200 kHz and 24 dB less at 51 kHz. A 0-100 m daytime net indicated an even lower daytime density of Metridia lucens. The only remaining significant copepod scatterer was Centropages typicus (1.0-1.4 mm length) in concentrations up to 100/m^3 above 40 m at night and also present at 40/m^3 in the daytime 0-100 m net. Their smaller size would make their contribution comparable to that of Metridia lucens and in any case negligible compared to observed VBS's.

We conclude that neither euphausiid nor copepod scattering accounts for observed VBS's above 60 m depth. However, the nearly identical shapes of the 51 and 200 kHz profiles suggest both arise from the same biological component. The approximate equivalence in scattering strengths at 51 and 200 kHz indicates they are relatively large "geometric" targets. These scatterers also displayed a "patchy" echogram appearance suggestive of active aggregation. We suggest these shallow echoes arise from fish which normally avoid small sampling nets.

Between 70 and 170 m , 51 kHz scattering levels (Figure 4) are limited by ambient noise resulting in a relatively ragged response after noise cancellation. Below 170 m, 51 kHz VBS's increase with depth but without a well defined peak at 190 m as observed at 200 kHz. The 200 kHz maxima between 120 and 200 m apparently correspond to the cloud-like structures on the echograms. These maxima vary from section to section with maximum levels of about -55 dB. The 200 kHz VBS's exceed those at 51 kHz by as much as 16 dB supporting a "Rayleigh" scattering mechanism at 200 kHz.

Net data showed that deep copepod distributions differed markedly from those in the upper 60 m mixed layer. The dominant species was Calanus finmarchicus stage V (C.F. V) (2.17 mm length tightly distributed, with theoretical TS's of -125 db at 51 kHz and -101 dB at 200 kHz). Calanus hyperboreus stage IV (C.H. IV) (3.3 mm, TS -114 dB at 51 kHz and -91 dB at 200 kHz) was intermixed with C.F. V in a 10-25% relative proportion with the highest proportions below 220 m. Typical total copepod densities were of the order of 1000/m^3 but highly variable. The maximum sampled density of C.F. V was 2000/m^3 and 500/m^3 for C.H. IV. Both occurred in a 220-230 m sample at station 21. At these densities the combined theoretical VBS would be about -63 dB at 200 kHz. This was roughly 8 dB lower than the highest observed backscattering levels at 200 kHz. However maximum copepod densities may well be underestimated in our sparse direct sampling data base. Maximum daytime euphausiid densities were about $0.2/m^3$ (dominately Meganyctiphanes norvegica below 200 m). Theoretical target strengths between -68 and -70 dB at 51 kHz and -66 to -68 dB at 200 kHz indicate that euphausiid scattering could be significant at 51 kHz below 200 m and would also explain the smaller difference between 51 and 200 kHz VBS's in this depth range. An unknown contribution from ground fish may also be important in reducing the difference. However most ground fish are within several metres of bottom during daylight hours.

CONCLUSIONS

False color multifrequency acoustic imagery is a promising rapid survey tool for delineating contrasting marine biological populations. An imagery capability in real time would assist the execution of marine biological surveys. Wide frequency diversity in acoustic sounding is ideally required since components of a mixed population can be separated only if they differ in their basic scattering modes over the range of observational frequencies.

ACKNOWLEDGEMENTS

We thank the Defence Research Establishment, Atlantic (DREA) for making available their barge acoustic calibration facility and Drs. A. Herman and R. Shotton for constructive comments on this manuscript.

REFERENCES

Anderson, V.C., 1950, Sound scattering from a fluid sphere, J. Acoust. Soc. Am., 22:246.

Clay, C.S. and Medwin, H., 1977, "Acoustical Oceanography Principles and Applications", John Wiley, New York.

Greenlaw, C.F., 1979, Acoustical estimation of zooplankton populations, Limnol. Oceanogr., 24:226.

Greenlaw, C.F., 1977, Backscattering spectra of preserved zooplankton, J. Acoust. Soc. Am., 62:44.

Greenlaw, C.F. and Pearcy, W.G., 1985, Acoustical patchiness of mesopelagic micronekton, J. Marine Res., 43:163.

Greenlaw, C.F. and Johnson, R.K., 1982, Physical and acoustical properties of zooplankton, J. Acoust. Soc. Am., 72:1706.

Holliday, D.V. and Pieper, R.E., 1980, Volume scattering strengths and zooplankton distributions at acoustic frequencies between 0.5 and 3 MHz, J. Acoust. Soc. Am., 67:135.

Johnson, R.K., 1977, Sound scattering from a fluid sphere revisited, J. Acoust. Soc. Am., 61:375.

Pieper, R.E., 1979, Euphausiid distribution and biomass determined acoustically at 102 kHz, Deep-Sea Res., 26:687.

Richter, K.E., 1985, Acoustic scattering at 1.2 MHz from individual zooplankters and copepod populations, Deep-Sea Res., 32:149.

Sameoto, D.D., 1982, Zooplankton and micronekton abundance in acoustic scattering layers on the Nova Scotian Slope, Can. J. Fish. Aquat. Sci., 39:760.

Sameoto, D.D., Jaroszynski, L.O. and Fraser, W.B., 1980, BIONESS, a new design in multiple net zooplankton samplers, Can. J. Fish. Aquat. Sci., 37:722.

STUDY OF THE PSEUDO - LAMB WAVE S_o GENERATED IN THIN CYLINDRICAL SHELLS INSONIFIED BY SHORT ULTRASONIC PULSES IN WATER

Maryline Talmant and Gérard Quentin

G.P.S. Tour 23 - Université Paris 7 - 2 Place Jussieu
75251 Paris Cedex 05 - France

ABSTRACT

We studied the backscattering of short ultrasonic pulses by very thin shells filled with air or water. The shells are such that $0.96 < b/a < 0.99$. The domain of the dimensionless frequency factor ka which is studied is $20 < ka < 300$. We observed two circumferential waves, one, with a high velocity, that we have identified as the pseudo-Lamb wave S_0, and one with slow velocity, the nature of which is not easy to specify. The signal analysis has been performed both in the time and frequency domains for shells filled with air and only in the time domain for shells filled with water.

NOTATION

a(b) = external (internal) radius of the cylindrical shell
e = thickness of the shell
d = e/2 = half thickness of the shell
a' = (a + b)/2 = mean radius of the shell
c = velocity of sound in water
$c_L(c_T)$ = velocity of longitudinal (transverse) waves inside the material constituent of the shell
ρ = density of the material constituent of the shell
$c_G(c'_G)$ = group velocity of the first (second) circumferential wave circumnavigating the shell
θ(θ') = angle of incidence generating the first (second) circumferential wave.

INTRODUCTION

The study of acoustic backscattering by elastic bodies has been very fruitful for many years. Initially, the theoretical model has been presented for bodies with simple geometry and with high symmetry as the sphere or the cylinder, where the method of separation of variables can be used. In theoretical studies, the first step is generally to describe the steady state of the scattered pressure in the far field, versus the dimensionless factor ka ; this function, normalized by the incident pressure is labelled "form function of the body". Doolittle and Überall[1] were the first to present the form function of one shell insonified by a plane wave as a Rayleigh

Table 1. Geometrical parameters of the five shells

index of the shell	1	2	3	4	5
e(mm)	0.12	0.18	0.24	0.29	0.40
b(mm)	10	10	10	10	10
a(mm)	10.12	10.18	10.24	10.29	10.40
b/a	0.988	0.982	0.976	0.972	0.961

series of normal modes of vibration. Two complementary approaches are used : the normal modes theory and the circumferential waves theory. The first approach shows the resonant character of the limited body. The second approach permits an isolation of the contributions of the different waves excited by the scattering process : the reflected waves, the transmitted waves and the circumferential waves. The relationship between these two approaches has been etablished[2,3].

The first experimental works have shown the circumferential feature of one or two waves generated when an ultrasonic wave is scattered from a thin cylindrical shell[4,5,6,7]. The first wave has a large velocity of the order of 5500 m/s ; the second one is slower with a velocity of the order of 2000 m/s. These two waves are compared with the Lamb waves inside a plate. All authors consider the quick wave as being the same as the first symmetrical Lamb mode S_o inside a plate[8,9,10]. The validity of this analysis becomes increasingly good when the ratio b/a tends towards 1. The physical nature of the slow wave is rather more controversial and the fact that it has a specific domain of generation has received various interpretations[8,9]. Our experimental study follows the work of Fekih and Quentin[11]. We have studied the backscattering by thin cylindrical shells with a ratio b/a greater than 0.96 in a quite large range of values of the dimensionless frequency parameter extending from 20 to 300.

EXPERIMENTAL SET - UP

The sound velocity inside the water placed in the tank is c = 1480 m/s. Five circular cylindrical shells have been studied. All are made of duraluminum (c_L = 6370 m/s, c_T = 3130 m/s and ρ = 2800 kg/m^3) ; their common length is 20 centimeters and their common internal diameter is 20 millimeters. Their only difference is the thickness of the shell (see table 1). The shells are filled with air and then with water. The first set of experiments allows the study of the circumferential waves circumnavigating the shell. The second set of experiments allows also the study of the waves transmitted inside the water filling the shell.

A pulse generator delivering short pulses (200 nanoseconds) is connected to the ultrasonic transducer acting as both transmitter and receiver. Three transducers have been used with center frequencies respectively equal to 1, 2.25 and 5 megahertz. Their frequency spectrum is wide band with respective bandwidth of 0.5 to 1.5 megahertz ; 1.3 to 3.5 megahertz and 3 to 7 megahertz within 20 decibels. The dimensionless frequency range which can be easily studied extends therefore from 20 to 300. The received signal is amplified through a wide band amplifier and then, displayed on a sampling oscilloscope when we study the waveform in the time domain. For the studies performed in the frequency domain the oscilloscope is replaced by an analog spectrum analyzer. The shell is placed in the far field of the transducer and the width of the ultrasonic beam at the position of the shell is larger than the shell's diameter 2a. The axis of this beam is perpendicular to the axis of the shell.

EXPERIMENTAL RESULTS

Time domain (shells filled with air)

One typical plot of the time domain signal is shown on Fig.1. The first echo corresponds to the specular reflexion on the front face of the shell. It is followed by oscillations due to multiple reflections inside the thickness of the shell[11]. After the first echo, we notice two sets of echoes labelled E_i and E'_i. The successive echoes of each set are regularly spaced in time and have a decreasing amplitude. The two sets of echoes are distinguishable not only by the arrival time of the first echo (respectively τ_1 and τ'_1) and by the periodicity of the echoes (respectively τ and τ') but also by the frequency content and the time width of the echoes. The amplitude of each of these two sets of echoes is smaller than that of the specular echo by more than one order of magnitude. On Fig.1 the amplitude of echo E'_1 is larger than that of echo E_1. This ratio of amplitudes inverts oneself in the high frequency regime (5 MHz transducer).

Other experiments performed in the bistatic geometry where the receiving transducer is different from the transmitting one (Fig.2) have shown the circumferential nature of the two waves giving rise to the sets of echoes E_i and E'_i. Each of these waves is generated at a given angle of incidence θ. In the bistatic geometry the axis of the two ultrasonic beams are tilted at an angle θ with respect to the normal to the shell and we can choose this angle in order to enhance a given set of echoes. The observation of the arrival time of the first echo shows that it varies proportionnaly to the angular path γ accomplished by the wave circumnavigating the shell. The circumferential wave propagates around the shell and radiates inside water under an angle equal to its angle of incidence. In the backscattering geometry at normal incidence, $\gamma = 2\pi - 2\delta$ and these two paths are possible for the same kind of wave circumnavigating clockwise or counterclockwise around the shell. The corresponding echoes interfere constructively because their arrival time is exactly the same.

In the backscattering experiments described in this paper the time τ_1 (respectively τ'_1) corresponds to a circumnavigation of $(2\pi - 2\delta)$ radians[1] around the shell and the time τ (respectively τ') to a circumnavigation of 2π radians. Therefore the periodicities τ and τ' are related respectively to the group velocities c_G and c'_G through the relationship :[8]

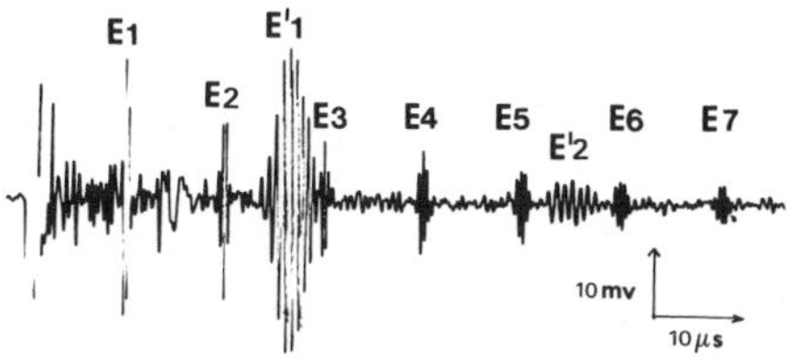

Figure 1 : Experimental signal backscattered from a duraluminum shell with b/a = 0.976 and an insonifying center frequency of 2.25 MHz.

$$\tau^{(')} = 2\pi a' / c_G^{(')} \qquad (1)$$

These periodicities are measured by superposition of echoes using an oscilloscope with double time base and digital delay. The bandwith of the transducer being large, the velocity $c_G^{(')}$ corresponds to a mean group velocity near the center frequency of the superposed echoes. On all five shells the values of the measured periodicities have quite the same values :

$\tau \sim 12\ \mu s$ $\qquad c_G \sim 5400$ m/s

$\tau' \sim 34\ \mu s$ $\qquad c_G' \sim 1900$ m/s

The values are in good agreement with the results of Ryan[6] and Dragonette[7,8] On Fig.3 we have plotted the phase and group velocities of the first symmetric S_o and antisymmetric A_o Lamb waves in a plate in air with a half thickness d, where abcissa is in units of $k_T d$ as usually[12]. The velocity of the first wave c_G corresponds quite exactly to that of the Lamb wave S_o. This circumferential wave can consequently be considered as a pseudo-Lamb wave S_o. The velocity c_G' of the second wave does not correspond within experimental errors to that of the A_o Lamb wave for one plate.

Frequency analysis (shells filled with air)

On Fig.4 are plotted the frequency spectra of the total backscattered signal obtained for shell 5 insonified successively by the three transducer These spectra are not normalized with respect to the frequency response of each transducer. Nevertheless in the center frequency region of each transducer where the transmitted amplitude varies slowly with frequency, the measured backscattering spectrum may be considered as differing very little from the "form function" in the far field multiplied by one constant. The spectra plotted on Fig.4 exhibit sudden amplitude variations for specific values of ka which can be divided into two series depending on their ka periodicity (δka or (δka)'), their amplitude and their observation domain. The comparison of the spectra of echoes E_i' and of the frequency domain of generation of the oscillations with periodicity (δka)' shows that they are due to the same phenomenon : the propagation of the "slow" wave. Similarly for the echoes E_i and the periodicity δka.

The oscillations with periodicity δka corresponding to the pseudo-Lamb wave S_o are observed for almost the whole range of frequencies in the backscattering spectrum except in the highest frequency domain where they disappear for frequencies higher than a critical frequency $(ka)_c$ which is a

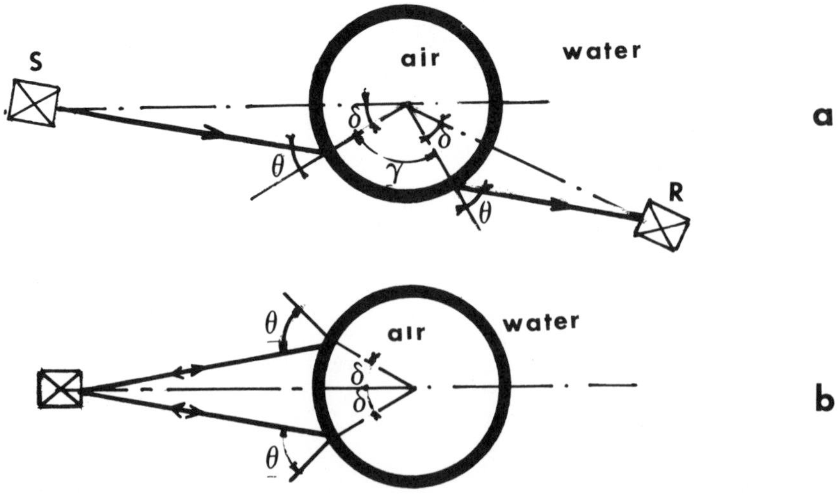

Fig.2 (a) Bistatic geometry (b) Backscattering geometry

decreasing function of the shell's thickness (Table 2). The oscillations due to the second wave ("slow" wave) do exist only in a limited range of values of ka. We label D_{ka} and D_{fe} the ranges where this wave is experimentally observed respectively in function of the dimensionless parameter ka or of the other parameter fe (MHz x mm) commonly used for plates. Table 2 shows that D_{fe} is almost a constant whatever the shell studied ; at the opposite D_{ka} is a decreasing function of the thickness of the shell. The range D_{ka} corresponding to the shell with b/a = 0.982 is in rather good agreement with the work of Dragonette[8] who has shown that the wave that he calls A_o is generated in the range D_{ka} extending from 55 to 90 for one shell with b/a = 0.98. The range D_{fe} measured on this same shell is in rather good agreement with the theoretical calculations of Breitenbach et al[9].

Group velocity of the two waves observed experimentally

We have used in the beginning of this work the overlapping of echoes method to deduce the group velocity of circumferential waves from time domain measurements. It is also possible, when the wave has a dispersion sufficiently small to deduce this velocity from the backscattering spectra. One can show[8] that, for non dispersive waves the difference δka between two maxima or two minima, corresponding to the same wave, is related to the group velocity (equal to the phase velocity) through the relationship

$$\delta ka = c_G/c \qquad (2)$$

The values plotted on Fig.3 are deduced from this relationship. For the pseudo-Lamb wave S_o, the periodicity δka varies very little and this wave can be considered as a non-dispersive one in the frequency range studied. In addition the comparison with the value of the group velocity of the Lamb-wave S_o in a plate exhibits a very good agreement confirming the identification. For the second wave (the one with a slow velocity) the situation is different because the periodicity(δka) varies (noticeably) even in the narrow range of frequencies where this wave is observed. In this dispersive case the

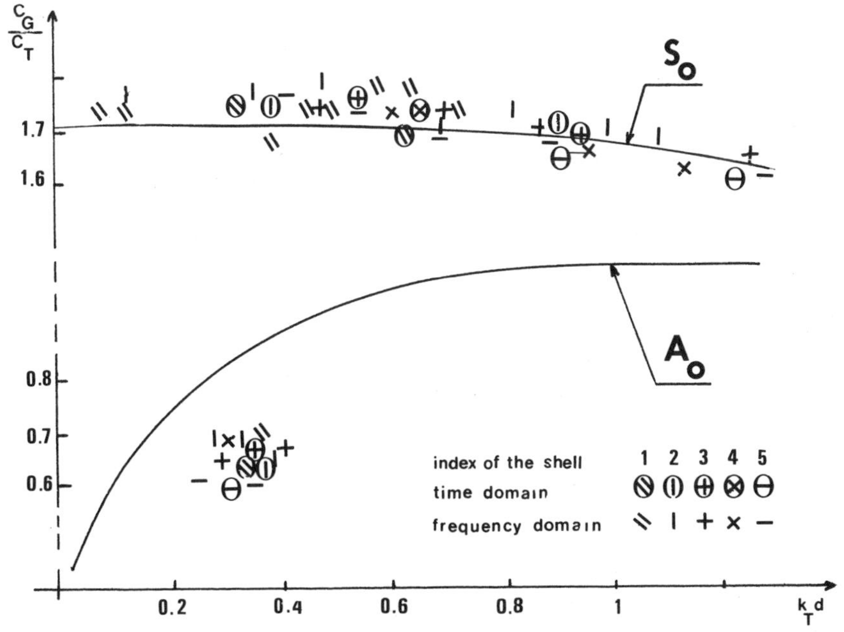

Fig.3 Group velocity curves of A_o and S_o Lamb modes

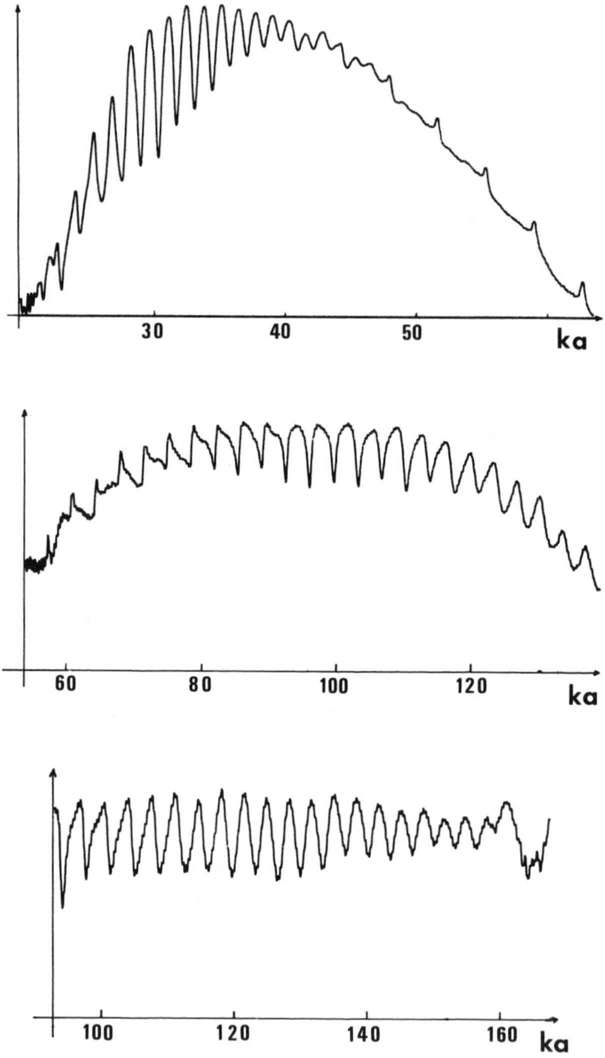

Fig.4 Backscattered spectrum from the shell n°5.

Table 2. Cut-off frequency for the pseudo-Lamb wave S_o Frequency range of observation of the second wave in units of ka or fe

index of the shell	1	2	3	4	5
$(ka)_c$	330	280	250	210	170
D_{ka}	80-130	60-90	40-70	35-50	25-40
D_{fe} (MHz x mm)	0.25-0.35	0.25-0.37	0.25-0.4	0.25-0.35	0.2-04

relationship given above is not valid and must be replaced by another one where the order of the resonance must be known.

"Attenuation" of the two circumferential waves

The circumferential waves, when propagating, radiate inside water. From the ratio of the amplitude of the frequency spectra of two successive echoes E_i and E_{i+1} due to the pseudo-Lamb wave S_o, or E'_i and E'_{i+1}, due to the second wave, we deduce for each wave its coefficient α of "attenuation" through radiation inside water during one travel equal to the circumference of the shell. Fig. 5 shows the curves obtained for the two waves. For the pseudo-Lamb wave S_o this "attenuation" which is low at low frequencies increase very sharply when ka > 80 until values attaining 4 dB/radian. In addition the range of ka where this rapid increase is observed is an increasing function of the ratio b/a. This explains why on Fig.2 the oscillations of S_o type disappear for ka $\sim$ 170 (shell with b/a = 0.961) and only for ka $\sim$ 330 for one thinner shell (b/a $\sim$ 0.988).

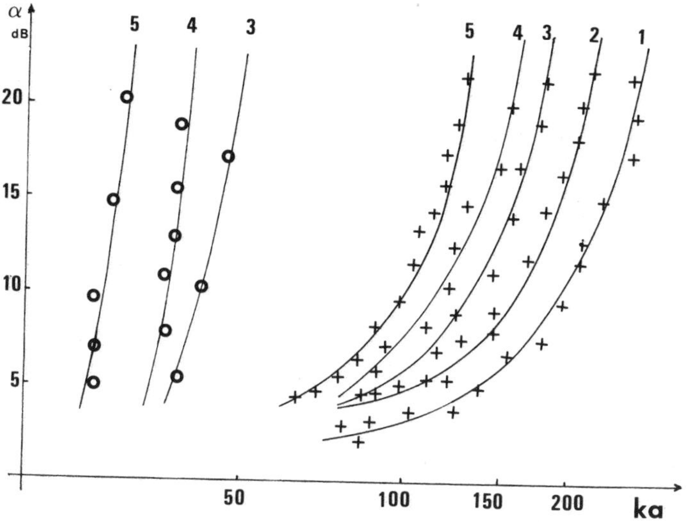

Fig.5 Reradiation coefficient of the two circumferential waves versus ka (+: "S_o" ; O: the "slow" wave). The numbers design the index of each shell.

Shells filled with water

The time domain signal is much more complicated than for the same shell filled with air. We observed, in addition to the case of air-filling, the waves transmitted into internal water. We identified again the pseudo-Lamb wave S_0 and verified that the value of its "attenuation" coefficient is twice as large as for air-filling confirming that this "attenuation" is purely a consequence of radiation inside water. The identification of the various experimental echoes is on the way.

CONCLUSION

These experimental results show that there is one simple acoustic signature for thin shells filled with air. In addition to the specular echo this signature is given by the echoes of two circumferential waves : the pseudo-Lamb wave S_0 and one second wave probably dispersive and generated in a narrow frequency range. We have studied the variations with frequency of the group velocity and of the attenuation due to radiation is water. Recent works[13] show that the specific nature of the wave generated in shells is complicated. The fact that these waves are generated only in a limited frequency range is an important criterion for target identification[14]. We have studied these ranges for the two waves observed in the frequency interval extending from ka = 20 to ka = 300.

REFERENCES

1. R.D. Doolittle and H. Uberall, Sound scattering by elastic cylindrical shells, J. Acoust. Soc. Am. 39 : 272 (1966).
2. H. Uberall, L.R. Dragonette, and L. Flax, Relation between creeping waves and normal modes of vibration of a curved body, J. Acoust. Soc. Am. 61 : 711 (1977).
3. A. Derem, Relation entre la formation des ondes de surface et l'apparition de résonances dans la diffusion acoustique, Revue du CETHEDEC. 58 : 43 (1979).
4. R.E. Bunney, R.R. Goodman, and S.W. Marshall, Rayleigh and Lamb waves on cylinders, J. Acoust. Soc. Am. 46 : 1223 (1969).
5. C.W. Horton and M.V. Mechler, Circumferential waves in a thin-walled air-filled cylinder in a water medium, J. Acoust. Soc. Am. 51 : 295 (1972)
6. W.W. Ryan, Acoustical reflections from aluminum cylindrical shells immersed in water, J. Acoust. Soc. Am. 64 : 1159 (1978).
7. W.G. Neubauer and L.R. Dragonette, Observation of waves radiated from circular cylinders caused by an incident pulse, J. Acoust. Soc. Am. 48 : 1135 (1970).
8. L.R. Dragonette, Evaluation of the relative importance of circumferential or creeping waves in the acoustic scattering from rigid and elastic solid cylinders and from cylindrical shells, Nav. Res. Lab. Rep. NRL 8216 (1978).
9. E.D. Breitenback, H. Uberall, and K.B. Yoo, Resonant acoustic scattering from elastic cylindrical shells, J. Acoust. Soc. Am. 74:1267(1983)
10. G. Maze, J. Ripoche, A. Derem, and J.L. Rousselot, Diffusion d'une onde ultrasonore par des tubes remplis d'air immergés dans l'eau, Acustica. 55 : 69 (1984).
11. M. Fekih and G. Quentin, Scattering of short ultrasonic pulses by thin cylindrical shells : generation of guided waves inside the shell, Physics Letters. 96 : 379 (1983).
12. I.A. Viktorov, "Rayleigh and Lamb waves", Plenum Press, New York (1967)
13. J.L. Rousselot, Comportement acoustique d'un tube cylindrique mince en basse fréquence, Acustica. 58 : 291 (1985).
14. S.K. Numrich, N.H. Dale, and L.R. Dragonette, Generation and exploitation of plate waves in submerged, air filled shells, Advances in fluid-structure interaction 78 (1984).

SCATTERING OF AN ACOUSTIC WAVE BY A TRANSMISSION-LOSS TILE

R.J. Brind

Admiralty Research Establishment
Portland, Dorset, U.K. DT5 2JS

ABSTRACT

The interaction of an plane acoustic wave in water with an infinite, doubly periodic transmission-loss tile is investigated theoretically. The tile consists of an array of voids in a lossy rubber fluid loaded layer; the plane wave is incident at an arbitrary angle. A doubly periodic Green's Function is introduced and shown to lead to an elastodynamic representation for the scattered field in terms of an integral over the surface of a void in one cell only. The integral representation can serve as the basis for approximations, or the scattering problem can be reduced to an integral equation which is solved numerically. The reflected and transmitted fields are shown to consist of various spectral orders. Numerical methods for solving the integral equation are investigated.

INTRODUCTION

Recently, a transmission-loss material has been developed for a variety of acoustic applications. It is manufactured in the form of tiles which consist of a regular array of voids in a rubber plate. There is a need for accurate modelling of this material, in order that a physical understanding of its operation can be gained, and as an aid to the development of means of measuring its transmission-loss performance at lower frequencies. Previous modelling work at A.R.E. has used an approach based on the effective medium approximation, which assumes the scatterers (the air cavities in the tile) are small, randomly distributed and that there is no interaction between them. Here we develop an alternative treatment that retains the doubly periodic structure and includes multiple scattering effects.

Our approach is to aim to solve the scattering problem exactly for a fixed periodic structure. Radlinski (1980) has presented a theoretical treatment of the scattering of acoustic waves in water by multiple gratings of compliant tubes. He also considered the scattering of normally incident waves by a grating in a viscoelastic layer by the method of partial domains. Fokkema (1981) has analysed the two-dimensional scattering of elastic waves by a periodic interface between two media using an integral equation formulation. Wirgin (1984) has analysed the three-dimensional scattering of scalar waves by hard or soft eggcrate surfaces using the Rayleigh hypothesis.

In this paper, we assume the tile is composed of homogeneous linearly elastic material and that the solution in the absence of the cavities is known. Damping in the rubber is modelled by taking the elastic constants to be complex. By the introdution of singular solutions, Green's Functions, corresponding to arrays of sources of acoustic waves and longitudinal and transverse bulk waves, we are able to derive an integral representation for the scattered field in terms of an integral over the surface of the void in one cell only. An integral equation for the solution to the scattering problem is derived. A numerical technique for its solution is developed for the corresponding integral equation in the simpler problem of the scattering by an isolated void in rubber.

FORMULATION OF THE PROBLEM

The tile consists of linearly elastic material and lies in $-h/2 \leqslant x' \leqslant h/2$ with water above and below. It has a periodic structure with periods d_1 and d_2 in the x_1 and x_2 direction respectively, see Fig.1. An harmonic time dependence $\exp(-i\omega t)$ is assumed where ω is the radian frequency.

The equation of motion in the solid requires that the displacement fields have a potential representation

$$\underline{u} = \nabla \phi + \nabla \times \underline{\psi} \tag{1}$$

where

$$(\nabla^2 + k_L^2)\,\phi = 0 \qquad k_L = \omega/c_L \qquad c_L^2 = (\lambda + 2\mu)/\rho_s \tag{2}$$

$$(\nabla^2 + k_T^2)\,\underline{\psi} = \underline{0} \qquad k_T = \omega/c_T \qquad c_T^2 = \mu/\rho_s \tag{3}$$

and ρ_s is the density of the rubber, λ and μ the Lame constants.

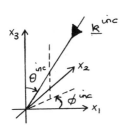

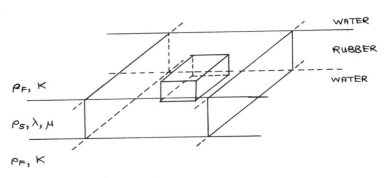

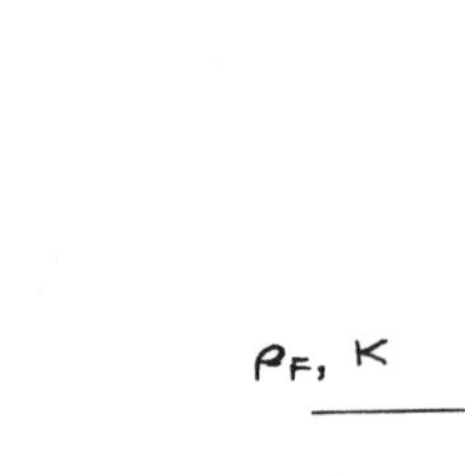

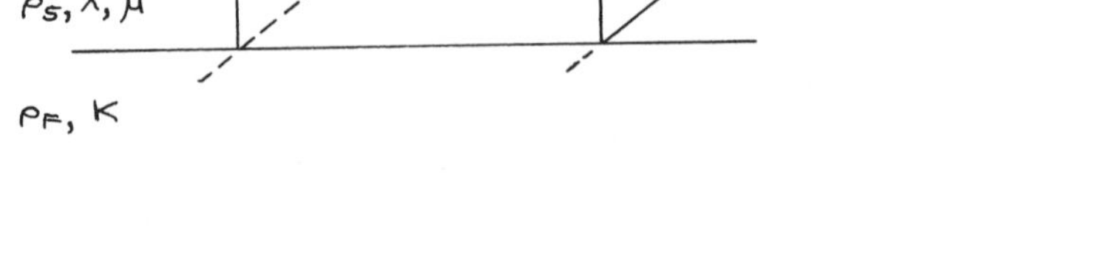

Similarly the displacement in the fluid has a representation

$$\underline{u} = \nabla \chi \tag{4}$$

where

$$(\nabla^2 + k_F^2)\,\chi = 0 \qquad k_F = \omega / c_F \qquad c_F^2 = K / \rho_F \tag{5}$$

and ρ_F is the density of water, K its bulk modulus.

The scattering problem is posed as follows. A plane wave is incident on the tile. In the absence of the voids, planes waves are reflected and transmitted into the water and are set up in the rubber layer. The angles of propagation, which can be complex, are determined by Snell's law. When the voids are present some of the energy of the incident wave is scattered into higher spectral orders. The incident field is taken to be the field that would occur in the absence of the voids. The scattered field

(a) satisfies the equations of motion in the solid and the fluid and the interface conditions,

(b) consists of outgoing waves at $x = -\infty, \infty$

(c) negates the surface traction of the incident field on the surface of the voids.

If the incident wave strikes the tile from a direction specified by spherical polar angles θ^{inc} and ϕ^{inc} as shown in Fig.1, the form of the exciting field and the periodicity of the file enforce a quasi-periodicity in the scattered field such that

$$\underline{u}^{sc}(x)\,\exp(-ik_1^{inc}x_1 - ik_2^{inc}x_2) \qquad k_{1,2}^{inc} = -k_F \sin\theta^{inc} \begin{smallmatrix}\cos\\ \sin\end{smallmatrix} \phi^{inc} \tag{6}$$

is periodic in x_1 and x_2 with periods d_1 and d_2 respectively.

GREEN'S FUNCTIONS FOR THE TILE

Consider an array of sources of W-waves in free-space (where W=L,Tk,F) at positions $\underline{x}'_{mn} = \underline{x}' + \hat{\underline{x}}_1 md_1 + \hat{\underline{x}}_2 nd_2$ where $m,n = -\infty, \infty$ with phase factors $\exp(-ik_1 md_1 - ik_2 nd_2)$, i.e.,opposite to that of the incident wave. It can be shown from a manipulation of the Fourier transform of the free-space point source potential,

$$\begin{aligned} g^W &= \exp\{ik_W|\underline{x}-\underline{x}'|\} / 4\pi|\underline{x}-\underline{x}'| \\ &= \iint d\xi_1 d\xi_2 \, \frac{i\exp\{i\xi_1(x_1-x_1') + i\xi_2(x_2-x_2') + i\eta^W|x_3-x_3'|\}}{8\pi^2\eta^W} \end{aligned} \tag{7}$$

where

$$\eta^W = (k_W^2 - \xi_1^2 - \xi_2^2)^{1/2} \qquad Re(\eta) > 0 \qquad Im(\eta) > 0 \tag{8}$$

making use of the generalised function result

$$\sum_{m=-\infty}^{\infty} \exp\{im(\xi_1 - k_1)d_1\} = \sum_{p=-\infty}^{\infty} \delta(\xi_1 - k_1 - 2\pi p/d_1)/d_1 \tag{9}$$

that the array of sources has an expansion

$$G^W = \sum_{p,q=-\infty}^{\infty} G_{pq}^W(\underline{x}, \underline{x}') \tag{10}$$

where

$$G_{pq}^W = \frac{i\exp\{-ik_1^p(x_1-x_1') - ik_2^q(x_2-x_2') + ik_{W3}^{pq}|x_3-x_3'|\}}{2d_1d_2k_{W3}^{pq}} \tag{11}$$

$$k_{W3}^{pq} = (k_W^2 - k_1^{p2} - k_2^{q2})^{1/2} \qquad (12$$

$$k_i^p = k_i^{inc} + 2\pi p/d_i \qquad (13$$

This array of sources does not satisfy the equations of motion and the boundary conditions for the fluid loaded plate. However, we will show th non-singular terms can be added so that the appropiate conditions for eac type of wave source are satisfied.

Consider first an array of L-sources, putting $x_1'=x_2'=0$ without loss o generality. Since all field variables derived from each term in the expansion of the potential retain its spatial variation in the 1-2 plan we can treat each pq term individually. The 1-2 axes can be rotated that the term is a function of x_1^{pq} only, i.e., the problem is two-dimensional. Thus the field from an array of L-sources is

$$\underline{U}^L = \sum_{p,q=-\infty}^{\infty} \{ \nabla \Phi_{pq}^L + \nabla \times \Psi_{pq}^L \hat{x}_2^{pq} \} \qquad |x_3| \leqslant h/2 \qquad (14$$

$$\underline{U}^L = \sum_{p,q=-\infty}^{\infty} \nabla \chi_{pq}^L \qquad |x_3| \geqslant h/2 \qquad (15$$

We suppose that the field consists of symmetric and antisymmetric longitudinal and transverse standing waves in the solid, and of outgoing plane waves in each of the fluid half-spaces. We write, omitting a fact $\exp(ik_1^{pq} x_1^{pq})$ everywhere,

$$\Phi_{pq}^L = G_{pq}^L + A_{pq}^L \cos(k_{L3}^{pq} x_3) + B_{pq}^L i \sin(k_{L3}^{pq} x_3) \qquad (16$$

$$\Psi_{pq}^L = C_{pq}^L i \sin(k_{T3}^{pq} x_3) + D_{pq}^L \cos(k_{T3}^{pq} x_3) \qquad (17$$

and

$$\chi_{pq}^L = (E_{pq}^L + F_{pq}^L) \exp(ik_{F3}^{pq} x_3) \qquad x_3 \geqslant h/2 \qquad (18$$

$$= (E_{pq}^L - F_{pq}^L) \exp(-ik_{F3}^{pq} x_3) \qquad x_3 \leqslant -h/2 \qquad (19$$

The interface conditions are continuity of surface traction and normal displacement. The two conditions on σ_{23} are satisfied identically. The remaining six continuity conditions are sufficient to determine the six remaining constants. The details are not given here.

It should be emphasised that the Green's Functions themselves are no two-dimensional, as they are the sum of terms with different planes of symmetry. The potentials for an array of F-sources in the water and Tk-sources in the solid are calculated in a similar way.

DERIVATION OF INTEGRAL REPRESENTATIONS

An application of the Elastodynamic Green's theorem to one cell of t tile yields the usual integral representation for the compressional and shear potentials of the scattered field in the solid.

$$\rho_s \omega^2 \phi^{sc}(\underline{x}') = -\int_{\partial V_s} ds_j \{ u_i^{sc} \sigma_{ij}^L(\underline{x},\underline{x}') - \sigma_{ij}^{sc} U_i^L(\underline{x},\underline{x}') \} \qquad (20$$

$$\rho_s \omega^2 \psi_k^{sc}(\underline{x}') = -\int_{\partial V_s} ds_j \{ u_i^{sc} \sigma_{ij}^{Tk}(\underline{x},\underline{x}') - \sigma_{ij}^{sc} U_i^{Tk}(\underline{x},\underline{x}') \} \qquad (21$$

where $\underline{ds}$ is the outward normal on the surface ∂V_s enclosing the solid in o cell, $\underline{u}^W$ and $\underline{\underline{\sigma}}^W$ are the singular solutions for a phased array of W-sources in the tile without voids. The periodicity factor of the Green's Functions is opposite to that of the scattered field, and hence the integrals over the sides between cells cancel. An application of Green's Theorem to the fluid volume, with $\underline{x}'$ still in the tile, produces

$$0 = \int_{\partial V_{F\pm}} ds \left\{ u_n^{sc} P^W(\underline{x},\underline{x}') - p^{sc} U_n^W(\underline{x},\underline{x}') \right\} \qquad W = L, Tk \tag{22}$$

where $\partial V_{F\pm}$ is the surface between the solid and water at $x_3 = \pm h/2$. The periodicity of the Green's Functions means that the contribution from the sides of the fluid volumes above and below a cell is zero. The radiation condition ensures that the contribution from infinity is zero also. Adding Eq. (20) and Eq. (21) to Eq. (22) and making use of the continuity of normal displacement and surface traction (which in this case is normal to the surface), we obtain

$$\rho_S \omega^2 \phi^{sc}(x') = -\int_{\partial V} ds_j \left\{ u_i^{sc} \sigma_{ij}^L(\underline{x},\underline{x}') - \sigma_{ij}^{sc} U_i^L(\underline{x},\underline{x}') \right\} \tag{23}$$

$$\rho_S \omega^2 \psi_k^{sc}(\underline{x}') = -\int_{\partial V} ds_j \left\{ u_i^{sc} \sigma_{ij}^{Tk}(\underline{x},\underline{x}') - \sigma_{ij}^{sc} U_i^{Tk}(\underline{x},\underline{x}') \right\} \tag{24}$$

Using the Green's Functions corresponding to an array of F-sources in the fluid, we derive a similar representation for the field in the fluid

$$\rho_F \omega^2 \chi^{sc}(\underline{x}') = -\int_{\partial V} ds_j \left\{ u_i^{sc} \sigma_{ij}^F(\underline{x},\underline{x}') - \sigma_{ij}^{sc} U_i^F(\underline{x},\underline{x}') \right\} \tag{25}$$

INTEGRAL EQUATIONS AND NUMERICAL SOLUTION

Since the singularities of the doubly periodic Green's Functions about the source point are the same as those of the free-space Green's Function, integral equations for this problem can be derived in a similar manner to those for the exterior T-problem of elastodynamics considered by Jones (1984). The following equations are obtained by forming the integral representations for the displacement and surface traction from Eq. (23) and Eq. (24) and allowing the observation point $\underline{x}'$ to approach the surface of the void. The displacement representation yields a singular integral equation of the second kind

$$\frac{1}{2} u_i^{sc}(\underline{x}') + \int_{\partial V} ds_q\, u_p^{sc}\, \Gamma_{i,k}\left(\sigma_{pq}^L, \sigma_{pq}^{Tk}\right) = \int_{\partial V} ds_q\, \sigma_{pq}^{sc}\, \Gamma_{i,k}\left(U_p^L, U_p^{Tk}\right) \tag{26}$$

where $\Gamma_{i,k}$ is a first order differential operator

$$\rho_S \omega^2 \Gamma_{i,k}(\phi, \psi^k) = \frac{\partial \phi}{\partial x_i'} + \varepsilon_{imk} \frac{\partial \psi^k}{\partial x_m'} \tag{27}$$

The traction representation yields an integro-differential equation of the first kind

$$n_j \Sigma_{ij,k}\left(\int_{\partial V} ds_q\, u_p^{sc} \sigma_{pq}^L, \int_{\partial V} ds_q\, u_p^{sc} \sigma_{pq}^{Tk} \right) \tag{28}$$

$$= \frac{1}{2} n_j \sigma_{ij}^{sc} + n_j \Sigma_{ij,k}\left(\int_{\partial V} ds_q \sigma_{pq}^{sc} U_p^L, \int_{\partial V} ds_q \sigma_{pq}^{sc} U_p^{Tk}\right)$$

where $\Sigma_{ij,k}$ is a second order differential operator

$$\rho_s \omega^2 \Sigma_{ij,k}(\phi, \psi^k) = \left\{\lambda \delta_{ij} \frac{\partial^2}{\partial x'_\ell \partial x'_\ell} + 2\mu \frac{\partial^2}{\partial x'_i \partial x'_j}\right\} \phi \tag{29}$$

$$+ \mu \left\{\varepsilon_{i\ell k} \frac{\partial^2}{\partial x'_j \partial x'_\ell} + \varepsilon_{j\ell k} \frac{\partial^2}{\partial x'_i \partial x'_\ell}\right\} \psi^k$$

The integral equation of the second kind is to be preferred as it is likely to be better conditioned numerically. Numerical difficulties also occur at certain frequencies that correspond to eigenfrequencies of the complementary interior problem, and Jones (1984) has shown for the isolated scatterer problem that these can be overcome by taking a linear combination of the two equations. However, the frequencies of immediate interest in the transmission-loss tile problem are low enough that difficulties in numerical solution arising from non-uniqueness are not anticipated.

A numerical technique for the solution of the integral equation has been implemented, initially for the isolated scatterer problem. The surface of the void in Fig.1 is split up into rectangular flat patches S_j ,on each of which the scattered displacement is assumed to be a constant vector $\underline{v}_j$. The constant vectors are determined by satisfying the integral equation at the centres of the patches $\underline{x}'_j$. Jones (1985) has presented transformations which convert the highly singular integrals over the patches to contour integrals around the boundary curves C_j . For our geometry in the isolated scatterer case, these transformations simplify to

$$\int_{S_1} ds_q v_p \Sigma_{pq,k}\left(\frac{\partial g^L}{\partial x'_i}, \varepsilon_{imk} \frac{\partial g^T}{\partial x'_m}\right)$$

$$= \int_{S_1} ds_q \left\{v_q \frac{\partial (g^T - g^L)}{\partial x_i} - v_i \frac{\partial g^T}{\partial x_q}\right\} \tag{30}$$

$$+ \int_{C_1} d\ell_p \, \varepsilon_{pqr} \left\{\frac{2}{k_T^2} v_r \frac{\partial^2}{\partial x_i \partial x_q}(g_L - g_T) + \delta_{ir} v_q g^T\right\}$$

and

$$\int_{S_1} ds_q \sigma_{pq}^{sc} \left\{\frac{\partial^2}{\partial x'_i \partial x_p}(g^L - g^T) + k_T^2 \delta_{ip} g^T\right\}$$

$$= \int_{S_1} ds_q \sigma_{pq}^{sc} \left\{n_i n_p (k_L^2 g^L - k_T^2 g^T) + \delta_{ip} g^T k_T^2\right\} \tag{31}$$

$$+ \int_{C_1} d\ell_r n_q \sigma_{pq}^{sc} \left\{\varepsilon_{rim} n_m \frac{\partial}{\partial x_p} - \varepsilon_{rmp} n_i \frac{\partial}{\partial x_m}\right\}(g^L - g^T)$$

The remaining surface integrals involve at most an integrable singularity. The integrals have been evaluated numerically using Gauss-Legendre quadrature schemes. The integral equation was approximated by a matrix equation, which was solved with the aid of a NAG Library subroutine.

DISCUSSION

We have presented a method of deriving an integral equation formulation of this scattering problem. The integration is over the surface of a void in one cell only, but multiple scattering effects have been included by the introduction of the doubly periodic Green's Functions. The kernel of the integral equation is strongly singular, but the equation can be solved numerically using Boundary Integral Equation methods developed for isolated scatterers. A numerical method for solving the integral equation has been implemented and applied to the isolated scatterer problem. It is hoped to apply the method to the integral equation for the transmission-loss tile problem in the near future.

The Green's Functions have been calculated in closed form for a single layer of rubber. Multi-layered plates can be treated but it may then not be feasible to write down analytic expressions for the Green's Functions.

REFERENCES

Fokkema, J. T., 1981, Reflection and Transmission of Elastic Waves by the Spatially Periodic Interface between Two Solids, Wave Motion, 3(2):33.

Jones, D. S., 1985, Boundary Integrals in Elastodynamics, I.M.A. J. of Appl. Maths., 34:83.

Radlinski, R. P., 1980, Multiple Scattering from Gratings of Compliant Tubes in Fluid and in a Viscoelastic Layer Immersed in Fluid, in "Acoustic, Electromagnetic and Elastic Wave Scattering - Focus on the T-Matrix Approach", V .K. Varadan and V. V. Varadan, eds., Pergamon, New York.

Wirgin, A., 1984, Scattering of Sound from Hard and Soft Eggcrate Surfaces, J. Acoust. Soc. Amer., 75(2):340.

FREQUENCY DEPENDENCE OF THE INTERACTION OF ULTRASOUND WITH SUSPENDED SEDIMENT PARTICLES

A.S.Schaafsma and A.J.Wolthuis

Delft Hydraulics Laboratory
P.O. Box 177
2600 MH Delft, The Netherlands

ABSTRACT

Measurements of the ultrasonic attenuation in a wide frequency range, caused by sand and clay particles as well as glass spheres in suspension, are presented. It is found that the behaviour of the attenuation as a function of frequency depends strongly on both particle size and shape. The experimental results agree well with recently published theoretical calculations in a qualitative sense. The measurement of partial concentrations of clay and sand particles in a mixed suspension, is envisaged as an application of the present multifrequency approach.

INTRODUCTION

The development of new measuring methods for suspended sediment transport has formed part of the research program of the Delft Hydraulics Laboratory for many years. An ultrasonic method to measure continuously the local sand transport was developed (Jansen, 1979). Several prototype instruments were built for use during the construction of the storm surge barrier in the mouth of the Eastern Scheldt tidal estuary (Schaafsma and der Kinderen, 1985). The best results for this specific field situation, were obtained with the following narrow beam double bistatic scattering system. Continuous ultrasound of a fixed frequency was emitted by a single source. The sound intensity scattered under an angle of 120 degrees with respect to the forward direction was measured along two soundpaths of different length. The two signals obtained, enabled a correction for the attenuation loss, so that relatively high sand concentrations, up to about 5 kg/m^3, could still be measured. Further, by choosing a sound frequency of 4.5 MHz, equal sensitivity (within 10 %) was achieved for sand sizes between about 80 μm and 300 μm. This choice also reduced the sensitivity for the type of silt particles present in the estuary, with at least a factor of 10, as compared to that for sand.

The Eastern Scheldt closure works being finished, the interest has turned to problems related to silt. The work presented in this paper is related to two kinds of field situations, which are found in the Dutch estuaries and along the coast, and can be characterized in the following, somewhat idealized way.

In the first situation, both silt, of size between about 1 μm and 50μm, and sand particles, of size from 50 μm to about 1 mm, are present in suspension in varying mass ratios. A measuring method is required which is able to determine the (partial) mass concentrations of the sand and the silt seperately. Preferably there should be some size resolution within the silt and sand size ranges.

In the second type of situation, only silt particles are in suspension. Depending on the salinity of the water and the history of the suspension, part or all of the silt particles may have flocculated. A measuring method should be able to determine the mass concentration of the silt independent of the degree of flocculation. Prefarably the size distribution of the flocs should be measurable with some resolution.

The problem is : how to meet these requirements ? Optical extinction methods are in use for turbidity measurements. It is well known however that these optical methods suffer from a strong particle size dependence. The reason for this is that the wavelength of light is small compared to all relevant particle sizes. Therefore optical methods operate in the geometric range as regards the scattering process. The principal advantage of ultrasonic methods is that the wavelength of sound in water can be varied over a wide range. The sound wavelength of the above mentioned ultrasonic sand transport meter for example, was deliberately chosen of the same order of magnitude as the relevant particle sizes, around 150 µm, to minimize the grain size dependence. For both larger sizes (geometric scattering region) and smaller sizes (Rayleigh scattering region), the sensitivity of the instrument decreases strongly.

The present work concerns the feasibility to extend this approach to broader particle size distributions, including both finer clay particles and sand. The idea is to use the frequency dependence of scattering and attenuation of ultrasound to discriminate between different particle sizes. This idea is not new. For example Flammer (1962), showed that the frequency dependence of the attenuation in the range of 2.5 to 25 MHz can be used to infer particle size distibutions of sand particles with sizes between about 60 µm and 1 mm.

In our problem, the presence of finer clay particles required the use of frequencies above 25 MHz. Therefore a laboratory set-up has been realized to measure attenuation and scattering under some selected angles in the frequency range of 1 to 100 MHz, using bursts of ultrasound.

EXPERIMENTAL

The experiments were carried out in a vertical perspex cylinder with an inner diameter of 10 cm. The suspension was recirculated using a slurry pump. Six different sets of broadband ultrasonic transducers, covered the frequency range of about 0.7 MHz to 100 MHz. The transducers were placed in two measuring sections on top of each other and arranged in appropriate configurations for attenuation and scattering measurements. A third measuring section contained a movable suction pipe, with an inner diameter of 5 mm, in order to take samples which were used to obtain estimates of the particle concentration.

The choice was made to use short bursts of sine waves, of some fixed frequency, as the excitation signal for the sound emitting transducers. The repetition frequency of the bursts was about 1 kHz. In this way the entire frequency range could be sampled at an arbritrary number of discrete values.

The detection method was still in development: attenuation measurements could be performed correctly, but scattering measurements were not yet possible. For frequencies below about 20 MHz the average amplitude of a single burst was determined by amplitude demodulation and integration. Subsequently, averaging was performed over typically 2000 bursts. At higher frequencies a selective detection method was envisaged but not yet implemented. Therefore the pulse heights were read from an oscilloscope screen, which increases the experimental error in this range, also since averaging is not possible.

Table 1. Data regarding particle size and concentration of the suspensions for which the attenuation was measured (see Fig. 2. and Fig. 3.). The effective diameters d_{16}, d_{50} and d_{84} correspond to the 16%, 50% and 84% values respectively, of the cumulative size distributions by weight or by volume. $C_{m,n}$ is the nominal mass concentration of the suspension as it was prepared before the measurement. $C_{m,s}$ is the mass concentration as measured by suction at the end of the measurement.

particles	sieve size fraction (μm)	optical size analysis d_{16} (μm)	d_{50}	d_{84}	$\frac{d_{84}}{d_{16}}$	concentration $C_{m,n}$ (kg/m^3)	$C_{m,s}$	$\frac{C_{m,s}}{C_{m,n}}$
sand ρ =2650 kg/m^3	180-212	171	207	263	1.54	5.00	5.07	1.01
	90-106	88	115	143	1.63	2.00	1.72	0.86
glass spheres ρ=2900 kg/m^3	180-212	167<all d<216			<1.29	5.00	4.45	0.89
	90-106	80	95	120	1.50	2.00	1.60	0.80
clay								
China ρ=2660 kg/m^3		d < 50				0.75	0.69	0.92
Westerwald ρ=2890 kg/m^3		d < 50				0.75	0.57	0.76

The sound attenuation constant of the water itself increases with the square of the frequency, reaching a value of about 2000 db/m at f=100 MHz. This had the following two consequences for the experimental set-up.

In the first place, the sound path, i.e. the distance between the emitting and receiving transducers, was chosen smaller for the higher frequencies. The distances used were: 10 cm for the transducers with a centre frequency of 1.0, 2.25, 7.5 and 30 MHz; 5 cm and 2.7 cm for the 50 MHz and 100 MHz transducer pair respectively.

In the second place, in order to avoid a significant influence of the temperature dependence of the attenuation constant of the water, the water temperature was stabilized within 0.02^{o}C. All measurements reported here were carried out at a temperature of 20.0^{o}C.

The measurement procedure for attenuation measurements was to perform a reference measurement in water without particles first. Subsequently the suspension of particles in water was measured. The difference between the two measurements was the attenuation due the particles. The concentration was chosen for each suspension in such a way, that the accuracy of the measurement was optimized for the higher frequency range, 50 MHz to 100 MHz. The proportionality between the attenuation constant and the particle concentration was verified for a few types of suspensions and further it was assumed to be valid.

RESULTS AND DISCUSSION

Characterization of particles and suspensions

For the present study we have chosen some suspensions of particles in the size ranges of sand as well as that of silt. Starting from natural sand and commercially available leadglass spheres, relatively narrow size fractions were obtained by repeated sieving. The clays were not treated further.

An independent measurement of the size distribution was performed using a commercial particle sizer (Malvern) based on light diffraction: the results are collected in Table 1. The size distribution of the 180-212 μm glass spheres was so narrow, that all effective sizes were within one size interval of the particle sizer.

The size determination of the clay particles was not well possible with the optical diffraction method. The reason was that a considerable fraction of these particles is smaller than the fundamental lower limit of this method, which is about 5 μm.

Therefore we used scanning electron microscopy (SEM), to arrive at the following semi-quantitative judgement. Both clays mainly consist of platelike particles of a thickness of a few micrometers, as illustrated by the photomicrograph shown in Fig. 1. The dominant dimensions of the plates are somewhat higher for China clay, ~ 10 μm, than for Westerwald clay, ~ 5 μm. The China clay definitely contains a higher number of very fine, ~ 1 μm, particles.

The suspensions were prepared at the desired nominal concentrations listed in Table 1. At the end of each attenuation mesurement, the concentration also was determined from a sample taken with the suction method. This latter concentration was always lower than, or equal to, the nominal one. From this and other evidence, we concluded that some fraction of the particles was not in suspension, but remained settled at some places in the circuit. However, the concentration as measured by suction reproduced well, within 5 % in all cases, and was considered to be the best estimate of the actual concentration during the attenuation measurement. There was one exception in this respect: for Westerwald clay we found a scatter of about 50 % in the measured concentrations.

Fig. 1. Particles of China clay, as seen by the scanning electron microscope.

Suspensions of sand and glass spheres

Fig. 2 shows the result of the attenuation measurements for the 180-212 μm and the 90-106 μm sieve fractions of sand and glass spheres.

We should note that the attenuation data are presented here as: $\alpha/1000C_v$, where α is the attenuation constant of the particles in db/m, and C_v is the volume concentration of the particles in suspension as measured by suction (see Table 1.).

The difference between the smoothly increasing attenuation for the sand particles and the oscillatory behaviour for the glass spheres is striking. For sand similar results were published by Flammer (1962). However the present authors are not aware of any published data for spheres, which show the marked oscillations.

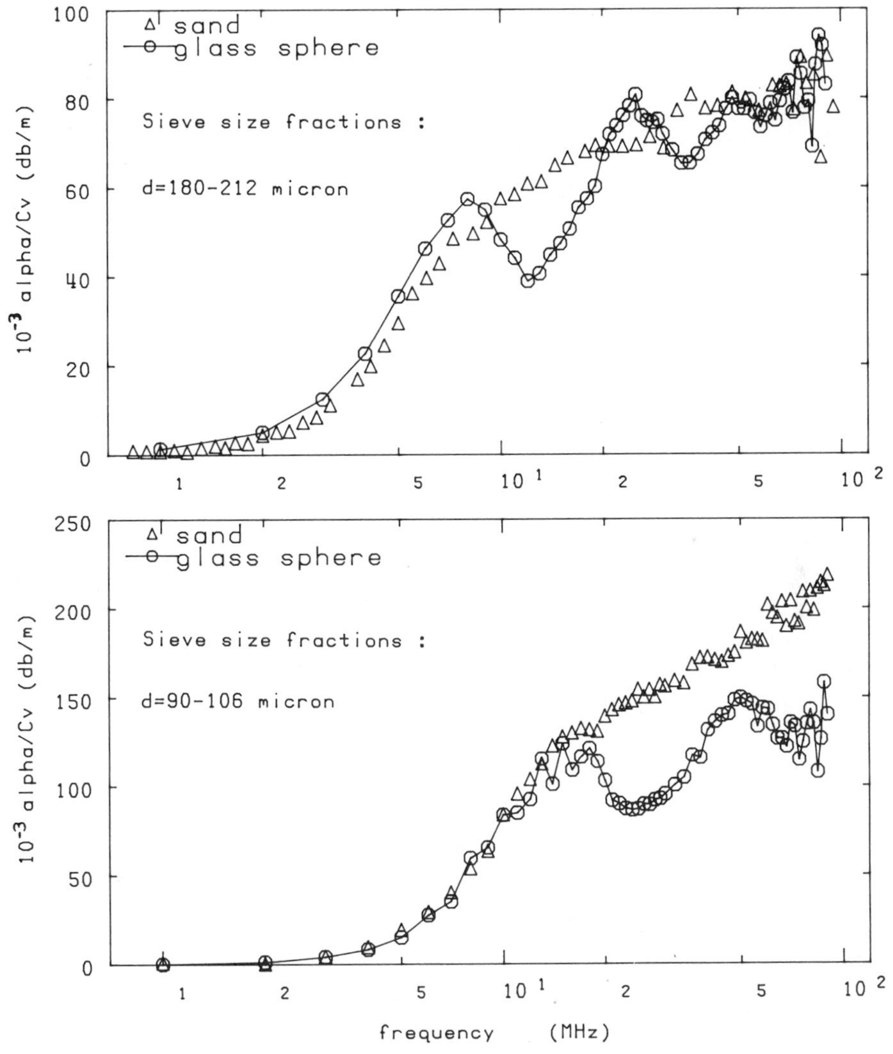

Fig. 2. Measured frequency dependence of the attenuation due to sand particles and leadglass spheres of approximately the same size, for two values of the size (upper and lower figure). The attenuation is normalized in such a way, that it applies to a volume concentration C_v equal to 0.001 in all cases.

Another noteworthy aspect of the result shown in Fig. 2 is that the frequency dependence is almost the same for sand and glass spheres in the lower frequency range, i.e below the first maximum of the glass sphere curve.

Suspensions of clay

The results for the two different kinds of clay, China and Westerwald are shown in Fig. 3. For the Westerwald clay we attach value only to the shape of the curve, but not to the absolute values, because of the problem with the concentration mentioned above. As expected, the attenuation start to increase at much higher frequencies compared with the 200 μm and 100 μm sized particles. Also, for China clay, the absolute value of the attenuation per unit of concentration is considerably higher for f > 80 MHz. In principle this provides a means to distinguish between silt and sand sized particles.

Comparison with theory

For a comparison with theoretical results it is convenient to express the measured attenuation in terms of an effective cross section σ per particle, normalized in some way e.g. by dividing by the particle's projected cross section πa^2, a being its effective radius. Let α be the attenuation constant in db/m and C_v the volume concentration of particles in m^3/m^3. It is straightforward to show that,

$$\frac{\sigma}{\pi a^2} = (4/3)\,\frac{a\alpha}{4.343\,C_v}$$

In Fig. 4 the measurements for narrow size fractions of sand and glass spheres are presented this way as a function of ka, where k is the wavenumber for sound waves in water. It was assumed that all particles of a given suspension have the same size. The value of the effective diameter d or radius a was taken equal to the value of d_{50}, from the optical

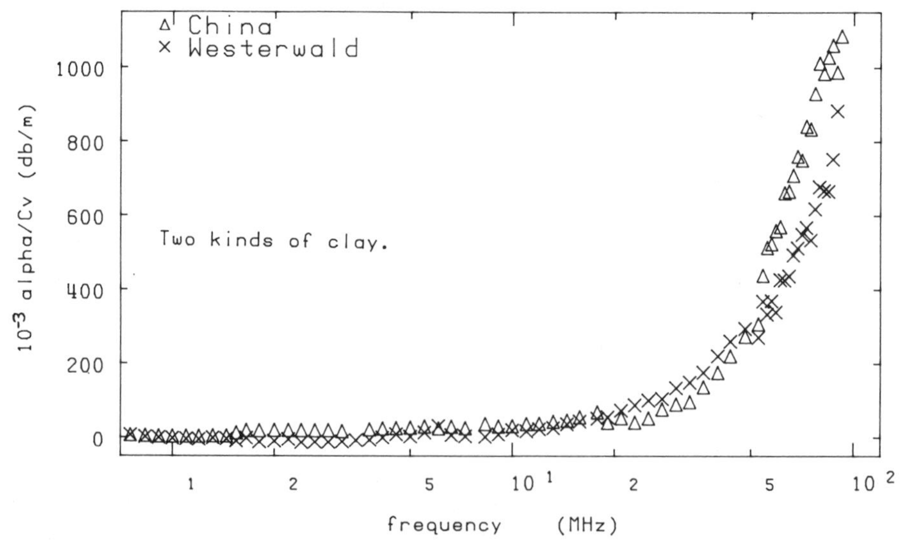

Fig. 3. Measured frequency dependence of the attenuation due to clay particles, pertaining to a concentration by volume of 0.001.

diffraction size measurement given in Table 1. For the 180-212 μm size fraction of glass spheres, the average sieve diameter, d=196 μm, was used.

Also shown in Fig. 4 are two theoretical results for the attenuation due to scattering by spherical particles. The smoothly increasing curve corresponds to the well known rigid movable sphere model. The other curve is the calculation for elastic quartz spheres as given by Hay and Mercer (1985). This latter curve shows an oscillatory behaviour, which is remarkably similar to that found for the glass spheres. It should be noted however that the elastic properties of leadglass spheres are likely to be different from that of quartz. Therefore only a qualitative value should be attributed to the agreement between theory and experiment in the case of the glass spheres.

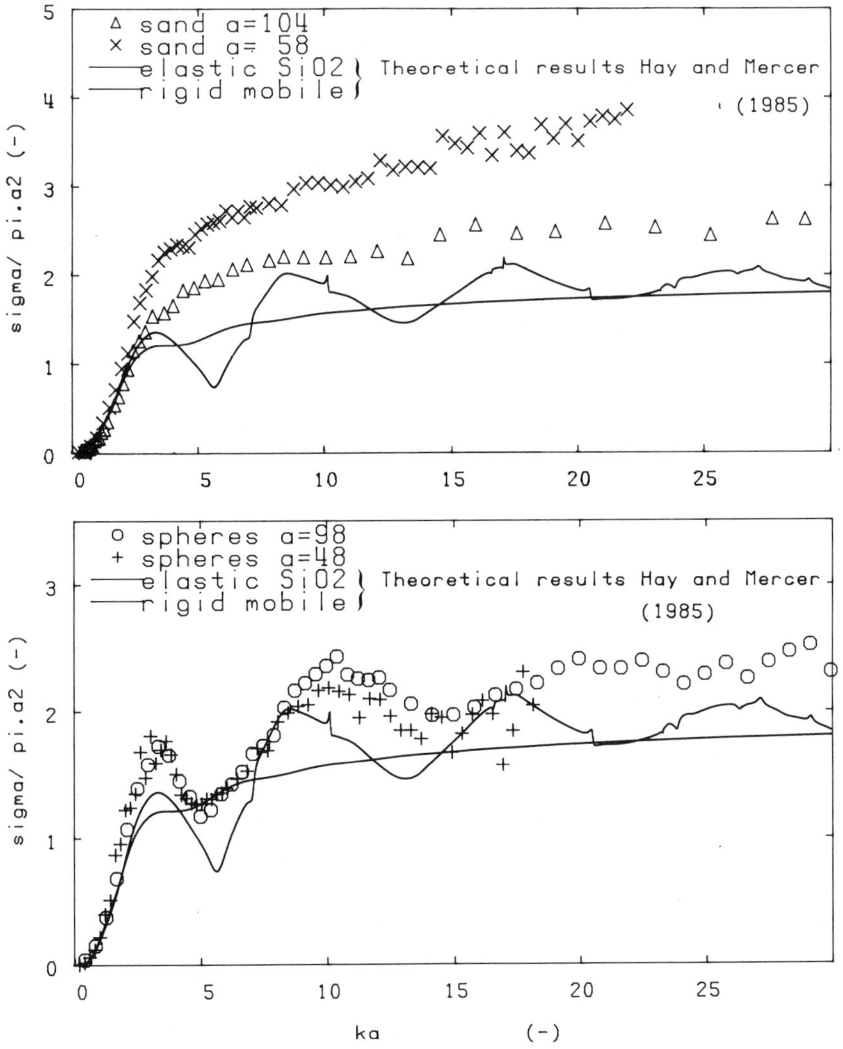

Fig. 4. Normalized attenuation cross section $\sigma/\pi a^2$ as measured for sand (upper figure) and leadglass spheres (lower figure), compared with the theoretical attenuation due to scattering for elastic quartz spheres, which are the oscillating curves, and for rigid movable spheres, as calculated by Hay and Mercer (1985).

There is also a qualitative agreement between the behaviour of the attenuation of the sand grains and that predicted by the rigid movable sphere model.

As regards the quantitative discrepancy between theory and experiment for values of ka $\gtrsim$ 3 we remark the following. The approximation to use a single effective size, in the normalization procedure of the data, may be too crude. Instead a distribution of sizes should be used. Also other effects could contribute to this discrepancy. However it seems advisable to collect additional, and complementary, experimental data prior to a further elaboration of this point. For example, experiments with sand particles which have an increasing degree of sphericity, could be valuable in this respect.

ACKNOWLEDGEMENTS

The authors wish to thank Dr. A.E. Hay for making the results of his and Mercer's theoretical calculations (Hay and Mercer, 1985) available to them.

REFERENCES

Jansen, R.H.J., 1979, An ultrasonic doppler scatterometer for measuring suspended sand transport, in: "Proc. Ultrasonics Int. 1979 Conf., Graz, Austria", pp. 366-371, IPC Science and Technology Press, Guildford, U.K.

Schaafsma, A.S. and der Kinderen, W.J.G.J., 1985, Ultrasonic instruments for the continuous measurement of suspended sand transport, in: "Proc. IAHR Symp. on Measuring Techniques in Hydraulic Research", Balkema Publ., Rotterdam.

Flammer, G.H., 1962, Ultrasonic measurement of suspended sediment, Geological Survey Bulletin 1141-A, US Government Printing Office, Washington.

Hay, A.E. and Mercer, D.G., 1985, On the theory of sound scattering and viscous absorption in aqueous suspensions at medium and short wavelengths, J. Acoust. Soc. Am., 78:1761.

SOUND SCATTERING IN AQUEOUS SUSPENSIONS OF SAND:

COMPARISON OF THEORY AND EXPERIMENT

J. Sheng and A. E. Hay

Department of Physics and Newfoundland Institute for Cold Ocean Science
Memorial University of Newfoundland
St. John's, Newfoundland, Canada, A1B 3X7

ABSTRACT

Theoretical estimates of the form factor and the linear attenuation coefficient are compared with the available data for dilute aqueous suspensions of sand. As far as the existing data are concerned, the measured attenuation coefficients are the most useful. The available data on scattered intensities are too few for the comparison with theory to be conclusive. From the comparisons with the attenuation data it appears that a spherical model is a reasonable approximation. Three theoretical models are considered, in which the spherical scatterer is assumed to be either elastic, or completely rigid, or both rigid and immovable. The rigid movable model provides the best fit to the data. The comparatively poor agreement with the results from the elastic model indicates that resonance excitation does not occur, probably because natural sand grains are irregularly shaped and inhomogeneous in composition. The rigid immovable model fits the data the least well, indicating that the inertia of the particles is important. Approximate expressions for the form factor and attenuation coefficient have also been constructed, based on the so-called high-pass model introduced by Johnson (1977). The high-pass model provides a fit to the data which is as good as the rigid movable case.

INTRODUCTION

The purpose of this paper is to compare the available experimental data on attenuation and scattering of sound by sand grains in water with different theoretical models in which the sand grain is approximated by a homogeneous solid sphere. The objective is to determine first whether the homogeneous sphere approximation is useful, and second which of the spherical models, if any, is the most suitable. The problem is part of the more general one of interpreting the relationship between the scattered acoustic signal, and particle concentration and size, that is associated with the recent application of acoustic remote sensing techniques to sediment transport studies in the ocean.

Our approach is to use the phase shift formalism in the partial wave expansion of the scattered pressure field, following Faran (1951; see also Hay and Mercer, 1985). The theoretical estimates are made for an inviscid, non heat-conducting fluid and a homogeneous spherical scatterer. The scatterer is assumed to be either elastic, or rigid but movable, or both rigid and immovable. By elastic we mean that shear and compression waves may propagate within the material, and that the incident wave can induce displacements of the scatterer's center of mass. In a rigid scatterer no sound propagation occurs. An immovable scatterer is infinitely dense. Comparisons are also made with a modified form of the so-called high-pass model introduced by Johnson (1977).

The data sources are Flammer (1962), Jansen (1977), Clarke et al. (1984), and Schaafsma and der Kinderen (1985). Flammer measured the additional attenuation caused by the suspended sediment at the concentration $M = 2.67\ kg/m^3$. The sand particles were sieved into 17 size fractions in the 44 to 1000 μm diameter range. Both Jansen and Schaafsma and der Kinderen used bistatic systems and measured the scattered intensity, as a function of sand concentration and size, for a scattering angle of 120°. The attenuation coefficient is obtained from the dependence of the scattered signal on concentration. Jansen's measurements were made at 8 MHz, in the concentration range 0.1 to 30 kg/m^3, and for four sieve fractions in the 50 to 280 μm range. Schaafsma and der Kinderen used 4.5 MHz systems and natural sand size distributions in the 50 to 100 μm range. The measurements were made at concentrations less than 5 kg/m^3. Clarke et al. measured the backscattered intensity at 3 MHz for three size fractions in the 30 to 300 μm range. The concentrations were less than 0.1 kg/m^3.

THEORY

In a narrow beam bistatic or monostatic system the mean square scattered pressure can be shown to be given by

$$\hat{p}_s^2 = S^2 H_o^2 \exp[-4\alpha_s r_o] \tag{1}$$

where S is a constant depending on the geometry of the system, the attenuation in the ambient fluid, and the sound pressure level of the transmitted pulse. The distance from each transducer to the detected volume is r_o, and α_s is the additional attenuation due to scattering from the particles. H_o depends upon the magnitude of the form factor as well as the size distribution and concentration of the suspended particles, which can be written as

$$H_o^2 = \frac{3M}{4\pi\rho_o'} \left[\int_0^\infty |f_\infty|^2 a^2 n(a)da \Big/ \int_0^\infty a^3 n(a)da \right] \tag{2}$$

where ρ_o' is the grain density of each scatterer, n(a) the size spectral density, M the mass concentration, and f_∞ the form factor, which can be written as (Faran, 1951)

$$f_\infty(\theta) = -\frac{2}{k_c a} \sum_{n=0}^{\infty} (2n+1)i \sin\eta_n \exp[-i\eta_n]P_n(\cos\theta) \tag{3}$$

Here θ is the scattering angle, a the particle radius, k_c the incident wavenumber, η_n the phase shift, and P_n the Legendre polynomial of order

n. For particles of uniform size, the attenuation coefficient due to scattering is given by

$$\frac{\rho_o'\alpha_s}{k_c M} = \frac{3}{4}\frac{\mathrm{Im}[f_\infty(0)]}{(k_c a)^2} \tag{4}$$

The magnitude of the form factor for the high-pass model is given by (Sheng, 1986)

$$|f_\infty| = \frac{K_f(k_c a)^2}{1+K_f(k_c a)^2} \tag{5}$$

where $K_f = (2/3)|\gamma_\kappa+\gamma_\rho\cos\theta|$, and γ_κ and γ_ρ are the usual compressibility and density contrasts in the Rayleigh range (e.g. Hay, 1983). The attenuation coefficient is

$$\frac{\rho_o'\alpha_s}{k_c M} = \frac{K_\alpha(k_c a)^3}{1+(4/3)K_\alpha(k_c a)^4+\beta(k_c a)^2} \tag{6}$$

where $K_\alpha = (\gamma_\kappa^2 + \gamma_\rho^2)/6$ and β is an adjustable constant ≥ 1. In this paper the value of β is chosen as 1.

RESULTS

Attenuation coefficient

The attenuation coefficients measured by Flammer for a mixture of Missouri River sand and blasting sand (M. S. mixture) are shown in Fig. 1. The data for particle diameters greater than 500 μm have not been used since greater uncertainties in the measured values for those fractions were reported (Flammer, 1962). The solid lines in Figs. 1a and b represent the theoretical results for the elastic and rigid movable cases, respectively. It can be seen that the rigid movable case fits the data better than the elastic case for $k_c a < 10$ and both cases fit the data very well in the geometric region. The extrema in the theoretical curve for the elastic case near $k_c a = 5.7$, 8.2, 13.3, and 17.1 in Fig. 1a are associated with Rayleigh wave resonances (Hay and Mercer, 1985). Since these features do not appear in Flammer's data, it is concluded that resonance excitation does not occur for these sand grains.

Since there is little difference between the rigid movable and rigid immovable cases for very large $k_c a$, the comparisons for these cases are considered only for $k_c a \leq 5$, and are shown in Fig. 2. It is obvious that the rigid movable case provides the better fit to the data. The agreement is quite good for $k_c a < O(2)$. For larger $k_c a$ the measurements are larger than the theoretical values. This is also illustrated by Fig. 3a, in which the calculated estimates in the rigid movable case for uniform particle size are plotted against the measurements. In this Figure we have non-dimensionalized α_s by a/ϵ, where $\epsilon = M/\rho_o'$ is the volume concentration of suspended particles. This form permits the high and low $k_c a$ data to be distinguished in scatter diagrams like those in Fig. 3. It can be seen that for $a\alpha_s/\epsilon > O(0.4)$, corresponding to $k_c a > 3$ (Hay and Mercer, 1985), the points are below the straight line. The

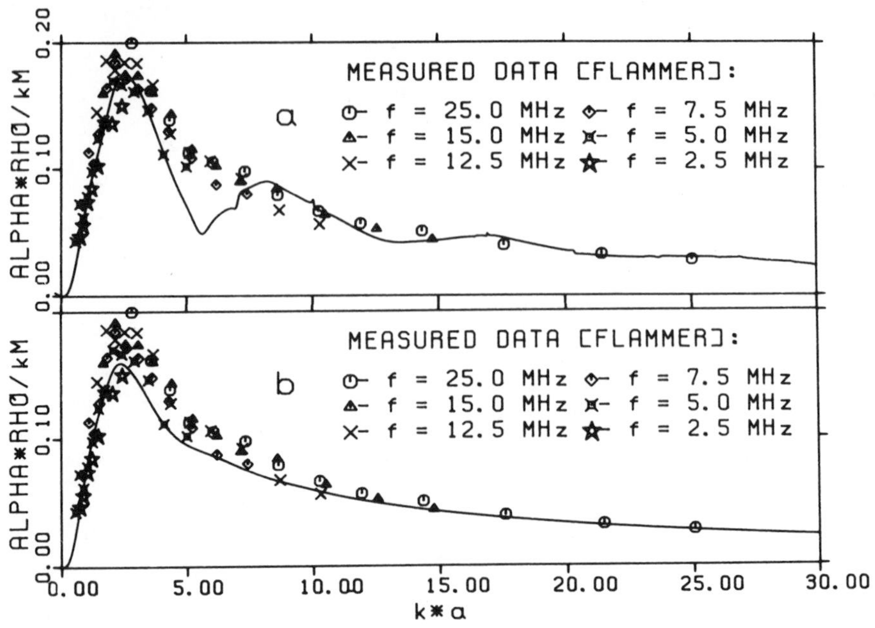

Fig. 1. Comparison of measured and computed values of $\rho_0'\alpha_s/k_c$ M using Flammer's data. The solid lines are the theoretical results for: (a) elastic case; (b) rigid movable case.

same effect is evident in Jansen's and Schaafsma and der Kinderen's data (Fig. 3b). These three data sets are therefore consistent. There is good agreement with the rigid movable model at low $k_c a$, but the measurements exceed the calculated values for larger $k_c a$. A probable explanation for the higher attenuation as $k_c a$ increases is additional scattering from irregularities in grain shape. This is observed in the attenuation of light by irregular particles (Reagan and Herman, 1980).

Further evidence for the importance of irregularities is provided by measurements of attenuation made by Flammer in suspensions of glass beads as well as the M. S. mixture, at the same frequencies. The comparison between the data and theoretical estimates in the rigid movable case is shown in Fig. 4. The rigid-movable model fits the glass bead data very well (Fig. 4a) although the data have a barely perceptible minimum near the Rayleigh wave resonance at $k_c a = 5.7$. In contrast, the data for the M. S. mixture is again larger than predicted for $2 < k_c a < 15$ (Fig. 4b).

Finally, returning to Fig. 2a, it can be seen that the high pass model (Eq. 6) provides a very reasonable fit to the data for $k_c a < 5$. This model may therefore prove to be a useful approximation to the attenuation of sand in aqueous suspensions by irregularly shaped particles like sand grains.

Form factor

No absolute measurements of scattered intensity are available. In order to compare measured and theoretical scattered intensities, therefore, the mean square voltage (V_s^2) at the receiver was normalized by its average value for each data set, and the theoretical estimates of

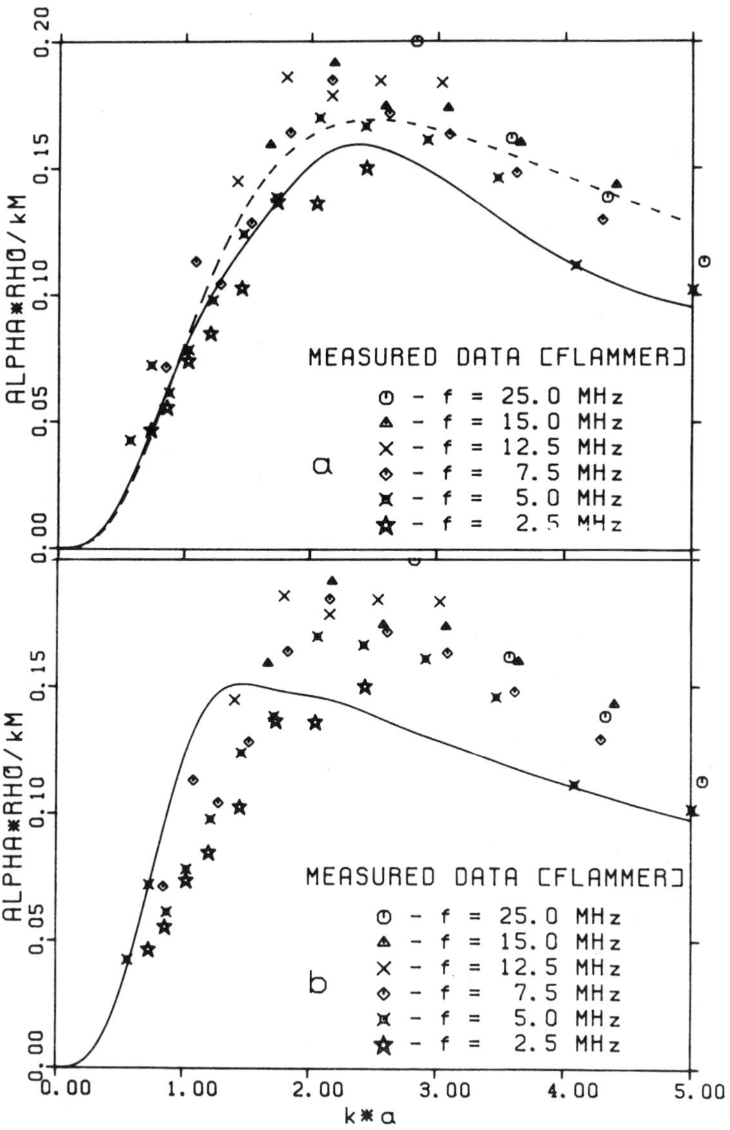

Fig. 2. Comparison of measured and computed values of $\rho_o' \alpha_s / k_c$ M using Flammer's data. (a) The solid line is the rigid movable case and the dashed line is the high-pass model; (b) the solid line is the rigid immovable case.

H_o^2 (Eq. 2) were normalized by the average values taken over all size fractions. The comparison between normalized values of H_o^2 and V_s^2 is shown in Fig. 5a for the rigid movable case assuming uniform particle size. Comparing this to the attenuation values in Figs. 3a and b, there is clearly larger scatter in the data. In fact, the correlation coefficient for Fig. 5a is only 0.12, as opposed to 0.98 and 0.97 for Figs. 3a and b, respectively, and is even worse for the other two physical models (Sheng, 1986). Some improvement in the correlation is achieved if we follow the suggestion of Clarke et al. (1984), and use the total scattering cross-section to compute the scattered intensity at any given angle. The idea is that the random orientations of the irregularly

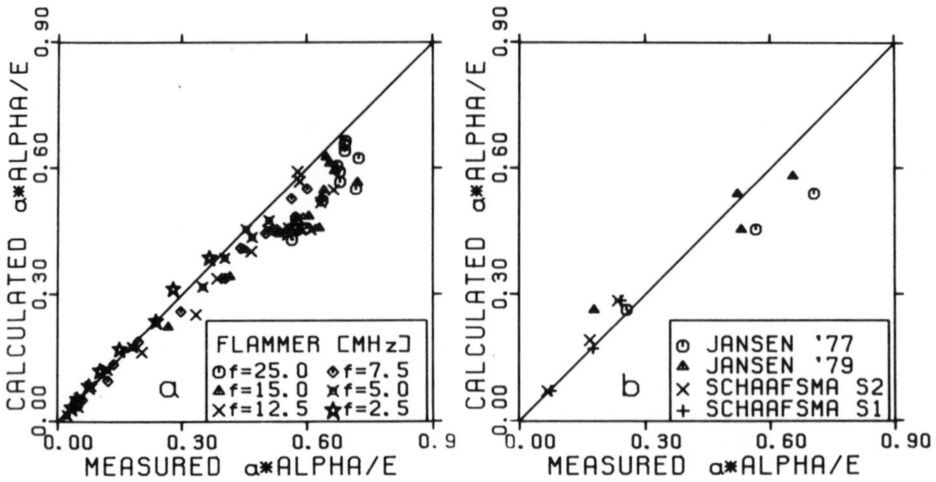

Fig. 3. Comparison between calculated and measured values of $a\alpha_{s}/\epsilon$, where the calculations are for the rigid movable case and uniform particle size. (a) Flammer's data. (b) Jansen's and Schaafsma and der Kinderen's data.

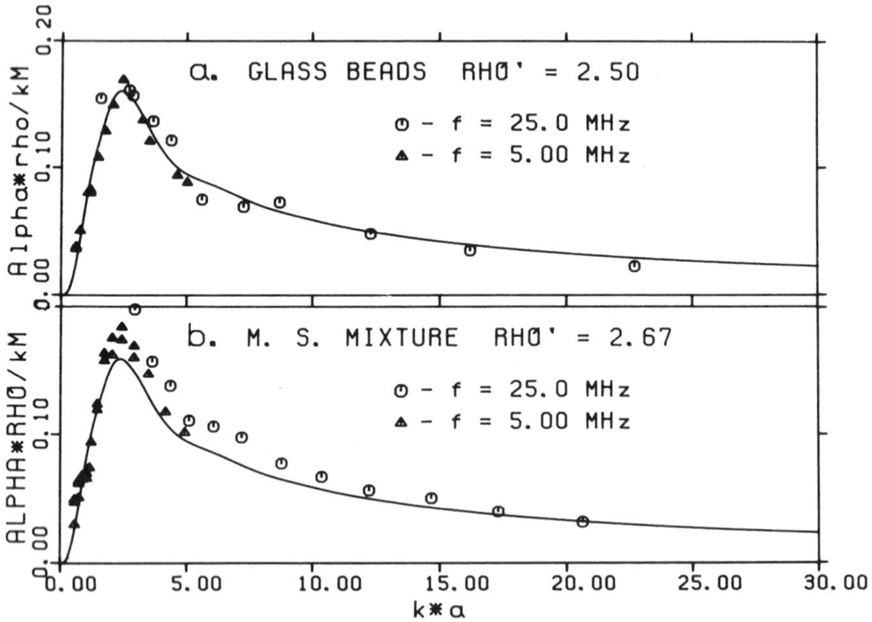

Fig. 4. Comparison of measured and computed values of $\rho_{0}'\alpha_{s}/k_{c}$ M using Flammer's data for: (a) glass beads; and (b) M. S. mixture. The solid lines are theoretical results for the rigid movable case and uniform particle size.

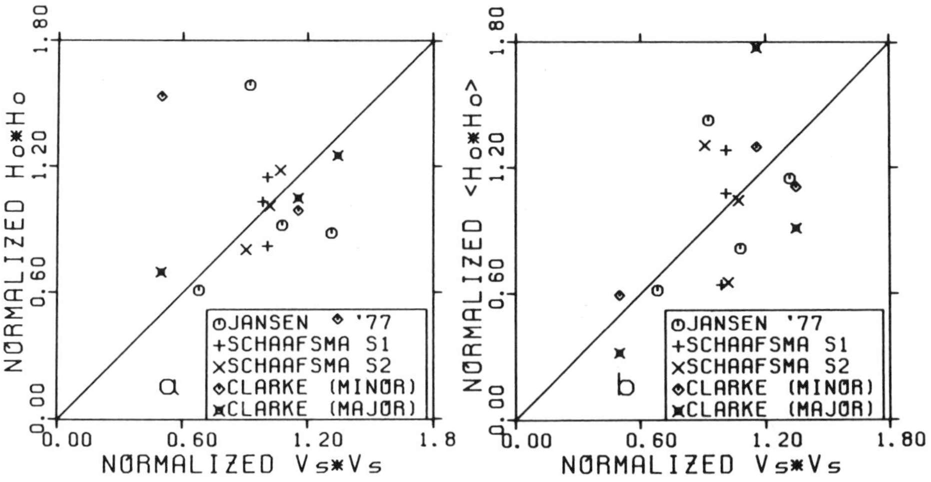

Fig. 5. (a) Normalized values of H_o^2 and V_s^2 for uniform size. (b) Normalized values of $<H_o^2>$ and V_s^2 for uniform particle size.

shaped sand grains will result on average in roughly isotropic scattering. Then $|f_\infty|^2$ would be proportional to the total scattering cross-section, normalized by πa^2. Assuming this proportionality, Eq. (2) was used to calculate $<H_o^2>$, where $<H_o^2>$ represents the value of H_o^2 for isotropic scattering. The normalized values of $<H_o^2>$ and V_s^2 are plotted in Fig. 5b for the rigid movable case assuming uniform particle size. The scatter is clearly reduced, and the cross correlation coefficient is increased to 0.55. This correlation, however, and those for the other cases (Sheng, 1986), are still too low to discriminate among the models. The effects of size distribution have also been investigated, but the correlation remains poor.

CONCLUSION

Comparisons between the theory for a homogeneous sphere and measured attenuation coefficients in aqueous suspensions of sand indicate that a spherical shape is a reasonable approximation. Furthermore, the fact the rigid movable case provides the best fit to the data suggests that the inertia of the particles is important, and at the same time that resonance excitation does not occur in natural sand grains. This is understandable, given the irregular shapes and inhomogeneous composition of sand grains. The finding supports a similar conclusion by Clarke et al. (1984) based on backscattered intensity measurements. The comparisons also indicate, however, that surface roughness is important for $k_c a > 2$.

Comparisons between theoretical and experimental scattered intensities exhibit a very high degree of scatter. There are a number of possible contributing factors. One of these is the normalization procedure, which would not contribute if absolute measurements had been made. A second is that it is difficult to make unambiguous estimates of the relative intensities of the scattered signals at different radii in some of the experiments. Thirdly, it is possible that in Jansen's experiments the particles adopted a preferred orientation because they were allowed to free fall through the detected volume. Regardless, a more comprehensive set of scattering measurements is needed.

REFERENCES

Clarke, T. L., Proni, J. R., and Craynock, J. F., 1984, A Simple Model for the Acoustic Cross Section of Sand Grains, J. Acoust. Soc. Amer., 76(5):1580.

Faran, J. J., Jr., 1951, Sound Scattering by Solid Cylinders and Spheres, J. Acoust. Soc. Amer., 23(4):405.

Flammer, G. H., 1962, Ultrasonic Measurement of Suspended Sediment, Geological Survey Bulletin 1141-A, US Government Printing Office, Washington.

Hay, A. E., 1983, On the Remote Acoustic Detection of Suspended Sediment at Long Wavelengths, J. Geophys. Res., 88:7525.

Hay, A. E., and Mercer, D. G., 1985, On the Theory of Sound Scattering and Viscous Absorption in Aqueous Suspensions at Medium and Short Wavelengths, J. Acoust. Soc. Amer., 78(5):1761.

Jansen, R. H. J., 1979, An Ultrasonic Doppler Scatterometer for Measuring Suspended Sand Transport, in: "Ultrasonic International 79, Conference Proceedings, Graz, Austria", IPC Science and Technology Press, Guildford, U.K.

Johnson, R. K., 1977, Sound Scattering From a Fluid Sphere Revisited, J. Acoust. Soc. Amer., 61(2):375.

Reagan, J. A., and Herman, B. M., 1980, "Light Scattering by Irregularly Shaped Particles Versus Spheres: What are some of the Problems presented in Remote Sensing of Atmospheric Aerosols?", in: "Light Scattering by Irregularly Shaped Particles", D. W. Schuerman, ed., Plenum Press.

Schaafsma, A. S., and der Kinderen, W. J. G. J., 1985, Ultrasonic Instruments for the Continuous Measurement of Suspended Sand Transport, in: "Proc. IAHR Symp. on Measuring Techniques in Hydraulic Research", Balkema Publ. Rotterdam.

Sheng, J., 1986, "Sound Scattering and Attenuation in Aqueous Suspensions of Sand: Comparison of Theory and Experiment", M.Sc. Thesis, Memorial University of Newfoundland.

A COMBINED SEISMIC REFLECTION PROFILER AND SIDESCAN SONAR SYSTEM FOR DEEP OCEAN GEOLOGICAL SURVEYS

D. J. Dodds* and G. B. J. Fader**

* GeoAcoustics Inc.
Box 772
Aurora, Ontario
Canada L4G 4J9

** Atlantic Geoscience Center
Bedford Institute of Oceanography
Box 1006
Dartmouth, Nova Scotia
Canada B2Y 4A2

ABSTRACT

Geological mapping of the deep ocean is necessary to permit exploitation of mineral resources and to establish jurisdiction under the Law of the Sea convention. A new deep water mapping system ("SeaMor") can be towed as much as 6 km below the surface and produces high resolution seismic reflection profiles and sidescan sonar images of the sea floor. The processed images are positioned in absolute geographical coordinates. A broadband (300 Hz - 10kHz) "boomer" source is used in the seismic profiler system to achieve high resolution and deep penetration (e.g. 2 m and 200 m) simultaneously, and it is expected that processed data will exhibit still higher resolution. The boomer is operated at 2 km depth by means of an active pressure compensation system and an unusually wide air gap, and the source behavior at depth can be predicted by a simple theoretical model. Images collected on a test cruise on the continental slope off Nova Scotia, Canada, reveal geological features not seen before in this environment, including iceberg furrows, pockmarks (gas escape craters), and slumped sediments. The resolution and broadband nature of the seismic system make it possible to distinguish between coherent internal stratification and incoherent backscattering from within a sediment unit. This facilitated the interpretation of phenomena such as slumping.

INTRODUCTION

Standard procedures conducted before continental shelf hydrocarbon drilling involve the collection and interpretation of high resolution seismic reflection profiles and sidescan sonograms to define seabed and subsurface geological hazards such as shallow gas, slumping, and faulting. A wide variety of systems have been developed to collect information on geological hazards and determine the stability characteristics of these relatively shallow water sites. Exploration has now been extended to greater depths on the adjacent continental slopes where shallow water systems cannot provide the resolution and penetration required for sediment mapping and characterization which is essential for safe drilling.

Deep ocean mapping is becoming important outside the hydrocarbon industry. The recent discovery of polymetallic sulphide deposits on deep ocean spreading ridges has created a need to map the distribution and thickness of these deposits for assessment of their potential as a resource. Also, under the Law of the Sea Convention the outer limit of jurisdiction of a signing member is determined both by the gradient of the seabed of the continental slope and by sediment thickness. Standard 3.5 kHz profilers do not easily penetrate high roughness seabeds or hard seabeds of gravel, sands, or disturbed sediments. Surface towed airgun systems have good penetration but inadequate resolution for these applications.

This paper describes a new deep water geological mapping system (SeaMor) for quantitative seabed assessment. It combines a seismic reflection profiler, an integrated sidescan sonar and a multisensor navigation capability. The system is designed for operation to 6 km depth. As illustrated in Fig. 1, the towed vehicle separates into a dead weight depressor and a trailing neutrally buoyant instrument carrier to enhance pitch and yaw stability during data collection. Up to 10 km of electromechanical tow cable may be used. The profiler uses a broadband electrodynamic source with an active pressure compensation system (presently designed for 2 km maximum depth). The profiler data is corrected for vertical motion of the vehicle, source level, spreading losses, and various other effects. The sidescan sonar operates at 27 to 30 kHz and has a 1.7° horizontal beam. This permits mapping of a swath 6 km or more in width. Sidescan data is processed to place each feature in the correct geographical position on the image, and to remove the effects of system geometry on the image intensity. The navigation capability includes an acoustic measurement of the position of the towed vehicle relative to the ship, a magnetic measurement of the heading of the vehicle, a pressure measurement to give vehicle depth, inertial measurement of short term translational and rota-

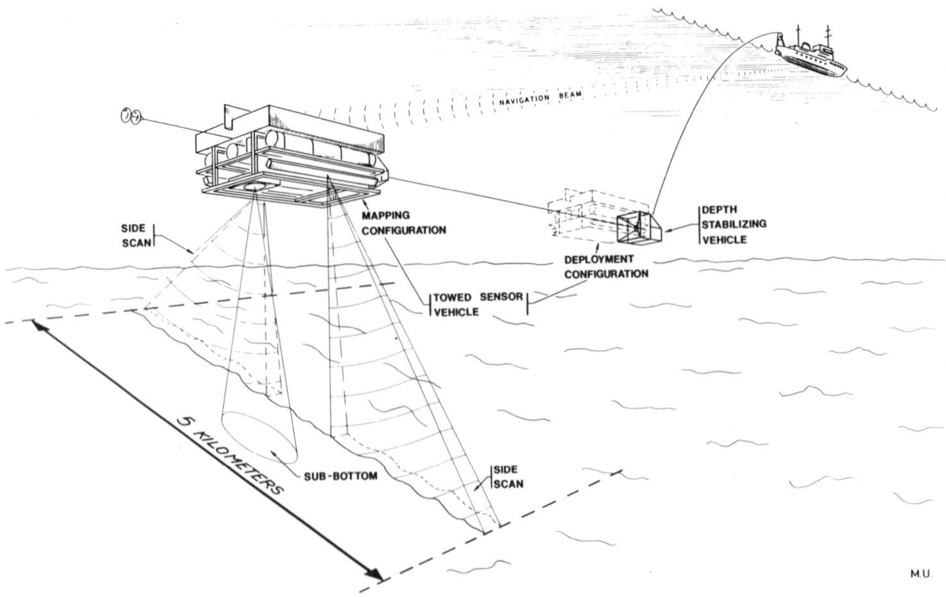

Fig. 1. SeaMor towed vehicle configuration.

tional motion, and software which integrates these inputs to derive parameters for the plotting of the sidescan and profiler data.

The emphasis in this paper is on the acoustic properties of sediments and the design of the seismic profiler component of the mapping system.

SEISMIC REFLECTION CHARACTER OF SURFICIAL SEDIMENTS

Through the process of interpretation, geologists translate the results of physical experiments into models of earth structure and processes. The reflection characteristics of sediments on which these interpretations are based are coherence, continuity, amplitude, spacing, relief, structure, and boundary relationships. For example, recent seafloor muds are typically characterized as acoustically transparent ponded sediments with weak continuous coherent reflectors. On the other hand, glacially deposited sediments can be described as medium to high intensity closely spaced conformable continuous coherent reflections with occasional hyperbolic reflections. These acoustic characteristics are widespread and are seen in sediments deposited by similar processes on the continental shelves on both sides of the Atlantic Ocean (e.g. Hovland, 1983, and King and Fader, 1986).

Interpretation of reflection characteristics of sediments provides clues to the processes that deposited or modified these sediments, but does not directly provide a quantified estimate of the sediment type or "lithology". A classification based on quantitative sea floor reflectivity measurements provides a close approximation of lithology, greatly increasing the power of seismic reflection interpretation to include lithology (Parrott et al, 1980).

ACOUSTIC MODELS OF SURFICIAL SEDIMENTS

Reflectivity

The complex structure of the sea floor can be approximated by simplified models. Fortunately, in most environments, the contacts between distinct sediment types are close to horizontal. This leads to a simple model of the sea floor as consisting of a sequence of horizontal layers of materials which are characterized acoustically by a sound speed and a density. The normal incidence reflection coefficient of the top of each layer is determined by ratio of its acoustic impedance (the product of sound speed and density) with that of the layer above. It has been shown that the density and sound speed of unconsolidated saturated sediments are determined quite precisely by the sediment porosity (Nafe and Drake, 1963). In some cases, a sediment unit has uniform porosity and small grains and fits the model well. Some sediment units have internal layering and again fit this model, although the acoustically defined internal layers are not distinct sediment types from the geological point of view. The reflection coefficients are usually small, except at the water-sediment interface, so that sound paths involving multiple reflections are relatively unimportant. When the sea floor fits this model, an omnidirectional sound source will generate an echo consisting of a replica of the incident signal for each layer.

We routinely measure and display an estimate of the reflection coefficient of the water-sediment interface, which helps in interpreting the type of sediment present (lithology) at the sea floor.

Absorption

In practice, echos from layer boundaries are not exact replicas of the incident signals. One reason is that sediments exhibit frequency dependent absorption of sound. Absorption increases with frequency; usually it is approximately proportional to frequency. Values of 0.01 dB/m for clay to 0.3 dB/m for coarse sand at 1 kHz are typical (Dodds, 1980).

Surface roughness

In many cases, the interfaces separating materials of different acoustic impedance are rough, and so do not fit the simple model. If the roughness is on a scale of 0.1 to 10 wavelengths, significant scattering of incident sound will occur. Roughness at the water-sediment interface is most significant, because the acoustic impedance contrast tends to be greatest there. An omnidirectional source will generate backscattered sound from such a seabed which will be superimposed on the echos from deeper layer boundaries and may mask some of them. Also, the apparent reflection coefficient of the seabed will be reduced due to loss of energy by scattering. On the other hand, the scattered energy contains statistical information on the seafloor roughness parameters (Dodds, 1984). We presently display a metric which quantifies this scattering as an aid to interpretation of sediment lithology.

Volume Scattering

Some sediment units have acoustic impedance variations which are not organized in layers. These may be due to porosity variations or to embedded objects such as pebbles or boulders which scatter sound energy from within the volume of the sediment unit. If a few large scattering features are present, each one produces a hyperbolic artifact on the seismic profile. If there are many scatterers, the returned echo contains an incoherent backscattered component. The effect increases rapidly with frequency, causing incident sound to be attenuated and masking echos from deeper layers. Again, this scattered energy does contain information about the character of the sediment (Dodds, 1984).

DESIGNING A SEISMIC PROFILER FOR DEEP OCEAN SURFICIAL SEDIMENTS

Impulsive Sources

It is important to be able to resolve layering in the sea floor. Layering may indicate a boundary between dissimilar sediment units; it may give clues to processes which have reworked the material of a unit; or it may help in differentiating one unit from another. Adjacent layer boundaries can be resolved only if the replica pulses reflected from them do not overlap. The incident pulse should therefore be as short as possible, which implies that the bandwidth of the signal should be as great as possible. But it is also desirable to obtain an image from the greatest possible depth below the seabed. To achieve penetration through sediments, the signal frequencies should be as low as possible. Reducing signal frequency reduces absorption in the sediments and reduces the level of masking energy from surface roughness and volume scatterers.

The desirablity of having low frequencies and a large bandwidth at the same time leads to the use of impulsive signals rather than gated sine waves in the profiler application.

The effect of scattered energy, which is to mask the reflections from layer boundaries, makes some directionality desirable. A directional source and receiver will favour near-vertical directions, which will include the normal incidence reflections from the layer boundaries, and will discriminate against scattered energy travelling at oblique angles. Too much directionality will cause the loss of normal incidence reflections when the survey system is not in a level attitude or when the layer boundaries are not exactly horizontal.

Impulsive Piston Source

The type of source used in the SeaMor profiler is an electromagnetically driven circular piston (boomer). As shown in Fig. 2, a flat coil is embedded in the body of the source and a circular plate (piston) is placed adjacent to it with an intervening air gap. A capacitor is discharged through the coil. The magnetic field of the coil induces eddy currents and acts on those currents to repel the plate. The SeaMor source piston is 0.5 m in diameter. A 60 uF capacitor charged to 6 kV delivers 1 kJ to the source. This produces a driving force on the order of 100 kPa with a duration of 0.3 ms. The resulting acoustic pulse has useful components from 300 Hz to 10 kHz.

The air gap is crucial to the operation of the source. Without the air gap, the high ambient pressure (20 MPa at 2 km depth) holds the piston against the coil. Relative motion is impossible and sound cannot be generated. Even with a gap containing air at ambient pressure, the motion of the piston is somewhat restricted since the gap behaves as a spring. If the air behaves as an ideal gas, the spring stiffness is inversely proportional to the gap volume and proportional to the air pressure (which must equal the ambient pressure). The magnetic driving force on the piston drops as the gap is increased (Hutchins, 1974), so that selection of the gap thickness is a tradeoff between driving force and spring stiffness. For operation at 2 km, a source designed for shallower operation was modified by increasing the gap. This compensates for the increase in stiffness of the air spring caused by the high ambient pressure, at the expense of reducing the driving force.

In order to maintain the air gap at ambient pressure, a combination of active and passive pressure compensation is used (Fig. 3). The air gap

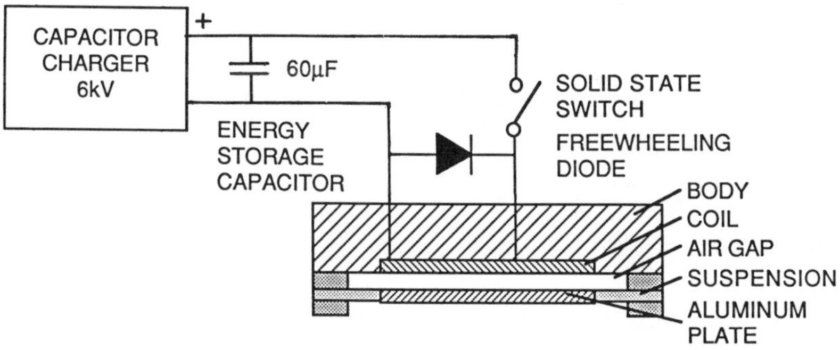

Fig. 2. A boomer seismic reflection source. The energy storage capacitor is discharged through the coil, producing a magnetic field. Eddy currents are induced in the adjacent plate, and the plate is repelled by the magnetic field of the coil. The plate moves outward as a piston, generating an acoustic output. The freewheeling diode, and the inductance of the coil, cause the current in the coil to continue flowing after the discharge of the capacitor. This slows the return of the plate to its initial position.

is connected to the passive pressure compensator, which is a lens shaped air reservoir with an elastic sheet forming one surface of the lens. Its volume is about four times the volume of the air gap in the boomer. The active compensator consists of a high pressure air reservoir and a valve system. The valves feed air to the air gap and the passive compensator as the depth increases, and release air as the depth decreases. The active compensator is disabled when the depth is greater than one quarter of the maximum expected tow depth. Below that depth, the passive compensator is compressed. The system thereby keeps the air gap at ambient pressure at all depths, yet does not consume air below one quarter of the maximum tow depth.

The high ambient pressure has a further effect in that once the piston is displaced, the air spring contains potential energy which must be dissipated. The combination of the compliance of the air gap and the mass-like radiation impedance of the piston produces a resonant system, which undergoes a damped oscillation referred to as "ringing". As the ambient pressure is increased, the energy stored in this system after firing the source increases, and so does its resonant frequency. An electrical analog of the resonant system is shown schematically in Fig. 4. Voltage e is the analog of the magnetic driving force on the piston and current i is the analog of the velocity of the piston. The radiation impedance of the piston can be represented approximately by the network L_w, C_w, R_1, R_2. Inductance L_p and capacitance C_g are analogs of the mass

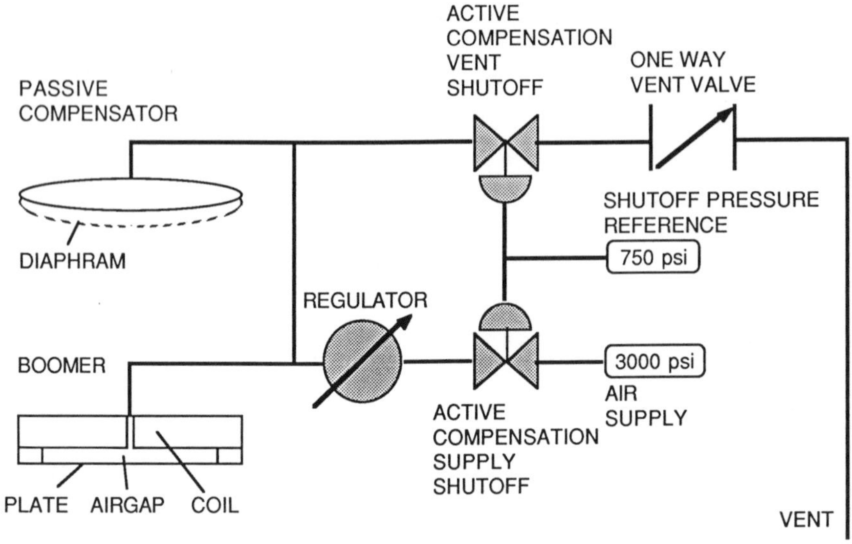

Fig. 3. The boomer pressure compensation system. The system maintains the air gap of the boomer at ambient pressure. During a descent to 500 m depth, air is fed from the air supply to the boomer and passive compensator. At 500 m the ambient pressure reaches the 750 psi reference pressure, and the active compensation supply and vent are shut off. As descent continues, the passive compensator diaphragm is deflected, reducing the system volume and maintaining it at ambient pressure. On ascent, the process is reversed. At 500 m the active compensation vent is opened, and the expanding air is released as the system is brought to the surface.

of the boomer plate and the compliance of the gap. The values of the circuit elements are as follows.

$$L_w = \rho d^3/3 \quad (1)$$

$$C_w = 1.2/[d\rho c^2] \quad (2)$$

$$R_1 = \pi d^2 \rho c/4 \quad (3)$$

$$R_2 = 0.441 R_1 \quad (4)$$

$$C_g = 1.4 v p \quad (5)$$

$$L_p = m \quad (6)$$

where d is the piston diameter, 0.50 m.
c is the speed of sound in water, 1450 m/s.
p is the ambient pressure.
ρ is the density of water, 1030 kg/m^3.
v is the volume of the air gap, $1.8 \times 10^{-3} m^3$.
m is the mass of the piston, 3.4 kg.

This model is based on several approximations. The analog of radiation impedance is an approximation to the theoretical radiation impedance of a circular piston in an infinite plane baffle. The air in the gap is treated as an ideal diatomic gas which undergoes adiabatic expansion and compression, and the behavior of the air spring has been linearized. It has been assumed that the components of the system, including the mounting of the boomer body, are rigid. Finally, the driving force has been assumed to be unaffected by motion of the plate.

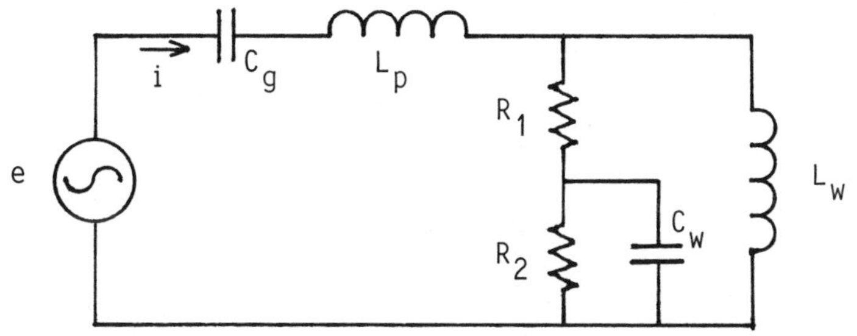

Fig. 4. An electrical analog of the mechanically resonant boomer. The voltage e and the current i are the analogs of the driving force and the piston velocity respectively. The components of the resonant system are the radiation impedance of the piston, the mass of the piston, and the the air spring formed by the gap between the boomer plate and coil. The network L_w, C_w, R_1, R_2 is an approximate analog of the radiation impedance. The elements L_p and C_g are analogs of the piston mass and air spring respectively. With typical values of the elements, the circuit has an underdamped response at a frequency of a few hundred Hertz.

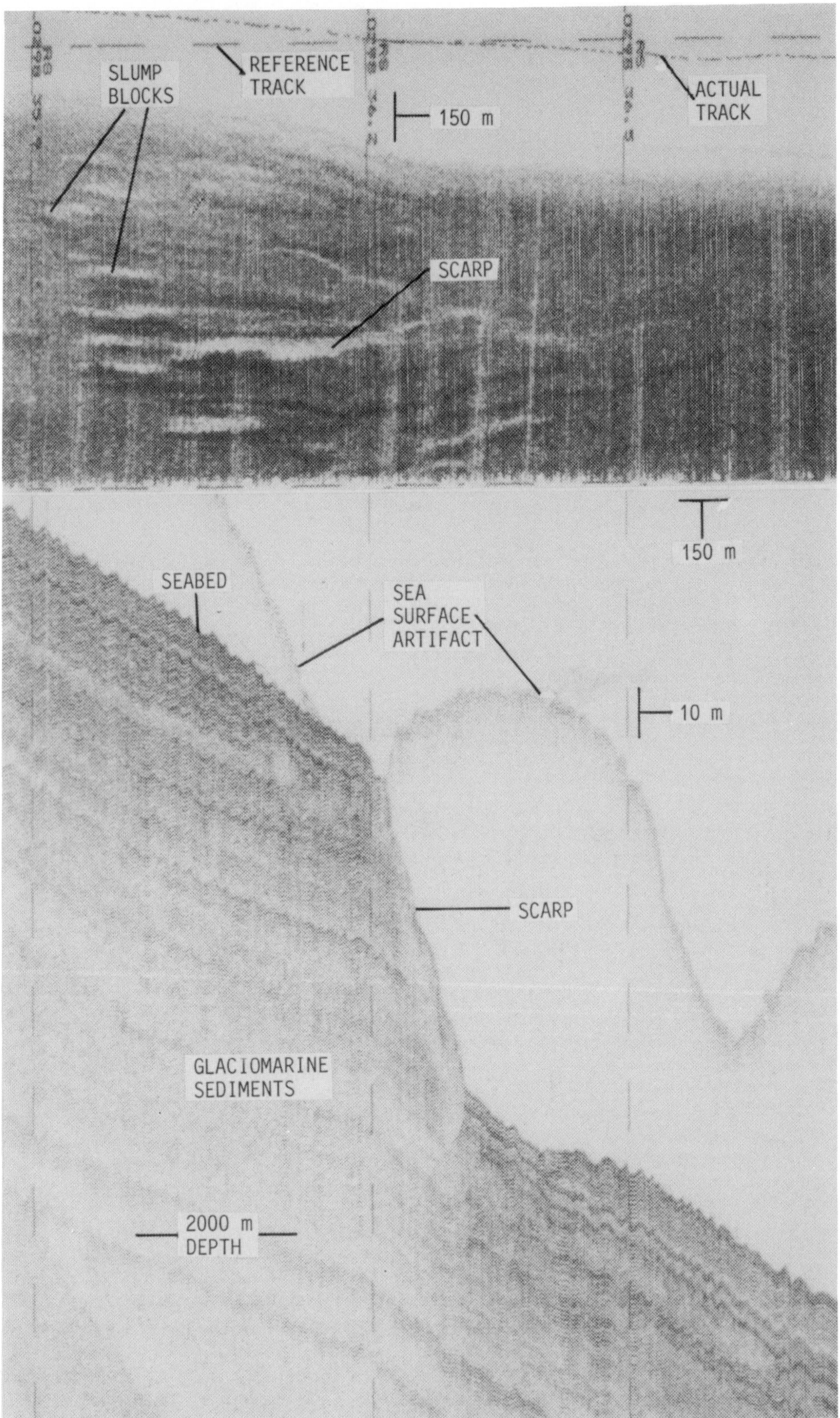
SLUMP BLOCKS
REFERENCE TRACK
150 m
ACTUAL TRACK
SCARP
150 m
SEABED
SEA SURFACE ARTIFACT
10 m
SCARP
GLACIOMARINE SEDIMENTS
2000 m DEPTH

Fig. 5. (Opposite page.) Processed high resolution seismic reflection profile and sidescan image from the continental slope off Nova Scotia, Canada. The water depth is 1850-1980 m. The sidescan image (right side only) appears at the top, with the profiler image below. The image shows a large (60 m) erosional scarp cut in glaciomarine seaward dipping sediments. Shallow reflections visible at the left of the profiler image terminate at the scarp, while deeper reflections (up to 140 m below the seabed to the left of the scarp) are continuous across the width of the image. The along-track and cross-track scales of the sidescan image are the same. The straight horizontal dashed line is the reference track, while the irregular line is the actual track of the towed vehicle. The sidescan sonar image has been processed so that each point on the image is presented in the correct geographical position. This requires corrections for slant range, yawing of the vehicle, and deviation of the actual track of the vehicle from a straight line reference track. In this image, the slant range correction is incomplete, and is based on a fixed bottom depth parameter. The vertical scale of the profiler image is exaggerated. Towed vehicle vertical motion effects on the order of 30 m have been removed from the profile, although the apparent roughness of the seabed (less than 2 m) is thought to be due to residual effects of the motion compensation. The profiler signal was received using a 3 m line array. Its amplitude has been corrected for source level and range, and the signal has been bandpass filtered to improve the signal to noise ratio. Deconvolution processing, which would further improve the resolution of the seismic profile, has not been implemented here.

The characteristic equation of the electrical analog can be calculated by standard methods (e.g. Kirwin and Grodzinsky, 1980) and its poles located. At an ambient pressure of 20 MPa, a pair of complex poles corresponding to a frequency of 500 Hz and a damping factor of 0.12 is found. At an ambient pressure of 11 MPa the frequency is 370 Hz and the damping factor is 0.09. Observations of the near field boomer output at 2 km depth (20 MPa pressure) gave a frequency of 800 Hz and a damping factor of 0.2; at 1100 m depth (11 MPa) the frequency was 600 Hz and the damping factor was 0.1. The discrepancies are thought to be due to the approximations inherent in the model. The absence of a baffle, the departure of the air from ideal behaviour at high pressures, and the compliance of the boomer body mounting will all increase the natural frequency of the actual system. The model can serve as a guide to the design of boomer sources for even greater depths.

The 800 Hz ringing frequency at 2 km depth is in the low end of the frequency range considered most useful for high resolution subbottom profiling. It is therefore desirable to attenuate the ringing so as to maximize the resolution. This is achieved by the application of a digital deconvolution operator to the received signal.

Seismic Receivers

SeaMor carries three seismic receivers. A 3 m single channel line array is towed behind the vehicle. The directionality of the streamer discriminates against off-axis surface and volume backscattered sound which might interfere with signals received from deeper horizons. It also discriminates against system noise, which is further reduced by the separation between the streamer and the vehicle. As well, two single element hydrophones are mounted on the vehicle. One is mounted directly below the source where it is used to monitor the near field source output. The near field output can be used to calculate the far field output which is used to normalize the energy of the seabed reflection to obtain seabed reflectivity measurements. The frequency and damping factor of the boomer ringing is also measured from the near field output and used in automatically calculating the coefficients of the deconvolution filter. The second single element hydrophone gives a seismic profile with higher resolution than can be achieved with the line array because variation in its vertical position is more precisely known, but it does not achieve the same penetration as the line array.

RESULTS

The sea trials and initial data collection by the system were conducted on the continental slope off Nova Scotia, Canada, adjacent to Sable Island Bank in water depths of 200-2500 m (Fader, 1984). A total of 85 hours of survey data were collected. An example of a processed record is shown in Fig. 5. The most conspicuous aspect of the seismic reflection data is the depth of penetration (up to 200m) and the degree of resolution (about 2 m). This is in contrast to previously collected 3.5 kHz data from the area which failed to penetrate the harder sea floor sediments.

Analysis of the data showed many geological features and relationships not previously observed in this environment. Interpretation of the sidescan sonograms showed that from 200-600 m water depth the seabed is covered with very old iceberg furrows. Downslope, from 500-1100 m, pockmarks (gas escape craters) cover the seabed with shallow cone-shaped depressions. Further down the slope the seismic profiles showed large areas of slumped sediments, characterized by truncated reflections, eroded sections, and scarps. The effects of slumping are also indicated on the sidescan image by arcuate scarps and hummocky relief. In the subsurface, the seismic profile shows a buried wedge shaped zone of incoherent reflections interbedded with a section of high intensity continuous coherent reflections. The wedge shaped body is interpreted as a "till tongue", a glacial transitional feature found at the lift off or floating margin of the continental ice sheet which extended across the shelf during the last glaciation at 60,000 y.B.P. (King and Fader, 1986). The ability to distinguish between coherent and incoherent reflections from these sediments is a characteristic of the broadband source used in this system. Through correlation with subsurface samples, these profiles will enable the time intervals for major instabilities to be determined for the deep water of the continental slope.

ACKNOWLEDGEMENTS

The development of this system was funded by the Government of Canada. The National Research Council, the Department of Energy, Mines, and Resources, the Department of National Defence, and the Department of Fisheries and Oceans administered the funding and provided the assistance and advice of their scientists and engineers. The support of these individuals and organizations is gratefully acknowledged. The SeaMor system was engineered and constructed by Huntec (70) Limited. The authors thank R. W. Hutchins, who was responsible for the boomer design and gave advice on modelling its response, and D. R. Parrott, who reviewed the manuscript.

REFERENCES

Dodds, D.J. 1984. Surface and volume backscattering of broadband acoustic pulses normally incident on the sea floor: Observations and models. Invited paper. Abstract only. Program of the 107th meeting, Journal of the Acoustical Society of America 75(Supp.1).

Dodds, D.J. 1980. Attenuation estimates from high resolution subbottom profiler echoes in W.A. Kuperman and F.B. Jensen. Bottom Interacting Ocean Acoustics. Plenum Press. New York.

Fader, G.B.J. 1984. C.S.S. Hudson, Cruise report no. 84-029. Atlantic Geoscience Center, Bedford Institute of Oceanography, Dartmouth, Nova Scotia, Canada.

Hovland, M. 1983. Elongated depressions associated with pockmarks in the western slope of the Norwegian Trench. Marine Geology 51:35-46.

Hutchins, R.W. 1974. Computer simulation model of a transiently excited underwater sound projector. Proceedings of Oceans '74, I.E.E.E. International Conference on Engineering in the Ocean Environment, Halifax, Nova Scotia.

King, L.H. and Fader, G.B. 1986. Wisconsinan glaciation of the Atlantic continental shelf - southeast Canada. Geological Survey of Canada, Bulletin 363, 72 pp.

Kirwin, G.J. and Grodzinsky, S.E. 1980. Basic Circuit Analysis. Houghton Mifflin Co. Boston.

Nafe, J.E. and Drake, C.L. 1963. Physical properties of marine sediments. in M.N. Hill, ed., The Sea. Vol. 3. John Wiley and Sons.

Parrott, D.R., Dodds, D.J., King, L.H., Simpkin, P.G. 1980. Measurement and evaluation of the acoustic reflectivity of the sea floor. Canadian Journal of Earth Sciences 17(6).

SIGNAL PROCESSING OF

OCEAN ACOUSTIC TOMOGRAPHY DATA

Duncan Sheldon and G. Clifford Carter

Naval Underwater Systems Center
New London, CT 06320, USA

ABSTRACT

The phase closure technique is applied to ocean acoustic tomography. This technique is widely used in radio astronomy to obtain images of remote distributed sources. Distinctions are made between the source imaging problem and the arrival time estimation problem. A signal processing procedure is described which is applicable if: (1) Mean signal travel times can be determined, and (2) Fluctuations in the signal travel times are directly related to the oceanic variations of interest.

INTRODUCTION

Ocean acoustic tomography[1] is a procedure for measuring the field of sound-speed fluctuations within a volume of ocean by means of acoustic transmissions. These transmissions are projected over many vertical "slices" of the ocean, each containing one source and usually one receiver. Within a slice the transmissions follow many diverse paths (described by a multipath ray diagram). Demonstrations of ocean acoustic tomography have involved stationary source and receiver moorings[2], and both ship-to-ship and mixed ship-to-mooring tomography have been proposed[3]. The purpose of the demonstrations was to verify that tomographic measurements of the ocean's temperature agree satisfactorily with traditional measurements and that eddy motion can be identified and tracked.

As emphasized in Reference 2, the demonstrated procedure depends critically on the resolution, identification, and stability of multipaths over long ranges. If multipaths cannot be identified, both analytically (by path) and experimentally (by measuring travel times), ocean tomography is not applicable. The multipaths must be stable over the time scale of mesoscale processes. However, oscillation in travel times due to tidal motion of the source mooring and tidal currents can be tolerated. These disturbances do not significantly affect differences in arrival times between paths and do not disorder the paths' spatial arrangement. The demonstrations[2] were carried out using (presumably) omnidirectional receivers, i.e., without recourse to an array to resolve path arrival directions.

The first demonstration that long-range multipaths could be unambiguously tracked over several months is described in Reference 4. Briefly stated, this work showed the large-scale oceanic variations that we want to observe (eddy motion, for example) are indeed the variations which principally affect the observable acoustic data. The question of observing shorter time-scale oceanic variabilities by means of tomography arises naturally from previous work. Studies of the effects of internal waves, tides, and currents on acoustic transmissions are listed in Reference 4.

The present work considers an invariant termed "phase closure" which can be applied to tomographic signal processing. It is expected that this invariant will be useful in tomographic investigations of oceanic variabilities such as internal waves. Phase closure can be implemented by using generalized cross-correlation methods. These methods include delay, square, and sum procedures, as well as direct cross-correlation (Figs. 1 and 2).

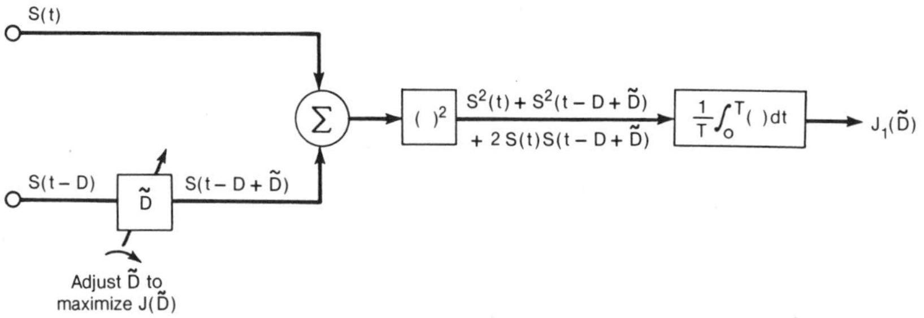

Figure 1.

Conceptual Delay, Sum, Square and Integrate Configuration

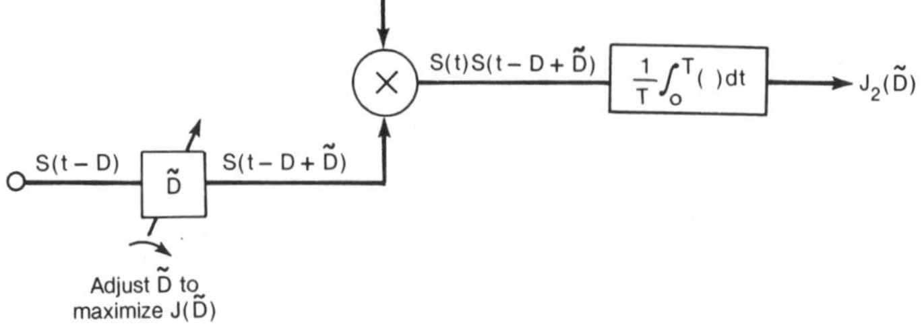

Figure 2.

Conceptual Cross-Correlator Configuration

In both cases the hypothesized delay, $\tilde{D}$, can be adjusted in order to maximize the configuration's output. In terms of time-delay-estimation performance the configurations are equivalent[5]. Present tomographic procedure[2] calls for determining the arrival times of all multipath signals by replica correlation. Phase closure offers a way to reduce the computational burden by making more effective use of generalized cross-correlations corresponding to zero time delay.

The technique of deriving corrected phase information from long-baseline interferometer data by summing the observed phase fluctuations around closed loops of baselines is widely used in radio astronomy[6-9]. In this field, phase closure is used to meliorate the effects of atmospheric turbulence and instrument errors and thereby contributes to obtaining maps of compact radio-frequency sources. Very long baselines (hundreds or thousands of kilometers) are used to achieve the high angular resolution necessary to map remote distributed sources. Ocean acoustic tomography sources are not distributed; the receiver separation distances are large in order to observe large volumes of ocean. Unlike radio astronomy, ocean acoustic tomography involves propagation paths that are neither essentially straight nor similarly curved; many quite different paths occur. Another distinction between long baseline interferometry and ocean tomography is that independent atmospheric disturbances act on the radio signals. The paths followed by the acoustic signals experience variations that are far from independent. Indeed, it is this lack of independence that necessitates the complicated process of inversion.

OCEAN TOMOGRAPHY

Among the substantitive (and interrelated) problems associated with ocean tomography are: (1) Inversion, (2) Ray path perturbations, and (3) Signal arrival time measurement and estimation. Inversion procedures are usually linearized, i.e., travel time perturbations due to changes in sound speed are assumed to occur along unperturbed ray paths. Atomic frequency standards are required to make arrival time measurements, and signal processing methods are employed to make the best estimates of arrival times. Long-range tomography demands low-frequency broadband sources[3]. The Ocean Tomography Group[2] used a 224 Hz carrier with a 20 Hz bandwidth. The present work considers the application of phase closure to obtain improved estimates of signal arrival times. This method requires a tonal source.

The processing methods and time scales of the demonstrated arrangement are of interest. High processing gain was achieved by transmitting a sequence of 127 digits whose period was nearly eight seconds. Each complete transmission consisted of twenty-four consecutive sequences lasting nearly 192 seconds. Coherent averaging of the twenty-four sequence receptions implied a theoretical travel time precision of 2 ms for a resolved multipath arrival. Transmissions were at hourly intervals on every third day over approximately one hundred calendar days. These transmission times were related to the various relevant time scales: months for mesoscale ocean variability and hours for tidal corrections and internal wave motion. The fact that long-range multipaths can be temporally tracked over several months suggests that it is appropriate to attempt to track multipath temporal and spatial perturbations over shorter times. Alternatively stated, if mean travel-times of multipaths can be established, perhaps their perturbations can also be measured and estimated. Processing economy in such an effort is essential.

PHASE CLOSURE

The principal signal processing problems in radio astronomy are resolution and coherence. High resolution, now about 10^{-4} arc-seconds, is accomplished by aperture synthesis over long baselines. Signal coherence over long distances is achieved by using closure techniques.

Consider the triad of widely separated radio telescopes A, B, and C. Assume they collect and process data from a single frequency source (e.g., a cosmic maser) and share an accurate time base. Each pair of receivers [(A,B), (B,C), and (C,A)] forms a radio interferometer; data from each receiver are continuously and coherently added to data from an opposing receiver. For example, data from Receiver A are added to both Receiver B data [Interferometer (A,B)] and Receiver C data [Interferometer (C,A)]. A radio interferometer is the exact analog of an optical interferometer. In the optical case the interference fringe pattern has a spatial distribution (e.g., over a screen); in the radio-frequency case the pattern appears along a time base. The nearly-sinusoidal variations of the radio interferometers' patterns are caused by the earth's rotation. (A fixed angular position of the earth corresponds to a fixed position on a radio interferometer's interference pattern, just as a fixed point on an optical interferometer's screen receives constant light intensity.) In the absence of disturbances, signals separated by long baselines can be used to obtain images of extended celestial sources.

The principal disturbances are instrument errors and delays caused by the earth's atmosphere. These delays introduce phase fluctuations in the signals arriving at the receivers and produce temporal shifts in the corresponding interference fringe patterns. However, the undesirable affects of the phase fluctuations can be removed by using the fact that the sum of the temporal shifts of the interference patterns around a closed baseline loop is zero. Applying the phase closure technique renders the radio signals coherent.

CLOSURE METHODS AND OCEAN TOMOGRAPHY

Closure methods can be applied to the arrival time estimation problem. An important distinction between existing ocean tomography procedures and closure procedures involves the use of tones rather than broadband signals. Consider a vertical hydrophone array receiving tomographic signals traveling through a non-stationary ocean (Fig. 3).

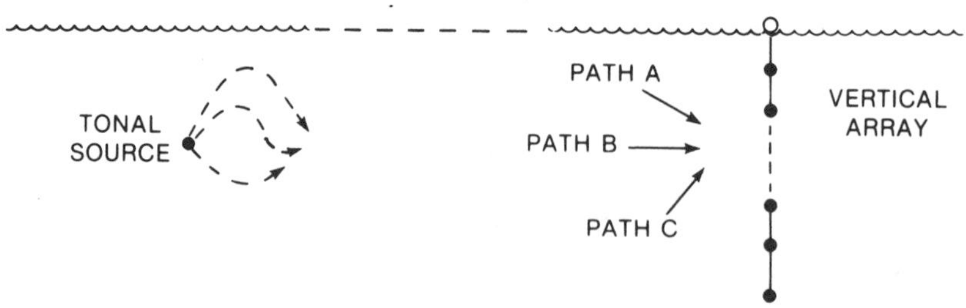

Figure 3.

Source, Paths, and Vertical Array

Assume the array can resolve Paths A, B, and C by their directionality, i.e., the paths lie in different beams of the array. The relevant travel times are defined in Tables I and II.

Table I

TRAVEL TIMES

Path	Travel Time	Mean Travel Time	Travel Time Perturbation
A	t_A	$\bar{t}_A$	δt_A
B	t_B	$\bar{t}_B$	δt_B
C	t_C	$\bar{t}_C$	δt_C

Table II

TRAVEL TIME DIFFERENCES

Paths	Interference Pattern	Travel Time Difference	Mean Travel Time Difference	Travel Time Difference Perturbation
A,B	I_{AB}	t_A-t_B	$\bar{\tau}_{AB}$	$\delta\tau_{AB}$
B,C	I_{BC}	t_B-t_C	$\bar{\tau}_{BC}$	$\delta\tau_{BC}$
C,A	I_{CA}	t_C-t_A	$\bar{\tau}_{CA}$	$\delta\tau_{CA}$

From these definitions we have

$$
\begin{aligned}
t_A-t_B &= (\bar{t}_A+\delta t_A) - (\bar{t}_B+\delta t_B) = \bar{\tau}_{AB} + \delta\tau_{AB} \\
t_B-t_C &= (\bar{t}_B+\delta t_B) - (\bar{t}_C+\delta t_C) = \bar{\tau}_{BC} + \delta\tau_{BC} \qquad (1) \\
t_C-t_A &= (\bar{t}_C+\delta t_C) - (\bar{t}_A+\delta t_A) = \bar{\tau}_{CA} + \delta\tau_{CA}
\end{aligned}
$$

Phase closure is based on the following travel-time difference identity:

$$(t_A-t_B) + (t_B-t_C) + (t_C-t_A) = 0 \qquad (2)$$

Since the above equations hold when all the delta quantities are zero,

$$\bar{\tau}_{AB} + \bar{\tau}_{BC} + \bar{\tau}_{CA} = 0 \qquad (3)$$

and

$$
\begin{aligned}
\delta t_A - \delta t_B &= \delta\tau_{AB} \\
\delta t_B - \delta t_C &= \delta\tau_{BC} \qquad (4) \\
\delta t_C - \delta t_A &= \delta\tau_{CA}
\end{aligned}
$$

where

$$\delta\tau_{AB} + \delta\tau_{BC} + \delta\tau_{CA} = 0 \tag{5}$$

We assume the $\bar{\tau}$'s and $\delta\tau$'s are known, and the δt's are unknown.

Three interference patterns (I_{AB}, I_{BC}, and I_{CA}) could be formed by processing the corresponding paths for variable $\tilde{D}$'s as shown in Figure 1. I_{AB} (for example) has a local maxima when $\tilde{D} = -(t_A - t_B)$. In general

$$\begin{aligned} \tilde{D}(A,B) &= -(t_A - t_B) \\ \tilde{D}(B,C) &= -(t_B - t_C) \\ \tilde{D}(C,A) &= -(t_C - t_A) \end{aligned} \tag{6}$$

correspond to local maxima in the patterns I_{AB}, I_{BC}, and I_{CA}. Ranges of $\tilde{D}$ values would be required to actually produce these interference patterns. However, the patterns are not required for every measurement. For processing economy, it is the values of the intensities (on the interference patterns) corresponding to

$$\begin{aligned} \tilde{D}(A,B) &= 0 \\ \tilde{D}(B,C) &= 0 \\ \tilde{D}(C,A) &= 0 \end{aligned} \tag{7}$$

which are of interest. These intensities (say J_{AB}, J_{BC}, and J_{CA}) are determined by $(t_A - t_B)$, $(t_B - t_C)$, and $(t_C - t_A)$. As in the case of the travel times, we let

$$\begin{aligned} J_{AB} &= \bar{J}_{AB} + \delta J_{AB} \\ J_{BC} &= \bar{J}_{BC} + \delta J_{BC} \\ J_{CA} &= \bar{J}_{CA} + \delta J_{CA} \end{aligned} \tag{8}$$

where the bar and delta again indicate mean and perturbation quantities. If the δJ's can be measured and related to the corresponding $\delta\tau$'s, and if one of the travel times [t_A, t_B, or t_C] can be measured directly (say t_A), an expedient procedure is available for estimating the remaining δt's (here δt_B and δt_C). The procedure follows directly from Equations (4) and (5). The above results can be extended to any number of multipaths [A, B, C, D, ...]. There will typically be more than ten multipaths, and a processing advantage can be gained by avoiding all but one replica correlation [or some equivalent direct travel-time calculation].

DISCUSSION

Phase closure has been related to estimating multipath perturbation times. The suggested procedure for achieving signal processing economy requires experimental verification. Beam resolution and interference pattern stability [for estimating the $\delta\tau$'s from the δJ's] are principal issues.

REFERENCES

1. Munk, W., and Wunsch, C., Ocean Acoustic Tomography: A Scheme for Large Scale Monitoring, Deep Sea Res., 26A, 1979, pp. 123-161.
2. Ocean Tomography Group, A Demonstration of Ocean Acoustic Tomography, Nature, 229, 9 September 1982, pp. 121-125.
3. Munk, W., and Wunsch, C., Observing the Ocean in the 1990s, Phil. Trans. R. So. Lond. A., 307, 1982, pp. 439-464.
4. Spiesberger, J. L., Spindel, R. C., and Metzger, K., Stability and Identification of Ocean Acoustic Multipaths, J. Acoust. Soc. Am., 67, No. 6, June 1980, pp. 2011-2017.
5. Carter, G. C., Time Delay Estimation for Passive Sonar Signal Processing, IEEE Trans. on Acous., Sp., and Sig. Proc., ASSP-29, No. 3, June 1981, pp. 463-470.
6. Jennison, R. C., A Phase Sensitive Interferometer Technique for the Measurement of the Fourier Transforms of Spacial Brightness Distributions of Small Angular Extent, R. Astro. Soc. Monthly Notices, 118, 1958, pp. 276-284.
7. Rodgers, A.E.E., et al., The Structure of Radio Sources 3C 273B and 3C 84 Deduced from the 'Closure' Phases and Visibility Amplitudes Observed with Three-Element Interferometers, Astrophys. J., 193, 15 October 1974, pp. 293-301.
8. Readhead, A.C.S., and Wilkinson, P. N., The Mapping of Compact Radio Sources from VLBI Data, Astrophys. J., 223, 1 July 1978, pp. 25-36.
9. Readhead, A.C.S., Radio Astronomy by Very-Long-Baseline Interferometry," Sci. Am., 246, June 1982, pp. 52-61.

SEISMIC PROFILING WITH A PARAMETRIC, SELF-DEMODULATED RICKER WAVELET

T. G. Muir, R. J. Wyber,* J. B. Lindberg, and L. A. Thompson

Applied Research Laboratories, The University of Texas at Austin
Austin, Texas 78713-8029, U.S.A.

ABSTRACT

The nonlinear self-demodulation of a burst of high frequency sound produces a highly directive, low frequency transient in the parametric array process. Ricker wavelets, of interest in exploration seismology, can be generated from the self-demodulation of a primary pulsed in a Gaussian envelope. The application of this process to sub-bottom profiling is explored in terms of resolution and depth of penetration. Sea trials conducted in a complicated geological area offshore Oahu, Hawaii, U.S.A., are discussed. It is shown that parametric operation can lead to greater depth of penetration in the sediments for systems compared on the basis of equal angular resolution and transducer size.

INTRODUCTION

The nonlinear parametric array has long been advocated for use in high resolution profiling of the ocean bottom. In the present paper, we examine this application from the viewpoint of seismic exploration, posing some fundamental questions on the real risks and payoffs of this approach. The method is illustrated through the presentation of tutorial material as well as the results of new theory and experiment.

Ricker Wavelet

The Ricker wavelet is a prominent waveform in seismology that was derived in a classic paper presented in 1951 (Ricker, 1952). Ricker developed the laws of propagation in an attempt to discover "the form and nature of the primary seismic disturbance which proceeds outward from the explosion of a charge of dynamite in the earth". Although Ricker's analysis was flawed by the assumption of dissipation proportional to the second power of frequency, his wavelet has survived as an almost universally used waveform in exploration seismology. It is especially useful in the display of seismic sections, due to its phase sensitive asymmetry and its high range resolution (see Fig. 1).

Parametric Generation of Ricker Wavelets

Some 20 years ago, Berktay (1965) introduced the transient concept of Westervelt's parametric transmitting array (Westervelt, 1963) with a model as follows:

$$p_s(t) = \left(1 + \frac{B}{2A}\right) \frac{P^2 S_o}{16\pi \rho_o c_o^4 \alpha R_o} \frac{\partial^2}{\partial t^2} f^2\left(t - \frac{r}{c_o}\right) , \qquad (1)$$

*On attachment from Royal Australian Naval Research Laboratory, 1983-1985.

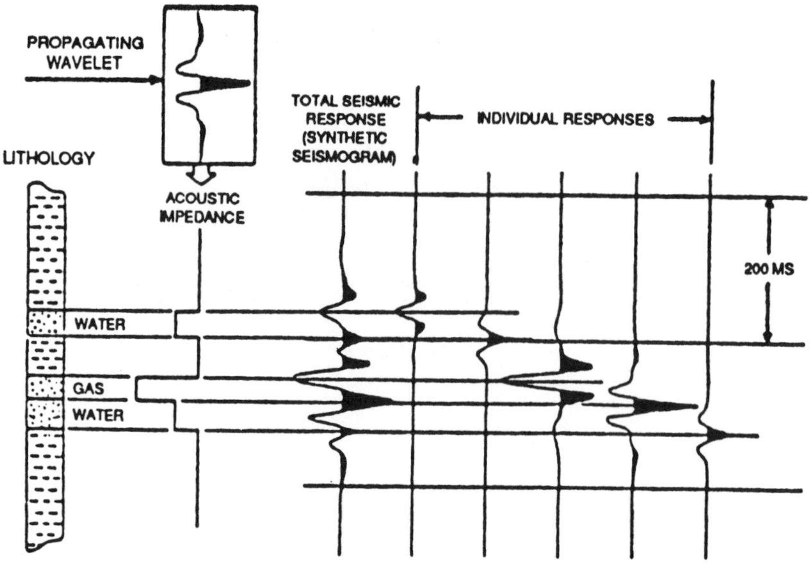

FIGURE 1
RELATIONSHIP BETWEEN LITHOLOGY, PROPAGATING RICKER WAVELET AND SEISMIC RESPONSE
(FROM NIEDEL, N., Tutorial Lecture)

where $p_s(t)$ = secondary waveform of the parametric transient at great distances, on the acoustic axis, B/A = parameter of nonlinearity, P = peak primary pressure amplitude, S_o= cross-sectional area of primary beam, ρ_o = static density, c_o = small signal sound speed, α = primary absorption coefficient, R_o = range to observation point, f = envelope function of the primary pulse, and t and r = time and space coordinates, respectively. This model has been invaluable in the development and understanding of nonlinear acoustics and its applications (Moffett, 1970; Mellen and Browning, 1970; Muir and Vestrheim, 1979). An example is illustrated in Fig. 2, which shows the parametric generation of a Ricker wavelet through the self-demodulation of a primary pulse transmitted in a Gaussian envelope.

One is attracted to the seismic potential of nonlinearly generated waveforms like that of Fig. 2, especially when it is considered that they are generated in a narrow beam, with no sidelobes. However, there is considerably more to this problem than meets the eye.

Although of pioneering significance, the Berktay model (1965) is limited to the following conditions.

(1) post-interaction ranges,
(2) shock free primaries (quasilinear conditions),
(3) idealized interaction volumes, devoid of diffraction effects, and
(4) the axial field.

In practice, these restrictions are almost always violated in the application of real parametric systems of practical interest.

The assessment of parametric seismic systems therefore requires the use of realistic models to account for complex effects. Despite considerable model development over the past 20 years, no single theory has been advanced that satisfies the practical demands of a realistic system. Assumptions lead to compromises with reality. These involve not only the restrictions mentioned above, but also many others, the most important of which is one involving narrowband primary radiations.

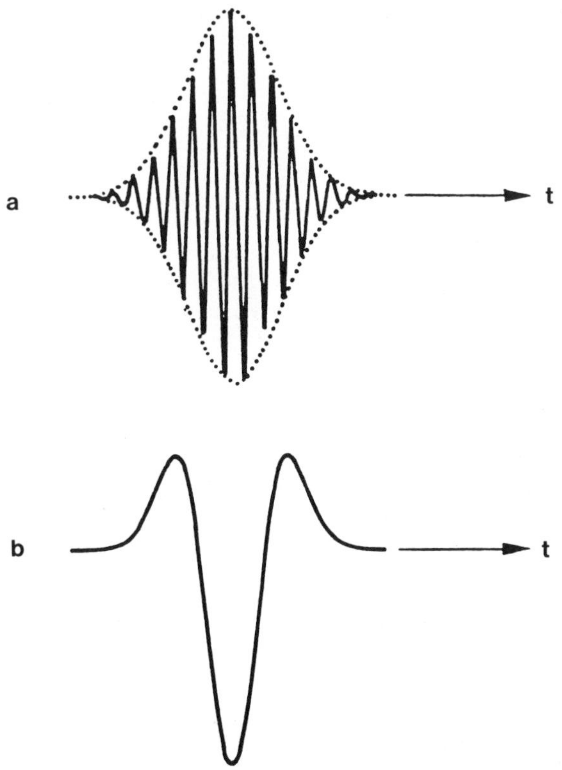

FIGURE 2
NONLINEAR DEMODULATION OF A PRIMARY PULSED IN A GAUSSIAN ENVELOPE;
(a) PRIMARY PULSE, (b) NONLINEARLY GENERATED RICKER WAVELET

In order to develop realistic models we are left with no other choice but to employ numerical physics in conjunction with computer capacity. Here, we outline the development of a family of models derived by Wyber (1985) that are not limited by the aforementioned restrictions. Basically, they utilize the Lighthill inhomogeneous wave equation, in the form

$$\nabla^2 p_2 - \frac{1}{c_o^2}\frac{\partial^2}{\partial t^2} p_2 = -\rho_o \frac{\partial q}{\partial t} \quad , \tag{2}$$

where q, the source strength density, is

$$q = \frac{\left(1 + \frac{1}{2}\frac{B}{A}\right)}{\rho_o^2 c_o^4} \frac{\partial}{\partial t} p^2 \quad . \tag{3}$$

Here, the superscript 2 denotes second order quantities, i.e., parametrically generated sound of both lower sideband (difference frequency) and upper sideband (harmonics).

In the quasilinear approximation the pressure p, on the right-hand side of Eq. (3), is taken as $p = p_1$, i.e., as the first order pressure. In the present family of models, we allow $p=p_1+p_2$, in order to include the effects of shock formation in the primaries. The models are formulated with convolution techniques so as to solve Eq. (2) for realistic geometries. Ample use of Fourier transform techniques is employed to provide for the treatment of bandwidth.

Brief functional descriptions of the models are as follows.

PC Model. A useful model can be implemented on a personal computer, although the result is limited by some simplifications. These include the assumption of uniform waves that are either collimated or tapered in intensity, transverse to the axis, as well as the limitation of accounting for only the second harmonic in shock formation. This means that diffraction effects are not fully accounted for and that the result is valid only up to shock parameters ($\sigma = [1 + B/2A] \in k$) of around 2.

Mainframe Model. When using a mainframe computer to do the computation, one may deal with all the harmonics, which removes the limitation on moderate shock parameters. Some simplification on interaction in the nearfield of the primaries is nonetheless required; however, most diffraction effects are accounted for.

Supercomputer Model. The incorporation of the exact source geometry in the calculation of nearfield interaction is enhanced by the use of great computer capacity, as well as parallel architecture and vector processing. McDonald and Kuperman (1986) have developed one such model for supercomputer use.

An example of one calculation, compared to the experiment, is shown in Fig. 3. The experimental parameters of the test sonar are discussed in the appendix, and elsewhere (Muir et al., 1980); but it should be stated here that this system produces transient signals in the 1-5 kHz band from the nonlinear demodulation of primary signals in the 10-15 kHz band. The data shown are for the sonar operating at shock parameter values near a value of 3, at a test range of 76 m. The asymmetry in the waveform (difference in heights of the positive peaks) is due to the existence of a nearfield component that develops as the first time derivative of the envelope squared. This component is superimposed on the farfield component that is derived from the second time derivative, as shown in Eq. (1). The peak pressure of the calculated waveform is some 3 dB higher than the measured waveform. This is due to the fact that the PC model used here treats only the second harmonic as a loss mechanism and therefore begins to lose accuracy at shock parameters in the neighborhood of 1 to 2. The mainframe model yields a result that agrees with experiment, within prevalent accuracy limits.

Seismic Profiling Concepts

There are several seismic exploration applications that could conceivably utilize the high angular resolution offered by the nonlinear, parametric source. These range from mining geophysics and engineering to the civil engineering of harbors, docks, and drilling sites, and eventually to deep strata geophysics for petroleum exploration.

Unterberger et al.(1980) have examined parametric generation in rock salt and have demonstrated the potential for identifying features and measuring distances in mines located in salt domes. This work is ongoing, and a system has recently been developed that will be used for further tests at frequencies of several kilohertz.

The use of high resolution parametric systems to identify and chart ore deposits on the ocean bottom has been explored by Kosalos (1978).

Each of these applications either has or could have its own unique system configuration, especially designed for performance and engineering economy. One such system is illustrated in Fig. 4, for the application of exploring and delineating structures in the surficial sediments. A parametric source is oriented downward in a single beam configuration. Echoes from the sub-bottom strata are received on elements of a vertical, towed line array. This array is processed to receive signals in the end-fire direction, which enables the suppression of water surface reflections as well as of own ship's noise.

Tests were conducted with one such system, described in the appendix, but before the results are discussed, it will be instructive to examine the problems of spatial resolution and depth of penetration.

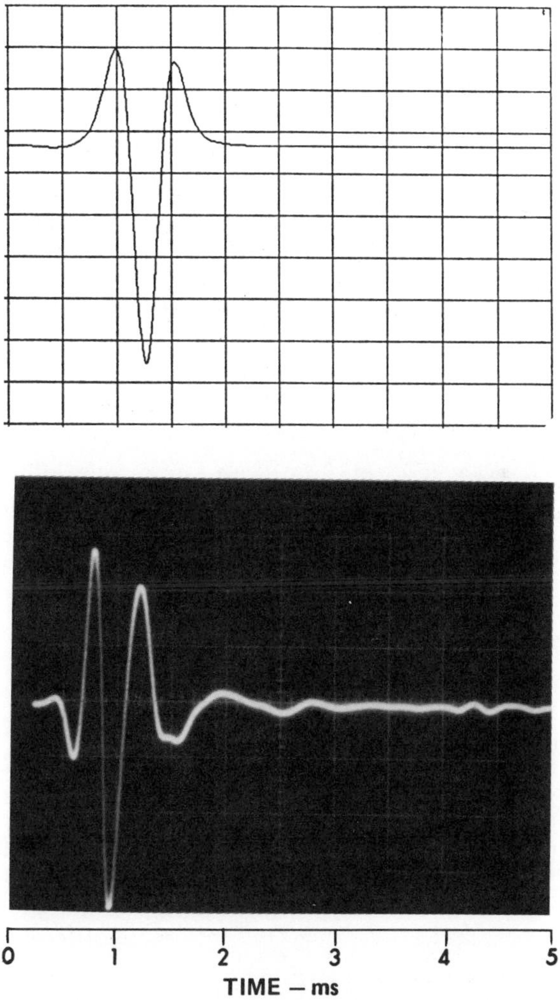

FIGURE 3
CALCULATED AND MEASURED SECONDARY PULSE
GENERATED BY A PARAMETRIC SONAR

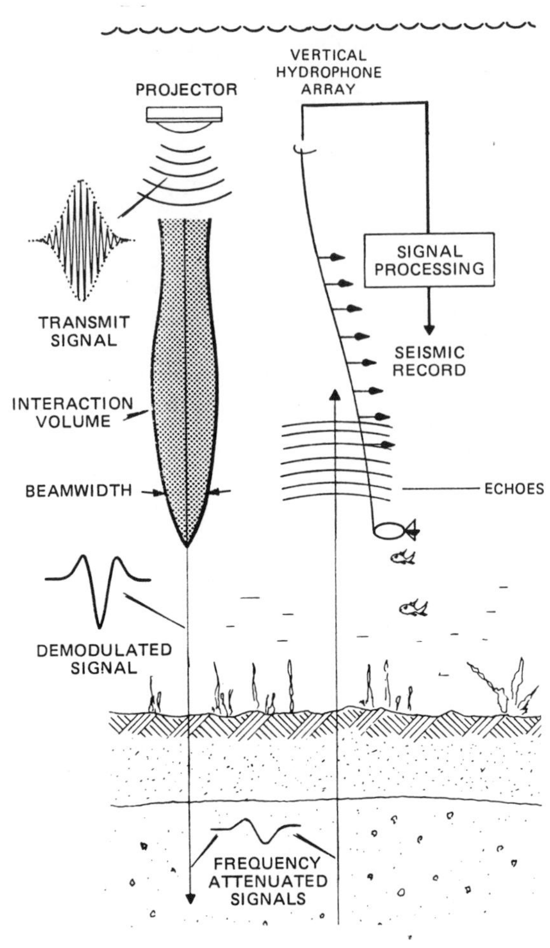

FIGURE 4
CONCEPT OF SEDIMENT PROFILING WITH A PARAMETRICALLY GENERATED RICKER WAVELET

Spatial Resolution

In high resolution acoustic systems, the lateral or angular resolution is normally specified by some parameter of the beam, i.e., its half-power width. As the beamwidth becomes large, the lateral resolution is often specified by the width of the first Fresnel zone on the plane of view, i.e., on a sub-surface stratum. These criteria are given as R sin θ_{HP} and $2\sqrt{R\lambda}$, respectively, where R is range, θ_{HP} is half-power beamwidth, and λ is acoustic wavelength.

The question arises as to what criterion governs nonlinear parametric profilers of practical interest. Some calculations on this problem are shown in Fig. 5, which plots the range at which the lateral width of a beam exceeds the width of the first Fresnel zone. This is done for three frequencies that are close to those desired in specific applications: 2.5 kHz for civil engineering work involving surficial sediments, 250 Hz for shallow seismic exploration, and 25 Hz for deep seismic applications. The ranges shown are waterpath equivalent ranges.

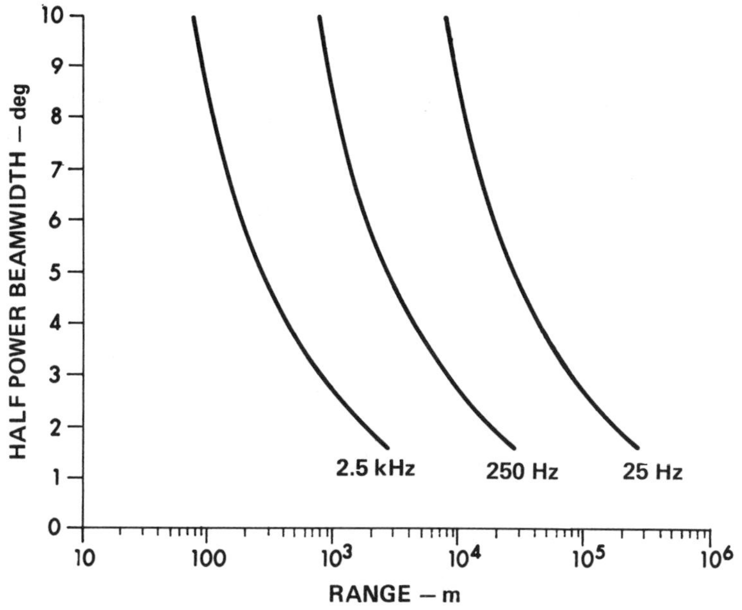

FIGURE 5
RANGE AT WHICH BEAMWIDTH RESOLUTION EXCEEDS FRESNEL ZONE RESOLUTION

Although the resolution that may be achieved in any particular system is a matter of design, limited by engineering economy, the curves do show the advantage of parametric systems that can operate at the smallest possible beamwidths. This appears to be the case for all the applications addressed here. On the other hand, even moderate sized beams appear to offer advantages, out to significant ranges into the bottom.

It should be remarked that the concept of Fresnel zone resolution is not always valid for all types of geological structures. Referring to Fig. 6, it is intuitive that a Fresnel zone system might adequately locate the major fault at A in the smooth upper section, and that it would undoubtedly fare worse at points B and C in the more complicated lower section, due to strong off-axis reflections within the large field of view. At point D, the broadbeam system would still view the off-axis discontinuities at B and C, even though they may well be outside the areal coverage of the first Fresnel zone. When displayed as a seismic section plot, the data would be fraught with the usual phantom parabolas. These can be removed with considerable seismic data processing, to reveal a more accurate sub-surface plot. It is nonetheless interesting to speculate on the improvements that might be gained by applying the equivalent amount of seismic data processing to raw data acquired with a narrow beam.

Depth of Penetration

The depth to which a sub-bottom profiling system can penetrate is often more important than its lateral resolution. It is therefore instructive to compare parametric and linear systems as to depth capability. This can be done without getting into system specific assumptions by comparing the required figure of merit (FOM) (Urick, 1967) as a function of depth of penetration. Recalling that a system's FOM may be equated to the allowable losses the system may sustain and still function, we have

$$FOM = 20 \log 2 (R_W + R_S) + 2\alpha_S R_S + CL + 2 ICL + RL \quad , \qquad (4)$$

where R_W is waterpath range, R_S is sediment penetration range, α_S is the sediment attenuation coefficient, CL is conversion loss, ICL is interface coherence loss, and RL is the reflection loss at the last detectable interface.

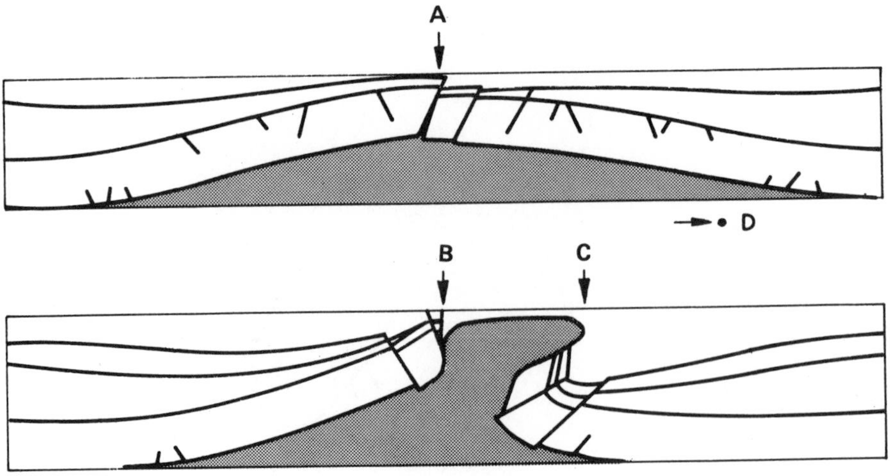

FIGURE 6
COMPARISON OF GEOLOGIC SECTIONS OF AN EMERGING SALT DOME
(From McQuillin et al., An Introduction to Seismic Interpretation)

Here, we will assume a waterpath of 1000 m, a sediment absorption that is linear with frequency but is a function of depth, i.e., $\alpha_s = k_p(R_s)f$, a conversion loss of 40 dB for parametric operation and 0 dB for linear operation, an interface coherence loss of 0.01 dB/m of sediment, and a reflection loss of 12 dB at a hypothetical planar stratum that we plan to detect. With this scenario, we may use the spreading law for image reflection, 20 log $2(R_w+R_s)$. For purposes of simplification, we will work with the center frequency component of the wideband Ricker wavelet signal.

It remains to define the attenuation coefficient, $k_p(R_s)$, which of course varies widely over the oceans. A generic geoacoustic attenuation curve has been advanced by Hamilton (1980) for a bottom made up of sand-silt, turbidites, and sedimentary and basalt rocks. A plot of this curve is reproduced in Fig. 7. Note that k_p has units in dB/m-kHz, and that it decreases with depth.

Due to the means of measurement, Fig. 7 presumably includes all attenuation mechanisms, among which are intrinsic absorption, transmission through reflectors (layered strata), reflector roughness and curvature, and scattering by inhomogeneities (Hamilton, 1980). The data of Fig. 7 are more appropriate for deep ocean sediments than for shallow ocean sediments, where attenuation near the surface is normally much higher.

In comparing depth penetration capability for systems utilizing sound beams, it is appropriate to compare those of equal beamwidth, and hence equal lateral resolution. Since size of the source is also a very real and very important constraint, we are obliged to also compare systems of compatible source size.

For the case of parametric systems operating vertically in the ocean at low frequencies, the effect of ocean acoustic absorption on beamwidth is quite small, due to low values of ocean acoustic absorption (Muir and Goldsberry, 1981). For this reason, the parametric beamwidth is usually very close to that of the primaries. If this is the case, the linear system that is equal in beamwidth and transducer size to a parametric system is one that operates at the primary frequency.

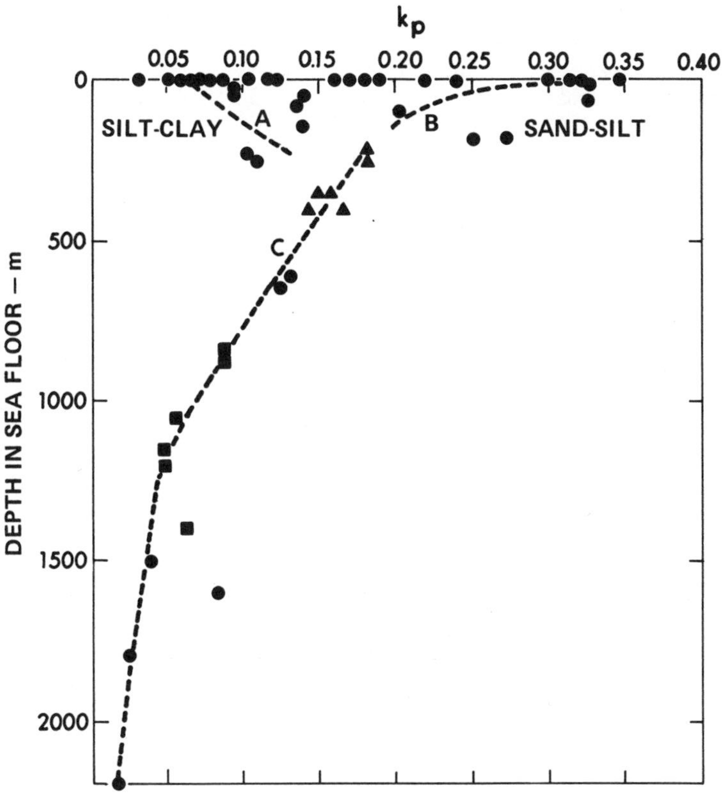

FIGURE 7
ATTENUATION OF COMPRESSIONAL WAVES versus DEPTH IN THE SEA FLOOR
(FROM HAMILTON, 1980)

For purposes of comparison, we choose a ratio of primary to secondary frequencies of 10, and compare the depth of penetration for parametric frequencies chosen for several general applications, profiling the surficial sediments, moderate depth seismic, and deep seismic. Some results are given in Fig. 8. In the upper part of the figure, we see that it requires less FOM (and hence less source level) for the 25 kHz primaries of a 2.5 kHz parametric system in the first few meters of bottom penetration. Beyond about 6 m into the sediment, the parametric system requires less FOM for any given depth of penetration. This means that despite a 40 dB conversion loss, the 2.5 kHz parametric system will out-perform an equivalent linear system of the same size and beamwidth.

In the center of Fig. 8, we see a comparison appropriate for a 250 Hz parametric profiler. Here the crossover point is around 70 m into the sediment; beyond 70 m the parametric system would be preferred due to the relative ease of achieving penetration capability.

The lower part of Fig. 8 shows a comparison appropriate for a 25 Hz parametric profiler. Here the crossover point is around 2000 m into the sediment. However, since the curves beyond this point are very similar, one must conclude that the nonlinear parametric approach may not be cost effective, since the linear 250 Hz system offers essentially the same depth capability. On the other hand, if performance capability at the greater depths is essential, it may be more easily achieved with the 25 Hz parametric system.

The trend toward less penetration payoff with decreasing operating frequencies is of course due to the decrease of the attenuation versus depth curve, shown in Fig. 7. As sound goes deeper into the sediments, the payoff for decreasing the frequency becomes less and less pronounced.

It should be remarked that the scenario used for Fig. 7 is quite idealized in that it involves sediments of fairly low absorptivity as well as a flat, planar reflecting surface.

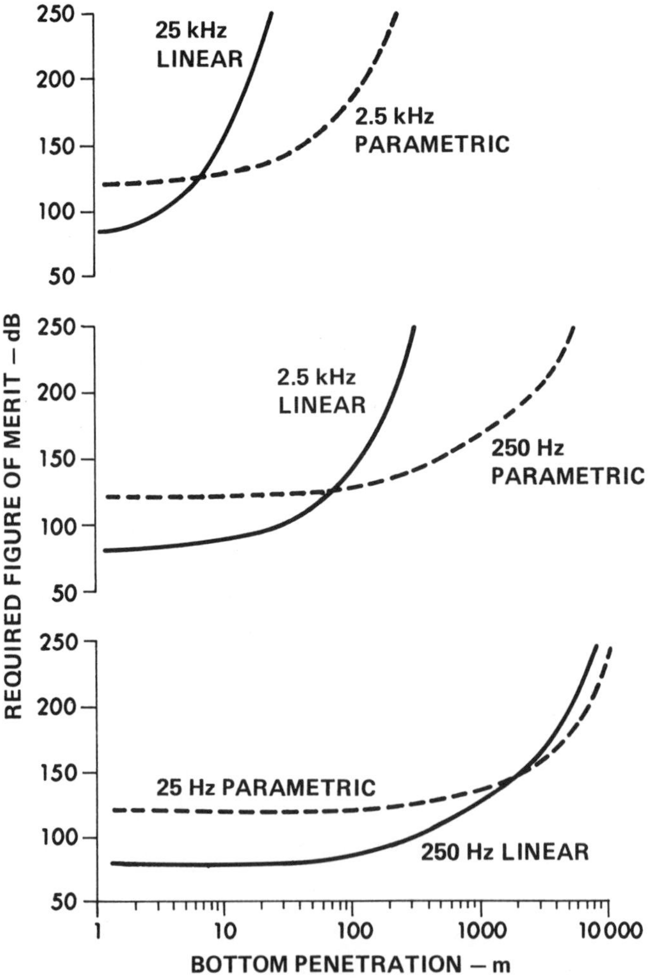

FIGURE 8
COMPARISON OF BOTTOM PENETRATION CAPABILITY

Experiments

The system of Fig. 9 was deployed from the R/V KAIMALINO in a water area offshore Oahu, Hawaii, in some bottom profiling measurements. R/V KAIMALINO is a SWATH ship that floats on two submerged pontoons. The parametric system (described in the appendix and elsewhere (Muir et al., 1980) was deployed through a deck center well in a 2.4 m diam spherical tow fish, stabilized by rope fairing lines streamed astern. The soundhead was pointed downward in the sphere and projected primary pulses centered at 13 kHz toward the bottom.

Primary source levels up to 246 dB re 1 μPa at 1 m (extrapolated from nearfield measurements made at 76 m) were used to generate Ricker wavelets at 195 dB source level, extrapolated from the same measurement range. The wavelets thus produced had a center frequency near 2.5 kHz with useful frequency content spanning the 1-5 kHz band. Echoes were received on an annular array of nine hydrophones located around the periphery of the circular disk sound source.

Echoes were also received on a vertical array of eight hydrophones, each processed separately. The processing involved coherent summation in a time delay beamformer to provide for a downward looking beam. This provided for some suppression of own ship's noise from the towing vessel as well as multiple reflections off the water surface. Signal processing gain of the stacked vertical array elements goes as 10 log n, where n = number of elements coherently summed.

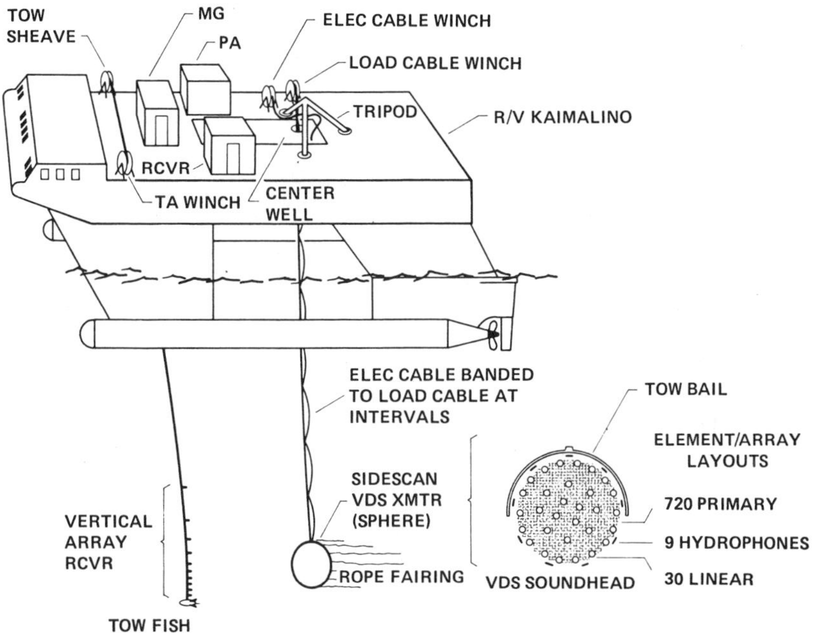

FIGURE 9
EXPERIMENTAL CONFIGURATION

Bottom geology at the test site is depicted in Fig. 10. This depiction was taken from a Hawaii Institute of Geophysics report by Tsutsui et al. (1986), which describes the geology in terms of a narrow shelf, a steep escarpment, and a moderate slope, before a final descent to the original submarine slope of Oahu Island. This slope is founded on bedrock created by extensive volcanism which terminated some 2.7 million years ago. Overlying volcanic bedrock is a Pleistocene formation that presents a prominent escarpment, leading shoreward to a shelf edge. A sediment cover from shore to shelf edge varies in thickness, and overlies a submerged reef platform cut by several sandfilled channels connecting to the beaches. A sand wedge 5-17 m in thickness is located along the shelf break.

As can be seen, the geology in this area is quite complicated. The sediment cover is a coarse calcareous sand rather than a soft sand-silt and there are no flat reflecting strata in the sub-bottom. As a result, the assumptions used in the example of Fig. 8 are not applicable and penetration depths are significantly reduced.

In the acquisition of sub-bottom profiling data, R/V KAIMALINO was driven along pre-determined lines, parallel and perpendicular to the beach, so as to sample various geological features in both the shallow sediments overlying the reef and the deeper sediments off the submarine slope. It should be remarked that own ship's noise problems were present during the tests, and this further reduced attainable penetration depths in the bottom.

Test Results

It is instructive to examine raw data received by the annular array, so as to compare the modes of operation prior to signal processing. This is done in Fig. 11, which compares test results acquired in only 17 m of water between the soundhead and the water-sediment interface. The undulations in the record of the sediment-water interface are due to vertical motion of the ship in ocean swell. Note that there are strong sub-bottom features to the right in each data set. These are thought to be the remnants of coral formations in the shape of plumes or plateaus. Although these occur at a very shallow depth in the sediment, they are better delineated by the parametric system than by the linear system of equal beamwidth, operating at the primary frequency. This is presumably due to the higher sediment attenuation at the higher frequency.

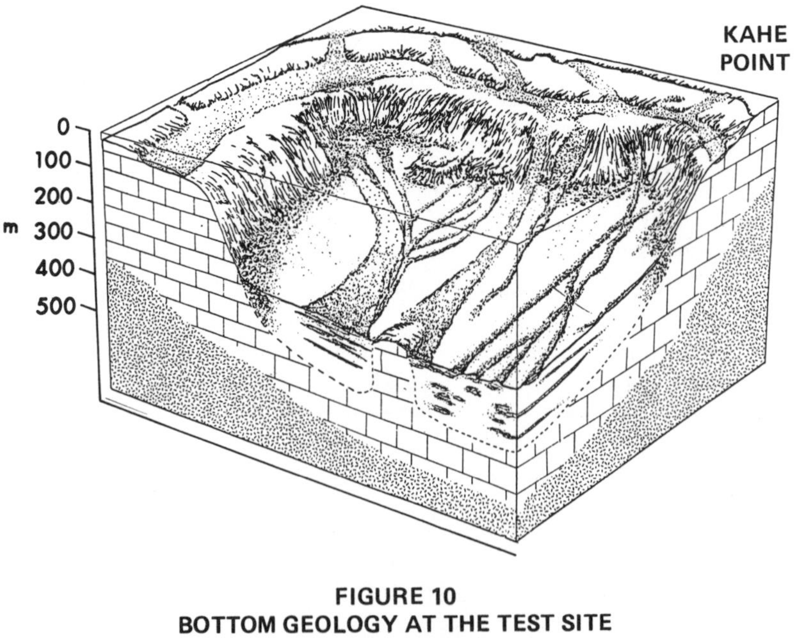

FIGURE 10
BOTTOM GEOLOGY AT THE TEST SITE

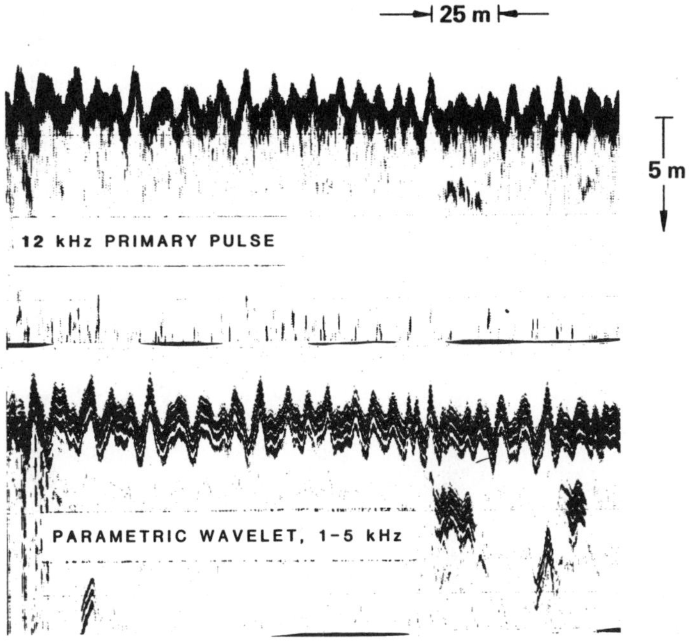

FIGURE 11
RAW DATA FROM PROFILING TESTS OVER AN AREA CONTAINING CORAL PLUMES

After processing, which includes removal of the effect of vertical motion of the ship, as well as coherent, end-fire summation of the output of the vertical array elements, we obtain the result shown in Fig. 12. The features in the surficial sediment are seen to be plateau-like structures, which could be remnants of a coral reef.

The banded nature of the data displayed in Fig. 12 is due to intensity rectification in plotting. The recorder is made to print during the positive-going excursion of the Ricker wavelet, so as to simulate the more conventional seismic plots in common use. One such plot is shown in Fig. 13. Here, the positive-going excursion of the transient signal is darkened, as in Fig. 1. The coral plumes or plateaus present in the right of Fig. 12 are here seen on an expanded scale, in greater detail (note the scale change). Also evident are vertical "streaks" that are individual traces containing intense low frequency oscillation; these have not yet been explained.

Some field data on a larger structure are shown in Fig. 14. Here, we see the cross-section of a submerged stream, with a sediment load in its bed. The width of the stream is approximately 350 m and the depth of the sedimentary bed is 15 m, which is displayed on an exaggerated scale. Geologists have noted significant offshore movement of sediments such as these during hurricane activity (Tsutsui et al., 1986). The linear system provided no penetration of this stream bed. Note the good delineation of the sides of the submarine canyon, despite their slope (approximately 45°).

Normally, narrowbeam profiling of sloped structures is difficult because most of the reflected energy misses the receiver on the return path (Kosalos, 1978). Profiling data off the submarine slope is shown in Fig. 15. Here, the sediment-water interface has an angle of about 5° with the horizontal, which is larger than the system beamwidth. This means that

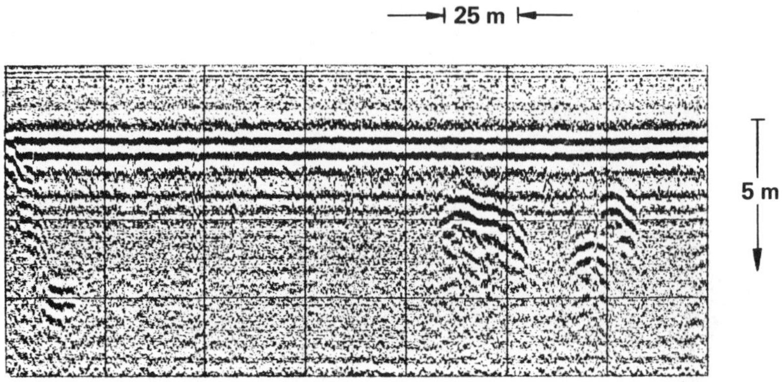

FIGURE 12
PROCESSED DATA FROM VERTICAL ARRAY; PARAMETRIC RICKER WAVELET

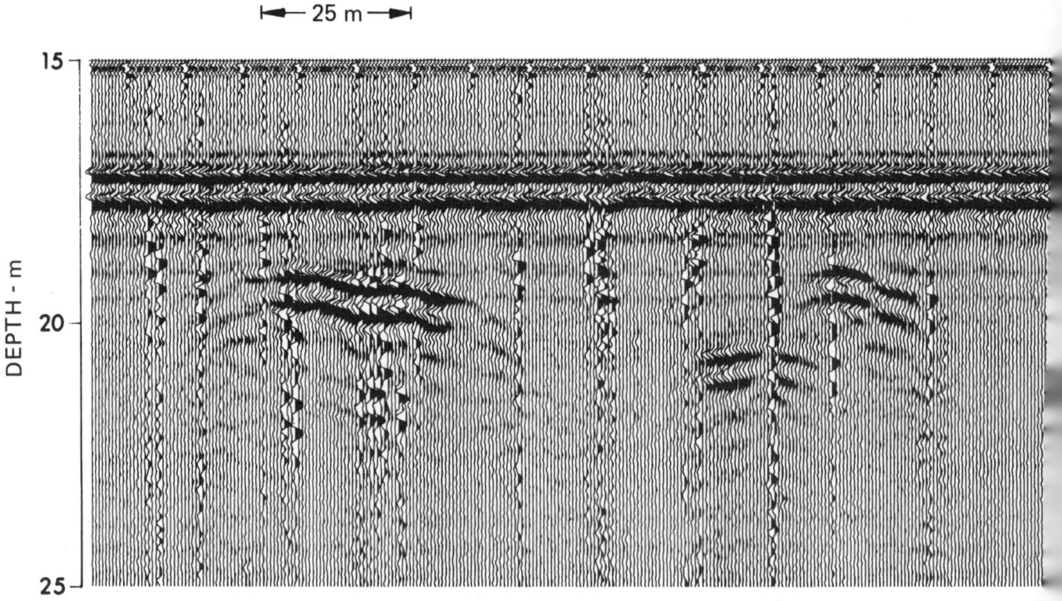

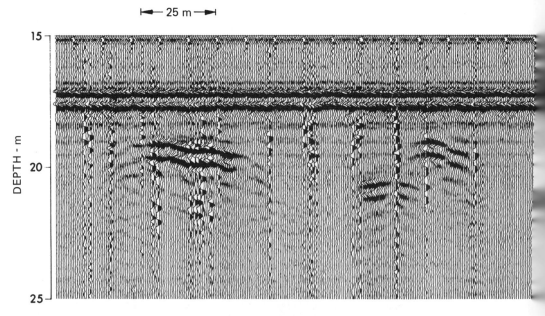

FIGURE 13
PROCESSED DATA OF THE RIGHT HALF OF FIGURE 12
DISPLAYED IN SEISMIC SECTION PLOT

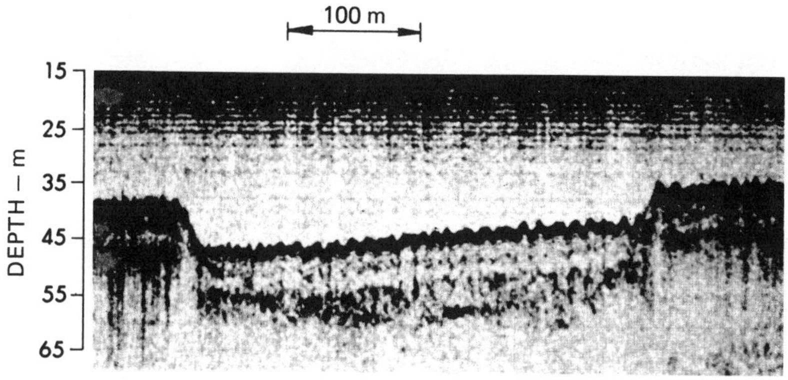

FIGURE 14
PARAMETRIC PROFILING OF A SUBMERGED STREAM

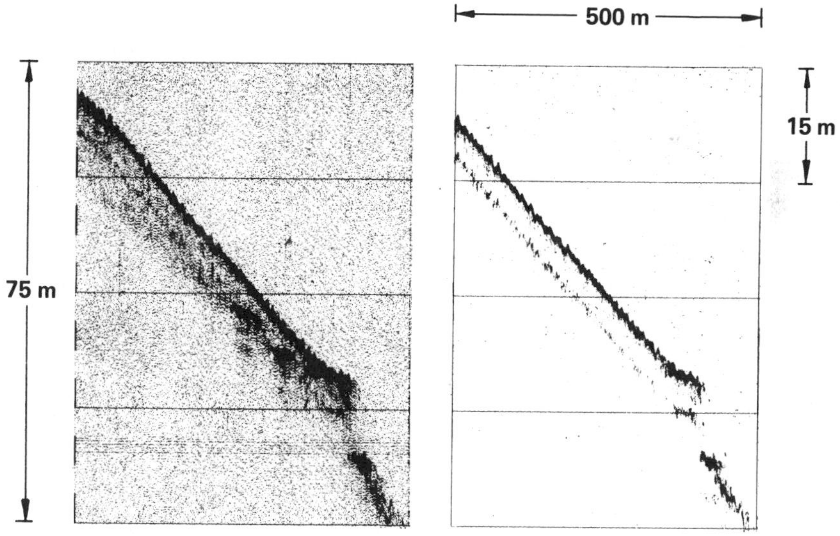

FIGURE 15
PROFILING THE SUBMARINE SLOPE OFF OAHU

most of the energy in the ~4° beam is returned to the surface at ~10° off the vertical, and therefore misses the soundhead. In this situtation, only the incoherent, rough surface component of the scattered field returns to the soundhead and provides a measurable signal. Profiling data are nonetheless acquired, as is evident in the figure. Here the parametric system portrays a sediment cover overlying bedrock that contains several sub-surface features near an escarpment. Although the linear system displays the exposed escarpment, it shows no real penetration of the sediment cover; it shows only a "ghost" beneath the interface that is due to an ocean surface reflection (path from receiver to surface to receiver).

SUMMARY AND CONCLUSIONS

Problems associated with sub-bottom profiling with narrowbeam systems have been examined. These include modeling and design, as well as signal processing and testing.

It was shown that operation of a profiler in a nonlinear, parametric mode provides deeper penetration at essentially the same resolution as that of the primary frequency sound normally radiated. This is due to high, frequency dependent attenuation in the surficial sediments that is higher than the conversion loss from primary to secondary sound.

APPENDIX: PARAMETRIC SOURCE

Since the research tool has been described elsewhere (Muir et al., 1980), it is only necessary to briefly review its parameters here. The source is 2.3 m in diameter and consists of an array of 720 elements for the projection of primary frequency sound in the 11-16 kHz band. Primary source levels for the generation of short, parametric transients are limited by the rise time of the source and by finite amplitude effects associated with shock formation at the primary frequency. The sound levels for parametric generation of a Ricker wavelet were measured with the present research tool in the nearfield of the primaries, well within the parametric growth region. The results are given below.

Table A-1. Parametric Transient Parameters for Present Research Tools

Range (R,m)	Relative Range $R/(a^2/\lambda)$	Primary Sound Level (dB re 1 μPa, rms)	Secondary Sound Level (dB re 1 μPa, rms)
16.8	0.7	243.4	187.5
25.6	1.0	247.8	191.4
76.2	3.0	246.3	194.6

The sound pressure levels are specified above as the equivalent rms value of a sinusoidal test signal substituted at the same peak positive to negative value as that of the measured signal. The primary signal contains nonlinear harmonic components that increase in content with increasing range. At 76 m, for example, the fundamental (at 13 kHz) had a measured sound pressure level of 239.6 dB while the fundamental plus harmonics measured 246.3 dB. Since the Ricker wavelet is asymmetric, the levels shown above represent maximum peak to trough amplitude. In the farfield, the primary to secondary conversion loss should approach 40 dB or so, depending on design and measurement conditions.

The measured test signal at the range of 76 m and its frequency spectrum are shown in Fig. A-1, below. It can be seen that the parametric Ricker wavelet has an overall duration of something less than 1 ms, and that the frequency spectrum peaks near 2.5 kHz, with useful frequency content spanning the 500 Hz to 5 kHz frequency band.

The beamwidth of the research tool operating in the parametric mode lies near 3-4°, as measured between the half-power points, over the aforementioned band. Some beam pattern measurements are shown below, in Fig. A-2.

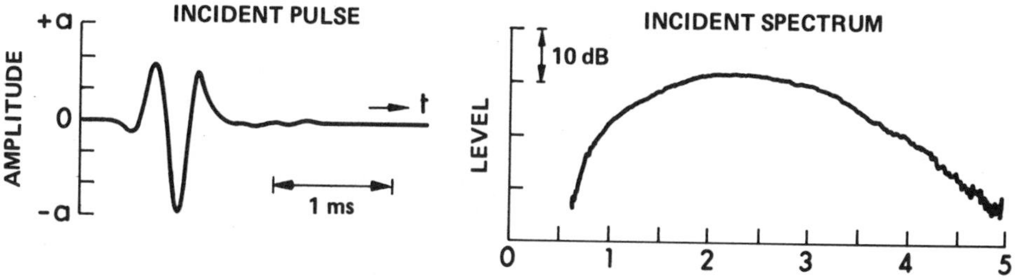

FIGURE A-1
MEASURED TEST SIGNAL

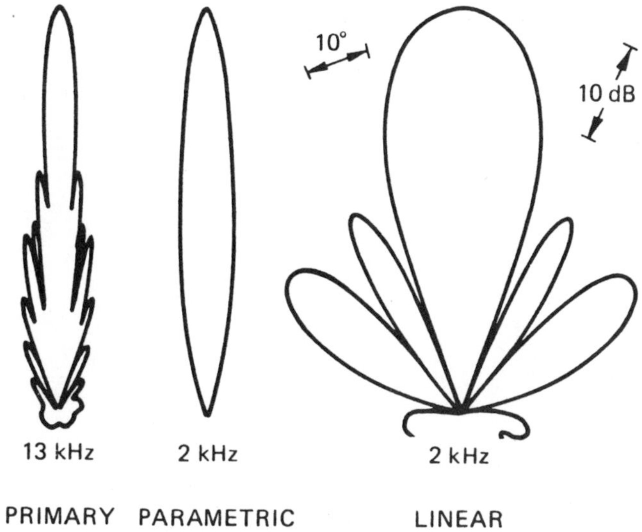

FIGURE A-2
BEAM PATTERNS

REFERENCES

Berktay, H. O., 1965, "Possible Exploitation of Nonlinear Acoustics in Underwater Transmitting Applications," J. Sound Vib. 2:435-461.

Hamilton, E. J., 1980, "Geoacoustic Modeling of the Sea Floor," J. Acoust. Soc. Am. 65(5):1313-1340.

Kosalos, J., 1978, "A Deep Towed 3 kHz Parametric Source," in Proceedings Joint USN/SEG Symposium, NORDA, Bay St. Louis, MS.

McDonald, B. Edward, and Kuperman, W. A., 1986, "Time Domain Formulation for Pulse Propagation, Including Nonlinear Behavior at a Caustic," submitted to J. Acoust. Soc. Am.

Mellen, R. H., and Browning, D. G., 1970, "Self-Demodulation of Acoustic Waves," in "Nonlinear Acoustics," Proceedings of the 2nd International Symposium on Nonlinear Acoustics, ARL:UT, Austin, Texas, T. G. Muir, ed., p. 57.

Moffett, M. B., 1970, "Large-Amplitude Pulse Propagation, A Transient Effect," in "Nonlinear Acoustics," Proceedings of the 2nd International Symposium on Nonlinear Acoustics, ARL:UT, Austin, Texas, T. G. Muir, ed., p. 143.

Muir, T. G., et al., 1980, "A Low-Frequency Parametric Research Tool for Ocean Acoustics," in "Bottom Interacting Ocean Acoustics," W. A. Kuperman and F. B. Jensen, eds., Plenum Press, New York, pp. 467-483.

Muir, T. G., and Goldsberry, T. G., 1981, "Signal Processing Aspects of Nonlinear Acoustics," in "Underwater Acoustics and Signal Processing," Proceedings NATO Advanced Study Institute, Copenhagen, L. Bjorno, ed., D. Reidel, Boston.

Muir, T. G., and Vestrheim, M., 1979, "Parametric Arrays in Air with Application to Atmospheric Sounding," in Proceedings of the 8th International Symposium on Nonlinear Acoustics, Journal de Physique, Paris, P. Alais and A. Zarembowitch, eds., pp. C8-89.

Ricker, Norman, 1952, "The Form and Laws of Propagation of Seismic Wavelets," Geophysics18(1):10-40.

Tsutsui, B., Campbell, J., Frisbee, J., and Coulbourn, W. T., 1986, "Storm Generated, Episodic Sediment Movements off Kahi Point, Oahu, Hawaii," Hawaii Inst. Geophys. Rept., in publication.

Unterberger, R. R., Wang, A. M., and Muir, T. G., 1980, "Nonlinear Sonar Probing of Salt," in "Proceedings of the 50th Meeting of the Society of Exploration Geophysics," Vol. III, 1795-1824.

Urick, R. J., 1967, "Principles of Underwater Sound for Engineers," McGraw-Hill, New York.

Westervelt, P. J., 1963, "Parametric Acoustic Array," J. Acoust. Soc. Am. 35: 535-537.

Wyber, R. J., 1985, personal communication.

REMOTE ACOUSTIC MAPPING OF A SUBMARINE SPRING PLUME

E. Colbourne and A. E. Hay

Department of Physics and Newfoundland Institute for Cold Ocean Science
Memorial University of Newfoundland
St. John's, Newfoundland, Canada, A1B 3X7

ABSTRACT

The results of a recent investigation of the submarine spring in Cambridge Fiord, Baffin Island, are presented. The brackish water plume rising from the spring was mapped using a Ross Laboratories 192 kHz acoustic sounder and a microwave positioning system. CTD profiles and acoustic backscatter data were acquired at different spatial positions relative to the plume axis. In addition, visual observations together with CTD and current measurements were made at the vent location 47 m below the surface using the submersible, PISCES IV. Digitally enhanced acoustic images are used to characterize the geometric properties of the plume, and to estimate vertical velocities within the plume. The acoustic backscatter intensities are compared to the fine structure amplitudes in the temperature and salinity fields.

INTRODUCTION

A buoyant plume has been observed rising from a submarine fresh water spring at a depth of 47 meters at the head of Cambridge Fiord, Baffin Island. This submarine spring results in the annual formation of a polynya in late March, which was discovered by the Royal Canadian Air Force in 1952 and also reported by Inuit hunters from nearby Pond Inlet. In an earlier study of the polynya, Sadler and Serson (1980) suggested that the spring may originate from a nearby lake whose water level decreased by about 2 meters during the winter, giving a discharge rate of approximately 0.14 m^3 s^{-1}. More recently it was shown that the plume could be detected using a high frequency acoustic sounder (Hay, 1984).

The purpose of this paper is to present preliminary results from a further investigation of the Cambridge Fiord submarine spring and plume in September 1985. This investigation involved more extensive acoustic sounding and CTD (Conductivity Temperature Depth) surveys conducted with the aid of a microwave positioning system to accurately map the horizontal extent of the plume. Also CTD and vertical current measurements were made at the vent using the submersible PISCES IV. Ambient currents and stratification were also measured. The objectives of the study were to characterize the plume's bulk properties and behavior, and to investigate the relationship between the lateral decay of the acoustic backscatter amplitudes and that of temperature and salinity fine structure in the spreading plume.

METHODS

The brackish water plume from the submarine spring was mapped from a 7 m launch using a 192 kHz Ross Laboratories acoustic sounder and a Motorola Miniranger microwave positioning system. The acoustic sounding transects spaced at approximately 10 m intervals were conducted by executing circular arcs around each shore transponder in the general area of the plume. A second mapping of the plume was made in conjunction with a grid of CTD stations while the launch was four-point moored. The analog acoustic backscatter signals and trigger pulse were recorded on a Racal Store 4 instrumentation tape recorder.

The acoustic sounder was a commercial prototype built by Ross Laboratories, Inc., and has been described elsewhere (Hay, 1983). The sounder has a nominal beamwidth of 2.3^{o} and was operated with a pulse length of 0.5 ms. The recordings were made at a tape speed of 38.1 cm/s, for which the recorder's frequency response was flat from 0 to 5 kHz, -3 dB at 6.2 kHz and -20 dB at 10 kHz. The rms tape noise level was approximately -43 dB at 3 kHz. Each analog tape was calibrated by recording the output of 3 dc levels from the recorder calibrator output. The analog signal was digitized on a HP model 5451B Fourier Analyzer at an effective rate of 20 kHz by using a playback tape speed of 19.1 cm/sec and a real time sampling rate of 10 kHz. The recorded trigger pulse was used to activate the 4096-point digitizing window. The acoustic data was processed on a VAX 11/785 and a HP-1000 computer system. False color images of the acoustic data were obtained on a Norpak VDP-11 image processing system. The images were enhanced by both thresholding and contrast stretching.

The CTD measurements were obtained using a Guildline model 8770 portable CTD system. The time interval between successive pressure, conductivity, and temperature records was 190 ms. Typical ascent and descent rates were of the order of 0.3 m/s giving a spatial resolution of about 10 cm. The vertical current measurements were made with a Neil Brown Instruments Systems DRCM-2 acoustic current meter. This current meter was mounted on the remote manipulator arm of PISCES IV. Measurements were made while the submersible was sitting on the bottom and the current meter maintained in a horizontal orientation using externally mounted spirit levels.

The CTD digital data processing was done on a VAX 11/785 computer. The pressure time series was smoothed by computing a sliding 15-point least squares fit to a second order polynomial. This eliminated the small amplitude random changes in the pressure arising from the low resolution of the portable system (0.12 dbars) and noise in the least significant bit of the pressure word. Temperature and salinity were then computed using the UNESCO 1978 practical salinity scale. The rms fine structure fluctuations in temperature and salinity were computed by high pass filtering the data. A filter of approximately 1.5 m in length with a cutoff at about 1.0 m was used.

Acoustic data profiles were selected by averaging 5 consecutive transmissions parallel to the descending CTD package. The acoustic backscatter signals were corrected for spreading and attenuation by calculating $v' = v_{rms} F$ where

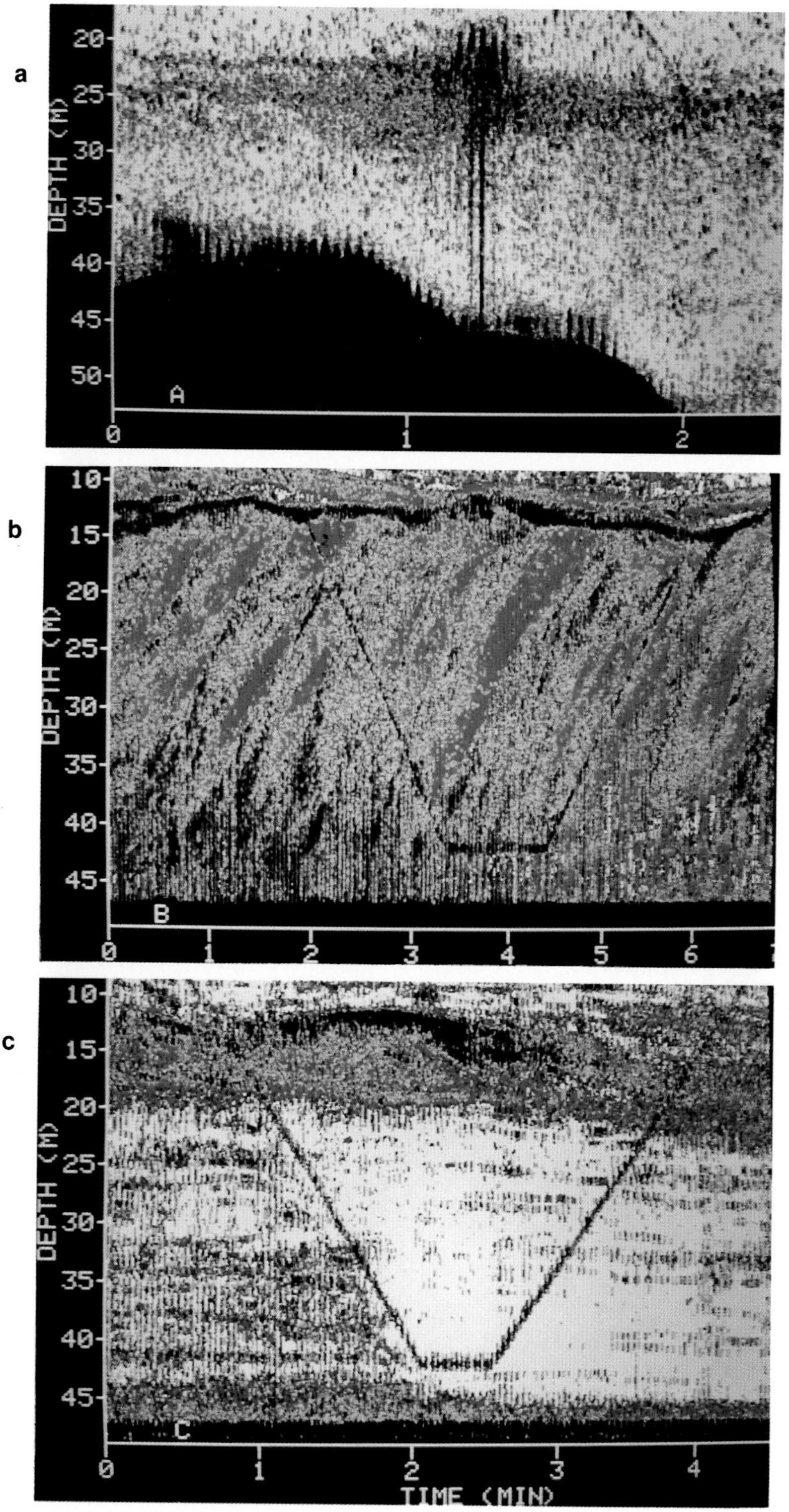

Fig. 1. False color, digitally enhanced acoustic images of the submarine spring plume obtained (a) while steaming directly over the rising plume (horizontal scale is approximately 75 m per minute), (b) while moored over the vent, and (c) while moored over the spreading plume.

$$F = \frac{r}{r'} \exp[2\alpha(r-r')] \qquad (1)$$

and v_{rms} is the measured root mean square backscatter amplitude, α the attenuation coefficient and r' is 25 m, half the total water depth. The attenuation coefficient at 0^oC and 31.0 ppt salinity for 192 kHz was calculated to be $5.97 \times 10^{-3}\ m^{-1}$ (Clay and Medwin, 1977 p. 98).

RESULTS

Figure 1a shows a false color image of an acoustic transect acquired while steaming over the vent. These images are made from 430 x 590 pixels with red representing high signal level while blue represents the lowest level. The narrow approximately vertical rising plume overshoots its height of neutral buoyancy due to its vertical momentum, reaches it maximum height of rise at 17 m depth and then falls back to spread radially outward along isopycnal surfaces at depths of 20 to 30 m. The amplitude of the acoustic backscatter decreases with radial distance from the plume axis. The vent is located at a depth of 47 m. Observations made from within the submersible indicate that the near bottom zone of high amplitude backscatter is due to euphausiids. The periodic nature of this zone and the appearance of two rising plumes are due to the launch rolling in about 1 to 2 meter seas. Acoustic images obtained on consecutive days show that the maximum height of rise increased by 5 m. This is believed to be related to changes in ambient stratification associated with a reduction in the speed of up-inlet winds.

A typical acoustic map of the spreading plume in horizontal cross section is shown in Figure 2. This map was constructed from 8 consecutive acoustic transects averaged over 1 m in the vertical centred at 23 m depth. The contours represent different levels of backscatter signal. The contouring interval is 0.1 volts with the highest level contour at 0.5 volts around the plume's center, on a relative scale of 0 to 0.65 volts. The average background level of approximately 0.1 volt has been removed. The along-shore extent of the plume at this depth is approximately 80 m and the off-shore extent about 45 m.

Figure 1b shows a depth versus time false color image of the plume obtained while the launch was four-point moored over the vent. This image shows discrete scattering structures, similar to those observed previously (Hay, 1984), rising from the bottom at constant speed and then decelerating above 20 m depth. Maximum ascent rates calculated from the slope of the straight line trajectories of these structures are about 32 cm/s in the bottom 20 m. Time series of horizontal and vertical velocity components obtained from PISCES show maximum vertical speeds near 28 cm/s. These measured speeds compare very well to those calculated from the acoustic sounding records. This image was obtained one day after the acoustic transect shown in Figure 1a. The previously mentioned increase of about 5 m in the maximum height of rise is clearly seen by comparing Figs. 1a and b.

The echo from the CTD instrument package can be seen in Figure 1b as oblique traces during ascent and descent and a horizontal trace at about 42 m where the CTD was held at constant depth for approximately 1 minute. Temperature, salinity and density (σ_t) profiles corresponding to the CTD descent are shown in Figure 3. The large amplitude fluctuations and fine structure associated with the acoustic scattering structures are clearly visible in this profile. Similar CTD profiles obtained away from the plume do not exhibit these gravitationally unstable features. Minimum salinities in this profile are about 29.4 at 42 m depth. However, time

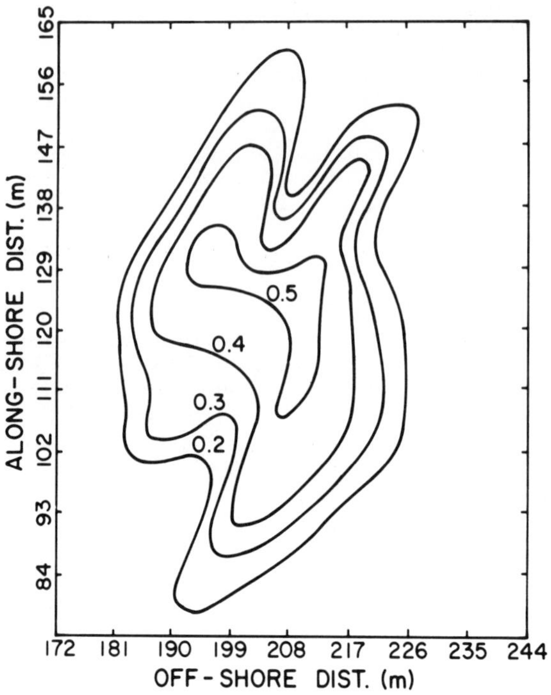

Fig. 2. Contours of acoustic backscatter showing a horizontal cross section of the spreading plume.

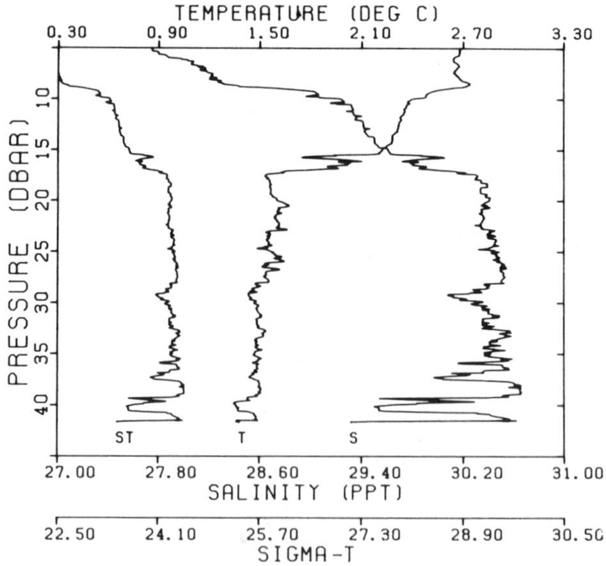

Fig. 3. Typical temperature (T), salinity (S) and density anomaly (ST) profiles corresponding to the image in Figure 1b.

series of temperature and salinity obtained from PISCES IV at the vent exhibit large fluctuations, particularly in salinity, with salinities of less than 10 recorded. The fluctuations in temperature are smaller than those in salinity, quite unlike previous studies when the reverse was true. Time series of CTD measurements at 21.5 m depth, however, show fluctuations in temperature and salinity of comparable amplitude.

The rms fluctuations in temperature and salinity are shown in Figure 4a and 4b for the CTD cast shown in Figure 1b through the rising plume. The corresponding block averaged acoustic backscatter signal is shown in Figure 5d. The increased rms fluctuations in salinity below 30 m depth are associated with a large increase in the acoustic backscatter.

The image shown in Figure 1c was obtained with the launch four-point moored over the spreading plume. The descending CTD probe enters the top of the plume at 12 m depth and emerges at 20 m depth. The horizontally coherent structures typical of biological scatterers seen below 25 m depth in the first 1.5 minutes of the image disappear as the CTD descends. The relatively scatter free region after the CTD has passed is indicative of an avoidance response outside the spreading plume.

The temperature, salinity and density anomaly profiles corresponding to the descending CTD cast in Figure 1c are shown in Figure 5. Large amplitude fluctuations in temperature and salinity occur where the CTD enters the plume at 12 m depth. Outside the 12 to 20 m depth range the density profile remains relatively stable. Artificial fluctuations in the salinity and density profiles may be present due to the time constant mismatch of the temperature and conductivity sensors and the physical separation of the sensors.

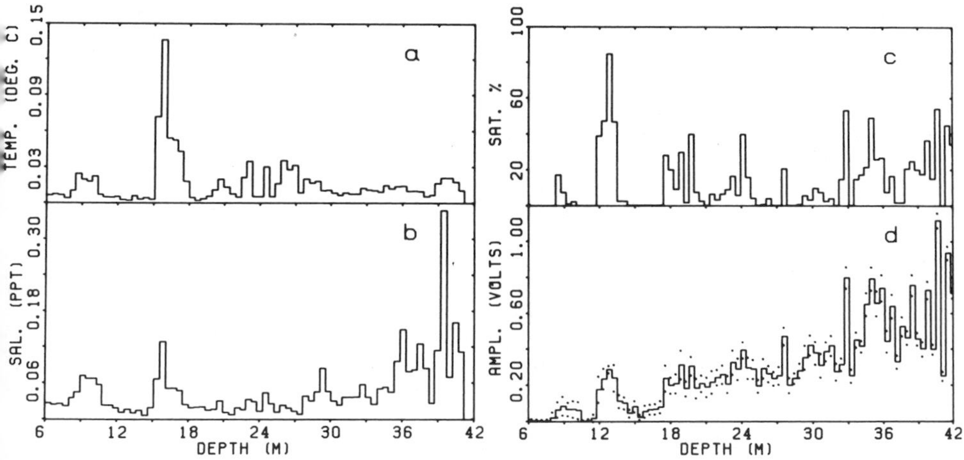

Fig. 4. Profiles of temperature, salinity and acoustic backscatter corresponding to Figure 1b. (a) rms fluctuations in Temperature. (b) rms fluctuations in Salinity. (c) Percentage of acoustic transmissions that saturated the detector. (d) Block averaged acoustic backscatter and standard deviations obtained from each block average.

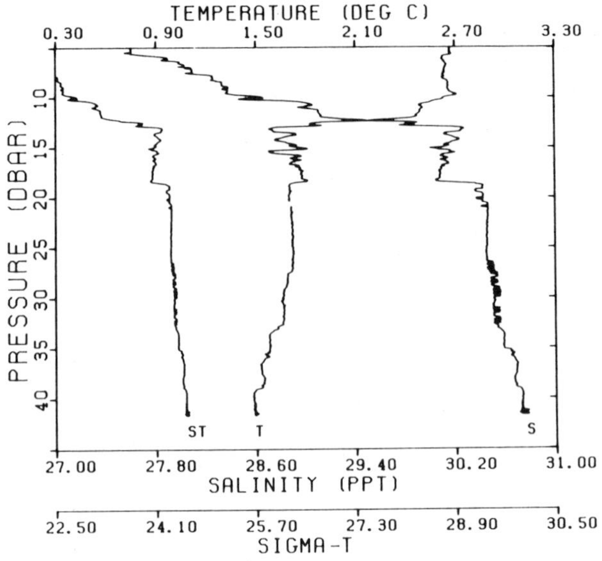

Fig. 5. Typical temperature, salinity and density profiles through the spreading plume corresponding to Figure 1c.

The rms fluctuations in temperature and salinity are shown in Figure 6a and 6b for a CTD profile through the spreading plume. The corresponding block averaged acoustic backscatter signal is shown in Figure 6d. The increased acoustic backscatter beginning at approximately 12 meters depth is clearly associated with the large increase in the rms values of temperature and salinity at the same depth. Cross correlation coefficients of 0.76 and 0.45 were obtained between the acoustic backscatter signal and the rms fluctuations in temperature and salinity respectively.

The visual and photographic observations at the vent show that the bottom in the discharge zone consists of cobbly material and rocks. These are kept free of sediment and detritus by the discharge which prevents such material from accumulating locally. The discharge itself has no initial momentum, and the discharge region is of large horizontal extent (several m on a side). The water escaping from the bottom is clear, and appears to be free of gas bubbles or sediment. Large variations in optical refractive index, with apparent spatial scales at least as small as 1 mm, are produced as the discharge mixes with the surrounding sea water. If the observed acoustic backscatter is caused by turbulent microstructure, then the refractive index fluctuations are expected to have spatial scales equal to one half the acoustic wavelength (Tatarski, 1961), about 0.4 cm at 192 kHz. Such scales are certainly present, and this provides additional support for a scattering mechanism associated with turbulence. Other candidate mechanisms, such as biota, suspended sediments or gas bubbles must be considered, but these seem unlikely. In particular, the avoidance response exhibited towards the descending CTD instrument package by biological organisms, which is evident below the spreading plume, does not appear to affect the backscatter from the plume itself.

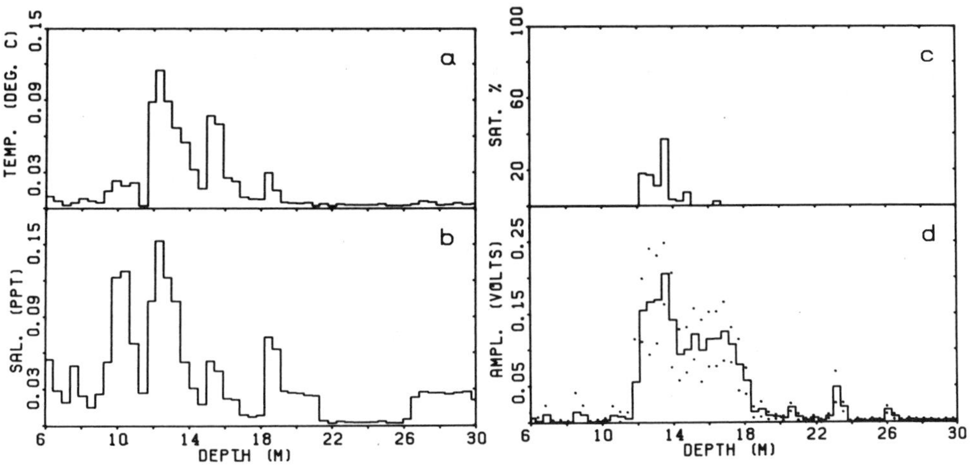

Fig. 6. Profiles of temperature and salinity fluctuations and acoustic backscatter through the spreading plume. See also Fig. 4.

SUMMARY

These results further demonstrate the advantages of using acoustic remote sensing techniques to detect and characterize buoyant plumes in the ocean. Augmented now by the submersible observations and more detailed acoustic soundings and CTD data, a body of evidence is being accumulated which points to the backscatter mechanism being acoustic refractive index fluctuations caused by turbulent mixing. Further investigations, however, including turbulence microstructure measurements, multiple frequency soundings and biological samples, are needed.

REFERENCES

Clay, C. S., and H. Medwin, 1977, "Acoustical Oceanography", John Wiley, New York.

Hay, A. E., 1984, Remote Acoustic Imaging of the Plume from a Submarine Spring in an Arctic Fiord, Science, 255:1154.

Hay, A. E., 1983, On the Remote Acoustic Detection of Suspended Sediment at Long Wavelengths, J. Geophys. Res., 88:7525.

Sadler, H. and H. Serson, 1980, An Unusual Polynya in an Arctic Fiord, in: "Fiord Oceanography", H. J. Freeland, D. M. Farmer, C. D. Levings, Eds., Plenum, New York.

Tatarski, V., 1961, "Wave Propagation in a Turbulent Medium", Dover, New York.

ACOUSTIC REMOTE SENSING OF THE WAVEHEIGHT DIRECTIONAL SPECTRUM OF SURFACE GRAVITY WAVES

Steven H. Hill

Department of Oceanography
University of British Columbia
Vancouver, B.C. Canada V6T 1W5

INTRODUCTION

Knowledge of the directional waveheight spectrum of surface gravity waves on water could be of great importance in improving our understanding of such phenomena as: the generation and propagation of wind waves; air-sea interactions; coastal erosion processes; and others. Engineering studies of the effects of waves on man-made structures such as port facilities, ships, drilling rigs, etc. would be greatly aided by the ability to make precise, reliable measurements of the waveheight directional spectrum. Such measurements are difficult to obtain; there are only a few instruments able to make them routinely, and all these instruments actually float on the surface, thus affecting the measurements by their presence, as well as being subjected to the many and varied mechanical stresses present at the sea surface. Thus the development of an instrument able to make reliable measurements of the surface geometry while remote from the surface would be of considerable scientific and practical importance.

Although some study of acoustic scattering from the surface had gone on previously, Eckart(1953) was the first to apply rigourous mathematical analysis to the problem of extracting information about the surface geometry from the study of such scattered acoustic energy. He reached one fundamental conclusion: that much more information about the sea surface would be revealed by long wave scattering (i.e. acoustic wavelengths much greater than the rms waveheight) than by short wave scattering. This is unfortunate, since it is much more efficient to produce short wavelength acoustic radiation using existing transducer technology than it is to produce long wavelength radiation. Following Eckart, many other workers have made contributions, some of those contributions having been reviewed by Fortuin(1970). Investigations have continued up to the present, including, among others, experimental work using forward scattering by Roderick and Cron(1970), and Williams(1973), who also gives a concise review of related work. Considerable purely theoretical analysis of the problem has also been carried out, for example: Harper and Labianca(1975) developed a general perturbation theory for long wave scatter from a moving rough surface; and Labianca(1980) has applied inverse scattering methods to show how the directional waveheight spectrum might be estimated using a bistatic arrangement.

See page 809 for Abstract.

The method to be described in this paper does not build on any previous work in acoustics, but rather is an adaptation of a technique which has been developed for measuring ocean waves and surface currents using HF RADAR. This work has been described by Barrick(1977), Barrick and Lipa(1977), and Lipa and Barrick(1986), among others. Adaptation of the electromagnetic scattering analysis to the acoustic case has resulted in both complications and simplifications, as will be seen below.

In this paper I will outline in a general way the theory of the technique, and will describe the instrument system we are using. A substantial amount of work developing models of the expected results has been done, and I will present the results of some of that work, together with some preliminary acoustic data from the instrument. Finally, I will outline the future research plans for the system.

THEORY

Full development of the theory used in this technique would be too lengthy to include in this paper, but I will present an outline of the development to give at least a basic understanding of the method and its underlying assumptions. The geometry of the experiment is shown in Figure 1 below. Acoustic energy is incident on the surface from below at the co-elevation angle θ, and is then scattered in many directions by the rough surface.

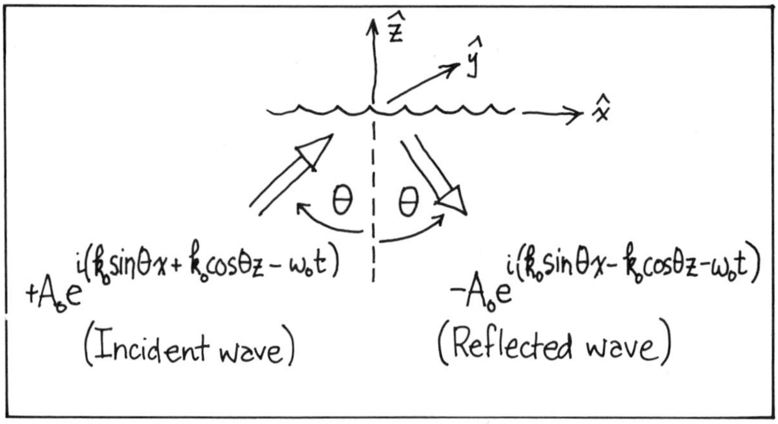

Figure 1. Geometry of the scattering problem.

We first expand the surface height as a real variable in a two dimensional exponential Fourier series:

$$z = \eta(\underline{r},t) = \sum_{\underline{k},\omega} H(\underline{k},\omega)\, e^{i(\underline{k}\cdot\underline{r} - \omega t)} \tag{1}$$

where $z = \eta(\underline{r},t)$ is the surface height above the mean, $H(\underline{k},\omega)$ are the Fourier coefficients for the surface height, $\underline{k}$ is the two-dimensional spatial wavenumber, and $\underline{r}=(x,y)$ is the horizontal position vector. We also use the deep water dispersion relation for gravity waves:

$$\omega^2 = gk \; ; \quad k = |\underline{k}| \, , \quad g = \text{acceleration of gravity} \tag{2}$$

The acoustic wavefield within the water is also expressed as a series expansion with two parts. For the first part, we assume that we are in the far-field, so that the acoustic waves incident on the surface can be taken to be plane waves, which can be described in terms of the carrier frequency and the angle θ. We assume without loss of generality that the plane of incidence is the x-z plane. Then we combine the incident wave with the wave which would be reflected from a perfectly flat pressure-release surface, and we assume that the remaining term is due to the acoustic energy reflected from the surface roughness. Thus the expression for the acoustic pressure is:

$$A(\underline{r},z,t) = 2iA_0 \sin(k_0 \cos\theta z) e^{i(k_0 \sin\theta x - \omega_0 t)} + \sum_{\underline{k},\omega} A(\underline{k},\omega) e^{i(\underline{k}\cdot\underline{r} - k_z z - (\omega+\omega_0)t)} \tag{3}$$

Now applying the wave equation to (3) and assuming that $\omega \ll \omega_0$, we have:

$$k_z = \begin{cases} \sqrt{k_0^2 - k^2} & ; \ k^2 < k_0^2 \\ i\sqrt{k^2 - k_0^2} & ; \ k^2 > k_0^2 \end{cases} \tag{4}$$

Finally, we use the boundary condition for the pressure release surface:

$$A(\underline{r}, \eta(\underline{r},t), t) = 0 \tag{5}$$

We now apply a perturbation expansion to the boundary condition (5) for the acoustic pressure. The following terms are considered to be first-order quantities in the expansion:

$$k_0\eta, \quad k_z\eta, \quad \partial\eta/\partial x, \quad \partial\eta/\partial y$$

The perturbation expansion allows us to derive equations relating the first and second order acoustic Fourier coefficients to the first and second order surface height Fourier coefficients. By manipulation of these relations, and using (3), (4), and work by Weber and Barrick(1977) on the nonlinear theory for surface gravity waves, we can derive relationships relating the Doppler spectrum of the acoustic backscatter to the waveheight directional spectrum of the surface, for both first and second orders. These relationships are most easily expressed when quantities are normalized using the following scheme:

1. Wavevectors are normalized by dividing by the Bragg wavenumber, equal to $2*k_0*\sin\theta$, where k_0 is the wavenumber of the incident acoustic energy.

2. Frequencies are normalized by dividing by the Bragg frequency, which is given by:

$$\omega_B = \sqrt{2 g k_0 \sin\theta} \quad ; \quad \therefore \ \eta = \omega/\omega_B$$

3. Doppler spectra are made dimensionless by multiplication by ω_B.

4. Waveheight directional spectra $S(\underline{k})$ are normalized as follows:

$$Z(\underline{k}) = (2k_0 \sin\theta)^4 S(\underline{k})$$

Then for the first order, we have:

$$\underset{\sim}{\sigma}_1(\theta,\eta) = \pi/2 \cot^4\theta \sum_{m=\pm 1} Z(m\hat{x})\delta(\eta+m) \tag{6}$$

Thus the form of the first order Doppler spectrum should be two delta functions at $\eta = \pm 1$, the height of which should be equal to the magnitude of the normalized waveheight directional spectrum for $|\underline{K}| = 1$. This is the well-known Bragg reflection.

For the second order, we have:

$$\sigma_2(\theta,\eta) = \frac{\pi}{2}\cot^4\theta \sum_{m,m'=\pm 1} \int_{-\infty}^{\infty} K\,dK \int_{-\pi}^{\pi} d\varphi \,\sin^2\theta\, |\gamma_T|^2 Z(m\underline{K}) Z(m'\underline{K}') \,\delta(\eta - m\sqrt{K} - m'\sqrt{K'}) \tag{7}$$

The term γ_T is the coupling coefficient, which describes both the hydrodynamic and acoustic second order interactions at the surface. $\underline{K}$ and $\underline{K}'$ are normalized wave-vectors, where $\underline{K} = (K,\varphi)$ in polar co-ordinates. In addition, we have the Bragg constraint:

$$\underline{K} + \underline{K}' = -\hat{x} \tag{8}$$

By the use of (8) and the frequency constraint implicit in the delta function in (7), the integral in (7) can be simplified somewhat. It should be noted that the expressions given in (6) and (7) are the acoustic Doppler spectra which would be measured by a narrow-beam antenna looking in the direction given by $\hat{x}$.

THE MEASUREMENT SYSTEM

A side view of the instrument package, which has been deployed for the greater portion of the past year in 50 m of water in Patricia Bay, B.C., is shown in Figure 2 on the next page. The receivers are nine ITC 1001 omni-directional hydrophones arranged in a square 3 by 3 array on the top of the package. Pre-amplifiers for these provide a 60 dB gain at the nominal operating frequency of 400 Hz. To the left is the projector, an Argo Technology Model 201 low frequency omnidirectional source, with resonant frequency of 2 kHz. The projector is driven by a high power audio amplifier. The electronic components are housed in a large spherical pressure case. Power supply and communication to and from the package is through an armoured cable containing two copper power conductors and two optical fibers. The package is connected to a shore-based data collection and control facility housed approximately 1.5 km away in the Institute of Ocean Sciences (IOS), near Sidney, B.C.

After pre-amplification, received signals from each transducer pass through a programmable amplifier with gain programmable from 0 to 60 dB, and then to a synchronous quadrature demodulator. The programmable oscillator used in the demodulator also produces the driving signal for the projector. The resulting 18 signals then pass through a bank of programmable low-pass filters, and into a 12-bit analog to digital conversion system. All system components and functions are software controllable from the surface. There are two CompuPro MC68000 systems used to run the system, one in the instrument package, controlling the sampling sequence and data storage and transmission to the surface, and the other at the surface passing operating software and system parameters to the subsurface unit, and handling data reception and storage.

A typical sampling sequence proceeds as follows. First the projector is enabled, and a short sequence of cycles (typically 7) of the carrier frequency is transmitted. After some preset waiting period, the receivers are enabled, and at some later point digitization of data begins. Data points will continue to be digitized, separated in time by the number of cycles in the transmitted pulse, until a preset time is reached. All waiting periods, delays, etc. are specified in terms of the cycle time of the carrier frequency. This measurement cycle is repeated (typically at 4 Hz) until a specified number of pulses have been transmitted.

Naturally, the frequency resolution of the system is limited by the sampling scheme chosen. A fundamental limitation to the frequency resolu-

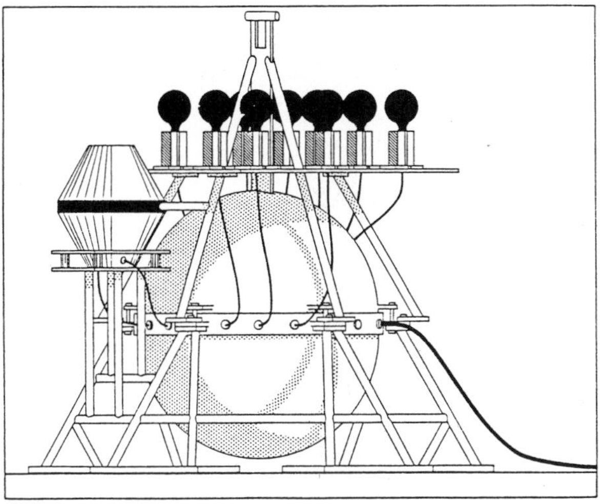

Fig. 2 Side view of Instrument package

tion is the short pulse length. Another contributing factor unique to the acoustic case is the effect of variations in the co-elevation angle, again due to the finite pulse width, but now because it is not an impulse. As we have seen in the previous section, the Bragg frequency is a function of the co-elevation angle, but the co-elevation angle is different for different parts of the transmitted pulse, and this results in an effective smearing of the returned Doppler spectrum. This effect is worse when the co-elevation angle is small. It is estimated that the combination of these two effects for a 17 ms pulse incident at 45 degrees will result in a "smearing width" of about 0.13 Hz.

EXTRACTION OF THE WAVEHEIGHT DIRECTIONAL SPECTRUM FROM THE DOPPLER SPECTRUM

Given any set of 4 receiving hydrophones arranged in a square (with our 3 by 3 array we have 4 of such sets), we can, by proper manipulation of the signals from these transducers, simulate an antenna with a beam pattern given by:

$$r(\theta,\psi,\phi) = \sin^2\theta \cos^4\left[\frac{\psi-\phi}{2}\right] \tag{9}$$

provided that the separation between the transducers is much less than the wavelength of the transmitted acoustic waves. The antenna is "tuned" for a specific co-elevation angle θ. The angle ψ is the angle at which the antenna is aimed, and the angle ϕ is the angle from which energy is being received. Then we can define the output of our system as:

$$\tilde{\sigma}(\eta,\theta,\psi) = \frac{\sin^2\theta}{2\pi}\int_{-\pi}^{\pi}\cos^4\left[\frac{\psi-\phi}{2}\right]\sigma(\eta,\theta,\phi)\,d\phi \tag{10}$$

where $\tilde{\sigma}(\eta,\theta,\psi)$ is the measured acoustic Doppler spectrum, and $\sigma(\eta,\theta,\phi)$ is the narrow beam Doppler spectrum (both first and second order) defined in (6) and (7) above. Now we can write the $\cos^4$ term as an exact truncated Fourier series:

$$\cos^4\left[\frac{\psi-\phi}{2}\right] = \sum_{n=-2}^{2} a_n f_n(\psi) f_n(\phi) \tag{11}$$

where
$$f_n(\phi)=\begin{cases}\cos n\phi & ; n=0,1,2\\ \sin(-n\phi) & ; n=-1,-2\end{cases} \quad \text{and} \quad a_n=\begin{cases}\cdot 125 ; & n=\pm 2\\ \cdot 5 ; & n=\pm 1\\ \cdot 375 ; & n=0\end{cases}$$

Now if we write our measured spectrum $\tilde{\sigma}$ as a truncated Fourier Series:

$$\tilde{\sigma}(\eta,\theta,\psi) = \frac{1}{2\pi}\sum_{n=-2}^{2} B_n(\eta) f_n(\psi) \tag{12}$$

then, by definition:

$$B_n(\eta) = \frac{2}{\epsilon_n}\int_{-\pi}^{\pi}\tilde{\sigma}(\eta,\theta,\psi) f_n(\psi)\, d\psi \quad \text{where} \quad \epsilon_n=\begin{cases}2 ; & n=0\\ 1 ; & n=\pm 1,\pm 2\end{cases} \tag{13}$$

and by inspection of (11) and (12) we can write:

$$B_n(\eta) = \sin^2\theta\, a_n \int_{-\pi}^{\pi} \sigma(\eta,\theta,\phi) f_n(\phi)\, d\phi \tag{14}$$

The B_n defined in (13) are the data products we use, and relation (14) allows us to get the waveheight directional spectrum from the B_n, since we have analytical expressions for the narrow-band Doppler spectra, given in (6) and (7) above.

MODELLING OF THE DOPPLER SPECTRA, AND PRELIMINARY MEASUREMENTS

As a first investigation into the expected form of the Doppler spectra, a model of the response of an omnidirectional sensor to various surface wave conditions was developed, using the expressions for the first and second order narrow-beam Doppler spectra shown in (6) and (7), and an antenna output given by:

$$\tilde{\sigma}(\eta,\theta) = \frac{1}{2\pi}\int_{-\pi}^{\pi}\sigma(\eta,\theta,\phi)\, d\phi$$

For the directional wave spectra, the Phillips non-directional spectrum shown below was used, with a directional factor proportional to $\cos^4$. Then the directional wave spectrum is given by:

$$S(k,\phi) = f(k)g(\theta) ; \text{ where } f(k) = \begin{cases}\frac{\cdot 005}{k^4} & ; k > k_c = u^2/g\\ 0 & ; k \leq k_c\end{cases}$$
$$g(\theta) = \frac{4}{3\pi}\cos^4\left[\frac{\phi^*-\theta}{2}\right]$$

ϕ^* is the principal direction of wave propagation, and 'u' is the windspeed. Figure 3 on the next page shows the effect of increasing windspeed on the spectrum. Note that the theoretical spectrum has been convolved with a Gaussian "smearing function" with -20 dB width of 0.1 Hz, to account for the finite frequency resolution of the system. Increasing windspeed has two effects: the energy in the second order spectrum increases, and the separation between the first and second orders decreases, so that eventually the first order peak will be lost in the second order. Because the area in which the instrument is deployed is strongly asymmetric with respect to depth and fetch, I then used a more complex spectral model, the JONSWAP non-directional spectrum and the Pierson-Neumann-James directional distribution. This model was successfully used by Seymour(1977) to predict wave spectra in areas of restricted fetch. An asymmetrical omnidirectional Doppler spectrum is predicted for some wind directions. The predicted spectrum for southwest winds is shown in Figure 4 on the next page, together with a measured spectrum taken when winds were blowing strongly from that direction. Note that the asymmetry is reproduced, and that the first and second order portions of the measured spectrum appear to be "smeared" together.

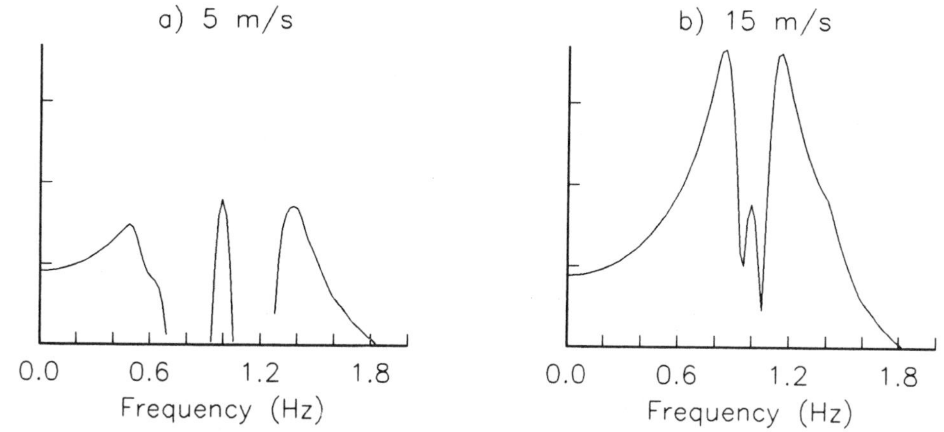

Figure 3. Effect of windspeed on omnidirectional spectra. Windspeed: a) = 5 $m.s^{-1}$; b) = 15 $m.s^{-1}$. These spectra are symmetric, so only the positive Doppler side is shown. Vertical axis is log of spectral power.

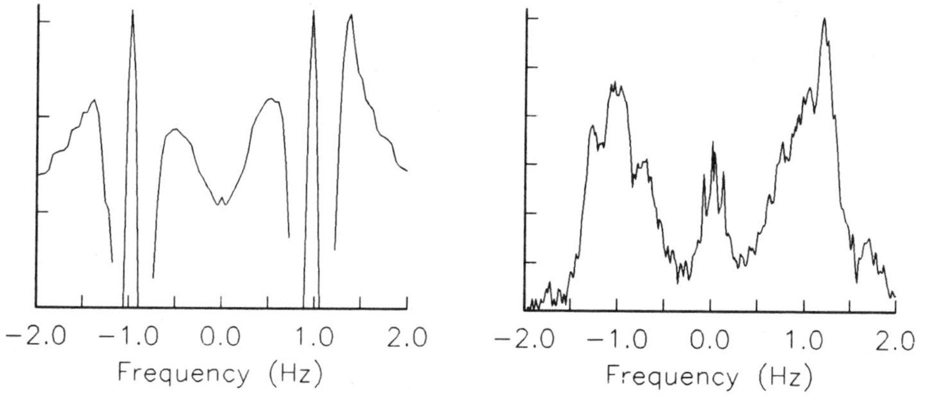

Figure 4. Predicted omnidirectional spectrum using JONSWAP model for southwest winds on the left, and a measured spectrum for the same wind direction on the right. Vertical axis is log of spectral power.

Finally, model B_n functions, as given in (14) above were computed, so as to gain some feeling for the expected variation in the form of these functions with changing wind direction. Figure 5 shows the computed B_n for $n = 0, \pm 1$ for southwest winds together with the same functions calculated from measured data. Although the measured functions are noisy, they show that the general shape of the computed functions is reproduced by the data. Despite the fact that these data are very preliminary, we can deduce some useful information about the surface wave spectrum (i.e. the rough principal direction of the waves) from them.

FUTURE PLANS

As well as making some improvements to the instrument to increase the signal-to-noise ratio, we will be replacing the projector with a unit capable of efficient operation at frequencies of as low as 50 Hz in the fall of this year. This should allow us to increase the separation of first

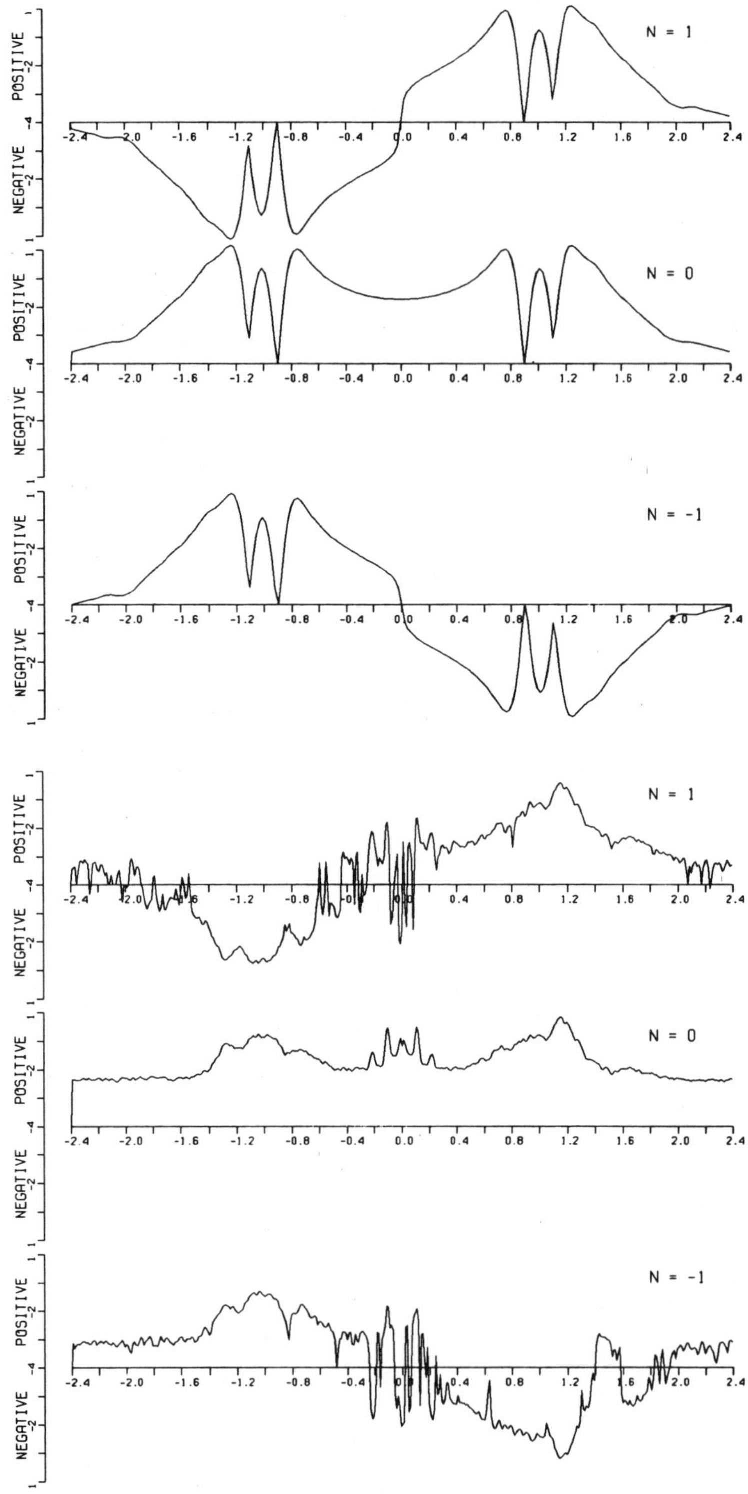

Figure 5. Top: computed b_n functions for n=0, ± 1 for southwest winds. Bottom: same functions calculated from measured data with similar winds.

and second order energy in the spectrum, making the extraction of waveheight directional spectral information much easier. After testing of the new configuration in Patricia Bay, we then will be deploying the instrument off the west coast of Vancouver Island, probably in the spring of 1987.

REFERENCES

Barrick, D.E. 1977. The ocean waveheight non-directional spectrum from inversion of the HF sea-echo Doppler spectrum. Remote Sens. Environment. 6. 201-227.

Barrick, D.E. and B.J. Lipa. 1979. A compact transportable HF radar system for directional coastal wave field measurements. in **Ocean Wave Climate,** edited by M.D. Earle and A. Malahoff. Plenum Press, New York.

Eckart, C. 1953. The scattering of sound from the sea surface. J. Acoust. Soc. Am. 25(3). 566-570.

Fortuin, L. 1970. Survey of literature on reflection and scattering of sound waves at the sea surface. J. Acoust. Soc. Am. 47(5). 1209-1228.

Harper, E.Y. and Labianca, F.M. 1975. Scattering of sound from a point source by a rough surface progressing over an isovelocity ocean. J. Acoust. Soc. Am. 58(2). 349-364.

Labianca, F.M. 1980. Estimation of ocean-surface directional-frequency spectra: the inverse problem. J. Acoust. Soc. Am. 67(5). 1567-1577.

Lipa, B.J. and D.E. Barrick. 1986. Extraction of sea state from HF radar sea echo: Mathematical theory and modelling. Radio Science 21(1). 81-100.

Roderick, W.I. and B.F. Cron. 1970. Frequency spectra of forward-scattered sound from the ocean surface. J. Acoust. Soc. Am. 48(3). 759-766.

Seymour, R.J. 1977. Estimating wave generation on restricted fetches. J. Waterway, Port, Coastal and Ocean Division, Proc. Am. Soc. Civil Eng. 103(WW2). 251-264.

Weber, B.L. and D.E. Barrick. 1977. On the nonlinear theory for gravity waves on the ocean's surface. Part 1: Derivations. J. Phys. Oceanogr. 7(1). 3-10.

Williams, R.G. 1973. Estimating ocean wind wave spectra by means of underwater sound. J. Acoust. Soc. Am. 53(3). 910-920.

ACKNOWLEDGEMENTS

I would like to thank the Science Council of B.C. and the National Engineering and Science Research Council for their support. Partial support for this project is from the Panel of Energy Research and Development, Project No. 67132. Instrument development and deployment has been carried out by Arctic Sciences Ltd. of Sidney, B.C., supported by a DSS-UP contract. David Lemon of Arctic Sciences Ltd. and Dr. David Farmer of IOS have made helpful suggestions and criticisms.

A PATTERN RECOGNITION APPROACH TO REMOTE ACOUSTIC BOTTOM CHARACTERIZATION

Thomas L. Clarke and John R. Proni

Ocean Acoustics Division-AOML-NOAA
4301 Rickenbacker Causeway
Miami, FL 33149

ABSTRACT

The possibility of extracting useful bottom information from reflected pulse waveforms at customary echo-sounding frequencies has been demonstrated experimentally. The one-dimensional nature of the sediment property continuum should also enable the use of remotely measured acoustical sedimentary properties to predict navigationally important mechanical characteristics.

A convenient mathematical model has been developed to assess the effects of bottom roughness and material properties on bottom echo shape. The physical basis of the model is explained and model output is presented.

The ability of this model to easily generate sample echoes from a wide range of bottom types permits a pattern recognition approach to be taken to the problem of extracting information from the echo signals. An adaptive algorithm can be "trained" using model generated echoes in the same way speech recognition systems are "trained". The linear discriminant algorithm can be trained to distinguish mud from gravel, but has difficulty with mud versus fine sand. A commercially available voice recognition system conversely has difficulty distinguishing sand from gravel. More sophisticated algorithms will be needed for general bottom discrimination.

INTRODUCTION

The possibility of extracting useful bottom information from reflected acoustic pulse waveforms has been demonstrated by many experimenters. Dodds (1984) bases bottom type discrimination on analysis of frequencies in the range 1 to 10 kHz generated by an impulsive sound source. Meng and Guan (1984) utilize higher frequencies in the 100 kHz range to establish statistical discriminators. Reliable discrimination in this higher range offers the possibility of adding remote acoustical bottom characterization to a conventional echo-sounder.

The ability to remotely determine bottom types acoustically depends on the acoustical properties of the sediments. In particular, the bulk sound speed, the bulk attenuation, and the bulk scattering properties are the major factors affecting echo formation. While these acoustic properties have intrinsic interest for many applications, the navigationally significant quantities such as sediment shear strength are mechanical. It is the mechanical properties that determine the ability of the bottom to impede the motion of shipping. Fortunately, marine sediments tend to form a one-dimensional continuum (Bachman, 1985). That is, acoustic properties are well correlated with sediment grain size which is in turn correlated with mechanical properties. Thus, measurements of bulk acoustic properties provide a means to effectively estimate the strength of sediments.

THE MODEL

A convenient mathematical model has been developed (Clarke et al, 1986) to assess the effects of bottom characteristics such as surface roughness and material properties on bottom echo shape. This model incorporates the effects of scattering from surface roughness, and of volume scattering from within the volume of the sediment.

Standard echo-sounding geometry is assumed in the model. The acoustic transducer which emits sound of wavelength λ is located distance D (the depth) from the sea bed. The bulk acoustic impedance of the sea bed enters the model through the bulk reflection coefficient R for sound incident on the bed from water. The roughness of the bed is described by an RMS height variation h and a correlation length L. This statistical roughness reduces the magnitude of the coherent bottom echo component and scatters energy into a reverberant tail whose decay time is determined by a combination of sea-bed slope, h/L, depth D, and transducer

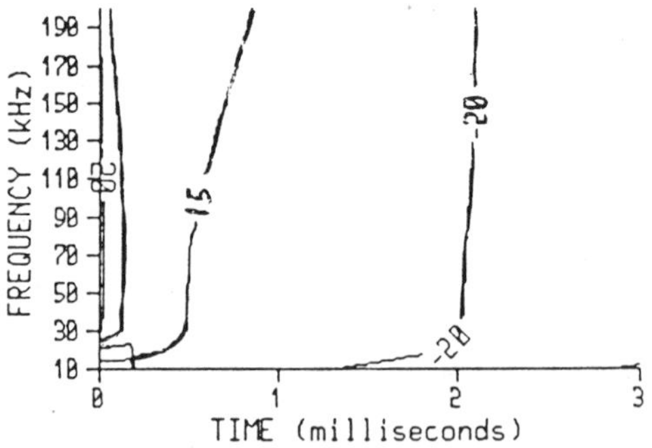

Figure 1: Sonogram for sandy bottom (400 microns) and surface roughness 1 cm and correlation length 10 cm. Contour lines are intensity in dB relative to perfect reflection.

beam angle. The factor by which the coherent echo is reduced is $\exp(-\phi^2)$ where $\phi=4\pi/\lambda$ is the nondimensionalized surface roughness. For roughness much greater than a wavelength, $\phi>>1$, almost all the echo energy is reverberant.

The fraction (1-R) of sound that penetrates the sea bed is scattered from the granularity of the sediment and is also rapidly absorbed by viscous losses (Biot absorption). The sound scattered from beneath the sea bed is highly frequency dependent. The OAD model uses Rayleigh theory which predicts a fourth power dependence on frequency for the volume scattering by the sediment. Empirical values are used for the Biot absorption. Jackson et. al. (1986) present a similar model that uses a somewhat different surface scattering model and empirical values for volume scattering and absorption.

Figure 1 shows model results in the form of contour plots of acoustic intensity versus time and frequency. Time is measured in milliseconds from the first echo return. The case shown is for a sandy bottom (400 micron diameter) and for h=1 cm and L=10 cm. These parameters would correspond to a gently rippled bottom. For low frequencies, less than 20 kHz, the .2 msec transmitted waveform is mirrored in the contours of the echo waveform. The mirrored pulse rapidly diminishes above 20 kHz as ϕ increases, and the echo becomes a reverberant tail approximately 1 msec long.

Figure 2 shows model output for 200 micron sand and a smooth bed, h=1 cm. The echo for this case is very similar to that for the coarse sand at high frequencies where volume scattering dominates. The mirrored transmitted pulse is visible at all frequencies because of the reduced roughness and only drops off a little at 200 kHz; surface scattering effects are negligible at all but the highest frequencies.

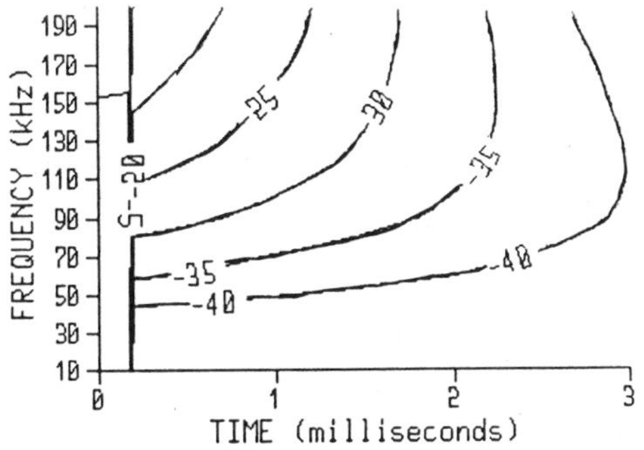

Figure 2 Sonogram for fine sandy bottom (200 microns) and surface roughness 1 cm and correlation length 10 cm. Contour lines are intensity in dB relative to perfect reflection.

Figure 3 shows model output for a smooth bottom and muddy sediment, 50 micron grain diameter. At low frequencies, the smooth sand and mud echoes are nearly identical since volume scattering is small. The big differences between the echoes from the two sediments are in volume scattering reverberant tails. The granularity of the sand produces a large scattering tail. It is this tail that is the basis for discrimination of bottom types.

RECOGNITION TRAINING

The ability of this model to easily generate sample echoes from a wide range of bottom types permits a pattern recognition approach to be taken to the problem of extracting information from the echo signals (Duda and Hart, 1973). In general terms, an algorithm for discrimination between echoes will be based upon some (possibly vector) parameter P. Starting with some initial parameter Value P_0, the pattern recognition algorithm is "trained" to regognize the echoes by determining a succession of values of the Parameter $\hat{P}_i$, i=1..., based on the success of the algorithm on the ith echo.

Perhaps the simplest discrimination algorithm uses linear discrimination. That is $\hat{X}$ is the vector of echo data, the parameter vector P, then the data is classified according to the sign of $\hat{X}.\hat{P} = \geq X(f,t)P(f,t)$ where f is the frequency of the echo and t is the time at which the echo amplitude is measured.

The training algorithm for linear discrimination is simple. Start with an initial value $\hat{P}_0$, then for the set of data $\{\hat{X}\}$, let $\hat{P}_i=\hat{P}_i+\hat{X}_i$ if the echo is misclassified, that is if the sign of $\hat{X}.\hat{P}$ is wrong. If the set of echoes can be correctly classified on the basis of a linear discriminant, then it can be shown that this algorithm converges to a parameter P that correctly classifies the echoes.

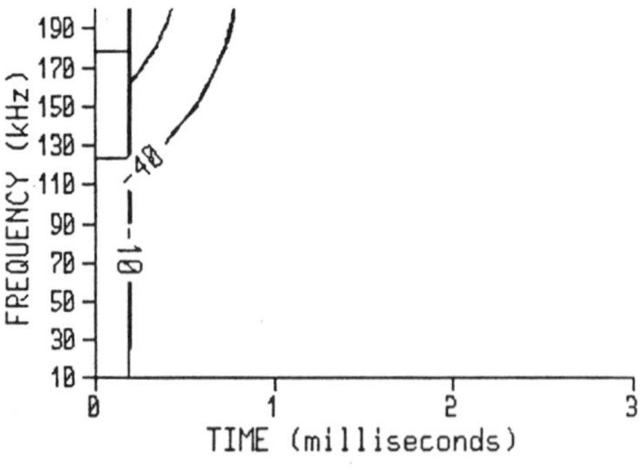

Figure 3 Sonogram for muddy bottom (50 microns) and surface roughness 1 mm and correlation length 10 cm. Contour lines are intensity in dB relative to perfect reflection.

For the echo-sounding case a computer program for testing recognition was written in which the echo vector $\hat{X}$ had thirty components. These were samples of echoes at the five different frequencies 10, 20, 40, 80, and 160 kHz at 6 different delay times beginning with the beginning of the echo waveform and equally spaced from 0 to 1 msec. This 30-vector was then used to train a linear discriminant algorithm. The set of training data were generated by combining preceding model produced predictions with random values of surface roughness, h, ranging from 0 to 10 cm, and with random noise added to the waveforms to produce a realistic test.

The linear discriminant algorithm converged when the criterion was correct classification of mud versus gravel. The parameter vector $\hat{P}$ was essentially zero except for 200 usec delay and 80 kHz and 160 kHz. The discriminant was approximately

$$\hat{X}.\hat{P} = X(160kHz,200\mu sec)-0.25X(80kHz,200\mu sec).$$

This agrees generally with the conclusions of Meng and Guan (1984).

When the criterion was correct discrimination between fine sand and mud, however, the aogorithm failed. This indicates that it is not possible to discriminate linearly between these two sediment cases for all possible combinations of bottom roughness in the presence of noise.

To test the possibility of using special purpose pattern recognition hardware to perform the discrimination, the model output was translated into the audio band by inputting calculated amplitudes to a digital to analog convertor at a clock rate reduced by a factor of 32. For a 10 kHz to 200 kHz sonogram, the resulting "audio echo" will be in the range 300 Hz to 6000 Hz. Model echoes for various sediments played in this fashion have sounds distinguishable by the human ear.

These echoes were then played into an Interstate Voice Products SRB-LC voice recognition circuit board installed in a personal computer. The software supplied with the board was utilized to "train" the board to recognize three classes of echoes: gravel, sand, and mud, corresponding to figures 1, 2 and 3. When echoes having different roughnesses were presented to the SRB-LC board, recognition was largely successful for the sand and mud cases, although discrimination between sand and gravel was not as reliable. This is, nevertheless, a remarkable performance for a small circuit board (14 ICs with supporting discrete circuitry) mounted in a garden variety personal computer.

CONCLUSIONS

An adaptive algorithm can be "trained" using model generated echoes in the same way speech recognition systems are "trained". The linear discriminant algorithm can be trained to distinguish mud from gravel, but has difficulty with mud versus fine sand. A commercial speech recognition circuit board was able to reliably discriminate sand from mud but had more difficulty with gravel.

These results suggest that a more complex algorithm will be needed to remotely recognize the full range of ocean bottom types. The simplicity of the voice recognition board is encouraging, however, so that is should be possible to incorporate bottom recognition capability into echo-sounders. As suggested by Clarke et. al. (1985) an acoustical definition of bottom depth may be possible through the use of specification for echo detection based on realistic bottom echoes from a variety of bottoms. The ability of the voice recognition board to discriminate various bottoms gives support to this suggestion.

REFERENCES

Bachman, R.T., 1985: "Acoustic and Physical Property Relationships in Marine Sediment," J. Acoust. Soc. Am., 78, 616-621.

Clarke, T.L., J. Proni, S. Alper, and L. Huff, 1985: "Definition of 'Ocean Bottom' and 'Ocean Bottom Depth'", Proceedings IEEE Oceans '85, 1212-1216.

Clarke, T.L., J. Proni, D. Seem and J. Tsai, 1986: "Joint CGS-AOML Acoustical Echo-Formation Research I: Literature Search and Initial Modelling Results," NOAA Tech Memo AOML-ERL.

Dodd, D.J., 1984: "Surface and Volume Backscattering of Broadband Acoustic Pulses Normally Incident on the Seafloor: Observations and Models," J. Acoust. Soc. Am., 75, Suppl 1, S29.

Duda, Richard O., and Peter E. Hart, 1973. Pattern Classification and Scene Analysis, Wiley, New York, 482p.

Jackson, D.R., D.P. Winebrenner, and A. Ishimaru, 1986, "Application of the Composite Roughness Model to High-Frequency Bottom Backscattering," J. Acoust. Soc. Am., 79(5), 1410-1422.

Meng, J. and D. Guan, 1984: "Acoustical Method for Remote Sensing of Seafloor Sediment Types," Institute for Acoustic Research, Peking, unpublished manuscript.

COMPARATIVE NUMERICAL STUDY OF VLF SIGNAL PROPAGATION CHARACTERISTICS FOR OCEAN BOTTOM AND MARINE BOREHOLE ARRAYS

G. J. Tango, H. B. Ali and M. F. Werby

Naval Ocean Research and Development Activity
NSTL, Mississippi 39529-5004 USA

ABSTRACT

Preliminary results of a numerical investigation of VLF propagation to ocean-bottom and buried sensors are presented. Using an environmental model representing deep ocean basin conditions, both single and multifrequency seismo-acoustic propagation are simulated by means of a new Fast-Field Reflectivity Program. Factors determining correct environmental model selection by using numerical results for sensors (hydrophone and geophone) in vertical and horizontal arrays are discussed. Some implications for VLF propagation of (sub)bottom mode trapping, P to S wave conversion, and multiple reverberations are considered. It is shown that additional studies are required to better identify both signal and noise mechanisms in long-distance VLF propagation.

INTRODUCTION

The behavior of seismo-acoustic propagation, particularly that induced by waterborne signals, is of considerable interest in a variety of fields that range from commercial oil exploration to ocean reconnaissance. Until recently the underwater acoustic community has, in large part, confined its attention to waterborne/shallow sediment propagation, and has neglected deeper subbottom effects. Regarding the latter, some evidence suggests that an increase in both SNR and understanding of VLF propagation phenomena may be overlooked if subbottom propagation is not considered. This paper discusses the preliminary results of an examination of one aspect of this problem: the relative reception responses of ocean bottom and subbottom sensors.

Numerous seismic propagation experiments using diverse sensor deployments in deep land and marine boreholes have been used to determine shallow sedimentary and crustal structure (Stephen, 1977). However, despite improved environmental data obtained from these measurements, the precise number and comparative magnitudes of environmental, experimental and wave-theoretical factors controlling long-range seismo-acoustic propagation, both within and from deep sea bottom, remain unclear. To examine some of the possible causes for this complexity, with reference to propagation, this paper reviews initial results of numerical investigations, using a new Fast-Field Reflectivity Program (Schmidt and Tango, 1986; Tango and Schmidt, 1985). Both single and multifrequency propagation simulations are considered for a representative deep-sea hemipelagic sediment velocity/depth profile. In the following, we examine factors governing correct environmental model selection and parameterization, and corresponding numerical results for hydrophones and geophones in vertical and horizontal arrays.

ENVIRONMENTAL MODEL

The environmental model used (Fig. 1) is based on regional U.S. Geological Survey and Deep Sea Drilling Project results, which represent deep ocean basin conditions typical of the Mediterranean Sea and the Indian Ocean. The water column is approximately 5000 m and is characterized by a shallow, weak, velocity inversion, followed by a gradual, positive, sound-speed gradient approaching the seabottom. Near-bottom zones of complex stratigraphy were retained rather than arbitrarily smoothed or deleted. Discretization using a spatial sampling of one-quarter shear wavelength was used to step-wise represent the indicated semicontinuous P and S velocity gradients. As in many deep-sea environments, geoacoustic parameters characterizing shallow sediment layers are known only within a wide range of uncertainty. Thus, the precise values of S wave velocity, and both P and S wave attenuation as a function of depth within the first 0.5 km of semiconsolidated sediments, are treated as weakly differentiated zones of approximately mud(stone) properties. Deeper basement levels below borehole total depth are derived from marine ophiolite cross sections that typify young ocean crust. Regional data and reported empirical parameterizations (Hamilton, 1980) were used to assign ranges of minimum and maximum P and S wave velocity and attenuation.

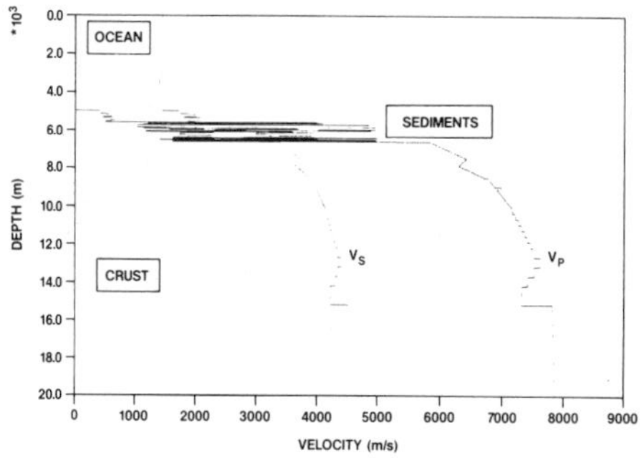

Figure 1. Young ophiolite velocity/depth profile used in numerical modeling.

For deep-water propagation, particularly to a shallow (near-surface) receiver location, the fine scale details of the elastic parameter depth distributions are of secondary importance in the VLF regime (5–25 Hz). In fact, such features, in general, are seismically unobservable at these frequencies. As a result, it is sometimes permissible in single frequency calculations of VLF reflection or transmission loss from thick sediments to simulate the net loss from shear rigidity by a two- to four-fold increase in compressional attenuation (Tunnell and Tango, 1985). However, for nearbottom or subbottom sensor arrays, serious errors can arise from neglect of the acoustic impedance microstructure within the first few P wavelengths of depth below the sea floor. This error occurs because both the number and the amplitude of compressional diving-head waves and Scholte-Rayleigh waves are governed by the near-bottom (particularly S wave) velocity and attenuation gradients. Detailed amplitude and phase behavior of in-bottom or bottom-returned signals is likewise controlled by the complexity of subbottom stratigraphy, as measurable by spatial autocorrelation and power spectral distributions. Thus, if multifrequency sources and near- or in-bottom propagation are of interest, then it may be insufficient to examine only the sensitivity of plane wave reflection coefficients and single frequency transmission loss. Since many individual small stratigraphic features are seismically unobservable at 10 Hz, the largest changes in layer thickness, depth, or elastic parameter value must be determined for the net layer complex, which would leave the recorded multifrequency signal unchanged for the case of on-bottom or in-bottom receivers.

Experimental Parameters for Numerical Modeling

A center frequency of 10 Hz is used in all CW and pulse calculations presented (the input pulse has a smooth power distribution from 0 to 25 Hz). The propagation behavior in this frequency band is critical for exploration and crustal seismic surveys, as well as in general studies of meteorologic and oceanographic background noise (Wagstaff, 1981). Because of the considerable interest in the relative performances of (omnidirectional) hydrophones and (dipolar) triaxial geophones (Brocher et al., 1982) as sensors of both waterborne and bottomborne signals, both synthetic normal stress (pressure), and vertical and horizontal particle velocity responses were considered. Because of the vertical depth dependence of the significant losses expected from interlayer reverberation, mode conversion, and/or scattering at the deeper bottom strata, a synthetic vertical seismic array through the bottom supplements the horizontal ocean-bottom hydrophone/geophone array.

Numerical Results

Fig. 2 shows transmission loss (in normal stress) plotted as range-depth contours, which result from a source located 100 m below the free surface. The loss levels were computed using the ''SAFARI'' FFP algorithm with 4096 discrete range points and 200 depth points. In the figure, low and high transmission loss zones are represented, respectively, by light and dark shading. Horizontal ranges extend from 0 to 50 km and vertical depth from 0 to 6000 m. The figure qualitatively suggests that at most ranges, the first 1000 m or so of bottom produces a higher loss/lower return than the water column at comparable ranges. In fact, examination of two individual transmission loss curves (Fig. 3) from receiver depths of 4973 m (sea floor) and 6629 m (basement) reveals basement levels of the order of 20 dB below sea-floor levels. Fig. 2 suggests one or more instances of lower-loss energy channelling at depths corresponding to elastic waveguides, defined by depth-adjacent high velocity layers. In contrast to the shallow-water case (generally shallow water depth of <10 P wavelengths), refracted and reflected normal mode propagation through the deep ocean dominates near-grazing seismic contributions from bottom interaction, for this environment (see Fig. 4).

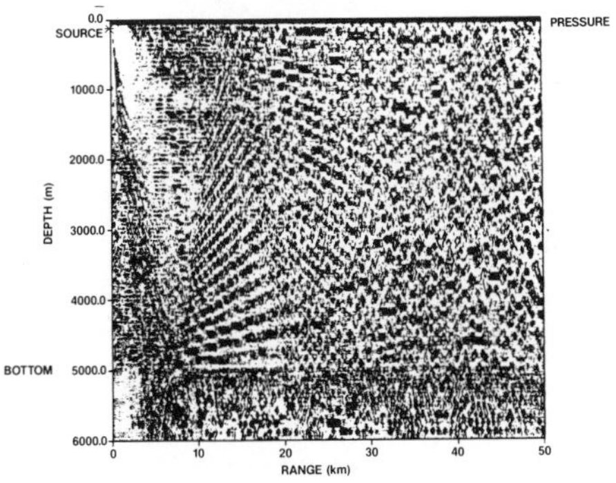

Figure 2. Range-depth contoured transmission loss (f= 10 Hz, Zs = 100 m) for deep ocean environment.

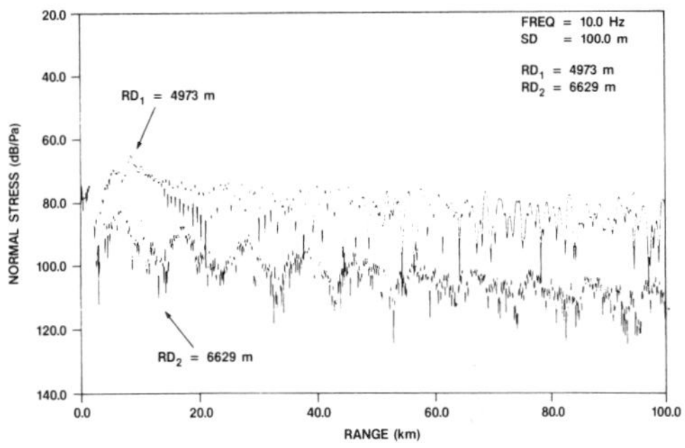

Figure 3. Comparative transmission loss (pressure) for seafloor (4973 m) and subbottom (6629 m) receivers.

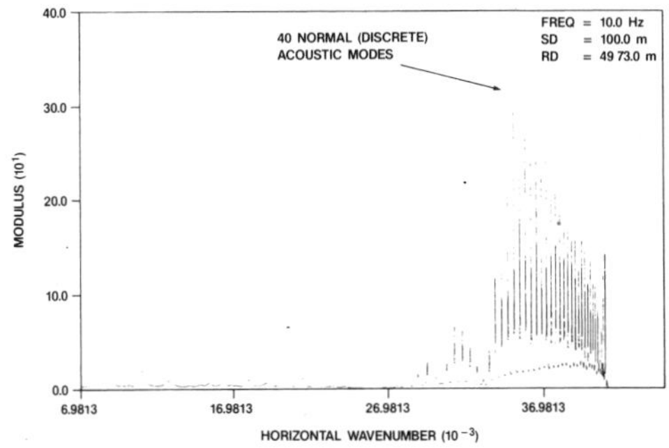

Figure 4. Full wave-theoretical Green's impulse response function, f= 10 Hz, source depth = 100 m, receiver depth = 4973 m (seafloor) computed by SAFARI FFP.

Fig. 5 represents (normal stress) transmission loss as a function of depth below the sea floor, through the sediment stratigraphy, to the basaltic basement. Here, the source-to-receiver separation ("offset") is 0 km. The depth parameter was sampled at 35-m increments, which represents approximately 0.25 of a compressional wavelength ($\lambda_P/4$). Although an overall increase in loss level with depth is evident (roughly following geometric spreading loss), this trend is strongly modulated by many localized oscillations of either sign, with transmission loss deviations up to 25 dB, especially in the shallowest sediments. This oscillatory behavior with depth of complex pressure and displacement has been previously noted (Gupta, 1965; Roden, 1968) as due to wavelength-dependent wave interference phenomena, which occur in the subbottom no less than in the water column. However, as shown by exploration borehole studies, signal loss/depth behavior cannot be accurately defined without a sufficiently dense spatial sampling of sensors, since complex subbottom acoustic impedance distributions will generally act as a complex bandpass filter to further complicate signal amplitude and phase behavior of both normal modes and body waves. The dangers of interpolative errors with respect to this distribution can be seen by comparing the previous loss/depth curve (for $\Delta z = \lambda_P/4$) with that in Fig. 6, which is based on a depth sampling interval of $\lambda_S/4$ (~ 1 m).

Further complications in this transmission loss/depth pattern versus range result from intrinsic changes in mode number and intermodal energy distribution, as well as from offset-dependent reflection coefficient behavior and other wave types generated by interaction with local layering (Spencer et al., 1977). Thus, conclusions concerning the depth-dependent signal reception (amplitude) response of either hydrophone or geophone sensors, notably in complexly stratified sea-floor intervals, can be erroneous if based on limited undersampled field measurements. To achieve interpolative sampling accuracy in modeling signal loss with depth, it is generally necessary to (1) impose arbitrary smoothing of single-frequency data during reception, based on assumed propagation mechanisms for "signal" and "noise"; (2) average experimental depth/range measurements, based on one or more interreceiver spacings and array apertures; and/or (3) consider only homogeneous or sufficiently simple geological environments (with a minimum of near-bottom noise-generating features). A judicious combination of these three options is probably the most realistic approach, and is widely employed in borehole seismic array assessments (e.g., for determining in situ properties such as attenuation) by using vertical seismic profiling (Hardage, 1984; Spencer et al., 1977).

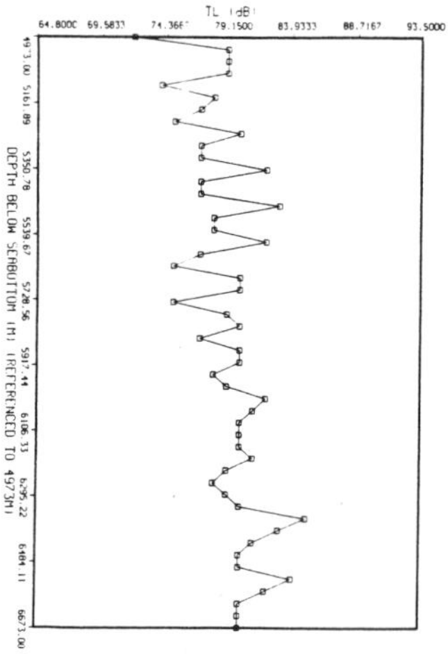

Figure 5. Transmission loss versus depth below seafloor (pressure) (f = 10 Hz).

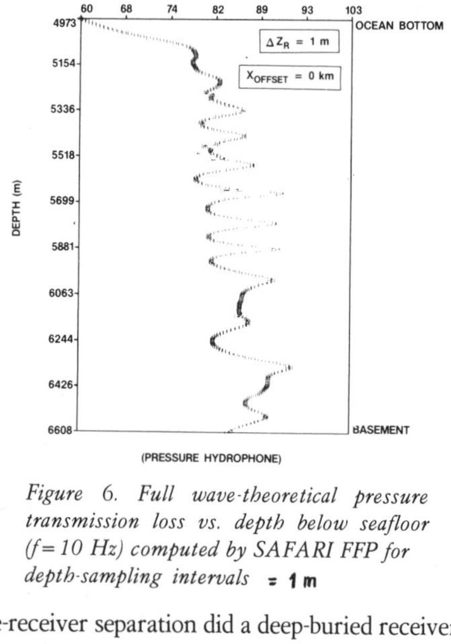

Figure 6. Full wave-theoretical pressure transmission loss vs. depth below seafloor (f = 10 Hz) computed by SAFARI FFP for depth-sampling intervals = 1 m

For this geology, it was found that at no source-receiver separation did a deep-buried receiver (either hydrophone or geophone) provide any clear or consistent signal loss minimization for bottom-based receivers (Tango and Ali, in preparation), arising from lithologic amplification due to both offset-dependent reflection coefficient and wavelength-dependent wavefield interference. Regarding these effects, deeper-penetrating propagation through several complex stratigraphic zones of comparatively high specific attenuation will work to significantly complicate, if not decrease, net signal versus depth. Thus, any buried sensor signal enhancement, if it exists as such, for the environment considered must be found at shallower depths, where it may still be true that SNR gain can be achieved by reducing noise level. For sources between 0 and 100 km, a burial depth of anywhere up to approximately 50–300 m below bottom is dictated for this environment, given the e-folding (exponential decay) depths for direct-arriving (10 Hz) normal modes, as shown by examining the Green's function in horizontal wavenumber as a function of depth below sea floor (Fig. 7).

An important conclusion is that, although for a given frequency (depth), a number of peaks and nulls exist in signal amplitude as a function of depth (frequency), for a realistically stratified ocean bottom it would be fortuitous, if not impossible, to experimentally localize and exploit these, solely on the basis of a limited number of single-frequency signal measurements. Viewed in a multifrequency context, signal spectral amplitude and phase behavior with depth is clearly a function of the number and the complexity of the "ringing tail" or coda of secondary arrivals following the direct wave (which are dependent upon the number, thickness, and acoutic impedance variability versus depth of the bottom layers).

Some of these complexities are further shown in Fig. 8, the synthetic seismogram corresponding to the levels predicted for hypothetical hydrophones emplaced 100 m below bottom (a depth, as previously indicated, at which minimal net signal loss occurs for 10 Hz). The results are here shown in velocity-normalized or reduced time format, which has the advantage of effectively compressing a large time series length into a graphically smaller time window. Fig. 8 represents the complete wavefield in a phase velocity window of 100–10000 m/sec. The direct water wave arrival at particular near-field ranges shows the effects of interference with subsequent interbed multiple reflections, P-S conversions generating a caustic, and head waves from the first few subbottom layers (thereafter separating in apparent velocity and decaying smoothly to a small fraction of its original amplitude). As McPherson and Frisk (1980) have shown for shallow water, the interbed seismic phases in the subbottom generally decay more rapidly (and in more complex fashion) in range than the direct water arrival. For this environmental model, deeper crustal arrivals, if present, are well below ($\sim$10–30 dB) direct water wave amplitudes at all pertinent ranges, as shown schematically in Fig. 9. Free ocean surface interactions give rise to several orders ($N > 5$) of nonnegligible water-bottom reflected multiples, persisting at and beyond ranges in excess of 50 km. The first few water-bottom multiples repeatedly interact with the bottom at delay times equal to integral multiples of the two-way travel time, and thereby give rise each to its own complex coda following the first arrivals. Each multiple-generated coda shows the same source-

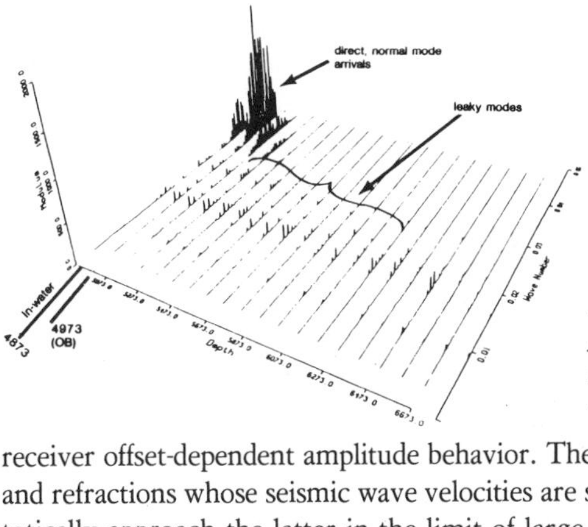

Figure 7. Full wave-theoretic Green's impulse response functions in horizontal wavenumber ($f = 10$ Hz) computed by SAFARI FFP for multiple subbottom receiver depths.

receiver offset-dependent amplitude behavior. These coda essentially consist of subbottom caustics and refractions whose seismic wave velocities are sufficiently close to that of water, so as to asymptotically approach the latter in the limit of large offsets. These water-bottom multiple-generated complexes clearly constitute important contributions to net signal strength as measured by both

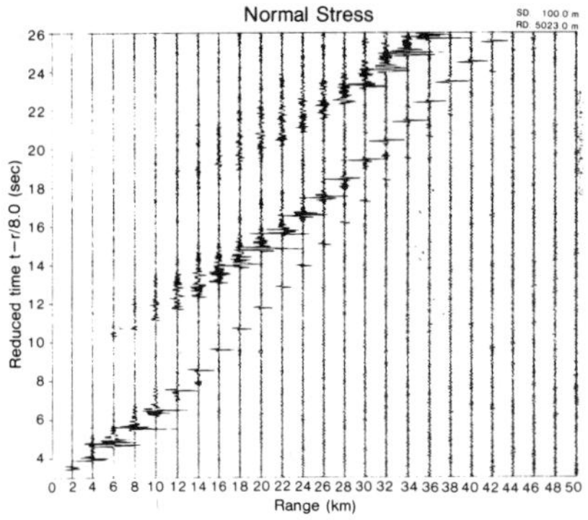

Figure 8. Complete reflectivity synthetic seismogram in normal stress (pressure) ($f_{peak}=10$, bandwidth = 0–25 Hz, source depth = 100 m, receiver depth = 5023 m = 50 m below bottom) computed by SAFARI for 25 offsets (x = 0–50 km).

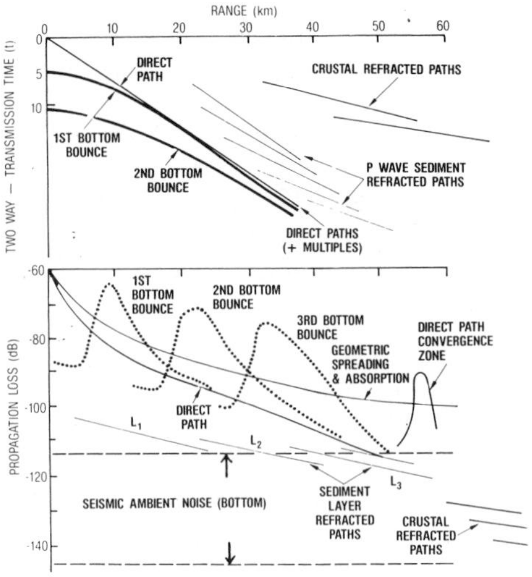

Figure 9. Wavefield schematic of time/range response, and pressure/range components.

buried hydrophone and geophone receivers, even at subbottom depths corresponding to a considerable fraction of a P wavelength for individual ranges up to ~20 km.

To identify the geologic layers giving rise to particular arrivals and the subsequent propagational behavior of these arrivals, both exploration and deep crustal seismic reconnaissance have underscored the importance of supplementing, where possible, the time-range information of a horizontal seismic profile with time-depth data from a borehole-measured vertical seismic profile. Fig. 10 shows the synthetic vertical seismic profile (SVSP) response in vertical particle velocity at a single, near-vertical incidence, source-receiver offset range. In addition to the amplitude-decaying direct water wave arrival, strong reverberant complexes are graphically seen generated from several alternating high- and low-velocity subbottom layers, which correspond to components of the water-bottom multiple-generated complexes above. Comparison of hydrophone and geophone responses shows that a considerable fraction of upcoming arrivals comprise P-S converted modes, in addition to interbed P multiples despite the small incident angles subtended. As source-receiver offset separation increases (not shown), the net SVSP-recorded amplitudes of both ocean bottom and interbed multiples increase sharply, and eventually (for x > 10 km) exceed the amplitude of the direct water wave as well as singly reflected waves. Despite a theoretically expected lower maximum frequency content due to increased attenuation from increasingly oblique raypaths through the bottom, the reverberation, conversion, and refracted arrivals still maintain a "noisy" high-frequency interference character that is indicative of the complicated nearbottom stratigraphy of their generation.

Because only a single-point impulsive source located many wavelengths above the sea floor is used here (thus precluding significant seismic interface wave generation), the observed noise on the synthetic seismograms is a combination of (1) signal-generated noise (with the same apparent velocity, but differing amplitude and phase versus frequency behavior than the source pulse) due to bottom layer interaction, and (2) numeric temporal wraparound aliasing, endemic to reflectivity-type synthetic seismic modeling algorithms. The wraparound largely comprises multiply reflected arrivals that fall outside the maximum computational window allowed by the frequency sampling interval selected, and thus alias at times before the first arrival (carrying over to mix with true arrivals). Although true deep-sea ocean-bottom seismo-acoustic ambient noise generally shows far more complex and varying statistics than simply delayed-damped-interfering signal replica-

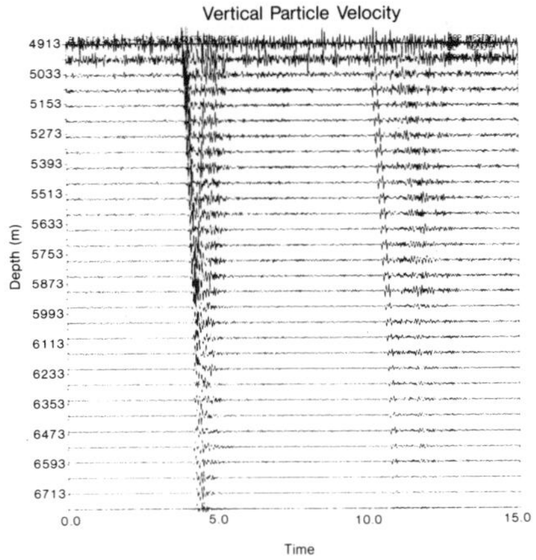

Figure 10. Synthetic vertical seismic profile (vertical particle velocity), ocean bottom to 2000 m subbottom depth.

tions, the range-depth amplitude behavior of signal aliasing could be taken to roughly approximate one possible kind of signal-generated noise, as might be found in a very noisy reverberant environment, due (e.g.) to rough-bottom topography. The variability in effectiveness of signal-generated noise suppression due to locating pressure and horizontal particle velocity sensor arrays in the (sub)bottom is seen by comparing pressure and particle velocity SVSPs (not shown). In both cases, the greatest noise reduction occurs only within the shallowest subbottom depths, as might be expected from the locally high values of P and S wave attenuation used to characterize the shallowest bottom depths. A somewhat lesser problem with near-vertically propagating arrivals arises on the hydrophone as compared to the geophone vertical array. However, as is also reasonably expected, it is not clear that this reduction in signal-generated noise is any less variable than the depth-dependent amplitude decrease of the signal. Even below a subbottom depth approximating an e-folding decay depth for the direct water arrival (or, more precisely, its direct wave root (Stephen and Bolmer, 1984)), both reverberant and P-S converted signal-generated codas are generated by the large-amplitude downgoing noise in the first few hundred meters of subbottom. It may be that an extremely variable site-dependent mixture of ambient and signal-generated noise could partially explain some of the often-conflicting results of field experiments seeking to assess the relative S/N merits of vertical versus horizontal arrays of diverse sensors (using only a gross or incomplete statistic).

DISCUSSION

The above (preliminary) simulations suggest that whereas direct wave/normal mode signal amplitudes decrease almost steadily in depth independently of geology, the net signal level, defined in terms of both the direct arrival and its signal-generated coda, can exhibit highly complex geology-dependent amplitude and phase behavior in both range and depth. More quantatitive and precise prescriptions on optimal sensor locations and types must be preceded by use of enhanced array-signal processing capabilities, together with more advanced propagation modeling algorithms. In particular, to assess experimentally controlled reception of net signal and noise requires computation of statistical figures of merit from synthetic seismic algorithm outputs used as inputs to a selection of candidate array/signal processing techniques and implementations, for both specific arrival types and net arrivals from specific layers in more realistic (range dependent) geologies. Recent analysis has shown the viability of supplementing Fast-Field Reflectivity results with those of finite difference, asymptotic ray, and generalized ray-reflectivity algorithms (Ursin et al., 1986). Such supplementary modeling capabilities are necessary because the results of different combinations of signal/array processing will be markedly different according to the number and kinds of arrivals considered. This difference is due, not only to differing apparent velocity and frequency power levels, but also to widely different wave velocities, directions, and particle motions—all of which are clearly functions of both the type and the path of seismo-acoustic arrivals. The failure of many investigations to address these questions is probably responsible for results of horizontal and vertical arrays showing no clear VLF reception advantage, since here both signal and noise may overlap in apparent velocity, frequency, and/or propagation mechanism and path.

These provisional results do not suffice for particular recommendations or site-independent general conclusions, but suggest possible sources of conflicting results of some previous marine bottom and borehole reception experiments. It is probable that any net signal enhancement, via ambient noise level reduction, to be gained by lowering receivers into deep-sea subbottom boreholes may be evident only (1) when noise components decrease with depth more rapidly than signal amplitude decreases (a local function of absolute noise level and geology); (2) if noise and signal spectral characteristics are sufficiently simple (smooth), are stationary in range and depth, and do not overlap in terms of frequency, wavenumber, azimuth, arrival type and path; (3) if frequency-dependent lithologic layering is sufficiently undifferentiated in range and depth, or otherwise permits resonant effects to be exploited or removed; and (4) if all possible "signal" and "noise" discriminants are dynamically employed in array/signal processing for specific reception objectives. Complementary numerical and experimental investigations are currently continuing to further assess these factors for generic environmental and experimental conditions.

REFERENCES

Brocher, T. M., Iwatake, B. T., Gettrust, J. F., Sutton, G. H., and Frazer, L. N., 1982. Comparison of the S/N ratios of low frequency hydrophones and geophones as a function of ocean depth, *Bull. Seismo. Soc. Am.*, 71(5): 1649–1659.

Gupta, I. N., 1965. Standing-wave phenomena in short-period seismic noise, *Geophysics*, 30(6): 223–245.

Hamilton, E. L., 1980. Geoacoustic modelling of the seafloor, *J. Acoust. Soc. Am.*, 68(5): 1313–1340.

Hardage, B. A., 1984. *Vertical Seismic Profiling*. Geophysical Press, Amsterdam and London.

Lewis, B. T. R. and McClain, J., 1977. Converted shear waves as seen on ocean bottom seismometers and surface hydrophones, *Bull. Seismo. Soc. Am.*, 67: 1291–1302.

MacPherson, M. K. and Frisk, G. V., 1980. The contribution of normal modes in the bottom to the acoustic field in the ocean, *J. Acoust. Soc. Am.*, 68(3): 929–940.

Roden, R. B., 1968. Seismic experiments with vertical arrays in boreholes, *Geophysics*, 33(2): 270–285.

Schmidt, H. and Tango, G. J., 1986. Efficient global matrix approach to computation of synthetic seismograms, *Geophys. J. Roy. Astro. Soc.*, 84: 331–359.

Spencer, T. W., Edwards, C. M., and Sonnad, J. R., 1977. Seismic wave attenuation in non-resolvable cyclic stratification, *Geophysics*, 42; 939–949.

Stephen, R. A., 1977. *The Oblique Seismic Experiment: Theory and Practice*, Ph.D. Dissertation, Department of Geophysics, Cambridge University, UK.

Stephen, R. A. and Bolmer, S., 1984. The "direct wave root" in marine seismology, *Bull. Seismo. Soc. Am.*, 75(3): 455–461.

Tango, G. J. and Schmidt, H., 1985. Exact full wave-theoretic synthetic VSPs using a new fast direct-global-matrix solution, Paper BHG3b, in: *SEG '85 Abstracts*, pp. 65–67.

Tunnell, T. and Tango, G. J., 1985. Predicted energy partitioning of bottom-interacting VLF seismo-acoustic energy, paper, in: *Proceedings, NATO-ASI Conference on VLF Seismo-Acoustic Propagation*, La Spezia, Italy, 18–23 June 1985, Plenum Press, New York.

Ursin, B., Stephen, R., Schmidt, H., Tango, G., and Arnsten, B., 1986. Comparison of asymtotic ray, finite difference and discrete wavenumber algorithms for vertical seismic profiling synthesis, paper presented at the *USN-SEG Symposium on High-Resolution Marine Seismology*, 27 February, 1986, NSTL, Mississippi

Wagstaff, R. A., 1981. Low frequency ambient noise in the deep sound channel—the missing component, *J. Acoust. Soc. Am.*, 69: 1009–1004.

APPLICATION OF ACOUSTIC WAVES AND ELECTRICAL CONDUCTIVITY FOR THE DETERMINATION OF MECHANICAL CHARACTERISTICS OF MARINE SEDIMENTS

Jean-Paul Longuemard* and Jean-Michel Daupleix†

*Ecole Centrale de Paris, Grande Voie des Vignes, 92295 - Chatenay-Malabry, France

†Université de Perpignan, Avenue de Villeneuve, 66025 - Perpignan, France

ABSTRACT

Marine sediments permit the propagation of two types of acoustic waves:

- compressional waves of speed (C_L)

$$C_L = \left(\frac{\lambda + 2\mu}{\rho}\right)^{1/2}$$

- shear waves of speed C_T

$$C_T = \left(\frac{\mu}{\rho}\right)^{1/2}$$

where λ is Lamé's coefficient, μ is the rigidity modulus and ρ is the wet density or wet unit weight.

The mechanical characteristics of marine sediments are dependent on λ and μ. The determination of these parameters is possible if the wet density is known. The authors propose the determination of ρ by the measurement of an electrical current I in circulation between two probes ($\rho = f(I)$). In this case, the calculation of the values of λ and μ with the speeds C_L and C_T is possible:

$$\mu = f_1\ (C_T,\ I \text{ or } \rho)$$

$$\lambda = f_2\ (C_L,\ C_T,\ I \text{ or } \rho)$$

The oedometric modulus E' is in relation with the compressional index (C_c):

$$E' = \frac{1 + e}{C_c} \cdot \frac{\sigma\Delta}{\log(1 + \frac{\Delta\sigma}{\sigma})}$$

where e is the void ratio, σ is the constraint, and $\Delta\sigma$ is the variation of the constraint.

For a sediment, the relation stress-strain is:

$\sigma = E'.\varepsilon$

This relation is identical with $\varepsilon = f(E)$ used for an elastic solid. For a visco elastic solid, (clay), $E' = E.f(\nu)$. The comparison between values of C_c calculated with acoustic parameters and the values obtained by the oedometric method are given.

RESUME

Les sédiments des fonds marins permettent la propagation de deux types d'ondes acoustiques :

- Longitudinales de célérité (C_L) :

$$C_L = \left(\frac{\lambda + 2\mu}{\rho} \right)^{1/2}$$

- Transversales de célérité (C_T) :

$$C_T = \left(\frac{\mu}{\rho} \right)^{1/2}$$

où :

λ est le coefficient de Lamé
μ est le module de rigidité
ρ est la masse volumique.

Le comportement mécanique des sédiments est calculable à travers λ, μ et le coefficient de Poisson, qui est peu variable d'un sédiment à un autre. Les auteurs proposent le calcul de λ et de μ à travers C_L et C_T en supposant connue la valeur de la masse volumique. Cette dernière est déterminée à partir de données concernant la résistivité électrique des argiles, soit $\rho = \nu(I)$. Dans ces conditions :

$$\mu = f_1(C_T, I \text{ ou } \rho)$$
$$\lambda = f_2(C_L, C_T, I \text{ ou } \rho)$$

Or, le module oedométrique E' est relié avec l'indice de compression C_c et l'indice des vides e par :

$$E' = \frac{1 + e}{C_c} \; \frac{\Delta\sigma}{\log\left(1 + \frac{\Delta\sigma}{\sigma}\right)}$$

où σ est la contrainte,
$\Delta\sigma$ est l'accroissement de contrainte.

E' est également fonction du module de Young $E' = E \,.\, f(\nu)$. Une comparaison entre les valeurs calculées de E d'après les mesures acoustiques et E' obtenues par mesures oedométriques est fournie .

THEORETICAL ANALYSIS

Lamé's constants λ and μ are related to the geotechnical parameters and acoustic waves by two equations concerning marine sediments:

- Geotechnical equation

It is in relation with the oedometric modulus E'. The stress-strain equation for a visco-elastic sediment is :

$$(\lambda + 2\mu)\, f(\nu).\varepsilon = \sigma \tag{1}$$

where :
- ε is the deformation
- ν is the Poisson coefficient
- E' is the oedometric modulus.

- Acoustic equation

$$(\lambda + \mu)\, \overrightarrow{\text{grad}}\, (\text{div}\, \vec{\phi}) + \mu\Delta\, \vec{\phi} + \vec{F} = 0 \tag{2}$$

where :
ϕ is the particule displacement vector

$$\vec{F} = \frac{\rho.\ \partial^2 \phi}{\partial t^2}$$

Two acoustic waves result from this equation :

- compressional waves

$$(\lambda + 2\mu)\Delta\, \psi_L = \rho \frac{\partial^2 \psi_L}{\partial t^2} \tag{3}$$

- shear waves

$$\mu\Delta\, \psi_T = \frac{\partial^2 \psi_T}{\partial t^2} \tag{4}$$

In this equation :
ρ is the density
$\vec{\phi} = \vec{\phi}_L + \vec{\phi}_T$ and

$\vec{\psi}_L = \text{div}\, \vec{\phi}$, $\vec{\psi}_T = \text{rot}\, \vec{\phi}$

The celerity of these waves is, respectively :

$$C_L = \left(\frac{\lambda + 2\mu}{\rho} \right)^{1/2} \tag{5}$$

$$C_T = \left(\frac{\mu}{\rho} \right)^{1/2} \tag{6}$$

The comparison between the stress-strain relation and λ and μ is difficult to calculate because the oedometric modulus is not constant ; it is necessary to use the Poisson coefficient and the void ratio :

$$E' = \frac{1 + e}{C_c} \cdot \frac{\Delta\sigma}{\log\left(1 + \frac{\Delta\sigma}{\sigma}\right)} \tag{7}$$

where C_c is the compressional index,
e is the void ratio

For $\Delta\sigma < \sigma$ this equation becomes :

$$E' = \frac{1 + e}{C_c} \cdot 2{,}3\, \sigma \tag{8}$$

Table 1. EXPERIMENTAL DATA

ρ Kg_m^{-3}	C_L ms^{-1}	C_T ms^{-1}	μ N	λ N	ν	$f(\nu)$	E' ϕ (C_L, C_T)	e	C_c	E' $\Phi(e$ & $C_c)$
1400	1460	2	$3 . 10^3$	$3 . 10^9$						
1450	1420	6	$0,52 . 10^5$	$2,9 . 10^9$	0,499	0,002	$0,6 . 10^7$	2,5	0,7	$0,8 . 10^7$
1500	1480	10	$1,5 . 10^5$	$3,28 . 10^9$	0,4988	0,0035	$1,1 . 10^7$	2,2	0,5	$1,12 . 10^7$
1550	1500	17	$4,5 . 10^5$	$3,4 . 10^9$	0,4984	0,0045	$1,5 . 10^7$	2	0,45	$1,23 . 10^7$
1600	1522	23	$8,4 . 10^5$	$3,7 . 10^9$	0,498	0,006	$2,1 . 10^7$	1,8	0,33	$1,49 . 10^7$
1650	1550	35	$20 . 10^5$	$3,96 . 10^9$	0,4975	0,007	$2,7 . 10^7$	1,7	0,3	$1,58 . 10^7$
1700	1580	40	$2,7 . 10^6$	$4,24 . 10^9$	0,4975	0,007	$2,9 . 10^7$	1,6	0,25	$1,83 . 10^7$
1750	1613	45	$3,54 . 10^6$	$4,54 . 10^9$	0,4985	0,008	$3 . 10^7$	1,5	0,22	$2,1 . 10^7$
1800	1650	52	$4,8 . 10^6$	$4,89 . 10^9$)	0,4965	0,007	$4 . 10^7$	1,4	0,17	$2,49 . 10^7$
1850	1690	100	$1,85 . 10^7$	$5,26 . 10^9$)		0,009	$4,6 . 10^7$	1,3	0,12	$3,38 . 10^7$
1900	1733	115	$2,5 . 10^7$	$5,6 . 10^9$	0,496	0,011	$6 . 10^7$	1,1	0,08	$4,63 . 10^7$
1950	1780	125	$3,04 . 10^7$	$6,14 . 10^9$	0,4955	0,017	$1 . 10^8$	1	0,05	$0,7 . 10^8$
2000	1830	140	$3,92 . 10^7$	$6,66 . 10^9$	0,495	0,027	$1,8 . 10^8$	0,7	0,03	$1,3 . 10^8$
2050	2000	200	$8,2 . 10^7$	$8,2 . 10^9$	0,4949			0,55	<0,03	

The utilisation of the Poisson coefficient ν is also possible :

$$E.f(\nu)\varepsilon = \varepsilon.E'$$

The ν' values, for a clay, are between 0,495 and 0,499. The result of this is:

$$0{,}002 < f(\nu) < 0{,}027$$

The determination of λ and μ is important. It is necessary to know these parameters in order to calculate the marine sediment comportment. This transaction is possible if the density is specified. The authors use the relation between the wet density and the electrical current that can circulate in the clay to calculate the wet density. The experimental data obtained in situ and in artificial clay permits the following proposition (Fig. 1-2):

$$\rho = 10^3 . (1 + KI)$$

Where :

ρ is expressed in $Kg.m^{-3}$

K is a constant dependent on the experimental apparatus

I is the electrical current.

Under these conditions, it is possible to make a value table of C_T, C_L, $\rho=f(I)$ and λ and μ then, to compare E' to C_c.

EXPERIMENTAL DATA

λ and μ are calculated after the celerity C_T, C_L and the electrical current have been measured. The values obtained by this method are listed on the Figs. 2 and 3 enclosed 1. These values grow with the density. The celecity of the shear waves is very dependent on the wet density while the compressional waves grow less with this parameter. The harmonious combination of acoustic and electrical results allows the evaluation of Lamé's constant. If we consider that the Poisson coefficient is fixed or varies little, it is possible to determine an evaluation of the oedometric modulus :

The corresponding data between the two methods is described in Table 1. E' is determined by :

- C_L, C_T and Poisson coefficient,
- ρ , C_c and σ .

The second result is obtained with $\sigma < 5.10^5$ Pascals and $\Delta\sigma < 2.10^5$ Pascals. Under these conditions, the values of E' are between $1{,}3 . 10^8$ Pascals for sand and $0{,}8 . 10^7$ Pascals for mud or clay.

It is equally possible to give a direct correlation between C_c and C_L on E' and C_L. This method is very interesting for clay or satured sediments ; in this case, the relation is described in Figure 6.

CONCLUSION

Acoustic measures are useful for the prevision of geotechnical sediment parameters. It is possible to find E' or C_c if we know C_L, C_T and ρ or the electrical current ($\rho = f(I)$). It is equally possible to use a direct

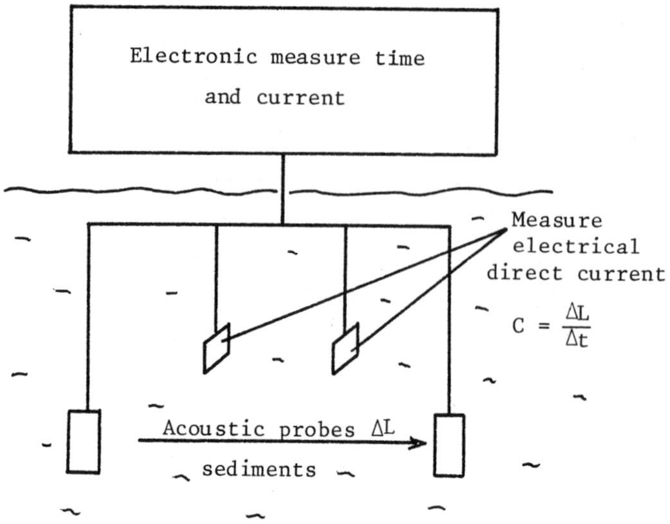

Figure 1. EXPERIMENTAL APPARATUS

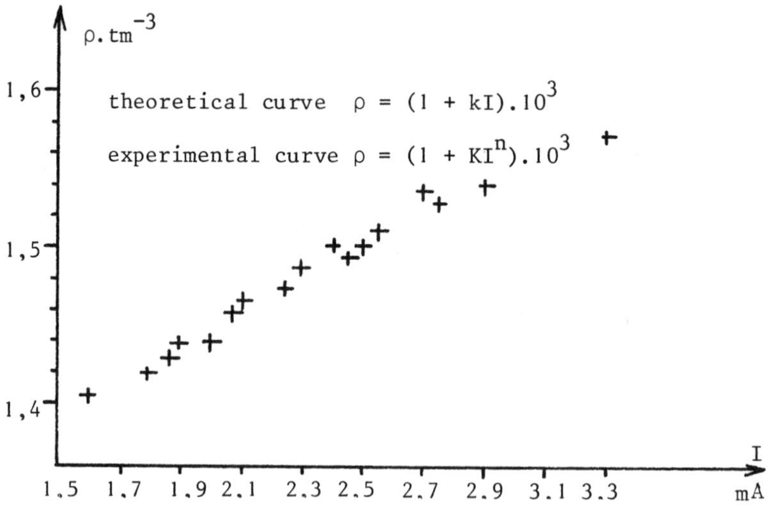

Figure 2. CORRELATION BETWEEN ρ and I FOR A CLAY

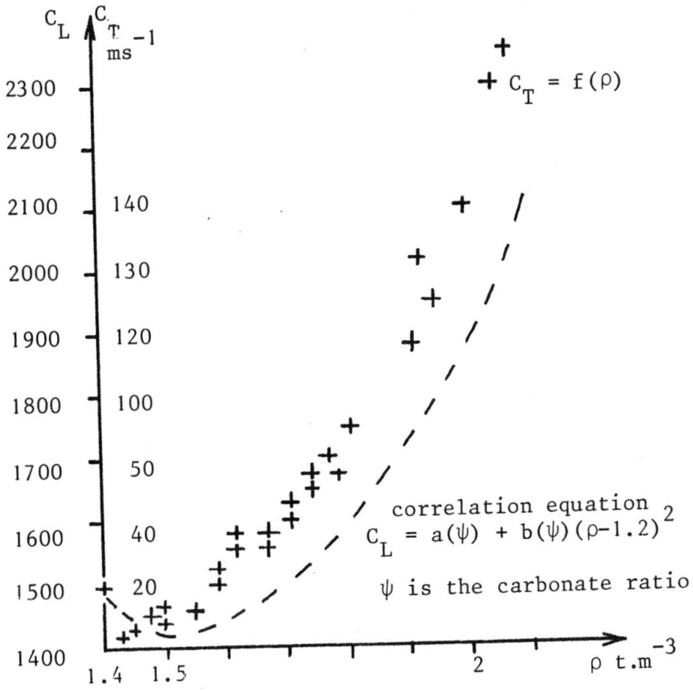

Figure 3. RELATION BETWEEN ρ. C_L AND C_T

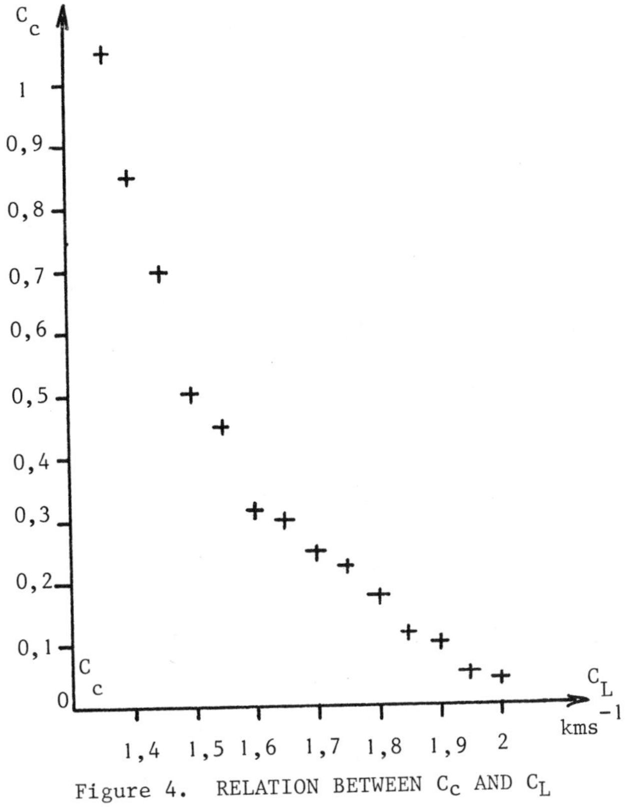

Figure 4. RELATION BETWEEN C_c AND C_L

correlation between the acoustic properties of marine sediments and oedometric index:

$$CL = 1400 + \frac{77,5}{(C_o + 0,02)^{0,5}}$$

REFERENCES

- J. COSTET & G. SANGLERAT : "Cours pratique de mécanique des sols", t.1, Dunod, Paris 1975.

- E.L. HAMILTON : "Sound Velocity : relations in sea floor, sediments and rocks" - J. Acoustical Soc. amer., Vol. 63 (2), pp. 366-377 1974 February.

- E.L. HAMILTON : "Prediction of deep sea sediments properties state of the art - Deep sea sediments physical and mechanical properties" Marine Sciences 2, Plenum Press, New-York 1974, pp. 1-43.

- C.C. LEROY, J.-M. DAUPLEIX & J.-P. LONGUEMARD : "Relationships between the acoustical characteristics of deep sea sediments and their physical environment" - Symp. on Ocean seismo-acoustic, Congrès NATO - La Spezia, Italy, 1985 June.

- J.-P. LONGUEMARD & J. MOUSSIESSIE : "Relation célérité des ondes de compressions - masse volumique de sédiments marins saturés" - Rev. du Cethedec, 20th year, 76, 1983.

- J.-P. LONGUEMARD : "Influence de l'indice de compression des sédiments marins sur la célérité d'ondes ultrasonores de fréquences comprises entre 15 et 100 KHz" - Rev. du Cethedec, 19th year, 71, 1982.

- Th. PLAIRE : "Effets d'un champ électrique sur des argiles : application à la diminution de l'adhérence de sédiments argileux à des parois métalliques", thèse de Docteur de l'Université des Sciences de Toulouse, à soutenir devant la Commission de Perpignan.

- R.D. STOLL : "Acoustic waves in ocean sediments, geophysics vol. 42, N°4, pp. 715-725, 1977.

THE ESTIMATION OF DENSITY, P-WAVE, AND S-WAVE SPEEDS OF THE TOP-MOST LAYERS OF SEDIMENTS

S. Levy, J. Cabrera, K. Stinson,
D. Oldenburg and *R. Chapman

IT&A and *Department of National Defence
#241 7080 River Road
Richmond, B.C. V6X 1X5

ABSTRACT

Reflection arrivals from the water bottom are analysed and estimates of the elastic parameters in the top-most layers of sediments are calculated. The amplitude of the reflection coefficient from an interface is a function of the angle of incidence of the incoming source energy at the reflecting interface, and the contrasts in the density, and P- and S-wave speeds across the interface. A nonlinear parametric inversion has been developed to yield estimates of these elastic parameters from the amplitude offset information contained in pre-critical reflections. The inversion method is applied to synthetic data and also to data from the Cascadia Basin.

INTRODUCTION

In this paper we evaluate the viability of a remote sensing system through which the elastic parameters of the ocean bottom sediments can be estimated. Our method relies on acoustic probing of the ocean floor with an experiment in which reflection arrivals from the water bottom at various source-receiver offsets are analysed to yield estimates of the density ρ, the P-wave speed α, and the S-wave speed β in the top sedimentary layers.

The basic experiment is shown in Figure 1. We assume a point source excitation in the water at a depth z_s and a series of omnidirectional hydrophones at depth z_r. In the physical experiment it may be that a single source is repeatedly moved and fired into a single fixed receiver, or that the source is fixed and the receiver is moved. Irrespective of the manner in which the data are acquired, the geometry shown in Figure 1 is applicable so long as the earth is a one-dimensional medium.

The data consist of a set of seismograms in the time-offset (t-x) space from which reflections from successively deeper sedimentary layers can often be identified. The amplitude and phase characteristics of the reflected arrivals are a function of the source-receiver offset and of the physical properties of the medium. In all of the work carried out here it will be assumed that the frequency content of the recorded waves is such that within the experimental model environment we observe plane rather than spherical-wave fronts. The theoretical relationship for the reflection coefficient as a function of offset is therefore determined by the plane-wave reflection coefficients for waves propagating in acoustic or elastic media.

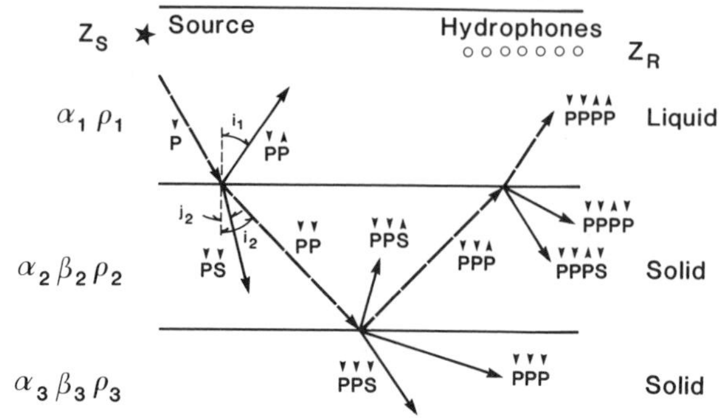

Figure 1: Ray path geometry of plane waves interacting at a liquid-solid and solid-solid interfaces. The "↓" denotes a downward travelling plane wave and the "↑" denotes an upgoing wave. For example, the compressional wave used to find the layer parameters a_3, β_3 and ρ_3 is denoted as PPPP ↓↓↑↑ (written as PP in the text) and its travel path is depicted by the bold line. The angles i_1, i_2 and j_2 are respectively the angle of incidence of the initial P wave in the water layer and the refracted angles of the P and S waves in the top layer of sediments.

The primary goal of this paper is to show how precritical amplitude information as a function of offset can be inverted to yield estimates of the density and P and S-wave speeds of the medium. Computation of the desired parameters involves an iterative approach in which the parameter set (α, β, ρ) is estimated for one layer at a time. Starting with a knowledge of the density and sound speed in the water column, we show how the theoretical relationship defining the plane-wave reflection for an acoustic-elastic interface can be used to determine the values of the material parameters (α_2, β_2, ρ_2) which best reproduce the observed reflection amplitudes. The material parameters in deeper layers are estimated by using an approximate amplitude formula corresponding to a reflection from an elastic-elastic interface.

Our inversion approach assumes a one-dimensional earth model in which the elastic properties for all layers overlying the target reflecting interface are known. Using the known layer parameters, forward ray tracing modelling gives the ray path with the associated reflection and transmission coefficients at shallower interfaces as well as the incident angles at the target interface for all measurement stations. The latter information is utilized in the estimation of the observed reflection strength and the inversion proceeds. Assuming trial values of the P-wave and S-wave speeds in the target medium, the density which gives the smallest RMS misfit between the calculated and observed amplitudes can be evaluated. The resultant misfit error between the observed and calculated data for this trial triplet of P-wave speed, S-wave speed and density is then tabulated. Repeating this process over the allowed space of P-wave and S-wave speeds in the target medium generates a surface of misfit error values. The minimum error position on this surface is selected as the best estimate of the P-wave and S-wave speeds, with corresponding density.

In the next section we outline the inversion method in detail. We then illustrate the approach with a synthetic example, and then a real data example.

Successful application of the inversion technique to real data requires that reliable estimates of the reflection amplitudes be obtained over a wide range of offsets. In field experiments however, the reflection sequence of sedimentary layers is blurred by convolution with the source signature and its ghost. In addition, there may be some variability in shot size. Recorded field data will therefore have to be processed to remove these effects so that true amplitudes can be recovered. A processing flow designed to carry this out is presented, and the subsequent amplitude - offset inversion is applied to data recorded by R. Chapman of DREP in the Cascadia Basin.

INVERSION PROCEDURE

The geometry of the assumed experiment is given in Figure 1. We re-emphasize that the earth model is assumed to be composed of planar interfaces and that the density and seismic wave speeds are constant within each layer. It is also assumed that for the frequency band of interest the travel path from the source to the ocean bottom interface is several wavelengths in length so that the impinging wave may be adequately approximated by a plane wave. Under these conditions ray theory can be used to trace the propagation path for the seismic energy, and plane-wave reflection and transmission coefficients can be computed for each boundary. The plane-wave reflection coefficient at a boundary separating the lth and $(l+1)^{th}$ layer can be written as

$$R = f(i, \alpha_l, \beta_l, \rho_l, \alpha_{l+1}, \beta_{l+1}, \rho_{l+1}) \tag{1}$$

where "i" is the angle of incidence of the impinging wave and "f" is a known function. The functional form of f will depend upon whether the boundary is an acoustic-elastic or an elastic-elastic interface.

Given $(\alpha, \beta, \rho)_l$, (the properties of the shallower medium), and a knowledge of the reflected amplitudes R_k, k=1...,N at N different angles of incidence i_k, we want to find the material parameters of the deeper layer, $(\alpha, \beta, \rho)_{l+1}$. The values of these parameters are those which minimize the RMS error E between the observed reflections and those computed from the theoretical formula in (1):

$$E = \left(\frac{1}{N} \sum_{k=1}^{N} (R_k^0 - R_k^c)^2 \right)^{1/2} \tag{2}$$

where the superscripts 0 and c denote observed and computed responses respectively.

Although at each layer interface there are three parameters to be determined, it is more convenient to find the globally smallest misfit by specifying trial values of the P and S wave velocities in the lower medium and then to use a least squares approach to find the density which minimizes (2) for that choice of velocities. Carrying this out for the range of possible P and S wave velocity values constitutes a suite of misfit values that is searched to find a minimum. This global search yields the $(\alpha, \beta, \rho)_{l+1}$ which best reproduces the observed reflection amplitudes.

We now turn our attention to the details of this estimation procedure. Two cases must be considered. For the ocean-sediment interface the function 'f' in equation (1) is given by the acoustic-elastic plane wave reflection coefficient formula. For subsequent layers we will use an approximate formula for elastic-elastic interfaces.

Inversion of P-Wave Reflection Amplitudes for the Acoustic-Elastic Case

Considering a horizontal interface, the $\overset{\downarrow\uparrow}{PP}$ plane-wave reflection coefficient is given by (Brekhovskikh, 1980)

$$\overset{\downarrow\uparrow}{PP} = \frac{\dfrac{\rho_2}{\rho_1}\dfrac{\alpha_2}{\alpha_1}\dfrac{\cos^2 2j_2}{\cos i_2} + \dfrac{\rho_2}{\rho_1}\dfrac{\alpha_2}{\alpha_1}\dfrac{\beta_2}{\alpha_2}\dfrac{\sin^2 2j_2}{\cos j_2} - \dfrac{1}{\cos i_1}}{\dfrac{\rho_2}{\rho_1}\dfrac{\alpha_2}{\alpha_1}\dfrac{\cos^2 2j_2}{\cos i_2} + \dfrac{\rho_2}{\rho_1}\dfrac{\alpha_2}{\alpha_1}\dfrac{\beta_2}{\alpha_2}\dfrac{\sin^2 2j_2}{\cos j_2} + \dfrac{1}{\cos i_1}}, \tag{3}$$

where i_1, i_2 and j_2 are the angles from the vertical of the incident P wave, the transmitted P wave, and the transmitted S wave respectively (see Figure 1).

By defining:

$$m = \rho_2/\rho_1$$

$$x = \frac{\alpha_2 \cos^2 2j_2}{\alpha_1 \cos i_2} \qquad y = \frac{\alpha_2 \beta_2 \sin^2 2j_2}{\alpha_1 \alpha_2 \cos j_2} \qquad z = \frac{1}{\cos i_1}$$

we obtain

$$\overset{\downarrow\uparrow}{PP} = R = \frac{mx + my - z}{mx + my + z} . \tag{4}$$

Equation (4) may be rearranged to yield

$$m(1 - R)\ (x + y) = (1 + R)\ z . \tag{5}$$

If x and y are specified then equation (5) involves only a single unknown m.

Suppose that we have a number of observed reflection amplitudes R_k corresponding to N different angles of incidence i_k, k=1,..,N. That is

$$m(1 - R_k)\ (x_k + y_k) = (1 + R_k)\ z_k \qquad k=1,..,N \qquad (6)$$

Equations (6) are easily solved using least squares to yield a value of m and the misfit to the observed reflection amplitudes can be evaluated from equation (2).

To find a global solution we perform a systematic search of the solution space by solving equations (6) after specifying particular values of the P- and S-wave velocities in an allowed range $\alpha_{2(min)} \le \alpha_2 \le \alpha_{2(max)}$ and $\beta_{2(min)} \le \beta_2 \le \beta_{2(max)}$. An error matrix $\{\epsilon_{js}\}$ corresponding to the values obtained from equation (2) for a P-wave velocity α_{2j} and an S- wave velocity β_{2s} is scanned to find the minimum value. This determines the global best fit solution.

In this work we have grey scaled the negative of the values of $\{\epsilon_{js}\}$ and refer the resultant plot as an error surface. This error surface provides considerable insight about how well the velocity parameters are determined by our inversion method.

Inversion of P-Wave Reflection Amplitudes for the Elastic-Elastic Case

We assume that the acoustic-acoustic inversion procedure has been applied to the water bottom reflections to determine the material parameters of the first layer of sediments, and that we now want to find the elastic properties of deeper layers. The inversion proceeds in the following manner:

(1) Knowing the P-wave velocity in the top-most layer of sediments and the travel time to the second reflecting interface, we calculate the thickness of the first layer of sediments. A ray tracing algorithm is then used to determine the angles of all the involved wave modes generated along the path of the observed compressional waves.

(2) With the ray paths known, we compute the transmission coefficients of compressional waves at the ocean bottom interface.

(3) Using the information found in Step 2 and the observed PP amplitudes for the target interface, we estimate the corrected reflection coefficients at this interface. Note that in addition to the correction for transmission loss, the correction for geometrical spreading is also done at this step.

(4) To invert these amplitudes, we follow a procedure which is similar to that the presented for the acoustic-elastic inverstion. With the assumption that density, and the P and S wave velocities of the second layer of sediments are not 'too different' from those of the first layer, the observed amplitudes from an interface separating these layers are given by the expression (Aki and Richards, 1980)

$$\overset{\downarrow\uparrow}{PP} = \frac{1}{2}\,(1 - 4\beta^2p^2)\,\frac{\Delta\rho}{\rho} + \frac{1}{2\cos^2 i}\,\frac{\Delta\alpha}{\alpha} - 4\beta^2p^2\,\frac{\Delta\beta}{\beta} \qquad (7)$$

where

$$\Delta\rho = \rho_3 - \rho_2$$

$$\rho = \frac{\rho_2 + \rho_3}{2}$$

$$\Delta\alpha = \alpha_3 - \alpha_2$$

$$\alpha = \frac{\alpha_2 + \alpha_3}{2}$$

$$\Delta\beta = \beta_3 - \beta_2$$

$$\beta = \frac{\beta_3 + \beta_2}{2}$$

(Note from Figure 1 that layer 1 is the water layer, layer 2 is the first sediments layer, etc.)

$$p = \text{ray parameter} = \frac{\sin i_1}{\alpha_1},$$

and the angle 'i' is computed via

$$\frac{\Delta\alpha}{\alpha} \tan i = i_2 - i_1$$

with

$$\frac{\sin i_1}{\alpha_1} = \frac{\sin i_2}{\alpha_2}.$$

Equation (7) gives a good approximation to the reflection coefficient when the quantities $\frac{\Delta\rho}{\rho}$, $\frac{\Delta\alpha}{\alpha}$ and $\frac{\Delta\beta}{\beta}$ are small and 'i' is not close to 90°.

We now define

$$x = \frac{1}{2}(1 - 4\beta^2 p^2)$$

$$b = \frac{1}{2\cos^2 i}\frac{\Delta\alpha}{\alpha} - 4\beta^2 p^2 \frac{\Delta\beta}{\beta}.$$

Hence equation (7) reads as

$$\overset{\downarrow\uparrow}{PP} = R = x \frac{\Delta\rho}{\rho} + b \; . \tag{8}$$

Equation (8) is a linear equation in $\Delta\rho/\rho$. Given N amplitude observations R_k $k=1,\ldots N$ and corresponding values of x_k and b_k, least squares may be used to solve (8) to find the best relative density contrast. We obtain

$$\frac{\Delta\rho}{\rho} = \frac{\sum_{k=1}^{N} (R_k - b_k)\, x_k}{\sum_{k=1}^{N} x_k^2} \tag{9}$$

from which we find (using the definitions given in (7))

$$\rho_3 = \rho_2 \left(\frac{2 + \Delta\rho/\rho}{2 - \Delta\rho/\rho} \right) . \tag{10}$$

For any selected trial values of P-wave and S-wave velocities (α_3, β_3) in the second layer of sediments, we can therefore compute the density ρ_3 which provides a best fit to the observed reflection amplitudes.

For the estimated (α_3, β_3, ρ_3) values, we use expression (7) to compute reflection amplitudes and find the RMS error between these and the observed data according to equation (2). As in the acoustic-elastic inversion, the RMS error surface is calculated over the range of physically realistic P and S wave velocities. This RMS error surface is searched, and the set of quantities (α_3, β_3, ρ_3) which gives the least RMS error constitutes our best estimate of the P and S wave velocities and the density of the second layer of sediments.

Material properties for underlying layers may be then found by repeating the above sequence of steps.

SYNTHETIC EXAMPLE

We shall now apply the techniques of the last section to a synthetic example. Using the reflectivity method of Fuchs and Muller (1971), we have calculated a set of synthetic seismograms for the model in Figure 2. The synthetic seismograms are given in Figure 3. We have measured a set of reflection amplitudes corresponding to the first and second interfaces. These amplitudes, adjusted for geometrical spreading and transmission losses, are given in Table I. Although there is a third interface in the synthetic model, we will not be analysing that here as reflection amplitudes from that interface could be reasonably obtained only for the four smallest offsets.

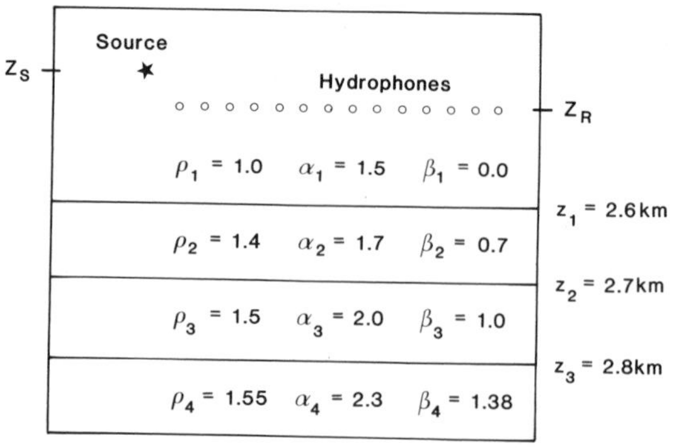

Figure 2: Earth model used to generate the synthetic seismograms shown in Figure 3. The source (*) is placed at a depth Z_s = 0.19 km. The receivers (o) are at a depth Z_R = 0.50 km. The nearest offset receiver is at x_1 = 0.5 km; and receiver spacing is Δx = 0.50 km. The explosive point source consists of a Ricker wavelet whose center frequency is 40 Hz. The depths of the layer interfaces (measured from the ocean surface) are given by z_i, i = 1,2,3 on the right. The densities ρ_i (gm/cc), P-wave speeds α_i (km/sec) and S-wave speeds β_i (km/sec) are displayed for each layer.

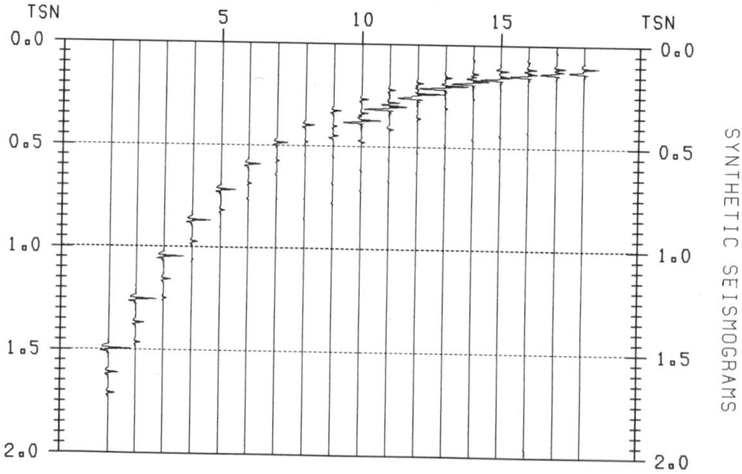

Figure 3: Synthetic seismograms corresponding to the earth model in Figure 2. The seismograms represent the vertical component of the displacement and have been reduced with a reduction velocity of 1.7 km/sec and a reduction time of 1.4 secs. Frequency band is 10 to 100 Hz.

Applying the acoustic-elastic inversion method to the water-bottom reflections, we obtained the estimates α_2 = 1.7 km/sec, β_2 = 0.68 km/sec, and ρ_2 = 1.31 gm/cc. The least squares error surface for this analysis is shown in Figure 4. The global RMS minimum error is concentrated in the region β_2/α_2 = 0.4 and α_2 = 1.7 km/sec. The fact that the global minimum is very localized provides confidence in the estimated parameters. That confidence is enhanced by the good fit to the observed amplitudes of those calculated using the estimated parameters (see Figure 5).

Table I: Amplitude-offset data obtained from the first two reflection events in Figure 3. The amplitudes have been corrected for geometrical spreading and transmission losses.

OFFSET	WATER-BOTTOM	SECOND LAYER
0.500000	0.191978	0.0945
1.000000	0.187096	0.0848
1.500000	0.179959	0.0743
2.000000	0.170498	0.0626
2.500000	0.157705	0.0518
3.000000	0.145613	0.0440
3.500000	0.138290	0.0411
4.000000	0.130329	0.0482
4.500000	0.120497	0.1159
5.000000	0.118206	0.1446
5.500000	0.121216	

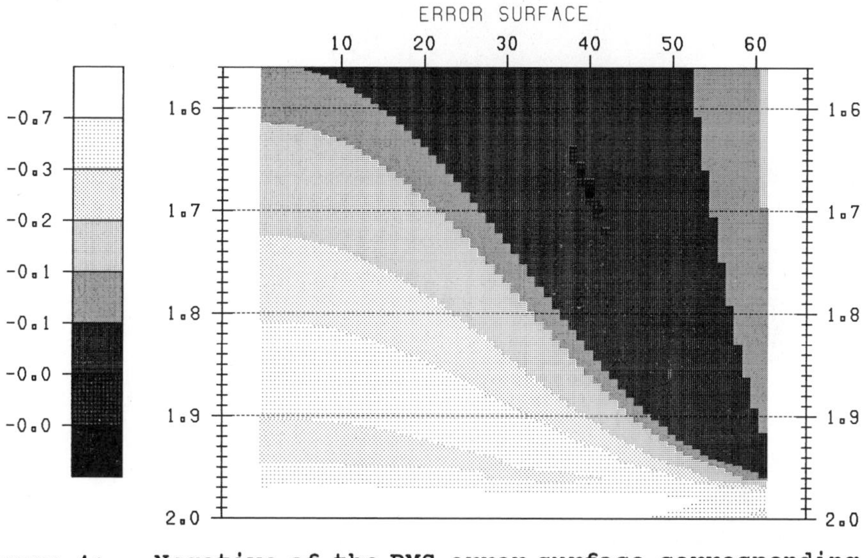

Figure 4: Negative of the RMS error surface corresponding to the data of reflector 1 in Table I. The horizontal axis denotes velocity ratios β/α times 10^{-2}; the vertical axis represents P-wave velocity α.

With the estimates for (α_2, β_2, ρ_2) we can carry out the elastic-elastic inversion of reflection amplitudes from the second interface. The final results are α_3 = 1.95 km/sec, β_3 = 0.9 km/sec, and ρ_3 = 1.38 gm/cc, and the corresponding error surface is displayed in Figure 6. The decrease in localization of the global minimum of the error should be considered to represent inversion results with a larger uncertainty than the results for the first layer. This is corroborated by the poorer fit of the calculated results to the observed ones (Figure 7). To summarize then, we tabulate the inversion results in Table II and conclude that our inversion technique has yielded earth parameters which are in reasonable agreement with the true ones.

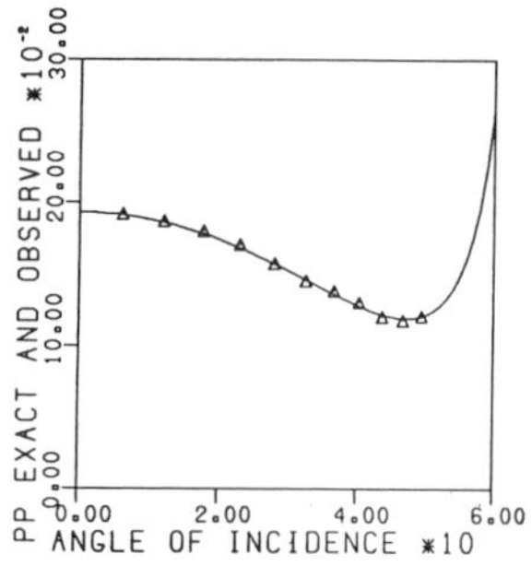

Figure 5: Calculated (continuous curve) and observed (triangles) reflection strengths for the water-bottom interface in the model of Figure 2 (acoustic-elastic inversion).

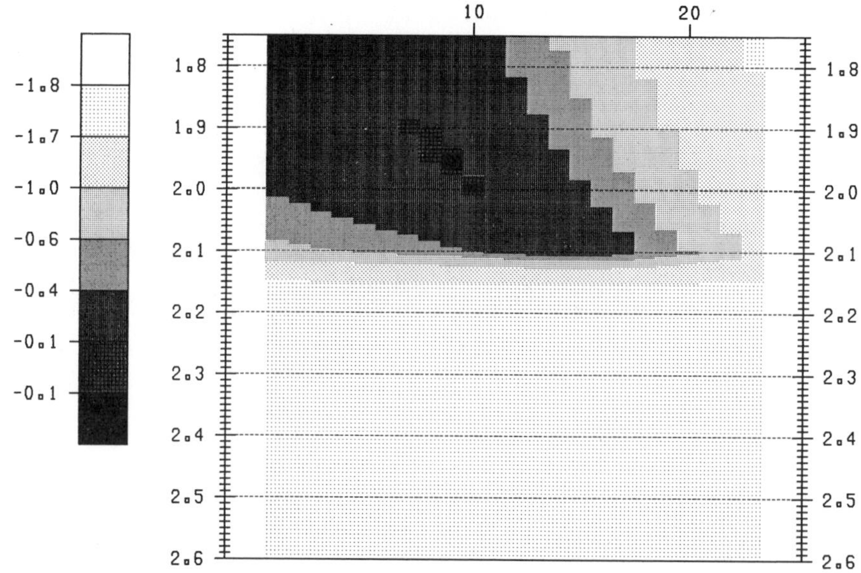

Figure 6: Negative of the RMS error surface corresponding to the data of reflector 2 in Table I. The first trace on the left corresponds to a S-wave velocity of 0.5 km/sec; the last trace on the right is for ß = 1.6 km/sec. The vertical axis denotes P-wave velocity.

FIELD DATA EXAMPLE

To further evaluate the proposed inversion algorithms we have analysed a set of reflection seismograms acquired by Dr. R. Chapman in the Cascadia Basin. These data correspond to an experiment in which a boat steamed away from the receiver location triggering off explosions at pre-determined offsets (ranging from 0.6 to 26 km). The direct arrivals and water bottom reflection are respectively displayed in Figures 8(a) and (b). Cursory inspection of these data reveals some amplitude imbalances which may be due to differential recording gain, insufficient temporal sampling rate, and differential source strength. As well, for all but the nearest offsets the source ghost interferes with the direct arrival.

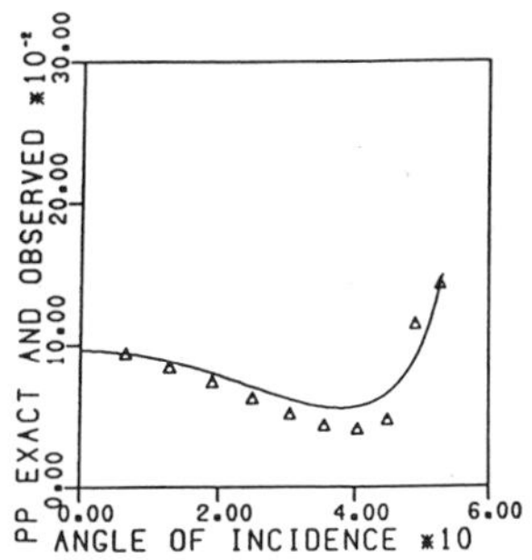

Figure 7: Calculated (continuous curve) and observed (triangles) reflection strengths for the second interface in the model of Figure 2 (elastic-elastic inversion).

Table II: Velocity-density inversion results obtained by inverting amplitude-offset data for the synthetic data shown in Figure 3.

REFLECTOR #1: ACOUSTIC-ELASTIC INVERSION

PARAMETER	TRUE	RECOVERED
α_2 (km/sec)	1.70	1.70
β_2 (km/sec)	0.70	0.68
ρ_2 (gm/cc)	1.40	1.31

REFLECTOR #2: ELASTIC-ELASTIC INVERSION

PARAMETER	TRUE	RECOVERED
α_3 (km/sec)	2.00	1.95
β_3 (km/sec)	1.00	0.90
ρ_3 (gm/cc)	1.50	1.38

Before we can proceed with the inversion, it is important to carry out pre-inversion processing to ensure that the data comply with the basic assumptions which underly the proposed inversion. These assumptions are:

(a) To calculate the incidence angle at the water bottom, we assume that the exact locations of the source and each of the receivers are known. The water bottom is assumed to be a planar interface whose depth and dip angle are known.

(b) For the application of a geometrical spreading correction we assume that the sound speed throughout the water column is a constant.

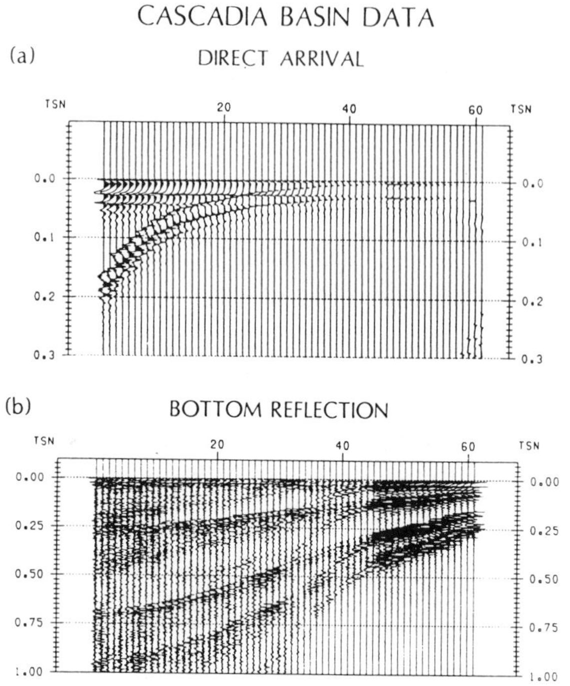

Figure 8: Direct arrival and source ghosts recorded in the Cascadia Basin experiment are shown in (a). The data have been flattened to the onset of the direct arrival. Reflections from the sediments are shown in panel (b). The data have been flattened to the onset of the water-bottom reflection.

(c) For the purpose of estimating the absolute reflection strength, we assume that the direct arrival at the near offset hydrophones (after the geometrical spreading correction has been applied) constitutes a reasonable approximation of the source function. The absolute reflection strength may then be calculated by dividing the peak amplitude of the water bottom reflection by that of the direct arrival. Another possible approach (which will be illustrated here), is to estimate the source signature for each shot and then to obtain an estimate of the reflection amplitudes after source signature deconvolution.

(d) It is assumed that both the source and the receivers produce or record a consistent signature irrespective of the direction at which the wave is leaving the source or arriving at the receivers (i.e. both source and receivers are omnidirectional).

The pre-inversion processing necessary to ensure that the data comply with the basic assumptions which underly the inversion will be dependent upon the data. Presented here, in Figure 9, is the processing flow that was used to analyse the Cascadia Basin data.

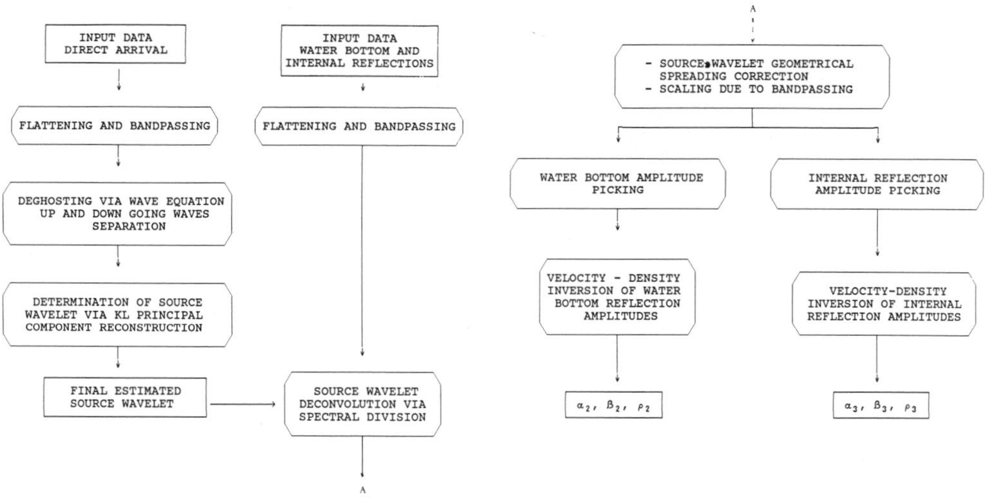

Figure 9: Flow chart for processing the field data shown in Figure 8.

The first task is to estimate the source wavelet for each shot, which has been accomplished through the following steps:

(1) The data are bandpassed to attenuate the effects of high frequencies noise. These data are shown in Figure 10(a), datumed to the direct arrival.

(2) In order to separate the direct arrival from its ghost we use a wave equation separation technique (DREP Report #DSS 06SB.97708 - Wave Equation Deconvolution - 1985). The computational procedure involves considering the problem as if the seismic source was directly below the receiver. In this configuration, the direct arrival would be an up-going wave, while the ghost is a down-going wave. Using the fact that the pressure at the surface of the water is zero, it is possible to separate the observed pressure field into up- and down-going wave components, thereby separating the ghost from the direct arrival. Estimates of the direct arrivals produced in this way are shown in Figure 10(b).

(3) The estimated source wavelets obtained from Step 2 are then subjected to a Karhunen-Loeve analysis which reveals a substantial common component for the analysed set (see Figure 10(c)) with only a relatively small residual. We conclude that both the direct arrivals displayed in Figure 10(b) and their most common parts (Figure 10(c)) are probably reasonable representations of the source signatures and either could be used as such.

(4) With the estimated source signatures we are now in a position to deconvolve the data. Our objectives are two fold: (a) we wish to enhance event definition, remove the bubble pulse signature and isolate the target reflections, and (b) we wish to scale reflection events on each of the seismograms by the corresponding source wavelets.

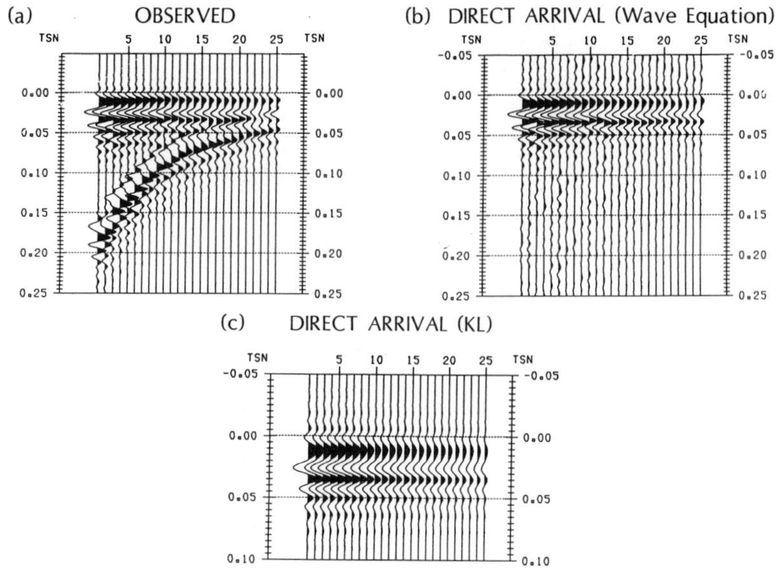

Figure 10: **Direct arrivals and source ghosts recorded in the offset range .66 km to 5.0 km are given in 10(a) (the data are limited to the frequency band 10-80 Hertz). The estimated direct arrivals obtained through the wave equation deghosting algorithm are in (b). The Karhunen-Loeve analysis of the direct arrivals in (b) resulted in the set of first principal component wavelets presented in panel (c). These wavelets will be used to carry out a signature deconvolution of the data in Figure 8(b).**

The source deconvolution was performed using spectral division with the deconvolved data at each recording station obtained by inverse Fourier transforming the quantity

$$D(\omega) = \frac{X(\omega) \cdot W^*(\omega)}{W(\omega) \cdot W^*(\omega) + \eta} . \tag{11}$$

In (11), $X(\omega)$ is the Fourier representation of the recorded seismogram, $W(\omega)$ is the Fourier representation of the corresponding source signature, $()^*$ denotes the complex conjugate and η is a stabilization parameter. Since at each recording station we have used a source signature which has been estimated from the analysis of the direct arrivals, the above deconvolution accomplishes preliminary data scaling as well as spiking in the same step. Geometrical spreading and transmission loss effects are still present in the data and must still be accounted for (please refer to Figure 9).

Narrowing our attention to the window of interest (see Figure 11(a)), we apply the signature deconvolution using the wavelets of Figure 10(c) to obtain the data of Figure 11(b). Inspection of the water bottom reflection on the deconvolved data reveals a significant decrease in signature duration. As well, the reflection event around 0.1 seconds (presumed to correspond to the second layer of sediments), is now clearer and can be tracked across the whole window of interest. The inversion then proceeds as discussed earlier, to obtain first the material parameters of the first-layer water bottom sediments, and then to obtain the material parameters corresponding to the layer underneath it.

The resultant parameter estimates (Table III) are very reasonable for both layers. In the calculated and observed reflection responses for the two reflectors (Figure 12), we note the increased scatter of the reflection amplitude estimates for the internal reflection event.

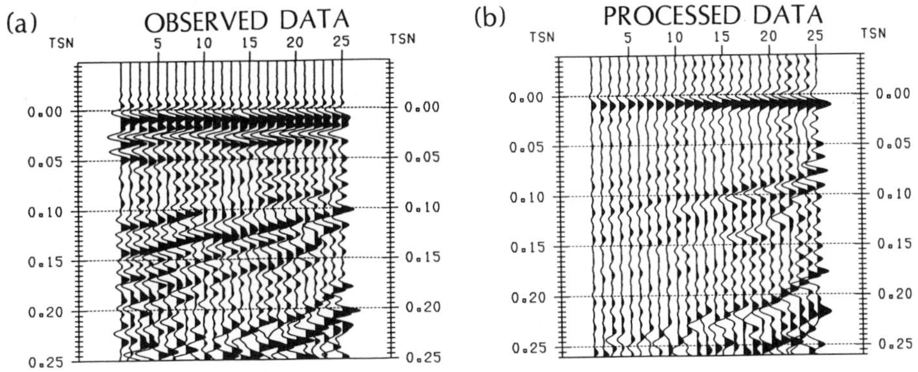

Figure 11: A narrow data window containing the events to be analysed in the amplitude-offset inversion are shown in panel (a). Of particular interest is the event at zero-time (water-bottom reflection) and the event at about 0.1 seconds. Panel (b) shows the data in 11(a) after the bubble pulse signatures have been removed through application of signature deconvolution.

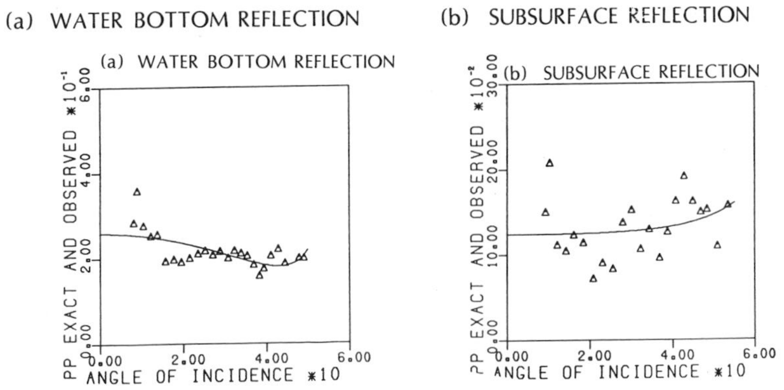

Figure 12: The calculated and observed reflection response for the water bottom reflection are shown in 12(a). The calculated and observed reflection response for the internal reflection are displayed in 12(b).

DISCUSSION

We have developed a method to estimate the elastic parameters of the water bottom sediments from amplitude-offset measurements. For each layer there are three parameters to be determined and the best set of parameters is found using a least squares approach. The accuracy with which the true parameters can be recovered depends strongly upon the quality of data. For synthetic models where amplitude-offset information was well determined, we showed that elastic parameters in the top two layers could be well recovered. For field data however, there are complications which can arise if the initial assumptions about the physical model and the data have been violated. Difficulties in obtaining estimates of the reflection amplitude also exist because of our inexact knowledge of the source wavelet. In practice, field data will need to be processed before reflector strengths can be determined and, as a consequence, there will exist larger uncertainties about the reflection amplitudes. This uncertainty may decrease the quality of the inversion results.

Table III: Velocity-density inversion results obtained by inverting amplitude-offset data shown in Figure 11(b).

REFLECTOR #1: ACOUSTIC-ELASTIC INVERSION

PARAMETER	RECOVERED
α_2 (km/sec)	1.82
β_2 (km/sec)	0.82
ρ_2 (gm/cc)	1.40

REFLECTOR #2: ELASTIC-ELASTIC INVERSION

PARAMETER	RECOVERED
α_3 (km/sec)	1.90
β_3 (km/sec)	0.75
ρ_3 (gm/cc)	1.72

Nevertheless, since both the P- and S-wave speeds in the sediments indicate strong dependency on the relative (trace to trace) reflection response variations and are fairly insensitive to errors in absolute reflection magnitudes we have found that the wave-speed parameters can in general be estimated with a fairly high confidence bound. The density on the other hand should be treated with caution since it is strongly affected by the estimated absolute reflection strengths. We have compared the P-wave speed estimated through the amplitude offset inversion to that obtained from the inversion of travel time measurements and the agreement within reasonable error bound is good.

Finally, we are encouraged by the results obtained in this work and feel that the amplitude-offset inversion presented here is a viable geophysical tool that should be used for the estimation of the elastic parameters of the water-bottom sediments. We recommend however, that future experiment design should prefer a setup of a single source radiating into a linear array of hydrophones. With such a field configuration, the task of reflection scaling is considerably simpler than the procedure followed in this work.

REFERENCES

Aki, K., and Richards, P., 1980, Quantitative Seismology - Theory and Methods. San Francisco: W.H. Freeman and Co.

Brekhovskikh, L.M., 1980, Waves in Layered Media. New York: Academic Press.

Fuchs, K., and Muller, G., 1971, Computation of Synthetic Seismograms with the Reflectivity Method and Comparison of Observations: Geop. J. R. Ast. Soc., V. 23, 417-433.

DREP Report DSS #06SB.97708, Wave Equation Deconvolution, 1985. Prepared by Inverse Theory and Applications, Inc.

ETL REFLECTIVITY AS A FEATURE FOR

RECOGNITION OF TOP-LAYER MARINE SEDIMENTS

Jinsheng Meng* and Dinghua Guan

Institute of Acoustics, Academia Sinica, P.O. Box 2712
Beijing, China

*now at Department of Engineering Physics, Technical University of Nova Scotia, P.O. Box 1000, Halifax, N.S Canada, B3J 2X4

ABSTRACT

For the sake of improving acoustic recognition of top-layer sediments of the sea bottom, ETL reflectivity is proposed in this paper. As a result of eliminating the influence of surface roughness and volume inhomogeneities of the seabed, ETL reflectivity correlates closely with top-layer sediment. The results of sea-going experiments show that ETL reflectivity is a better feature for recognition of sediment than the conventional ones measured by the average amplitude or energy of bottom echoes.

INTRODUCTION

The acoustic reflectivity of the sea bottom has been found to correlate significantly with sediment porosity and mean grain size, so it can be used as a feature for recognition of marine sediments (Breslau, 1965; Danbom, 1976; Bell et al., 1974; MacIsaac et al., 1977). The bottom reflectivity is usually measured by the average amplitude of the bottom echoes. In certain cases, it is calculated by the total energy of the echo instead.

It is well known that the surface of the sea floor is, generally speaking, rough for a sound wave of the frequency band most in use, and there are a variety of layers and scatterers within the sediment. Consequently, the conventional reflectivity of the sea bottom depends not only on the impedance of the top-layer sediment but also on the roughness and the inhomogeneities of the seabed, as well as the geometry of the measurement. As a result, the success rate for recognition of marine sediments by the conventional reflectivity method is often poor.

Therefore, it would be useful to propose a reflectivity which is independent of both the roughness and inhomogeneities of bottom. Such a reflectivity would, of course, correlate more closely with the impedance of top-layer sediment, and be a better feature than the conventional ones measured simply by the echo amplitude or energy.

CONVENTIONAL REFLECTIVITY

Let us discuss a general case in measuring the reflectivity of seabottom with both surface roughness and volume inhomogeneities.

A set of bottom echoes, $\{P_i(t),\ i=1,2,\ldots N\}$, is obtained by sequential transmissions of wide-beam, short duration soundpulses, $P_0(t)$, normally incident on the sea floor. Both transducer and hydrophone are mounted at the hull of a ship under-way. The incident wave is reflected and scattered at both the surface and at inhomogeneities within the volume of the sea bottom. For this reason, a bottom echo, P(t), can be divided into three components -- one comes from the volume of seabed, l(t) (the subbottom component), and other two come from the irregular surface, r(t) (the specular component, i.e. the coherent one) and s(t) (the scattered component, i.e. the incoherent one).

The ensemble average values of the sound energy of the bottom echoes and their three components over the set of echoes are denoted by E, E_r, E_s, and E_l, respectively, i.e.

$$
\begin{aligned}
E &= \left\langle \int_0^T P_i^2(t)dt \right\rangle , \\
E_r &= \int_0^T r^2(t)dt , \\
E_s &= \left\langle \int_0^T s_i^2(t)dt \right\rangle , \\
E_l &= \left\langle \int_0^T l_i^2(t)dt \right\rangle .
\end{aligned}
\qquad (1)
$$

where, integral duration T should be long enough to include almost all of the scattered component coming from the rough surface of the sea floor, and <.> denotes ensemble averaging over the set of echoes. Owing to the independence of the three components, we have

$$E = E_r+E_s+E_l \quad .$$

The acoustic reflectivity is usually determined by the average amplitude of echoes according to the following equation

$$V_a = 2H\frac{\langle A\rangle}{A_0} , \qquad (2)$$

where, A_0 is the amplitude of source pulse at a distance of one meter from the transducer, and $\langle A\rangle$ is the average amplitude of the echoes coming from the sea bottom at H meters depth.

Another way to calculate the reflectivity is by using the total echo energy, E, i.e.

$$V_e = 2H\left(\frac{E}{E_0}\right)^{1/2} , \qquad (3)$$

where,

$$E_O = \int_0^{T_0} P_O^2(t)dt ,$$

is the sound energy of source pulse, T_O being the duration of the incident pulse.

In the case of two half-infinite homogeneous media with a plane interface, the reflectivity at the interface is the so called Rayleigh reflectivity which depends only upon the acoustic impedances of the media. If an ideal bottom consists of homogeneous sediment which has an impedance the same as that of the top-most layer sediment of the real bottom, and if this ideal bottom has a smooth surface, then the reflectivity of this ideal bottom is the Rayleigh reflectivity, and it should be a very good feature for the top-layer sediment.

Unfortunately, neither the amplitude reflectivity nor the energy reflectivity is equivalent to the Rayleigh reflectivity. The amplitude one is smaller than the Rayleigh one unless the sea floor is quite smooth for the wavelength of the incident sound. As to energy reflectivity, if the sediment within the depth to which the incident wave can penetrate is homogeneous, it is about the same as the Rayleigh reflectivity under certain restrictions (Berry, 1973; Brekhovskikh, 1974). For conventional frequencies in underwater acoustics, however, the inhomogeneities of the subbottom play a non-negligible part in the formation of the bottom echo. It is well known that the echo from a sea bottom having a soft sediment on its top part does contain a subbottom component, so that the reflectivity measured by the total echo energy may be greater than the Rayleigh one.

Therefore, it is not unimportant to find a "better" reflectivity in the sense of improving the acoustic recognition of top-layer marine sediments.

ETL REFLECTIVITY

In this paper, we suggest a reflectivity defined by the following equation

$$R_{etl} = 2H\left(\frac{E_r+E_s}{E_0}\right)^{1/2}. \qquad (4)$$

According to the above definition, this reflectivity is determined by the energy of two components coming from the Top-Layer of sediment. So it can be referred to as ETL reflectivity.

For the purpose of calculating the ETL reflectivity, three correlation coefficients are employed, i.e. the correlation coefficient between adjacent echoes

$$R_m = \frac{\langle\int_0^T P_i(t)P_{i+1}(t)dt\rangle}{\langle\int_0^T P_i^2(t)dt\rangle} , \tag{5}$$

the correlation coefficient of echoes with the source pulse

$$R_s = \frac{\langle\int_0^T P_0(t)P_i(t)dt\rangle}{[\int_0^T P_0^2(t)dt\langle\int_0^T P_i^2(t)dt\rangle]^{1/2}} , \tag{6}$$

and the correlation coefficient of subbottom components

$$\rho_1 = \frac{\langle\int_0^T l_i(t)l_{i+1}(t)dt\rangle}{\langle\int_0^T l_i^2(t)dt\rangle} . \tag{7}$$

Some deductions yield the proportions of the three component energies in the total echo energy, i.e.

$$\frac{E_s}{E} = 1-R_m/\rho_1+(1/\rho_1-1)R_s^2 , \tag{8}$$

$$\frac{E_1}{E} = (R_m-R_s^2)/\rho_1 , \tag{9}$$

and

$$\frac{E_r+E_s}{E} = 1-(R_m-R_s^2)/\rho_1 . \tag{10}$$

Combining Eq. (4) with Eq. (10), we get

$$R_{etl} = 2H[(1 - \frac{R_m-R_s^2}{\rho_1})\frac{\langle\int_0^T P_i^2(t)dt\rangle}{\int_0^T P_0^2(t)dt}]^{1/2} . \tag{11}$$

In addition, ρ_1 has to be known before calculating of R_{etl} by Eq. (11). In the first instance, the minimum of ρ_1 can be determined as follows. It is obvious that the energy of the scattered component, E_s, can not be negative. So, by Eq. (8), we get

$$\rho_{min} = \frac{R_m-R_s^2}{1-R_s^2} . \tag{12}$$

Then the possible maximum proportion of subbottom component energy in the total echo energy, $(E_1/E)_{max}$, can be determined by Eq. (9) and Eq. (12). If $(E_1/E)_{max}$ is not great, say less than 10%, the sediment inhomogeneities can be ignored. This seems to be a criterion that determines whether or not the sea bottom can be referred to as homogeneous for the frequency band of the incident wave. In this case, the ETL reflectivity is about the same as the energy one. If $(E_1/E)_{max}$ is not very small, the coherence of the echo tail can be used as an estimate of ρ_1. In such a case, the ETL reflectivity will be significantly different from the energy one, and can make a contribution to the acoustic recognition of marine sediments.

Table 1. The Results of Sea-Going Experiments

Station	P(%)	$M_d(\phi)$	V_a	R_{etl}
C	50.3	5.60	0.22	0.28
L	50.3	5.35	0.26	0.28
I	50.0	5.80	0.27	0.29
H	61.2	5.75	0.15	0.22
O	61.2	5.75	0.12	0.18
B	41.2	4.15	0.36	0.39
D	41.2	3.23	0.40	0.40

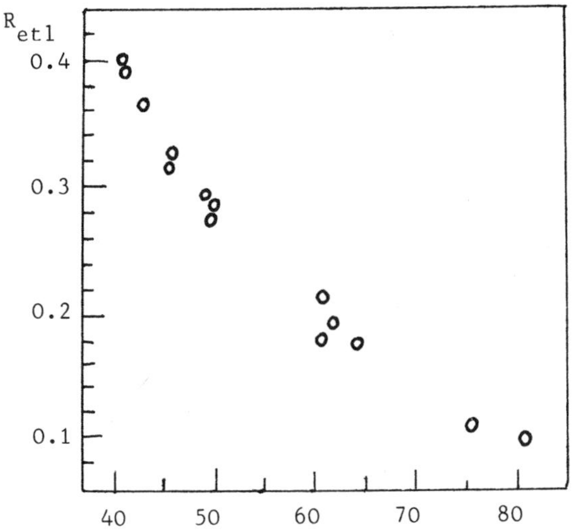

Figure 1: ETL Reflectivity versus Sediment Porosity

RESULTS OF EXPERIMENTS

The experiments were conducted at the South Yellow Sea. Table 1 gives some of the results. The sediment porosity, P, at some stations, for example stations C, L, and I, are about the same. It can be found from the table that the ETL reflectivities are coincident, but the amplitude ones, V_a, are relatively varied. Figure 1 shows the correlation of the ETL reflectivity with sediment porosity. The results of these experiments indicate that the ETL reflectivity is a better feature than the conventional energy and amplitude reflectivities for the recognition of the top-layer sediments of sea bottom.

REFERENCES

Bell, D.L. and Porter, W.J., 1974, Remote sediment classification potential of reflected acoustic signal, in: "Physics of Sound in Marine Sediments", L. Hampton, ed., Plenum, New York.

Berry, M.V., 1973, The statistical properties of echoes diffracted from rough surface, Phil. Trans. Roy. Soc., Ser. A. 273-611.

Brekhovskikh, L.M., 1974, "Ocean Acoustics", Science Press, Moscow (in Russian).

Breslau, L., 1965, Classification of sea floor sediments with a shipborne acoustical system, Le Petrol et la Mer, 132:1-9.

Danbom, S.H., 1976, Sediment classification by seismic reflectivity, OCEAN 76, 16D1-6.

MacIsaac, P.R. and Dunsiger, A.D., 1977, Ocean sediment properties using acoustic sensing, Proc. of 4th Int. Conf. on POAC '77, 1074-1086.

INVERSION OF OCEAN SUBBOTTOM REFLECTION DATA

David J Thomson

Defence Research Establishment Pacific
FMO Victoria, B. C., Canada V0S 1B0

ABSTRACT

An important inverse problem in underwater acoustics is the reconstruction of the ocean subbottom structure (e.g., the density and sound speed profiles) from aperture-limited and bandlimited measurements of bottom-reflected sound. Knowledge of these subbottom properties is needed for accurate simulation of low-frequency sound propagation using sophisticated computer models. This paper extends an approximate inverse solution method (Candel et al., 1980; Thomson, 1984) based on the scattering of plane waves by a stratified medium to accommodate point source geometry for precritical receiver offsets. In addition, a nonlinear renormalization of the Riccati equation for the plane wave reflection coefficient as used by Jaggard and Kim (1985) is shown to improve the accuracy of the reconstructions. To illustrate the method, numerical inversions are presented for both plane-wave and point-source synthetic reflection data for two geoacoustic models.

INTRODUCTION

For a proper assessment of acoustic propagation conditions in the ocean, account must be made for the sound waves that interact with the ocean bottom. The traditional measure of this interaction is the bottom reflection loss, $-10 \log_{10}|R|^2$, where R is the (complex) plane wave reflection coefficient and depends on frequency and grazing angle. Experimentally, $|R|^2$ is inferred from measurements of the propagation loss along bottom-interacting paths which are interpreted using a specular reflection model. Because low-frequency sound waves can penetrate the sediments and undergo refraction within the subbottom, this interpretation can lead to inaccurate, even negative values of bottom loss (Santaniello et al., 1979).

A typical low-frequency acoustic reflectivity experiment performed in the deep ocean uses explosive charges detonated at increasing ranges (offsets) from a fixed receiver system. The analysis and interpretation of the shot waveforms reflected and refracted by the ocean bottom is described by Dicus (1976). The processed traces are fitted to simple geoacoustic models of the bottom sediments. In contrast, direct methods for reconstructing a layered medium from its plane-wave reflection response have been developed. A summary of formal solutions to this so-called inverse problem is reviewed by Newton (1981). Bube and Burridge (1983) give an extensive theoretical and numerical treatment of the one-dimensional problem of reflection seismology for normally incident plane waves. In this case only the impedance profile can be recovered. Separate determination of both the density and the sound speed profiles requires at least two plane-wave reflection seismograms for different angles of incidence (Candel et al., 1980; Coen, 1981; Raz, 1981).

Point-source reflection data first must be processed into signals with plane-wave amplitudes and phases. Several techniques have been proposed for effecting this reduction, e.g., Hankel transform inversion (Frisk et al., 1980), slant stacking (Chapman, 1981), and towed-array beamforming (Kuperman et al., 1985). Each of these methods requires the coherent field to be known along a finite range aperture. When the source/receiver offset is of the order of the distance between the source and the reflecting layers, however, asymptotic evaluation of the Hankel transform by the method of stationary phase is suitable (Fuchs, 1971). In this case, the plane-wave reflection coefficient at each range is related to the measured field response by simple geometrical amplitude and phase factors.

In this paper, the stationary phase approximation is used to simplify the integral representation for the reflected field above the ocean bottom so that with simple preprocessing, the noniterative inverse method of Candel et al. (1980) can accommodate point source reflection data. Moreover, reformulation of their method in terms of a Riccati equation for R allows the more accurate nonlinear approximation described by Jaggard and Kim (1985) to be used. The improved accuracy is illustrated using the normal-incidence plane wave response of a model comprising large sound speed variations. The simultaneous recovery of the density and the sound speed profiles is demonstrated using both plane-wave and point-source reflection data for a model representative of deep ocean sediments.

THEORY

Mathematical Model

The acoustic problem to be considered is based on the following mathematical model. A layered inhomogeneous liquid with density $\rho(z)$ and sound speed $c(z)$ is considered to occupy the region $0 < z < H$ of a cylindrical coordinate system (r, θ, z), z positive downwards. Both $\rho(z)$ and $c(z)$ are assumed to have piecewise continuous variation with depth. The ocean and basement regions are taken to be homogeneous liquid half-spaces characterized by the constant density and sound speed pairs ρ_o, c_o for $z < 0$ and ρ_1, c_1 for $z > H$. Absorption within each region can be accommodated by allowing the wavenumbers $k_o = \omega/c_o$, $k_1 = \omega/c_1$ and $k(z) = \omega/c(z)$ assume complex values. If $p(r,z)\exp(-i\omega t)$ represents the field in $z < 0$ due to a harmonic point source located at $r = 0$, $z = -z_o$, then the pressure $p(r,z)$ has the Hankel transform representation

$$p(r,z) = S_-^{-1}\exp(ik_oS_-) + (i/2)\int_{-\infty}^{\infty}\eta_o^{-1}\exp[i\eta_o(z+z_o)]R(\xi)H_o^{(1)}(\xi r)\xi d\xi, \quad (1)$$

where ξ and $\eta_o = (k_o^2 - \xi^2)^{1/2}$ are the horizontal and vertical wavenumber components for $z < 0$, $S_- = [r^2 + (z-z_o)^2]^{1/2}$ is the slant distance from the source to the receiver, and $H_o^{(1)}$ is the Hankel function of the first kind of order zero. The first term in Eq. (1) corresponds to the direct arrival. The second term corresponds to waves which have interacted with the bottom. $R(\xi)$ denotes the plane wave reflection coefficient associated with the region $z > 0$.

Stationary Phase Approximation

The Hankel transform in Eq. (1) can be inverted for the plane wave reflection coefficient R provided p is known over a suitable range aperture (Frisk et al., 1980). However, if $z, z_0 \gg H$ and if the source/receiver geometry is such that the dominant contribution to the integral occurs for precritical reflection angles, then the wavenumber integral may be evaluated asymptotically by the method of stationary phase (Fuchs, 1971). These conditions arise in deep water reflectivity measurements. Application of the method reduces Eq. (1) to

$$p(r,z) = S_-^{-1}\exp(ik_oS_-) + R(\xi_o)\, S_+^{-1}\exp(ik_oS_+), \tag{2}$$

where $S_+ = [r^2 + (z+z_0)^2]^{1/2}$ is the slant distance from the source to its mirror image about the plane z = 0. The point of stationary phase, $\xi_o = k_o\cos\theta_o$, occurs at the specular angle θ_o determined from $\tan\theta_o = (z+z_0)/r$. When frequency domain deconvolution processing is applied to the time domain equivalent of Eq. (1), the ratio of the reflected to direct fields is obtained (Santaniello et al., 1979). For precritical grazing angles, where Eq. (2) applies, scaling the amplitude by S_+/S_- and modulating the phase by $\exp[ik_o(S_- - S_+)]$ recovers the plane wave reflection coefficient R.

Linearization of the Riccati Equation

For precritical grazing angles, R can be interpreted everywhere within $0 < z < H$ as the local ratio of upgoing to downgoing wave fields that satisfies the nonlinear Riccati equation (e.g., Bregman et al., 1985)

$$dR/dz = -2i\eta R + \gamma(1 - R^2), \tag{3}$$

subject to the initial condition R(H) = 0. Here, $\eta = [k^2(z) - \xi^2]^{1/2}$ is the vertical wavenumber at depth z and $\gamma = (2Y)^{-1}(dY/dz)$ denotes one-half the logarithmic derivative of the "vertical admittance", $Y = (\eta/k)(\rho c)^{-1}$. When $|R| \ll 1$, Eq. (3) can be linearized by neglecting the term in R^2. This corresponds to the situation where the amplitudes of the reflected waves are much smaller than the amplitude of the incident wave, and is equivalent to the application of the forward scattering approximation used by Candel et al. (1980). In this case, Eq. (3) can be integrated analytically and evaluated at z = 0 to give

$$R(\xi) = -\int_0^H \gamma(s)\, \exp[2i\int_0^s \eta(t)dt]\, ds = -\int_0^{\zeta(H)} \gamma(\zeta)\, \exp[2\pi if\zeta/c_o]\, d\zeta, \tag{4}$$

where the last equality is a result of defining the new depth coordinate, $\zeta = (2/k_o)\int_0^z \eta(t)dt$. Recently, Jaggard and Kim (1985) relaxed the restriction $|R| \ll 1$ by making use of a nonlinear approximation to the Riccati equation. They applied the transformation

$$R^*(\xi) = \tanh^{-1}R(\xi) = (1/2)\log\left\{ [1 + R(\xi)]\, [1 - R(\xi)]^{-1}\right\}, \tag{5}$$

together with the approximation $R^*(\xi) \cong R(\xi)/[1 - R^2(\xi)]$ to Eq. (3) and obtained a solution to the Riccati equation equivalent to Eq. (4) but with R replaced by R*.

Inversion Formulas

Eq. (4) is recognized as a Fourier transform for R (or R*). Inversion of this transform recovers γ from which Y can be obtained by integration. The unique determination of both $\rho(z)$ and c(z) requires reflection responses for two probing directions θ_1 and θ_2. With $\theta_2 > \theta_1$, the set of four first order equations required for the simultaneous inversion of density and sound speed are given by (Candel et al., 1980; Thomson, 1984)

$$dY_1/d\zeta_1 = -(2/c_o)\, r_1\, Y_1; \qquad dz/d\zeta_1 = (2\rho c_o Y_1)^{-1}, \tag{6}$$

$$dY_2/d\zeta_1 = -(2/c_o)\, r_2\, Y_2^2/Y_1; \qquad d\zeta_2/d\zeta_1 = Y_2/Y_1, \tag{7}$$

subject to initial conditions that depend on the known values ρ_o, c_o and θ_o in $z < 0$. The subscripts 1 and 2 denote quantities that correspond to the two grazing angles. At each integration depth z, the density and sound speed can be determined using

$$\rho^2 c_o^2 = [\cos^2\theta_1 - \cos^2\theta_2]\, [Y_2^2 - Y_1^2]^{-1}, \tag{8}$$

$$c^2/c_o^2 = [Y_2^2\cos^2\theta_1 - Y_1^2\cos^2\theta_2]\, [Y_2^2 - Y_1^2]^{-1}. \tag{9}$$

In Eq. (6) and Eq. (7), $r(t) = r(\zeta/c_o)$ denotes the plane-wave time response obtained from the reflection coefficient $R(\xi) = R(2\pi f\cos\theta/c_o)$ by inverse Fourier transformation with respect to frequency.

NUMERICAL EXAMPLES

In this section, numerical results based on a computer realization of the inversion formulas of Eqs. (6) - (9) are presented for simulated data. Two geoacoustic models were used in the simulations. Model 1 was adapted from Fig. 3 of Bregman et al. (1985) and includes a low sound speed layer and large sound speed contrasts but no density variation. Model 2 was derived from the analysis of Chapman et al. (1983) of shot waveforms taken in deep water over the Alaskan Abyssal Plain. It contains both sound speed and density discontinuities and a region with a constant ($1\ s^{-1}$) sound speed gradient. Plane-wave (bandlimited) impulse responses are inverted for both models. In addition, point-source seismograms for model 2 are inverted after reduction to plane-wave responses via frequency-domain deconvolution processing and application of the asymptotic amplitude and phase corrections.

Plane-Wave Results

For plane wave excitation, the simulations were performed in the following way. At each grazing angle, the complex frequency response R(f) was computed by integrating Eq. (3) numerically at discrete frequencies f_k separated by $\Delta f = 0.125$ Hz. The complex reflection coefficients were filtered using a sinc x_k function with $x_k = f_k\pi/90$ and then zero-padded to give a complex sequence length of 1024. The time domain response, r(t), was computed every $\Delta t = 1/(N\Delta f)$ s using an inverse fast Fourier transform (FFT) and multiplied by Δf to approximate the analytical Fourier transform result. With r_1 and r_2 computed for grazing angles θ_1 and θ_2, recovery of c(z) and $\rho(z)$ proceeded according to Eq. (6) - (9). Finally, the reconstructed profiles were compared to the sound speed and density profiles initially used to generate the synthetic impulse responses.

In Fig. 1, the reconstructed sound speed profile is compared to the initial profile for Model 1. Since the density profile is known for this model, only a single bandlimited (0 - 90 Hz) normally incident plane wave impulse response was used for the inversion. Fig. 1(a) shows the comparison when the forward scattering approximation of Eq. (5) was used. Although the agreement is very good, the departures are observed to increase with depth, a feature common to Born-like methods which do not account for subbottom multiples (e.g., Lahlou et al., 1983). The large sound speed contrasts for this model give rise to significant subbottom

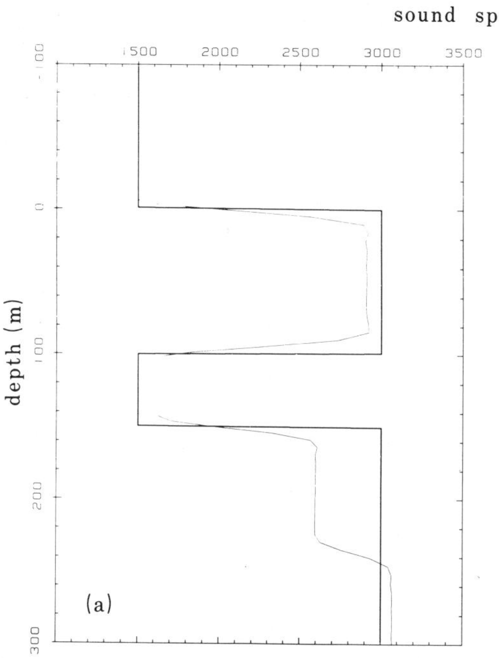

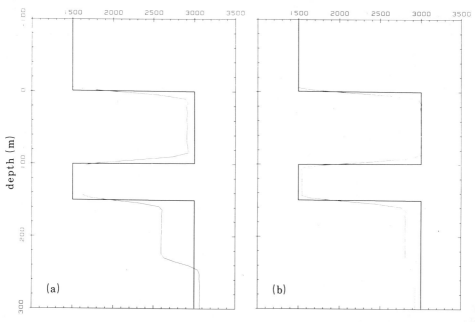

Fig. 1. Reconstructed sound speed profile for Model 1 (a) before and (b) after $\tanh^{-1}R$ transformation. The inversion is based on a single bandlimited (0 - 90 Hz) sinc-filtered plane wave impulse response at normal incidence.

reverberation, the effects of which are evident in the reconstructed curve near z = 250 m. In Fig. 1(b), the results are seen to be improved when the reflection coefficient is first transformed according to the inverse hyperbolic tangent relation of Eq. (5). In this case, even the effects of the multiple arrivals are observed to be reduced considerably.

In Fig. 2, the reconstructed sound speed and density profiles are compared to the initial profiles for the Alaskan Abyssal Plain model. The reconstructed profiles are based on the simultaneous integration of Eq. (6) - (9) for two plane wave impulse responses at the oblique grazing angles of 45° and 60°. The reflection coefficients of both impulse responses were subjected to the $\tanh^{-1}$ transformation and sinc-filtering in the frequency band 0 - 90 Hz prior to the inversion. The agreement is seen to be excellent, especially for the density variations. For this model, the multipath amplitudes are small and their effects are not evident for the depths shown. Higher frequencies would have been needed to resolve the small decrease in sound speed at z = 0 m. It is worthwhile remarking that the presence of the low frequencies in the bandlimited impulse responses used in both models enabled the recovery of the trends of the profiles, including the sound speed gradient for Model 2.

Point Source Results

To examine the accuracy of the asymptotic result given in Eq. (2), point source seismograms were generated for Model 2 and processed into plane wave impulse responses prior to inversion. The simulation was

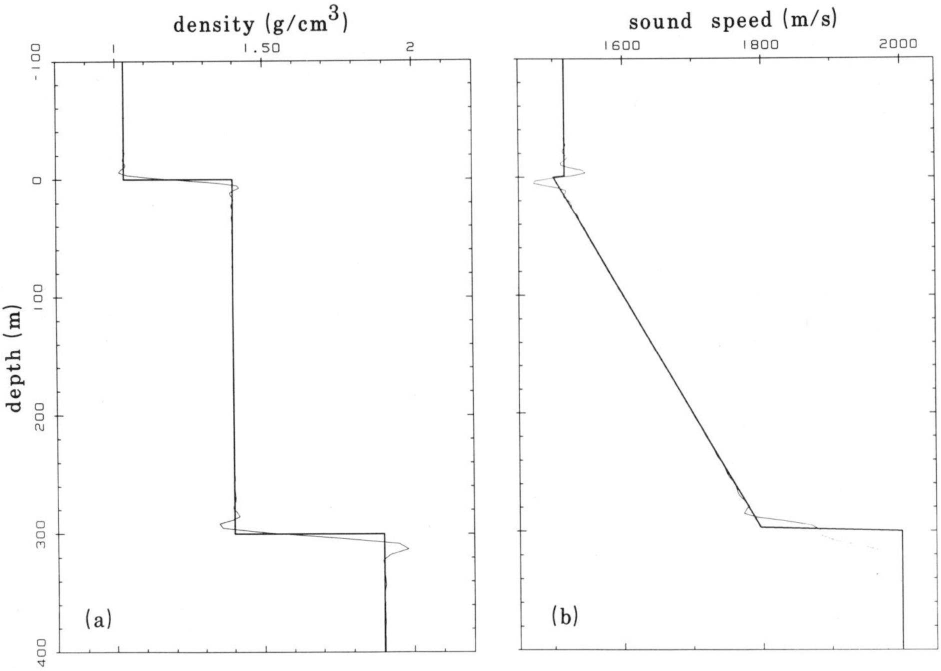

Fig. 2. Reconstructed profiles of (a) density and (b) sound speed for Model 2. The simultaneous inversion is based on two bandlimited (0 - 90 Hz), R*-transformed, sinc-filtered plane wave impulse responses at grazing angles of 45° and 60°.

carried out in the following way. Synthetic seismograms were generated at several ranges using a recently developed computer code described by Schmidt and Tango (1986). These data were then processed into ocean bottom impulse responses by frequency domain deconvolution methods (e.g., Santaniello et al., 1979). Specifically, the ocean bottom impulse response H(f) was determined at each source/receiver offset according to

$$H(f) = \left\{ Y(f)\, X^*(f) \,/\, [\, |X(f)|^2 + A \,] \right\} G(f), \tag{10}$$

where X*(f) is the complex conjugate spectrum of the direct path arrival, Y(f) is the corresponding spectrum of the bottom interacting arrival, A is a suitably determined "white noise" constant, and G(f) is a filter function used to exclude frequencies outside the band of interest. Both X and Y were determined by Fourier transformation of appropriate time-windowed intervals of each synthetic seismogram. For the results in this paper, A = 0 and G was chosen to be a sinc function over the band 0 - 90 Hz. Finally, for each source/receiver location, the plane wave impulse response was estimated from Eq. (10) according to the asymptotic result of Eq. (2), i.e.,

$$R(f) \cong (S_+/S_-) \exp[ik_0(S_- - S_+)]\, H(f). \tag{11}$$

Fig. 3 gives the comparison between the initial and reconstructed profiles based on the application of Eq. (10) and Eq. (11) to synthetic point source seismograms computed for Model 2. The source and receiver were located 3600 m and 3225 m respectively above the ocean bottom interface. The results are shown for simultaneous inversion of two processed

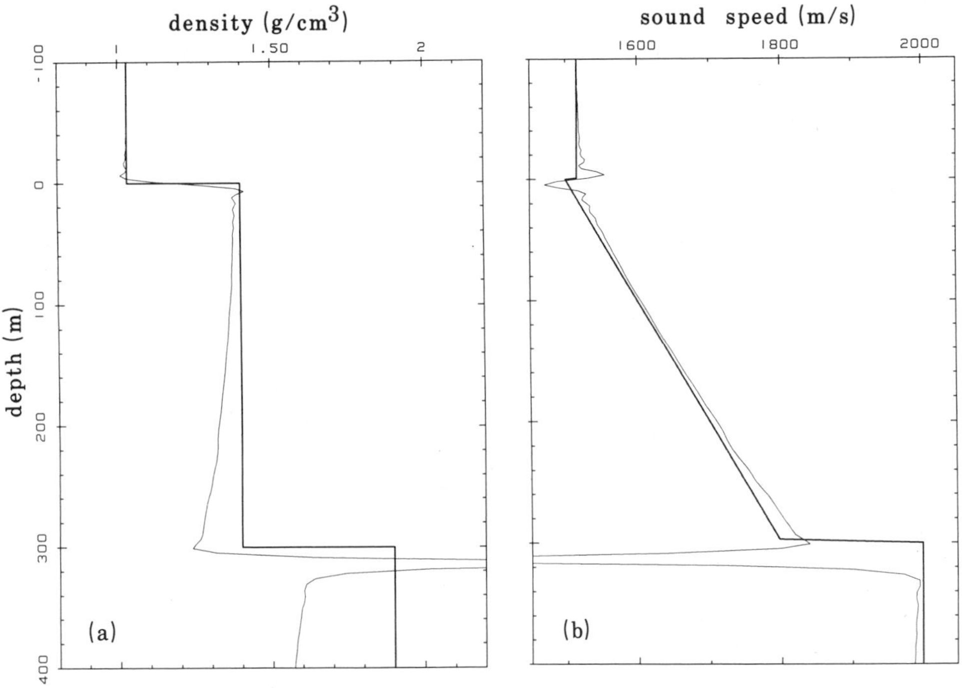

Fig. 3. Reconstructed profiles of (a) density and (b) sound speed for Model 2. The simultaneous inversion is based on two bandlimited (0 - 90 Hz), R*-transformed, sinc-filtered point source reflection responses at offsets of 3500 m and 6000 m.

seismograms at horizontal offsets of 3500 m and 6000 m. The source waveform consisted of a triangular pulse of duration 10 ms. In spite of the rather large offsets, very good reconstructions are observed down to a depth of 300 m. The departures evident at greater depths are due in part to the presence of low-frequency, signal-like artifacts in the synthetic seismograms. As discussed by Chapman and Orcutt (1985), these artifacts arise from sampling and/or truncation effects associated with the numerical integration of the wavenumber representation of Eq. (2) resulting in ghost arrivals time-coincident with the reflected waves.

Two remarks should be made about the application of this method to real data. First, the very low frequency information needed for recovering the profile trends is usually missing from recorded data. In this case, the data must be processed differently in order to recover the trends, e.g., Lahlou et al. (1983) describe a method based on measuring peak amplitudes. Second, the reconstructions in this paper are based on minimal data only. For real data, the signal-to-noise ratio could be improved by inverting many impulse responses simultaneously (Bregman et al., 1985).

SUMMARY

The results presented in this paper demonstrate that the density and sound speed profiles of a layered ocean subbottom can be reconstructed from two bandlimited plane-wave impulse responses at different precritical grazing angles. Point-source reflection data simulated for a model of the

Alaskan Abyssal Plain were reduced to plane-wave responses by applying frequency-domain deconvolution processing followed by simple (asymptotic) amplitude and phase adjustments. A nonlinear transformation of the reflection coefficient was shown to improve the accuracy of the reconstructions.

REFERENCES

Bregman, N. D., C. H. Chapman, and R. C. Bailey, 1985, A noniterative procedure for inverting plane-wave reflection data at several angles of incidence using the Riccati equation, Geophys. Prospect., 33:185.

Bube, K. P. and Burridge, R., 1983, The one-dimensional inverse problem of reflection seismology, SIAM Rev., 25:497.

Candel, S. M., DeFillipi, F., and Launay, A., 1980, Determination of the inhomogeneous structure of a medium from its plane wave reflection response, Part I: a numerical analysis of the direct problem, and Part II: a numerical approximation, J. Sound Vib., 68:571.

Chapman, C. H., 1981, Generalized Radon transforms and slant stacks, Geophys. J. R. astr. Soc., 66:445.

Chapman, C. H. and Orcutt, J. A., 1985, The computation of body wave synthetic seismograms in laterally homogeneous media, Rev. Geophys., 23:105.

Chapman, N. R., Zelt, C. A., and Busch, A. E., 1983, Geoacoustic modelling of deep ocean abyssal plains, in: "Acoustics and the Sea-Bed", N. G. Pace, ed., Bath University Press, Bath.

Coen, S., 1981, Density and compressibility profiles of a layered acoustic medium from precritical incidence data, Geophysics, 46:1244.

Dicus, R. L., 1976, Preliminary investigations of the ocean bottom impulse response at low frequencies, U.S. Naval Oceanographic Office Tech. Note TN 6130-4-76.

Frisk, G. V., Oppenheim, A. V., and Martinez, D. R., 1980, A technique for measuring the plane-wave reflection coefficient of the ocean bottom, J. Acoust. Soc. Am., 68:602.

Fuchs, K., 1971, The method of stationary phase applied to the reflection of spherical waves from transition zones with arbitrary depth-dependent elastic moduli and density, Z. Geophysik, 37:89.

Jaggard, D. L. and Kim, Y., 1985, Accurate one-dimensional inverse scattering using a nonlinear renormalization technique, J. Opt. Soc. Am. A, 2:1922.

Kuperman, W. A., Werby, M. F., Gilbert, K. E., and Tango, G. J., 1985, Beam forming on bottom-interacting tow-ship noise, IEEE J. Oceanic Eng., OE-10:290.

Lahlou, M., Cohen, J. K., and N. Bleistein, 1983, Highly accurate inversion methods for three-dimensional stratified media, SIAM J. Appl. Math., 43:726.

Newton, R. G., 1981, Inversion of reflection data for layered media: A review of exact methods, Geophys. J. R. astr. Soc., 65:191.

Raz, S., 1981, Direct reconstruction of velocity and density profiles from scattered field data, Geophysics, 46:832.

Santaniello, S. R., DiNapoli, F. R., Dullea, R. K., and Herstein, P. D., 1979, Studies on the interaction of low-frequency acoustic signals with the ocean bottom, Geophysics, 44:1922.

Schmidt, H. and G. Tango, 1986, Efficient global matrix approach to the computation of synthetic seismograms, Geophys. J. R. astr. Soc., 84:331.

Thomson, D. J., 1984, An inverse method for reconstructing the density and sound speed profiles of a layered ocean bottom, IEEE J. Ocean. Eng., OE-9:18.

AN INVERSE METHOD FOR OBTAINING THE ATTENUATION PROFILE AND SMALL VARIATIONS IN THE SOUND SPEED AND DENSITY PROFILES OF THE OCEAN BOTTOM

Subramaniam D. Rajan and George V. Frisk

Woods Hole Oceanographic Institution
Woods Hole, MA 02543

ABSTRACT

An inverse method is presented for the determination of the compressional wave speed, compressional wave attenuation, and density as a function of depth for a horizontally stratified ocean bottom. It is based on a perturbation technique for which the required input information is the plane-wave reflection coefficient of the bottom as a function of incident angle at a fixed frequency. The reflection coefficient is related to variations of the acoustic properties about known reference values through a nonlinear integral equation which is then linearized using the Born approximation. An acceptable stable solution of the integral equation is obtained using a priori constraints on the solution. Resolution of the solution obtained is studied using the resolving power theory of Backus and Gilbert[1]. Examples of inversions using synthetic data are presented using noise-free and noisy data. Results obtained with noise-free data show good agreement between true and reconstructed profiles. Inversions performed with noisy data yield stable, acceptable results.

INTRODUCTION

The compressional wave attenuation profile of marine sediments is an important acoustic parameter that affects sound propagation in the ocean. Methods for obtaining this parameter from field measurements have been reported in the literature[2]. These methods in general follow the seismic reflection/refraction type experiment using a broadband source. Assumptions that are made in determining the compressional wave attenuation profile from such experiments are that the attenuation is linearly dependent on frequency, and that the dispersion of the compressional wave speed is small and can be ignored. However, the linear dependence of attenuation on frequency has not been proved conclusively[2] especially in the low-frequency region (50-500Hz) of interest to us. Also visco-elastic models for marine sediments (e.g. the standard linear model[3]) indicate that for highly permeable marine sediments like coarse sand, the dispersion of compressional waves can be substantial. In view of these facts, we propose a monochromatic source technique for determining the attenuation profile using the plane-wave reflection coefficient as a function of horizontal wavenumber (or incident angle) as input data. The reflection coefficient can be obtained from measurements of the point source field as a function of range[4]. In solving the inverse prolem we assume that good estimates of the compressional wave speed and density profiles are available. Typically these initial estimates are obtained using schemes that treat the sediment as lossless and therefore will have small errors[5]. The inverse method proposed yields the corrections to the sound speed and density profiles in addition to obtaining the attenuation profile.

FORMULATION OF THE INVERSE PROBLEM

In formulating the inverse problem, we make the assumptions that the marine sediments are horizontally stratified and can be modelled as a fluid at the frequencies of interest to us (50-500Hz). The acoustic parameters that are of interest are the compressional wave speed, compressional wave attenuation, and density, all being functions of depth only. The model of marine sediments we consider is shown in Fig. 1(a). Here region I represents the ocean, region II the sediment layer, and region III the subbottom. Although the theory can be developed for the case where the acoustic properties in regions II and III are the unknowns to be determined[6], for simplicity we assume that the acoustic parameters in regions I and III are known and the only unknowns are the acoustic parameters in the sediment layer. Further we assume that an initial guess of the sound speed and density in region II is available. Figure 1(b) represents this initial guess model or the background model for the ocean bottom. In Fig. 1, C_0 and ρ_0 represent the constant sound speed and density in the water column. Similarly C_2, α_2, and ρ_2 are the constant sound speed, attenuation and density in the subbottom.

Consider the model in Figure 1(a). Let a plane wave of unit amplitude and frequency ω be incident on the interface at $z = 0$ at an angle θ, with respect to the

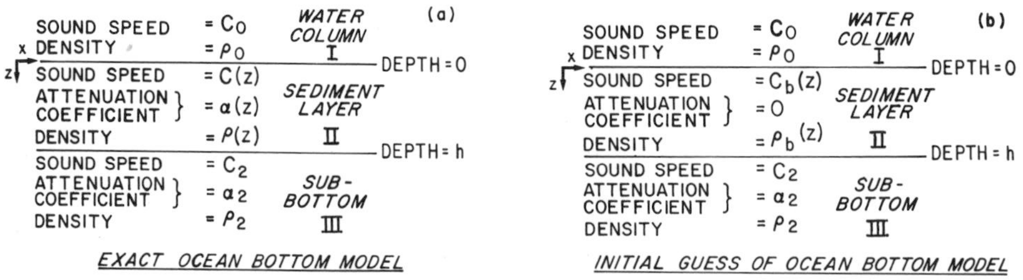

Fig. 1. The (a) exact and (b) guess models for the ocean bottom.

vertical. The field in region I is the sum of the incident and reflected waves and is

$$P_0(z) = \exp\left[i(K_0^2 - k_x^2)^{1/2} z\right] + R(k_x)\exp\left[-i(K_0^2 - k_x^2)^{1/2} z\right], \quad z \le 0, \tag{1}$$

where $P_0(z)$ is the pressure field, $R(k_x)$ is the plane -wave reflection coefficient, $k_x = K_0 \sin\theta$ is the horizontal wavenumber, and $K_0 = \omega / C_0$. Since the subbottom is a half-space, only an outgoing transmitted field $P_2(z)$ exists in this region and is

$$P_2(z) = T(k_x)\exp\left[i(K_2^2 - k_x^2)^{1/2} z\right], \quad z \ge h, \tag{2}$$

where $T(k_x)$ is the transmission coefficient.

In the sediment layer, the pressure field is obtained by solving the equation

$$\frac{d^2 P(z)}{dz^2} + \rho(z)\left[\frac{1}{\rho(z)}\right]' \frac{dP(z)}{dz} + \left[K^2(z) - k_x^2\right] P(z) = 0, \quad 0 \le z \le h, \tag{3}$$

where

$$K(z) = \frac{\omega}{C(z)} + i\alpha(z), \tag{4}$$

and $C(z), \rho(z)$, and $\alpha(z)$ are the sound speed, density, and attenuation respectively. The boundary conditions to be satisfied are the continuity of pressure and normal particle velocity at the interfaces $z = 0$ and $z = h$. This differential equation can be transformed into a Schrodinger type equation by defining a new function $v(z) = \rho^{-1/2}(z)P(z)$. Substituting for $P(z)$ in Eq. (3), we obtain

$$\frac{d^2v(z)}{dz^2} + [K^2(z) + \mu(z) - k_x^2]v(z) = 0, \tag{5}$$

where

$$\mu(z) = \frac{\rho^{1/2}(z)}{2}\left[\frac{\rho'(z)}{\rho^{3/2}(z)}\right]'. \tag{6}$$

Here the primes denote the derivative with respect to z.

For the background model we have similar expressions for the pressure field in the three regions.

$$P_{0b}(z) = \exp[i(K_0^2 - k_x^2)^{1/2}z] + R_b(k_x)\exp[-i(K_0^2 - k_x^2)^{1/2}z], \quad z \le 0, \tag{7}$$

$$\frac{d^2v_b(z)}{dz^2} + [K_b^2(z) + \mu_b(z) - k_x^2]v_b(z) = 0, \quad 0 \le z \le h, \tag{8}$$

$$P_{2b}(z) = T_b(k_x)\exp[i(K_2^2 - k_x^2)^{1/2}z], \quad z \ge h, \tag{9}$$

where

$$v_b(z) = \rho_b^{-1/2}(z)P_b(z), \tag{10}$$

$$\mu_b(z) = \frac{\rho_b^{1/2}(z)}{2}\left[\frac{\rho_b'(z)}{\rho^{3/2}(z)}\right]', \tag{11}$$

and

$$K_b(z) = \frac{\omega}{C_b(z)}. \tag{12}$$

Here $P_{0b}(z)$, $P_b(z)$, and $P_{2b}(z)$ are the pressure fields in the three regions, $R_b(k_x)$ and $T_b(k_x)$ are the reflection and transmission coefficients for the background model, and $C_b(z)$ and $\rho_b(z)$ are the sound speed and the density in the sediment layer. The attenuation in the sediment layer for the background model is zero. We now express $K(z) = K_b(z) + \delta K(z) + i\alpha(z)$ and $\mu(z) = \mu_b(z) + r(z)$. The quantities $\delta K(z)$ and $i\alpha(z)$ are small real and imaginary perturbations around $K_b(z)$ since $C(z)$ is close to the background value $C_b(z)$ and the attenuation $\alpha(z)$ is small compared to $K_b(z)$ for marine sediments at the frequencies of interest to us. Similarly $r(z)$ is small compared to $\mu_b(z)$ since $\rho_b(z)$ is close to $\rho(z)$. Substituting this expression for $K(z)$ and $\mu(z)$ into Eq. (5), we obtain

$$\frac{d^2v(z)}{dz^2} + [K_b^2(z) + \mu_b(z) - k_x^2]v(z) = -\{2K_b(z)[\delta K(z) + i\alpha(z)] + r(z)\}v(z). \tag{13}$$

We now multiply Eq. (13) by $v_b(z)$ and Eq. (8) by $v(z)$, take their difference and integrate both sides between the limits 0 and h to obtain

$$v_b(z)v'(z) - v(z)v_b'(z)|_0^h = \int_0^h -\{2K_b(z)[\delta K(z) + i\alpha(z)] + r(z)\}v(z)v_b(z)dz. \tag{14}$$

Assuming that the densities in the true model and the background model are continuous up to their second derivatives, the left hand side of Eq. (14) reduces to

$$[\rho_b(z)\rho(z)]^{-1/2}[P_b(z)P'(z) - P(z)P_b'(z)]|_0^h. \tag{15}$$

Making use of Eqs. (1), (2), (7), and (9) we obtain

$$\frac{2i(K_0^2 - k_x^2)^{1/2}[R_b(k_x) - R(k_x)]}{\rho_0} = \int_0^h \{2K_b(z)[\delta K(z) + i\alpha(z)] + r(z)\}v_b(z)v(z)dz. \tag{16}$$

This equation cannot be solved exactly for the unknowns $\delta K(z), \alpha(z)$, and $r(z)$ because $v(z)$ depends on the unknown. The Born approximation is now applied to this equation, i.e. we set $v(z) = v_b(z)$. Then we have

$$\frac{2i(K_0^2 - k_x^2)^{1/2}[R_b(k_x) - R(k_x)]}{\rho_0} = \int_0^h \left\{ \frac{2K_b(z)[\delta K(z) + i\alpha(z)] + r(z)}{\rho_b(z)} \right\} P_b^2(k_x, z)dz. \tag{17}$$

We now have an integral equation which relates the plane-wave reflection coefficient $R(k_x)$ to the unknowns. The imaginary part of the solution to Eq. (17) yields $2K_b(z)\alpha(z)$ from which the attenuation $\alpha(z)$ is obtained. The real part of the solution yields $2K_b(z)\delta K(z) + r(z)$, and we must separate these two terms in order to determine $C(z)$ and $\rho(z)$. To do this we follow the approach of Stickler[7] and perform the experiment at two frequencies ω_1 and ω_2. Let the real parts of the unknowns determined at the two frequencies be $d_1(z)$ and $d_2(z)$. Expressing $K_b(z)$ and $\delta K(z)$ in terms of the frequency and the sound speed we have

$$d_1(z) = \frac{2\omega_1^2[C(z) - C_b(z)]}{C_b^3(z)} + r(z), \tag{18}$$

$$d_2(z) = \frac{2\omega_2^2[C(z) - C_b(z)]}{C_b^3(z)} + r(z). \tag{19}$$

Solving this pair of equations, we obtain $C(z)$ and $r(z)$. Knowing $r(z)$ and $\mu_b(z)$ we compute $\mu(z)$ which is then used in Eq. (6) to solve for $\rho(z)$.

Region of validity of the Born approximation

An important consideration in obtaining the solution to this integral equation is to determine the region where the Born approximation is valid. For a perturbation $\alpha(z)$ in the attenuation coefficent it can be shown[6] that the Born approximation is valid when

$$\|\alpha\|\|K_b\|h|G(z, z'; k_x)|_{max} << 1, \tag{20}$$

where $\|.\|$ represents the norm defined by

$$\|f\| = \left\{ \int_0^h [f^2(z)]dz \right\}^{1/2}, \tag{21}$$

and $G(z, z'; k_x)$ is the solution to the equation

$$\frac{d^2G(z, z'; k_x)}{dz^2} + [K_b^2(z) - k_x^2]G(z, z'; k_x) = \delta(z - z'), \;\; 0 < z, z' < h. \tag{22}$$

The region of validity of the Born approximation therefore depends on the magnitude of the perturbation, the extent of the perturbation, and the magnitude of the Green's function $G(z, z'; k_x)$. For a given perturbation and depth h the optimum aperture will correspond to k_x values for which $|G(z, z'; k_x)|_{max}$ is small. It can be shown[6] that this region corresponds to angles of incidence which are pre-critical with respect to the subbottom.

SOLUTION OF THE INTEGRAL EQUATION

We now describe the method for solving Eq. (17). To make the analysis simpler we consider the case where the only unknown is the compressional wave attenuation $\alpha(z)$. We therefore make $C(z) = C_b(z)$ and $\rho(z) = \rho_b(z)$. With this assumption $\delta K(z) = 0$ and $r(z) = 0$ and the integral equation is then

$$\frac{(K_0^2 - k_x^2)^{1/2}[R_b(k_x) - R(k_x)]}{\rho_0} = \int_0^h \frac{K_b(z)}{\rho_b(z)} \alpha(z) P_b^2(k_x, z)dz. \tag{23}$$

From field measurements the plane-wave reflection coefficient $R(k_x)$ is obtained. The plane-wave reflection coefficient $R_b(k_x)$ is computed for the known background model. Reflection coefficients from field measurements are normally available only at a discrete set of horizontal wavenumbers[4] $[R(k_{x_n})], n = 1, \cdots, N$. Equation(23) then takes the form

$$d_n = \int_0^h \alpha(z) G_n(z) dz, \; n = 1, \cdots, N, \tag{24}$$

where

$$G_n(z) = \frac{K_b(z)}{\rho_b(z)} P_b^2[(k_{x_n}), z], \tag{25}$$

and

$$d_n = [K_0^2 - (k_{x_n})^2]^{1/2} [R_b(k_{x_n}) - R(k_{x_n})]. \tag{26}$$

Equation (24) is a Fredholm integral equation of the first kind. Two important problems one encounters in solving equations of this type are those of non-uniqueness and instability of the solution. Methods for solving this class of integral equation or equivalently the matrix equation $\boldsymbol{Ax} = \boldsymbol{y}$ have been studied by many investigators and different approaches are available. We use the method originally proposed by Phillips[8] and later modified by Twomey[9]. This method falls within the general category of regularization methods and is based on adding constraints to the solution. Specifically we assume that the solution is in some sense smooth. A measure of smoothness $S[\alpha(z)]$ is defined and we look for the solution $\alpha(z)$ which minimises $S[\alpha(z)]$ subject to Eq. (24) being satisfied. The measure of smoothness $S[\alpha(z)]$ is defined as

$$S[\alpha(z)] = \int_0^h \left[\frac{d^2 \alpha(z)}{dz^2} \right]^2 dz. \tag{27}$$

To solve Eq. (24) we first reduce it to a matrix equation by using an appropriate quadrature scheme.

$$d_n = \sum_{m=1}^{M} G_{nm} \alpha_m \quad n = 1, \cdots, N, \tag{28}$$

or equivalently

$$\boldsymbol{d} = \boldsymbol{G\alpha}. \tag{29}$$

The representation of the quadratic measure for smoothness in discrete form is

$$S(\boldsymbol{\alpha}) = \sum_m (\alpha_m - 2\alpha_{m-1} + \alpha_{m+1})^2. \tag{30}$$

This measure can be written as $\boldsymbol{\alpha}^T \boldsymbol{H} \boldsymbol{\alpha}$ where the superscript refers to the transpose. The matrix $\boldsymbol{H}$ is

$$\boldsymbol{H} = \begin{bmatrix} 1 & -2 & 1 & . & . & . & . & . & . & . & . \\ -2 & 5 & -4 & 1 & . & . & . & . & . & . & . \\ 1 & -4 & 6 & -4 & 1 & . & . & . & . & . & . \\ . & 1 & -4 & 6 & -4 & 1 & . & . & . & . & . \\ . & . & . & . & . & . & . & . & . & . & . \\ . & . & . & . & . & . & . & . & . & . & . \\ . & . & . & . & . & . & . & . & 1 & -2 & 1 \end{bmatrix}. \tag{31}$$

The problem is therefore to find an $\boldsymbol{\alpha}$ that minimizes $S(\boldsymbol{\alpha})$ subject to Eq. (29) being satisfied. Using λ as the Lagrange multiplier, we need to solve for the $\boldsymbol{\alpha}$ that minimizes

$$\lambda \left[(\boldsymbol{d} - \boldsymbol{G\alpha})^T (\boldsymbol{d} - \boldsymbol{G\alpha}) \right] + \boldsymbol{\alpha}^T \boldsymbol{H} \boldsymbol{\alpha}. \tag{32}$$

Differentiating with respect to each element of $\boldsymbol{\alpha}$ and equating it to zero, we obtain

$$\boldsymbol{\alpha} = (\boldsymbol{G}\boldsymbol{G}^T + \lambda \boldsymbol{H})^{-1} \boldsymbol{G}^T \boldsymbol{d}. \tag{33}$$

By imposing the smoothness constraint as described above, we have in effect chosen out of many possible solutions which satisfy the data one which meets the smoothness requirement. We see that the non-uniqueness inherent in the problem has been overcome by redefining what is an acceptable solution.

To obtain Eq. (17), we linearized the nonlinear equation around the background model by using the Born approximation. If the background model is not close to the true model, an iterative process is adopted wherein the solution obtained at one step of the iteration is used as the background model to generate a new estimate of the unknown.

Resolution

Having obtained a solution by the procedure described above, we need to determine how close these estimates are to the true values. Backus and Gilbert[1] have proposed a method for determining the resolution obtainable when the unknown is determined from a finite set of data. We start with the equation

$$d_n = \int_0^h \alpha(z) G_n(z) dz, \quad n = 1, \cdots, N. \tag{34}$$

Let $\hat{\alpha}(z_0) = \sum_{n=1}^{N} a_n(z_0) d_n$, where $\hat{\alpha}$ is an estimate of α obtained from the data. Substituting for d_n from Eq. (34) we have

$$\hat{\alpha}(z_0) = \int_0^h \alpha(z) \sum_{n=1}^{N} a_n(z_0) G_n(z) dz, \tag{35}$$

$$\hat{\alpha}(z_0) = \int_0^h \alpha(z) A(z, z_0) dz, \tag{36}$$

where

$$A(z, z_0) = \sum_{n=1}^{N} a_n(z_0) G_n(z). \tag{37}$$

If $A(z, z_0) = \delta(z - z_0)$ then $\hat{\alpha}(z_0) = \alpha(z)$, which is the exact solution. However with a finite set of data it is not possible to obtain the exact solution and the estimate of $\alpha(z)$ will be a local average of the exact values. The averaging kernel $A(z, z_0)$ is obtained by determining the coefficients $a_n(z_0)$ which make $A(z, z_0)$ approach a delta function in a least squared sense. The width of the averaging kernel obtained is a measure of the resolution possible. The resolution measure used in this paper is the resolution width $RL(z_0)$ defined by,

$$RL(z_0) = h \left(\frac{\int_0^h |A(z, z_0)|^2 dz_0}{\int_0^h |A(z_0, z_0)|^2 dz_0} \right). \tag{38}$$

RECONSTRUCTION OF AN ATTENUATION PROFILE

We apply the procedure desribed here to reconstruct an attenuation profile in the sediment layer using synthetic data. For this purpose we assume the model for the bottom shown in Fig. 2, where the exact attenuation profile in the sediment layer is shown in Fig. 2(b). The sound speed and density profiles for the sediment layer are assumed known, and the only unknown is the attenuation profile. The background model consists of the model in Fig. 2(a) with $\alpha(z)$ in the sediment layer equal to zero. The quantities $R(k_x)$, $R_b(k_x)$, and $P_b(k_x, z)$ are computed using a propagator matrix method[6]. In all of these examples, we compute the reflection coefficient for a discrete set of pre-critical angles (20 values of k_x) corresponding to the optimum aperture for the Born approximation.

The result of the reconstruction using noise-free data is shown in Fig. 2. Seven iterations were performed to obtain this result. We note that the reconstructed profile is in good agreement with the exact profile. The resolution of this result was computed using the method described earlier and the resolution widths at the various depths are shown in Fig. 3.

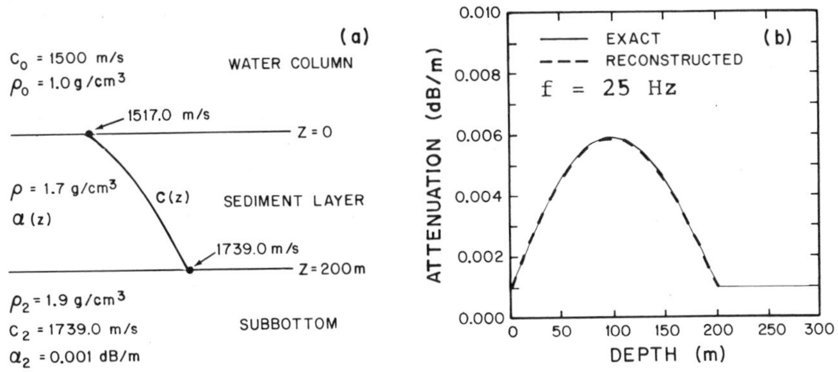

Fig. 2. The (a) bottom model and the (b) exact and reconstructed attenuation profiles.

We now add zero mean Gaussian random noise to the real and imaginary part of the input data. For a given signal to noise ratio (SNR) the variance σ_d^2 is determined from

$$SNR = 10 \log_{10} \left[\frac{I_s}{\sigma_d^2} \right], \tag{39}$$

where

$$I_s = \frac{1}{N} \sum_{n=1}^{N} |(R(k_{x_n})|^2 . \tag{40}$$

The reconstruction obtained for $SNR = 30dB$ is shown in Fig. 4. We note that even with the addition of noise the inversion algorithm remains stable and the reconstructed result retains the general characteristics of the exact profile, though appearing noisy.

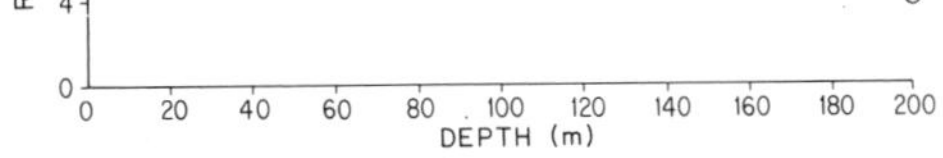

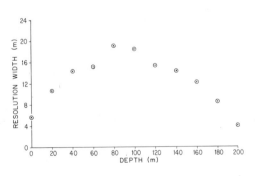

Fig. 3. Resolution width vs depth.

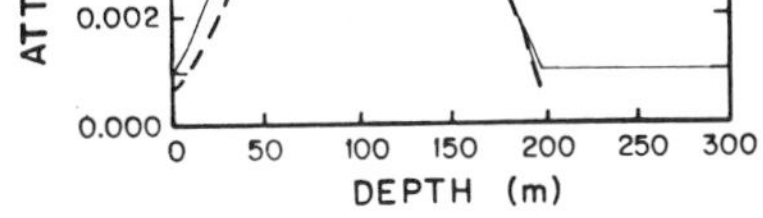

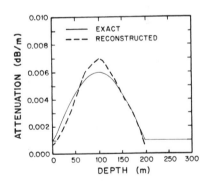

Fig. 4. Reconstruction using noisy data.

RECONSTRUCTION OF SOUND SPEED AND ATTENUATION PROFILES

We now demonstrate the simultaneous reconstruction of sound speed and attenuation profiles. Fig. 5(a) gives the exact, reconstructed, and background profiles for the sound speed while Fig. 5(b) shows the reconstructed and exact profiles of the attenuation. The background attenuation profile was assumed zero. Again good agreement between the exact and reconstructed profiles is observed.

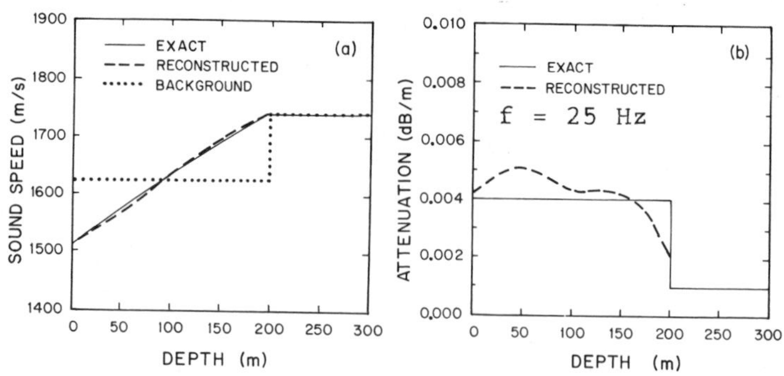

Fig. 5. Exact and reconstructed (a) sound speed and (b) attenuation profiles.

CONCLUSIONS

We have demonstrated using synthetic data that a perturbation method can be used to obtain the acoustic parameters of the ocean bottom given the plane-wave reflection coefficient at a discrete set of angles in the precritical region. Work is now in progress to determine the plane-wave reflection coefficient from field measurements which will then be used in the perturbation method described in this paper to obtain the acoustic parameters of the bottom.

REFERENCES

1. G. Backus and F. Gilbert, The resolving power of gross earth data, Geophys. J. R. Astr. Soc.,16:169, (1968).

2. R. D. Stoll, Marine sediment acoustics, J. Acoust. Soc. Am., 77:1789, (1985).

3. Keiiti Aki and P. G. Richards, " Quantitative Seismology: Theory and Methods, Vol. I ", Freeman, San Francisco, (1980).

4. G. V. Frisk, A. V. Oppenheim, and D. R. Martinez, A technique for measuring the plane-wave reflection coefficient of the ocean bottom, J. Acoust. Soc. Am., 68:602, (1980).

5. A. Merab, " Reconstruction of ocean bottom velocity profiles from monochromatic scattering data," Sc.D. Thesis, M.I.T./W.H.O.I. Joint Program, Cambridge, MA and Woods Hole, MA (August 1986).

6. Subramaniam D. Rajan, " An inverse method for obtaining the attenuation profile and small variations in the sound speed and density profiles of the ocean bottom, " Ph.D Thesis, M.I.T./W.H.O.I. Joint Program, Cambridge, MA and Woods Hole, MA (May 1985).

7. D. C. Stickler, Inverse scattering in a stratified medium, J. Acoust.Soc. Am., 74:994, (1983).

8. D. L. Phillips, A technique for numerical solution of Fredholm integral equations of the first kind, J. Assoc. Compt. Mach., 9:84, (1962).

9. S. Twomey, On the numerical solution of the Fredholm integral equation of the first kind by the inversion of the linear system produced by quadrature, J. Assoc. Compt. Mach.,10:97, (1963).

A PERTURBATIVE INVERSE METHOD FOR THE DETERMINATION OF GEOACOUSTIC PARAMETERS IN SHALLOW WATER

James F. Lynch, Subramaniam D. Rajan, and George V. Frisk

Woods Hole Oceanographic Institution
Woods Hole, MA 02543

ABSTRACT

An inverse method is described for obtaining the acoustic properties of a horizontally stratified bottom in shallow water from measurements of the magnitude and phase versus range of the pressure field due to a CW point source. These data are numerically Hankel transformed to obtain the depth-dependent Green's function versus horizontal wavenumber. The Green's function contains prominent peaks at horizontal wavenumbers corresponding to the eigenvalues for any trapped and virtual modes excited in the waveguide. The eigenvalues are then used as input data to a perturbative inverse scheme. The method is applied to the problem of determining the compressional wave speed as a function of depth and is demonstrated for experimental data. The results are compared with those obtained using an iteration of forward models method.

INTRODUCTION

A technique for acoustically characterizing horizontally stratified shallow water waveguides has been described in the literature[1]. The method consists of measuring the magnitude and phase versus range of the pressure field due to a CW point source and numerically Hankel transforming these data to obtain the depth-dependent Green's function versus horizontal wavenumber. In the context of normal mode theory, the Green's function contains prominent peaks at horizontal wavenumbers corresponding to the eigenvalues for any trapped and virtual modes excited in the waveguide. The positions and magnitudes of these modal peaks are sensitive to the acoustic properties of the bottom, and this sensitivity is the basis for our proposed inverse method for determining geoacoustic models in shallow water.

THEORY

Consider a water column of thickness h, characterized by a constant density and a sound speed $c(z)$, and bounded at the surface and bottom by horizontally stratified media. Then the spatial part of the acoustic pressure p, due to a point source at $r = 0$ and $z = z_o$ with harmonic time dependence $e^{-i\omega t}$, satisfies the inhomogeneous Helmholtz equation

$$\left[\frac{1}{r}\frac{\partial}{\partial r}\left(r\frac{\partial}{\partial r}\right) + \frac{\partial^2}{\partial z^2} + k^2(z)\right] p(r; z, z_o) = -2\left[\delta(r)/r\right]\delta(z - z_o), \tag{1}$$

where the wavenumber $k(z) = \omega/c(z)$. The solution of Eq. (1) can be expressed as the zero-order Hankel transform of the depth-dependent Green's function $g(k_r)$[2,3]

$$p(r) = \int_0^\infty g(k_r)\, J_o(k_r r)\, k_r dk_r, \tag{2}$$

where r is the horizontal range, k_r is the horizontal wavenumber, and J_0 is the Bessel function of order zero. Note that $p(r)$ and $g(k_r)$ form a conjugate transform pair, where

$$g(k_r) = \int_0^\infty p(r) J_0(k_r r)\, r dr, \tag{3}$$

The Green's function satisfies

$$\left[\frac{d^2}{dz^2} + k^2(z) - k_r^2\right] g(k_r; z, z_o) = -2\delta(z - z_o), \tag{4}$$

along with the impedance boundary conditions at the surface and bottom. These boundary conditions can be expressed in terms of the plane-wave reflection coefficients $R_S(k_r)$ and $R_B(k_r)$ of the surface and bottom, respectively. Although both $p(r)$ and $g(k_r)$ depend on the source and receiver depths, z_o and z, these depths will be viewed as parameters in the problem, while r and k_r will act as the conjugate transform variables.

In general, the Green's function is characterized by a finite-valued continuum and a discrete set of resonances occurring at the poles of $g(k_r)$ which, in conventional modal methods, give rise to the continuous and discrete portions of the normal mode spectrum, respectively. Specifically, the modal representation of the field is given by

$$p(r) = i\pi \sum_n a_n \phi_n^*(z_o)\phi_n(z) H_o^{(1)}(k_n r) + I(r), \tag{5}$$

where the eigenfunctions ϕ_n and eigenvalues k_n satisfy the equation

$$\left[\frac{d^2}{dz^2} + k^2(z) - k_n^2\right] \phi_n(z) = 0 \tag{6}$$

along with impedance boundary conditions at the interfaces. Here $H_o^{(1)}$ is the zero order Hankel function of the first kind, a_n is the modal normalization constant, and the asterisk denotes complex conjugate. The modal field can be obtained from the Hankel transform representation by treating the original path of integration along the real k_r-axis in Eq. (2) as a portion of a closed contour in the complex k_r-plane. The calculation then involves the determination of the contributions of poles enclosed by the contour and branch cuts along the contour. The discrete sum in Eq. (5), corresponding to modes perfectly trapped in the waveguide, arises from the residues at the poles k_n along the real axis. In general, there is also a continuum contribution $I(r)$ which arises from branch line integrals and sometimes, depending upon the choice of branch cuts, from the contribution of virtual modes with complex eigenvalues. Typically, the trapped modal sum in Eq. (5) dominates the long-range behavior of the field, whereas the continuum contribution is only significant at short range.

Thus, the modal representation is intimately connected with the analytic properties of the Green's function. Therefore, knowledge of even gross features of $g(k_r)$, for example the number of resonances and their positions in horizontal wavenumber, provides information about the nature of modal propagation in the waveguide. Furthermore, the modal characteristics are directly related to the acoustic properties of the bottom. As a result, the basic principle underlying the inverse method discussed here is first to obtain an estimate of $g(k_r)$ from measurements of $p(r)$ by numerically performing the Hankel transform in Eq. (3). The modal features of $g(k_r)$ are then used to determine a geoacoustic model.

In this paper, we will concentrate on the case of an isovelocity water column with sound speed c and a pressure-release surface with $R_s = -1$, although we emphasize that our methods are not restricted to this simplified example. Under these circumstances, the Green's function is given by[1]

$$g(k_r) = \frac{e^{i\gamma z_-} - e^{i\gamma z_+} + R_B(k_r)e^{2i\gamma h}\left[e^{-i\gamma z_+} - e^{-i\gamma z_-}\right]}{-i\gamma\left[1 + R_B(k_r)e^{2i\gamma h}\right]} \tag{7}$$

where $z_+ = z + z_o, z_- = | z - z_o |$,

$$k_r^2 + \gamma^2 = k^2 = (\omega/c)^2, \tag{8}$$

and γ is the vertical wavenumber. The denominator of Eq. (7) when set equal to zero is the well known characteristic equation governing the modal propagation in an isovelocity water column overlying an arbitrary horizontally stratified bottom:

$$1 + R_B(k_r)e^{2i\gamma h} = 0. \tag{9}$$

EXPERIMENTAL METHOD

The current WHOI method involves an experiment that consists of towing a two frequency CW source at fixed depth away from two moored receivers over an aperture which extends from zero range to several kilometers. The experimental configuration for obtaining the data processed in this paper is shown in Fig. 1. The receivers quadrature demodulate the signal, that is, they remove the harmonic time dependence by beating the signal down to 0 Hz, and digitally record the real and imaginary parts (quadrature components) of the spatial part of the acoustic field $p(r)$. In order to avoid aliasing errors in the discrete Hankel transform, the field must be sample at least every half-wavelength in range, corresponding to the fact that g decays exponentially for $k_r > k$ and $z, z_o < h^1$. For the frequencies of interest, we can obtain this sampling rate for drift rates of a half knot or less. The Hankel transform is performed using the Fourier-Bessel series[1], which is an exact algorithm for k_r- limited fields. Fourier-Bessel series is slow but reliable, in that among the exact Hankel transform algorithms its associated errors are best understood.

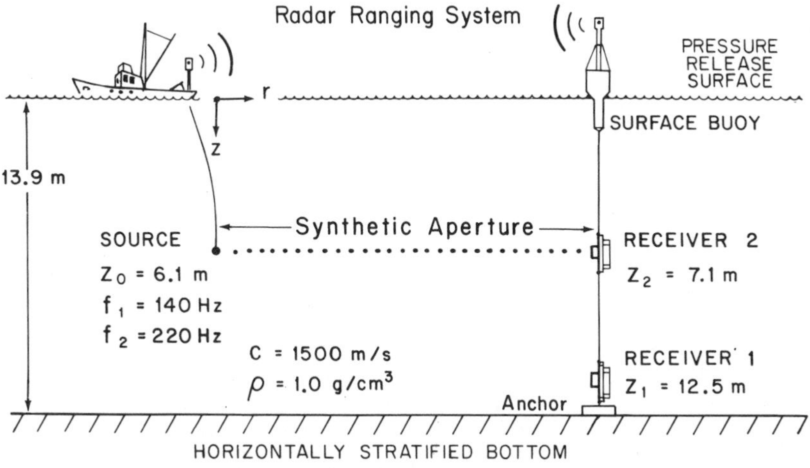

Fig. 1. Typical shallow water experimental setup.

STANDARD INVERSION TECHNIQUE - FORWARD MODEL ITERATION

Before discussing our perturbative inverse technique, we will look at a more standard inverse method, iteration of forward models, the eventual intent being to compare the results of this type of inverse to our perturbative method.

The forward model iteration inversion procedure used at WHOI[4] consists of assuming a geoacoustic model, computing the Green's function, and calculating the mean-square difference between the theoretical and experimental Green's functions for each frequency/receiver depth combination. The bottom parameters are then varied and this procedure is repeated until the total mean-square difference at all four frequency/receiver depth pairs is minimized. Comparisons are also made between the measured and theoretically computed pressure field magnitudes, so that a consistent, balanced fit is obtained in both domains.

In computing the theoretical Green's functions required for the forward modeling inversion procedure, we use the SAFARI program[5,6], where we assume a fluid bottom model consisting of a sequence of isovelocity, constant density layers with absorption. Sound velocity gradients are approximated by a series of thin isovelocity layers with a thickness that is a small fraction ($\approx 1/7$) of a wavelength. We also introduce a compressional wave attenuation $\alpha = 0.1$ dB/λ in order to reduce the theoretical Green's function peaks to a finite height. The primary advantage of SAFARI in our application is its speed, since we generally require Green's function computations for hundreds of bottom models in our forward modeling inversion procedure.

PERTURBATIVE INVERSE TECHNIQUE

For the inverse technique[7] we will describe how one uses the location of the trapped mode eigenvalues as data to invert for the difference $\delta c(z)$ between the actual sound speed profile of the bottom, $c(z)$, and that associated with a plausible background model. Specifically, one locates the mode peaks in $g(k_r)$ as calculated by a finite aperture approximation to Eq. (3), subtracts the mode peak locations found in the background model to form δk_m for each of the modes, and uses these as input to an integral equation for $\delta c(z)$, viz.

$$\delta k_m = \frac{1}{k_m^{(o)}} \int_0^\infty \rho_o^{-1}(z)(\phi_m^{(o)})^2(z) k^{(o)^2}(z) \frac{\delta c(z)}{c_o(z)} dz. \tag{10}$$

In Eq. (10), $k_m^{(o)}$ and $\phi_m^{(o)}$ are the eigenvalues and eigenvectors for the background model, $\rho_o(z)$ and $c_o(z)$ are the density and sound speed profiles for the background model, and $k^{(o)}(z) = \omega/c_o(z)$. Equation (10) is obtained by a straightforward application of the first order perturbation theory, and is readily extended to find attenuation and density profiles as well. Equation (10) is a Fredholm integral equation of the first kind, and in general will not give a unique answer for $\delta c(z)$ unless constrained. The two types of constraints we have employed in our solutions of Eq. (10) are the minimum norm solution constraint (used in the singular value decomposition[8] and spectral expansion methods[9]) and the smoothness constraint (used in the regularization method[10]). Since the solutions obtained are reasonably independent of the constraint imposed, we feel confident in the interchanging solution methods arbitrarily and still trusting the answers.

One distinct advantage of using perturbative linear inverse theory to obtain geoacoustic models of the bottom is the simplicity with which one can perform an error analysis for the bottom model obtained by computing its resolution (which is directly related to the bias) and variance. How one obtains the resolution estimate associated with the bottom model generated is described in a companion paper to this one by Rajan and Frisk[11] and so will not be described further here. The variance of the estimate for δc (the error bars) can be expressed in a simple form in the context of the singular value decomposition inversion. First, we note that by using a numerical integration (quadrature) scheme, we can write Eq. (10) in a matrix form

$$\mathbf{d} = \mathbf{G}\mathbf{m} \tag{11}$$

where $\mathbf{m}$ is the desired bottom model ($\delta c(z)$ in this case), $\mathbf{d}$ is the data (the difference in modal eigenvalues), and $\mathbf{G}$ is the kernel. By using the singular value decomposition to break $\mathbf{G}$ to into a product of three matrices[12],

$$\mathbf{G} = \mathbf{U}\mathbf{\Lambda}\mathbf{V}^{\mathbf{T}}, \tag{12}$$

it can be shown that under certain reasonable assumptions the solution error covariance matrix $\mathbf{C_e}$ can be expressed as [13]

$$\mathbf{C_e} = \sigma_\nu^2 \mathbf{V}\mathbf{\Lambda}^{-2}\mathbf{V}^{\mathbf{T}}. \tag{13}$$

The matrices $\mathbf{V}$ and $\mathbf{\Lambda}$ are obtained directly from the singular value decomposition in Eq. (12). The σ_ν^2 factor is obtained by estimating the errors in the modal eigenvalues due to experimental errors or limitations. In the Hankel transform synthetic aperture array experiment, error in the modal eigenvalues can arise from 1) the finite aperture of the array, 2) range-variable bathymetry effects, 3) rough surface scattering, 4) ranging errors, and so on. The finite aperture L of the array limits the accuracy in determining the modal eigenvalues to π/L; efforts are currently being made with high resolution spectral estimation techniques and other methods to circumvent this classical limitation. Efforts to remove range variable bathymetry effects and quantify the other eigenvalue errors mentioned are also being pursued by the authors, though space forbids their description here.

DATA ANALYSIS AND INVERSION RESULTS

The pressure fields measured for a sandy bottom over a range aperture of 1325 m in Nantucket Sound, Massachusetts in May 1984 are shown in Figs. 2 and 3. The magnitudes exhibit strong spatial interference patterns which are particularly evident at 220 Hz. These patterns arise due to the coherent combination of two normal modes of comparable strength at 220 Hz, and one dominant and one weak mode at 140 Hz. The decay with range of the weaker mode at 140 Hz is also apparent in the data. We note that the water depth changes from 13.9 m to 14.6 m at a range of about 600 m. This change in bathymetry has a pronounced effect on the 220 Hz modal cycle distance, which changes from about 125 m to 150 m. The results of the forward model iteration inversion procedure are shown in the theoretical curves for $p(r)$ in Figs. 2 and 3. Excellent agreement is obtained for the 220 Hz data, but for the 140 Hz data, only fair agreement results. Considering the amount of computation invested in such

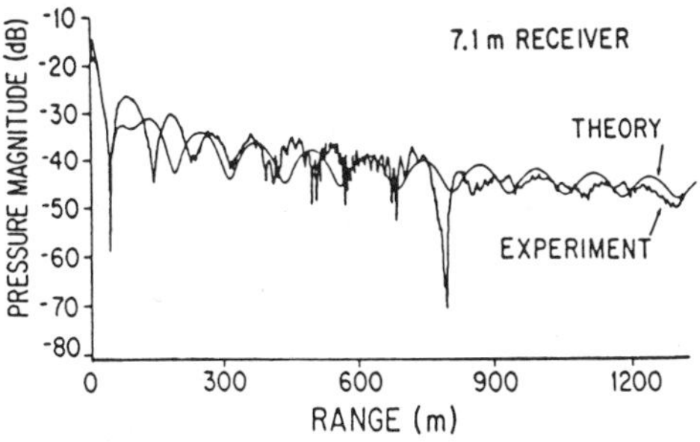

Fig. 2. Comparison of theoretical and experimental pressure field magnitudes and phases at 140 Hz, using forward model iteration.

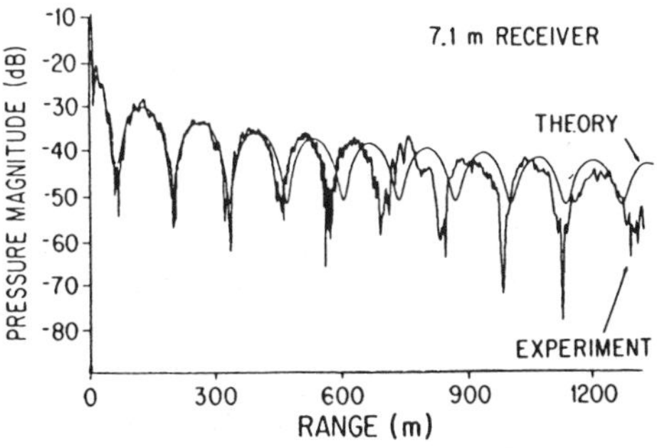

Fig. 3. Comparison of theoretical and experimental pressure field magnitudes and phases at 220 Hz, using forward model iteration.

a calculation (several hundred iterations to the SAFARI fast field program, which gives both $g(k_r)$ and $p(r)$), this result is a bit of a disappointment. The likely reason for such a failure is the inherent simplicity of the bottom models we iterated, where only one or two parameters were varied at a time. A more general approach in which the bottom is not constrained to a few isovelocity or simple gradient layers is required - inverse theory is one candidate, which we next examine.

The pressure field generated by using the perturbative inverse scheme is shown in Fig. 4 for the 140 Hz data. The agreement between theory and measured data is excellent both at 140 Hz and at 220 Hz (which is not shown, since the level of agreement is comparable to our Fig. 3 result). It is of interest to note that this agreement was obtained after only a few iterations of an inversion program which used a crude 1820 m/s Pekeris sediment model as a background. This demonstrates both the efficiency and robustness of the approach as compared to standard forward modeling.

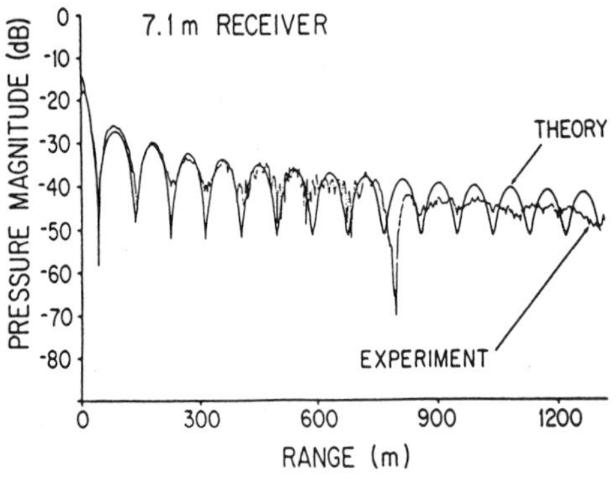

Fig. 4. Comparison of theoretical and experimental pressure field magnitudes at 140 Hz, using perturbative inverse technique.

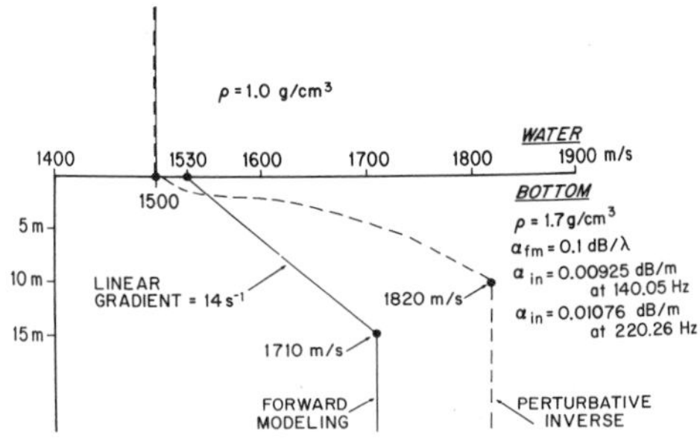

Fig. 5. Geoacoustic models of the bottom produced by 1) forward modeling and 2) perturbative inverse.

The geoacoustic models of the bottom produced by the two techniques are shown in Fig. 5. The high near surface gradients produced by both models are consistent with the results of Hamilton for sandy bottoms. The perturbative inverse technique, however, also produces a continuously variable gradient profile (until constrained at 10 m sediment depth) which allows it to fit data which seemingly can't be handled by simpler bottom models, such as the constant gradient layer model used in our forward approach.

As mentioned, another advantage to perturbative inverse techniques is the ease with one can obtain error estimates for the bottom model obtained. Specifically, one can estimate the resolution kernel, $R(z_o, z)$, and the RMS error in the sound speed, δm^{RMS}. These quantities are shown in Figs. 6 and 7 respectively. The resolution kernels shown in Fig. 6 are a measure of how well one can discriminate bottom features in the vertical direction as a function of how deep into the sediment one looks. For instance, if one looks at 15 m depth (which is 1.1 m into the sediment, since the water is 13.9 m deep), one obtains a resolution kernel peak width of 1.6 m, meaning that one can only resolve features 1.6 m in vertical extent or greater. (An alternate way of looking at this is that the bottom sound speed estimate one obtains at 15 m represents an average of the neighboring sediments over 1.6 m). As one goes deeper into the sediment, the resolution kernels increase in width, finally reaching an asymptotic 8.8 m peak width at about 24 m (10.1 m into the sediment). The RMS error in the sound speed estimate, shown in Fig. 7, also shows a degradation in the quality of the sound speed estimte with increasing depth. Particularly noticeable is the sharp increase in the variance at six meters depth into the sediment. This is caused by a sharp decrease in the amount of acoustic energy sampling the bottom past six meters deep, i.e. the trapped modes used for our inversion only show significant strength down to that level. This is also the reason for the resolution kernel growing wider with increasing depth.

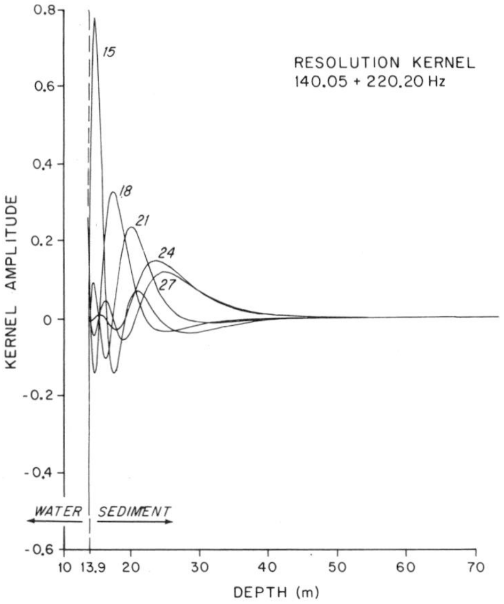

Fig. 6. Resolution kernels generated using both 140 Hz and 220 Hz data in inversion.

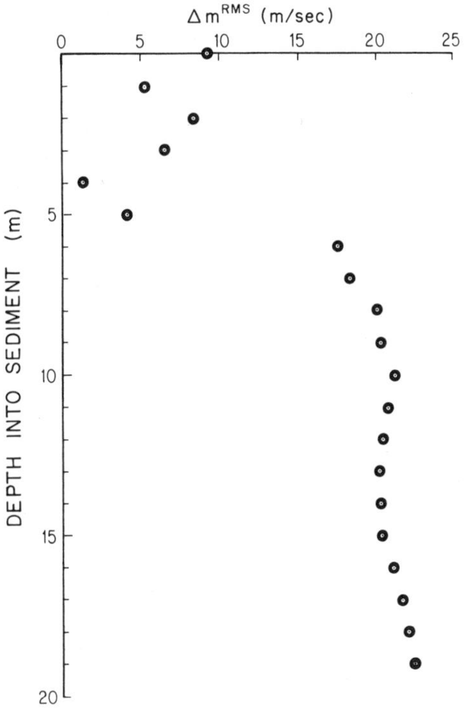

Fig. 7. Variance vs. depth for perturbative inverse bottom model.

CONCLUSIONS

A perturbative inverse method for obtaining the sound velocity profile of the bottom using the trapped mode eigenvalues measured in shallow water has been demonstrated. It is seen that the method can generate physically reasonable bottom models quickly (few iterations are needed) and robustly (it is not overly sensitive to the background model used). Moreover, the method is such that error estimates for the bottom model obtained (viz., the resolution and variance) are readily generated. Standard iteration of forward models inversion seem to be slow, inefficient, and generally restricted to much simpler models when compared to the perturbative inverse scheme. Research is currently being pursued to extend the perturbative inverse scheme to include compressional wave and shear attenuation profiles, shear speed profiles, and density profiles, as well as to incorporate different types of acoustic data as input.

ACKNOWLEDGEMENTS

The authors would like to gratefully acknowledge all the engineering, marine, and programming support personnel who made our shallow water experiments both possible and successful. The support of ONR codes 220 and 1125 UA is also gratefully acknowledged. This paper is WHOI Contribution No. 6230.

REFERENCES

1. G.V. Frisk and J.F. Lynch, Shallow water waveguide characterization using Hankel transform, J. Acoust. Soc. Am., 76:204, (1984).

2. G.V. Frisk, A.V. Oppenheim, and D.R. Martinez, A technique for measuring the plane-wave reflection coefficient of the ocean bottom, J. Acoust. Soc. Am., 68:602, (1980).

3. F.R. DiNapoli and R.L. Deavenport, Theoretical and numerical Green's function field solution in a plane multi-layered medium, J. Acoust. Soc. Am., 68:602, (1980).

4. G.V. Frisk, J.F. Lynch, and J.A. Doutt, The determination of geoacoustic models in shallow water , paper presented at the symposium on Ocean Seismo-acoustics, SACLANT ASW Research Centre, La Spezia, Italy, (June 1985).

5. H. Schmidt and F.B. Jensen, A full wave solution for propagation in multilayered viscoelastic media with application to Gaussian beam reflection of fluid-solid interfaces, J. Acoust. Soc. Am., 77:813, (1985).

6. H. Schmidt and F.B. Jensen, Efficient numerical solution technique for wave propagation in horizontally stratified environments, Report SM-173, SACLANT ASW Research Centre, La Spezia, Italy, (1984).

7. S.D. Rajan, J.F. Lynch, and G.V. Frisk, Perturbative inversion schemes to obtain bottom acoustic parameters in shallow water, submitted to J. Acoust. Soc. Am.

8. R.A. Wiggins, The general linear inverse problem: implication of surface waves and free oscillations for the earth structure, Rev. Geophysical and Space Physics, 10:251 (1972).

9. R.L. Parker, Understanding inverse theory, Ann. Rev. Earth Planet. Sc., 5:35, (1977).

10. S. Twomey, On the numerical solution of Fredholm integral equation of the first kind by inversion of the linear system produced by quadrature, J. Acoust. Compt. Mach., 10:97, (1963).

11. S.D. Rajan and G.V. Frisk, An inverse method for obtaining the attenuation profile and small variations in the sound speed and density profiles of the ocean bottom paper presented at the I.C.A. Congress Symposium on Underwater Acoustics, Halifax, Canada, (July 1986).

12. C. Lanczos, Linear differential operators, Van Nostrand, New York, (1961).

13. Keiiti Aki and P.G. Richards, Quantitative Seismology: Theory and methods, Vol. II, Foreman, San Francisco, (1980).

A SYNTHETIC APERTURE-ARRAY TECHNIQUE FOR FAST APPROXIMATE GEOBOTTOM RECONNAISSANCE

M. F. Werby, G. J. Tango, and H. B. Ali

Naval Ocean Research and Development Activity
NSTL, Mississippi 39529–5004 USA

ABSTRACT

In the absence of detailed geological sampling data or of exact inversion algorithms for seismo-acoustic profiling data, an approximate and efficient method for identifying geoacoustic properties of shallow water environments would be a valuable tool for exploration and survey applications. In this paper, we outline results of an ongoing effort at NORDA in developing and testing numerical algorithms for estimating sea-bottom properties. Use is made of mode discrimination using synthetic aperture simulation. The first phase of the present method consisted of the development and implementation of a simple linear towed-array beamforming of bottom interacting modes from ship self-noise. The second stage consisted of analyzing more complicated layered bottoms. The most recent effort comprises examination of other spectral analysis techniques for modal decomposition by beamforming. This paper reviews the salient features and theoretical formulation of mode detection by arrays, and presents further simulated results for multilayered coastal-shelf environmental models.

INTRODUCTION

Rapid mapping of sea-bottom geophysical properties over wide, shallow ocean regions is of considerable interest to marine geotechnical, exploration, and naval research communities. Direct geologic core sampling and conventional high-resolution remote surveying methods for determining these properties are generally both costly and time consuming, the latter being further restricted in its abilities to resolve the shallowest layers (Brocher and Ewing, 1986). Thus, in the absence of a robust general-applications direct inversion algorithm, it is clearly desirable to seek possibly less-precise, but nonetheless effective methods to complement existing techniques, as well as to employ previously unused data sources.

This paper discusses the further implementation of one such method (Kuperman et al., 1984), based on beamforming bottom-interacting pressure fields from dedicated or towship self-noise acoustic sources, using a towed or synthetic aperture array. The method rests on the fact that the primary geophysical sea-floor properties of a bottom-limited environment (including characteristic compressional and shear velocity, attenuation, and layer thickness) determine the behavior of waterborne acoustic signals propagating as normal modes. Thus, in principle, an appropriate measure of these bottom-interacting signals can indicate the type and properties of the local sea floor beneath the array.

Because near-field acoustic self-noise generated by a towship propulsion system can strongly interfere with, or even dominate, all other modal contributions, considerable interest has been directed toward suppressing near-field towship acoustic noise by directly identifying and separating plane wave signal components on the basis (e.g.) of mode arrival angle, amplitude and phase (Rupe and Lunde, 1982). The approach taken here, by contrast, is to attempt beamforming to enhance the ship source signal for detecting and identifying bottom-interacting normal modes (of low order), thereby determining the local bottom geoacoustic properties.

THEORETICAL BASIS

The present method rests on three fundamental concepts. The first (Fig. 1) is the inverse Fourier-Bessel transform pair relation between the seismo-acoustic wavefield in range, $P(r)$, as seen by a horizontal sensor array, and the depth-dependent Green's function, $G(k_r)$, the impulse response of the total ocean/bottom waveguide:

$$P(r) = \int_0^{\infty} G(k_r)\, J_0(k_r r)\, k_r\, dk_r$$

$$G(k_r) = \int_0^{\infty} P(r)\, J_0(k_r r)\, r dr$$

With a $1/r^{1/2}$ correction for amplitude loss due to geometrical spreading, $G(k_r)$ can be approximately obtained as an inverse Fourier transform of $P(r)$.

Williams (1968), DiNapoli (1977), Kuperman et al. (1984), and others have pointed out the operational analogies between the beamformed output of a horizontal equispaced line array and the inverse FFT (Ziomek, 1985). That is, the discrete inverse Fourier transform, inverting pressure in range to impulse response kernel in wavenumber/incident angle, and an FFT beamformer algorithm operating from range to angle domains have precisely the same mathematical form. If in the Green's function above, we use the outgoing wave approximations, viz,

$$J_0(k_r r) = \tfrac{1}{2}[H_0^{(1)}(k_r r) + H_0^{(2)}(k_r r)] \approx \tfrac{1}{2}H_0^{(1)}(k_r r),$$

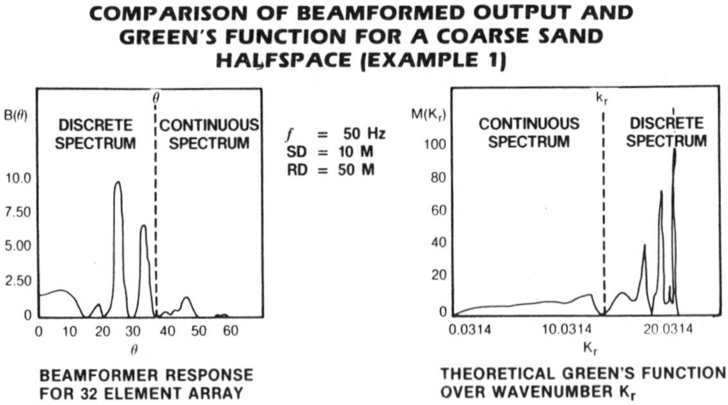

$$|B(\theta)| = C|G(K_{r_m})$$
$$C = \left[\Delta K_r(K_{r_m})^{1/2}\right]/N$$
$$K_{r_m} = K/\cos\theta_m$$
$$\theta_c = \cos^{-1}(V_{P_1}/V_{P_2})$$

Figure 1.

and the asymptotic form

$$H_o^{(1)}(k_r r) \sim \left(\frac{2}{\pi k_r r}\right)^{1/2} e^{i(k_r r - \pi/4)}$$

and then compare with the beamformer response (expressed as a discrete Fourier Transform; Kuperman et al., 1985)

$$B(\theta_m) = \sum_{n=0}^{N-1} W_n P(r_n) \exp^{-ikr_n \cos\theta_m}$$

yields the result that

$$B(\theta_m) = |C_m| \, |G(k_{rm})| \, .$$

This relation states that the beamformer output is directly proportional to the Green's impulse response function of the bottom expressed in incident angle space (Fig. 1, comprising Example 1, a coarse sand half-space). The above relation is rigorous (Brigham, 1974), given linearity and the above constraints, and represents the second basic concept of the present method. Physically, this means that the peak amplitudes of the propagating (discrete) modes in the Green's function correspond directly to the angular values of maximum beamformer output. Thus, in principle, estimation of sea-floor geoacoustic parameters by direct inversion of acoustic field data is immediate, since knowledge of the incident mode angles is equivalent to knowledge of the compressional and/or shear velocities of one or more bottom layers, via Snell's law. Simple geometrical optics is the third relationship needed in the present method. Werby and Tango (1985) and Ioup et al. (1986) have outlined some of the environmental and experimental limitations of this technique.

DISCUSSION OF NUMERICAL RESULTS

Unless otherwise stated, beamformer response is computed for source frequencies of 25 and 50 Hz, with corresponding array-element spacings of 30 m and 15 m, respectively (where $c = f\lambda = 1500$ m/sec). As shown in Example 1 above, the present array consists of only 64 elements towed directly behind the source vessel. The source is taken as a single point radiator at a constant depth of 10 m, an approximation that is valid for low frequencies and azimuthal angles near zero. A constant water depth of 100 m overlies a range-independent bottom consisting of one, two, or three layers. Significantly different shallow-water geomodels, i.e., those of coarse sand, silt-clay/sand/basalt and silt-clay/sand/limestone, are examined in terms of their discrimination using beamformed modal responses. The geoacoustic parameterizations used have been previously published (Jensen and Kuperman, 1980), and are representative of sea bottoms encountered in shallow coastal shelf regions of some Canadian and Mediterranean areas. Because of the importance of the complete wavefield comprising discrete, leaky, and evanescent modes in shallow-water environments (where source and/or receiver are located near the bottom), a full wave-theoretic (FFP) solution was used to generate synthetic pressure fields for the following simulations (Schmidt and Tango, 1986), although this requirement is not fundamental for the method.

The following numerical experiments were chosen to illustrate the concept of selectively processing synthetic aperture data over small adjacent segments at progressively farther source-array offset distances (Frisk and Lynch, 1984). The motivation behind this approach (Fig. 2) is the fact that as the central elements of the receiving array are moved farther from the CW source, the number of effective ray transitions (i.e., received by the array) through the various bottom layers increases; therefore, the cumulative mode amplitude loss from bottom attenuation is increased. Thus, one expects at moderate ranges to begin to see modes from the deepest-accessible

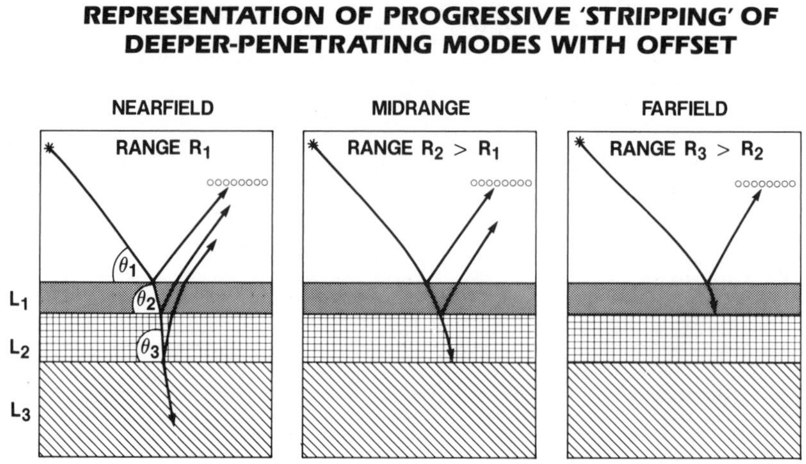

Figure 2.

basement layer extinguished first, followed by modes of the next layer at larger offsets, and so on, until only the mode(s) from the top layer remain. This technique enables one to estimate the critical angles and P and S velocities for each layer in turn by forming synthetic apertures as a function of offset. With these estimates and with additional geoacoustic knowledge, a first estimate of the properties of a shallow sea-floor geobottom can be obtained. With such initial bottom models, other more refined inversion methods, such as the perturbative techniques under development by Frisk and Lynch, can be used to obtain a more refined seafloor description.

In the following, all beamformer outputs are corrected for cylindrical spreading and renormalized to a linear scale, suppressing Gibb's oscillations and sidelobes from finite array aperture, thereby enhancing definition of mode arrival angle.

Fig. 3a–d illustrates Example 2 (silt-clay/sand/basalt), which was examined at 50 Hz. Fig. 3a shows beamformed results for apertures just behind the source. This figure includes several large amplitude discrete trapped modes, as well as lesser virtual modes (from all layers) between 10° and 35°. The synthetic data is processed so that beamformed results are examined at adjacent offsets corresponding to successive multiples of total water depth (expressed in terms of P-wavelength). The first few offsets are generally sufficient to identify which (decaying) modes are leaky. Only a select number from 30 offsets (0–7 km) are shown. Fig. 3a exhibits a high-angle response maxima at about 75°; the persistence of the response over appreciable offsets (not shown here) beyond the near-field identifies it as corresponding to a compressional velocity critical angle of the deepest basement layer (basalt). Fig. 3b can be similarly used to determine the critical angle of the basement shear velocity (52°). Fig. 3c must then show the last attenuating modes from the intermediate layer (at an offset of 52 wavelengths), occurring at about 33°, that correspond to coarse sand. Finally, Fig. 3d illustrates the sole remaining mode at about 8° from the topmost silt-clay layer (since, by the final offset of 116 wavelengths, all other deeper modes have been attenuated). This same array-offset processing sequence is repeated below for 25 Hz, in Figs. 4a–d. Although the compressional mode from the basement is no longer seen (due to the larger relative strengths of the remaining modes), the bottom shear arrival still occurs at about 50°. The compressional cutoff mode from the middle layer is now seen at about 28°, and at about 9° for the top layer; these agree satisfactorily with the previous mode angle locations at 50 Hz. Comparison of these extracted values with the critical angles in Table 2 illustrates that the method is fairly reliable in this case, using only simple visual interpretation. More precision would be obtained with additional beamformed responses computed in a narrow band of adjacent frequencies if progressive interlayer refraction effects are corrected for.

We now consider Example 3, comprising the same shallow-most layer geologies, with limestone replacing the previous basalt basement. In the same manner as previously described,

Table 1. Input geology and inversion estimate.

Layer No.	Geology	V_p	V_s
	Example 1		
Layer 1	Clay-Silt	1515	100
Layer 2	Sand	1800	600
Layer 3	Basalt	5250	2500
	Example 2		
Layer 1	Clay-Silt	1550	200
Layer 2	Sand-Silt	1650	200
Layer 3	Limestone	3500	1750

Table 2. Theoretically expected critical angles.

Layer 1	Example 1 8.1°	Silt-Clay	Example 2 11.3°	Silt-Clay
Layer 2	33.6°	Coarse-Sand	24.8°	Silt-Sand
Basement	Shear 53.1° Compressional 73.4°	Basalt	31.0° 64.6°	Limestone

one can determine that the bottom-most layer for 50 Hz (Fig. 5a–d) has a compressional velocity, but has a weak critical angle at about 60°, and a shear velocity corresponding to about 30°. Note here that at 25 Hz (Fig. 6a–d) only the compressional cutoff mode is seen here at 65°. The middle layer has a critical angle of about 23° and the topmost layer about 11°. Once again, these results agree well with the known values in Table 2.

Note that we have not determined either P or S wave attenuations or density in the preceding analysis, since current small-aperture array resolution is insufficient to accurately ascertain the change in mode signature width as a function of bottom absorption. More fundamentally, although bottom attenuation is, in principle, determinable from mode widths in a Pekeris waveguide, this is no longer true for a multilayered bottom. Specific attenuation estimates in many geographic areas may be obtained from known empirical relations (Hamilton, 1980) in terms of P and/or S wave velocities.

CONCLUSIONS

A chronic problem with many extant seismic profiling or iterative inversion schemes is the lack of an initial bottom model estimate for the shallowest sea-floor layers. The present approximate direct inversion technique can be considered, even in its present simple form, as a potential adjunct and initial input to more sophisticated inversion methods. Extensive environmental simulations to date have suggested that, for many cases of canonic velocity/depth structure in shallow water, analysis of beamformed response versus incident angle can be further aided by systematically incorporating known geologic associations and delimiting *a priori* expected layer structures. The present technique is thus complementary to standard acoustic subbottom profiling, which provides direct assessments of sediment thickness and areal distribution (however, with little direct information on sediment properties). All necessary array data acquisition and processing operations can, in principle, be made in real-time by a ship of opportunity by using its own self-noise or a dedicated CW source. Despite the apparent complexity of some beamformed mode responses (e.g., those of higher frequency, incorporating multiple higher-order modes from several bottom layers), systematic and potentially diagnostic Green's function mode responses clearly exist for many representative bottom types, which may be further enhanced through larger synthetic apertures or, alternatively, through more sophisticated beamforming and/or signal proc-

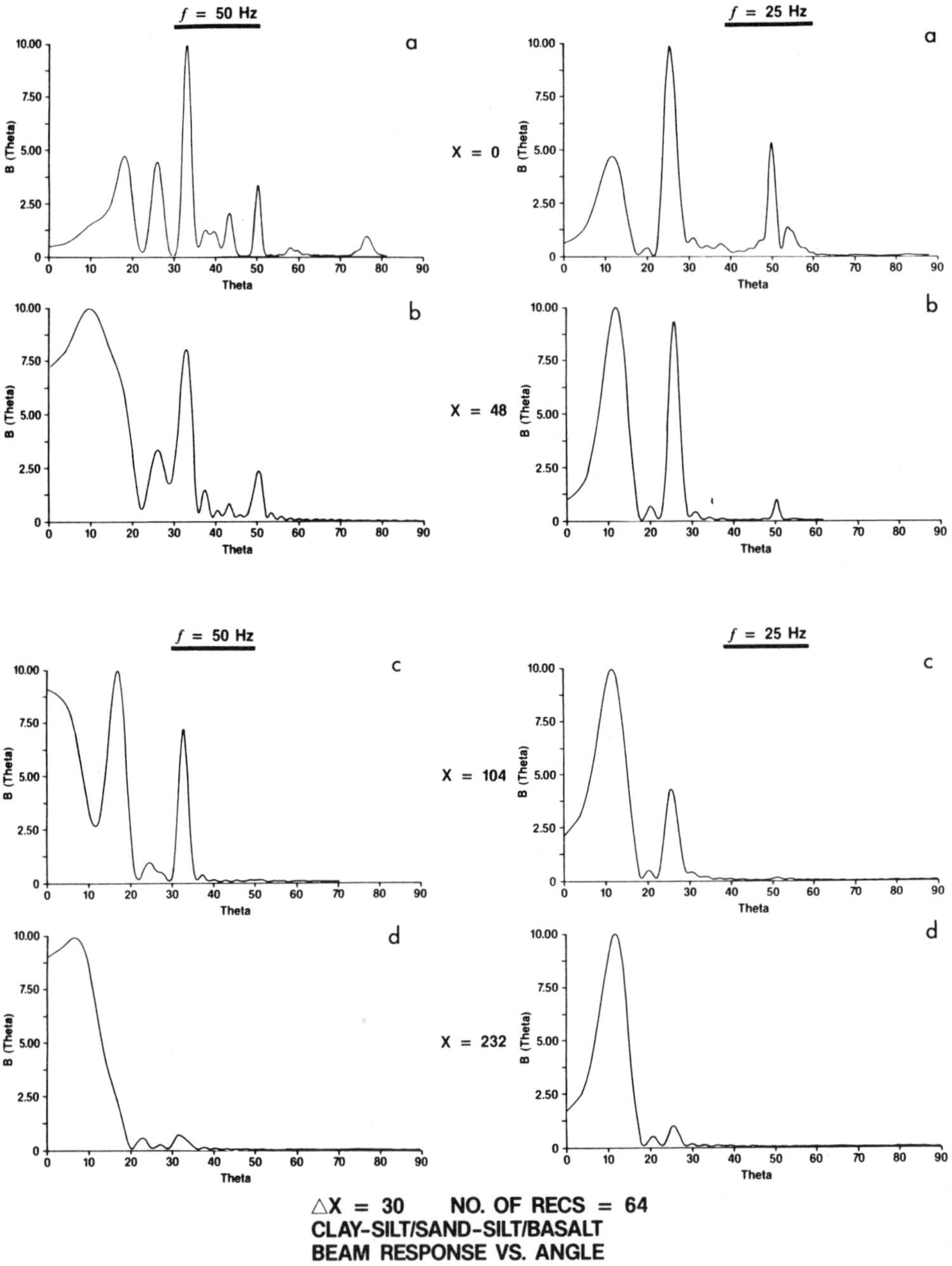

Figure 3.

Figure 4.

essing techniques (Burg, 1968; Capon, 1969; Pisarenko, 1973). An example of the latter is the use of a deconvolution algorithm postprocessor, which incorporates additional constraints on the statistical properties of the mode signal, such as nonnegativity of its spectral autocorrelation function (Ioup et al., 1986), to increase resolution by removing truncation effects of a finite array aperture. These and other additional processing methods may allow improved mode-angle estimation using fewer array sensors. Use of multifrequency beamforming can, in principle, yield results similar to those of the related "plane wave decomposition" technique in exploration seismology.

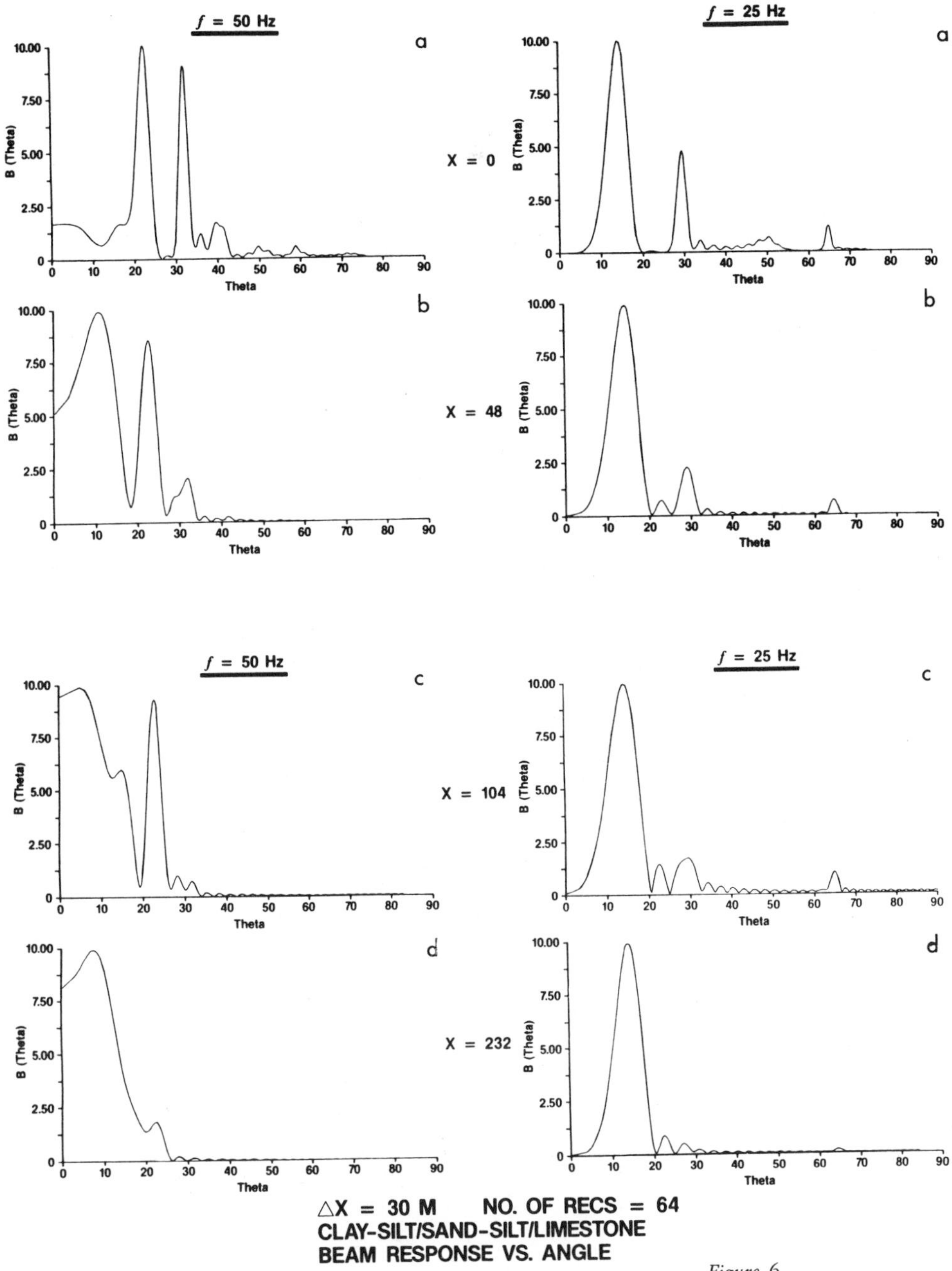

Figure 5.

Figure 6.

More fundamentally, strict mathematical justification for the present technique can be given only if the ocean bottom is vertically inhomogeneous in a discrete-step sense and is range-independent. However, recent numerical experiments using the parabolic equation algorithm for more complicated geobottom environments (including interfacial roughness, continuous vertical velocity gradients, and lateral inhomogeneities) suggest that the present technique has more general validity. The judicious use of all experimental free parameters (e.g., number and type of array receivers, spatial weights and subarray localization, and multiple array offsets), with improved signal-array processing and data display will, it is hoped, substantiate and extend the applicability of this method.

REFERENCES

Brigham, E. O., 1974. *The Fast Fourier Transform*, Englewood Cliffs, New Jersey, Prentice Hall.

Brocher, T. M. and Ewing, J. I., 1986. A comparison of high-resolution seismic methods for determining seabed velocities in shallow water, *J. Acoust. Soc. Am.*, 79(2): 286–298.

Burg, J. P., 1968. A new (maximum entropy) spectral analysis technique for time series data, paper presented at the *NATO Advanced Study Institute on Signal Processing in Underwater Acoustics*, Enschede, Netherlands.

Capon, J., 1969. High-resolution frequency-wavenumber spectrum analysis, *Proc. IEEE*, 57: 1408–1418.

DiNapoli, F. R., 1977. *The Inverse Fast Field Program: An Application to the Determination of the Acoustic Parameters of the Deep Ocean Bottom*, NUSC Tech. Memo. 77160, New London, Connecticut.

Frisk, G. V. and Lynch, J. F., 1984. Shallow water waveguide characterization using the Hankel transform, *J. Acoust. Soc. Am.*, 76: 205–216.

Frisk, G. V., Lynch, J. F., Tango, G. J., and Werby, M. F., 1986. Mapping seafloor geoacoustic properties in shallow water from monochromatic pressure field data: two methods based on Hankel transform inversion, OTC Proceedings Paper 5276, *1986 Offshore Technology Conference*, Houston, Texas, May 5–8, 1986.

Hamilton, E. L., 1980. Geoacoustic modelling of the seafloor, *J. Acoust. Soc. Am.*, 68: 1313–1340.

Ioup, G. E., Ramaswamy, M., Tango, G. J., and Werby, M. F. 1986. Applications of constrained-iterative-deconvolution (CID) to seismic transfer function estimation and underwater acoustic environmental beamforming (in preparation).

Jensen, F. B. and Kuperman, W. A., 1983. Optimum frequency of propagation in shallow-water environments, *J. Acoust. Soc. Am.*, 73: 813–819.

Kuperman, W. A., Werby, M. F., and Gilbert, K. E., 1984. Towed array response to ship noise: a nearfield propagation problem, paper presented at the *NATO-ASI Conf. Underwater Acoustics and Signal Processing*, Luneberg, W. Germany, 1 August 1984, in: G. Urban (ed)., *Adaptive Signal Processing*, Dordrecht and Boston, D. Reidel, 1985, pp. 39–46.

Kuperman, W. A., Werby, M. F., Gilbert, K. E., and Tango, G. J., 1985. Beamforming on bottom-interacting tow-ship noise, *IEEE J. Oceanic Engin.*, OE-10(3): 290–298.

Pisarenko, V. F., 1973. The retrieval of harmonics from a covariance function, *Geophys. J. R. Astro. Soc.*, 33: 347–366.

Rupe, U. E. and Lunde, E. B., 1982. *A Comparison of Signal Processing Algorithms to Suppress Tow-Vessel Noise in a Towed Array, With Results from a Shallow-Water Field Trial*, SACLANTCEN Memo SM-158, La Spezia, Italy.

Schmidt, H. and Tango, G., 1986. Efficient global matrix approach to the computation of synthetic seismograms, *Geophys. J. R. Astro. Soc.*, 84: 331–359.

Watson, G. N., 1966. *Theory of Bessel Functions*, Cambridge, University Press.

Werby, M. F. and Tango, G. J., 1985. Characterization of average geoacoustic bottom properties from expected propagation behaviour at very low frequencies (VLF), using a towed array simulation, paper presented at the *NATO-ASI Conf., VLF Underwater Seismo-Acoustics*, La Spezia, Italy, 16–23 June 1985, in: T. Akal and J. Berkson (eds.), *Proceedings* volume (to appear).

Williams, J. R., 1968. Fast beamforming algorithm (based on the FFT), *J. Acoust. Soc. Am.*, 44(5): 563–564.

Ziomek, R., 1985. *Underwater Acoustics: A Linear Systems Approach*, Academic Press, Orlando, Florida.

BURIED TARGET REFLECTIVITY EVALUATION THROUGH AN ARRAY OF SENSORS

R. Carbó Fité and C. Ranz Guerra

Instituto de Acústica C.S.I.C., Serrano, 144
28006 Madrid, Spain

ABSTRACT

The problem of locating and characterizing targets buried in the sea bottom was studied in a previous paper (Porc. Inst. Acoustics, Vol. 7, PT.3, 1985, pp.169-177). In that paper a bottom reflectivity map was first obtained by acoustic insonification; a frequency post-processing analysis was done afterward. That method is very time consuming because of the requirement to carefully sound around the buried target.

In this paper we tried to avoid that time consuming process by using a vertical linear array of sensors located in line with the emitter. The set of echoes coming from the bottom and the target, reach each one of the hydrophones with delays being a function of their depth. The sum of all the echoes detected by each of the sensors, gives a signal in which its spectral density determines, in frequency space, the presence as well as the position and dimensions of the target.

SPECTRAL DENSITY OF THE ECHO

Figure 1, shows the N element array and the emitter, both located along the same vertical axis. When the emitter-receiver system is over a region with no target, only the bottom surface produces an echo. In such a case the signal picked up by each element only differs, in delay and amplitude, from the radiated signal:

$$X_n(t) = r \frac{X(t - t_0 - n\tau)}{c_0(t_0 + n\tau)} \qquad (1)$$

$X(t)$ being the normalized emitted signal, c_0 the sound velocity in water, r the bottom reflection coefficient and t_0 the time that the sound pulse takes to reach the first hydrophone element after being reflected from the bottom. The spectral density of the echo captured by any array element is proportional to the spectral density $Y(\omega)$, of the radiated signal:

$$Y_n(\omega) = Y(\omega)\ [r/c_0(t_0 + n\tau)]^2\ . \qquad (2)$$

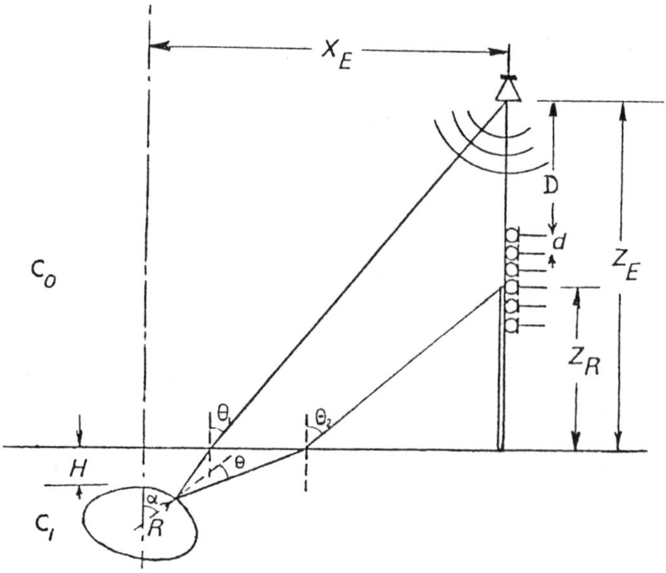

Figure 1: Experimental geometry.

The whole array gives a signal that is the sum of those coming from each individual sensor:

$$X_R(t) = \sum_{n=0}^{N} X_n(t) \quad . \tag{3}$$

Its spectral density reaches a maximum value at frequencies where $\omega.\tau$ is an even multiple of π.

$$Y_R(\omega) = Y(\omega) \sum_{n,m=0}^{N} \frac{r^2 \cos(m-n)\,\omega\tau}{c_0^2(t_0 + n\tau)\,(t_0 + m\tau)} \quad . \tag{4}$$

The ratio $G(\omega) = Y_R(\omega)/Y(\omega)$ is given by Figure 2 with peaks clearly defined, equally spaced and with identical levels.

The presence of a buried target produces a new echo $Y_n(t)$. For each element

$$Y_n(t) = \frac{X(t - t_0 - n\tau - \nu_n)}{c_0(t_0 + n\tau + \nu_n)}\,(1 - r_n^2).e^{-\mu h_n}.\sigma_n \tag{5}$$

where μ is the absorption coefficient of the sand bottom, h_n is the acoustic path of the signal through the bottom, σ_n is the target strength of the target which depends on the angle of incidence of the sound beam on it and hence is also function of the element "n" considered. Finally $(1 - r_n^2)$ represents the transmission losses in the water-sand and sand-water paths.[1]

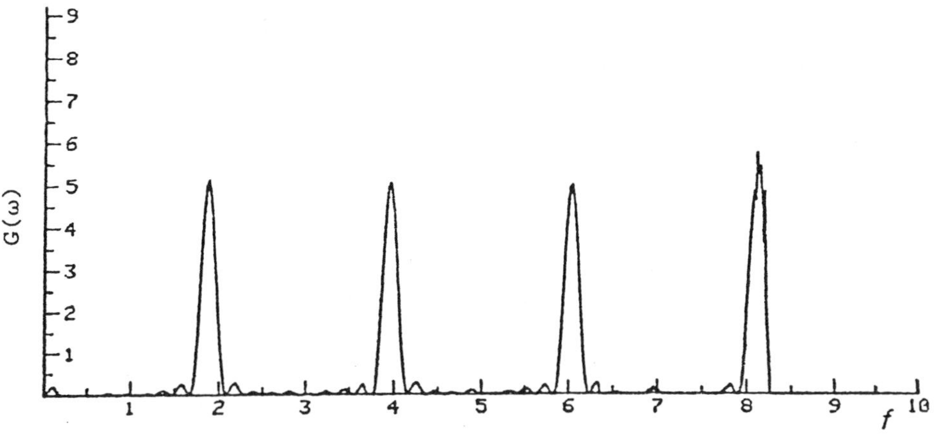

Figure 2: Relative Spectral Density of the bottom surface echo.

Each array element receives a reflected signal that is the sum of those produced by the bottom surface and the buried target, so

$$S_n(t) = X_n(t) + Y_n(t) = A_n X(t-t_0-n\tau) + B_n X(t-t_0-n\tau-\nu_n) \quad (6)$$

Then, as before, its spectral density, would be

$$Y_n(\omega) = Y(\omega) \cdot (A_n^2 + B_n^2 + 2A_n B_n \cos[\omega\nu_n]) \quad (7)$$

In this instance the ratio $G(\omega)$ is a cosine function of period ν_n with maxima and minima such that A_n and B_n can be evaluated.

When the target has finite dimensions, the delays ν_n are a function of the buried object shape; consequently $G(\omega)$ will show a period that is a function of the element location. ν_n depends on the distance from the buried object to the vertical of the array-emitter system, X_E, and also on the buried depth H as well as on the curvature radius of the target over the isonified zone.

From the geometry shown in Figure 1, the dependence is of the following form

$$\nu_n = \nu(o) - (D + nd)\frac{\xi}{c_o} \quad (8)$$

where $\nu(o)$ would be the delay found when the position of both emitter and the "n" hydrophone coincide:

$$\nu(o) = \frac{2Z_E}{c_o \cos\theta_1} \quad \frac{2[H + R(1-\cos\alpha)]}{c_1 \cos\alpha} \quad . \quad (9)$$

The angles α and θ_1 are given by

$$\sin\theta_1 = \frac{c_1}{c_o}\sin\alpha, \quad \tan\theta_1 = \frac{X_E - (H+R)\tan\alpha}{Z_E} \quad . \quad (10)$$

The parameter ξ is nondimensional and can be determined by

$$\xi = \frac{1}{\cos\theta_2} - 1 \tag{11}$$

where $\sin\theta_2 = \frac{c_1}{c_0} \sin(\alpha+\theta)$ and α and θ can be computed through

$$X_E = [H+R(1 - \cos\alpha)] \tan(\alpha-\theta) + Z_E\tan\theta_1 + R\sin\alpha \tag{12}$$

$$[H+R(1-\cos\alpha)] [\tan(\alpha+\theta)-\tan(\alpha-\theta)] = (Z_E-D-nd)\tan\theta_2-Z_E\tan\theta_1 \quad . \tag{13}$$

Generally ξ depends on the hydrophone location (D+nd), so it will be represented by ξ_n; and the delay ν_n is not a linear function of the "n" hydrophone coordinates.

As an experimental result, Figure 3, shows the ν_n values obtained from a buried sphere, as in functions of Z_E, X_E, H and R, for each of the N elements of an array. It can be seen, that when the horizontal distance from the target decreases, the variation of ν_n also decreases.

The ratio of the spectral densities is, as we have shown above, a cosine function with a growing period when the depth of the element increases (Z_E = 1.7 m, X_E = .5 m, H = .06 m, R = .15 m) Figure 4. The experimental values are represented in Figure 3.

The whole array signal (buried target present) is

$$S_R(t) = \sum_{n=0}^{N} S_n (t) = \sum_{n=0}^{N} A_n X(t-t_0-n\tau) + \sum_{n=0}^{N} B_n X(t-t_0-n\tau-\nu_n) \quad . \tag{14}$$

and the spectral density[2]

$$Y_R(\omega) = Y(\omega)\left[\sum_{n,m=0}^{N} A_n A_m \cos (m-n)\omega\tau + \sum_{n,m=0}^{N} A_n B_m \cos[\omega[(n-m)\tau+\nu_n]] + \sum_{n,m=0}^{N} B_n B_m \cos[\omega(\nu_n-\nu_m)]\right] . \tag{15}$$

is composed of three terms: 1) the spectral density of the bottom surface signal with peaks at frequencies which are multiples of c/d; 2) the spectral density coming from the interaction of the bottom surface reflection and the target surface reflection; 3) the spectral density that is also a consequence of the object, and it varies with the difference between the delays, of the target reflected signal, picked up by any two elements of the array.

DEEP WATER

When the bottom is far away from the array, $Z_E \gg D+Nd$, the dependence of the delay ν_n with the hydrophone depth (D+nd) becomes linear, and the acoustic wave attenuation, corresponding to any of the array sensors, is identical; so

$A_n = A_0$ and $B_n = B_0 \quad \forall n$.

In a given position (X_E, Z_E) of the acoustic system, the three terms of the spectral density of the whole echo captured by the array are:

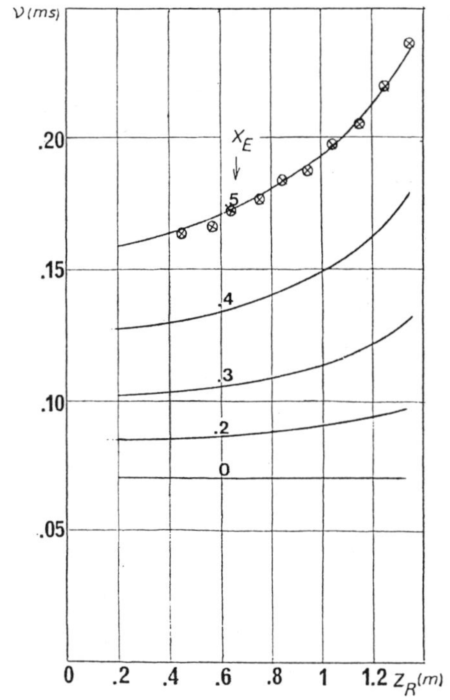

Figure 3: Delays between the bottom reflected and target reflected signals when the depth of the array element increases.

i)

$$\sum_{n,m=0}^{N} A_n A_m \cos\omega(m-n)\tau = NA_0^2 \frac{\sin^2(N+1)\frac{\omega\tau}{2}}{\sin^2 \frac{\omega\tau}{2}} , \qquad (16)$$

that is a spectral density in which the maxima are separated by $\Delta\omega_1 = \pi/\tau$, where $\Delta\omega_1$ depends only on the separation distance between the elements of the array, ii)

$$\sum_{n,m=0}^{N} A_n B_m \cos\omega[(m-n)\tau+\nu_n] = \qquad (17)$$

$$NA_0B_0\cos\omega[\nu_0+\xi(\frac{H}{C} + N\frac{\tau}{2})] \frac{\sin[(N+1)(1+\xi)\frac{\omega\tau}{2}]\ \sin[(N+1)\frac{\omega\tau}{2}]}{\sin(1+\xi)\frac{\omega\tau}{2}\ \sin\frac{\omega\tau}{2}} ,$$

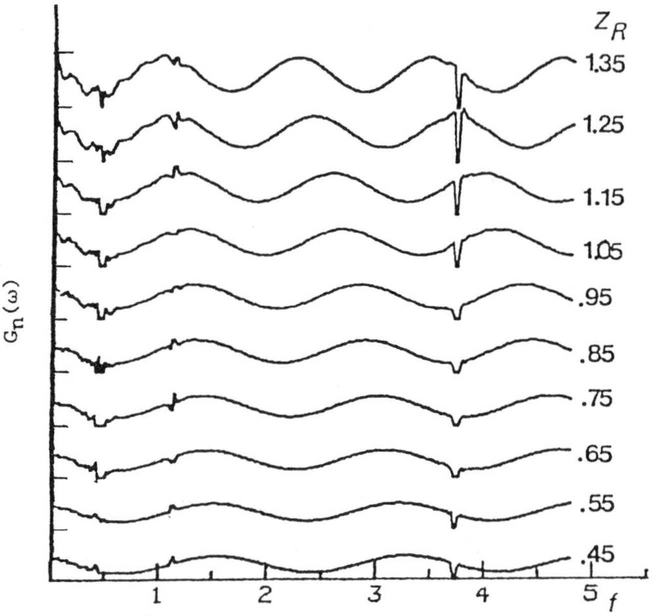

Figure 4: Relative spectral density of the echo coming from a buried sphere. Each curve corresponds to an individual element.

and iii)

$$\sum_{n,m=0}^{N} B_nB_m\cos\omega(\nu_m-\nu_n) = NB_0^2 \frac{\sin^2 (N+1)\xi\ \frac{\omega\tau}{2}}{\sin^2\xi\ \frac{\omega\tau}{2}} . \qquad (18)$$

There are two types of "peaks" in the second term of the spectral density: those that coincide with those present in the first term of the sum, and those whose separation distance is

$$\Delta\omega_2 = \pi/\tau(1+\xi) . \qquad (19)$$

The amplitude of both is modulated by $\cos\omega\nu_{N/2}$. We have to recall that $\nu_{N/2}$ is a function of the geometry of the array (D,d) of the parameter ξ and of the delay $\nu(o)$; that for a given position (X_E, Z_E) depends on the burial depth, H, and the target radius, R.

The third term of the spectral density shows an identical structure to the first one, but where the distance between maxima is

$$\Delta\omega_3 = \pi/\tau\xi \quad . \tag{20}$$

ξ varies with the square of the distance X_E, so $\Delta\omega_3$ increases rapidly when coming closer to the vertical of the target.

Referring to the experimental results of Figure 5, the parameter ξ is much less than 1 (ξ=0.075) and so the third term of the sum will only be of importance at very high frequencies. The "peaks" of the second term all coincide with those of the first and the modulating effect of $\cos\omega\nu_{N/2}$ appears very clearly.

If the buried object presents a plane parallel surface to the bottom (the curvature radius goes to infinity), the delay ν_n is constant (ν=H/c) and then the variation of the third term in $\Upsilon_B(\omega)$ vanishes. The peaks at frequencies which are multiples of c/d, periodically vary with ν. See Figure 6.

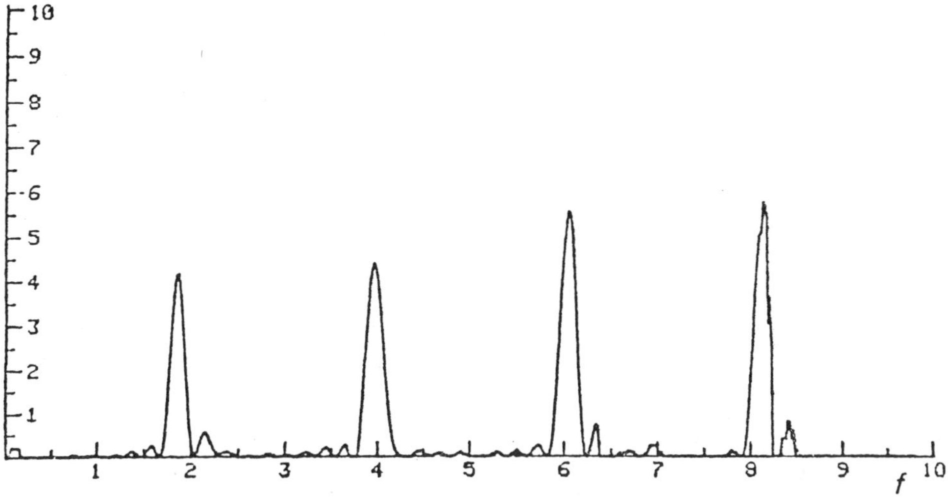

Figure 5: Relative spectral density of the echo from buried sphere.

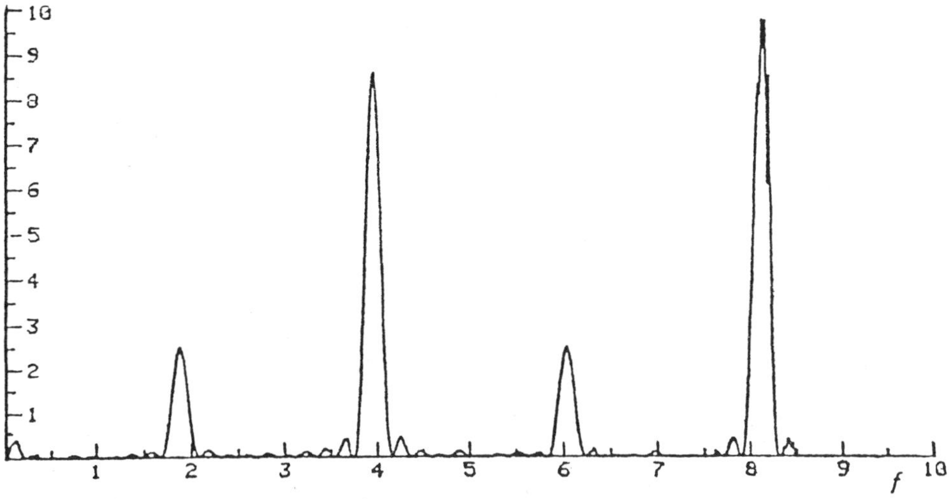

Figure 6: Relative spectral density of the echo from a plane parallel buried target.

CONCLUSION

$G(\omega)$ is the ratio between two spectral densities: a) the one corresponding to the radiated signal captured by one of the hydrophones of the array and b) the one corresponding to the whole signal reflected by a flat bottom where a target was buried. $G(\omega)$ allows information of:

- The presence of a target; $G(\omega)$ varies and some "peaks" of variable amplitude come into sight.

- The buried situation (X_E, H) by measuring the separation distance between peaks (those corresponding to the three terms of $G(\omega)$: $\Delta\omega_1$, $\Delta\omega_2$, $\Delta\omega_3$).

- The target dimensions (R) by computing the modulation period of the "peaks", $\nu_{N/2}$.

REFERENCES

1. R.W. G. Haslett, "Acoustic Echoes from Targets Under Water". Underwater Acoustics, R.W.B. Stephens, Ed., Wiley-Interscience, London, pp. 129-197 (1970).
2. L.J. Ziomek, "Underwater Acoustics: A Linear Systems Theory Approach" Academic Press, New York, pp. 153-176 (1985).

EFLECTIVITY OF A LAYER WITH ARBITRARY PROFILES OF DENSITY

ND SOUND VELOCITY

P. Cobo, C. Ranz and R. Carbó

Instituto de Acústica
Serrano 144, 28006 Madrid, SPAIN

BSTRACT

The sea bottom acoustic reflection direct problem has been solved by n upwards integration, from the basement to the upper sediment layer, of n acoustic impedance Riccati equation. The starting data are the water nd basement impedance, as well as the arbitrary density and sound elocity profiles of the sediments. The bottom reflection response will e the result of the whole process.

The theoretical model results are compared with those from the xperiments. The experimental model refers to a two layer bottom (sand nd gravel) made in our laboratory.

NTRODUCTION

When a sound beam strikes a sea bottom some of the energy is eflected upwards. This echo comes from all interfaces: water-sediment, ediment-sediment, and sediment-basement. All contributions build the eflection response.

When the density-velocity profiles, as well as the water and asement impedance are known, we are able to synthetize the reflection esponse (direct problem). When, on the other hand, the reflection esponse is deconvolved to extract the velocity-density profiles, this is alled the inverse problem.

The geoacoustic sea bottom model that we assume, Fig. 1, consists in n inhomogeneous layering of sediments sandwiched between two homogeneous emi-infinite media: water and basement. The water and basement mpedances, Zw and Zb, and the arbitrary profiles, c(x), (sound velocity) nd $\rho(x)$ (density) in the sediments, are the input data for the problem.

We try to obtain the reflection response of a system (in the requency domain) to an acoustic isonification. By hypothesis, the sea ottom is in the source far field in order to assume a plane wave front; nly normal incidence is considered. Both hypotheses simplify the athematical model, mainly avoiding the transverse wave problem. In rief, we seek the normal incidence plane wave reflection response.

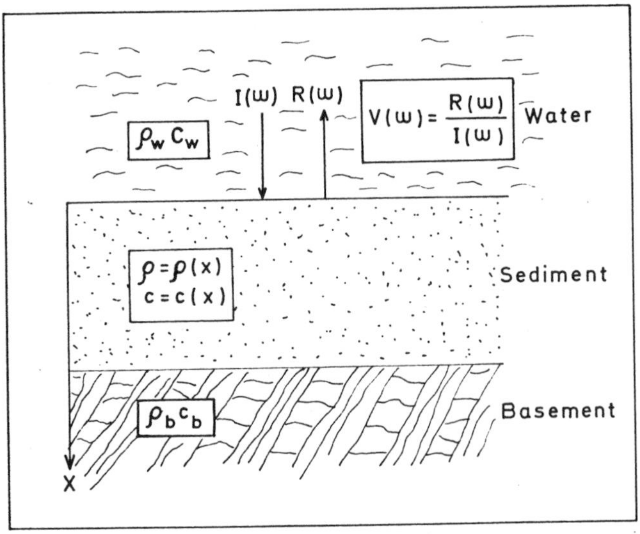

Figure 1: Geoacoustic sea bottom model.

A very general outlook on the direct problem and some exact solutions in the case of particular profiles of sound velocity and density, can be found in Brekhovskikh (1980). Gupta (1965) solved the problem by dividing the inhomogeneous layer into a number of homogeneous sublayers. Similarly, the well known Goupillaud model in Geophysics (Robinson, 1984), considers layers of identical "acoustical thickness" (the two-way travel time) and applies the z transform to get the synthetic seismogram. Hawker and Foreman (1978) numerically integrated an acoustic pressure wave equation.

In this paper, a Riccati equation for the acoustic impedance is integrated numerically from the basement (initial condition) up to the interface water-sediment, to get the input impedance of the system. Once this input impedance is known, together with that of the water, the evaluation of the reflection coefficient is straight forward.

THEORETICAL MODEL

Let us consider a one-dimensional inhomogeneous medium in the x axis direction, with a thickness d, and located between two homogeneous media. From one of these media a plane wave sound beam at normal incidence (x axis) strikes the separation surface. The equations to consider are

$$\nabla p(x,t) = -\rho(x)\frac{\partial v(x,t)}{\partial t}$$

and (1)

$$\frac{\partial p(x,t)}{\partial t} = -p(x)\, c^2(x)\, \nabla . v(x,t).$$

The first is the linearized Euler equation, while the second is the linearized continuity equation. Assuming harmonic time dependence (or considering the Fourier transforms of Eq. 1), we obtain

$$\frac{dZ}{dx} = i\,\omega\,\rho(x)\left[\frac{Z^2}{\rho^2(x)\;c^2(x)} - 1\right]$$

Riccati equation

$$Z^* = \sqrt{\mu}\;\; Z/\rho_b\; c_b$$

$$X^* = \omega\; x\; /\sqrt{\mu}\;\; c_b$$

Scaling

$$A(x) = \rho_b\; c_b^{\;2}/\rho(x)\;\; c^2(x)$$

$$B(x) = \mu\;\rho(x)/\rho_b$$

$$Z^* = R^* + i\; X^*$$

$$F(x,\; R^*,\; X^*) = -2\; A(x)\;\; R^*\; X^*$$

$$G(x,\; R^*,\; X^*) = A(x)\,(R^{*2} - X^{*2}) - B(x)$$

Functions defined

$$dR^*/dx^* = F(x,\; R^*,\; X^*)$$

$$dX^*/dx^* = G(x,\; R^*,\; X^*)$$

System to be integrated

$$R^{*(o)}_{n-1} = R^*_{n+1} - 2h\; F(x_n,\; R^*_n,\; X^*_n)$$

$$X^{*(o)}_{n-1} = X^*_{n+1} - 2h\; G(x_n,\; R^*_n,\; X^*_n)$$

Predictor

$$R^{*(i)}_{(n-1} = R^*_n - (h/2)\left[F(x_n,\; R^*_n,\; X^*_n) + F(x_{n-1},\; R^{*(i-1)}_{n-1}\;,\; X^{*(i-1)}_{n-1})\right]$$

$$X^{*(i)}_{n-1} = X^*_n - (h/2)\left[G(x_n,\; R^*_n,\; X^*_n) + G(x_{n-1},\; R^{*(i-1)}_{n-1}\;,\; X^{*(i-1)}_{n-1})\right]$$

Corrector

Figure 2. Sketch of the numerical algorithm

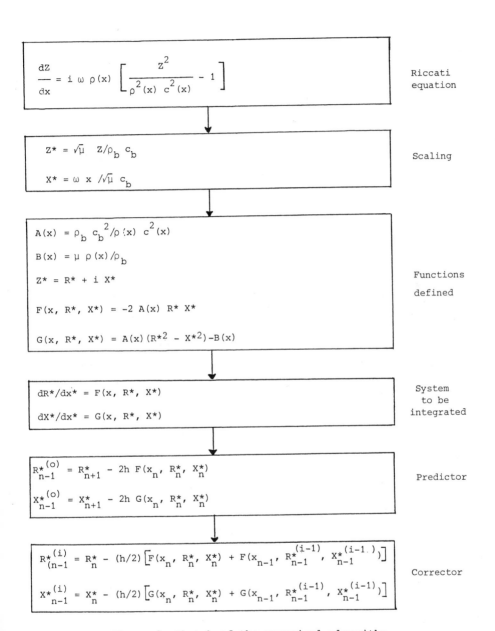

$$P(x,t) = P_1(x)\, e^{i\omega t}$$
$$v(x,t) = v_1(x)\, e^{i\omega t} \qquad (2)$$

and

$$\frac{\partial p_1}{\partial x} = -\, i\, \omega\, \rho(x)\, v_{1x}$$
$$\frac{\partial V_{1x}}{\partial x} = -\, \frac{i\, \omega}{\rho(x)\, c^2(x)}\, P_1\,, \qquad (3)$$

v_{1x} being the component of v_1 in the x axis direction. By defining $Z = \frac{P_1}{v_{1x}} = \frac{P(x,t)}{v_x(x,t)}$, and after some simple mathematical transformation, we can finally get

$$\frac{dZ}{dx} = i\, \omega\, \rho(x) \left[\frac{Z^2}{\rho^2(x)\, c^2(x)} - 1 \right], \qquad (4)$$

a Riccati differential equation for the acoustic impedance in an one-dimensional inhomogeneous medium.

Equation 4 is scaled and split in both real and imaginary parts. The resulting equation system is numerically integrated from the basement upwards by a second order predictor-corrector algorithm. This numerical algorithm is sketched in Figure 2. The reason for introducing the scale factor, μ, is the following: by using a second order predictor-corrector as an integration formula, the truncation error is of the order of h^3 (McCracken and Dorn (1965)), h being the integration step. Due to the scale change

$$h^* = \frac{2\pi f}{\sqrt{\mu}\, c_b}\, h \qquad (5)$$

where c_b is the sound velocity in the basement; so μ can be used to compensate the increase of h^* with frequency, keeping then a control on the truncation error. In this case we choose

$$\mu = 1 + \left(\frac{2\pi f d}{c_b} \right)^2 \qquad (6)$$

where d is the sediment layer thickness. For high frequencies h^* tends to 1/N, N being the integration step number, N = d/h.

Not very much can be said in a general sense about the stability of the algorithm for any type of functions $\rho(x)$ and c(x). Nevertheless, the numerical solution is checked against some well known solutions for particular cases; for instance, Fig. 3 shows the comparison of the numerical and exact solutions for a layer with a linear gradient of velocity (Officer, 1958).

Once the input impedance of the water-sediment interface is obtained, and the water impedance being known, the reflection coefficient simply is written as

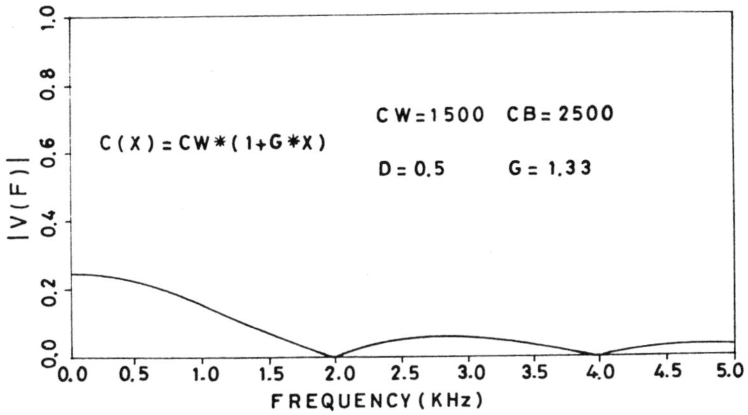

Figure 3: Numerical and exact solutions for a layer with linear gradient of velocity (both curves overlap, to the resolution of the diagram).

$$V(f) = \frac{Z_{in} - Z_w}{Z_{in} + Z_w} = \frac{Z^*_{in} + \sqrt{\mu}\,(Z_w/Z_b)}{Z^*_{in} + \sqrt{\mu}\,(Z_w/Z_b)} \qquad (7)$$

RESULTS

A wide series of profiles were introduced, as a subprogram, into our algorithm. As an example, Fig. 4 shows the comparison of four combinations of linear, quadratic and exponential profiles:

$$c(x) = \begin{cases} c_0\ (1 + b_1\ x)^p \\ c_0\ \exp\ (b_1\ x) \end{cases} \qquad \rho(x) = \begin{cases} \rho_0\ (1+b_2\ x)^q \\ \rho_0\ \exp\ (b_2\ x) \end{cases} .$$

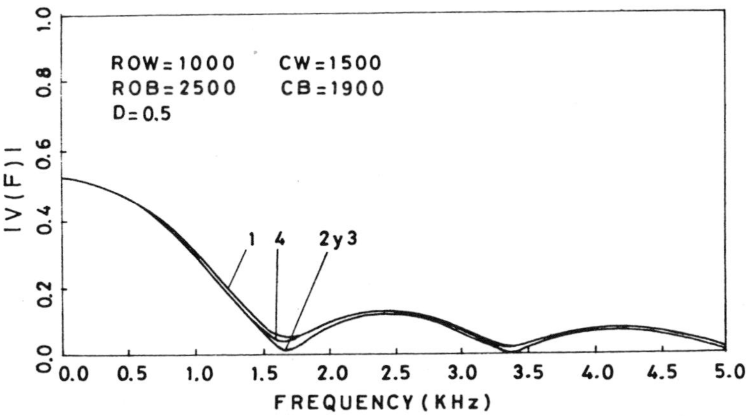

Figure 4: Numerical solutions for four different profiles in a layer, all profiles having identical limiting values.

Table I presents the b_1, b_2, p and q values corresponding to those results of Figure 4. In all cases the limiting values were identical: $c_0 = c_w = 1500$ m/s, $c = c(d) = 1900$ m/s, $\rho_0 = \rho_w = 1000$ Kg/m³ and $\rho(d) = \rho_b = 2500$ Kg/m³.

Table I: Values of the profile parameters used in Figure 4.

NUMBER	B1	P	B2	Q
1	0.533	1	3.000	1
2	0.251	2	1.160	2
3	0.470	EXP	1.830	EXP
4	0.470	EXP	3.000	1

It would be most interesting to check the model results against experience. In order to do this we made an artificial bottom in our experimental water tank. It consisted of two homogeneous layers: a layer of gravel, 25 cm thick, and, overlying it, a layer of fine sand 20 cm thick. The reflection response of such a layered medium was experimentally obtained by irradiating the bottom with a 35 kHz resonance frequency directional sonar. Figure 5 shows the experimental and theoretical results along with the value of all parameters used in the models, in MKS units

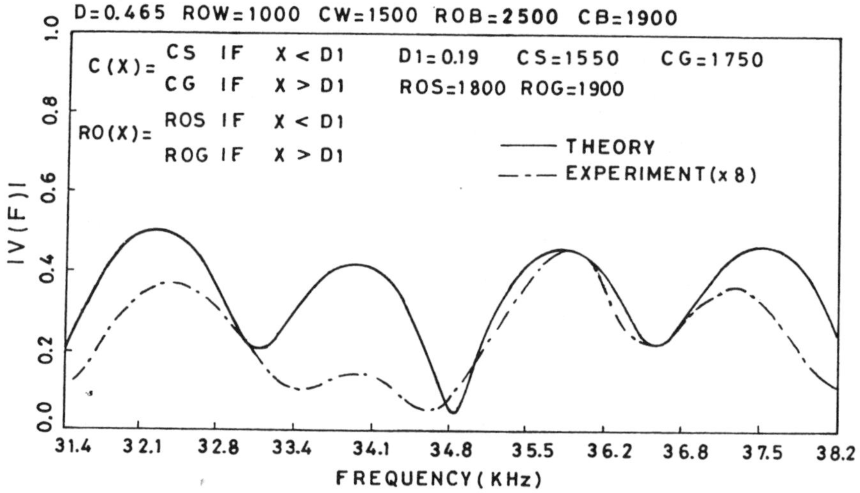

Figure 5: Numerical and experimental results for a two layer artificial sediment.

As it can be seen, the location of all maxima and minima are very close; we think the differences found in the levels, ought be thought as due to the absorption and to those other factors present in the experiment but not in the model: inhomogeneties, variable layer thickness, beamwidth of the source, etc.

CONCLUSIONS

We present a new approach in the calculation of the normal incidence plane wave reflection response based in a numerical integration of an acoustic impedance Riccati equation.

We think that this method affords the possibility of working with any sound velocity and density profiles, those profiles being either analytical functions or experimental data. So, this formulation of the problem seems to be very much appropriate to solve the inverse problem because it will bring us to the resolution of an integral equation for the acoustic impedance. That is the target with which the authors are, at this moment, deeply involved.

ACKNOWLEDGEMENT

The research described herein could not have been completed without the financial support of the Comision Asesora de Investigacion Cientifica y Técnica (Program 2517-83).

REFERENCES

Brekhovskikh, L.M., 1980, "Waves in layered media", Academic Press, New York.

Gupta, R., 1966, "Reflection of sound waves from transition layers", J. Acoust. Soc. Am., 39(2), 255-260.

Hawker, K.E. and Foreman, T.L., 1978, "A plane wave reflection loss model based on numerical integration", J. Acoust. Soc. Am., 68(5), 1313-1340.

McCracken, D.D. and Dorn, W.S., 1965. "Numerical methods and FORTRAN programming", John Wiley and Sons Inc., New York.

Officer, C.B. 1958, "Introduction to the theory of sound transmission", McGraw Hill, New York.

Robinson, E.A., 1984, "Seismic inversion and deconvolution", Geophysical Press, London.

OUND SPEED PROFILE INVERSION IN THE OCEAN

Linda Boden and John A. DeSanto

Center For Wave Phenomena
Colorado School of Mines
Golden, Colorado 80401

NTRODUCTION

An inversion technique has been developed to recover the depth ependent sound speed profile in the upper ocean waveguide using ynthetically generated acoustic data. This is a direct technique in which sound speed profile correction is explicitly written as an integral over he scattered field data.

These results are derived starting from a Fourier-Bessel representation f the scattered field. The method uses a Born approximation on the depth ependent Green's function, a WKB representation for the wave functions, and he far field range solution. The kernel resulting from these pproximations is evaluated asymptotically and a linear relation between the cattered data and the profile correction is the final result.

This linearity allows the data and the profile correction to be written s a quasi-Fourier transform pair. As a result, the fast Fourier transform an be used to calculate the profile correction. As will be illustrated, a mall perturbation in the upper ocean can be recovered from a first guess inear input profile, if the background profile slope is assumed known.

HEORY

A brief overview of the scattered data derivation and the relation etween this data and a profile correction are presented in this section. he details can be found in Boden (1985).

The expression for the scattered data is derived for non-arctic rofiles with a single turning point, and the receiver is located above the ource in depth.

The specific evaluation of the kernel (mentioned above) differs epending on the depth location in the waveguide relative to the source and eceiver positions. In effect, different layers are created in the ocean as result of the asymptotic analysis. The derivations of the scattered data nd the profile correction are discussed for the uppermost asymptotic layer between the surface and the receiver) only. This layer is designated as egion 1.

See page 810 for Abstract.

Derivation of the Scattered Field Representation: Region 1

The behavior of the acoustic velocity potential field, $\phi(r,k,z,z_s)$, for an angle independent point source is described by the inhomogeneous Helmholtz equation

$$\nabla^2\phi(r,k,z,z_s) + k^2n^2(z)\phi = -\delta(r)\delta(z-z_s)/2\pi r \tag{1}$$

where ∇^2 is the cylindrical Laplacian, z_s and z are the source and field positions respectively, $k = \omega/c_0$ is the scalar wavenumber, $n(z) = c_0/c(z)$ is the index of refraction and c_0 is chosen to be the sound speed at the channel axis. The source and field positions are located in the medium separated by a horizontal distance, r. Although the propagation is in range, the sound speed depends only on depth; therefore, Eq. (1) has a separable solution which can be written as a Hankel contour integral (Deavenport, 1966),

$$\phi(r,k,z,z_s) = \frac{k^2}{4\pi}\int_C H_0^{(1)}(kr\beta)F(z,z_s,k,\beta)\beta d\beta \quad , \tag{2}$$

which is identically equal to the Fourier-Bessel representation. The separation parameter is $\beta = n(z)\sin\theta(z)$, where the angle is measured from the vertical.

The depth dependent solution satisfies an ordinary differential equation (Ahluwalia and Keller, 1977), and with the introduction of the travel length coordinate

$$\tau(z) = \int_0^z f(z',\beta)dz' \tag{3}$$

where

$$f(z',\beta) = \left[n^2(z')-\beta^2\right]^{1/2} \tag{4}$$

this differential equation is written as

$$\frac{d^2}{d\tau^2}F(\tau,\tau_s,\beta) + \frac{1}{f^2(z,\beta)}\frac{df}{dz}\frac{dF}{d\tau} + k^2F = -\frac{\delta(\tau-\tau_s)}{f(z_s,\beta)} \tag{5}$$

where

$$\frac{1}{f^2(z,\beta)}\frac{df}{dz} = \frac{\frac{d}{dz}n^2(z)}{2[n^2(z)-\beta^2]^{3/2}} \quad . \tag{6}$$

Equation (6) contains the profile perturbation $\varepsilon(z) = d(n^2(z))/dz$, which is independent of our separation parameter, β.

Choosing the WKB-Green's function which satisfies

$$\frac{d^2}{d\tau^2}G(\tau,\tau_s,\beta) + k^2G = -\delta(\tau-\tau_s) \tag{7}$$

nd using Green's Theorem on Eq. (5) and Eq. (7), the depth dependent solution is written as an integral equation in coordinate space

$$F(\tau,\tau_s,\beta) = G(\tau,\tau_s,\beta) \left[f(z_s,\beta) \right]^{-1}$$

$$+ \int_0^{z_b} G(\tau,\tau',\beta) \left[\frac{1}{2f^2(z',\beta)} \varepsilon(z') \frac{d}{dz'} F(\tau',\tau_s) \right] dz' \quad . \tag{8}$$

he first term represents the incident field and the integral term epresents the scattered field. Equation (8) is solved in Born pproximation by setting $F' \sim G'/f(z_s,\beta)$.

Combining the depth dependent solution with the far field range olution leads to the scattered data representation as the total field, ϕ, inus the incident field, ϕ_I, the first term in Eq. (8). The result is

$$D(k,r,z,z_s) = \phi - \phi_I = A(k,r) \int_0^{z_b} K(k,r,z',z,z_s)\, \varepsilon(z')dz' \tag{9}$$

here

$$A(k,r) = k^{3/2}\exp(-i\pi/4)/\ (4\pi^{3/2}(2r)^{1/2}) \tag{10}$$

nd K is the kernel to be evaluated asymptotically. The transform pair iven by Eq. (9) between the scattered data and the profile correction is he main result of this work.

In the upper region K becomes

$$K(k,r,z,z',z_s) = - \left[\frac{i}{4k} \right] \mathrm{sgn}(z'-z_s) \int_C \frac{\beta^{1/2}\exp\left[ik\Phi_1(\beta)\right]}{f(z_s,\beta)\ f^2(z',\beta)} d\beta \tag{11}$$

here the phase, Φ_1, is given by

$$\Phi_1(\beta) = r\beta + \int_{z'}^{z} \left[n^2(z'') - \beta^2 \right]^{1/2} dz'' + \int_{z'}^{z_s} \left[n^2(z'') - \beta^2 \right]^{1/2} dz'' \quad . \tag{12}$$

he contour is distorted about that branch point with the lowest index of efraction value, (DeSanto, 1984). The leading order branch point in the hase occurs at the variable index of refraction, n(z'). Evaluation of the hase at this branch point results in wavelike solutions in depth. These hase integrals are evaluated by setting $\beta \sim n(z')$ (or $\sin\theta(z) \sim 1$) and inearizing the integrands. The amplitude terms are also evaluated near $\beta =$ (z'). The analytic portions of the integrand are taken out from under the integral and the remaining singular portion is evaluated along the teepest descent path. The scattered data representation in region 1 ecomes

$$D_1(k,r,z,z_s) = \int_0^z B(z',z_s)\exp\left[ikP(r,z,z_s,z') \right] \varepsilon(z')dz' \qquad (1$$

where B and P are the amplitude and phase terms resulting from th linearization and asymptotics. Note that the amplitude factor, A(k,r), an (-i/4k) have been swept into the data. The amplitude and phase are given b

$$B(z',z_s) = -\frac{2\pi i}{3} \frac{1}{\sqrt{n(z')}} \frac{1}{\sqrt{n^2(z_s)-n^2(z')}} \qquad (14$$

and

$$P(r,z,z_s,z') = rn(z') + 2/3 \left[\varepsilon(z') \right]^{1/2} \cdot \left[(z_s-z')^{3/2} + (z-z')^{3/2} \right] . \qquad (15$$

Equation (13) is the integral representation used to generate the syntheti data for any given depth dependent profile.

<u>Inverse Equation: Region 1</u>

In the previous section a linear relation between the scattered data $D_1(k)$, and the profile correction, $\varepsilon(z')$, was obtained. The goal is t write $D_1(k)$ and $\varepsilon(z')$ as a quasi-Fourier transform pair. If the profile i assumed known everywhere except region 1, then the inversion can b performed in region 1 alone by subtracting the contributions of the othe regions from the total scattered data.

The transform variables are $P(z')$ and k. Equation (13) is rewritten a

$$D_1(k) = \int_{P(o)}^{P(z)} \frac{B(P(z'))}{|dP/dz'|} \exp\left[ikP(z') \right] \varepsilon(P(z'))dP \quad . \qquad (16$$

Multiplying by $\exp[-ikP(\bar{z}')]$, the profile correction is given by

$$\varepsilon(P(\bar{z}')) = \frac{|dP/dz'|_{z'=\bar{z}'}}{B(P(\bar{z}'))} \int_{-\infty}^{\infty} D_1(k)e^{-ikP(\bar{z}')}dk \quad . \qquad (17$$

Now that the scattered data and the profile correction are written as a quasi-Fourier transform pair, the inversion procedure is straightforward The data are generated using Eq. (16) for some true profile over th bandwidth necessary for a stable inversion. The true data are input int Eq. (17) and the profile correction is calculated as a function of th phase. It only remains to solve for $\varepsilon(\bar{z}')$ using relationships betwee $\varepsilon(P(\bar{z}'))$, $\varepsilon(\bar{z}')$ and the phase derivative. No iteration is necessary t recover the depth dependent sound speed profile.

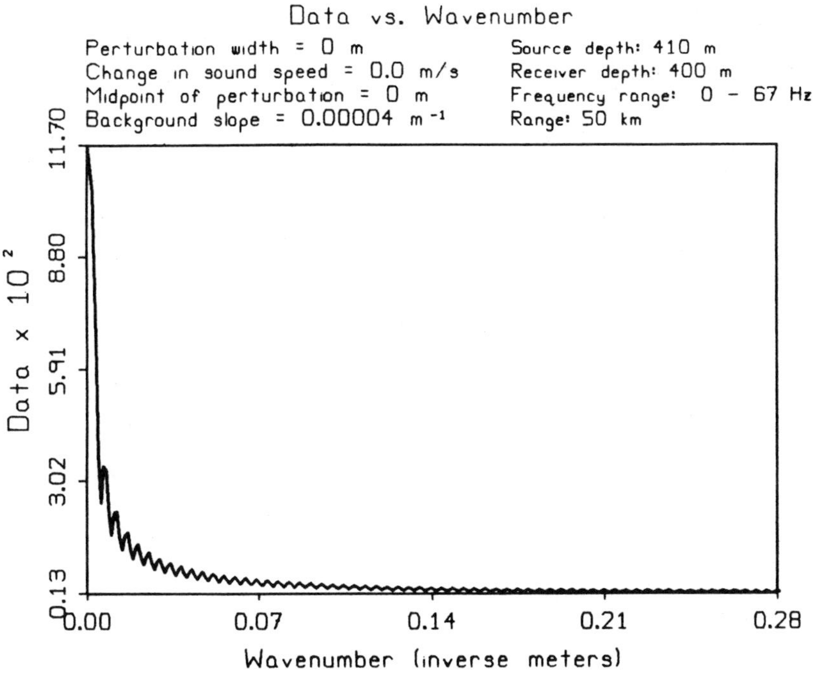

Fig. 1a. Spectrum of data generated for linear input profile.

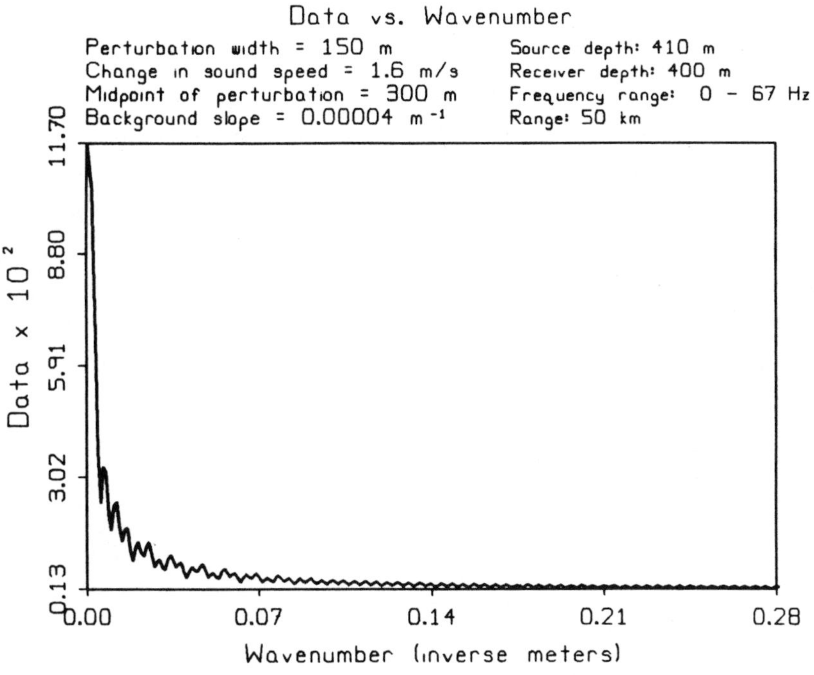

Fig. 1b. Spectrum of data generated for profile with a small perturbation.

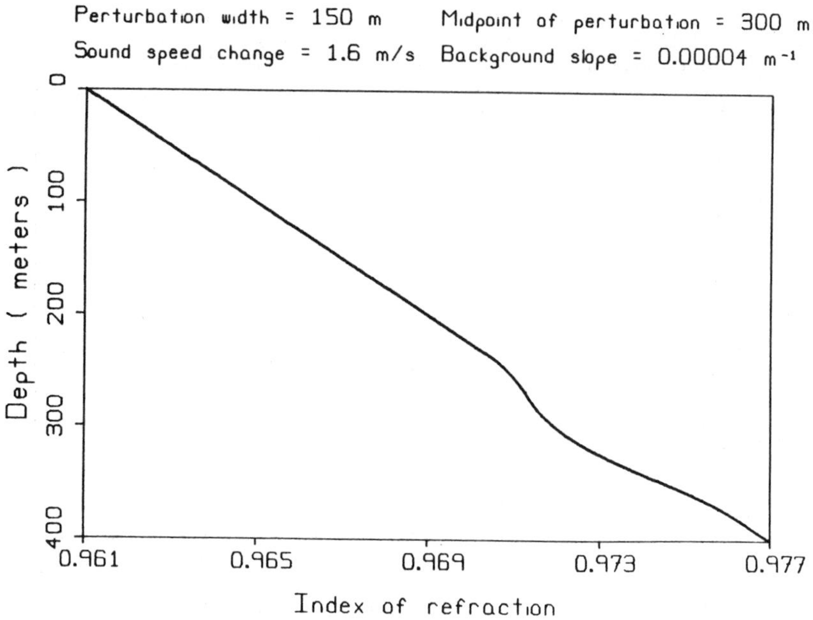

Fig. 2a. True input index of refraction.

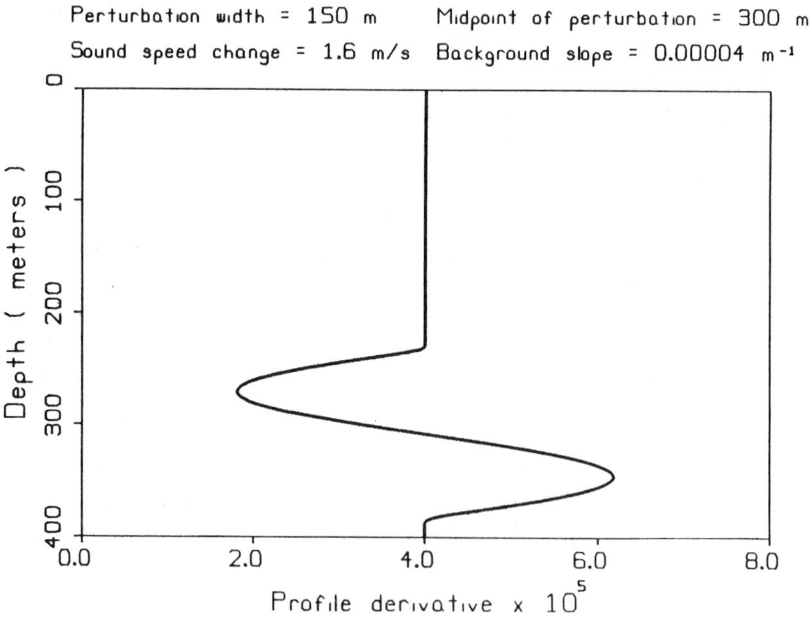

Fig. 2b. True input slope.

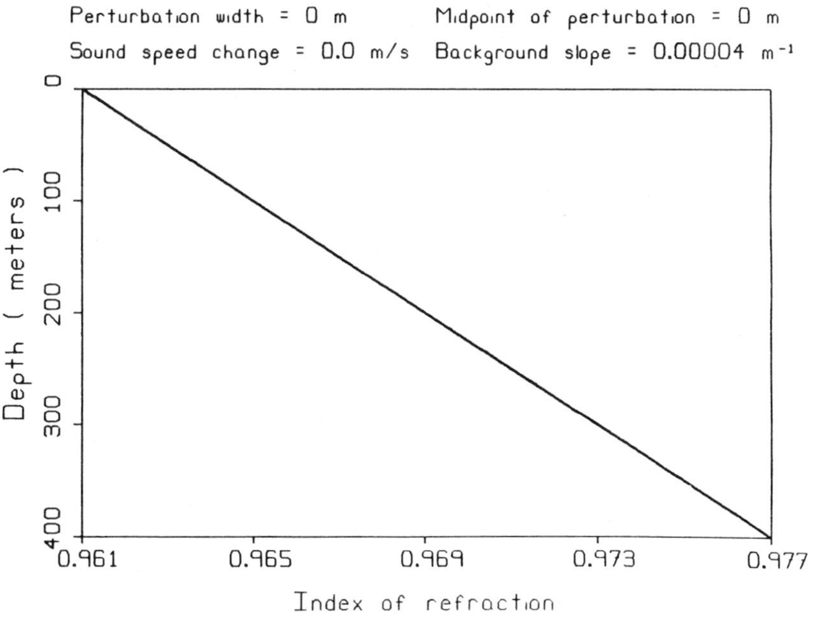

Fig. 3a. Guess input index of refraction.

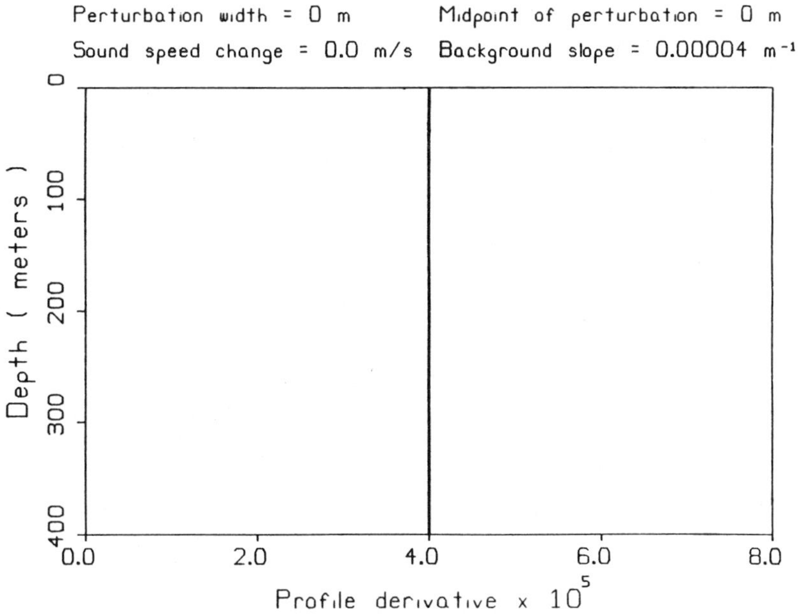

Fig. 3b. Guess input slope.

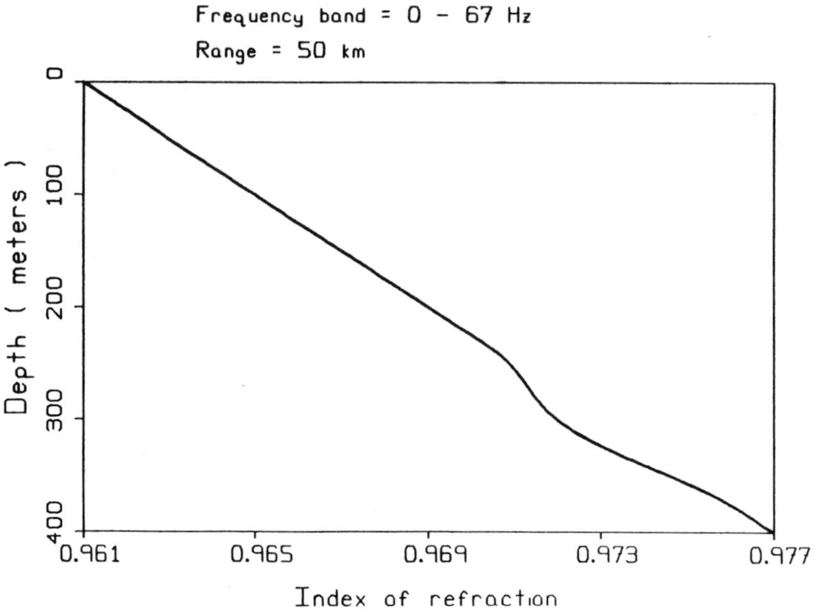

Fig. 4a. Output index of refraction.

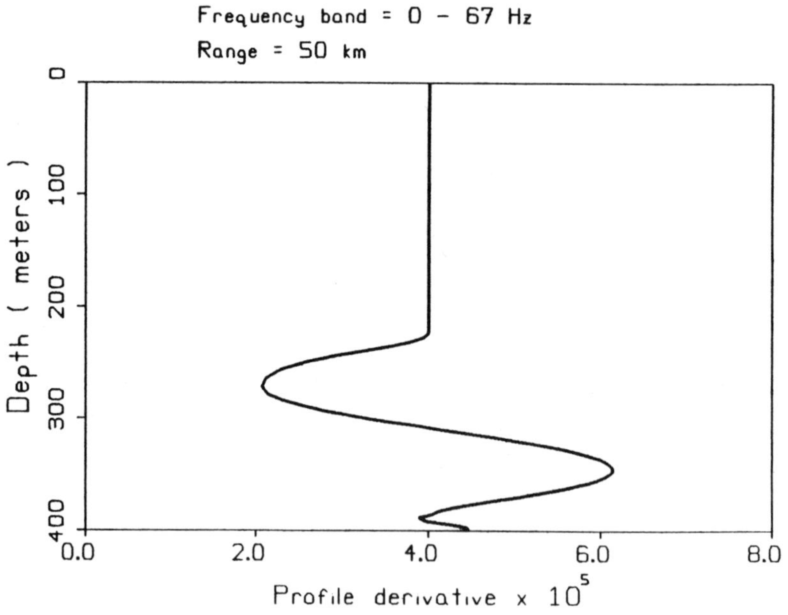

Fig. 4b. Output slope.

NUMERICAL RESULTS

The synthetic scattered data can be calculated for any arbitrary input profile. However, the inversion algorithm cannot differentiate two different depths having the same phase value; therefore, in order to perform an inversion the phase must be monotonic. The phase monotonicity depends exclusively on the specific input profile and the source-receiver offset in range.

An example of the spectrum of the data (region 1) for a linear input profile (see Fig. 3a and Fig. 3b) is illustrated in Figure 1a. The data vary smoothly and exhibit a regular decrease in amplitude as the wavenumber is increased. Figure 1b illustrates the spectrum when a small sound speed perturbation (of a type illustrated in Fig. 2a and Fig. 2b) is introduced into the profile. The behavior of the data clearly indicates the presence of the perturbation.

To test the inversion algorithm, the data in Figure 1b is used as input data in Eq. (17). Figure 2a illustrates the true index of refraction and Figure 2b the true profile slope used to generate this data. A linear guess profile, (Fig. 3a, index of refraction and Fig. 3b, profile slope) is used as the input guess profile in the inversion. The output profile and slope are illustrated in Figures 4a and 4b respectively.

The output profile and slope are nearly identical to the true profile and slope. This is illustrated especially well if the true and output slopes are compared.

CONCLUSIONS

A direct inversion technique which relates a profile correction and the scattered acoustic field data as a quasi-Fourier transform pair was discussed in this paper. The calculation of the profile correction is efficiently accomplished via an FFT. No iteration is required to reconstruct the depth dependent sound speed profile in the upper ocean waveguide from this profile correction. Small perturbations about the main profile trend are recovered very accurately if the background slope is assumed known.

REFERENCES

Ahluwalia, D.S. and Keller, J.B., 1977, Exact and asymptotic representations of the sound field in a stratified ocean, **in**: "Wave Propagation and Underwater Acoustics," J.B. Keller and J.S. Papadakis, eds., Springer, New York.

Boden, L.R., 1985, Sound Speed Profile Inversion In The Ocean, M.Sc. Thesis in Geophysics, Colorado School of Mines, Golden, Colorado.

Deavenport, R.L., 1966, A normal mode theory of an underwater acoustic duct by means of Green's function: **Radio Science,** Vol. 1, p. 709-724.

DeSanto, J.A., 1984, Oceanic sound speed profile inversion: **IEEE J. of Oceanic Engr.**, vol. OE-9, p. 12-17.

A TECHNIQUE FOR MODELING THE 3-D SOUND SPEED DISTRIBUTION AND STEADY STATE WATER VELOCITY FIELD FROM HYDROPHONE DATA

P. D. Young

Advanced Systems Development Center
Lockheed-California Company
Box 551 (Dept. 78-55, Bldg. 360, Plant B-6)
Burbank, California, USA 91520

ABSTRACT

This paper discusses a method presently under development which may be used to map the sound speed distribution and the steady state water velocity field of a region of ocean on the basis of hydrophone data collected on a number of discrete sources with known positions and spectral emission characteristics. The paper falls into two parts, the first being a presentation of the mathematical basis of the method, and the second being a discussion of implementation techniques. The method is most directly applicable to underwater environments where the acoustic energy is partly confined by an irregular bottom topography.

MATHEMATICAL DEVELOPMENT

The equations on which this treatment is based are

$$\frac{\partial \rho}{\partial t} + \sum_{i=1}^{3} \frac{\partial(\rho V_i)}{\partial x_i} = Q \tag{1}$$

and

$$\frac{\partial(\rho V_i)}{\partial t} + \sum_{j=1}^{3} \frac{\partial(\rho V_i V_j)}{\partial x_j} = -\frac{\partial P}{\partial x_i} + F_i \tag{2}$$

which respectively describe conservation of mass and conservation of fluid momentum; ρ is the fluid density, $\underset{\sim}{V}$ is the water velocity, Q gives the rate of mass density introduction to the system, P is the pressure field, and $\underset{\sim}{F}$ is the external force acting on the system. These equations neglect viscosity and heat conduction effects, but make no assumptions otherwise. Making the definitions

$$\rho = \rho_o + \rho_o \kappa p \tag{3}$$

$$\underset{\sim}{V} = \underset{\sim}{u} + \underset{\sim}{v} \tag{4}$$

$$Q = Q_o + q \tag{5}$$

$$P = P_o + p \quad , \tag{6}$$

where ρ_o, Q_o, and P_o are the time-averaged fluid density, source, and pressure field distributions respectively, κ is the isothermal compressibility, $\underset{\sim}{u}$ is the steady state component of the water velocity field, and $\underset{\sim}{v}$, q, and p are the time-varying components of the water velocity field, the source distribution, and the pressure field respectively, and the definitions

$$\zeta = \rho_o \kappa \tag{7}$$

(the reciprocal of the square of the sound speed) and

$$\underset{\sim}{w} = \rho_o \underset{\sim}{v} \quad , \tag{8}$$

and assuming a time dependence of $\exp(-i\omega t)$, setting $\underset{\sim}{F}$ to zero to reflect the assumption that no external forces act on the system, and discarding terms relating to nonlinear interactions of p and $\underset{\sim}{v}$ with themselves and each other, we get the hydrostatic equations

$$\nabla \cdot (\rho_o \underset{\sim}{u}) = Q_o \tag{9}$$

$$\nabla \cdot (\rho_o u_i \underset{\sim}{u}) = -\frac{\partial P_o}{\partial x_i} \quad , \; i = 1 \text{ to } 3 \tag{10}$$

and the equations in the harmonically varying distributions

$$-i\omega \underset{\sim}{w} - i\omega\zeta p \underset{\sim}{u} + \underset{\sim}{X} + \underset{\sim}{Y} + \underset{\sim}{Z} = -\nabla p \tag{11}$$

$$\nabla^2 p + \zeta\omega^2 p = i\omega q - 2\nabla\cdot\underset{\sim}{X} - \nabla\cdot\underset{\sim}{Z} \quad , \tag{12}$$

where

$$X_i = \nabla\cdot(u_i \underset{\sim}{w}) \tag{13}$$

$$Y_i = \nabla\cdot(\underset{\sim}{u} w_i) \tag{14}$$

$$Z_i = \nabla\cdot(\zeta p u_i \underset{\sim}{u}) \quad ; \tag{15}$$

Eq. (11) derives directly from Eq. (2) with the indicated substitutions, while Eq. (12) is reached by differentiating Eq. (1) with respect to time, taking the divergence of Eq. (2), and eliminating the common term from between them. These equations are solved by a multiorder expansion technique, in which we make the substitutions

$$\zeta = \zeta_o + \varepsilon\zeta_1 \tag{16}$$

$$\underset{\sim}{u} = \varepsilon \underset{\sim}{u}_1 \tag{17}$$

$$p = \sum_{n=0}^{\infty} \varepsilon^n p_n \tag{18}$$

$$\underset{\sim}{w} = \sum_{n=0}^{\infty} \varepsilon^n \underset{\sim}{w}_n \tag{19}$$

and separate the terms according to powers of ε; for Eq. (11) this gives

$$-i\omega \underset{\sim}{w}_o = -\nabla p_o \tag{20}$$

$$-i\omega \underset{\sim}{w}_1 = i\omega\zeta_0 \underset{\sim}{u}_1 p_0 - \underset{\sim}{X}_0 - \underset{\sim}{Y}_0 - \nabla p_1 \tag{21}$$

$$-i\omega \underset{\sim}{w}_2 = i\omega\zeta_0 \underset{\sim}{u}_1 p_1 + i\omega\zeta_1 \underset{\sim}{u}_1 p_0 - \underset{\sim}{X}_1 - \underset{\sim}{Y}_1 - \underset{\sim}{Z}_0^{(0)} - \nabla p_2 \tag{22}$$

and, for $n \geq 3$,

$$-i\omega \underset{\sim}{w}_n = i\omega\zeta_0 \underset{\sim}{u}_1 p_{n-1} + i\omega\zeta_1 \underset{\sim}{u}_1 p_{n-2}$$

$$- \underset{\sim}{X}_{n-1} - \underset{\sim}{Y}_{n-1} - \underset{\sim}{Z}_{n-2}^{(0)} - \underset{\sim}{Z}_{n-3}^{(1)} - \nabla p_n \quad , \tag{23}$$

while for Eq. (12) it gives

$$\nabla^2 p_0 + \zeta_0 \omega^2 p_0 = i\omega q \tag{24}$$

$$\nabla^2 p_1 + \zeta_0 \omega^2 p_1 = -\zeta_1 \omega^2 p_0 - 2\nabla\cdot\underset{\sim}{X}_0 \tag{25}$$

$$\nabla^2 p_2 + \zeta_0 \omega^2 p_2 = -\zeta_1 \omega^2 p_1 - 2\nabla\cdot\underset{\sim}{X}_1 - \nabla\cdot\underset{\sim}{Z}_0^{(0)} \tag{26}$$

and, for $n \geq 3$,

$$\nabla^2 p_n + \zeta_0 \omega^2 p_n = -\zeta_1 \omega^2 p_{n-1} - 2\nabla\cdot\underset{\sim}{X}_{n-1} - \nabla\cdot\underset{\sim}{Z}_{n-2}^{(0)} - \nabla\cdot\underset{\sim}{Z}_{n-3}^{(1)} \quad , \tag{27}$$

where

$$X_{ni} = \nabla\cdot(u_{1i}\underset{\sim}{w}_n) \tag{28}$$

$$Y_{ni} = \nabla\cdot(\underset{\sim}{u}_1 w_{ni}) \tag{29}$$

$$Z_{ni}^{(0)} = \nabla\cdot(\zeta_1 p_n u_{1i} \underset{\sim}{u}_1) \tag{30}$$

$$Z_{ni}^{(1)} = \nabla\cdot(\zeta_1 p_n u_{1i} \underset{\sim}{u}_1) \quad . \tag{31}$$

Note that the right side of the wave equation in p_n for any n above zero is always available through assorted differential operations on solutions to this equation for lower orders; given the solution to the equation

$$\nabla^2 g(\underset{\sim}{x},\underset{\sim}{x}') + \zeta_0(\underset{\sim}{x})\omega^2 g(\underset{\sim}{x},\underset{\sim}{x}') = \delta(\underset{\sim}{x}-\underset{\sim}{x}') \tag{32}$$

for the boundary conditions appropriate to the specific problem under consideration, one can in principle solve for p_n and $\underset{\sim}{w}_n$ up to any order.

If the source in the zero order wave equation is taken to be a delta function peaking at $\underset{\sim}{x}'$, then we have

$$p_0(\underset{\sim}{x},\underset{\sim}{x}') = i\omega g(\underset{\sim}{x},\underset{\sim}{x}') \tag{33}$$

and, from Eq. (20),

$$\underset{\sim}{w}_0 = -i(\nabla p_0)/\omega \quad , \tag{34}$$

from which we can calculate $\underset{\sim}{X}_0$; if f_1 is the right side of Eq. (25),

$$f_1(\underset{\sim}{x},\underset{\sim}{x}') = -\zeta_1(\underset{\sim}{x})\omega^2 p_0(\underset{\sim}{x},\underset{\sim}{x}') - 2\nabla\cdot\underset{\sim}{X}_0(\underset{\sim}{x},\underset{\sim}{x}') \quad , \tag{35}$$

then we have for p,

$$p_1(\underset{\sim}{x},\underset{\sim}{x}') = \int g(\underset{\sim}{x},\underset{\sim}{x}'')f_1(\underset{\sim}{x}'',\underset{\sim}{x}')d\underset{\sim}{x}'' \quad . \tag{36}$$

One can extend this procedure to bootstrap to $p_n(\underset{\sim}{x},\underset{\sim}{x}')$ and $\underset{\sim}{w}_n(\underset{\sim}{x},\underset{\sim}{x}')$ for arbitrarily great n; if from these results we construct the functions G and $\underset{\sim}{H}$,

$$G(\underset{\sim}{x},\underset{\sim}{x}') = \sum_{n=0}^{\infty} p_n(\underset{\sim}{x},\underset{\sim}{x}') \tag{37}$$

$$\underset{\sim}{H}(\underset{\sim}{x},\underset{\sim}{x}') = \sum_{n=0}^{\infty} \underset{\sim}{w}_n(\underset{\sim}{x},\underset{\sim}{x}') \tag{38}$$

(where for calculational convenience ε is set equal to unity), then we have

$$p(\underset{\sim}{x}) = \int G(\underset{\sim}{x},\underset{\sim}{x}')q(\underset{\sim}{x}')d\underset{\sim}{x}' \tag{39}$$

and

$$\underset{\sim}{w}(\underset{\sim}{x}) = \int \underset{\sim}{H}(\underset{\sim}{x},\underset{\sim}{x}')q(\underset{\sim}{x}')d\underset{\sim}{x}' \quad . \tag{40}$$

In practice, unless $|\zeta_1|$ and $|\underset{\sim}{u}_1|$ are sufficiently small, there is the possibility that the series G and $\underset{\sim}{H}$ will not effectively converge; conversely, if $|\zeta_1|$ and $|\underset{\sim}{u}_1|$ are small enough, then calculations past the first order will be unnecessary. If ζ_0 is properly chosen, then ζ_1 need differ from it by no more than a percent in any practical situation, and circumstances where the steady state water velocity is as great as a percent of the speed of sound in water are extremely rare; calculations to the first order should be sufficient for an adequate approximation to the pressure field. For data inversion purposes we require p_1 in the form

$$p_1(\underset{\sim}{x},\underset{\sim}{x}') = \int h_1(\underset{\sim}{x},\underset{\sim}{x}';\underset{\sim}{x}'')\zeta_1(\underset{\sim}{x}'')d\underset{\sim}{x}'' + \int \underset{\sim}{h}_2(\underset{\sim}{x},\underset{\sim}{x}';\underset{\sim}{x}'')\cdot\underset{\sim}{u}_1(\underset{\sim}{x}'')d\underset{\sim}{x}'' \quad , \tag{41}$$

whereas substitution of Eq. (35) into Eq. (36) gives

$$p_1(\underset{\sim}{x},\underset{\sim}{x}') = -\omega^2 \int g(\underset{\sim}{x},\underset{\sim}{x}'')p_0(\underset{\sim}{x}'',\underset{\sim}{x}')\zeta_1(\underset{\sim}{x}'')d\underset{\sim}{x}'' - 2\int g(\underset{\sim}{x},\underset{\sim}{x}'')\nabla\cdot\underset{\sim}{X}_0(\underset{\sim}{x}'',\underset{\sim}{x}')d\underset{\sim}{x}'' \quad ; \tag{42}$$

the first term of Eq. (42) already falls into the desired form, and the second term can be rearranged to do so. Substitution of the identity

$$u_{1i}(\underset{\sim}{x}) = \int u_{1i}(\underset{\sim}{x}')\ \delta(\underset{\sim}{x}-\underset{\sim}{x}')\ d\underset{\sim}{x}' \tag{43}$$

into the expression for $\underset{\sim}{X}_0$ in this term immediately gives

$$\int g(\underset{\sim}{x},\underset{\sim}{x}'')\nabla\cdot\underset{\sim}{X}_0(\underset{\sim}{x}'',\underset{\sim}{x}')d\underset{\sim}{x}'' = \int \underset{\sim}{I}(\underset{\sim}{x},\underset{\sim}{x}';\underset{\sim}{x}'')\cdot\underset{\sim}{u}_1(\underset{\sim}{x}'')d\underset{\sim}{x}'' \tag{44}$$

$$\underset{\sim}{I}(\underset{\sim}{x},\underset{\sim}{x}';\underset{\sim}{x}'') = \int g(\underset{\sim}{x},\underset{\sim}{y})\nabla[\nabla\cdot\{\delta(\underset{\sim}{y}-\underset{\sim}{x}'')\underset{\sim}{w}_0(\underset{\sim}{y},\underset{\sim}{x}')\}]d\underset{\sim}{y} \quad , \tag{45}$$

which may be evaluated to give

$$\mathrm{I}(\underset{\sim}{x},\underset{\sim}{x}';\underset{\sim}{x}'') = \nabla''\{\nabla''g(\underset{\sim}{x},\underset{\sim}{x}'')\cdot\underset{\sim}{w}_o(\underset{\sim}{x}'',\underset{\sim}{x}')\} \quad . \tag{46}$$

Accordingly, we have

$$h_1(\underset{\sim}{x},\underset{\sim}{x}';\underset{\sim}{x}'') = -\omega^2 g(\underset{\sim}{x},\underset{\sim}{x}'')p_0(\underset{\sim}{x}'',\underset{\sim}{x}') \tag{47}$$

$$\underset{\sim}{h}_2(\underset{\sim}{x},\underset{\sim}{x}';\underset{\sim}{x}'') = -2\nabla''\{\nabla''g(\underset{\sim}{x},\underset{\sim}{x}'')\cdot\underset{\sim}{w}_o(\underset{\sim}{x}'',\underset{\sim}{x}')\} \quad . \tag{48}$$

Given a known source distribution $q(\underset{\sim}{x})$ and a number of pressure field measurements

$$\gamma_i = p(\underset{\sim}{x}_i) \quad , \; i = 1 \text{ to } N \quad , \tag{49}$$

we have

$$\Delta\gamma_i = \int h_{1i}(\underset{\sim}{x})\zeta_1(\underset{\sim}{x})d\underset{\sim}{x} + \int \underset{\sim}{h}_{2i}(\underset{\sim}{x}) \; \underset{\sim}{u}_1(\underset{\sim}{x})d\underset{\sim}{x} \tag{50}$$

$$\Delta\gamma_i = \gamma_i - \int p_o(\underset{\sim}{x}_i,\underset{\sim}{y})q(\underset{\sim}{y})d\underset{\sim}{y} \tag{51}$$

$$h_{1i}(\underset{\sim}{x}) = \int h_1(\underset{\sim}{x}_i,\underset{\sim}{y};\underset{\sim}{x})q(\underset{\sim}{y})d\underset{\sim}{y} \tag{52}$$

$$\underset{\sim}{h}_{2i}(\underset{\sim}{x}) = \int \underset{\sim}{h}_2(\underset{\sim}{x}_i,\underset{\sim}{y};\underset{\sim}{x})q(\underset{\sim}{y})d\underset{\sim}{y} \quad , \tag{53}$$

a system of equations which may be solved by Backus-Gilbert linear data inversion methods to give solutions of the form

$$\zeta_1(\underset{\sim}{x}) = \sum_{i=1}^{N} a_i h_{1i}(\underset{\sim}{x}) \tag{54}$$

$$\underset{\sim}{u}_1(\underset{\sim}{x}) = \sum_{i=1}^{N} a_i \underset{\sim}{h}_{2i}(\underset{\sim}{x}) \quad , \tag{55}$$

where the constant coefficients a_i are functions of the data. The condition that there be no steady state sources or sinks within the volume of water under consideration may be incorporated into the solution by use of the hydrostatic mass conservation relation Eq. (9); if we have a linearly independent basis set $\{\phi_i(x)\}$ whose elements satisfy the boundary conditions required for the pressure field, then we can express this equation in the desired integral form by writing the integral expressions

$$J_i = \int \phi_i(\underset{\sim}{x})\nabla\cdot\{\rho_0(\underset{\sim}{x})\underset{\sim}{u}_1(\underset{\sim}{x})\}d\underset{\sim}{x} \tag{56}$$

and requiring that every J_i be equal to zero. Use of the identity Eq. (43) quickly transforms these integrals into

$$J_i = -\int \underset{\sim}{u}_1(\underset{\sim}{x})\cdot\{\rho_0(\underset{\sim}{x})\nabla\phi_i(\underset{\sim}{x})\}d\underset{\sim}{x} \quad , \tag{57}$$

which may be processed along with Eqs. (50) on the same footing, with the J_i viewed as data whose values are equal to zero.

IMPLEMENTATION TECHNIQUES

The practical utility of this procedure is contingent on the existence of an expression for g in a form which expedites the indicated mathematical operations; the form found most suitable for this purpose is

$$g(\underset{\sim}{x},\underset{\sim}{x}') = \sum_{i=1}^{M} \sum_{j=1}^{M} c_{ij}\, \psi_i(\underset{\sim}{x})\psi_j(\underset{\sim}{x}') \quad , \tag{58}$$

where the c_{ij} are constant coefficients (which may be complex) and the ψ_i are selected elements of the set of term functions of a 3-D Fourier series expansion defined over a rectilinear volume large enough to enclose the water volume under study. As a basis set, the set of Fourier series term functions has the computational advantages that it is closed under the operations of multiplication, differentiation, and integration; if the elements of its subset are properly chosen, this subset will also be closed under differentiation and integration, and the errors resulting from its lack of closure under multiplication will be small. It is helpful in the selection process that an upper limit is set by the resolving power of the measured data on the detail that this working basis set will be called upon to represent, providing a guide for the estimation of the optimum size and composition of the set. Given g in this form, the required mathematical operations on g are reduced to a series of matrix operations, easily implemented in software.

The procedure by which g is derived starts with the definition of a basis set $\{\phi_i\}$

$$\phi_i(\underset{\sim}{x}) = \sum_{j=1}^{M} a_{ij}\psi_j(\underset{\sim}{x}) \tag{59}$$

such that the boundary conditions for the region being modeled are satisfied by all elements of this basis set. We suppose that g is expandable in the form

$$g(\underset{\sim}{x},\underset{\sim}{x}') = \sum_{i=1}^{N} B_i(\underset{\sim}{x}')\phi_i(\underset{\sim}{x}) \quad , \tag{60}$$

and substitute this into the differential equation for g to get

$$\sum_{i=1}^{N} B_i(\underset{\sim}{x}')\{D\phi_i(\underset{\sim}{x})\} = \delta(\underset{\sim}{x}-\underset{\sim}{x}') \tag{61}$$

$$D\phi_i(\underset{\sim}{x}) = \nabla^2\phi_i(\underset{\sim}{x}) + \zeta_o(\underset{\sim}{x})\omega^2\phi_i(\underset{\sim}{x}) \quad ; \tag{62}$$

multiplication of both sides by $\phi_j^*(\underset{\sim}{x})$ followed by integration over $\underset{\sim}{x}$ gives

$$\sum_{i=1}^{N} B_i(\underset{\sim}{x}') \langle\phi_j, D\phi_i\rangle = \phi_j^*(\underset{\sim}{x}') \tag{63}$$

$$\langle f,g\rangle = \int f^*(\underset{\sim}{x})g(\underset{\sim}{x})d\underset{\sim}{x} \quad , \tag{64}$$

a matrix equation with the solution

$$B_i(x') = \sum_{j=1}^{N} d_{ij}\phi_j^*(\underset{\sim}{x}') \quad , \tag{65}$$

where the coefficient matrix $\{d_{ij}\}$ is the inverse of the inner product matrix $\{\langle\phi_j, D\phi_i\rangle\}$, from which we have

$$g(\underset{\sim}{x},\underset{\sim}{x}') = \sum_{i=1}^{N} \sum_{j=1}^{N} d_{ij} \phi_i(\underset{\sim}{x})\phi_j^*(\underset{\sim}{x}') \quad . \tag{66}$$

Substitution of Eq. (59) into Eq. (66) then gives a solution in the form of Eq. (58).

In solving for the ϕ_i, the volume of water in question is modeled as an irregular solid polygon whose surface is made up of planar triangular and trapezoidal sections, on each of which each of the ϕ_i is required to satisfy some specified boundary condition to an acceptable degree of approximation. To this end, let $\phi(\underset{\sim}{x})$ be a representative element of $\{\phi_i\}$, assumed to have the form

$$\phi(\underset{\sim}{x}) = \sum_{i=1}^{M} a_i \psi_i(\underset{\sim}{x}) \quad , \tag{67}$$

and let $F_{jk}\{f(\underset{\sim}{x})\}$ represent coefficient k of a 2-D Fourier series analysis of the evaluation of the 3-D function $f(\underset{\sim}{x})$ on the surface of section j; the boundary condition that $\phi(\underset{\sim}{x})=0$ over section j (a condition appropriate to a pressure release surface) may then be represented by the system of equations

$$0 = \sum_{i=1}^{M} a_i F_{jk}\{\psi_i(\underset{\sim}{x})\} \quad , \tag{68}$$

the condition that $\hat{n}\cdot\nabla\phi(\underset{\sim}{x})=0$ over section j, where $\hat{n}$ is the unit normal vector to the section (appropriate to a hard, nonporous ocean bottom) is expressed as

$$0 = \sum_{i=1}^{M} a_i F_{jk}\{\hat{n}\cdot\nabla\psi_i(\underset{\sim}{x})\} \quad , \tag{69}$$

and the condition that $\hat{n}\cdot\nabla\phi(\underset{\sim}{x})=\alpha\phi(\underset{\sim}{x})$ over section j, where α is some constant characteristic of the section (appropriate to a partially absorbing ocean bottom) takes the form

$$0 = \sum_{i=1}^{M} a_i F_{jk}\{\hat{n}\cdot\nabla\psi_i(\underset{\sim}{x})-\alpha\psi_i(\underset{\sim}{x})\} \quad . \tag{70}$$

It is helpful that, if the ψ_i are 3-D Fourier series term functions, then analytical expressions exist for all of these 2-D Fourier series integrals, eliminating the need for numerical integration. Situations in which ϕ is required to fit some specified distribution over the surface of the section may be dealt with by doing a 2-D Fourier series analysis of the desired surface distribution and matching the resulting series coefficients to the corresponding coefficients for ϕ; if S_k is the value of coefficient k for the analysis of the desired surface distribution over section j, then we have

$$S_k = \sum_{i=1}^{M} a_i F_{jk}\{\psi_i(\underset{\sim}{x})\} \quad . \tag{71}$$

The collected equations in the a_i for all of the sections constitute a least-squares problem, which may be solved by various matrix methods to yield a collection of linearly independent solutions which satisfy the specified boundary conditions to good approximation, if not exactly.

A NEW FORM OF THE WAVE EQUATION FOR SOUND IN A GENERAL LAYERED FLUID

O. A. Godin

P.P. Shirshov Institute of Oceanology
the USSR Academy of Sciences
Moscow, V-218, USSR

ABSTRACT

By introduction of a new vertical coordinate, the wave equation in a layered medium is transformed into the reduced wave (Helmholtz) equation. In the new form of the wave equation, effects of gravity as well as of density and of mean current stratification appear only in the effective wave number and in the transformation of the variable used. Starting from this equation, a number of new results are obtained. Some applications to three-dimensionally varying media are considered. The equation of sound propagation in an ocean with arbitrarily slow mean currents is obtained and the corresponding parabolic approximation is discussed.

1. INTRODUCTION

Sound fields in the ocean are usually studied theoretically on the basis of the wave equation

$$p_{tt} - c^2(\mathbf{r})\Delta p = 0, \tag{1}$$

which reduces for monochromatic waves to the Helmholtz (reduced wave) equation

$$\Delta p + k^2 p = 0, \quad k(\mathbf{r}) = \omega/c(\mathbf{r}), \tag{2}$$

where p is acoustic pressure, c is sound velocity, t is time, k is wave number, and $\mathbf{r} = (x, y, z)$ designates the space coordinates. We shall use letters in subscript to designate differentiation. It is supposed that all medium parameters are independent of time. Eqs. (1) and (2) with the boundary conditions on p and its normal derivative continuity are valid when the density ρ of the medium is constant, its velocity $\mathbf{u}$ in the absence of sound equals zero and the action of gravity is neglected.

When nonuniformity of the density is taken into account, the wave equation has the form

$$\Delta p - \nabla(\ln\rho)\cdot\nabla p + c^{-2}p_{tt} = 0. \tag{3}$$

For monochromatic waves it is usually convenient to use this equation in another form (Brekhovskikh, 1980)

$$F + [k^2 + \Delta\rho/2\rho - 3(\Delta\rho/2\rho)^2]F = 0, \quad F= p/\rho^{1/2}. \tag{4}$$

Eqs. (3) and (4) contain derivatives of the density in their coefficients. This imposes some restrictions on the permissible dependence of $\rho(\mathbf{r})$, which are not met in many practical problems. When the density of the medium is allowed to vary on small scale intervals, large and rapidly varying coefficients appear in the equations. This prevents us from using many known analytical and numerical techniques designed to solve differential equations.

The purpose of the present paper is threefold. The first one is to find equations governing elastic wave propagation in a moving inhomogeneous compressible fluid in a uniform gravitational field, and cast the equations in a common form, which doesn't contain derivatives of the medium's parameters. Advantages of such a form of the wave equation for a medium at rest and without gravity were illustrated by a number of examples (Godin, 1985b). We manage to solve the stated problem in the case of a layered medium, where parameters depend on the vertical coordinate z only, for waves with harmonic dependence on coordinates x, y and time t. This wave equation is the Helmholtz one with some effective wave number dependence on the earth's gravity and stratifications of the density and the velocity of the current.

The second purpose is to utilize an analogy of the wave equation obtained with well-studied Eq. (2) to generalize some results proved earlier for immovable media with g=constant.

The third purpose consists in development of an approximate wave equation for sound in a three-dimensionally inhomogeneous medium with slow currents. We shall study this equation under conditions typical of oceanic environments.

The whole presentation is within the scope of linear acoustics.

2. DERIVATION OF THE EQUATION FOR ACOUSTIC-GRAVITY WAVES

The motion of a fluid is described by hydrodynamic equations, that is the Euler, continuity and state equations:

$$\partial\tilde{\mathbf{v}}/\partial t + (\tilde{\mathbf{v}}\cdot\nabla)\tilde{\mathbf{v}} = -\nabla\tilde{p}/\tilde{\rho} + \mathbf{f}/\tilde{\rho}, \tag{5}$$

$$\partial\tilde{\rho}/\partial t + \nabla\cdot(\tilde{\rho}\,\tilde{\mathbf{v}}) = 0, \tag{6}$$

$$\tilde{p} = \tilde{p}(\tilde{\rho}, S). \tag{7}$$

Here $\mathbf{f}=-\tilde{\rho}\, g\mathbf{v}\, z$ is the volume density of the gravity force, g=const, $\tilde{p}=p_0 + p$ is pressure, $\tilde{\mathbf{v}} = \mathbf{u}+\mathbf{v}$ is the particle velocity, $\tilde{\rho} = \rho+\rho'$ is the density. p, $\mathbf{v}$ and ρ' are additions to appropriate quantities due to a sound wave. In a multicomponent medium (for example, in sea water) $\tilde{p}$ in Eq. (7) depends also on concentrations of components. We shall neglect the irreversible processes of admixture, diffusion and heat transfer. Then the entropy and concentrations of components are constants in every particle, and $\tilde{p}$ is a single-valued function of $\tilde{\rho}$ in it. Hence,

$$(\partial/\partial t + \tilde{\mathbf{v}}\cdot\nabla)\tilde{p} = \tilde{c}^2\,(\partial/\partial t + \tilde{\mathbf{v}}\cdot\nabla)\rho, \quad \tilde{c}^2 = (\partial\tilde{p}/\partial\tilde{\rho})_S. \tag{8}$$

By the process of linearization we obtain from Eqs. (5), (6) and (8) the set of linear differential equations for acoustic quantities p, $\mathbf{v}$ and ρ':

$$d\mathbf{v}/dt + (\mathbf{v}\cdot\nabla)\mathbf{u} + \nabla p/\rho - \rho'\nabla p_0/\rho^2 = 0, \quad (9)$$

$$(\mathbf{v}\cdot\nabla)p_0 + dp/dt = (\tilde{c}^2 - c^2)\mathbf{u}\cdot\nabla\rho + c^2[\mathbf{v}\cdot\nabla\rho + d\rho'/dt], \quad (10)$$

$$d\rho'/dt = \rho'\nabla\cdot\mathbf{u} + \nabla\cdot(\rho\mathbf{v}) = 0, \quad (11)$$

where

$$d/dt = \partial/\partial t + \mathbf{u}\cdot\nabla. \quad (12)$$

Let's suppose now, that current is horizontal and the medium is layered: $p_0=p_0(z)$, $\mathbf{u}=\mathbf{u}(z)$, $\rho=\rho(z)$, $c=c(z)$. Then $(\mathbf{u}\cdot\nabla)\mathbf{u}=0$ and $\nabla p_0=-\rho g \mathbf{v} z$. It is convenient to consider an arbitrary wave field as a superposition of elementary waves harmonically dependent on time and horizontal coordinates. Using the designations

$$h(\omega, \mathbf{q}, z) = \iiint_{-\infty}^{+\infty} dx dy d\omega\, h(\mathbf{r},t)\, e^{-i\mathbf{r}\cdot\mathbf{q}+i\omega t}, \quad \mathbf{q} = (q_1, q_2, 0) \quad (13)$$

for spectral quantities, where h stands for p, **v** or ρ', for the harmonic waves we obtain, instead of Eqs. (9)-(11),

$$-i\omega U\mathbf{v} + w\mathbf{u}_z = -\nabla p/\rho - g\rho'\nabla z/\rho, \quad (14)$$

$$\rho g w + i\omega U p = (i\omega U\rho' - w\rho_z)c^2, \quad (15)$$

$$w\rho_z + \rho(i\mathbf{q}\cdot\mathbf{v} + w_z) = i\omega U\rho', \quad (16)$$

where w is the z component of **v** and $U = 1-\mathbf{q}\cdot\mathbf{u}/\omega$. This set of equations easily reduces to 2 scalar ordinary differential equations of the first order:

$$p_z + gc^{-2}p = i\rho(\omega^2U^2 + g\rho_z/\rho + g^2c^{-2})w/\omega U, \quad (17)$$

$$w_z - (gc^{-2} + U_z/U)w = i(\omega^2U^2 - q^2c^2)p/\omega U\rho c^2. \quad (18)$$

They are supplemented by the following boundary conditions

$$[w/U]_S = 0, \quad [p - i\rho g w/\omega U]_S = 0. \quad (19)$$

Here [F] means a jump of a function F at any surface S, and S in this application is a horizontal plane z=const. One can obtain the boundary conditions using Eqs. (14), (17) and (18). The conditions mean that the pressure and vertical component are equal for particles lying on different sides of S.

Using Eqs. (17) and (18) it is not difficult to derive one equation of the second order in closed form. For example, substituting w from Eq. (17) in Eq. (18), one obtains

$$F_{zz}+\{k^2U^2-q^2+0.5(\ln L)_{zz}-[0.5(\ln L)_z]^2\}F=0, \quad F = p[\rho(\omega^2U^2-N^2)]^{-1/2} \quad (20)$$

where

$$L=\rho(\omega^2U^2-N^2)\exp\left[2\int_{z_0}^{z} gc^{-2}\,dz\right], \quad z_0=\text{const}, \quad N=[-g\rho_z/\rho + g/c^2]^{1/2} \quad (21)$$

is the Vaisala frequency. For the particular case of the atmosphere, considered as an ideal gas with γ standing for the ratio of specific heats at constant stress and constant strain, taking into account the relation (Tatarskii 1979)

$$(\ln p_0)_z = -\gamma\, g/c^2, \tag{22}$$

one can cast L in the form

$$L = \rho(\omega^2U^2 - N^2)[p_0(z_0)/p_0(z)]^{2/\gamma}. \tag{23}$$

Then Eq. (20) coincides with Ostashev's (1984) result.

To obtain a wave equation without derivatives of the medium's parameters in the coefficients it is convenient to introduce new dependent variables

$$W = w/U, \qquad s = f(z)\cdot(p - i\rho g W/\omega), \tag{24}$$

which are continuous throughout the medium. Here f is any continuous function. Then Eqs. (17) and (18) become

$$s_z + s\,(gq^2/\omega^2U^2 - f_z/f) = i\rho f(\omega U^2 - g^2q^2/\omega^3U^2)W, \tag{25}$$

$$W_z - (gq^2/\omega^2U^2)W = (i\omega/\rho f)(c^{-2} - q^2/\omega^2U^2)s. \tag{26}$$

Further transformations are simplified greatly if

$$f(z) = \exp\Big[gq^2\omega^{-2}\int_{z_0}^{z} U^{-2}\,dz\Big]. \tag{27}$$

Then from Eqs. (25) and (26) we easily obtain

$$s_{zz} - [\ln \rho\, f^2(\omega^2U^2 - g^2q^2/\omega^2U^2)]_z s_z + (k^2U^2 - q^2)\,(1 - g^2q^2/\omega^4U^4)s = 0. \tag{28}$$

Now to eliminate the derivatives from the medium parameters in the coefficients of the wave equation, by analogy, with (Godin, 1985b) it is enough to introduce a new independent variable

$$\zeta(z) = \rho_0^{-1}\int_{z_0}^{z} \rho f^2(U^2 - g^2q^2/\omega^4U^2)\,dz, \quad \rho_0 = \text{constant} \tag{29}$$

In the new variables Eq. (28) and the boundary conditions (19) become

$$s_{\zeta\zeta} + (k^2U^2 - q^2)\,(\rho_0/\rho U^2 f^2)^2\,(1 - g^2q^2/\omega^4U^4)^{-1}s = 0, \quad [s]_S = 0, \quad [s_\zeta]_S = 0. \tag{30}$$

If the quantity $U^2 - g^2q^2/\omega^4U^2$ has a fixed sign in the region considered, the transformation of the variable (29) is reversible, and a single-valued inverse function $z(\zeta)$ exists. In the case of a medium at rest U = 1 and this condition is always fulfilled.

Let's turn to the derivative of the wave equation for acoustic-gravity waves with arbitrary dependence on **r** and t. The quantity

$$G = dp/dt - \rho g w, \tag{31}$$

is the rate of the pressure change in a fluid particle (but not at a fixed point **r** = const). In an harmonic wave $d/dt = -i\omega U$, $q^2 = -\Delta_\perp$, $-i\omega U_z = \mathbf{u}_z \cdot \nabla$, and according to Eq. (24) $G = -i\omega U s/f$. (Here $\Delta_\perp = \nabla_\perp^2$, $\nabla_\perp = (\partial/\partial x, \partial/\partial y, 0)$) . In this notation, after some transformations, one can cast Eq. (28) in the form

$$(d^4/dt^4 + g^2\Delta_\perp)\{(d^2/dt^2)[c^{-2}\, d^2G/dt^2 - \rho\nabla\cdot(\nabla G/\rho)] - N^2\Delta_\perp G + \rho(\rho^{-1}\mathbf{u}_z\cdot\nabla)_z\, dG/dt\} + 4(d^3/dt^3)(\mathbf{u}_z\cdot\nabla)(d^2G/dt^2 - \mathbf{u}_z\cdot\nabla dG/dt + g\Delta_\perp G) = 0 \tag{32}$$

Eq. (32) is linear and applicable to every spectral component of G. Coefficients of the equation do not depend on ω or **q**. Hence, G(**r**, t) with arbitrary dependence on **r** and t must obey Eq. (32). Quite analogously one obtains from Eq. (20)

$$(d^2/dt^2 + N^2)[(d^2/dt^2)(\Delta p - c^{-2}d^2p/dt^2 - 2g\, c^{-3}c_z p - \rho_z\rho^{-1}p_z) + N^2\,\Delta_\perp p] = 2(d^2/dt^2)(\mathbf{u}_z\cdot\nabla d/dt + NN_z)(p_z + g\, c^{-2}p). \tag{33}$$

Eq. (32) is of the eighth order in d/dt and Eq. (33) is of the sixth order. They become simpler in a number of particular cases. For example, in the case of uniform flow (**u** = const) we have

$$(d^2/dt^2)\,[(\rho c^2)^{-1}\, d^2G/dt^2 - \nabla\cdot(\nabla G/\rho)] - \rho^{-1}N^2\Delta_\perp G = 0. \tag{34}$$

It is worth noticing that in some cases a good choice of the dependent variable leads to simpler equations as compared to the consequences of Eqs. (32) or (33). For example, in an incompressible fluid, the equation for G is of the sixth order in d/dt but the vertical displacement, Y, of particles in the wave which is connected with the velocity by the relation w = dY/dt, obeys a much simpler equation

$$\nabla\cdot(\rho d^2\nabla Y/dt^2) = g\rho_z\nabla_\perp Y \tag{35}$$

resulting from Eqs. (17 and (18).

Of course, one can derive Eqs. (32) and (33) without use of the spectral representation of the field in intermediate transformations. We shall illustrate this fact supposing **u** = 0 for simplicity. Up to the end of the section, the medium will be supposed not to be layered. After elimination of ρ' and the horizontal components of **v**, Eqs. (9) - (11) reduce to

$$(\rho c^2)^{-1}G_{tt} - \nabla\cdot(\rho^{-1}\nabla_\perp G) + \hat{A}_1 w = 0, \quad \hat{A}_1 = \partial^3/\partial t^2\partial z - g\nabla\cdot(\rho^{-1}\nabla_\perp\rho), \tag{36}$$

$$\rho^{-1}G_z + (g/\rho c^2)G + \hat{A}_2 w = 0, \quad \hat{A}_2 = \partial^2/\partial t^2 + g\,\partial/\partial z. \tag{37}$$

Applying operator $\hat{A}_2$ to Eq. (36) and $\hat{A}_1$ to Eq. (37) and subtracting the results, one finds

$$\{(g\partial/\partial z + \partial^2/\partial t^2)\,[(\rho c^2)^{-1}\partial^2/\partial t^2 - \nabla\cdot(\rho^{-1}\nabla_\perp)] - [\partial^3/\partial t^2\partial z - g\nabla\cdot(\rho^{-1}\nabla_\perp\rho)]\cdot(\rho^{-1}\partial z + g/\rho c^2)\}G + g^2\mathcal{R}\, w = 0, \tag{38}$$

where $\mathcal{R}$ is the commutator:

$$\mathcal{R} = [\partial/\partial z,\ \nabla\cdot(\rho^{-1}\nabla_{\perp}\rho)] \tag{39}$$

If $\mathcal{R} = 0$, expression (38) transforms into a closed equation on function G. Particularly this happens for an arbitrary dependence $c = c(\mathbf{r})$, if $\rho = \rho(z)$ or $\rho = \rho(x, y)$. For a layered medium ($\rho = \rho(z)$, $c = c(z)$) Eq. (38) is consistent with Eq. (32)

In a recent book (Boyles, 1984) the equation

$$\Delta p - \rho^{-1}\nabla\rho\cdot\nabla p - c^{-2}p_{tt} = \rho^{-1}\nabla p_0\cdot(\rho^{-1}c^{-4}\nabla p_0 + \rho\nabla(\rho c^2)^{-1})p \tag{40}$$

was offered for acoustic-gravity waves in a three-dimensionally homogeneous medium at rest. Eq. (40) is derived from the expression

$$c^{-2}[p_{tt} - \rho^{-1}\nabla p_0\cdot\nabla p + \rho'(\nabla p_0/\rho)^2] + \rho^{-1}\nabla\rho\cdot\nabla p - \rho'\rho^{-2}\nabla\rho\cdot\nabla p_0 =$$
$$\Delta p - \nabla\cdot(\rho'\rho^{-1}\nabla p_0) \tag{41}$$

by the substitution

$$\rho'(\mathbf{r}, t) = c^{-2}(\mathbf{r}, t)\ p(\mathbf{r}, t). \tag{42}$$

The expression (41) is correct and coincides with our formula (50), if we set $\mathbf{u} = 0$. But Eq. (40) is an erroneous one. The cause is that small changes in density and pressure are related by the expression $d\tilde{\rho} = d\tilde{p}/c^2$ in a fixed fluid particle. Eq. (42) is incorrect since, due to the earth's gravity, the acoustic pressure p changes not only owing to a compression of the particle, but also owing to its displacement at the point with a different value of $p_0(\mathbf{r})$. One should use complicated Eq. (10) instead of Eq. (42). This prevents us from eliminating ρ' from (41) and from deriving a wave equation for acoustic-gravity waves in a general three-dimensionally varying medium.

3. THE ACOUSTIC WAVE EQUATION FOR A MOVING MEDIUM

Here we shall consider in some detail elastic waves with a frequency high enough to disregard the action of gravity on the wave (but not on formation of initial dependencies $c(\mathbf{r})$, $\rho(\mathbf{r})$ and $\mathbf{u}(\mathbf{r})$). In the following it will become evident that in the ocean this assumption is valid over all the acoustic range of frequencies. The medium is supposed to be layered. When $g = 0$, the transformation of variables (29), the wave equation and the boundary conditions (30) become

$$\zeta(z) = \rho_0^{-1}\int_{z_0}^{z} \rho U^2\, dz, \tag{43}$$

$$p_{\zeta\zeta} + (\rho_0/\rho U^2)^2\,(k^2U^2 - q^2)p = 0, \tag{44}$$

$$[p]_S = 0, \quad [p_\zeta]_S = 0. \tag{45}$$

Note that $\zeta(z)$ is strictly an increasing function. Fulfillment of the boundary conditions is guaranteed by Eq. (44) itself. That is why this equation describes sound propagation in an arbitrary layered media with effective piecewise continuous parameters. It is an Helmholtz equation with effective wave number

$$k_{ef} = \rho_0(k^2U^2 - q^2)^{1/2}/\rho U^2, \tag{46}$$

which is bounded unless U = 0. U = 0 when the projection of the phase velocity of the wave on the direction of current equals u. In this case a special resonance interaction between sound and current takes place (Fabricant, 1976). This effect is described by a singularity in Eq. (44).

For currents in the ocean, the Mach number $M = u/c << 1$. This enables us to simplify the wave equation by neglecting second and higher powers of u. Then letting g = 0, we obtain from Eqs. (20) and (44)

$$F_{zz}+\{k^2 - q^2 + 0.5(\ln\rho)_{zz} - [0.5(\ln\rho)_z]^2 - 2k^2m - m_{zz}\}F = 0, \quad F = e^m\rho^{-1/2}p, \tag{47}$$

$$p_{\zeta\zeta} + (\rho_0/\rho)^2[k^2 - q^2 + 2m(k^2 - 2q^2)]p = 0, \tag{48}$$

where

$$m(z) = 1 - U = \mathbf{q}\cdot\mathbf{u}(z)/\omega. \tag{49}$$

Now we consider the possibilities of finding the exact solutions of Eqs. (47 and (48) in terms of known special functions. With respect to Eqs. (20) and (44) with g = 0, it is only the case c = const, ρ = const, $\mathbf{u} = \mathbf{a}_1 z + \mathbf{a}_0$, where an exact solution is known for a smooth profile of $\mathbf{u}(z)$ and arbitrary $\mathbf{q}$.

If q = 0, the current doesn't affect sound propagation. If $\mathbf{q} \neq 0$ and ω are fixed, for arbitrary dependencies of two of the quantities from c(z), ρ(z) and m(z), one can find the third one so that Eqs. (47) and (48) will have elementry solutions. For example, the choice

$$k^2 = (a\rho^2/\rho_0^2 + q^2 + 4mq^2)/(1 + 2m), \tag{50}$$

leads to an exact solution $p = \exp(\pm ia\zeta)$ for arbitrary ρ(z) and $\mathbf{u}(z)$. It is of more interest, however, to search for exact solutions when stratification of all parameters ρ, c, and $\mathbf{u}$ is given independently. Let c = const, ρ = const. We are interested in those profiles $\mathbf{u}(z)$ independent of ω and $\mathbf{q}$ for which one can solve Eq. (47) exactly. To find such profiles, is a more complicated problem than to search for profiles, k(z), admitting exact solutions for ρ = const, $\mathbf{u} = 0$ (Brekhovskikh, 1980, Ch. 3), as the coefficient of Eq. (47) contains m and m_{zz} simultaneously. Many profiles, k(z), admitting exact soultions of the Helmholtz equation in terms of confluent hypergeometric functions were found in (Godin, 1980). Testing functions $m(z) \sim k^2(z)$ wiht k being taken from this paper, we find the following two families of profiles leading to the exact solutions of Eq. (47).

1) $$\mathbf{u} = \mathbf{a}_0 + \mathbf{a}_1(z + z_1) + \mathbf{a}_2(z + z_1)^2, \tag{51}$$

where $\mathbf{a}_j$ and z_1 are arbitrary constants, j = 0, 1, 2. For $a_2 \neq 0$ the solutions are expressed in terms of Weber parabolic cylinder functions (Abramovitz and Stegun, 1964, Ch. 19). For $a_2 = 0$, Eq. (47) becomes simpler and its solutions can be expressed in terms of Airy functions (Abramovitz and Stegun, 1964, Ch. 10).

2) $$\mathbf{u} = \mathbf{a}_0 + \mathbf{a}_1\exp(bz) + \mathbf{a}_2\exp(2bz) \tag{52}$$

In this case one can express soultions of Eq. (47) in terms of confluent hypergeometric functions (Abramovitz and Stegun, 1964, Ch. 10). For $a_1 = 0$ they reduce to Bessel functions. The last case was studied in some detail by Chunchuzov (1985).

Note that for currents, (51) and (52) one can obtain exact expressions for the sound field in some inhomogeneous media also. Eg. for **u**(z) (51) the solutions can be expressed in tems of Weber parabolic cylinder funtions, if

$$\rho(z) = \rho_0 \exp[d_0 + d_1(z + z_1) + d_2(z + z_1)^2], \tag{53}$$

where d_j are arbitrary constants. For **u**(z) (52) one can find the exact solutions in terms of confluent hypergeometric functions, if

$$\rho(z) = \rho_0 \exp[d_0 + d_1 \exp(bz) + d_2 z]. \tag{54}$$

The restriction c = const is also not obligatory. For instance, if **u** and ρ are defined by Eqs. (52) and (54) with $d_1 = 0$, $a_1 = 0$, the Eq. (47) has solutions in terms of Weber parabolic cylinder functions, if $c(z) = (b_1 z + b_0)^{-1/2}$ and $b_{1,2}$ are arbitrary.

One can also obtain all the indicated exact solutions by departing from Eq. (48). In difference from Eq. (47), Eq. (48) doesn't contain derivatives of m, but the coordinate ζ depends on **q**. In the factor before p in Eq. (48) the term $2m(k^2 - 2q^2)$ is small due to M<<1. The influence of the current on $\zeta(z)$ is relatively small too. The is why substituting $m(\zeta)$ for $m(z(\zeta))$ leads to a small discrepancy, which under some conditions is of the same order of magnitude as the disregarded terms $O(m^2)$. Then assuming c and ρ to be constants for simplicity one can transmit to Eq. (48) all the numerous exact solutions found for the case of a medium at rest (see Brekhovskikh, 1980; DeSanto, 1979; Godin, 1980). More precisely, if an exact general solution of an equation $f_{zz} + a[Q(z) - D]f = 0$ is known for all a and D, then one can find the sound pressure exactly (in the sense indicated earlier) in a medium with $m(z) = \text{const}\cdot Q(z)$ for arbitary ω.

Eq. (43) shows that $(\zeta - z)/z \sim m$, when $\rho = \rho_0$. The substitution $m(\zeta)$ for $m(z(\zeta))$ gives rise to an error of the order of k^2m^2z/l in the coefficient of Eq.(48) as long as $\zeta - z \le l$, where l is a scale of space variation for **u**(z). In the case kl >>1 one can disregard the error as long as mz/l << 1. In the opposite case of low frequencies, when $kl \le 1$, the condition of applicability of the substitution may be written as kmz << 1. It means that the phases of a wave at horizons $\zeta(z)$ and z are close.

In the acoustics of a medium at rest the notion of impedance is used widely. Let's generalize this notion to include sound waves, harmonically depending on horizontal coordinates and time, in moving layered media. We define impedance as

$$R = i\omega\rho_0 p/p_\zeta. \tag{55}$$

R doesn't depend on x, y, t, or the amplitude of a wave, and according to Eq. (45), is continuous at boundaries. Hence R possesses all the leading properties of the impedance in a medium at rest. Due to Eq. (14) in the last case $p_\zeta/i\omega\rho_0 = w$. That is why for u = 0, R is identical with the common definition (Brekhovskikh, 1980) of the impedance. For a plane wave with angle of incidence θ in a homogeneous uniformly moving medium Eq. (55) gives the expression $R = c/\cos\theta(1 + M\sin\theta)$, which coincides with the impedance introduced by Steinmetz and Singh (1972) and independently by Lyamshev (1981) for discretely-layered liquids.

In numerical simulations of wave propagation in continuously-layered media the Riccati differential equation for the impedance is very convenient (Krasnushkin, 1980). It is not too difficult to obtain an analogous equation in a moving medium too. By differentiating definition (55) with respect to ζ and using Eq. (44) we find

$$R_\zeta = i\omega\rho_0[1 - (k^2U^2 - q^2)\rho^{-2}\omega^{-2}U^{-4}R^2]. \tag{56}$$

In the usual coordinates this equation has the form

$$R_z = i\omega\rho U^2[1 - (k^2U^2 - q^2)\rho^{-2}\omega^{-2}U^{-4}R^2]. \tag{57}$$

Starting from the new form (44) of the wave equation, a number of results were recently obtained for different acoustical problems. It is beyond the scope of this paper to present all the applications of Eq. (44). One can find detailed presentations in (Godin, 1985a,b; 1986). Here we just mention some of the results.

- The plane wave transmission coefficient symmetry with respect to inversion of the path of the wave in moving layered absorbing media is stated.
- The excitation coefficient of a lateral wave by a point source situated above a layered half-space with piecewise smooth parameters is found in terms of the field in the half-space induced by a plane wave incident under critical angle of total reflection.
- The Riccati equation (56) enabled us to improve the known method of successive approximations (Brekhovskikh, 1980, Sec. 25.5) for calculation of the plane-wave reflection coefficient from thick layers.
- The uniform (with respect to the angle of incidence) asymptotic expression is obtained for the reflection coefficient of a plane wave incident on a thin (compared to wavelength) layer with arbitrary stratification of density and velocities of sound and the mean currents.
- The reciprocity principle and the flow reversion theorem are generalized to include acoustic-gravity waves in layered media. It is shown also that one can prove the reciprocity principle in a moving medium as well (in some cases, at least) by adequate choice of a physical quantity to characterize the wave field. Hence the reciprocity principle and the flow reversion theorem (stated for different physical quantities) can be valid simultaneously.

4. APPROXIMATE WAVE EQUATION FOR SOUND IN A THREE-DIMENSIONALLY INHOMOGENEOUS MEDIUM WITH SLOW CURRENTS

In the ocean, sound propagation is often affected significantly by flows caused by tidal currents and mesoscale eddies. In the atmosphere, wind is often the dominant factor in the formation of an acoustic field of a given source. In both cases the flow velocity is small compared to the sound velocity. As strong currents in the ocean are usually associated with mesoscale eddies and fronts, discrepancies between a medium and its layered models becomes significant. In ocean acoustics, the effects of currents have been studied mainly by the ray method (see eg., Newhall et al., 1980; Itzikowitz et al., 1983; Polyanskaya, 1985). Recently some papers have appeared (Ostashev, 1984; Lan and Tappert, 1985; Grigor'jeva and Yavor, 1986), where more accurate and more widely applicable wave approaches were used, but study proceeds from different model wave equations. In this section, we shall derive the equation of sound propagation in a three-dimensionally inhomogeneous medium with arbitrarily slow currents, systematically, departing from the exact equations of hydrodynamics. We shall be interested mainly in oceanic sound propagation, but the results obtained are applicable to the acoustics of the atmosphere as well.

Let's assume provisionally that flow is incompressible and medium paramaters are constant along particle trajectories in unperturbed flow, that is

$$\nabla\cdot\mathbf{u} = 0, \quad d\rho_0/dt = 0, \quad d\rho/dt = 0, \quad dc/dt = 0. \tag{58}$$

Then eliminating $d\rho'/dt$ from Eqs. (10) and (11) we obtain

$$c^2\nabla\cdot\mathbf{v} + (\mathbf{v}\cdot\nabla)p_0 + dp/dt = 0. \tag{59}$$

Eqs. (9) and (59), after some transformations, give

$$(d/dt)[(\rho c^2)^{-1}d^2p/dt^2 - \nabla\cdot(\nabla p/\rho)] + 2\nabla\cdot[\rho^{-1}(\nabla p\cdot\nabla)\mathbf{u}]$$
$$+ \{(\rho c^2)^{-1}\nabla p_0\cdot d^2v/dt^2 + (d/dt)\nabla\cdot(\rho^{-2}\rho'\nabla p_0) - (\partial \mathbf{u}/\partial x_i)\nabla(\rho^{-2}\rho'\partial p_0/\partial x_i)$$
$$- \nabla\cdot[\rho^{-2}\rho'(\nabla p_0\cdot\nabla)\mathbf{u}] + \nabla\cdot[(\mathbf{v}\cdot\nabla)(\mathbf{u}\cdot\nabla)\mathbf{u}] + (\partial \mathbf{u}/\partial x_i)\cdot\nabla u_j(\partial v_i/\partial x_j) +$$
$$2(\partial \mathbf{u}/\partial x_i)\cdot\nabla(\mathbf{v}\cdot\nabla)u_i\} = 0. \qquad (60)$$

Summation over repeated indices i, j = 1, 2, 3 is assumed here and below.

Consider a layered medium with a horizontal current: $u_3 = 0$. In this case the conditions (58) are fulfilled. Disregarding the gravity force we have $\nabla p_0 = 0$. Then all the terms in the braces in Eq. (60) equal zero and we obtain the exact wave equation

$$(d/dt)[(\rho c^2)^{-1}d^2p/dt^2 - \nabla\cdot(\nabla p/\rho)] + 2(\mathbf{u}_z\cdot\nabla)(p_z/\rho) = 0. \qquad (61)$$

For the case of current with constant direction, an analogous equation was derived by Goldstein (1976).

Note that in a general medium under the condition g = 0, according to the Euler equation, ∇p_0 , and hence the quantity in braces in Eq. (60), is proportional to M^2. Preserving in Eq. (60) only terms linear with respect to M, we obtain an (approximate) equation of sound propagation in a three-dimensionally varying medium with slow currents.

$$[(\rho c^2)^{-1}p_{tt} - \nabla\cdot(\nabla p/\rho)]_t + 2(\rho c^2)^{-1}\mathbf{u}\cdot\nabla p_{tt} + 2\nabla\cdot[\rho^{-1}(\nabla p\cdot\nabla)\mathbf{u}] = 0. \quad (62)$$

In fact Eq. (62) doesn't contain second derivatives of **u**. Using the condition of incompressibility, one can cast the last term in the form $2(\partial u_j/\partial x_i)\partial(\rho^{-1}\partial p/\partial x_i)/\partial x_j$.

Deriving Eq. (62), all terms $O((kl)^{-1}M^2A)$ were discarded, where $A = k^3cp/\rho$; k, p are typical values of corresponding quantities for a wave. During propagation, a wave "averages" small-scale variations of the medium with l less than a wavelength. That is why, in estimates, one should take $kl \geq 1$. The terms connected with current in Eq. (62) are of the order of MA and MA/kl. For slow currents and arbitrary medium inhomogeneities, they are large compared to the discarded ones.

When g = 0, one can show that condititons (58) are not necessary for Eq. (62) to be valid, as giving the conditions up leads to additional terms arising of the order of M^2 in Eq. (60).

Let us use expression (60) to derive an equation for sound propagation in the ocean. The ratio j of vertical and horizontal space scales of c, ρ and **u** variability are small. Typically (Brekhovskikh and Lysanov, 1982; Newhall et al., 1980) $j \leq 10^{-2}$ and $M \leq 10^{-3}$. It follows from anisotropy of current and its incompressibility that $u_3 < ju$, $(\mathbf{u}\cdot\nabla)\mathbf{u}$ - $jM^2c^2/l << g$. Consequently one should use the hydrostatic expression $\nabla p_0 = - \rho g \nabla z$ for the gradient of the unperturbed pressure.

In Eq. (60) the sum of terms containing p_0 has an estimate $Ag/\omega^2 l$. Other quantities are estimated as in the case g = 0. Because $g/M\omega c$ - 1/f(Hz) <<1 for acoustic frequencies, one can disregard the effect of gravity compared to the effects of currents. This results in Eq. (62) again. Taking into account the anisotropy of the medium, this equation may be simplified further without changing the order of magnitude of the discarded quantity. Neglecting quantities $O(jAM(kl)^{-1})$ we obtain finally

$$[(\rho c^2)^{-1}p_{tt} - \nabla\cdot(\nabla p/\rho)]t + 2(\rho c^2)^{-1}\mathbf{u}\cdot\nabla p_{tt} + 2\rho^{-1}\mathbf{u}\cdot\nabla p_z = 0. \qquad (63)$$

Here **u** stands for horizontal components of the current velocity.

An analysis of terms, discarded in Eq. (60) during derivation of Eq. (63), shows that at low frequencies the prominent error arises from neglecting gravity. But with f increasing, quantities proportional to M^2 discarded in the first term in the left-hand side of Eq. (60) become more significant. At intermediate frequencies, errors caused by the difference between the last terms in Eq. (62) and Eq. (63) can contribute. If one takes $l \approx 100$ (m) for definiteness, then the orders of magnitude of the errors, caused by gravity and proportional to M^2 terms, become equal at $f = (g/l)^{1/2}/2\pi M \approx 50$ (Hz) and other sources of error are insignificant at all frequencies.

Now compare our Eqs. (62) and (63) with those obtained earlier. Let the medium be an ideal gas. Deflections from the state $p_0 = \text{const}$, $\rho = \text{const}$ are small and caused by the deflection T_1 of temperature from $T_0 = \text{const}$. Then for a monochromatic sound wave, preserving in Eq. (62) only those terms which are linear with respect to T_1, we obtain

$$p + \omega c_0^2 = 2i\omega^{-1}(\partial^2/\partial x_j \partial x_k)(u_j \partial p/\partial x_k) - (\partial/\partial x_j)(T_1 T_0^{-1} \partial p/\partial x_j), \qquad (64)$$

where c_0 is sound velocity for $T = T_0$. Eq. (64) coincides with the A.S. Monin equation (Tatarskii, 1971), which is widely used in the theory of sound scattering by atmospheric turbulence. Note that departing from Eq. (62) (in difference from Eq. (64)) one can consider sound scattering by turbulent current in the ocean or in the atmosphere wtih inhomogeneous mean parameters.

In a recent paper (Robertson et al., 1985) under very limiting simplifying assumptions ($c = c(z)$, $\rho = \text{const}$, $g = 0$, $u_3 = 0$, current is directed from a sound source to a receiver) a propagation equation for monochromatic sound in an ocean with slow currents was derived. The result of this paper coincides with the corresponding consequence of Eq. (63).

The A.M. Obukhov (1943) equation

$$d^2F_t/dt^2 = c^2\Delta F_t + \rho^{-1}\nabla p_0 \cdot \nabla F_t + c^2 \Delta \mathbf{u} \cdot \nabla F = 0, \qquad p = -dF/dt, \qquad (65)$$

is rather often employed in atmospheric acoustics. In deriving Eq. (65), in addition to $M \ll 1$, assumptions were made of smooth variations of c and ρ ($kl \gg 1$) and of entropy being constant throughout the medium. In the ocean, the last assumption is knowingly violated. In Eq. (65) some terms of the order of M^2 are preserved when others are neglected. In the case $u = 0$, $g = 0$, Obukhov's equation (in difference from Eq. (62)) reduces to $p_{tt} = c^2\Delta(p/\rho)$ but not to the exact equation $p_{tt} = c^2\nabla\cdot(\nabla(p/\rho))$. It may be shown, however, that for $g = 0$ and in the limits of its applicability, Eq. (65) agrees with Eq. (62) obtained under more general and realistic assumptions.

Except for extemely oversimplified models, theoretical descriptions of sound propagation in the ocean can only be achieved using a computer. For the case of a medium at rest, a number of effective numerical techniques to calculate acoustic fields are known. Let's consider a possibility to use them for describing the influence of currents on monochromatic sound.

As was shown in Section 3, in a layered moving medium Eq. (61) for waves harmonically dependent on horizontal coordinates by transformation of independent variables reduces to the one dimensional Helmholtz equation. This makes it possible, after minor modifications, to use existing normal mode techniques of sound field calculation in the case of moving media. Existing programs for the field in a range-dependent waveguide (calculation by the adiabatic approximation) may be used in moving media in the same manner as in this approximation, the field is expressed through a number of solutions for layered waveguides (Brekhovskikh and Lysanov, 1982, Section 7.2).

Sound propagation in an ocean without currents is often studied by the parabolic equation method. Starting from Eq. (63) and following routine procedures (Brekhovskikh and Lysanov, 1982, Section 7.4) one can find the parabolic equation for sound moving in the ocean:

$$2ik_0F_r + F_{zz} + (k^2 - k_0^2 - 2k^2k_0\tilde{u}/\omega)F + 2k_0\tilde{u}_zF_z/\omega - ik_0F\rho_r/\rho - \rho_zF_z/\rho = 0, \qquad (66)$$

where k_0 = const, $r^2 = x^2 + y^2$, and $\tilde{u}$ is the projection of **u** on the direction of sound propagation. By assumption, the quantity $F = pr^{1/2}\exp(-ik_0r)$ is a slow function of r: $F_r \ll k_0F$. Deriving Eq. (66) we have neglected horizontal refraction. Eq. (66) agrees with 6 parabolic equations obtained by Robertson et al. (1985) and contains them as special cases.

In a layered medium Eqs. (63) and (66) have normal mode solutions. An analysis of these solutions shows that the parabolic Eq. (66) describes corrections to modal phase velocities c_n due to currents with relative error $b = 1 - \omega/k_0c_n$. The corresponding error in calculation for a mode phase under typical deep-water condidions, where $b < 0.03$, is small compared to the total change $O(Mk_0r)$ in the phase due to the current as well as to the error $O(b^2k_0r)$ of the parabolic approximation itself (Brekhovskikh and Lysanov, 1982, Section 7.4).

When ρ = const, and $\tilde{u} = 0$ the parabolic equation doesn't contain terms proportional to F_z. To eliminate such terms in the general case, new coordinates (r, ζ) are introduced, where

$$\zeta(r, z) = \rho_0^{-1}\int_{z_0}^{z} \rho(r, z)[1 - k_0\tilde{u}(r, z)/\omega]\, dz, \quad \rho_0 = \text{const.}, \quad z_0 = \text{const.} \qquad (67)$$

Neglecting terms proportional to jM, M^2 and ρ_r we obtain from (66)

$$2ik_0F_r + (k^2 - k_0^2 - 2k^2k_0\tilde{u}/\omega)F + \rho^2/\rho_0^{-2}(1 - 4k_0u/\omega)F_{\zeta\zeta} = 0, \qquad (68)$$

This parabolic equation (contrary to Eq. (66)) may be solved by an effective implicit finite-difference scheme studied by Lee et al. (1981). In the ocean, density inhomogeneities usually don't influence sound propagation appreciably. Taking $\rho_0 = \rho$ we find from Eq. (67)

$$\zeta - z = 2k_0\omega^{-1}\int_{z_0}^{z} u\, dz < Ml. \qquad (69)$$

Under typical conditions the difference between ζ and z is small: $\zeta - z \leq 1$ (m). Hence, in practical calculations, one is allowed to make no distinction between ζ and z.

ACKNOWLEDGEMENT

The author would like to thank Drs. V.V. Goncharov, V.M. Kurtepov, A.G. Voronovich and especially academician L.M. Brekhovskikh for many helpful discussions.

REFERENCES

Abramovitz, M., and Stegun, I.A., eds., 1964, "Handbook of Mathematical Functions", National Bureau of Standards.

Boyles, C.A., 1984, "Acoustic Waveguides. Applications to Oceanic Science", John Wiley and Sons, New York etc.

Brekhovskikh, L.M., 1980, "Waves In Layered Media", Academic Press, New York.

Brekhovskikh, L.M., and Lysanov, Yu., 1982, "Fundamentals of Ocean Acoustics", Springer-Verlag, Berlin etc.

Chunchuzov, I.P., 1985, "Field of a Point Sound Source in an Atmospherical Layer Close to the Earth", Sov. Phys. Acoust., 31: N 1.

DeSanto, J.A., 1979, "Theoretical Methods In Ocean Acoustics", in: "Ocean Acoustics", J.A. DeSanto, ed., Springer-Verlag, Berlin etc.

Fabricant, A.L., 1976, "Resonance Interaction of Sound Waves With Stratified Flow", Sov. Phys. Acoust., 22: N 1.

Godin, O.A., 1980, "On Reflection of Plane Waves From A Layered Half-Space", Dokl. AN SSSR, 255:1069. (In Russian)

Godin, O.A., 1985a, "Reciprocity Relations for Acoustic-Gravity Waves", in: "Wolny i Difraktzia 85", v.1, TGU Tbilisi. (In Russian)

Godin, O.A., 1985b, "On a Modification of the Wave Equation for a Layered Medium", Wave Motion, 7:515.

Godin, O.A., 1986, "A Modification of the Sound Propagation Equation For a Layered Medium", in: "Ocean Acoustics. 1984", Nauka, Moscow. (In Russian)

Goldstein, M.E., 1976, "Aeroacoustics", McGraw-Hill, New York etc.

Grigor'eva, N.S., and Yavor, M.I., 1986, "Method of Normal Modes for Acoustic Field Calculation in an Oceanic Waveguide Perturbed by a Current", Sov. Phys. Acoust., 32: N 1.

Itzikowitz, S., Jacobson, M.J.m and Siegmann, W.L., 1983 "Modeling of Long-Range Acoustic Transmission Through Cyclonic and Anticyclonic Eddies", J. Acoust. Soc. Amer., 73:1556.

Krasnushkin, P.E., 1980, "A Method of Recursive Impedance Calculation in Wave Problems in Elastic Media", Dokl. AN SSSR, 252:332. (In Russian)

Lan, N.P., and Tappert, F.D., 1985, "Parabolic Equation Modelling of the Effects of Ocean Currents on Sound Transmission", J. Acoust. Soc. Amer., 78:642.

Lee D., Botseas, G., and Papadakis, J.S., 1981, "Finite-Difference Soultion to the Parabolic Wave Equation", J. Acoust. Soc. Amer., 70:795.

Lyamshev, L.M., 1981., "On Definition of Impedance in Acoustics of Moving Media", Dokl. AN SSSR, 261:74. (In Russian)

Newhall, B.K., Jacobson, M.J., and Siegmann, W.L., 1980, "Effect of a Class of Random Currents on Acoustic Transmission in an Ocean with Linear Sound Speed", J. Acoust. Soc. Amer., 67:1997.

Obukhov, A.M., 1943, "On Sound Wave Propagation in a Vortex Flow", Dokl. AN SSSR, 39:46. (In Russian)

Ostashev, V.E., 1984, "Wave Description of Sound Propagation in a Startified Moving Atmosphere", Sov. Phys. Acoust., 30: N 4.

Polyanskaya, V.A., 1985, "On the Effect of a Currents Velocity Field on Sound Propagation in the Ocean", Sov. Phys. Acoust., 31: N 5.

Robertson, J.S., Siegmann, W.L., and Jacobson, M.J., 1985, "Current and Current Shear Effects in the Parabolic Approximation", J. Acoust. Soc. Amer., 77:1768.

Steinmetz, G.G., and Singh, J.J., 1972, "Reflection and Transmission of Acoustical Waves From a Layer with Space-Dependent Velocity", J.Acoust. Soc. Amer., 51:218.

Tatarskii, V.I., 1971, "The Effects of the Turbulent Atmosphere on Wave Propagation", Sect. 34, Jerusalem.

Tatarskii, V.I., 1979, "To the Theory of Sound Propagation in a Stratified Atmosphere", Izv. Atmospher. Ocean. Sci., 15: N 11.

PHASE SPACE METHODS AND PATH INTEGRATION: THE CONSTRUCTION OF NUMERICAL ALGORITHMS FOR COMPUTATIONAL ACOUSTICS

Louis Fishman

Department of Civil Engineering
The Catholic University of America
Washington, D.C. 20064 USA

ABSTRACT

Phase space, or "microscopic," methods and path (functional) integral representations provide the appropriate framework to extend homogeneous Fourier methods to inhomogeneous environments. Applied to the n-dimensional scalar Helmholtz equation, these techniques directly result in numerical algorithms appropriate for computational acoustics. Two complementary approaches to this representation and computation of the Helmholtz propagator are presented and examined numerically.

INTRODUCTION

The analysis and fast, accurate numerical computation of the n-dimensional classical physics wave equations are often quite difficult for rapidly changing, multidimensional environments extending over many wavelengths. For the most part, classical, "macroscopic" methods have resulted in direct wave field approximations (perturbation theory, ray-theory asymptotics, modal analysis, hybrid ray-mode methods), derivations of approximate wave equations (scaling analysis, field splitting techniques, formal operator expansions), and discrete numerical approximations (finite differences, finite elements, spectral methods). However, the "microscopic" phase space analysis developed over the past several decades by mathematicians studying linear partial differential equations and the global functional integral techniques pioneered by Wiener (Brownian motion) and Feynman (quantum mechanics) together provide the framework to both explicitly represent and, subsequently, directly compute the classical physics propagators. These phase space and path (functional) integral methods provide the appropriate framework to extend homogeneous Fourier methods to inhomogeneous environments.

PHASE SPACE AND PATH INTEGRAL CONSTRUCTIONS

For the scalar Helmholtz equation, there are two complementary approaches to this analysis and computation, as illustrated in Figure 1. The first is essentially a factorization/path integration/invariant imbedding approach. For transversely inhomogeneous environments, implying medium homogeneity with respect to a single distinguished direction, the n-

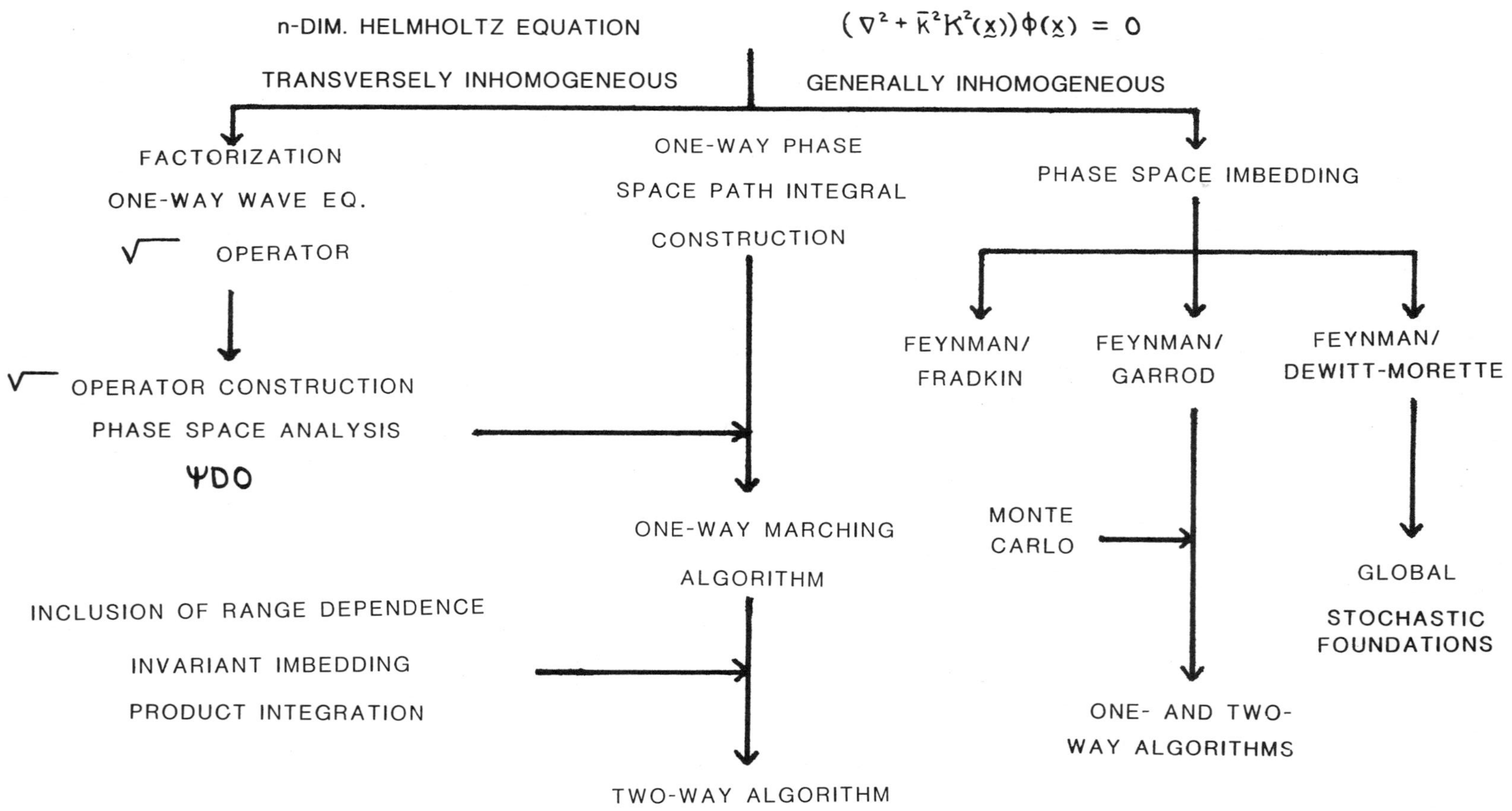

Fig. 1. Two complementary approaches to the analysis and computation of the n-dimensional scalar Helmholtz equation.

dimensional Helmholtz equation can be exactly factored into separate, physical forward and backward, one-way wave equations, following from spectral analysis.[1-5] The forward evolution (one-way) equation

$$(i/\bar{k})\partial_x \phi^+(x,\mathbf{x}_\perp) + (K^2(\mathbf{x}_\perp) + (1/\bar{k}^2)\nabla_\perp^2)^{1/2}\phi^+(x,\mathbf{x}_\perp) = 0 , \tag{1}$$

where $K(\mathbf{x})$ is the refractive index field and $\bar{k}$ is a reference wave number, is the formally exact wave equation for propagation in a transversely inhomogeneous half-space supplemented with appropriate outgoing wave radiation and initial-value conditions. While functions of a finite set of commuting self-adjoint operators can be defined through spectral theory, functions of noncommuting operators are represented by pseudo-differential operators.[2,5] The formal wave equation (1) is now written explicitly as a Weyl pseudo-differential equation in the form

$$(i/\bar{k})\partial_x \phi^+(x,\mathbf{x}_\perp) + (\bar{k}/2\pi)^{n-1}\int_{R^{2n-2}} d\mathbf{x}'_\perp\, d\mathbf{p}_\perp$$

$$\cdot\Omega_{\mathbf{B}}(\mathbf{p}_\perp,(\mathbf{x}_\perp + \mathbf{x}'_\perp)/2)\exp(i\bar{k}\mathbf{p}_\perp\cdot(\mathbf{x}_\perp - \mathbf{x}'_\perp))\,\phi^+(x,\mathbf{x}'_\perp) = 0. \tag{2}$$

In Eq.(2), the symbol $\Omega_{\mathbf{B}}(\mathbf{p},\mathbf{q})$ associated with the square root Helmholtz operator $\mathbf{B} = (K^2(\mathbf{q}) + (1/\bar{k}^2)\nabla_{\mathbf{q}}^2)^{1/2}$ satisfies the Weyl composition equation

$$\Omega_{\mathbf{B}^2}(\mathbf{p},\mathbf{q}) = K^2(\mathbf{q}) - \mathbf{p}^2 = (\bar{k}/\pi)^{2n-2}\int_{R^{4n-4}} d\mathbf{t}\,d\mathbf{x}\,d\mathbf{y}\,d\mathbf{z}\;\Omega_{\mathbf{B}}(\mathbf{t}+\mathbf{p},\ \mathbf{x}+\mathbf{q})$$

$$\cdot\Omega_{\mathbf{B}}(\mathbf{y}+\mathbf{p},\ \mathbf{z}+\mathbf{q})\exp(2i\bar{k}(\mathbf{x}\cdot\mathbf{y} - \mathbf{t}\cdot\mathbf{z})) \tag{3}$$

with $\Omega_{\mathbf{B}^2}(\mathbf{p},\mathbf{q})$ the symbol associated with the square of $\mathbf{B}$, $\mathbf{B}^2 = (K^2(\mathbf{q}) + (1/\bar{k}^2)\nabla_{\mathbf{q}}^2)$.[2,3,5] The generalized Fourier construction procedure for the square root Helmholtz operator can be summarized pictorially by the following correspondence diagram

$$\begin{array}{ccc} \mathbf{B}^2 & \Longleftrightarrow & \Omega_{\mathbf{B}^2} \\ \uparrow & & \Updownarrow \\ \mathbf{B} & \Longleftrightarrow & \Omega_{\mathbf{B}} \end{array}$$

where the arrows symbolize the one- and two-way mappings between the appropriate quantities.

Exact solutions of the Weyl composition equation (3) can be constructed in several cases.[6] For example, the symbol $\Omega_{\mathbf{B}}(p,q)$ for the two-dimensional ($n = 2$) quadratic medium, $K^2(q) = K_0^2 + w^2q^2$, is given by[6]

$$\Omega_{\mathbf{B}}(p,q) = -(\exp(i\pi/4)\varepsilon^{1/2}/\pi^{1/2})\int_0^\infty dt\, \exp(i(Yt + X\tanh t))$$

$$\cdot t^{-1/2}\,(iY\operatorname{sech} t + iX\operatorname{sech}^3 t - (\operatorname{sech} t)(\tanh t)) \tag{4}$$

with $X = (1/\varepsilon)(w^2q^2 - p^2)$, $Y = K_0^2/\varepsilon$, and $\varepsilon = w/\bar{k}$. Consistent with taking the square root of the indefinite Helmholtz operator, the corresponding symbols, generally, have both real and imaginary parts characterized by oscillatory behavior,[4,6] as illustrated in Figure 2. Nonuniform and uniform perturbation solutions corresponding to definite physical limits (frequency, propagation angle, field strength, field gradient) recover several known approximate wave theories (ordinary parabolic, range-refraction parabolic, Grandvuillemin-extended parabolic, half-space Born, rational linear) and systematically lead to several new full-wave, wide-angle approximations.[2-4,]

The exact pseudo-differential evolution equation (2) and, in general, the wide-angle extended parabolic approximate equations derived from the analysis of the composition equation[2-4,6] are singular integro-differential wave equations. Solution representations for such pseudo-differential equations can be directly expressed in terms of infinite-dimensional functional, or path, integrals,[7,8] following from the Markov property of the propagator. In an operator notation, then,

$$\exp(i\bar{k}\mathbf{B}x) = \lim_{N\to\infty} \prod_{j=1}^{N} \exp(i\bar{k}\mathbf{B}\Delta x_j) \tag{5}$$

where $\Delta x_j = x/N$, symbolically representing the propagator in terms of the infinitesimal propagator. As the operator symbol is not simply quadratic in $\tilde{p}$, the configuration space Feynman path integral formulation is not appropriate, necessitating the more general phase space construction.[4,7] This results in a parabolic-based (one-way) Hamiltonian phase space path integral representation of the propagator in the form[3,7]

$$G^{+}(x,\tilde{x}_{\perp}|0,\tilde{x}'_{\perp}) = \lim_{N\to\infty} \int_{R^{(n-1)(2N-1)}} \prod_{j=1}^{N-1} d\tilde{x}_{j\perp} \prod_{j=1}^{N} (\bar{k}/2\pi)^{n-1} d\tilde{p}_{j\perp}$$

$$\cdot\exp(i\bar{k}\sum_{j=1}^{N} (\tilde{p}_{j\perp}\cdot(\tilde{x}_{j\perp} - \tilde{x}_{j-1\perp}) + (x/N)\, H(\tilde{p}_{j\perp},\tilde{x}_{j\perp},\tilde{x}_{j-1\perp}))) \tag{6}$$

where

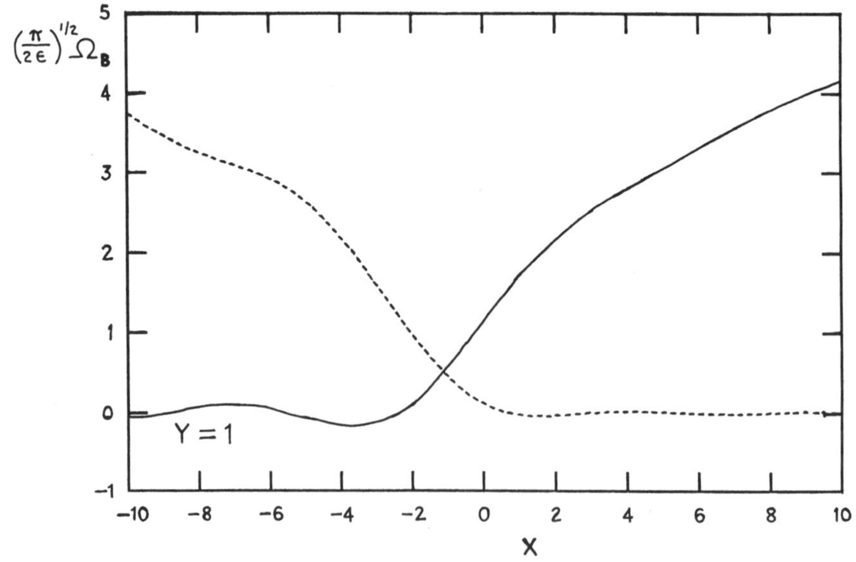

Fig. 2. The real (——) and imaginary (----) parts of the n = 2 quadratic medium symbol as a function of X for Y = 1.

$$H(\underset{\sim}{p},\underset{\sim}{q}'',\underset{\sim}{q}') = (\bar{k}/2\pi)^{n-1} \int_{R^{2n-2}} d\underset{\sim}{s}d\underset{\sim}{t}\ F(\underset{\sim}{q}'-\underset{\sim}{q}'',\underset{\sim}{s})$$

$$\cdot h_{\mathbf{B}}(\underset{\sim}{p},((\underset{\sim}{q}''+\underset{\sim}{q}')/2) - \underset{\sim}{t}) \exp(i\bar{k}\underset{\sim}{s}\cdot\underset{\sim}{t}). \qquad (7)$$

In Eq.(7), $F(\underset{\sim}{u},\underset{\sim}{v})$ and $h_{\mathbf{B}}(\underset{\sim}{p},\underset{\sim}{q})$ are related to the operator symbol $\Omega_{\mathbf{B}}(\underset{\sim}{p},\underset{\sim}{q})$ by

$$\hat{\Omega}_{\mathbf{B}}(\underset{\sim}{u},\underset{\sim}{v}) = F(\underset{\sim}{u},\underset{\sim}{v})\ \hat{h}_{\mathbf{B}}(\underset{\sim}{u},\underset{\sim}{v}) \qquad (8)$$

where $\hat{\Omega}_{\mathbf{B}}$ and $\hat{h}_{\mathbf{B}}$ are the corresponding Fourier transforms.[2,3,7]

The nonuniqueness of the lattice-approximation path integral representation is readily understood in terms of different discretizations, or quadratures, of the symbolic functional integral and corresponds to the representation of a given (fixed) operator by different operator-ordering, or pseudo-differential operator, schemes.[2,3,7,8] More fundamentally, in analogy with the Schrödinger equation for particle motion on a Riemannian space and the thermodynamic (Fokker-Planck) equation for particle diffusion, the algorithmic Helmholtz path integral construction reflects the stochastic nature of the integration.[4,9] Further, both the macroscopic and microscopic (infinitesimal) half-space propagators can be formally expressed as Fourier integral operators with complex phase.[4] The phase space path integral, thus, represents the macroscopic Fourier integral operator in terms of the N-fold application of the microscopic, or infinitesimal, Fourier integral operator in a manner which can be related to the global geometrical-optics construction of the macroscopic operator.[4,5]

The path integral formulation interprets the wave theory in terms of an infinitesimal propagator summed over all phase space paths. For the Helmholtz theory, the exact infinitesimal propagator is not, in general, given by the locally homogeneous medium propagator, as in the ordinary parabolic (Schrödinger) propagator construction.[8] The approximate extended parabolic wave theories then correspond to approximate infinitesimal propagators summed over the complete phase space. In retaining the "sum over all paths," diffraction, or full-wave, effects are incorporated.

For weakly range-dependent environments, range variability can be, at first, accommodated at the level of range updating, as in the case of the parabolic path integral.[1,8] For reflection/transmission from a planar interface separating two (different) transversely inhomogeneous acoustic half-spaces, the concept of reflection and transmission amplitudes generalizes to reflection (**r**) and transmission (**t**) operators. The reflection and transmission operators, which, when applied to the incident wave field at the interface, produce the initial values of the reflected and transmitted wave fields, are defined within the Weyl pseudo-differential operator framework and are explicitly determined by enforcing the well-known interface continuity conditions. The main result[10] is a composition equation of the form

$$\Omega_{\mathbf{B}_L}(\underset{\sim}{p},\underset{\sim}{q}) - \Omega_{\mathbf{B}_R}(\underset{\sim}{p},\underset{\sim}{q}) = (\bar{k}/\pi)^{2n-2} \int_{R^{4n-4}} d\underset{\sim}{t}d\underset{\sim}{x}d\underset{\sim}{y}d\underset{\sim}{z}\ (\Omega_{\mathbf{B}_L}(\underset{\sim}{t}+\underset{\sim}{p},\ \underset{\sim}{x}+\underset{\sim}{q}) +$$

$$\Omega_{\mathbf{B}_R}(\underset{\sim}{t}+\underset{\sim}{p},\ \underset{\sim}{x}+\underset{\sim}{q}))\ \Omega_{\mathbf{r}}(\underset{\sim}{y}+\underset{\sim}{p},\ \underset{\sim}{z}+\underset{\sim}{q}) \exp(2i\bar{k}(\underset{\sim}{x}\cdot\underset{\sim}{y} - \underset{\sim}{t}\cdot\underset{\sim}{z})) \qquad (9)$$

for the reflection operator symbol $\Omega_{\mathbf{r}}(\underset{\sim}{p},\underset{\sim}{q})$ and an analogous equation for the transmission operator symbol $\Omega_{\mathbf{t}}(\underset{\sim}{p},\underset{\sim}{q})$. The inclusion of a planar transition region of arbitrary length and inhomogeneity can be accomplished by factorization methods in conjunction with invariant imbedding.[4,11] Invariant imbedding constructs the initial-value system for the reflection and transmission operators associated with the transition region, transforming the Helmholtz boundary-value problem into an initial-value problem. A discretized formulation[11] provides the extension of Kennett's method[4,11] in reflection seismology. The resultant forward and backward wave fields propagating in the transversely inhomogeneous half-spaces are represented by the one-way path integrals, while, within the transition region, a formal path integral representation of the propagator can be expressed as a product integral.[8] This takes the form[4]

$$\mathbf{G} = \int_a^x \exp(i\bar{k}\underset{\approx}{\mathbf{H}}(s)ds) = \lim_{N \to \infty} \prod_{j=1}^{N} \exp(i\bar{k}\underset{\approx}{\mathbf{H}}(s_j)\Delta s_j) \tag{10}$$

where $s_j = a + (j-1/2)\Delta s_j$, $\Delta s_j = (x-a)/N$, a denotes the transition region boundary, $\underset{\approx}{\mathbf{H}}$ is the appropriate first-order Helmholtz equation matrix operator,[2,4] and with the product of exponential factors ordered from right (lower j) to left (higher j) reflecting the noncommutativity of the matrix operator $\underset{\approx}{\mathbf{H}}$ at different x. While product integration-based path integral constructions have been applied to the problems of nonrelativistic electron spin and the Dirac equation, such infinite products of matrices are, generally, only tractable in simple limiting cases.[4,8]

Rather than starting from a transversely inhomogeneous formulation and, subsequently, building in backscatter effects, the generalization of Fourier methods to arbitrary inhomogeneous environments and the construction of a dynamical basis for the Helmholtz equation can proceed, in the second approach, from the construction of truly global configuration space path integrals, which attempt to generalize, for example, the homogeneous half-space result[3,7]

$$G^+(x,\underset{\sim}{x}_\perp|0,\underset{\sim}{x}'_\perp) = \lim_{N \to \infty} \int_{R^{(n-1)(N-1)}} \prod_{j=1}^{N-1} d\underset{\sim}{x}_{j\perp} \, (i\pi x N^{(n-1)N/2}$$

$$\cdot(\bar{k}K_0/2\pi\delta_{(n-1)N+1})^{((n-1)N+1)/2} \, H^{(1)}_{((n-1)N+1)/2}(\bar{k}K_0\delta_{(n-1)N+1})) \tag{11}$$

where

$$\delta_{(n-1)N+1} = (N \sum_{j=1}^{N} (\underset{\sim}{x}_{j\perp} - \underset{\sim}{x}_{j-1\perp})^2 + x^2)^{1/2} \tag{12}$$

and $H^{(1)}_\nu(\xi)$ is the Hankel function. These elliptic-based (two-way) constructions, originating from the Fourier transform relationship between the Helmholtz and Schrödinger (parabolic) propagators, result in the approximate Feynman/Garrod path integral[3,7]

$$G(\underset{\sim}{x}|\underset{\sim}{x}') \cong (-1/2\bar{k}^2) \lim_{N \to \infty} \int_{R^{n(2N-1)}} \prod_{j=1}^{N-1} d\underset{\sim}{x}_j \prod_{j=1}^{N} (\bar{k}/2\pi)^n d\underset{\sim}{p}_j \frac{\exp(i\bar{k}S_N)}{(1/2 - \Sigma)} \tag{13}$$

where

$$S_N = \sum_{j=1}^{N} \underset{\sim}{p}_j \cdot (\underset{\sim}{x}_j - \underset{\sim}{x}_{j-1}) \tag{14}$$

corresponds to an appropriate discretized action and

$$\Sigma = (1/N) \sum_{j=1}^{N} (\underset{\sim}{p}_j^2/2 + V(\underset{\sim}{x}_j)) \tag{15}$$

plays a role analogous to an average energy with the identification $V(\underset{\sim}{x}) = (-1/2)(K^2(\underset{\sim}{x}) - 1)$. For a transversely inhomogeneous half-space, partial integration of Eq.(13) in conjunction with the reflection principle (or method of images) results in[3,7]

$$G^+(x,\underset{\sim}{x}_\perp|0,\underset{\sim}{x}'_\perp) \cong \lim_{N\to\infty} \int_{R^{(n-1)(2N-1)}} \prod_{j=1}^{N-1} d\underset{\sim}{x}_{j\perp} \prod_{j=1}^{N} (\bar{k}/2\pi)^{n-1} d\underset{\sim}{p}_{j\perp}$$

$$\cdot \exp(i\bar{k}(S_N + 2^{1/2}x(1/2 - \Sigma)^{1/2})) \tag{16}$$

with S_N and Σ taking on their appropriate forms in one-lower dimension. Formally reducing both the full- and transversely inhomogeneous half-space phase space Feynman/Garrod path integrals to configuration space path integrals[7] establishes the path functional character of the representation. Moreover, the approximate Feynman/Garrod path integral is exact in the homogeneous medium limit, incorporates significant backscatter information, and contains both the geometrical (ray) acoustic and ordinary parabolic approximations. This configuration space formulation for the two-way problem, initially based on a variational principle and phase space constructions, seeks to express the propagator in terms of a phase functional evaluated over an appropriate path space, as symbolically expressed in the Feynman/DeWitt-Morette representation.[3,7,9] The dynamical basis of the Helmholtz equation can, thus, be viewed in terms of a stochastic process embodying fixed "average energy" paths, or, alternatively, in terms of "free particle" motion.[3,7,9]

COMPUTATIONAL ALGORITHMS

Direct integration of the one-way phase space path integral provides the computational basis for the pseudo-differential wave equation (2). Choosing the standard ordering, $F(\underset{\sim}{u},\underset{\sim}{v}) = \exp(-i\bar{k}\underset{\sim}{u}\cdot\underset{\sim}{v}/2)$, in Eqs. (6), (7), and (8) results in a numerically more efficient post-point marching algorithm in the form

$$\phi^+(x+\Delta x,\underset{\sim}{x}_\perp) \cong \int_{R^{n-1}} d\underset{\sim}{p}_\perp \exp(i\bar{k}\underset{\sim}{p}_\perp\cdot\underset{\sim}{x}_\perp)\,(\exp(i\bar{k}\Delta x h_{\mathbf{B}}(\underset{\sim}{p}_\perp,\underset{\sim}{x}_\perp))\,\hat{\phi}^+(x,\underset{\sim}{p}_\perp)) \tag{17}$$

where $\hat{\phi}^+$ is the Fourier-transformed wave field and

$$h_{\mathbf{B}}(\underset{\sim}{p}_\perp,\underset{\sim}{x}_\perp) = (\bar{k}/\pi)^{n-1} \int_{R^{2n-2}} d\underset{\sim}{s}\,d\underset{\sim}{t}\;\Omega_{\mathbf{B}}(\underset{\sim}{s},\underset{\sim}{t}) \exp(-2i\bar{k}(\underset{\sim}{x}_\perp - \underset{\sim}{t})\cdot(\underset{\sim}{p}_\perp - \underset{\sim}{s})). \tag{18}$$

This marching algorithm provides the generalization of the Tappert/Hardin split-step FFT algorithm[1] to the full one-way (factored Helmholtz) wave equation. For a two-dimensional model ocean/bottom propagation environment with a perfectly reflecting ocean surface, the Fourier transform of the wav field in Eq.(17) is replaced by a discrete fast sine transform and the inverse transform is evaluated by a rectangular rule integration, enabling th propagated wave field to be expressed in the matrix form

$$\phi^{+}(x+\Delta x,z_n) = \sum_m A_{nm}\,\hat{\phi}^{+}(x,p_m) \quad (19$$

for each depth point z_n. In Eq.(19), ϕ^{+} and $\hat{\phi}^{+}$ are column vectors and the matrix $\underset{\approx}{A}$ is defined by its matrix elements

$$A_{nm} = \eta\sin(\bar{k}p_m z_n + \bar{k}\Delta x h^{o}_{\mathbf{B}}(p_m,z_n))\ \exp(i\bar{k}\Delta x h^{e}_{\mathbf{B}}(p_m,z_n)) \quad (20$$

where $h^{e}_{\mathbf{B}}$ and $h^{o}_{\mathbf{B}}$ are the even and odd parts with respect to p of $h_{\mathbf{B}}(p,z)$ in Eq.(18) and η is an appropriate transform normalization constant.[1,4,12]

The principal idea underlying the practical implementation of the phas space marching algorithm is the construction of a small number of approximate operator symbols, which, when taken together, allow for wave field computations over a very wide range of model environments and propagation parameters. In conjunction with a study of exactly soluble cases of the Weyl composition equation,[6] high-frequency, real Weyl high-frequency, uniform high-frequency, and low-frequency approximate symbols have been constructed.[2-4,6] Of particular significance is the fact that the manner of marching the radiation field is independent of the medium and any approximation to the square root Helmholtz operator, resulting in a modular code architecture and highly versatile propagation program. Moreover, the propagation models constructed and computed through the code correspond to singular integro-differential equation as well as partial differential equation approximations to the one-way wave equation. Indeed, this numerical algorithm represents one of the very few attempts to compute directly with pseudo-differential and Fourier integral operators. For the two-dimensiona case, the range-incrementing procedure is just a sequence of matrix multiplications, and, thus, ideally suited for computers which provide either a vector or a parallel pipe type of operation. Phase space filtering reduces both the size of the matrix multiplication and the number of matrix element initially computed, in particular, reducing the total range-incrementing computational time by almost an order of magnitude for typical model calculations.[4]

Extensive numerical calculations with the filtered one-way phase space marching algorithm on ocean acoustic, seismological, and extreme model environments designed to establish the range of validity and manner of breakdown of the extended (approximate symbol) theories have been extremely promising.[4,12] A representative extreme model propagation experiment is illustrated in Figure 3, with the corresponding transmission loss curves computed for the filtered high-frequency[4,12] and reference Fast Field Program (FFP)[4,12] codes displayed in Figure 4. The speed and modest storage requirements of the filtered one-way algorithm indicate that range-dependent calculations over extended environments should be feasible with current supercomputer technology. Both range-updating and the numerical calculation of the reflected and transmitted fields from an interface should be possible over distances on the order of 10^4 wavelengths. Preliminary computations with range-dependent Munk-profile deep ocean environments, including propagation through extended shadow regions, compare well with adiabatic normal-mode calculations.

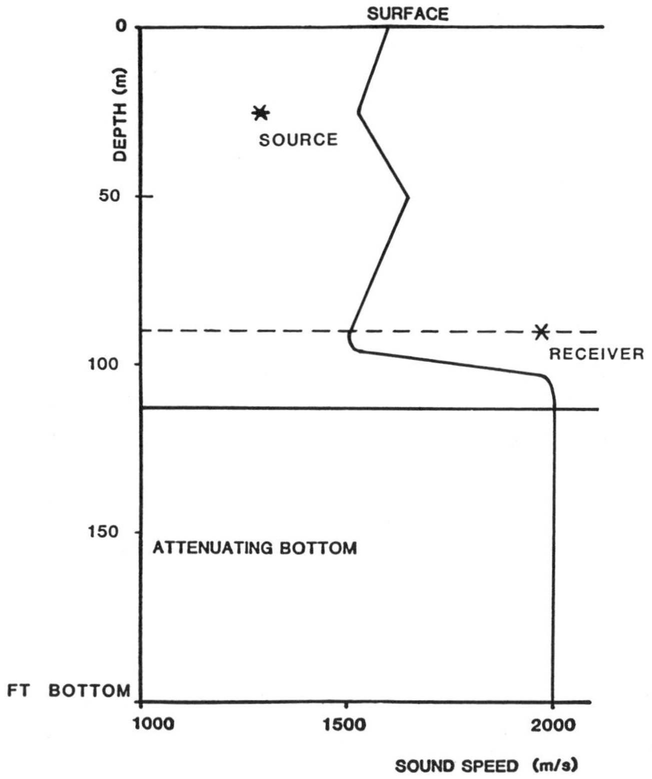

Fig. 3. Extreme model environment and propagation experiment.

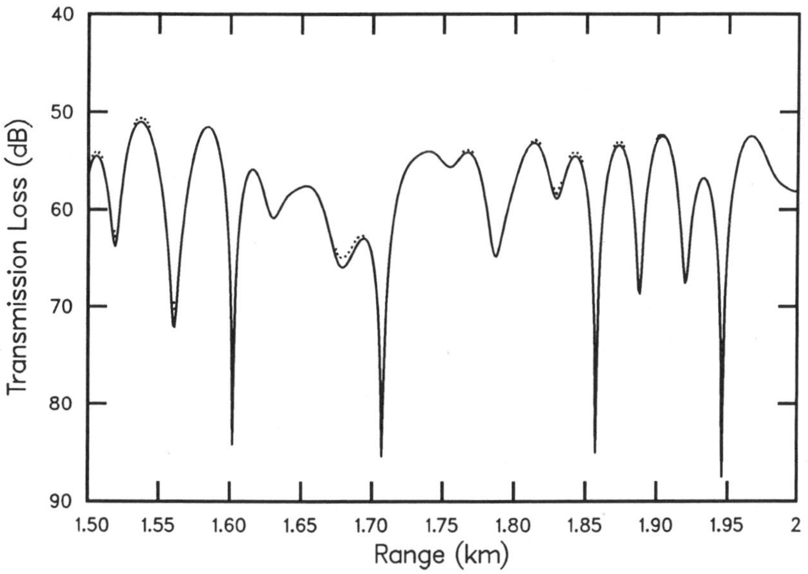

Fig. 4. Transmission loss (dB re 1 m) versus range (km) for the extreme model environment at 400 Hz. (——) High Frequency (80 degree filter) (....) FFP

Both the range-dependent and range-independent Feynman/Garrod path integral representations can be computed by standard Monte Carlo (statistical sampling) methods for the numerical evaluation of multiple integrals.[4] While numerically calculating Helmholtz wave fields as high (in principle, infinite)-dimensional integrals is quite distinct from the more traditional finite-difference and finite-element approaches, the Monte Carlo evaluation of functional integrals has been successfully applied in quantum mechanical, statistical mechanical, and quantum field theoretical calculations.[4] For the phase space representations of Eqs. (13) and (16) in two dimensions ($n = 2$), the modeling of realistic propagation experiments can involve the computation of thousand-dimensional oscillatory integrals. Correlated-sampling variance reduction techniques can dramatically improve the speed and accuracy of the algorithm.[4] Generally speaking, a large parallel processing capability should have a very favorable impact on the numerical computation of path integrals.[4]

ACKNOWLEDGMENTS

This work was supported under grants from the Office of Naval Research (N00014-85-K-0307) and the U.S. Army Research Office (DAAG 29-85-K-0002).

REFERENCES

1. J.A. Davis, D. White, and R.C. Cavanagh, "NORDA Parabolic Equation Workshop," NORDA tech. note 143, Naval Ocean Research and Development Activity, NSTL Station (1982).
2. L. Fishman and J.J. McCoy, Derivation and application of extended parabolic wave theories. Part I. The factorized Helmholtz equation, J. Math. Phys., 25 (2): 285 (1984).
3. L. Fishman and J.J. McCoy, Factorization, path integral representations, and the construction of direct and inverse wave propagation theories, IEEE Trans. Geosc. Rem. Sens., GE-22 (6): 682 (1984).
4. L. Fishman, J.J. McCoy, and S.C. Wales, Factorization and path integration of the Helmholtz equation: numerical algorithms, J. Acoust. Soc. Am., submitted for publication (1986).
5. M.E. Taylor, "Pseudodifferential Operators," Princeton University Press, Princeton (1981).
6. L. Fishman and A. Whitman, Exact and uniform perturbation solutions of the Helmholtz composition equation, J. Math. Phys., submitted for publication (1986).
7. L. Fishman and J.J. McCoy, Derivation and application of extended parabolic wave theories. Part II. Path integral representations, J. Math. Phys., 25 (2): 297 (1984).
8. L.S. Schulman, "Techniques and Applications of Path Integration," Wiley, New York (1981).
9. C. DeWitt-Morette, A. Maheshwari, and B. Nelson, Path integration in nonrelativistic quantum mechanics, Phys. Rep., 50 (5): March (1979).
10. J.J. McCoy, L. Fishman, and L.N. Frazer, Reflection and transmission at an interface separating transversely inhomogeneous acoustic half-spaces, Geophys. J. R. Astr. Soc., to appear (1986).
11. J.J. McCoy and L.N. Frazer, Propagation modelling based on wave field factorization and invariant imbedding, Geophys. J. R. Astr. Soc., to appear (1986).
12. L. Fishman and J.J. McCoy, A new class of propagation models based on a factorization of the Helmholtz equation, Geophys. J. R. Astr. Soc., 80: 439 (1985).

PREDICTING CONVERGENCE ZONE FORMATION IN THE DEEP OCEAN

John J. Hanrahan

Naval Underwater Systems Center
New London Laboratory, New London, CT 06320

ABSTRACT

Convergence zones are formed at discrete range intervals in the ocean provided that certain environmental conditions are satisfied. In order to evaluate the effect of these environmental features on the convergence zone path, the concept of a "depth excess" was developed. Depth excess is defined for a given location as the excess in water depth over that just required for a single ray from the surface to reach the convergence zone. By computing the average depth excess for a given month and location, it is possible to portray various ocean areas as a series of average depth excess contours. Then, by specifying a minimum acceptable depth excess for adequate zone formation, an examination of the monthly average depth excesses in various areas, along with the variability about the average, permits a realistic estimate of the reliability of useful convergence zone formation as a function of month and area.

INTRODUCTION

The objectives underlying this investigation were to

(a) examine the formation of convergence zones,
(b) determine the primary environmental influences on the occurrence of convergence zones, and
(c) develop a means for predicting the existence of convergence zones in the deep ocean.

The study produced a simple and direct technique - the depth excess technique for achieving these objectives which can be applied to any ocean. Depth excess is defined herein as the excess in water depth over that just required for a single ray from the surface to reach the convergence zone. Consequently, by computing the average depth excess for a given location and month, it is possible to portray any ocean area as a series of average depth-excess contours. If, then, for the purpose of adequate zone formation, a minimum acceptable depth excess is specified, an examination of the monthly average depth excesses in various areas, together with the variability about the average, permits a realistic estimate of the reliability of useful convergence-zone formation as a function of month and area.

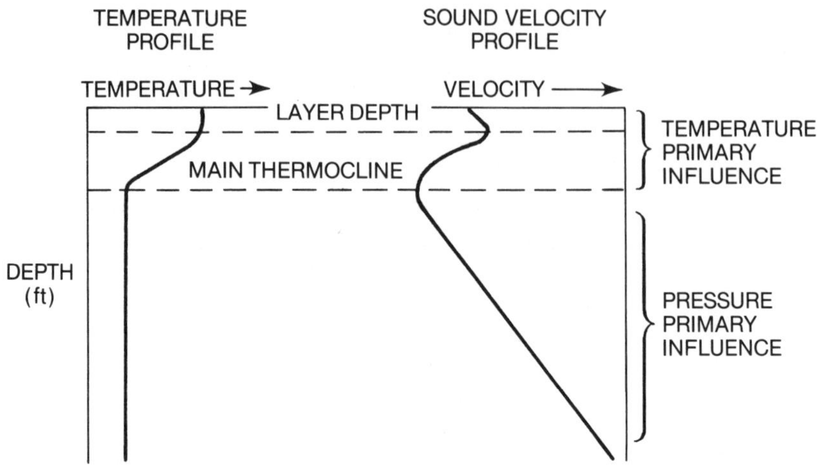

Figure 1: A typical deep water temperature and velocity profile.

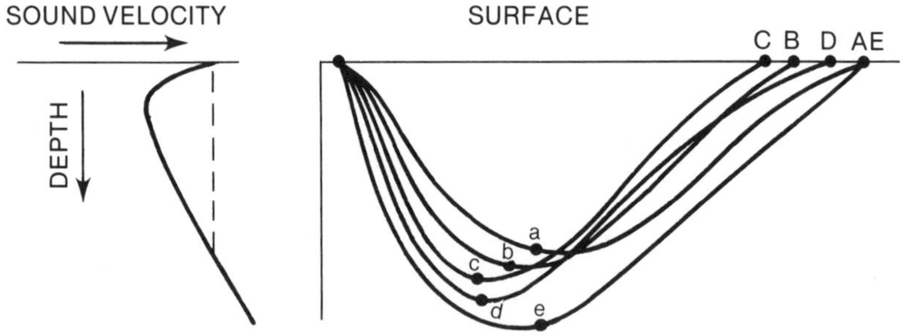

Figure 2: Formation of Convergence Zones by Rays that Escape Surface Channel.

COMPUTATION OF DEPTH EXCESS

Let us briefly review the mechanism that gives use to convergence zones. The basic thermal structure of deep ocean areas is essentially a three layer system as shown on Figure 1: a relatively warm and shallow surface layer in which most of the temporal and spatial changes in temperature occur; a transitional layer known as the main thermocline; and a very deep mass of water in which temperature decreases very slightly but uniformly with depth. As a consequence of the nature and stability of the deepest layer, a slightly positive velocity gradient usually exists below the thermocline.

Consider a sound ray leaving a source which is located near the surface. Assume that its inclination is steep enough to escape a surface layer, i.e., the beam pattern of the source in the vertical plane is wide enough for this assumption to obtain. The ray has a vertex velocity, that is, a velocity at which the sound ray becomes horizontal, which is given by Snell's Law as:

$$V_x = \frac{V_o}{\cos\theta}, \qquad (1)$$

where

V_x = the vertex velocity,

V_o = the sound velocity at the source,[1] and

θ = the angle of inclination of the ray relative to the surface.

If the bottom depth is great enough, the ray will travel in a downward direction until it reaches a depth at which the sound velocity corresponds to the ray's vertex velocity. At that depth, it will change direction and will be refracted upwards along the mirror image of its downward path until the ray finally reappears at the surface.

It may be seen from Eq. (1) that, as the inclination angle for a given V_o increases, the vertex velocity also increases. Hence, steeper rays must go to greater depths in order to reach their vertex velocities. The steeper rays will come to the surface at progressively shorter ranges until what we may call a focusing ray is reached. Progressively steeper angled rays will then come to the surface at longer and longer ranges. The double ray coverage defines a region popularly known as a convergence zone and is illustrated in Fig. 2. This cycle will continue: the sound will be bent downward and then upward to reappear at the surface at equally spaced intervals until the acoustic energy is dissipated. Thus, a series of discrete, annular zones is formed. The zone spacing will vary from approximately 35 miles at the equator to about 15 miles in higher latitudes. In some Arctic regions, where the positive gradient starts at the surface, the convergence zones widen and merge, giving continuous range coverage although not necessarily full depth coverage. (The latter is so because the rays reach a vertex velocity at comparatively shallow depths.)

By virtue of these focusing rays, the zones are also regions of high sound intensity. The one way spreading loss to the zone is 10 to 15 dB less than that expected under non-focusing conditions.

[1]In this report, the sound velocity at the source is taken to be equal to that at the surface. In practice, the source would be sufficiently close to the surface so that this procedure is permissible.

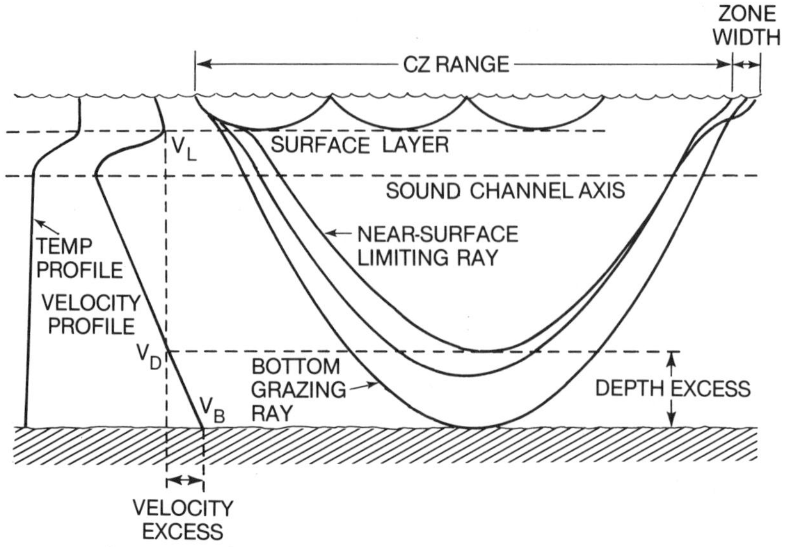

Figure 3: Depth Excess Defined for a Convergence Zone Path.

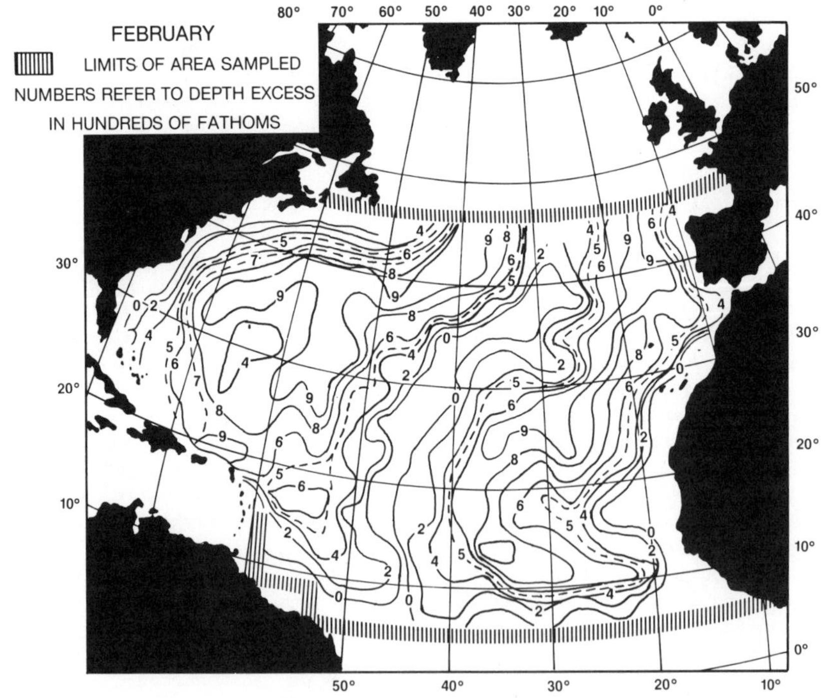

Figure 4: Depth Excess Contours for February.

The water depth is of vital importance to convergence-zone operation. It obviously must be sufficiently deep to permit some source rays to reach their vertex velocity as illustrated in Figure 3. The depth at which the vertex velocity is equal either to (1) the surface velocity, when there is no surface isothermal layer, or (2) the maximum velocity in a surface layer, is defined herein as the "limiting depth." This depth allows one ray from a source at the surface to reach the convergence zone. Therefore, depth excess for a given area is given by the equation:

$$\text{Depth Excess} = (\text{Bottom Depth for a Given Area}) - (\text{Limiting Depth for Same Area}) \quad (2)$$

If the depth excess does not exceed zero, all sound rays leaving the source will strike the bottom and no convergence zone will be formed. This condition is generally referred to as being "bottom-limited."

An equivalent expression in terms of sound velocity can be obtained by subtracting the vertex velocity from the bottom velocity.

The calculation of depth excess for a given area and month thus becomes a straightforward procedure. First, the bottom depth can be read from bathymetric charts. Second, the limiting depth is calculated by finding that depth on the velocity profile which yields a sound velocity equal to the mean surface velocity. Wilson's equation for sound velocity was used in these computations for relating temperature, salinity, and sound velocity (Ref 1), and third, the Depth Excess is determined quantitatively by taking the difference between bottom depth and limiting depth.

As an illustration, the North Atlantic was examined by finding the average depth excess for the month of February. A useful representation is presented in Figure 4 where contours are used to portray areas of equal depth excess. It is of interest to relate that the presence of a surface isothermal layer was found to decrease the depth excess by an amount equal to the depth of the layer. Thus, although winter is a favorable period for convergence-zone occurrence because of the lower surface temperatures encountered, the season's tempestuous weather results in very deep, well-mixed layers that partially negate the low surface temperatures.

COMPUTATION OF PROBABILITY OF ZONE OCCURRENCE

In order to interpret the depth excess contours in terms of probability of occurrence of a useful convergence zone, we require not only the average depth excesses but also (1) the depth excess required to form a useful zone and (2) the variability in depth excess about the average values.

Let us first consider the first item - the depth excess requirement. As was mentioned earlier, if the ocean depth is equal to the limiting depth, only one ray from a source at the surface will reach the convergence zone. Since a zone with reasonable width and intensity is required, it becomes obvious that one ray alone will not be sufficient and that, in fact, a "bundle" of rays is necessary. In this discussion, then, we define a "bundle requirement" as the extent in depth which is required beyond the limiting depth in order to form an adequate zone. The theoretical work of Pedersen and Keith suggests that a depth excess of 200 fathoms allows an amount of energy in the zone equal to about 50% of that potentially available if there was no depth restriction (Ref 2). However, because this study covered only one thermal profile, an exact bundle requirement to cover all situations is difficult to formulate at this time. Based on their work a 200-fathom figure appears to be reasonable and will be assumed for the purposes of this study. The results presented are

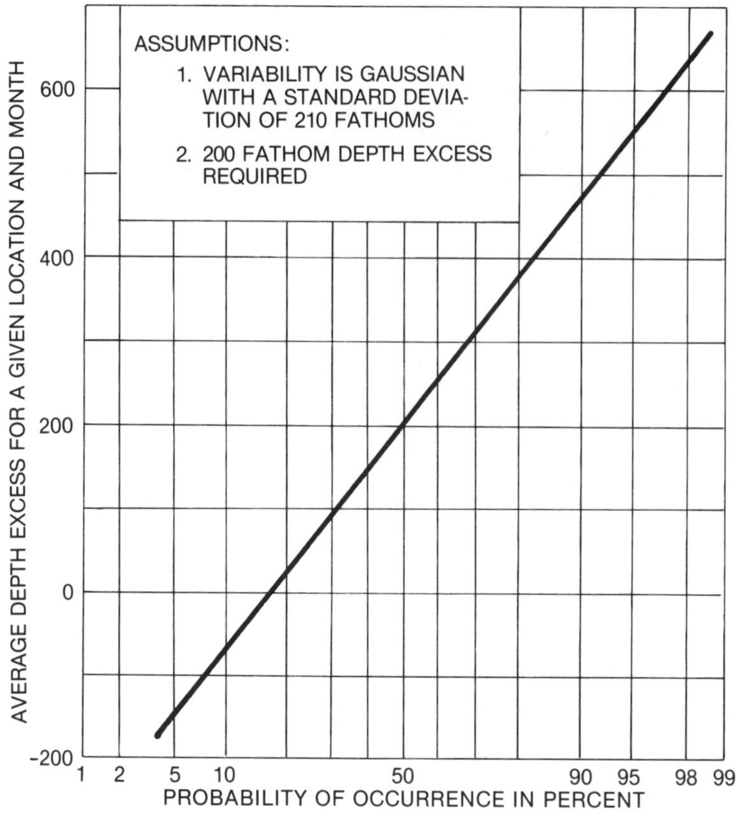

Figure 5: Probability of Occurrence of Useful Convergence Zone Path.

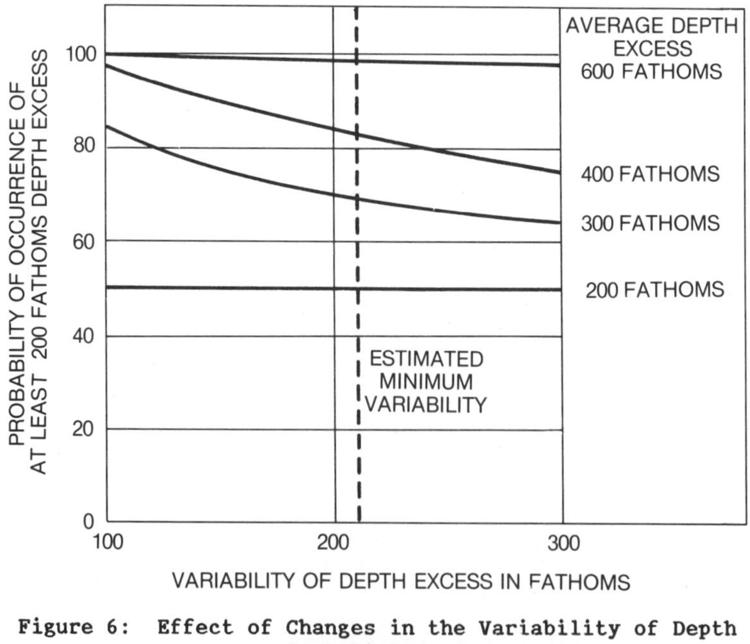

Figure 6: Effect of Changes in the Variability of Depth Excess as Probability of Occurrence.

sufficiently generalized to make possible a re-interpretation for any alternative bundle requirement.

To determine the degree of probability that the required bundle will be obtained, we must know the variability of the depth excess. The variability in depth excess from that computed from a mean surface temperature and a bottom-depth chart is probably to a great extent due to daily surface temperature variations and to errors in listed bottom depths.

The probable error between charted and actual depth is estimated as being within 5 to 6 per cent. This estimate is predicated on a comparison between the charted depths of the 6750 series of Hydrographic Office Bathymetric charts and some International Geophysical Year soundings.[3] The deep water areas of the ocean, that is, those greater than 1000 fathoms, average 2500 fathoms[4] and comprise about 90 per cent of the overall ocean area.[5] According to our estimates, the bottom-contour charts used in this study give bottom depths within a probable error of no less than 125 fathoms. Therefore, the standard deviation of the actual depths from the charted depths is estimated to be 185 fathoms.

The AMOS surface temperature measurement for a given day was then compared with the value obtained by interpolating between two successive monthly means as given in the World Atlas of Sea Surface Temperatures. As a result of that study, it was concluded that a standard deviation of $\pm 2^{\circ}$F from the assumed mean surface temperature could be expected on any given day. This figure of $\pm 2^{\circ}$F was estimated from the quartile spread of the deviations. What this means in terms of depth excess will now be shown.

Sound velocity, as was discussed earlier, depends primarily on pressure for depths below the main thermocline. The magnitude of the velocity-depth dependence is on the order of 0.0182 $\sec^{-1}$ which means that depth excess will change with surface temperature at the rate of roughly 50 fathoms per Fahrenheit degree increase in surface temperature. The depth excess would, therefore, be reduced by 100 fathoms for an increase of 2°F in surface temperature.

Since the variability of both the surface temperature, σ_T, and bottom charts, σ_{BC}, have been determined, the standard deviation of the depth excesses, σ_{DE}, may be found by the following equation:

$$\begin{aligned} \sigma_{DE}^2 &= \sigma_T^2 + \sigma_{BC}^2 \\ \sigma_{DE} &= \sqrt{(100)^2 + (185)^2} = 210 \text{ fathoms.} \end{aligned} \tag{3}$$

Under the assumptions (1) that the variability of depth excess is Gaussian with a standard deviation of 210 fathoms, and (2) that 200 fathoms of depth excess are required for a useful convergence-zone path to exist, a plot showing the probability of occurrence of a useful convergence-zone path versus average monthly depth excesses for a given month and area is given in Fig. 5. It may be seen that in order to insure that a depth excess of at least 200 fathoms occurs 84 per cent of the time, we must have an average depth excess of 410 fathoms.

Figure 6 is presented to illustrate the effect of changes in the variability of depth excess. It is believed that the calculated 210-fathom variability is a minimum and that other sources of uncertainty could increase it to as much as 300 fathoms. Generally speaking, the probability is not critically dependent upon changes in the standard deviation of the variability in depth excess. For example, if the variability should happen to be 300 fathoms instead of 210 fathoms for a given area which has an

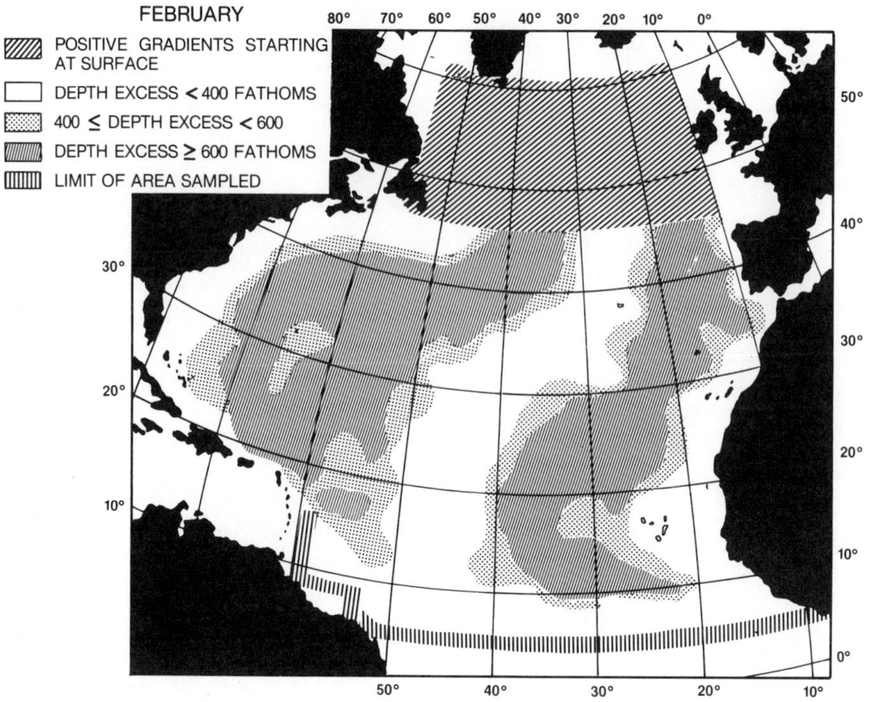

Figure 7: Areas where Depth Excess Exceeds 400 and 600 fathoms - February.

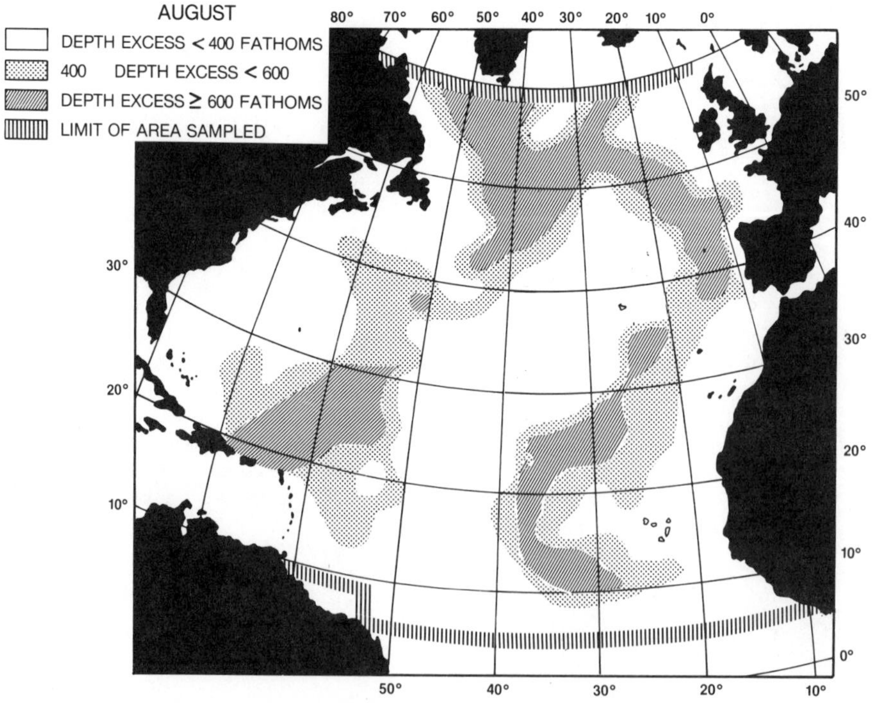

Figure 8: Areas where Depth Excess Exceeds 400 and 600 fathoms - August.

average depth excess of 400 fathoms, the probability may be seen to decrease from 83 to 75 per cent.

DISCUSSION OF AVERAGE DEPTH-EXCESS CONTOURS

The depth excess contours presented earlier for the North Atlantic in February may be represented in a simpler form (Figure 7) by arranging the ocean into the following three groupings: (1) areas where the average depth excess is less than 400 fathoms, shown as unshaded; (2) areas where the average depth excess is between 400 and 600 fathoms, represented by small dots; and (3) areas whose average depth excess is greater than 600 fathoms, indicated by cross-hatchings. According to Fig. 6, the cross-hatched areas are available for convergence-zone utilization at least 98 per cent of the time during the given month; the small dotted areas are open between 75 and 98 per cent; and the unshaded areas are probably of marginal usefulness.

The situation depicted here illustrates the near optimum case, since February is considered to be oceanographic winter, the ocean is characterized by minimum surface temperatures and exceptionally deep layers. As is expected from these considerations, and as shown in Fig. 7, convergence-zone is generally prevalent during this month. About the only area where coverage would be unavailable is in the vicinity of the Mid-Atlantic Ridge, as evidenced by the unshaded portion in the center of the chart. Areas above 46° north latitude were not sampled because positive gradients, starting at the surface, are prevalent there during the winter. Thus, in these regions, strong surface sound channels would exist, and convergence-zone formation would be more complex.

Worst case conditions occur in August which is considered to be oceanographic summer in the North Atlantic. The dominant reason for the relatively poor coverage in August is high surface temperature. From a consideration of comparative surface temperatures, Fig. 8 is probably indicative of the results that would be encountered in July and September also. This chart thus shows the lower limit to convergence-zone occurrence; and it may be seen that, although the performance is limited during the warm weather months, the North Atlantic is by no means devoid of coverage during this period.

CONCLUSIONS

Convergence zone propagation is recognized as one of the major paths by which energy can travel to long ranges in deep water. In order to predict the occurrence of convergence zones, the depth excess method discussed in this report has been developed, and charts can be prepared which delineate both the upper and lower limits on the occurrence of convergence zones. The conclusions which may be drawn are the following:

a. A convergence zone propagation path is controlled solely by environmental and physical conditions, specifically, the sea surface temperature and water depth.

b. The minimum average depth excess required for reliable occurrence of convergence zones is about 400 fathoms beyond the limiting depth defined in the text. This depth requirement allows for a proper ray bundle, variations in surface temperature, and probable errors in bottom-contour charts.

c. An average depth excess of 400 fathoms would correspond to an 83 per cent reliability for an estimated depth excess variability of 210 fathoms. The reliability would decrease to 75 per cent for a 300 fathom variability.

d. For the example evaluated for the North Atlantic, it was shown that the occurrence of convergence zones is strongly dependent upon seasons. The zones are most prevalent during the colder months. Although the summer period is generally poor, there remain large regions where convergence zones can still exist.

REFERENCES

1. W. D. Wilson, "Equation for the Speed of Sound in Sea Water," Journal of the Acoustical Society of America, Vol. 32, No. 10, Oct 1960, p. 1357.
2. M. A. Pedersen and A. J. Keith, Comparison of Experimental and Theoretical Sound Intensities for Convergence-Zone Transmission in 3100 Fathom Water, U. S. Navy Electronics Laboratory Report No. 738, 3 December 1956.
3. L. V. Worthington, Oceanographic Data from the R.R.S. Discovery II, International Geophysical Year Cruises One and Two, 1957, Woods Hole Oceanographic Institution Reference No. 58-30, June 1958.
4. The Application of Oceanography to Subsurface Warfare, Summary Technical Report of Division 6, NDRC, vol. 6A, 1946, ch. 5, p. 27.
5. Physics of Sound in the Sea, Summary Technical Report of Division 6, NDRC, vol. 8, ch. 5, p. 87.

A NEW APPROXIMATE SOLUTION VALID AT TURNING POINTS

Guan Ding-hua and Ling Hui

Institute of Acoustics, Academia Sinica
Beijing
People's Republic of China

ABSTRACT

A new approximate solution for a Helmholtz equation of one dimension is suggested. The solution is valid in cases both of having no turning point and of having only "linear" or "semi-linear" ones. The solution has the form of improved classical WKB solution with compensating functions. The solution is expressed by elementary functions.

I. INTRODUCTION

It is known that the Helmholtz equation of one dimension

$$\frac{d^2u}{dx^2} + q(x)u = 0 \tag{1}$$

is one of the basic equations in underwater acoustics. But its solutions can be expressed analytically only when q(x) is one of the several simple functions. The detailed discussions on analytically solving the equation (1) can be found in some books[1,2,3].

A few approximate solutions of equation (1) have been developed and accepted. The most widely used one is the traditional WKB solution[4]

$$u(x) \sim \begin{cases} \dfrac{C_-}{|q(x)|^{\frac{1}{4}}}\, e^{\pm\int^x |q(x)|^{\frac{1}{2}}dx}, & -q(x) \gg 1 \\[2ex] \dfrac{C_+}{(q(x))^{\frac{1}{4}}}\, e^{\pm i\int^x (q(x))^{\frac{1}{2}}dx}, & q(x) \gg 1 \end{cases} \tag{2}$$

The formula, however, is invalid near a turning point. In the vicinity of the turning point, the Airy function[4] or the numerical solution[5] can be used and joined to the WKB solutions at the ends of the vicinity. The former is analytic but unsatisfactory at the joining point and the latter is nonanalytic but satisfactory at the joining point.

For strictly solving the problem of turning points, there are GWKB solutions[4] expressed by Airy or Weber functions and other more strict solutions[4]. But all these solutions are very complicated. The efforts to simplify the solutions by replacing the special functions used in them with elementary functions have already had useful results. A few simplified solutions have been proposed[6,7] for several years.

In this paper we suggest a new approximate solution for the Helmholtz equation (1) derived through the Airy functions. This kind of solution has several characteristics: 1, it is valid even in an interval including a "linear" or "semi-linear" turning point; 2, it has the form of improved classical WKB solution with compensating functions; 3, it is expressed by elementary functions; and 4, its expression includes both q(x) and q'(x).

II. THE DERIVATION OF THE NEW SOLUTION

1. The Solutions of Equation $u'' + xu = 0$

From the theory of the Airy functions[3], we know that the solutions of the equation

$$u'' + xu = 0 \tag{3}$$

should be the linear combinations of the Airy functions Ai(-x) and Bi(-x). If we express the two independent solutions as

$$U_1(x) = \sqrt{\pi}\, e^{i\pi/4}[Ai(-x) - iBi(-X)] \tag{4}$$

$$U_2(x) = \sqrt{\pi}\, e^{-i\pi/4}[Ai(-x) + iBi(-x)] = U_1^*(x) \tag{5}$$

the asymptotic behaviors of them will be

$$U_1(x) \sim U_a(x) = \begin{cases} (-x)^{-\frac{1}{4}} \exp(-i\frac{\pi}{4} + \frac{2}{3}(-x)^{3/2}), & x \to -\infty \\ (x)^{-\frac{1}{4}} \exp(i\frac{2}{3} x^{3/2}), & x \to \infty \end{cases} \tag{6}$$

$$U_2(x) \sim U_a^*(x) \tag{7}$$

Taking into account that their Wronskian is a constant, we can express the solutions as

$$U_1(x) = \begin{cases} [-x+b_o(x)]^{-\frac{1}{4}} \exp[-i\frac{\pi}{4} - \int_o^x (-x)^{\frac{1}{2}}dx + ih_o(x)], & x < 0 \\ [x+a_o(x)]^{-\frac{1}{4}} \exp[-i\frac{\pi}{12} + i\int_o^x (x+a_o(x))^{\frac{1}{2}}dx], & x \geq 0 \end{cases} \tag{8}$$

$$U_2(x) = U_1^*(x) \tag{9}$$

where

$$a_o(x) = \frac{1}{\pi^2[Ai^2(-x)+Bi^2(-x)]^2} - x , \quad x > 0 \tag{10}$$

$$b_o(x) = \frac{e^{(8/3)(-x)^{3/2}}}{\pi^2[Ai^2(-x)+Bi^2(-x)]^2} + x , \quad x < 0 \tag{11}$$

$$h_o(x) = \text{arctg} \frac{Ai(-x)}{Bi(-x)} , \quad x < 0 \tag{12}$$

The analysis indicates that these functions

$$a_o(x) \to 0 , \quad \text{when } x \to \infty \tag{13}$$

$$b_o(x) \to 0 , \quad \text{when } x \to -\infty \tag{14}$$

$$h_o(x) \to 0 , \quad \text{when } x \to -\infty \tag{15}$$

2. The Solutions of Equation $u'' + k(x-x_0)u = 0$

When $q(x)$ of the equation (1) is a linear function

$$q(x) = k(x-x_0) \tag{16}$$

at the turning point x_0

$$\begin{cases} q(x_0) = 0 \\ q'(x_0) = q'(x) = k \neq 0 \\ q''(x_0) = q''(x) = 0 \end{cases} \tag{17}$$

the equation (1) has the form

$$u'' + k(x-x_0)u = 0 \tag{18}$$

From the equations (3), (8) and (9), we know that the two linear independent solutions for the equation (18) can be

$$u_I(x) = k^{-1/6} e^{i\pi/12} U_1[k^{1/3}(x-x_0)] \tag{19}$$

$$u_{II}(x) = u_I{}^*(x) \tag{20}$$

Since the integrations

$$\int_o^{k^{1/3}(x-x_0)} (-t)^{1/2} dt = s\int_o^x [-k(x-x_0]^{1/2} dx \tag{21}$$

$$\int_o^{k^{1/3}(x-x_0)} [t+a_o(t)]^{1/2} dt =$$

$$= s\int_o^x [k(x-x_0)+k^{2/3} a_o(k^{1/3}(x-x_0))]^{1/2} dx \tag{22}$$

where

$$s = \begin{cases} -1 , & q'(x_0) < 0 \\ +1 , & q'(x_0) > 0 \end{cases} \tag{23}$$

and the expression

$$k^{1/3}(x-x_o) = q(x)\,[q'(x)]^{-2/3} \tag{24}$$

the solutions' representations can be written as

$$u_I(x) = \begin{cases} \dfrac{e^{-i\pi/6}}{B_o(q,q')}\, e^{iH_o(q,q')} , & q(x) < 0 \\ \dfrac{1}{A_o(q,q')}\, e^{iG_o(q,q')} , & q(x) \geq 0 \end{cases} \tag{25}$$

$$u_{II}(x) = u_I{}^*(x) \tag{26}$$

where

$$A_o(q,q') = [q+q'^{2/3}a_o(qq'^{-2/3})]^{1/4} \tag{27}$$

$$B_o(q,q') = [-q+q'^{2/3}b_o(qq'^{-2/3})]^{1/4} \tag{28}$$

$$G_o(q,q') = s\int_{x_o}^{x} [A_o(q,q')]^2 dx \tag{29}$$

$$H_o(q,q') = is\int_{x_o}^{x} (-q)^{1/2} dx + h_o(qq'^{-2/3}) \tag{30}$$

Apparently, when x moves away from the turning point x_o, the solutions (25) and (26) will asymptotically reduce to their WKB forms (2).

3. The Elementary Functions' Approximation

Because of the complexity of the Airy functions, the representations (25) and (26) are not convenient for being used practically. For significantly diminishing the amount of calculation work, replacing the functions (10) to (12) by their elementary approximate functions is an effective method.

To accomplish the replacement, denote that

$$g_n(x) = \begin{cases} a_o(x) , & n=1 \\ b_o(-x^2) , & n=2 \\ h_o(x) , & n=3 \end{cases} \tag{31}$$

Assume that we can work out three substituting functions

$$f_n(x) = \begin{cases} a(x) , & n=1 \\ b_*(x) , & n=2 \\ h(x) , & n=3 \end{cases} \tag{32}$$

for functions in (31), that all of them are quotients of polynomials and that, in each of the quotients, the degree of the numerator is smaller than or, at least, equal to that of the denominator. On the assumption of the initial conditions

$$\left(\frac{d}{dx}\right)^{m_n} f_n(x)\Big|_{x=o} = \left(\frac{d}{dx}\right)^{m_n} g_n(x)\Big|_{x=o} ,$$

$$m_n = 0,1,\cdots,M_n \;, \quad n=1,2,3 \tag{33}$$

being satisfied, we may take the elementary functions a(x), b(x)=$b_*[(-x)^{\frac{1}{2}}]$ and h(x) as the expected approximate functions of $a_o(x)$, $b_o(x)$ and $h_o(x)$, respectively.

Three of the practical alternatives of the approximate functions are

$$a(x) = (2.5088251 + 2.6362805x + 1.4368776x^2 + 0.380284x^3)^{-1} \tag{34}$$

$$b(x) = (1.0 + 16.0x^3)[2.5088251 - 9.9521263x + 6.6902003x(-x)^{\frac{1}{2}} + 38.145223x^2 - 43.323546x^2(-x)^{\frac{1}{2}} - 114.53475x^3]^{-1} \;; \quad x < 0 \tag{35}$$

$$h(x) = (1.9098593 - 2.3028601x + 1.9373252x^2 - 1.323844x^3 + 0.7759298x^4)^{-1} \tag{36}$$

4. The Approximate Solutions of the Helmholtz Equation

After carefully examining the condition (17) and the solutions (25) and (26) of the equation (18), and the functions (34) to (36), we may see that the following predication is reasonably acceptable.

For equation (1)

$$\frac{d^2u}{dx^2} + q(x)u = 0 \tag{37}$$

if the known function q(x) has up to second derivatives and the point x_o is at least a "semi-linear" turning point or, definitely, at x_o

$$\begin{cases} q(x_o) = 0 \\ q'(x_o) \neq 0 \\ |q''(x_o)| \ll 1 \end{cases} \tag{38}$$

a pair of "independent" approximate solutions of the Helmholtz equation (37) can be expressed as

$$u_1(x) = \begin{cases} \dfrac{e^{-i\pi/6}}{[-q+B(q,q')]^{\frac{1}{4}}}\, e^{-s\int_{x_o}^{x}(-q)^{\frac{1}{2}}dx + iH(q,q')} \;, & x \in \{x\!: q<0\} \\[2ex] \dfrac{1}{[q+A(q,q')]^{\frac{1}{4}}}\, e^{is\int_{x_o}^{x}[q+A(q,q')]^{\frac{1}{2}}dx} \;, & x \in \{x\!: q\geq 0\} \end{cases} \tag{39}$$

$$u_2(x) = u_1{}^*(x) \tag{40}$$

where the three compensating functions

$$A(q,q') = q'^{2/3}\, a(qq'^{-2/3}) \tag{41}$$

$$B(q,q') = q'^{2/3}\, b(qq'^{-2/3}) \tag{42}$$

$$H(q,q') = h(qq'^{-2/3}) \tag{43}$$

and here the functions a(x), b(x) and h(x) are the ones given by the formulas (34) to (36).

It is obvious that, when the variable x is in an interval where

$$|qq'^{-2/3}| \gg 0 \tag{44}$$

the values of the three compensating functions will be relatively small enough so that the solutions (39) and (40) will reduce to the traditional WKB solutions (2).

III. THE NUMERICAL EXAMINATIONS

1. About the Elementary Approximate Functions a(x), b(x) and h(x)

For the formulas (8) and (9), when the functions $a_o(x)$, $b_o(x)$ and $h_o(x)$ are replaced by their elementary approximate substitutes a(x), b(x) and h(x), the amplitude and phase errors introduced in the oscillatory region can be expressed as

$$e_1(x) = [(x+a(x))^{-\frac{1}{4}}-(x+a_o(x))^{-\frac{1}{4}}]/(x+a_o(x))^{-\frac{1}{4}} \tag{45}$$

$$e_2(x) = (x+a(x))^{\frac{1}{2}}-(x+a_o(x))^{\frac{1}{2}} \tag{46}$$

and in the non-oscillatory region as

$$e_3(x) = [(-x+b(x))^{-\frac{1}{4}}-(-x+b_o(x))^{-\frac{1}{4}}]/(-x+b_o(x))^{-\frac{1}{4}} \tag{47}$$

$$e_4(x) = h(x)-h_o(x) \tag{48}$$

The tables of these error functions are as below.

Table 1.

x	$a_o(x)$	$e_1(x)$	$e_2(x)$
0.0	0.3986	0.0000	-0.0000
1.0	0.1437	0.0000	-0.0000
2.0	0.0613	0.0001	-0.0003
3.0	0.0315	0.0001	-0.0005
4.0	0.0186	0.0001	-0.0005
5.0	0.0122	0.0001	-0.0005
6.0	0.0086	0.0001	-0.0004

Table 2.

x	$b_o(x)$	$e_3(x)$	$h_o(x)$	$e_4(x)$
0.0	0.3986	0.0000	0.5236	0.0000
-0.5	-0.0762	0.0169	0.2648	0.0011
-1.0	-0.3308	0.0677	0.1116	0.0096
-1.5	-0.4126	0.0560	0.0382	0.0170
-2.0	-0.3849	0.0340	0.0106	0.0162
-2.5	-0.3282	0.0193	0.0024	0.0117
-3.0	-0.2815	0.0113	0.0005	0.0075
-3.5	-0.2493	0.0071	0.0001	0.0048
-4.0	-0.2269	0.0046	0.0000	0.0031

The above two tables indicate that the closeness of a(x) and h(x) to $a_o(x)$ and $h_o(x)$ are much better than that of b(x) to $b_o(x)$.

2. About the Approximate Solutions (39) and (40)

We have chosen the Weber equation

$$\frac{d^2u}{dx^2} + k_w(x^2-x_o^2)u = 0 \tag{49}$$

for examining the approximate solutions (39) and (40). Assume that $k_w = \pm 1$ and denote that the approximate solution, a linear combination of the solutions (39) and (40), for equation (49) as $u^{\pm}(x)$ and that the numerical solution for (49) as $S^{\pm}(x)$. Then, express the differences between the approximate and the numerical solutions as

$$e_c^{\pm}(x) = |u^{\pm}(x) - S^{\pm}(x)|/|S^{\pm}(x)| \tag{50}$$

when $k_w(x^2-x_o^2) \geqslant 0$, and

$$e_r^{\pm}(x) = [\mathrm{Re}\, u^{\pm}(x) - \mathrm{Re}\, S^{\pm}(x)]/\mathrm{Re}\, S^{\pm}(x) \tag{51}$$

$$e_i^{\pm}(x) = [\mathrm{Im}\, u^{\pm}(x) - \mathrm{Im}\, S^{\pm}(x)]/\mathrm{Im}\, S^{\pm}(x) \tag{52}$$

when $k_w(x^2-x_o^2) < 0$. Under the condition

$$\begin{cases} S^{\pm}(x_o) = u^{\pm}(x_o) \\ dS^{\pm}(x_o)/dx = du^{\pm}(x_o)/dx \end{cases} \tag{52}$$

the calculating results are as below.

Table 3. x_o=2.0

x	$u_r^-(x)$	$u_i^-(x)$	$e_c^-(x)$		$u_r^+(x)$	$u_i^+(x)$	$e_r^+(x)$	$e_i^+(x)$
0.5	-0.001	0.718	0.0231		1.640	-6.029	-0.0689	-0.0359
1.0	0.604	0.450	0.0068		0.766	-2.538	-0.1030	-0.0616
1.5	0.819	-1.389	0.0068		0.577	-1.298	-0.0954	-0.0546
2.0	0.706	-0.706	0.0000		0.706	-0.706	0.0000	
2.5	0.585	-1.253	0.0043	-0.0530	0.769	-0.091	0.0137	
3.0	0.890	-2.724	0.0975	-0.0382	0.422	0.506	0.0040	
3.5	2.500	-8.663	0.1055	-0.0040	-0.335	0.482	0.0096	
4.0	10.64	-38.46	0.0837	0.0102	-0.432	-0.317	0.0066	
x	$u_r^-(x)$	$u_i^-(x)$	$e_r^-(x)$	$e_i^-(x)$	$u_r^+(x)$	$u_i^+(x)$	$e_c^+(x)$	

Table 4. x_o=5.0

x	$u_r^-(x)$	$u_i^-(x)$	$e_c^-(x)$		$u_r^+(x)$	$u_i^+(x)$	$e_r^+(x)$	$e_i^+(x)$
3.0	0.163	-0.473	0.0042		35.18	-130.5	-0.0112	-0.0087
3.5	-0.528	-0.001	0.0040		5.671	-20.76	-0.0059	-0.0153
4.0	0.051	0.572	0.0060		1.259	-4.359	-0.0015	-0.0365
4.5	0.652	0.091	0.0028		0.521	-1.362	-0.0577	-0.0694
5.0	0.606	-0.606	0.0000		0.606	-0.606	0.0000	
5.5	0.529	-1.357	-0.0147	-0.0661	0.632	0.107	0.0036	
6.0	1.380	-4.705	0.0507	-0.0266	-0.028	0.545	0.0055	
6.5	7.524	-27.31	0.0333	-0.0051	-0.457	-0.176	0.0037	
x	$u_r^-(x)$	$u_i^-(x)$	$e_r^-(x)$	$e_i^-(x)$	$u_r^+(x)$	$u_i^+(x)$	$e_c^+(x)$	

These above two tables show that the smaller the value of $|q''(x_0)/q'(x_0)|$ is, the better the results of the representation (39) and (40) are. They also show that the difference between $b(x)$ and $b_0(x)$ causes the maximums of $|e_r^{\pm}(x)|$ and $|e_i^{\pm}(x)|$ not very small, so a better simple approximate function of $b_0(x)$ still needs to be found.

REFERENCES

1. D.Z. Wang and E.C. Shang, "Underwater Acoustics", Scientific Press, Beijing (1981).
2. L. Brekhovskikh, "Waves in Stratified Media", Scientific Press, Beijing (1960).
3. Z.X. Wang and D.R. Guo, "An Introduction of Special Functions", Scientific Press, Beijing (1965).
4. W.C. Qian et al., "The Theories of the Singular Perturbation and their Applications in Mechanics", Scientific Press, Beijing (1981).
5. H. Weinburg, Application of ray theory to acoustic propagation in horizontally stratified oceans, J.A.S.A., 58(1):97 (1975).
6. R.H. Zhang, About the regions near the turning points in underwater sound channels (I) — the normal mode theory, Acta Acustica, 5:28 (1980).
7. T.F. Gao, A modified WKB solution, unpublished, available from the Inst. of Acoustics, Beijing.

THE SOUND INTENSITIES NEAR THE CONJUGATE TURNING POINTS IN AN UNDERWATER SOUND CHANNEL

Zhang Renhe

Institute of Acoustics, Academia Sinica
P.O. Box 2712, Beijing, China

ABSTRACT

Following the work of Beilis (J. Acoust. Soc. Am. 74 p171, 1983), the author calculates ranges and acoustic intensities for deep caustics and turning points in an underwater sound channel. The acoustic intensity at such points is shown to be proportional to the cube root of frequency and inversely proportional to distance travelled. Results are compared with the normal mode and parabolic equation methods. - Ed.

INTRODUCTION

In the ocean sound channel, when the source is above the channel axis, there exists below the axis a conjugate depth which has the same sound velocity as the source depth. Conjugate turning-point convergence-zones with high intensities are formed in the vicinity of the conjugate turning points.

A little while ago, Beilis[1] used the normal-mode method to calculate transmission losses at the conjugate depth for the frequencies of 53 and 143 Hz. For the two frequencies, he has calculated 250 and 500 modes respectively, and his results show that the strong convergence-zones appear near the conjugate turning points. In this paper, we mainly discuss the fields in the vicinity of conjugate turning-points and give concise formulae for calculating the intensities.

RAY STRUCTURE NEAR THE CONJUGATE TURNING POINT

In Fig. 1 are shown a typical velocity profile in the Atlantic and a ray diagram with source at 100m depth, where T_0, T_1, T_2, .. are the conjugate turning points at 4000 m, i.e., the conjugate depth. Figure 2 is an enlarged ray diagram near the conjugate turning point T_5. In what follows, we shall analyze the ray structure near conjugate turning-points.

There are generally four types of rays in underwater sound channels. When the source is above the channel axis and the receiver is below it, the horizontal distances from the source to the receiver for the rays of four types may be written as:

$$
\left.
\begin{aligned}
r_n^{+-} &= (n+\tfrac{1}{2})S(\alpha_0)+L(z_1,\alpha_1)-L(z_2,\alpha_2)\\
r_n^{++} &= (n+\tfrac{1}{2})S(\alpha_0)+L(z_1,\alpha_1)+L(z_2,\alpha_2)\\
r_n^{-+} &= (n+\tfrac{1}{2})S(\alpha_0)-L(z_1,\alpha_1)+L(z_2,\alpha_2)\\
r_n^{--} &= (n+\tfrac{1}{2})S(\alpha_0)-L(z_1,\alpha_1)-L(z_2,\alpha_2)
\end{aligned}
\right\} \qquad (1)
$$

where n is a non-negative integer, α_1, α_2 and α_0 are the grazing angles of rays at the source depth z_1, receiver depth z_2 and axis z_0 respectively, $S(\alpha_0)$ is the cycle-distance, and the horizontal distances $L(z_1, \alpha_1)$ and $L(z_2, \alpha_2)$ are shown in Fig. 3. The horizontal distance of the n-th conjugate turning-point T_n is

$$r_n^{+-} = r_n^{++} = r_n^{-+} = r_n^{--} = (n+\tfrac{1}{2})S(\alpha_0^*)$$

where $\alpha_0^* = \arccos\left[c(z_0)/c(z_1)\right]$.

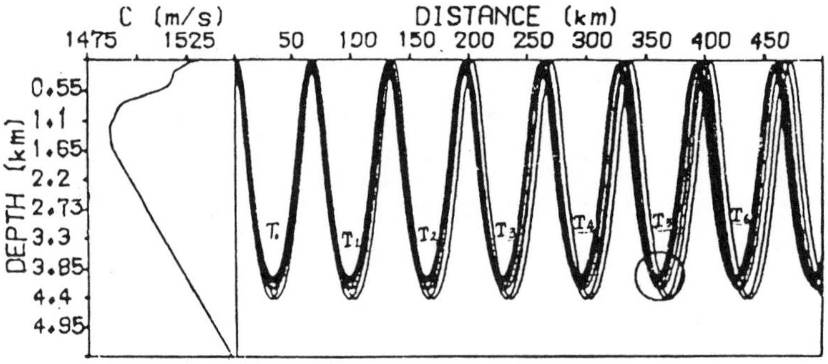

Figure 1: A velocity profile and a ray diagram.

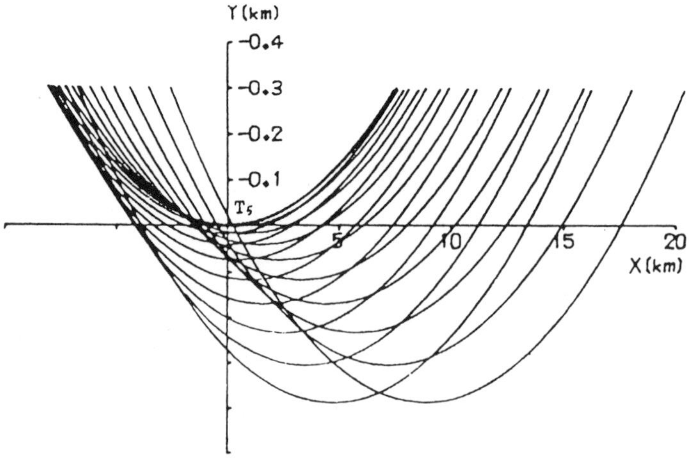

Figure 2: Enlarged ray diagram in the vicinity of T_5.

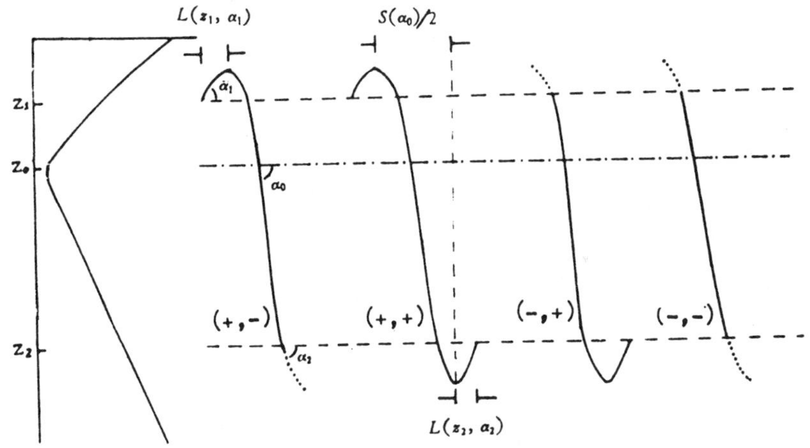

Figure 3: Rays of four types for n=0.

For the convenience of describing the behavior of rays near the conjugate turning-point, we introduce the reference coordinate system X-Y and make the origin coincide with the conjugate turning-point T_n, that is, $x=[r-(n+\frac{1}{2})S(\alpha_0^*)]$, $y=(z_2-\bar{z}_1)$ and $\bar{z}_1$ denotes the conjugate depth. For small angles α_1 and α_2, we can make use of the following approximate expressions:

$$L(z_1, \alpha_1)=\frac{\alpha_1}{a}, \qquad L(z_2, \alpha_2)=\frac{\alpha_2}{b}=\frac{\sqrt{\alpha_1^2-2by}}{b} \tag{2}$$

$$S(\alpha_0)=S(\alpha_0^*)+\frac{\alpha_1^2}{2}\dot{S}(\alpha_0^*)\cot\alpha_0^* \tag{3}$$

where $a=\left|(1/c)(dc/dz)\right|_{z_1}$, $b=\left|(1/c)(dc/dz)\right|_{z_2}$. Then we get

$$\left.\begin{aligned}
x_n^{+-} &= \frac{A}{2a}\alpha_1^2+\frac{\alpha_1}{a}-\frac{\sqrt{\alpha_1^2-2by}}{b}\\
x_n^{++} &= \frac{A}{2a}\alpha_1^2+\frac{\alpha_1}{a}+\frac{\sqrt{\alpha_1^2-2by}}{b}\\
x_n^{-+} &= \frac{A}{2a}\alpha_1^2-\frac{\alpha_1}{a}+\frac{\sqrt{\alpha_1^2-2by}}{b}\\
x_n^{--} &= \frac{A}{2a}\alpha_1^2-\frac{\alpha_1}{a}-\frac{\sqrt{\alpha_1^2-2by}}{b}
\end{aligned}\right\} \tag{4}$$

where $A=(n+\frac{1}{2})a\dot{S}(\alpha_0^*)\cot\alpha_0^*$, $\alpha_1 \geqslant 0$.

According to Eq. (4), the equations for caustics can be obtained from $(\partial x_n^{\pm\pm}/\partial\alpha_1)=0$. For definiteness, we suppose that $\dot{S}(\alpha_0^*)>0$, $a>b$. In this case, the (+,+) type rays cannot form a caustic, and the caustics formed by rays of the other three types are as follows:

$$(+,-)\text{type:}\quad x_n^{+-}=\frac{\alpha_1}{2a}\left[A\alpha_1+2-\frac{2a^2}{b^2(A\alpha_1+1)}\right] \qquad \alpha_1\geqslant 0,\quad y=\frac{\alpha_1^2}{2b}\left[1-\frac{a^2}{b^2(A\alpha_1+1)^2}\right] \tag{5}$$

$$(-,+)\text{type:}\quad x_n^{-+}=\frac{\alpha_1^2}{2a}\left[A\alpha_1-2-\frac{2a^2}{b^2(A\alpha_1-1)}\right] \qquad \frac{1}{A}>\alpha_1\geqslant 0,\quad y=\frac{\alpha_1^2}{2b}\left[1-\frac{a^2}{b^2(A\alpha_1-1)^2}\right] \tag{6}$$

$$(-,-)\text{type:}\quad x_n^{--}=\frac{\alpha_1}{2a}\left[A\alpha_1-2-\frac{2a^2}{b^2(A\alpha_1-1)}\right] \qquad \alpha_1>\frac{1}{A},\quad y=\frac{\alpha_1^2}{2b}\left[1-\frac{a^2}{b^2(A\alpha_1-1)^2}\right] \tag{7}$$

In Fig. 4 are shown the caustics corresponding to the ray diagram in Fig. 2, calculated by using (5), (6) and (7). It can be seen from Fig. 4 that the (+,-) and (-,-) type caustics intersect the conjugate depth (y=0). At the intersecting points, we have

$$(+,-)\text{type:}\quad \alpha_1=\frac{a-b}{bA},\qquad x_n^{+-}=-\frac{(a-b)^2}{2ab^2A} \tag{8}$$

$$(-,-)\text{type:}\quad \alpha_1=\frac{a+b}{bA},\qquad x_n^{--}=-\frac{(a+b)^2}{2ab^2A} \tag{9}$$

It can also be seen that the (+,-) type caustic has a turning point Q, and its position is

$$x_n^{+-}=-\frac{1}{2aA}\left[\left(\frac{a}{b}\right)^{2/3}-1\right]^2\left[2\left(\frac{a}{b}\right)^{2/3}+1\right],\qquad y=-\frac{1}{2bA^2}\left[\left(\frac{a}{b}\right)^{2/3}-1\right]^3 \tag{10}$$

Additionally, the (+,-) and (-,+) type caustics connect with each other and are tangent to the X-axis at the conjugate turning point.

INTENSITIES NEAR THE CAUSTICS AT THE CONJUGATE DEPTH

When the source is above the axis and the receiver is at or below the conjugate depth $\bar{z}_1$, by using the generalized phase-integral approximation[2] the field of a point source may be expressed as the sum of the following multipath integrals:

$$z_1<z_0<\bar{z}_1\leqslant z_2,\qquad P=\sum_{n=0}^{\infty}\left(P_n^{+-}+P_n^{++}+P_n^{-+}+P_n^{--}\right) \tag{11}$$

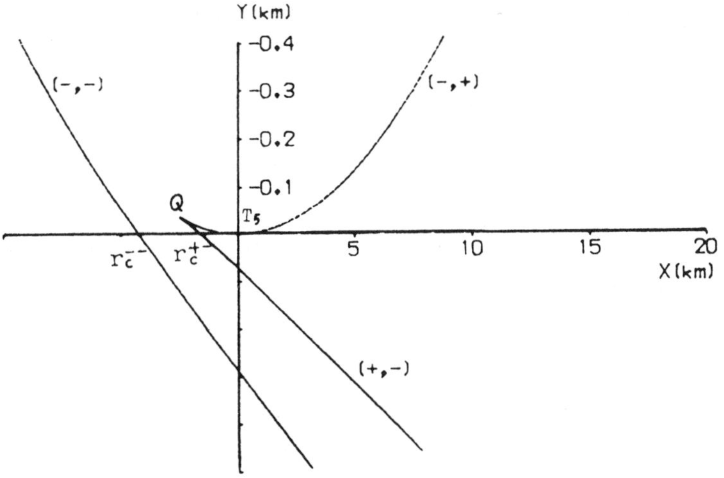

Figure 4: Caustics in the vicinity of T_5.

where

$$n \geq 1, \quad \left. \begin{aligned} P_n^{+-} &= \frac{e^{i\pi/4}}{\sqrt{2\pi r}} \int_{-K_1}^{K_1} \sqrt{\nu}\, q(z_1,\nu)\, q(z_2,\nu)\, e^{iW_n^{+-}(\nu)}\, d\nu \\ P_n^{++} &= \frac{e^{-i\pi/4}}{\sqrt{2\pi r}} \int_{-K_2}^{K_2} \sqrt{\nu}\, q(z_1,\nu)\, q(z_2,\nu)\, e^{iW_n^{++}(\nu)}\, d\nu \\ P_n^{-+} &= \frac{e^{i\pi/4}}{\sqrt{2\pi r}} \int_{-K_2}^{K_2} \sqrt{\nu}\, q(z_1,\nu)\, q(z_2,\nu)\, e^{iW_n^{-+}(\nu)}\, d\nu \\ P_n^{--} &= \frac{e^{i3\pi/4}}{\sqrt{2\pi r}} \int_{-K_0}^{K_0} \sqrt{\nu}\, q(z_1,\nu)\, q(z_2,\nu)\, e^{iW_n^{--}(\nu)}\, d\nu \end{aligned} \right\} \tag{12}$$

It should be pointed out that the limits of the multipath integrals are generally definite except for the direct wave P_0^{--}. The phase functions $W_n^{\pm\pm}(\nu)$ and the amplitude function $q(z,\nu)$ are given by

$$W_n^{\pm\pm}(\nu) = \nu r + (2n+1)\left[\int_\eta^\xi \sqrt{k^2(y)-\nu^2}\, dy - \frac{\pi}{2}\right] \pm \int_\eta^{z_1} \sqrt{k^2(y)-\nu^2}\, dy \pm \int_{z_2}^{\xi} \sqrt{k^2(y)-\nu^2}\, dy \tag{13}$$

and

$$q(z,\nu) = \begin{cases} \left\{Bd^{4/3} - Dd^{2/3}[k^2(z)-\nu^2] + [k^2(z)-\nu^2]^2\right\}^{-1/8} & |\nu| \leq k(z) \\ \sqrt{2}\, e^{-i\pi/4} \left\{Bd^{4/3} - Dd^{2/3}[k^2(z)-\nu^2] + 16[k^2(z)-\nu^2]^{-1/8}\right. & |\nu| > k(z) \end{cases} \tag{14}$$

In the equations mentioned-above, η and ξ are the upper and lower turning depths determined by $k(\eta)=k(\xi)=\nu$, B=2.152, D=1.619 and $d=|dk^2(z)/dz|$.

Suppose the receiver is at the conjugate depth, i.e., $z_2=\bar{z}_1$. We adopt the linear approximation for $k^2(z)$ in the vicinity of the source and conjugate depths:

$$\left.\begin{aligned} z\approx z_1, &\quad k^2(z)=k^2+2ak^2x(z-z_1) \\ z\approx \bar{z}_1, &\quad k^2(z)=k^2-2bk^2x(z-\bar{z}_1). \end{aligned}\right\} \tag{15}$$

Then, we have

$$\left.\begin{aligned} \int_{\eta}^{z_1}\sqrt{k^2(y)-\nu^2}\,dy=\frac{(k^2-\nu^2)^{3/2}}{3ak^2}\cong\frac{2^{3/2}}{3ak^{1/2}}(k-\nu)^{3/2} \\ \int_{z_2}^{\xi}\sqrt{k^2(y)-\nu^2}\,dy=\frac{(k^2-\nu^2)^{3/2}}{3bk^2}\cong\frac{2^{3/2}}{3bk^{1/2}}(k-\nu)^{3/2} \end{aligned}\right\} \tag{16}$$

where $k=k(z_1)=k(\bar{z}_1)$. Substituting (16) into (13), we get

$$\left.\begin{aligned} W_n^{+-}(\nu)&=\nu r+(2n+1)\left[\int_{\eta}^{\xi}\sqrt{k^2(y)-\nu^2}\,dy-\frac{\pi}{2}\right]-E_1(k-\nu)^{3/2} \\ W_n^{++}(\nu)&=\nu r+(2n+1)\left[\int_{\eta}^{\xi}\sqrt{k^2(y)-\nu^2}\,dy-\frac{\pi}{2}\right]+E_2(k-\nu)^{3/2} \\ W_n^{-+}(\nu)&=\nu r+(2n+1)\left[\int_{\eta}^{\xi}\sqrt{k^2(y)-\nu^2}\,dy-\frac{\pi}{2}\right]+E_1(k-\nu)^{3/2} \\ W_n^{--}(\nu)&=\nu r+(2n+1)\left[\int_{\eta}^{\xi}\sqrt{k^2(y)-\nu^2}\,dy-\frac{\pi}{2}\right]-E_2(k-\nu)^{3/2} \end{aligned}\right\} \tag{17}$$

where

$$E_1=2^{3/2}(a-b)/3abk^{1/2},\qquad E_2=2^{3/2}(a+b)/3abk^{1/2}. \tag{18}$$

According to the supposition that A>0 and a>b, only the (+,-) and (-,-) type caustics intersect the conjugate depth. So, we only calculate the values of P_n^{+-} and P_n^{--} near caustics at the conjugate depth. Differentiating (17), we get

$$\left.\begin{aligned} \dot{W}_n^{+-}(\nu)&=r-(n+\tfrac{1}{2})S(\nu)+\tfrac{3}{2}E_1(k-\nu)^{1/2} \\ \ddot{W}_n^{+-}(\nu)&=-(n+\tfrac{1}{2})\dot{S}(\nu)-\tfrac{3}{4}E_1(k-\nu)^{-1/2} \\ \dddot{W}_n^{+-}(\nu)&=-(n+\tfrac{1}{2})\ddot{S}(\nu)-\tfrac{3}{8}E_1(k-\nu)^{-3/2} \end{aligned}\right\} \tag{19}$$

where

$$S(\nu)=2\int_{\eta}^{\xi}\frac{\nu\,dy}{\sqrt{k^2(y)-\nu^2}}. \tag{20}$$

As the stationary-phase value ν_1 corresponding to the caustic makes $\dot{W}_n^{+-}(\nu_1)=\ddot{W}_n^{+-}(\nu_1)=0$, from the first two equations of (19) we then obtain

$$\left.\begin{aligned} &(k-\nu)^{1/2} \cong -3E_1/[4(n+\tfrac{1}{2})\dot{S}(k)] \\ &r_c^{+-} \cong (n+\tfrac{1}{2})S(k)-[(a-b)^2/2ab^2A] \end{aligned}\right\} \tag{21}$$

where r_c^{+-} is the horizontal distance of the intersection of the (+,-) type caustic and conjugate depth. It can be shown that Eq. (21) is equivalent to Eq. (8). Near the conjugate turning-point $(\nu_1 \approx k)$ the 2nd derivative $\ddot{S}(k)$ may be neglected, we than have

$$\dddot{W}_n^{+-}(\nu) \cong -\frac{3}{8}E_1(k-\nu_1)^{-3/2} \cong -[(2n+1)\dot{S}(k)]^3/9E_1^2 . \tag{22}$$

When $r \approx r_c^{+-}$, expanding $W_n^{+-}(\nu)$ as a power series in $(\nu-\nu_1)$ and using $\dot{W}_n^{+-}(\nu_1)=(r-r_c^{+-})$ and $\ddot{W}_n^{+-}(\nu_1)=0$, we get

$$W_n^{+-}(\nu)=W_n^{+-}(\nu_1)+(r-r_c^{+-})(\nu-\nu_1)+\frac{1}{6}\dddot{W}_n^{+-}(\nu_1)(\nu-\nu_1)^3+\ldots \tag{23}$$

Owing to the term $E_1(k-\nu)^{3/2}$ in Exp. (17), the convergence range of series (23) is

$$|\nu-\nu_1|<(k-\nu_1) . \tag{24}$$

Substituting (23) into (12), in the vicinity of the caustic near the conjugate depth, P_n^{+-} may be expressed as

$$P_n^{+-}=\sqrt{\frac{\nu_1}{2\pi r}}\,q(z_1,\nu_1)q(\bar{z}_1,\nu_1)e^{i\frac{\pi}{4}+W_n^{+-}(\nu_1)}\int_{-(k-\nu_1)}^{(k-\nu_1)} e^{i(r-r_c^{+-})\mu+i\frac{1}{6}\dddot{W}_n^{+-}(\nu_1)\mu^3}d\mu . \tag{25}$$

We introduce new variables:

$$\left.\begin{aligned} &s=\mu|\dddot{W}_n^{+-}(\nu_1)/2|^{1/3}\operatorname{sgn}[\dddot{W}_n^{+-}(\nu_1)] \\ &s_1=(k-\nu_1)|\dddot{W}_n^{+-}(\nu_1)/2|^{1/3} \\ &t=(r-r_c^{+-})|\dddot{W}_n^{+-}(\nu_1)/2|^{-1/3}\operatorname{sgn}[\dddot{W}_n^{+-}(\nu_1)] . \end{aligned}\right\} \tag{26}$$

Substituting (26) into (25) we get

$$P_n^{+-} \cong \frac{2^{5/6}\sqrt{\nu_1}}{\sqrt{\pi r}}\,q(z_1,\nu_1)q(\bar{z}_1,\nu_1)\,|\dddot{W}_n^{+-}(\nu_1)|^{-1/3}e^{i\frac{\pi}{4}+iW_n^{+-}(\nu_1)}\int_0^{s_1}\cos(\frac{s^3}{3}+st)ds, \tag{27}$$

When

$$s_1 >> 1, \tag{28}$$

the definite integral in Exp. (27) may be replaced by an infinite one, then, the intensity in the vicinity of the caustic at the conjugate depth can be simplified as

$$I_n^{+-} = |P_n^{+-}|^2 = \frac{2^{5/3}\nu_1}{r}\left[v(t)q(z_1,\nu_1)q(\bar{z}_1,\nu_1)\,|\dddot{W}_n^{+-}(\nu_1)|^{-1/3}\right]^2 \tag{29}$$

where $v(t)$ is the Airy function

$$v(t) = \frac{1}{\sqrt{\pi}}\int_0^\infty \cos\left(\frac{s^3}{3} + st\right)ds\ . \tag{30}$$

It is easy to show that when $s_1 > 1$, from (14) we can get

$$q(z_1,\nu_1) \cong q(\bar{z}_1,\nu_1) = (k^2 - \nu_1^2)^{-\frac{1}{4}}. \tag{31}$$

Substituting (22) and (31) into (29), we then obtain the simplified expression of the intensity in the vicinity of caustic as follows:

$$r \approx r_c^{+-}\ , \qquad I_n^{+-} = \frac{2^{5/3}\,v^2(t)\,k^{1/3}}{r}\left(\frac{ab}{a-b}\right)^{2/3}. \tag{32}$$

Notice that t=0 corresponds to the receiver on the caustic and $v(0)$ = 0.6293, t<0 corresponds to the receiver on the illuminated side of the caustic and that the Airy function has a maximum of 0.9493 when t=-1. Therefore, the intensity I_c^{+-} on the caustic and the maximum I_{max}^{+-} near the caustic are

$$r \approx r_c^{+-}\ , \qquad I_c^{+-} = \frac{1.26k^{1/3}}{r}\left(\frac{ab}{a-b}\right)^{2/3} \tag{33}$$

$$I_{max}^{+-} = \frac{2.86k^{1/3}}{r}\left(\frac{ab}{a-b}\right)^{2/3}. \tag{34}$$

For the multipath integral P_n^{--} , the intensity I_c^{--} on the caustic and the maximum intensity I_{max}^{--} can be similarly obtained:

$$r \approx r_c^{--}\ , \qquad I_c^{--} = \frac{1.26k^{1/3}}{r}\left(\frac{ab}{a+b}\right)^{2/3} \tag{35}$$

$$I_{max}^{--} = \frac{2.86k^{1/3}}{r}\left(\frac{ab}{a+b}\right)^{2/3}. \tag{36}$$

INTENSITIES AT THE CONJUGATE TURNING POINTS

When the receiver is just located at the conjugate turning point, i.e., $r=(n+\frac{1}{2})S(k)$, the stationary-phase value ν_1 is equal to k. Since $\ddot{W}_n^{\pm\pm}(k)=\infty$, the stationary-phase method can not be directly used to calculate the integrals (2). At the conjugate turning point, we express $W_n^{+-}(\nu)$ and $W_n^{-+}(\nu)$ as

$$\nu\approx k, \quad \left.\begin{aligned} W_n^{+-}(\nu)&=W_n^{+-}(k)-\tfrac{1}{4}(2n+1)\dot{S}(k)(\nu-k)^2-E_1(k-\nu)^{3/2} \\ W_n^{-+}(\nu)&=W_n^{-+}(k)-\tfrac{1}{4}(2n+1)\dot{S}(k)(\nu-k)^2+E_1(k-\nu)^{3/2}. \end{aligned}\right\} \tag{37}$$

We introduce the waveguide parameters ε_1 and ε_2 :

$$\varepsilon_1=(2E_1)^{4/3}/\pi^{1/3}(n+\tfrac{1}{2})|\dot{S}(k)|, \quad \varepsilon_2=(2E_2)^{4/3}/\pi^{1/3}(n+\tfrac{1}{2})|\dot{S}(k)|. \tag{38}$$

When

$$\varepsilon_1>1, \tag{39}$$

the phase functions $W_n^{+-}(\nu)$ and $W_n^{-+}(\nu)$ except the term $E_1(k-\nu)^{3/2}$ may be regarded as slowly-varing functions. Then, P_n^{+-} and P_n^{-+} at the conjugate turning point can be simplified as

$$\left.\begin{aligned} P_n^{+-}&=\frac{0.246k^{1/6}(ab)^{1/2}}{\sqrt{r}\;(a-b)^{2/3}}\,e^{-i\frac{\pi}{12}+iW_n^{+-}(k)} \\ P_n^{-+}&=\frac{0.246k^{1/6}(ab)^{1/2}}{\sqrt{r}\;(a-b)^{2/3}}\,e^{+i\frac{7\pi}{12}+iW_n^{-+}(k)}. \end{aligned}\right\} \tag{40}$$

Similarly, when

$$\varepsilon_2>1, \tag{41}$$

at the conjugate turning point we have

$$\left.\begin{aligned} P_n^{++}&=\frac{0.246k^{1/6}(ab)^{1/2}}{\sqrt{r}\;(a+b)^{2/3}}\,e^{i\frac{\pi}{12}+iW_n^{++}(k)} \\ P_n^{--}&=\frac{0.246k^{1/6}(ab)^{1/2}}{\sqrt{r}\;(a+b)^{2/3}}\,(2+e^{i\frac{\pi}{6}})\,e^{i\frac{\pi}{4}+iW_n^{--}(k)}. \end{aligned}\right\} \tag{42}$$

Finally, we obtain the intensity I_t at the conjugate turning point:

$$\begin{aligned} \varepsilon_1,\ \varepsilon_2>1, \quad I_t&=\left|P_n^{+-}+P_n^{++}+P_n^{-+}+P_n^{--}\right|^2 \\ &=\frac{0.0605abk^{1/3}}{r(a-b)^{4/3}}\left[1+(2+\sqrt{3})\left(\frac{a-b}{a+b}\right)^{2/3}\right]^2. \end{aligned} \tag{43}$$

SUMMARY

The analysis mentioned above shows that the intensities on the caustics and at the turning points located at the conjugate depth are proportional to the cube root of the frequency and inversely proportional to the distance.

In Fig. 5 are shown the transmission loss (TL) at 53 and 143 Hz calculated using several methods for the profile in Fig. 1, where TL in (a) and (c) were calculated using the normal mode method (NM), TL in (b) using the parabolic equation (PE) and dot-dash lines using Eq. (34). This figure shows that the lines of I_{max}^{+-} are consistent with the peak values of TL curves.

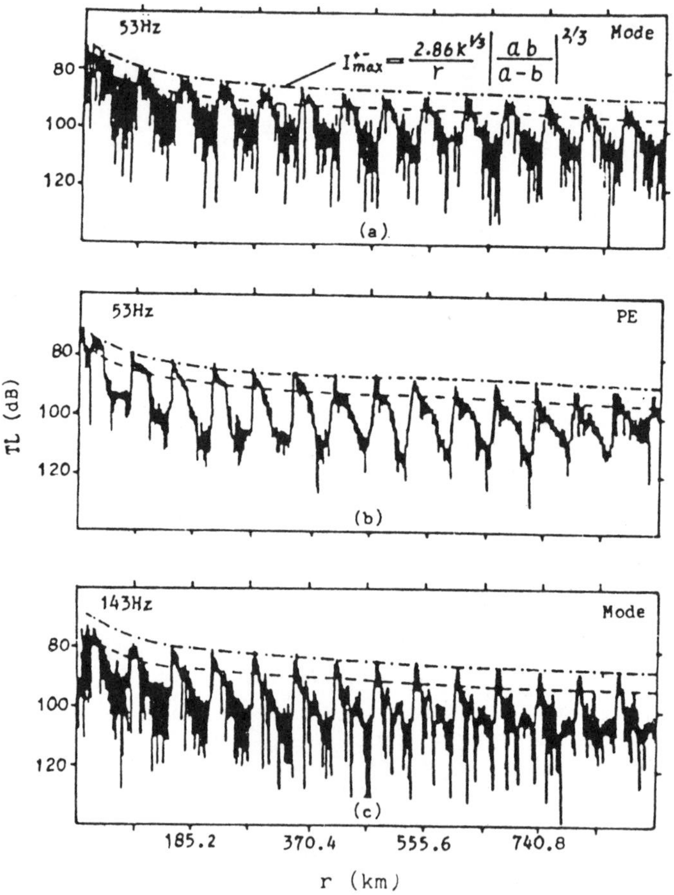

Figure 5: Transmission losses of 53 and 143 Hz at conjugate depth.

REFERENCES

1. A. Beilis, J. Acoust. Soc. Am., 74, 171 (1983).
2. Zhang Renhe, Chin. Phys., 1, 1064 (1981).

MODELING ACOUSTIC WAVES BY TRANSMUTATION METHODS

Robert P. Gilbert* and David H. Wood+†

*Department of Mathematical Science
The University of Delaware
Newark Delaware 19711

+Code 3332, New London Laboratory
Naval Underwater Systems Center
New London, CT 06320

†Department of Mathematics
The University of Rhode Island
Kingston, Rhode Island 02881

ABSTRACT

Our model of the ocean has two aspects: an idealized ocean, and some perturbations from that ideal. Acoustic waves in the idealized ocean can be treated analytically. The changes due to perturbations can be treated numerically and incorporated with the help of transmutation representations of the acoustic waves.

INTRODUCTION

The problem of computing a sound field in the ocean is greatly simplified by using a two-pronged approach involving 1) transmutation methods that analytically incorporate the major features of the ocean and its bounding surfaces, including the radiation condition and the average index of refraction; and 2) numerical methods to model the effects of the secondary, but important, detailed features of the ocean and its boundaries. In our dual approach, numerical techniques are used only to model the changes that result from perturbing the ocean from its analytical idealization.

In general, transmutation methods are used to transmute solutions of a simpler equation into solution of a more complicated equation. This technique has been extensively developed [1,2,3] and recently applied to the problem of acoustic waves in the ocean [4,5].

HELMHOLTZ EQUATIONS FOR IDEALIZED AND PERTURBED OCEANS

In an idealized ocean with a uniform index of refraction (whose nominal value can be taken to be unity), the acoustic waves are governed by the solutions v(z,r) of the idealized Helmholtz equation:

$$v(r,z) + v_{rr}(r,z) + 1/r\ v_r(r,z) + k^2 v(r,z) = 0, \qquad (1)$$

where k is the wavenumber of the acoustic wave. Since the index of refraction depends on the depth coordinate in the ocean, it is perturbed from its nominial value and becomes

$$n^2(z) = 1 + \epsilon s(z), \qquad (2)$$

where the parameter ϵ measures the strength of the perturbation. In this case, the acoustic waves are governed by the solutions u(z,r) of the perturbed Helmholtz equation:

$$u_{zz}(r,z) + u_{rr}(r,z) + 1/r\ u_r(r,z) + k^2 n^2(z)\ u(r,z) = 0. \qquad (3)$$

THE TRANSMUTATION REPRESENTATION OF ACOUSTIC WAVES

We use a transmutation, which in this case is an integral transformation with a kernel K(z,s), that transmutes the solutions of the idealized Helmholtz equation into solutions of the perturbed Helmholtz equationn. The transmutation we will use is

$$u(r,z) = v(r,z) + \int_b^z K(z,s)\ v(r,s)\ ds, \qquad (4)$$

where the kernel K(z,s) must satisfy some equations that we will present below.

All of the integrals in this paper are definite integrals, and their limits are always from b to z unless specifically shown otherwise. We will see that at z=b, both the value and the z derivative of v(r,z) are preserved. For this reason, it is usually convenient to take b to be the z coordinate of either the surface or bottom of the ocean, where boundary conditions have to be assigned--but for the time being, we do not say which boundary corresponds to z=b.

It can be shown [4,5] that K(z,s) must satisfy the hyperbolic partial differential equation

$$K_{zz} - K_{ss} + k^2(n^2(z) - 1)\ K = 0\ , \qquad (5)$$

and a condition on the diagonal of the s-z plane,

$$K(z,z) = -1/2 \int k^2(n^2(\zeta) - 1)\ d\zeta \qquad (6)$$

(see Figure 1).

THE KERNEL MUST SATISFY A "BOUNDARY" TYPE CONDITION

There is another condition [4,5] that the kernel K(z,s) must satisfy which is tied up with the boundary conditions on the acoustic wave. This condition is

$$K_s(z,b)\ v(r,b) - K(z,b)\ v_z(r,b) = 0\ . \qquad (7)$$

Let z=b be a boundary and let us now restrict our attention to the solutions v(r,z) of the idealized Helmholtz equation to those that satisfy an impedence condition at the boundary z=b, that is, we assume

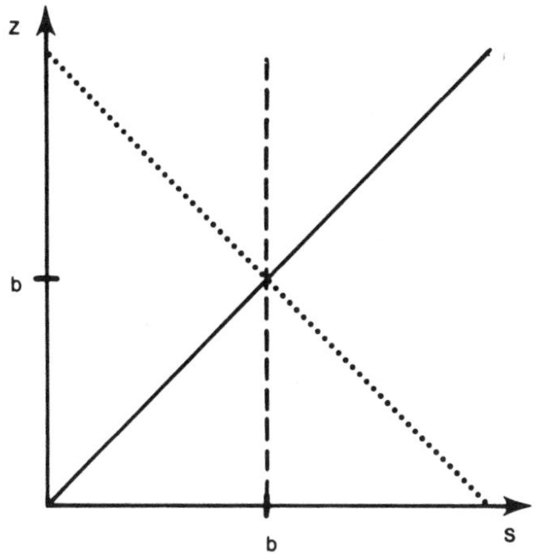

Figure 1: Diagram of the s-z plane

$$\alpha\, v(r,b) + \beta\, v_z(r,b) = 0 \,, \tag{8}$$

for some constants α and β. Using this information in (7), we see that the kernel must also satisfy an **impedence** type condition at s=b (not at z=b),

$$\alpha\, K(z,b) + \beta\, K_s(z,b) = 0. \tag{9}$$

The need to satisfy this condition is a serious impediment to finding the kernel. However, there are two cases in which (9) can be satisfied by using a more natural condition in its place. These cases are treated in the next two sections.

KERNEL CONDITION CORRESPONDING TO RIGID OCEAN BOTTOM

A rigid ocean bottom is modeled by assuming **that** the normal derivative is zero at the bottom, that is, by taking α=0 in (8) and therefore (9). One way to satisfy this is to impose the condition

$$K(z,-z+2b) = -1/2 \int (n^2(\varsigma) - 1)\, d\varsigma \tag{10}$$

on the other diagonal through the point (b,b)(see Figure 1). It can be shown that this makes K(z,s) a symmetric function of s about the line s=b. Because of this symmetry, it has a local maximum or minimum at s=b, so its derivative has to be zero there (we can show that the derivative exists and is continious).

KERNEL CONDITION CORRESPONDING TO USUAL OCEAN SURFACE

Using the condition that acoustic waves must be zero at the surface of the ocean is less restrictive than assuming that the bottom of the ocean is rigid. This surface condition is modeled by setting β=0 in (8) and therefore in (9). This is accomplished by forcing the kernel to be anti-symmetric about s=b by requiring that it satisfy a condition like (10) but with a change in sign,

$$K(z,-z+2b) = 1/2 \int (n^2(\varsigma) - 1)\, d\varsigma \,. \tag{11}$$

ANY BOUNDARY CONDITION AT z=b IS ALWAYS PRESERVED

To check boundary conditions, we recall the transmutation representation which is given by (4)

$$u(r,z) = v(r,z) + \int K(z,s)\, v(r,s)\, ds \,. \qquad (12)$$

Computing the derivative of this formula with respect to z, we obtain

$$u_z(r,z) = v_z(r,z) + K(z,z)\, v(r,z) + \int K_z(z,s)\, v(r,s)\, ds. \qquad (13)$$

At the boundary z=b, the first of these formulas gives

$$u(r,b) = v(r,b)$$

and, since K(b,b)=0, the second formula gives

$$u_z(r,b) = v_z(r,b) \,.$$

So, we see that the transmutation preserves values and derivatives at z=b. This means that z=b can conveniently be located where a known boundary condition has been assigned. In short, we say that the transmutation preserves boundary conditions at z=b.

ANY "OTHER" BOUNDARY CONDITION CAN NOT BE PRESERVED

Unfortunately, a boundary condition generally will not be preserved at an "other" boundary, at say, z=o (a small letter o to suggest "other"). At such a boundary, (12) gives

$$u(r,0) = v(r,o) + \int_b^o K(o,s)\, v(r,s)\, ds \qquad (14)$$

and (13) gives

$$u_z(r,o) = v_z(r,o) + K(o,o)\, v(r,o) + \int_b^o K(o,s)\, v(r,s)\, ds \qquad (15)$$

In general, these conditions will not simplify further because we cannot expect K(o,s) to be identically zero--infact, (6) shows that we cannot even expect K(o,s) to be zero at s=o.

This means that we are forced to accept an integral relationship between the boundary conditions of the idealized Helmholtz equation and those of the perturbed Helmholtz equation. There are several ways to cope with this situation but we mention only one here. When the parameter ϵ that measures the strength of the perturbation in (2) is small, we can expand (14) and (15) in power series in ϵ and find the first few terms. We are currently developing this technique [6].

EXISTENCE AND UNIQUENESS OF THE TRANSMUTATION KERNEL

Consider the system defined by (5), (6) and (10). We define a new coordinate system (ξ,η) with the following equations,

$$\xi = (z + s \;\; - 2z)/2 \qquad (16)$$

and

$$\eta = (z - s)/2 . \tag{17}$$

We define a new function $M(\xi,\eta)$ by the equation

$$M(\xi,\eta) = K(\xi+\eta+z_b, \xi-\eta+z_b) = K(z,s). \tag{18}$$

We can show that $M(\xi,\eta)$ satisfies the following partial differential equation and conditions

$$M_{\xi\eta}(\xi,\eta) + k^2(n^2(\xi+\eta+z_b) - 1)M(\xi,\eta) = 0 , \tag{19}$$

$$M(\xi,0) = -1/2 \int_0^{\xi} k^2(n^2(\zeta+b) - 1)\, d\zeta \text{ and} \tag{20}$$

$$M(0,\eta) = -1/2 \int_0^{\eta} k^2(n^2(\zeta+b) - 1)\, d\zeta . \tag{21}$$

These three equations form a Gorsat problem. Assuming that the index of refraction is continuous it is well known [7] that this problem has a unique solution.

ACKNOWLEDGEMENT

A portion of this research was performed while one of us (Gilbert) was on an Intergovernmental Personnel Act Mobility Assignment to the Naval Underwater Systems Center from the University of Delaware.

REFERENCES

1. R. W. Carroll, Transmutation and Operator Differential Equations, Mathematics Studies 37, North-Holland, New York 1982.
2. R. P. Gilbert, Function Theoretic Methods in Partial Differential Equations, Academic Press, New York, 1969.
3. I. N. Vekua, New Methods for Solving Elliptic Equations, J. Wiley, New York, 1967.
4. R. P. Gilbert and D. H. Wood, A transmutation approach to underwater sound propagation, Wave Motion (to appear).
5. M. D. Duston, R. P. Gilbert and D. H. Wood, A computation technique based on function theoretic representation, Proceedings of the 11th IMACS World Congress, North-Holland, New York, 1986 (to appear).
6. M. D. Duston, R. P. Gilbert, G. R. Verma and D. H. Wood, Perturbation of eigenfunctions using transformation (in preparation)
7. P. R. Garabedian, Partial Differential Equations, J. Wiley, New York, 1964.

BEYOND WHAT DISTANCE ARE FINITE-AMPLITUDE EFFECTS UNIMPORTANT ?

F. D. Cotaras,† D. T. Blackstock, and C. L. Morfey*

Applied Research Laboratories, The University of Texas at Austin
Austin, TX 78713-8029

ABSTRACT and INTRODUCTION

Summarized in this paper are the results of a numerical investigation of the effects of inhomogeneity, ordinary attenuation and dispersion, and nonlinear distortion on the propagation of finite-amplitude transients in a lossy stratified ocean. The results are obtained using a weak-shock propagation algorithm based on the equations of nonlinear geometrical acoustics (Cotaras, 1985). Reflections and caustics are avoided by careful selection of ray paths. Two explosion waveforms are considered: a weak shock with an exponentially decaying tail (hereinafter referred to as Pulse I) and a more realistic waveform that includes the first bubble pulse (hereinafter referred to as Pulse II). Numerical propagation of Pulse I along a 58.1 km path starting at a depth of 300 m (hereinafter referred to as the shallow source path) leads to the following conclusions. (1) The effect of inhomogeneity on nonlinear distortion is small. (2) Dispersion plays an important role in determining the arrival time of the pulse. (3) Neither nonlinear distortion nor ordinary attenuation (and dispersion) are paramount; both need to be included. The propagation of Pulse II is along a 23 km path starting from a depth of 4300 m (hereinafter referred to as the deep source path). For this pulse we consider two charge weights, 0.818 kg and 22.7 kg TNT, for each of which the energy spectrum of the received signal is calculated. The "true" energy spectrum, obtained by including finite-amplitude effects over the entire path, is compared with spectra obtained by neglecting finite-amplitude effects (1) entirely, (2) after the first 150 m, and (3) after the first 1100 m. Finite-amplitude effects are found to be of small consequence in the case of the 0.818 kg TNT explosion for frequencies below 6 kHz at distances beyond 1100 m. For the 22.7 kg explosion the corresponding quantities are 4 kHz and 1100 m.

NUMERICAL IMPLEMENTATION OF NONLINEAR GEOMETRICAL ACOUSTICS

The equations of nonlinear geometrical acoustics have been developed by Ostrovsky et al. (1975; 1976) and Pelinovsky et al. (1979) for the case in which losses are restricted to dissipation at shocks. (See also Cotaras and Blackstock (1986).) However, ordinary absorption, caused by viscosity and relaxation, and the associated dispersion may also be important because of their effects on the continuous section of the wave. For convenience we use the term "lossless ocean" to mean that shock losses are accounted for. "Lossy ocean" then implies the presence of ordinary losses and dispersion as well as shock losses.

---

†On temporary assignment from Defence Research Establishment Atlantic, PO Box 1012, Dartmouth, NS, B2Y 3Z7, Canada.

*Permanent address: Institute of Sound and Vibration Research, The University, Southampton, SO9 5NH, England.

A propagation algorithm based on nonlinear geometrical acoustics has been developed. For the algorithm, the key points of nonlinear geometrical acoustics are as follows. (1) The ray paths for finite-amplitude signals are the same as their small-signal counterparts. A computer ray model based on linear geometrical acoustics may therefore be used to calculate the ray paths; we chose MEDUSA (Foreman, 1983). (2) The equation governing the propagation of finite-amplitude signals along the ray paths reduces to the relations of weak-shock theory. (See, for example, Cotaras (1985), Chapter 4.) We therefore used the algorithm developed by Pestorius (1973), which implements the relations of weak-shock theory for plane waves in a pipe. Several modifications were, of course, required; for details, see Cotaras (1985, Chapter 5). Our algorithm accounts for the ordinary losses and dispersion of the ocean. Relaxation absorption, due to both boric acid and magnesium sulfate, is calculated from the equations of François and Garrision (1982); dispersion is included by using the relations given by Blackstock (1985).

RESULTS

The effect of ocean inhomogeneity on finite-amplitude distortion may be assessed by comparing the waveform for a homogeneous ocean with that for a stratified ocean. Examine Fig. 1(a). Along with the initial waveform (Pulse I at its reference range, 0.4 m), two other waveforms are shown. Waveform 2 results from numerically propagating the initial signal 58.1 km along the shallow source path through a lossless stratified ocean; waveform 1 is obtained by numerically propagating the same signal the same distance through a lossless homogeneous ocean. Note that the strong effect of geometrical spreading has been removed. The differences in amplitude and relative shock arrival time between the two resultant waveforms are clearly small (Morfey, 1984; Cotaras et al., 1984).

Regardless of inhomogeneity, however, Fig. 1(a) shows the importance of nonlinearity. Finite-amplitude effects cause the wave to attenuate and to stretch. The peak pressure in both waveforms 1 and 2 is about one-third that of the initial waveform. The decrease (about 10 dB) is due entirely to finite-amplitude effects. Stretching has two practical effects: (1) the shock arrives approximately 165 μs (three times the initial 1/e decay time) earlier than would be predicted using linear theory, and (2) the relative amount of low frequency energy is increased.

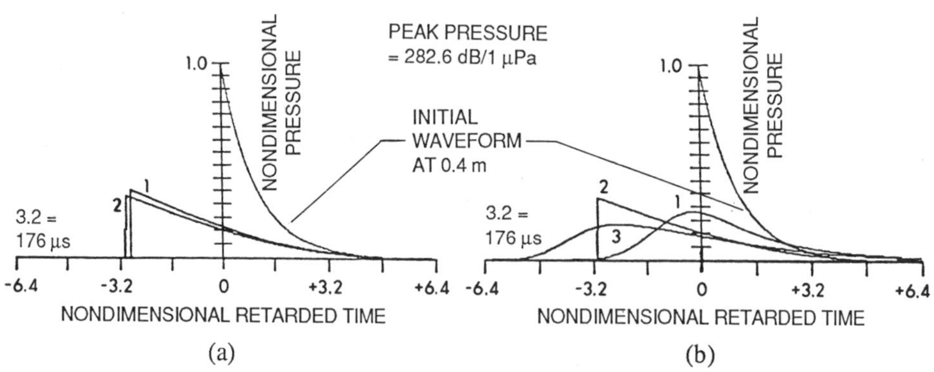

Fig. 1. (a) Effect of inhomogeneity on nonlinear distortion of Pulse I for a propagation distance of 58.1 km. Waveform 1, homogeneous ocean; waveform 2, stratified ocean. (b) Comparison of effects of nonlinearity and ordinary absorption (and dispersion) on Pulse I, for a propagation distance of 58.1 km. Waveform 1, ordinary attenuation and dispersion only; waveform 2, finite-amplitude effects only; waveform 3, finite-amplitude effects and ordinary attenuation and dispersion.

Combined Effects of Nonlinearity and Absorption

Throughout the propagation of a finite-amplitude wave, the effects of both nonlinearity and ordinary attenuation and dispersion are at play. To more clearly delineate the role of each, we examine them first separately and then together. Figure 1(b) shows four waveforms, one of which is the initial waveform, Pulse I at 0.4 m. The others are the waveforms at the end of the shallow source path. Included in the computations are ordinary attenuation and dispersion only (waveform 1), finite-amplitude effects only (waveform 2), and both finite-amplitude effects and ordinary attenuation and dispersion (waveform 3). It is clear from Fig. 1(b) that by itself neither finite-amplitude effects nor ordinary absorption can correctly account for the shape of the resulting waveform; both are required. The finite-amplitude effects try to maintain the initial shock while stretching and attenuating the wave. Ordinary absorption tries to round the initial shock while attenuating the wave.

It is interesting to examine the effect of dispersion on the position of the peak in the waveform. Consider the propagation, neglecting finite-amplitude effects, of Pulse I along the shallow source path. The waveforms in Figs. 2(a) and 2(b) are obtained at various positions along the shallow source path, from 0.4 m, the reference range, out to 58.1 km. (Waveforms at ranges between 0.4 m and 430 m are not shown because they are not resolved.) The calculations were made first without dispersion (Fig. 2(a)), and then with dispersion (Fig. 2(b)). At longer ranges the effect of dispersion is clear; it shifts the waveform forward. In Fig. 2(a) the peak pressure continuously moves backward; i.e., the effective propagation speed of the peak is less than c_0. The same is true of the waveforms in Fig. 2(b) for distances up to 8.6 km, but beyond that distance dispersion pulls the peak forward. By the time the wave has reached 58.1 km, the shift of the peak pressure due to dispersion is approximately 0.8 of the initial 1/e decay time. This would be a significant timeshift if one were trying to add signals coherently.

In Fig. 3 we again examine the effect of dispersion but this time with finite-amplitude effects included. In this case the waveforms at the lower ranges are resolved. Notice that in Fig. 3(a) the peak pressure moves forward until a range of 3300 m is reached and then moves backwards. In Fig. 3(b), however, the forward movement of the peak pressure is monotonic. The forward movement is at first due to finite-amplitude effects and in the end due to dispersion.

Figure 3(a) also enables us to examine the combined effects of nonlinearity and attenuation. As noted above, the peak pressure in Fig. 3(a) stops moving forward as the signal propagates beyond 3300 m. One may interpret this to indicate a change in the importance of finite-amplitude effects relative to that of attenuation. From 3300 m on, ordinary attenuation is the dominant mechanism of diminution, whereas for ranges less than 3300 m finite-amplitude effects are the principal mechanism. It is noted, however, that, judging by the peak pressure, the amount of attenuation beyond 3300 m indicated in Fig. 3(a) is less than that over the corresponding distance in Fig. 2(a). A possible

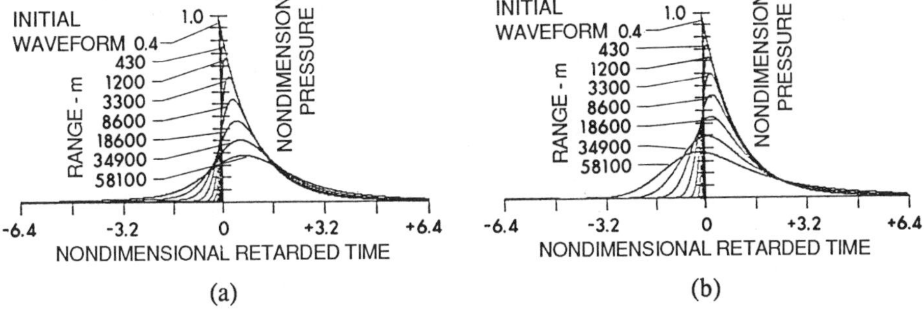

Fig. 2. Effect of dispersion on Pulse I at various ranges, finite-amplitude effects neglected. (a) Dispersion neglected, (b) dispersion included.

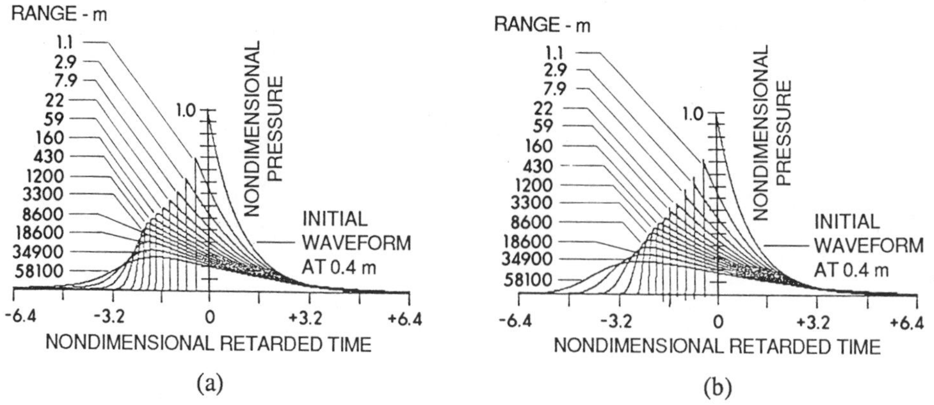

Fig. 3. Same as Fig. 2 but with finite-amplitude effects included.

explanation is as follows. Recall that the finite-amplitude stretching which has occurred up until 3300 m increases the relative amount of low frequency energy. The signal is therefore subject to attenuation at a lower rate than it would be if the stretching had not occurred. Thus, even though finite-amplitude effects are not dominant beyond 3300 m, their residual effect is noticeable. Although the transition point, 3300 m, applies only to this particular example, one can expect a similar behavior for other signals of similar initial shape.

Beyond What Distance Are Finite-Amplitude Effects Unimportant?

To answer the question posed in the title, we conducted a numerical experiment in which Pulse II signals from 0.818 kg and 22.7 kg TNT explosions were numerically propagated to a distance of 23 km along the deep source path. Attenuation and dispersion were accounted for over the entire 23 km, whereas nonlinear effects were accounted for as follows.

Case A: nonlinear effects neglected entirely,
Case B: nonlinear effects included only up to range 150 m,
Case C: nonlinear effects included only up to range 1.1 km,
Case D: nonlinear effects included for the entire 23 km.

Case D is used as the basis of comparison. Because the effective duration of Pulse II signals (shown in the inserts of Fig. 4) is much greater than that of Pulse I signals, time waveform resolution of the sort shown in Figs. 1-3 is not possible. Interesting results may, however, be found by comparing the spectra of the signals. The results are therefore presented in the form of energy spectrum plots. A few time waveforms are shown for clarity in interpreting changes in spectra.

The 0.818 kg TNT Explosion Results. The results from the 0.818 kg TNT explosion are presented in Figs. 4 and 5. The dotted curve in Fig. 4 is the initial energy spectrum at the reference range, 0.4 m. The dashed curve is the energy spectrum for Case A. The solid curve in the figures is the energy spectrum for Case D. Figure 5 shows a comparison of the spectra for Cases B and D and for Cases C and D. In all instances the effect of geometric spreading has been removed.

First examine Fig. 4. A comparison of the curves for Case D and Case A is itemized below.

(1) At the high frequency end (above 15 kHz) the solid curve is higher.
(2) In the middle range (approximately 1.5 - 15 kHz) the dashed curve is higher.
(3) At the low frequency end (below 1.5 kHz) the envelopes of the two curves are about the same.
(4) As the frequency decreases from about 400 Hz, the solid curve rises above the dashed curve.
(5) In the low frequency region the spectral peaks of the solid curve occur at slightly lower frequencies than the peaks of the dashed curve.

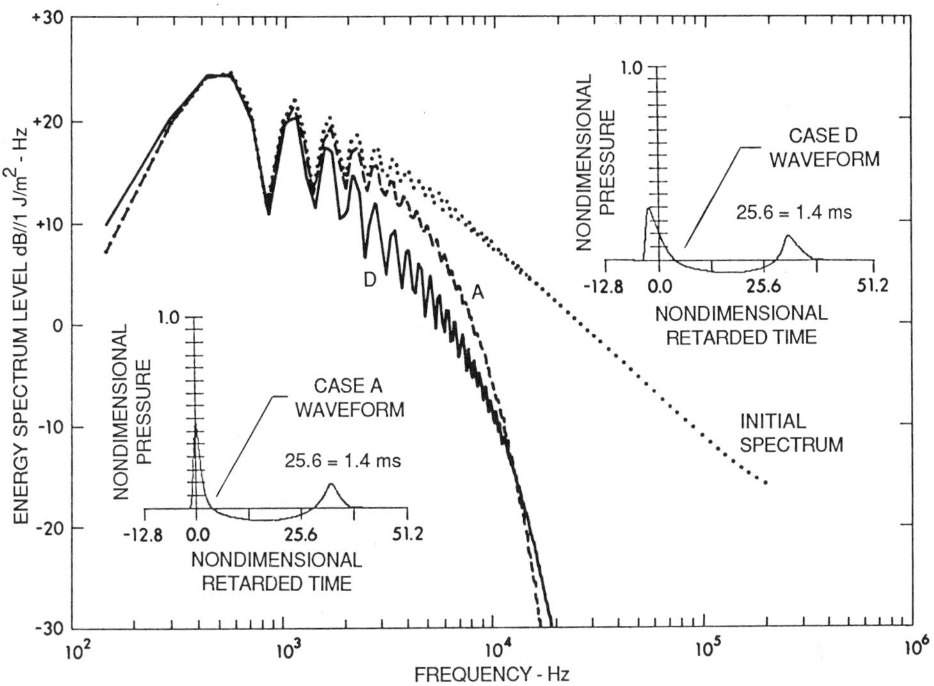

Fig. 4. Energy spectra of a 0.818 kg TNT explosion pulse (Pulse II) at the end of the deep source path. For Case A finite-amplitude effects are neglected, but for Case D they are included. For comparison the spectrum at the reference range (0.4 m) is also shown.

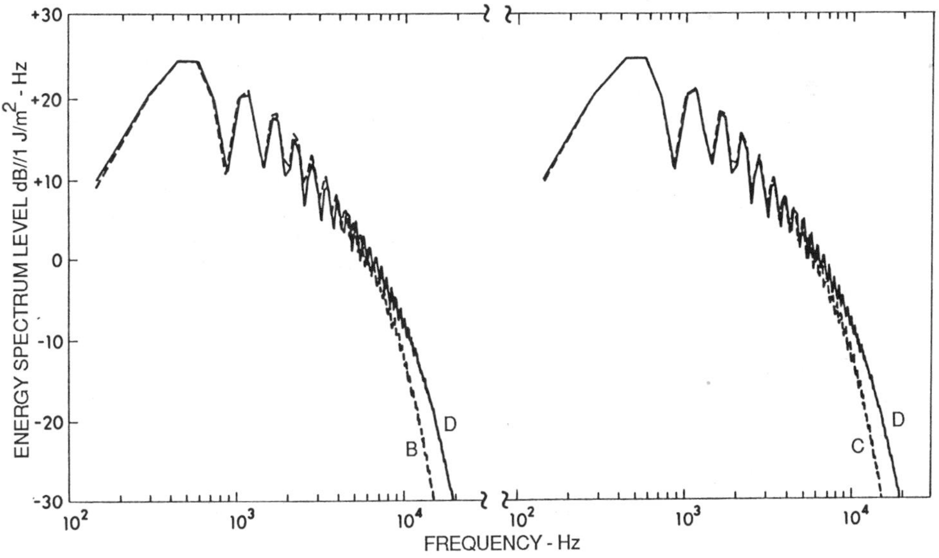

Fig. 5. Energy spectra of 0.818 kg TNT explosion pulse (Pulse II) at the end of the deep source path. Finite-amplitude effects are included in Case B over the first 150 m, in Case C over the first 1100 m, and in Case D over the entire 23 km.

The differences may be explained by considering the effects of nonlinear distortion. Compare the two time waveforms inserts in Fig. 4. Difference (1) is probably caused by the steepening of the compression part of the bubble pulse and the resistance of the shock to rounding. The same relative rise of the high frequency spectrum has been observed in propagation of intense noise in the atmosphere (Webster and Blackstock, 1978). Difference (2) is probably due to the increase in the decay time of the first peak (as the shock pulls ahead of the first zero). Differences (4) and (5) are probably due to the stretching of the time interval between the initial peak and the bubble pulse peak. The discussion of Fig. 5 is postponed for the moment.

The 22.7 kg TNT Explosion Results. The results from the 22.7 kg TNT explosion are presented in Figs. 6 and 7. As can be seen by comparing the initial energy spectra in Figs. 4 and 6, the energy spectra of the two explosions are very similar except that the larger explosion has an overall higher spectrum level and is shifted down in frequency. To avoid peak pressures too high to be handled correctly by weak-shock theory, we increased our reference range to 1.1 m (for the 22.7 kg pulse only). The choice of 1.1 m makes the peak pressure at the reference range the same for both explosions (282.6 dB//1 μPa). Because the only difference is the frequency shift of the spectrum, the nonlinear distortion of the two pulses is much the same (finite-amplitude effects scale with frequency). The only real difference is the effect of attenuation, which is less for the pulse from the larger charge because of its lower center frequency.

The general observations made about the solid and dashed curves in Fig. 6 are the same as those made about Fig. 4, except that the frequencies cited previously are higher. The differences between the solid and dashed curves of Fig. 6 may be summarized as follows.

(1) Above approximately 10 kHz the solid curve is higher.
(2) From about 500 Hz to 10 kHz the dashed line is higher.
(3) The solid curve is slightly higher from 50 to 500 Hz.
(4) Over the same frequency range as (3) the spectral peaks of the solid curve exhibit a slight downward shift.

The physical explanations for these differences are the same as those for the differences noted in Fig. 4.

Using Figs. 5 and 7, we now attempt to answer the question which is posed in the title by comparing the Case B and Case C energy spectra with that of Case D. For the 0.818 kg TNT explosive, the inclusion of finite-amplitude effects up to 150 m (Fig. 5, Case B) gives a 23 km spectrum that follows the Case D spectrum up to about 6 kHz. Less important effects are seen at the very low frequencies. The Case C spectrum also follows the Case D spectrum up to about 6 kHz. Since the spectra for Cases B and C are so similar, it is concluded that, for this charge weight, the finite-amplitude effects incurred between 150 and 1100 m are quite small. For the 22.7 kg TNT explosive, the Case B and Case C spectra follow the Case D spectrum (see Fig. 7) closely for frequencies below 4 kHz. It is therefore concluded that finite-amplitude effects can be neglected after a certain distance, and that the distance depends on both frequency and source strength. Ordinary absorption appears to dominate the propagation beyond that distance. Since the differences between the Case C and Case D spectra are approximately the same for both the 0.818 kg and 22.7 kg TNT explosions, it is also concluded that the differences between the two spectra due to absorption are small.

In summary the prevailing sentiment that "nonlinear effects are important only close to the source" is confirmed quantitatively. Notice, however, that "close to the source" is a relative restriction which depends not only on the charge weight but on frequency as well. Also note that the calculations are for rays that encounter neither reflections nor caustics.

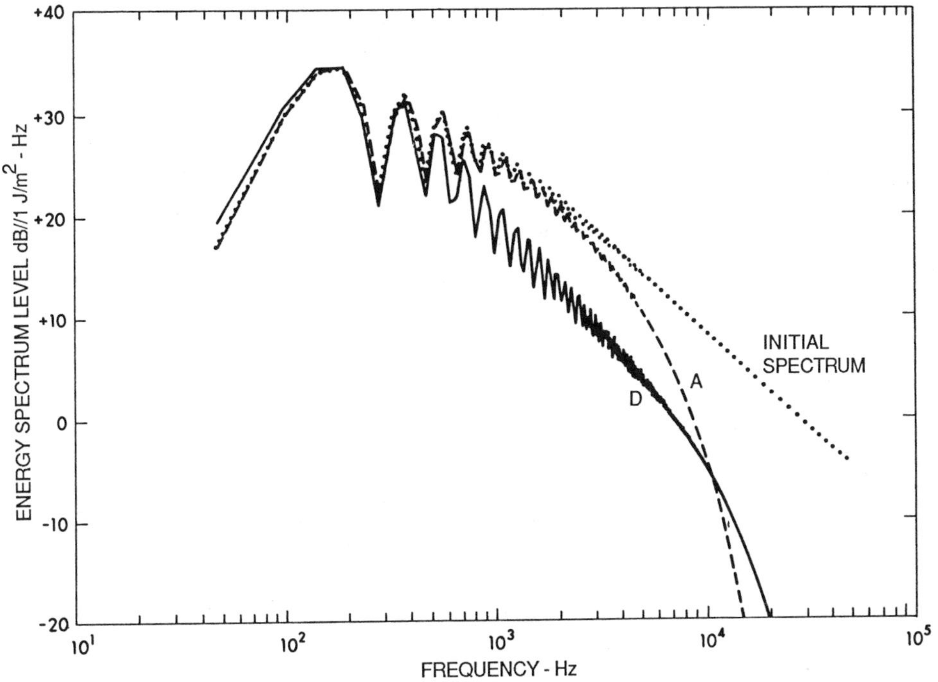

Fig. 6. Energy spectra of a 22.7 kg TNT explosion pulse (Pulse II) at the end of the deep source path. For Case A finite-amplitude effects are neglected, but for Case D they are included. For comparison the spectrum at the reference range (1.1 m) is also shown.

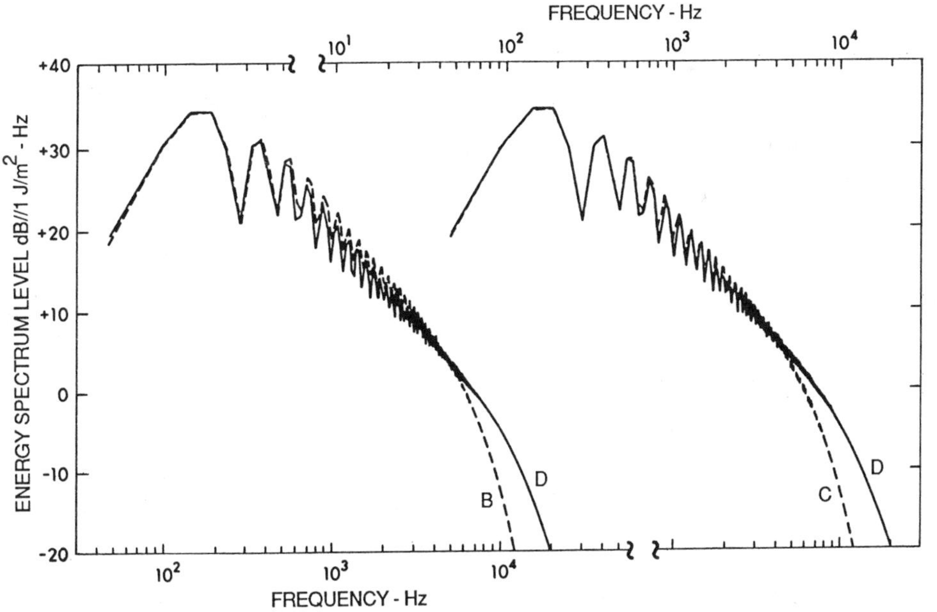

Fig. 7. Energy spectra of 22.7 kg TNT explosion pulse (Pulse II) at the end of the deep source path. Finite-amplitude effects are included in Case B over the first 150 m, in Case C over the first 1100 m, and in Case D over the entire 23 km.

ACKNOWLEDGMENT

The support of the U.S. Office of Naval Research is gratefully acknowledged.

REFERENCES

Blackstock, D. T., 1985, J. Acoust. Soc. Am. 77: 2050-2053.
Cotaras, F. D., 1985, Tech. Rept. ARL-TR-85-32, Applied Research Laboratories, The University of Texas at Austin (ADA 166 492).
Cotaras, F. D., Morfey, C. L., and Blackstock, D. T., 1984, J. Acoust. Soc. Am. 76: S39(A).
Cotaras, F. D., and Blackstock, D. T., 1986, Proc. 12th Intern. Cong. Acoust., Toronto, Canada.
Foreman, T. L., 1983, Tech. Rept. ARL-TR-83-41, Applied Research Laboratories, The University of Texas at Austin (ADA 137 202).
François, R. E., and Garrision, G. R., 1982, J. Acoust. Soc. Am. 72: 1879-1890.
Morfey,C. L., 1984, Tech. Rept. ARL-TR-84-11, Applied Research Laboratories, The University of Texas at Austin (ADA 145 079).
Ostrovsky, L. A., Pelinovsky, E. N., and Fridman, V. E., 1975, Proc. 6th Intern. Symp. Nonlinear Acoust., Moscow, USSR (Moscow University Press), Vol. 1, pp. 342-353.
Ostrovsky, L. A., Pelinovsky, E. N., and Fridman, V. E., 1976, Sov. Phys.-Acoust. 22: 516-520.
Pelinovsky, E. N., Petukhov, Yu. V., and Fridman, V. E., 1979, Izv., Acad. Sci. USSR, Atmos. Oceanic Phys. 15: 299-304.
Pestorius, F. M., 1973, Tech. Rept. ARL-TR-73-23, Applied Research Laboratories, The University of Texas at Austin (ADA 778 868).
Webster, D. A., and Blackstock, D. T., 1978, NASA CR 2992, Applied Research Laboratories, The University of Texas at Austin (N78-31876).

ATTENUATION OF LOW-FREQUENCY SOUND IN THE SEA: RECENT RESULTS

D. G. Browning† and R. H. Mellen*

†Naval Underwater Systems Center
New London, CT 06320

*Planning Systems Incorporated
New London, CT 06320

INTRODUCTION

Acoustic propagation loss is a critical factor in the design and performance of sonar systems. Absorption by the medium is generally the dominant attenuation mechanism in the sonar equation and is the subject of this paper.

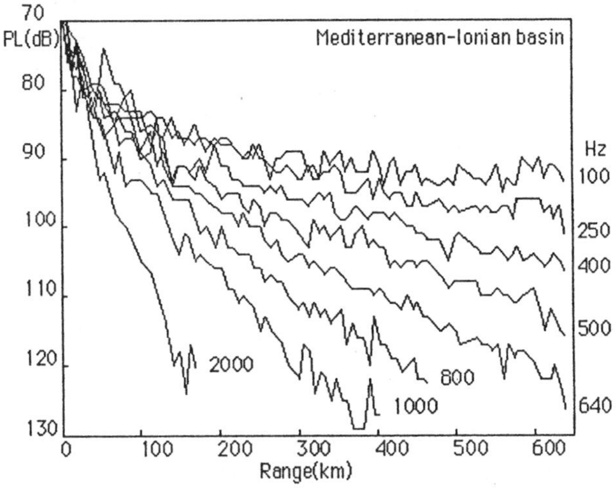

Fig. 1. Mediterranean propagation loss data.

Absorption tends to increase as the square of the frequency, which limits the maximum useable frequency for a given range. The situation is illustrated by the results of a recent sound-channel experiment in the Mediterranean Sea.[1] Figure 1 shows curves of propagation loss in dB//1m vs. range in kilometers at selected frequencies. The rapid increase in loss with increasing frequency is entirely due to absorption by the medium. The magnitude depends on a number of environmental factors, the most important of which has been found to be the pH value.

See page 810 for Abstract.

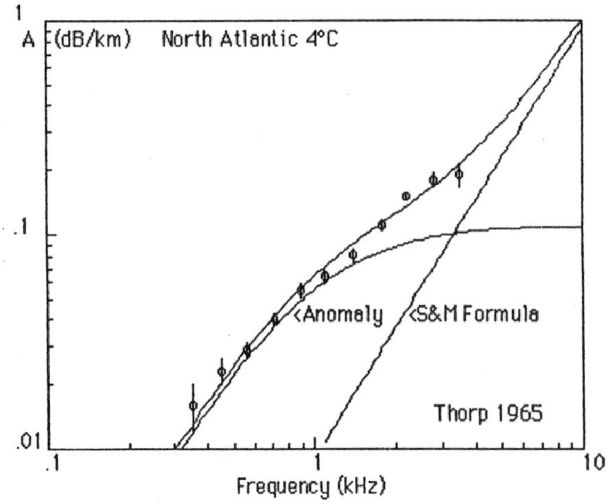

Fig. 2. Thorp's data and 2-component model.

In 1965, Thorp[2] carried out a sound-channel propagation experiment in the Bermuda-Eleuthera area over a 500 km track. Measurements were made near the sound-channel axis using explosive sources. The attenuation data shown in Figure 2 indicated a low-frequency anomaly. Comparison with the similar Mediterranean results of Leroy[3] and other data confirmed the conclusion that values below 1 kHz are an order of magnitude greater than predicted by the magnesium sulfate formula of Schulkin and Marsh.[4]

In a later paper, Thorp[5] modeled the anomaly by addition of a 1 kHz relaxation component to the S&M absorption formula. The magnesium sulfate relaxation frequency is 50 kHz or higher and, for frequencies below 10 kHz, the Thorp formula can be approximated as:

$$A = 0.009\ f^2 + 0.11\ f_r\ f^2/(f_r^2 + f^2)\ \text{dB/dm} \qquad (f_r = 1\ \text{kHz})$$

where f is the frequency in kHz and f_r is the relaxation frequency of the anomaly.

Based on the successful curve-fit, it appeared that the mechanism might be another chemical relaxation similar to magnesium sulfate but with a much lower relaxation frequency. T-jump relaxation frequency measurements by Yeager et. al.[6] indicated that the sea water constituent responsible to be boric acid. Resonator experiments by Simmons[7] succeeded in showing that the absorption also has the proper magnitude; however, the exact nature of the reaction remained uncertain. Later resonator experiments by Mellen et. al.[8] identified the reaction as an acid/base exchange between boric acid and carbonate, which Yeager et. al. had originally proposed. A relaxation of magnesium carbonate was also found to play a minor but significant part in sea water absorption and is included in the new model.

Concentrations of carbonate and borate increase approximately as 10^{pH} in the sea water range; therefore the absorption factors for both the boric acid/carbonate and magnesium carbonate relaxations will have the same dependence in the new model.

$$A = A_1(MgSO_4) + A_2(B(OH)_3) + A_3(MgCO_3)$$

$$A_n = a_n f^2 f_n / (f^2 + f_n^2)$$

$$a_1 = 0.5 \times 10^{-D(km)/20} \qquad f_1 = 50 \times 10^{T/60}$$

$$a_2 = 0.1 \times 10^{(pH-8)} \qquad f_2 = 0.9 \times 10^{T/70}$$

$$a_3 = 0.03 \times 10^{(pH-8)} \qquad f_3 = 4.5 \times 10^{T/30}$$

Atlantic 4°C pH 8.0
$A = 0.007f^2 + 0.1\,f^2/(1+f^2) + 0.18\,f^2/(6^2+f^2)$
N.Pacific 4°C pH 7.7
$A = 0.007f^2 + 0.05f^2/(1+f^2) + 0.09\,f^2/(6^2+f^2)$
Mediterranean 14°C pH 8.3
$A = 0.006f^2 + 0.26f^2/(1.4^2+f^2) + 0.78\,f^2/(12^2+f^2)$
Red Sea 22°C pH 8.2
$A = 0.005f^2 + 0.27f^2/(1.8^2+f^2) + 1.1f^2/(24^2+f^2)$
sub-Arctic -1.5°C pH 8.2
$A = 0.01f^2 + 0.14f^2/(0.85^2+f^2) + 0.2\,f^2/(4^2+f^2)$

Fig. 3. Simplified absorption formulae.

Although all three relaxations involved are now quite well understood, modeling absorption on purely chemical grounds is far too complex and the accuracy would be limited by the supporting laboratory measurements as well. Certain simplifications are possible because the range of the environmental parameters is so small. For a practical model, only factors critical to prediction within reasonable error limits need be retained.

The three relaxations involved have been measured in the laboratory, both independently and in combination in order to determine the reaction parameters. Results indicated that relaxation frequency depends only on temperature while magnitude depends only on pH and that all three components are simply additive.

The proposed simplified formula for the 3-relaxation model shown in the top box of Figure 3 is based on analysis of both laboratory and field data. The dominant feature of the model is the pH dependence of the boric acid and magnesium carbonate components. Values of pH in the World Ocean vary over a range of roughly 7.7 and 8.3, which corresponds to an absorption ratio of 4/1 at lower frequencies. The value pH=8.0 corresponds to Thorp's experiment and is used as a convenient reference value.

The S&M formula for magnesium sulfate has been modified according to the more recent experiments by Fisher and Simmons[9] and their pressure factor has been incorporated in the model as the depth term in D(km). Depth dependencies of the other relaxations are not yet known; however, measurements in both deep and shallow channels indicate that boric acid effects may be negligible. Magnesium carbonate effects may be greater but can be neglected because its contribution is so small. Salinity variations in the World Ocean are small and are neglected in order to simplify the model.

Specific values for several areas are shown in the bottom box.

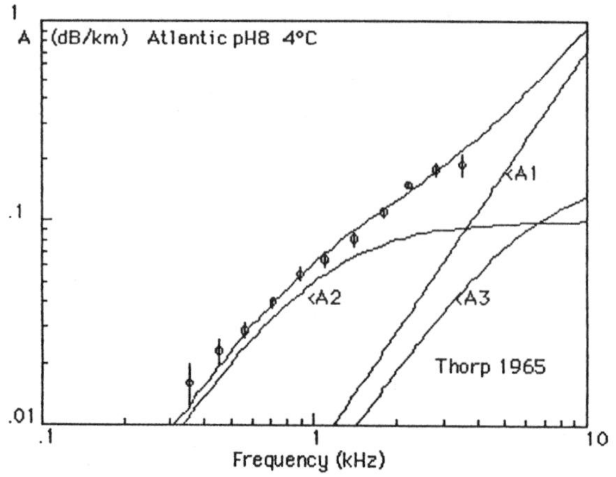

Fig. 4. Thorp's data and model.

Figure 4 shows Thorp's data compared to the 3-component model. The contributions of the components magnesium sulfate (A1), boric acid (A2) and magnesium carbonate (A3) are identified and the top curve is the sum. The fit to the data is equally as good as that of Figure 2.

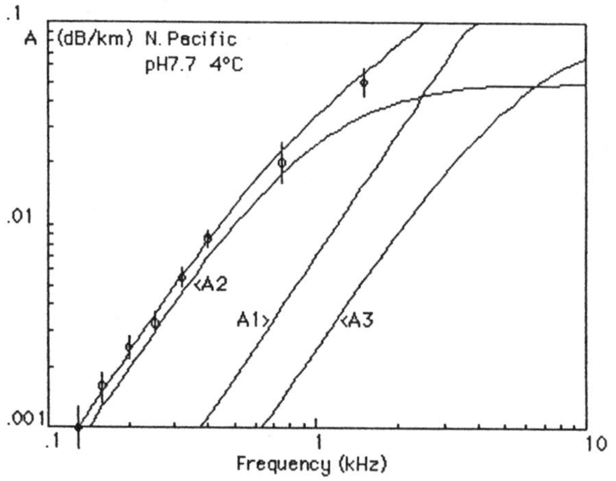

Fig. 5. North Pacific data and model.

In the North Pacific[10] case shown in Figure 5, the lower pH value reduces both the boric acid (A2) and the magnesium carbonate (A3) coefficients by a factor of two compared to Figure 4. Relaxation frequency depends only on temperature and remains the same. (Note the change in scale of the ordinate.)

Both cases are representative of ocean sound-channel measurements at mid-latitudes where the axis depth is roughly 1 km. In the North Pacific, the pH is uniformly lower than in the North Atlantic by approximately 0.3 units, accounting for the difference in the absorption spectra. This is not the case at higher latitudes where the axis rises to the surface where pH is uniformly high.

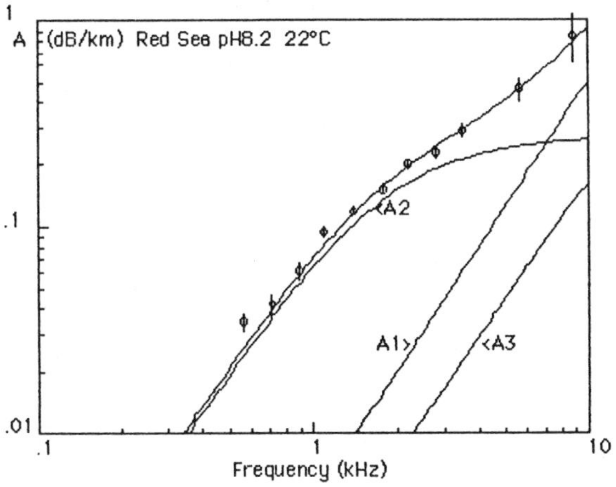

Fig. 6. Red Sea data and model.

The Red Sea[10] data were originally fitted with the 2-relaxation model and the results indicated a low-frequency relaxation at 1.5 kHz compared to the formula value 1.8 kHz. Figure 6 shows that an even better fit is achieved with the 3-component model using the latter value.

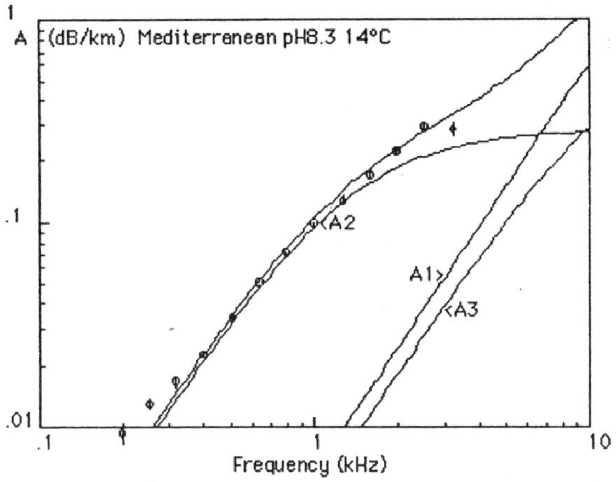

Fig. 7. Mediterranean data and model.

The Mediterranean experiment (Figure 1) was carried out in the deep Ionian basin between Sicily and Crete in July 1982 by SACLANTCEN.[1] In Figure 7 the attenuation data are compared to model using the measured value pH=8.3 in the calculation.

This is the only case so far in which the pH measurements were included in the experiment. All other values are estimates from archival data.

In both cases, the sound-channel axis depths are only 100-200m, which is typical of confined regions not subject to the circulation of the deeper cold water masses. In these regions, the pH tends to be uniformly high and the absorption coefficients are consequently greater than normal. However, the temperature is also greater, and the increased relaxation frequency reduces the differences at low frequencies.

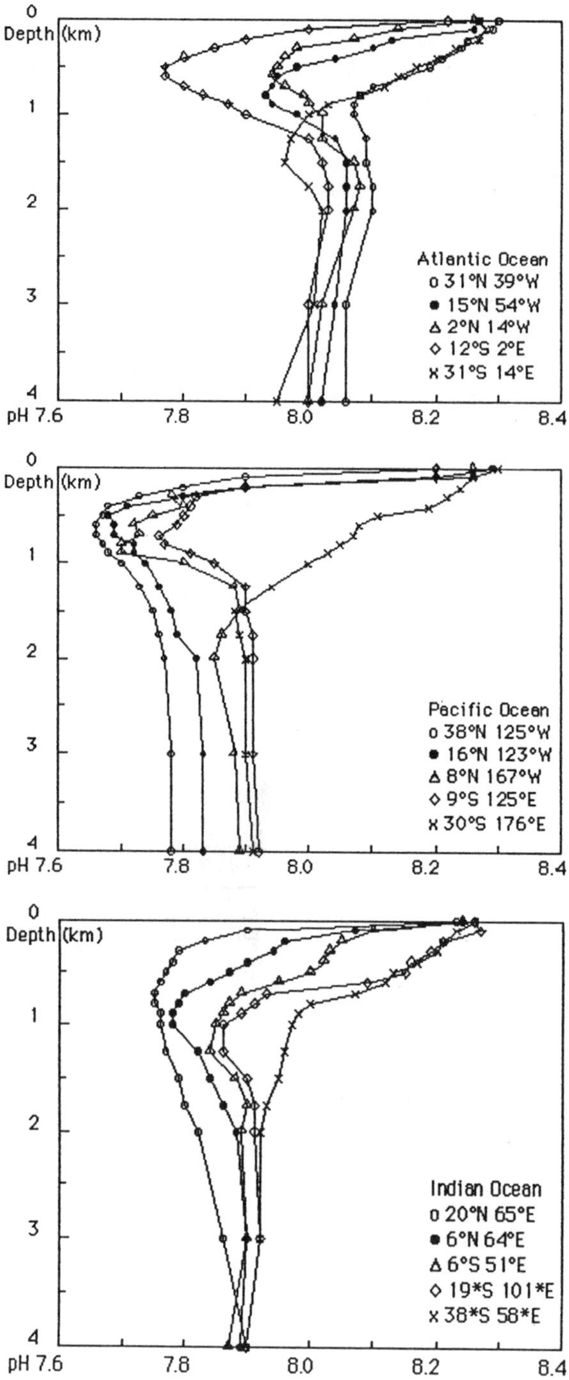

Fig. 8. PH profiles.

Figure 8 shows typical pH profiles for the Atlantic, Pacific and Indian Oceans from the GEOSECS[11] reports. Depth variability means that the net absorption will depend on the ray paths. However, for sound channel propagation, axial values give a reasonably accurate approximation.

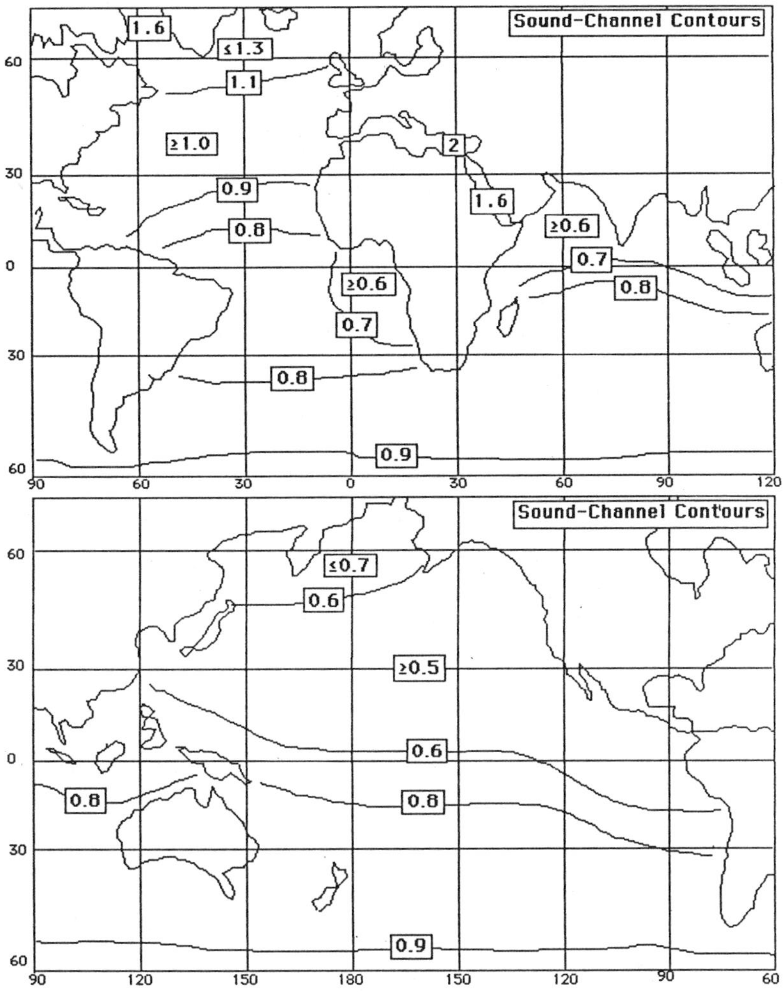

Fig. 9. K contours on the sound-channel axis.

After the pH dependence of the boric acid relaxation had been established by field measurements,[12] Lovett[13] published contour maps for the absorption coefficient on the sound-channel axis. The maps were based on pH contours at several depths reported in the World Ocean Atlas.[14] Figure 9 shows similar contours in terms of the factor $K=10^{(pH-8)}$ for the 3-component model. The contours have been modified by taking the GEOSECS data into account. Any discrepancies between the two have been resolved by adjusting the contours to minimize error. Considerable subjectivity is obviously involved; however, the new contours are entirely consistent with all available attenuation data and should give reasonably accurate predictions for pH dependence in regions where no data currently exist.

Note that the K values are multiplying factors for the absorption coefficients of both boric acid and magnesium carbonate but not that of magnesium sulfate. The net effect is that the full variability is realized only at very low frequencies and decreases rapidly with increasing frequency. Figure 10 shows the effective values K* vs. frequency for K values between 0.5 and 2, approximately represents the extremes to be expected under any sound-channel conditions including surface ducts.

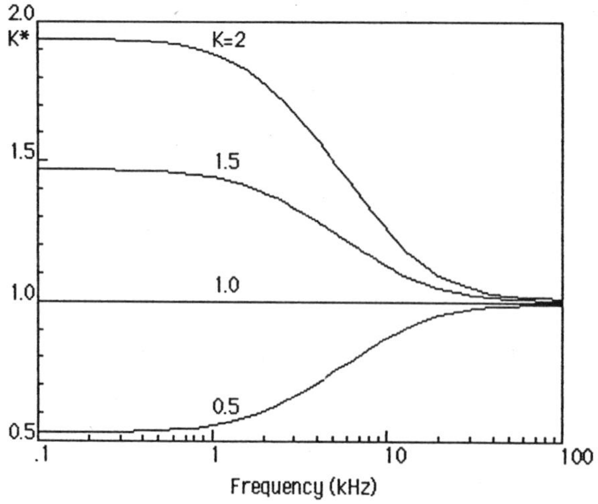

Fig. 10 Effective values K* vs. frequency.

REFERENCES

1. R. H. Mellen, T. Akal, E. H. Hug and D. G. Browning, Low-frequency sound attenuation in the Mediterranean Sea, JASA 78:S70 (1985).
2. W. H. Thorp, Deep ocean sound attenuation in the sub and low kilocycle-per-second region, JASA 38:648-654 (1965).
3. C. C. Leroy, Sound propagation in the Mediterranean Sea, in: "Underwater Acoustics," V. M. Albers, ed., Plenum (1967) Vol. 2, pp 203-241.
4. M. Schulkin and H. W. Marsh, Sound absorption in sea water, JASA 34:864-865 (1962).
5. W. H. Thorp, Analytic description of the low-frequency attenuation coefficient, JASA 42:270-271 (1967).
6. E. Yeager, F. H. Fisher, J. Miceli and R. Bressel, Origin of low-frequency sound absorption in sea water, JASA 53:1705-1707 (1973).
7. V. P. Simmons, "Investigation of the 1 kHz sound absorption anomaly in sea water," Ph.D. thesis, University of California, San Diego, CA (1975).
8. R. H. Mellen, D. G. Browning and V. P. Simmons, Investigation of chemical sound absorption in sea water by the resonator method, JASA Part I, 68:248-257 (1980); Part II, 69:1660-1662 (1981); Part III, 70:143-148 (1981); Part IV, 74:987-993 (1983).
9. F. H. Fisher and V. P. Simmons, Sound absorption in sea water, JASA 62:558-564 (1977).
10. "Attenuation of Low Frequency Sound in the Sea," NUSC Scientific and Engineering Studies, Volumes I & II (1981).
11. "GEOSECS Atlas, IDOE," Volume 1, Atlantic Ocean 1972-1973; Volume 3, Pacific Ocean 1973-1974; Volume 5, Indian Ocean 1977-1978.
12. R. H. Mellen and D. G. Browning, Variability of low-frequency sound absorption in the ocean: pH dependence, JASA 61:704-706 (1977).
13. J. R. Lovett, Geographic variation of low-frequency sound absorption in the Atlantic, Indian and Pacific Oceans, JASA 67:338-340 (1980).
14. "World Ocean Atlas," S. G. Gorshkov, ed., Pergamon Press, New York, Vol. 1, Pacific Ocean (1974); Vol. 2, Atlantic and Indian Oceans (1978).

ON THE CALCULATION OF ACOUSTIC INTENSITY FLUCTUATIONS CAUSED BY OCEAN CURRENTS

J.S. Robertson+, M.J. Jacobson*, and W.L. Siegmann*

+Department of Mathematics, U. S. Military Academy, West Point, NY, USA 10996-1788

*Department of Mathematical Sciences, Rensselaer Polytechnic Institute, Troy, NY, USA, 12180-3590

ABSTRACT

Ocean currents can cause significant and interesting effects on the intensity of underwater sound transmissions. We study this phenomenon via the parabolic approximation, beginning with conservation laws, and derive a family of equations, each of which is valid for different magnitudes of current speed, current gradient, and sound-speed variation. Numerical results indicate that some current structures can cause large variations in received intensity, and that substantial differences can occur in reciprocal transmissions. Current effects on intensity may be quite sensitive to the sound-speed distribution.

INTRODUCTION

Ocean currents cause interesting and significant effects on underwater sound. For example, in a time-independent ocean environment, currents cause reciprocity relations to fail. Based on ray theory, work has been done by the authors and others to model the influences of currents on acoustic transmissions. It has been shown, for instance, that certain current structures can cause large fluctuations in total-field intensity and per ray phase. These results are limited to high-frequency sound transmissions. To estimate current-induced effects for lower frequencies, a full-wave model should be used. One computational model is the parabolic approximation, which has been implemented using several algorithms, and is a particularly attractive method for efficiently generating transmission-loss calculations.

MODEL FORMULATION

In a complicated medium such as a moving ocean, it is not obvious how known parabolic equations should be modified to include current effects. For this reason, we systematically reformulated the governing time-dependent wave equation, starting from the conservation laws and state relations governing the ocean medium, and including medium motion (Robertson et al., 1985). If the current is assumed steady but non-uniform, additional terms occur in the wave equation. These terms depend on the current gradient and, as will be discussed below, can be significant in subsequent approximations.

We then assume that the sound source is time-harmonic and that the propagating wave is outgoing, thereby obtaining a reduced wave equation. After transforming coordinate systems, we invoke the far-field approximation, and proceed to generate a family of parabolic equations. Each of these equations depends on the relative sizes of three dimensionless parameters: a Mach number, the sound-speed deviation, and a shear number. The last parameter indicates the magnitude of current gradient, and also depends on source frequency. For example, a parabolic approximation appropriate for an isospeed sound channel, through which flows a steady depth-dependent current, is

$$2i\kappa_0\psi_r + \psi_{zz} - 2\kappa_0^2\left(\frac{u}{c_0}\right)\psi + \left(\frac{2}{c_0}\right)\left(\frac{du}{dz}\right)\psi_z = 0, \tag{1}$$

where κ_0 is a wave number, u is current speed, and ψ is an envelope of acoustic pressure in the far field. The reference sound speed c_0 in this isospeed case is equal to the sound speed c. Depth is indicated by z and range by r. It can be shown that range-dependent sound-speed profiles can be incorporated into these approximations in a straightforward way, provided that the horizontal gradient of sound speed is not large, which is often the case in many important ocean regions. For example, for a sound channel in which the sound-speed deviation is of the same order of magnitude as the Mach number, the appropriate equation is

$$2i\kappa_0\psi_r + \psi_{zz} + \kappa_0^2\,(n^2-1)\psi - 2\kappa_0^2\left(\frac{u}{c_0}\right)\psi + \left(\frac{2}{c_0}\right)\left(\frac{du}{dz}\right)\psi_z = 0, \tag{2}$$

where $n(r,z) = c_0/c$ is the index of refraction.

In Eqs. (1) and (2), the term which depends on the first derivative of current may or may not be retained, depending upon the size of the shear-number parameter. In the event that it is kept, the parabolic approximations are not in a "standard form"; that is, they cannot be solved directly with existing numerical implementations. However, it is possible to transform this family into related parabolic equations which <u>are</u> in a standard form. The transformed version of Eq. (2) is, for example:

$$2i\kappa_0\phi_r + \phi_{zz} + \kappa_0^2\,(n^2-1)\phi - 2\kappa_0^2\left(\frac{u}{c_0}\right)\phi - \frac{1}{c_0}\left(\frac{d^2u}{dz^2}\right)\phi - \frac{1}{c_0^2}\left(\frac{du}{dz}\right)^2\phi = 0, \tag{3}$$

where

$$\psi(z,r) = e^{-Mu(z)}\phi(z,r). \tag{4}$$

Note the appearance of two new terms in Eq. (3) which depend on the square of the derivative of current and the second derivative of current. The structure of Eq. (3) and other such equations suggests the use of an effective sound-speed profile (ESSP), which includes all sound-speed and current-related effects, and is used as the "actual" sound speed for numerical solution of the equations. For example, the ESSP corresponding to Eq. (3) is

$$\tilde{c} = c + u + \frac{1}{2\kappa_0^2\,c_0}\left(\frac{du}{dz}\right)^2 + \frac{1}{2\kappa_0^2}\left(\frac{d^2u}{dz^2}\right). \tag{5}$$

NUMERICAL RESULTS

To solve any of our parabolic equations numerically, we elect to use the IFD model developed by Lee and Botseas (1982). In the discussion below we consider several sound-speed profiles, together with current profiles, a

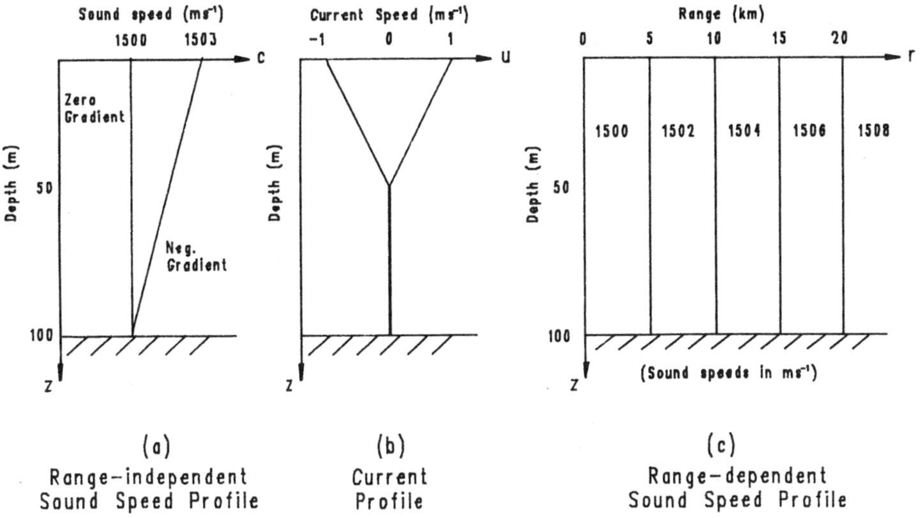

Fig. 1. Profiles of (a) range-independent sound speed, (b) current, and (c) range-dependent sound speed.

shown in Fig. 1. The zero-gradient and negative-gradient sound-speed profiles are sometimes range-independent as in Fig. 1(a), as are the current profiles in Fig. 1(b). The surface current may be either plus or minus 1 m/s. At other times, we employ a simple range-dependent sound-speed profile, Fig. 1(c), for which the horizontal gradient is both constant and small. Here, isopleths are vertical lines. We consider first the result of one calculation done in the isospeed channel, with surface current of magnitude 1 m/s. The source frequency is 200 Hz, for which it can be shown that current-gradient effects are negligible. The source and receiver depths are 25 m. In Figs. 2-6, the bottom acoustical properties are the same as in Robertson et al. (1985). Figure 2 shows a relative intensity in decibels versus range for three cases: no current present, a positive current in the source-receiver direction, and a negative current in the opposite direction. Several important current-related effects can be seen in the figure. When compared to the solid curve, representing intensity in the absence of any current, we see that a current with either direction can induce substantial variations in intensity. For example, with a positive current present, variations can exceed 10 dB over certain range intervals, such as those between 13 and 14 km. Similar behavior is seen for negative current. Current effects in both cases tend to increase with increasing range.

Intensity variations are highlighted in Fig. 3, which illustrates difference in relative intensity versus range for three cases: no current and positive current, no current and negative current, and positive and negative currents. Because source and receiver are at the same depth, this figure also illustrates one type of effect which may be seen in reciprocal transmissions (RTs). The intensity difference between positive and negative currents, indicated in Fig. 3 by the long-dashed curve, suggests that measurements of intensity variation between reciprocal source-receiver pairs may be very large. Near ranges 14 and 19 km, this difference attains a magnitude of nearly 20 dB. At other range intervals the difference is smaller, but significant. For example, between 10 and 12 km, the intensity difference is seen to generally be well over 4 dB. RT differences can also be significant in range-dependent channels. Using the sound-speed profile in Fig. 1(c), and the same source frequency, source-receiver depths, and current structure as above, the computed intensity difference between a source-receiver pair is shown in Fig. 4. Note that one effect of the range

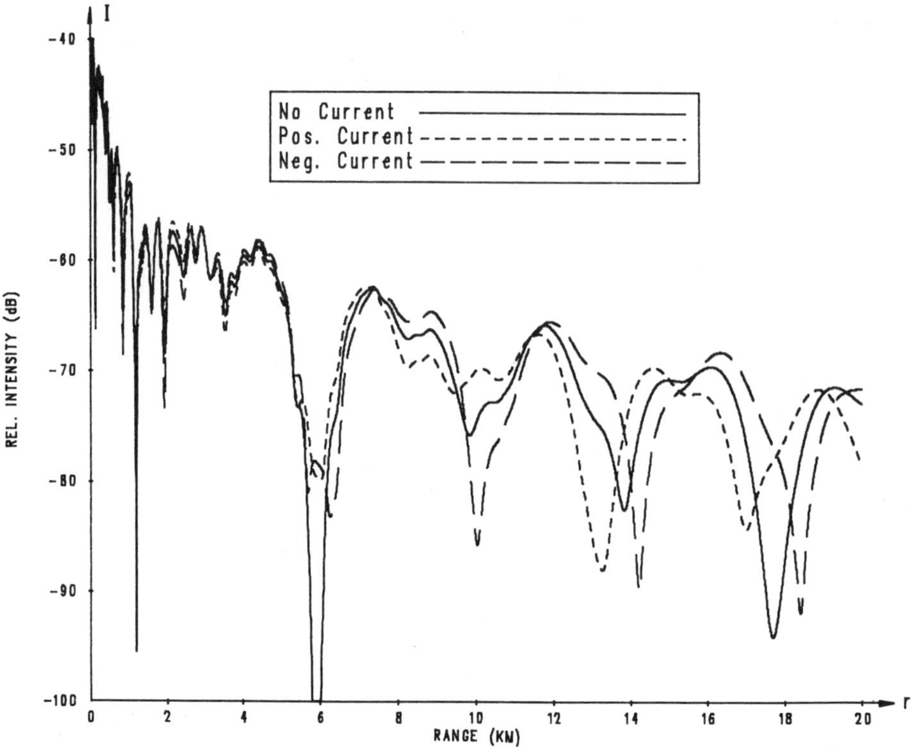

Fig. 2. Relative intensity versus range for three currents in an isospeed channel.

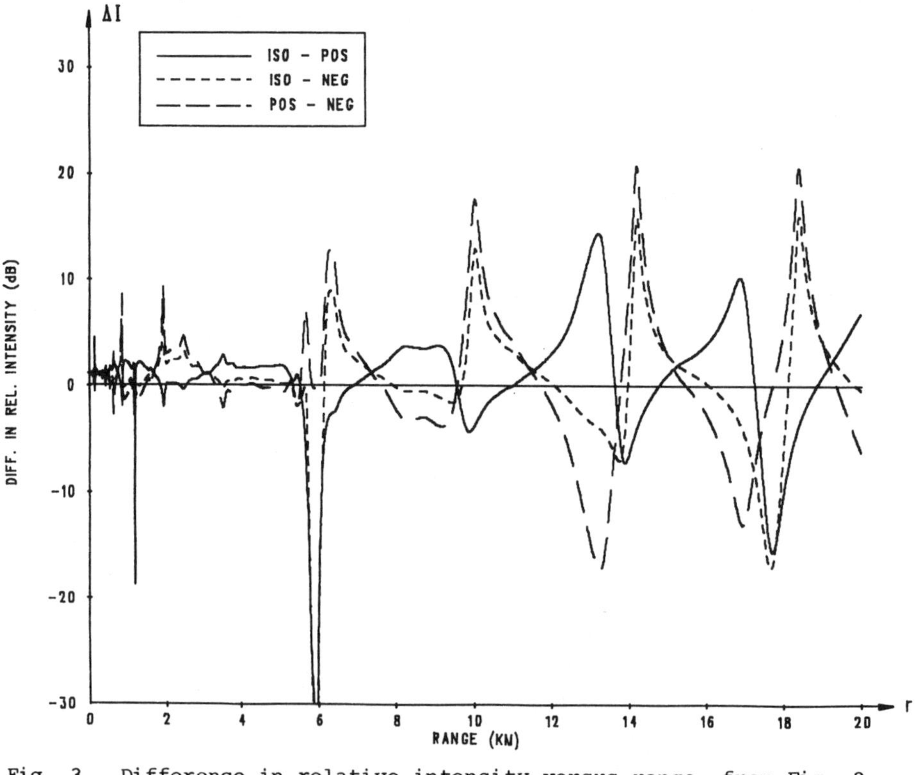

Fig. 3. Difference in relative intensity versus range, from Fig. 2. Oppositely-signed current effects are readily compared (long-dashed curve).

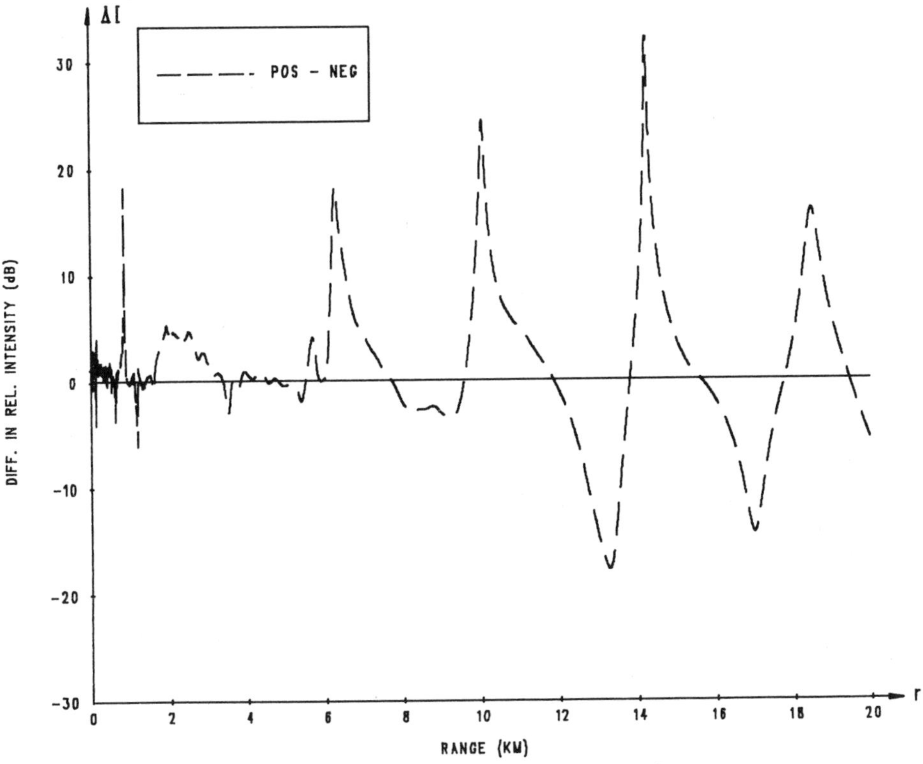

Fig. 4. Difference in relative intensity versus range, for current reversal. Reciprocal-transmission difference is shown for a range-dependent sound-speed profile.

variation in sound speed is to cause the relative intensity curve to shift toward the source, when compared to the analogous curve in Fig. 3. Maximum intensity values have been altered, also. For example, at a range 14 km, the peak difference is over 30 dB.

Another way to visualize the effects of reversing current direction is shown in Fig. 5. Here, differences in the two intensity functions are plotted as level curves in a portion of the depth-range plane. The difference is intensity for a positive current (in the source-receiver direction) minus intensity for a negative current, as in Fig. 1(b). The source frequency and source depth are again 200 Hz and 25 m, respectively. Contour intervals are 5 dB, with negative differences denoted by dotted curves. This figure illustrates the intensity differences that might be observed in a channel with tidal effects. At ranges larger than about 7 km, bottom attenuation has stripped away most higher modes, leaving a well-defined pattern of alternating intensity differences. Regions of large positive difference occur in finger-like patterns which alternately emanate from the channel surface and bottom. Similar structures are also seen for negative differences. In this example, a zone of very small differences extends in range across most of the channel at roughly mid-depth. Regions of maximum difference occur regularly above and below this zone. In contrast, the intensity-difference pattern in a negative gradient channel is noticeably different. Figure 6 shows the analogous level curves for the negative gradient profile of Fig. 1(a). The finger-like structures apparent in Fig. 5 have in some cases blended together in Fig. 6, leaving regions of high intensity difference located at many mid-depth points. The overall pattern is more complicated than the one present in the isospeed channel. Consequently, the intensity-difference pattern resulting from oppositely-signed currents

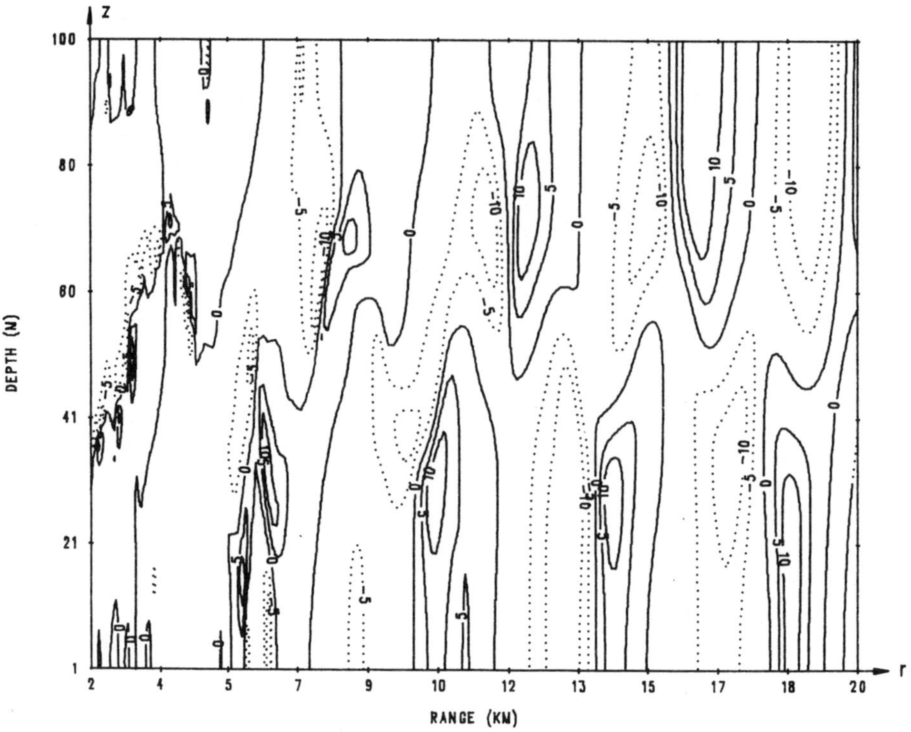

Fig. 5. Level curves of intensity difference in an isospeed channel, with current reversal.

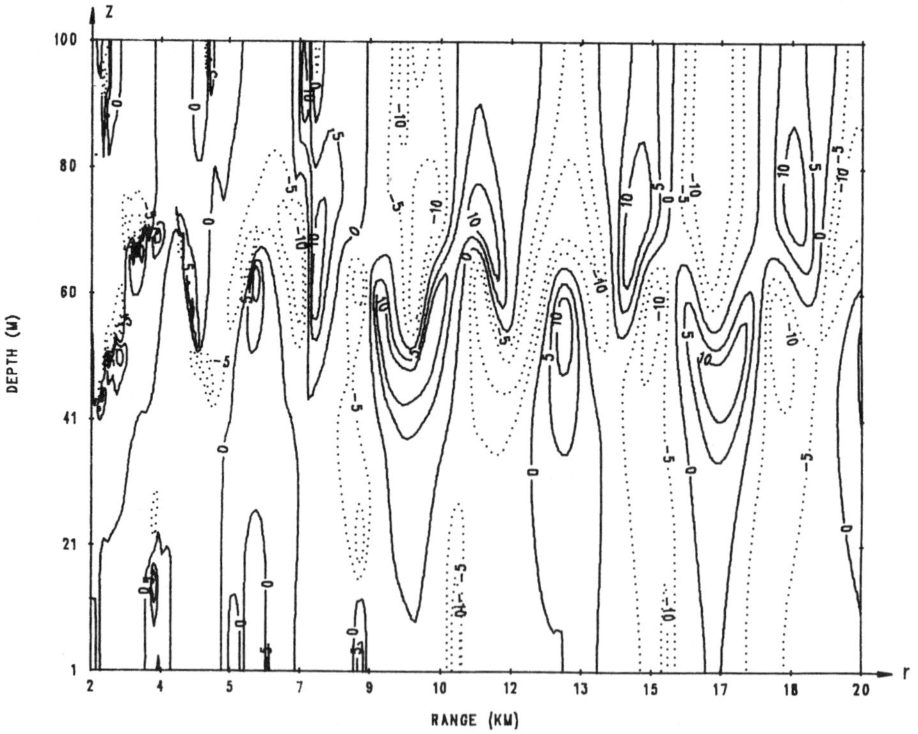

Fig. 6. Level curves of intensity difference for a negative gradient channel (see Fig. 1(a)), with current reversal.

appears to be very sensitive to these types of changes in the sound-speed profile.

For some current structures and lower source frequencies, the appropriate transformed parabolic equation will include new terms which depend on current concavity and (possibly) the square of current gradient. One type of current structure which may require additional terms is shown in Fig. 7(a). At the surface, the current speed is 1 m/s, and it decays to zero at the bottom. Note the appearance of several strong shear layers, particularly those at depths 35 and 60 m. The vertical shear structure seen here can be acoustically significant for sufficiently low source frequencies. For example, when the source frequency is 30 Hz, the ESSP is similiar to that given by Eq. (5) and is depicted in Fig. 7(b). In this example, concavity effects are significant, but shear effects can be neglected. Note that current concavity dominates the behavior of the ESSP. The current shear structure has introduced large rapid variations, one of which, at the depth 60 m, approaches 20 m/s in Fig. 7(b). For higher source frequencies, the magnitude of the variations decreases, yet may still be significant. We anticipate that this current structure can cause interesting acoustical effects.

In Fig. 8, we see one result of computations done with the current structure shown in Fig. 7(a). In order to observe concavity (or second derivative) effects, we solved the relevant parabolic equation, first with concavity included and then with concavity omitted. The root-mean-square difference of the intensities in the two cases, called J, was then calculated with a range averaging. Source and receiver depths are 25 m, and the source frequency is 100 Hz. The results for three different bottom types are shown. The rigid bottom is perfectly reflecting, while beneath the water column for the hard and soft bottoms is a second fluid layer with different sound speed and density. The hard bottom has larger discontinuities in these quantities than the soft bottom. For both hard and soft bottoms, a small amount of volume attenuation was introduced. As the bottom changes

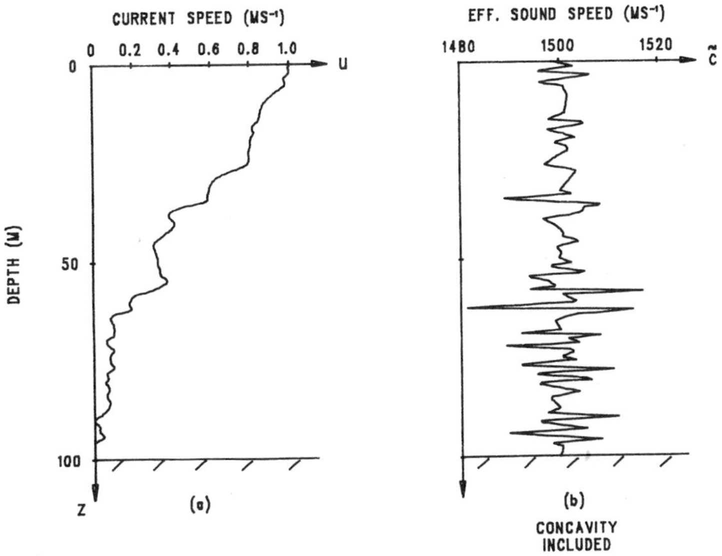

Fig. 7. Profiles of (a) a current with high shear, and (b) the effective sound speed.

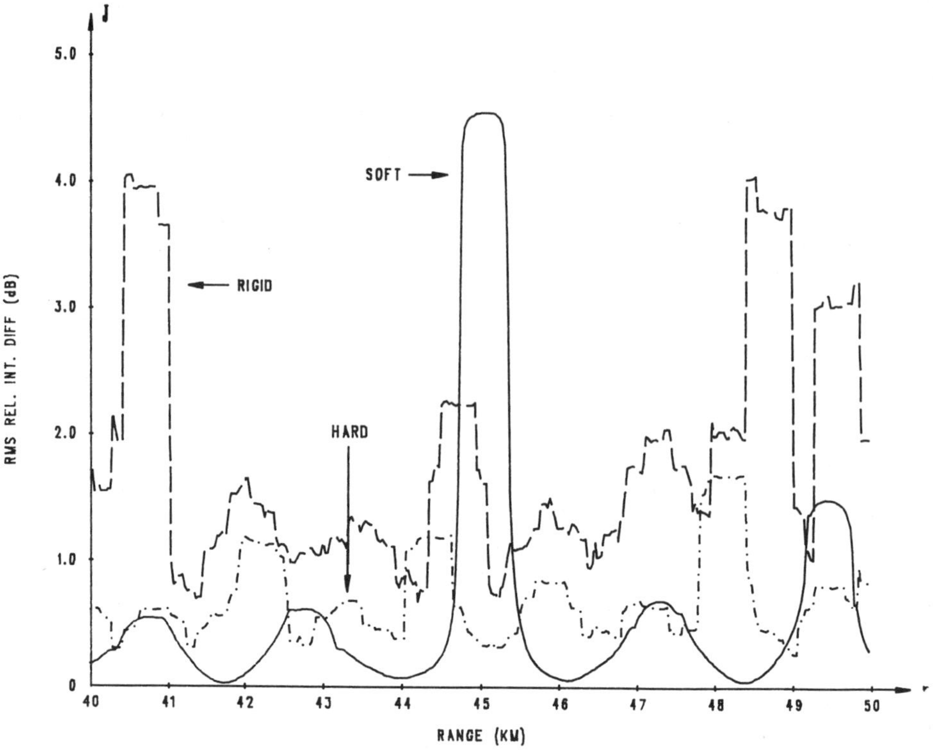

Fig. 8. RMS difference in intensity versus range for the current of Fig. 7(a) and for three bottom types.

from rigid to hard to soft, note that the overall values of J tend to decrease. However, the peak values may actually increase substantially. For example, at 45 km the soft bottom has a peak which is about 4 dB larger tha its overall value. Furthermore, the curves are smoother for both the hard and soft bottoms, since they attenuate higher modes more rapidly than the rigid bottom. These observations illustrate the strong dependence of concavity effects on bottom influences in underwater sound transmissions.

SUMMARY

We discuss a family of parabolic approximations, valid for depth- and range-dependent sound-speed profiles, which include effects caused by stead depth-dependent currents. These approximations permit examination of inten sity effects caused by currents for frequencies and environments where othe models may not be valid or convenient. Using a standard numerical implemen tation, we present the results of computations for several current and soun speed structures. They suggest that currents can cause significant intensi ty variations, principally by altering the effective sound-speed profile. Intensity differences arising from reciprocal transmissions are shown to be especially large. Also, current effects on intensity can be very sensitive to small changes in sound speed. Finally, the presence of current fine structure can introduce additional fluctuations in intensity predictions.

REFERENCES

Lee, D., and Botseas, G., 1982, IFD: An implicit finite difference com puter model for solving the parabolic equation, New London Lab., NUSC, New London, CT, TR 6659.

Robertson, J.S., Siegmann, W.L., and Jacobson, M.J., 1985, Current an current shear effects in the parabolic approximation for underwater sound channels, J. Acoust. Soc. Am., 77:1768.

QUICK NORMAL MODE TYPE STARTING FIELDS FOR PARABOLIC EQUATION MODELS

E. Richard Robinson, Hue B. Tran, and David H. Wood

Department of Mathematics
University of Rhode Island
Kingston, Rhode Island 02881 USA

Code 3332, New London Laboratory
Naval Underwater Systems Center
New London, CT 06320 USA

ABSTRACT

It is well known that Parabolic equation models require the user to provide a starting field. We present a method which directly generates the special combination of modes that gives the correct starting field. A comparison is made to the more customary Gaussian and normal mode starting fields.

INTRODUCTION

Before a parabolic equation model can be run the user must provide a starting field; i.e., values of the sound field as a function of depth at a fixed range. This is sometimes done without reference to the ocean sound speed profile of the area under consideration by using a Gaussian, sinc, or Bessel function. It is known [1], [5], however, that the correct starting field is given in terms of normal modes and, therefore, depends on the sound speed profile.

When we consider the normal modes expanded in Fourier series we find that the particular combination of modes needed to generate the starting field can be obtained by computing the -1/4 power of a matrix generated from the Fast Fourier Transform (FFT) of the sound speed profile.

To demonstrate the accuracy of our approach we exhibit graphs of various starting fields for a special sound speed profile.

THE ALGEBRAIC FORMULATION OF THE PERTURBED PROBLEM

If we assume that the ocean has a rigid bottom, and that the index of refraction varies from an average value of unity, then the normal modes for such an ocean must satisfy the vector differential equations (prime denotes differentiation with respect to depth, z)

$$\psi_\varepsilon''(z) + k^2\left(1+\varepsilon s(z)\right) \psi_\varepsilon(z) = \Lambda_\varepsilon \psi_\varepsilon(z)$$

$$\psi_\varepsilon(o) = 0 \ , \ \psi_\varepsilon'(h) = 0 \ , \qquad (1)$$

and, using superscript t to denote matrix transpose,

$$\int_0^h \psi_\varepsilon(z) \ \psi_\varepsilon^t(z) \ dz = I \ , \qquad (2)$$

where I is the identity matrix, ε is a parameter, and s(z) expresses the variation of the index of refraction from its nominal value of 1. The results corresponding to the special case $\varepsilon = 0$ are known [2]. The normal mode sum giving the sound field for the perturbed ocean is [2]

$$\psi_\varepsilon^t \ (z_0) \ H_0^{(1)} \left(\Lambda_\varepsilon^{1/2} r\right) \psi_\varepsilon(z) \ . \qquad (3)$$

We seek a matrix D (independent of z) such that

$$\psi_\varepsilon(z) = D\psi_0(z) \qquad (4)$$

and subject to the additional condition

$$\int_0^h \psi_\varepsilon(z) \ \psi_\varepsilon^t(z) \ dz = I \ ,$$

which implies that

$$DD^t = I \ .$$

The matrix D is made up of the coefficients of the Fourier series expansions of the perturbed modes in terms of the unperturbed modes. Substituting the relation (4) into (1) we obtain

$$D \ \psi_0''(z) + k^2 \left(1+\varepsilon \ s(z)\right) \ D\psi_0(z) = D\psi_0(z) \ . \qquad (5)$$

Using the value of $\psi_0''(z)$ from the unperturbed formulation, the above equation becomes

$$D\Lambda_0 \ \psi_0(z) + k^2 \left(1+\varepsilon \ s(z)\right) \ D \psi_0(z) = D \psi_0(z) \ . \qquad (6)$$

Multiplying (6) on the right by $\psi_0^t(z)$ and integrating from z = 0 to z = h we obtain, as a result of Eq. (2),

$$D\Lambda_0 + \varepsilon\, DA = \Lambda_\varepsilon D\ , \tag{7}$$

where the matrix A is given by

$$A = \int_0^h k^2\, s(z)\, \psi_0(z)\, \psi_0^t(z)\, dz\ . \tag{8}$$

Taking the transpose of (7) we have

$$D^t \Lambda_\varepsilon = \left(\Lambda_0 + \varepsilon\, A\right) D^t \tag{9}$$

and when this is multiplied from the right by D, we obtain our final equation governing D,

$$D^t \Lambda_\varepsilon\, D = \Lambda_0 + \varepsilon\, A\ . \tag{10}$$

Using the asymptotic form of the Hankel function [3] the desired perturbed starting field [1], [5] may be represented by

$$\sqrt{\frac{2}{\pi}}\ \psi_\varepsilon^t(z_0)\ \Lambda_\varepsilon^{-1/4}\ \psi_\varepsilon(z)\ . \tag{11}$$

Substituting (4) into (11), this starting field may be represented by

$$\sqrt{\frac{2}{\pi}}\ \psi_0^t \left(D^t\ \Lambda_\varepsilon^{-1/4}\ D\right)\Lambda_0\ . \tag{12}$$

It is now clear that by taking the (-1/4) root of (10) we obtain the combination ($D^t \Lambda_\varepsilon^{-1/4} D$) necessary for evaluation of the starting field (12).

RESULTS AND EXAMPLE

In the following example we consider results based on an idealized ocean sound speed profile. Comparison is made with starting fields generated by the SNAP model [4] and the Gaussian starting field for the case presented.

Example 1 - A Symmetric Sound Channel

The results presented in this example are based on an idealized ocean model with a symmetric sound channel. The sound speed profile is shown in Figure 1.

The function s(z) in this case is given by

$$s(z) = -2\cos\left(\frac{2\pi z}{h}\right)\ . \tag{13}$$

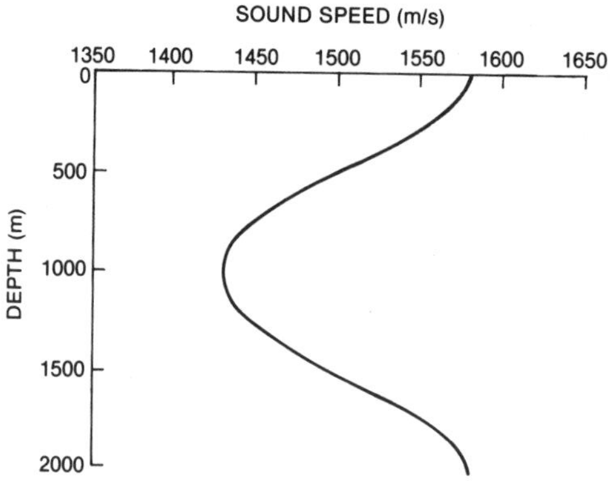

Figure 1. Ocean sound speed vs. depth for perturbation function s(z) = -2 cos (2π z/h, with ε = .05.

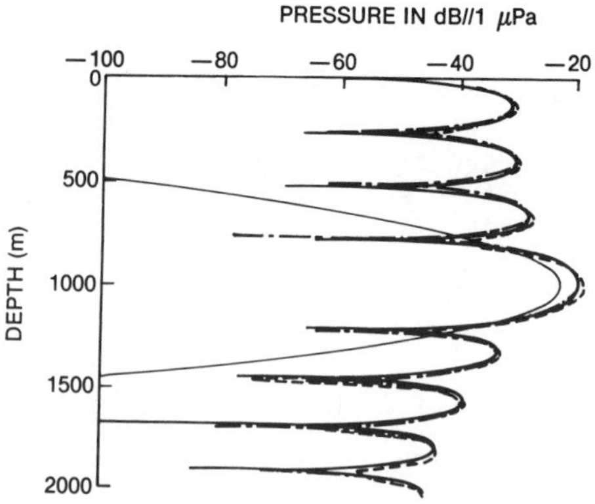

Figure 2. Heavy solid line represents pressure vs. depth for perturbation function s(z) = -2 cos (2π z/h), with frequency 3 Hz, propagating modes, source depth 1000m.

.... SNAP predicted starting field
---- unperturbed starting field
Gaussian starting field

In this case the entries of A are particularly simple:

if $|m-n| = 2$, then

$a_{mn} = k^2$,

if $|m+n-1| = 2$, then

$a_{mn} = -k^2$,

and for all other values of m and n,

$a_{mn} = 0$.

The starting fields for this perturbation function are shown in Figure 2 for a source depth of 1000m. The starting fields generated by SNAP, by a Gaussian starter and by the unperturbed modes($\varepsilon = 0$) are shown for comparison purposes.

CONCLUSION

It has been shown in our example that for a moderate perturbation of the ocean sound speed, i.e., $\varepsilon = .05$, the perturbed starting field differs significantly, at some depths, from both the unperturbed and Gaussian starting fields. The effects on the parabolic equation model loss versus range predictions in range dependent environments will be explored in a future paper.

REFERENCES

1. Wood, D. H. and J. S. Papadakis, Initial data for the parabolic equation, Proceedings of the Bottom-Interacting Ocean Acoustics Conference, Pleham Press, New York, 1980, 417-420.

2. Ahluwalia and Keller, Exact and asymptotic representations of the sound field in a stratified ocean, in Wave Propagation and Underwater Acoustics, Keller, J. B. and J. S. Papadakis, eds., Lecture notes in Physics, vol. 70, Springer Verlag, New York, 1977.

3. Gradshteyn, I. S. and I. M. Ryzhik, Table of Integrals, Series, and Products, A. Jeffrey, ed., Academic Press, New York, 1980.

4. Jensen, F. B. and M. C. Ferla, SNAP: The SACLANTCEN Normal-Mode Acoustic Propagation Model, SACLANTCEN Memorandum SM-121,15 January 1979, SACLANT ASW Research Center, La Spezia, Italy.

5. Robinson, E. R. and D. H. Wood, Ideal Starting Fields for Parabolic Equation Models (in preparation).

A POSTERIORI PHASE CORRECTIONS TO THE PARABOLIC EQUATION

David J. Thomson[§] and David H. Wood*

[§]Defence Research Establishment Pacific
FMO Victoria, B.C., Canada V0S 1B0

*Naval Underwater Systems Center
New London, CT, 06320 U.S.A.

ABSTRACT

A method is described for transforming numerical solutions of the Tappert and Hardin (1974) parabolic equation of ocean acoustics into solutions of the Helmholtz equation. For range-independent media, the phase errors inherent in parabolic equation predictions of oceanic waveguide sound propagation are removed exactly. The method is based on an integral transform established by DeSanto (1977). It is shown that the field satisfying the Helmholtz equation can be obtained from the Fourier transform of the field satisfying the parabolic equation via fast field program (FFP) techniques. Numerical examples are presented to illustrate this post-processing approach to parabolic phase error correction.

INTRODUCTION

At the 8th International Congress on Acoustics in 1974, Tappert and Hardin (1974) described a theoretical-numerical method based on a parabolic approximation to the wave equation for modeling low-frequency sound propagation in realistic ocean environments. Subsequent development of this standard parabolic equation (SPE) method is reviewed elsewhere (Tappert, 1977; Davis et al., 1982; Lee, 1984). Because the SPE is first-order in the range variable, it allows efficient numerical solution via noniterative marching techniques and is capable of modeling forward scattered propagation within range-dependent environments.

For range-independent environments, the validity of the parabolic approximation is generally tested by comparing SPE solutions to exact solutions of the Helmholtz equation (HE). These comparisons often display range dependent shifts in the interference patterns of intensity versus range curves. Methods for reducing these SPE range-phase errors have been proposed and implemented (Brock et al., 1977; DeSanto, 1977; DeSanto et al., 1978), but in many cases the SPE results are not improved significantly (Davis et al., 1982). One method for reducing the SPE range-phase errors is based on a stationary phase evaluation of an integral transform that connects the solution of the SPE to the solution of the HE for general range-varying media (DeSanto, 1977). An implementation of this method is described by DeSanto et al. (1978). It is worthwhile pointing out that this implementation requires twice the computing effort needed to solve the SPE itself and does not always reduce the range-phase errors significantly (Davis et al., 1982).

For the special case of range-independent waveguides, the integral relationship connecting the solutions of the SPE to the solutions of the HE simplifies considerably and can be demonstrated to be exact. This remarkable result was obtained previously by Polyanskii (1974) for two-dimensional waveguides. In this paper, a more accurate and efficient method of evaluating the exact integral transform is presented. The method is based on the fact that the Fourier transform of the SPE, with respect to the range coordinate, is connected to the Hankel transform of the HE via a nonlinear mapping between the respective horizontal wavenumbers. This mapping converts the SPE Green's function into the HE Green's function, from which the HE field can be recovered using well-known fast field program (FFP) methods (DiNapoli and Deavenport, 1980; Schmidt and Jensen, 1985).

THEORY

Let the region $z > 0$ of a cylindrical coordinate system (r,θ,z) be occupied by a medium with density $\rho(z)$ and sound speed $c(z)$. If $p(r,z)$ $\exp(-i\omega t)$ represents the acoustic field due to a point harmonic source located at $r = 0$, $z = z_o$, then for $r > 0$ the pressure p satisfies the scalar Helmholtz equation

$$r^{-1}\partial_r(rp_r) + \rho\partial_z(\rho^{-1}p_z) + k_o^2n^2p = 0, \tag{1}$$

where $n(z) = c_o/c(z)$ is the refractive index and $k_o = \omega/c_o$ is a reference wavenumber. Absorption can be accommodated by allowing $n(z)$ to become complex.

In the parabolic equation method, the pressure in Eq. (1) is approximated by

$$p(r,z) = \psi(r,z)\ (k_or)^{-1/2}\exp(ik_or), \tag{2}$$

where the ψ-field satisfies a parabolic wave equation. In particular, the SPE for ψ is given by (Tappert and Hardin, 1974)

$$\psi_r = i(k_o/2)(n^2 - 1)\psi + i(\rho/2k_o)\partial_z(\rho^{-1}\psi_z). \tag{3}$$

DeSanto (1977) has interpreted Eq. (2) as the first term in a stationary phase approximation of an exact integral transform relating p to ψ, i.e.,

$$p(r,z) = (2\pi i)^{-1/2}\int_0^\infty \psi(t,z)\ \exp[(ik_o/2t)(r^2 + t^2)]\ t^{-1}\ dt. \tag{4}$$

Rather than evaluating Eq. (4) directly, it is expedient to work in the Hankel transform domain. The Hankel transform of p, defined by

$$P_H(k,z) = \int_0^\infty p(r,z)\ J_o(kr)\ r\ dr, \tag{5}$$

determines the depth-dependent Green's function to the HE. The inverse transform is given by

$$p(r,z) = \int_0^\infty P_H(k,z)\ J_o(kr)\ k\ dk. \tag{6}$$

Application of the Hankel transform to Eq. (4) and subsequent manipulation of the right hand side leads to the result

$$P_H(k,z) = (i/2\pi)^{1/2} k_o^{-1} \Psi_F(s,z), \tag{7}$$

where Ψ_F denotes the Fourier transform of ψ,

$$\Psi_F(s,z) = \int_o^\infty \psi(t,z) \exp(-ist)\, dt. \tag{8}$$

In Eq. (7), the horizontal wavenumbers s and k are related via

$$k = k_o(1 + 2s/k_o)^{1/2}. \tag{9}$$

Eq. (7) is the basic theoretical result of this paper. It suggests an algorithm for converting the SPE Green's function $\Psi_F(s,z)$ into the HE Green's function $P_H(k,z)$ where the horizontal wavenumbers are mapped according to Eq. (9).

IMPLEMENTATION

According to Eq. (6), the solution to the HE can be obtained from a knowledge of its depth-dependent Green's function $P_H(k,z)$. The essence of the FFP method for evaluating an integral of this type is to manipulate it into the form of a finite Fourier integral so that the fast Fourier transform (FFT) can be invoked (DiNapoli and Deavenport, 1980; Schmidt and Jensen, 1985). The result of applying the FFP method to Eqs. (6) and (7) is the discrete complex series

$$Y_n = \sum_{m=0}^{N-1} X_m \exp(2\pi inm/N), \quad n = 0, 1, \ldots N-1 \tag{10}$$

where the complex input sequence is

$$X_m = (\Delta k/\pi k_o)\, k_m^{1/2}\, \Psi_F(s_m,z) \exp(im\Delta k r_{min}), \tag{11}$$

and the complex output sequence is

$$Y_n = r_n^{1/2}\, p(r_n,z) \exp(-ik_{min} r_n). \tag{12}$$

Here $r_n = r_{min} + n\Delta r$, $k_m = k_{min} + m\Delta k$ and $\Delta r \Delta k = 2\pi/N$.

A recipe for constructing $P_H(k,z)$ from $\Psi_F(s,z)$ and then recovering the HE field p(r,z) for each receiver depth of interest can be described in the following way. Solve Eq. (3) numerically on a range-depth grid with spacing Δt in range. Perform an N-point complex FFT to obtain Ψ_F at a spacing $\Delta s = 2\pi/(N\Delta t)$. Stretch the wavenumber axis using Eq. (9) and then interpolate Ψ_F onto a grid of spacing Δk and weight the spectrum according to Eq. (11) to obtain the HE Green's function $P_H(k,z)$. Finally, take an N-point inverse complex FFT to determine the solution to the HE on a grid of spacing $\Delta r = 2\pi/(N\Delta k)$.

Two remarks are in order. First, provided $\Delta t > 2\pi/k_o$, the mapping in Eq. (9) ensures that real k wavenumbers are obtained from real s wavenumbers. Second, a fast and accurate method of obtaining P_H on a uniform grid is to append zeroes to ψ before applying the FFT to obtain Ψ_F and then using linear interpolation in the wavenumber domain.

NUMERICAL EXAMPLES

To illustrate this post-processing procedure, Eq. (3) was solved numerically using the split-step Fourier algorithm of Tappert and Hardin (1974) for the environmental profiles given in Table 1. The bilinear sound speed profiles for models 1 and 2 in these examples correspond to test cases 2(b) and 2(c) respectively in the 1981 Parabolic Equation

TABLE 1. Environmental profiles*

Depth	Sound speed		Density	Attenuation
	model 1	model 2		
(m)	(m/s)	(m/s)	(g/cm^3)	(dB/m)
0	1500	1500	1.0	0.00
1000	1520	1520	1.0	0.00
1500	1744	1971	1.0	0.00
2048	1744	1971	1.0	0.05

*Each profile is a linear function of depth between given points.

Workshop (Davis et al., 1982). Each bilinear sound speed profile was terminated with an isovelocity half-space of the same sound speed as the bottom of the lower refracting layer. A small amount of absorption was introduced in the half-space to ensure that no energy entering the bottom was returned to the water column. For each example, the SPE was solved on a grid with a horizontal spacing of 25 m and a vertical spacing of 4 m for a reference sound speed of 1510 m/s. A source half-beamwidth of 30° for model 1 and 40° for model 2 was required to include all the significant refracted energy. All calculations were carried out at a frequency of 25 Hz for a source depth of 500 m and a receiver depth of 500 m.

For each model, the SPE and the reconstructed HE solutions are compared to a reference (exact) HE solution obtained using a code based on the FFP method (Schmidt and Jensen, 1985). In Fig. 1, the comparisons for model 1 are displayed in the form of propagation loss ($-20 \log_{10}|p|$) versus range curves. The upper panel shows the comparison for the SPE prediction. It is apparent that the SPE does not preserve the phases of the 22 modes propagating in this case. In the lower panel, the comparison for the processed SPE prediction shows that an excellent reconstruction has been achieved.

The corresponding comparisons for model 2 are shown in Fig. 2. For this example, the larger sound speed gradient of the lower refracting layer allows more modes to propagate. As a result, the effects of the range-phase errors associated with the SPE solution shown in the upper panel are more pronounced for this model than for model 1. The lower panel shows, however, that the processed SPE prediction again provides an excellent reconstruction of the reference FFP solution.

SUMMARY

A method has been presented for transforming numerical solutions to the standard parabolic equation of ocean acoustics (Tappert and Hardin, 1974) into solutions of the Helmholtz equation. The method is based on an integral relationship proposed by DeSanto (1977) that is exact for range-independent media. In the horizontal wavenumber domain, this connection becomes a local one and is given by Eqs. (7) and (9). In essence, these equations indicate how to map the SPE Green's function into the HE Green's function, after which the field satisfying the HE can be recovered using FFP techniques. The examples given demonstrate that an excellent estimate of the HE field can be provided efficiently and accurately using this postprocessing method.

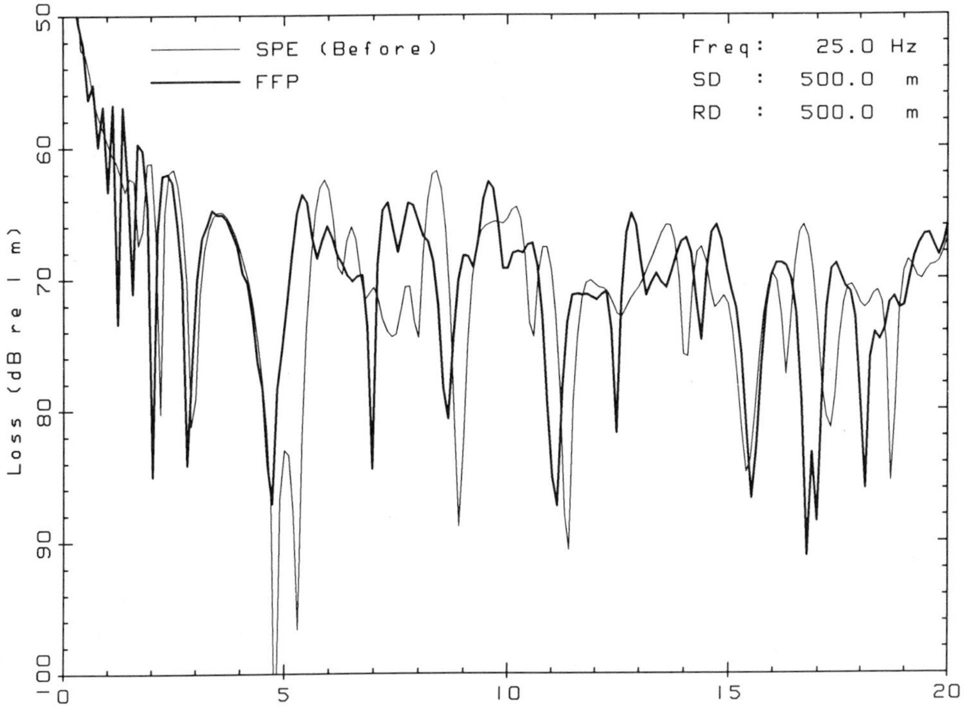

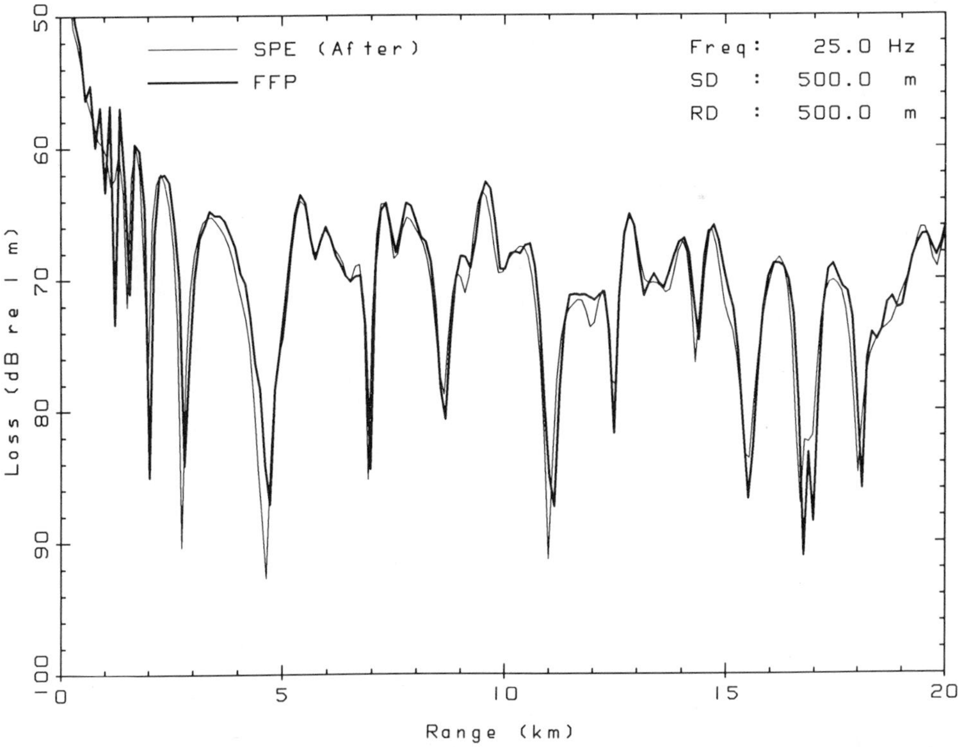

Fig. 1. Propagation losses versus range for model 1. The FFP (exact) prediction is compared to predictions for the SPE (upper panel) and processed SPE (lower panel) approximations.

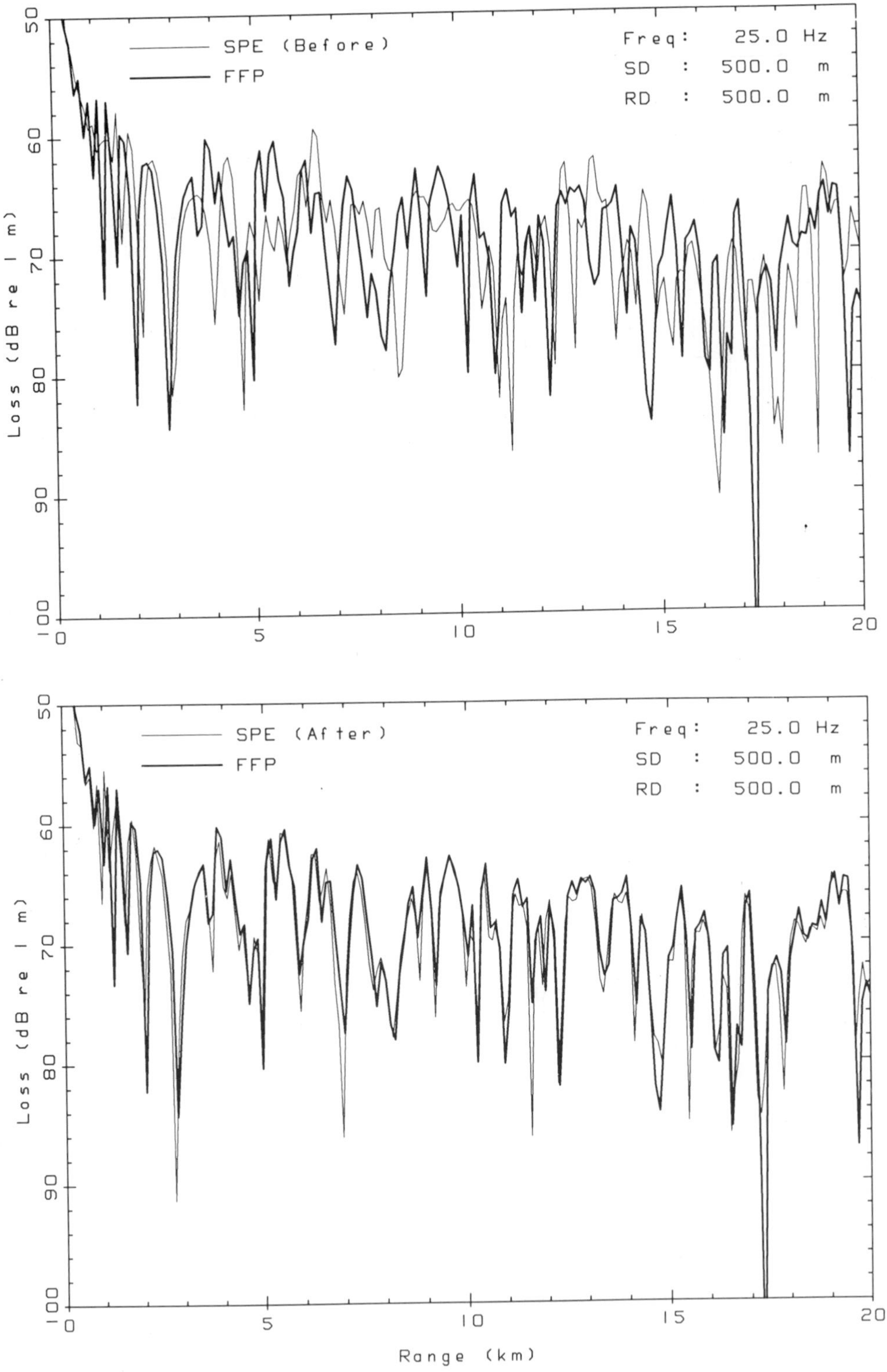

Fig. 2. Propagation losses versus range for model 2. The FFP (exact) prediction is compared to predictions for the SPE (upper panel) and processed SPE (lower panel) approximations.

REFERENCES

Brock, H. K., Buchal, R. N., and Spofford, C. W., 1977, Modifying the sound-speed profile to improve the accuracy of the parabolic-equation technique, J. Acoust. Soc. Am., 62:543.

Davis, J. A., White, D., and Cavanagh, R. C., 1982, NORDA Parabolic equation workshop, 31 March-3 April 1981, NORDA Tech. Note 143.

DeSanto, J. A., 1977, Relation between the solution of the Helmholtz and parabolic equation for sound propagation, J. Acoust. Soc. Am., 62:295.

DeSanto, J. A., Perkins, J. S., and Baer, R. N., 1978, A correction to the parabolic equation, J. Acoust. Soc. Am., 64:1664.

DiNapoli, F. R. and Deavenport, R. L., 1980, Theoretical and numerical Green's function field solution in a plane multilayered media, J. Acoust. Soc. Am., 67:92.

Lee, D., 1984, The state-of-the-art parabolic equation approximation as applied to underwater acoustic propagation with discussions on intensive computations, J. Acoust. Soc. Am. Suppl. 1, 76:S9; reprinted as NUSC Tech. Doc. 7247.

Polyanskii, E. A., 1974, Relationship between the solutions of the Helmholtz and Schrodinger equations, Sov. Phys. Acoust., 20:90.

Schmidt, H. and Jensen, F. B., 1985, Efficient numerical solution technique for wave propagation in horizontally stratified environments, Comp. & Maths. with Appls., 11:699.

Tappert, F. D., 1977, The parabolic approximation method, in: "Wave Propagation and Underwater Acoustics," J. B. Keller and J. S. Papadakis, eds., Springer-Verlag, New York.

Tappert, F. D. and Hardin, R. H., 1974, Computer simulation of long-range ocean acoustic propagation using the parabolic equation method, Proc. 8th Intern. Cong. on Acoustics, 2:452.

PERTURBATION MODELING FOR OCEAN SOUND PROPAGATION

Mark D. Duston, Ghasi R. Verma and David H. Wood

Department of Mathematics
The University of Rhode Island
Kingston, RI 02881, USA

Code 3332, New London Laboratory
Naval Underwater Systems Center
New London, CT 06320, USA

ABSTRACT

We assume that the speed of sound in the water and the bottom of the ocean is a function of only the depth, and not the range. We also assume that the ocean and its bottom eventually interface with a rigid halfspace. This problem can be solved by the method of normal modes, involving the eigenvalues and eigenfunctions of a depth dependent ordinary differential equation. Since the sound speed in this problem varies only a little from its average value, we exploit the fact that the eigenfunctions and eigenvalues are known when the sound speed is constant. We investigate the changes in these eigenvalues and eigenfunctions that result from changes in the depth dependent sound speed within the ocean and its bottom, using an algebraic formulation of the the effect of the perturbation.

INTRODUCTION

The speed of sound in the ocean is nearly uniform, but even these small variations in speed give many very important effects. We want to exploit the fact that if the sound speed were uniform, the ocean could be modeled using an equation for which we know [Ref. 1, pg. 34] the exact solution in a convenient form. In this paper, small changes in the equations governing an ocean with a uniform speed are thought of as "perturbations" and we seek the equations that govern only the changes in the resulting solutions. We expand the perturbed solutions as infinite "Fourier" series in the normal modes of the unperturbed problem. This gives us an eigenproblem for the matrix of Fourier coefficients. It has been shown that this result is consistant with the application of classical perturbation theory to the normal mode problem [Ref. 2].

THE IDEALIZED PROBLEM

Since we assume that in the idealized model of the ocean the sound speed is constant, the excess pressure p(r,z) in an inviscid, non-heat-conducting fluid is cylindrically symmetric and satisfies the Helmholtz equation [Ref. 1, pg. 19]

$$p_{rr} + \frac{1}{r} p_r + p_{zz} + k^2 p = 0. \tag{1}$$

In our previous work, [Ref. 3] we exhibited the solution of this equation by separation of variables with the boundary conditions

$$p(r,z) = 0 \quad \text{at } z = 0 \tag{2a}$$

and

$$p_z(r,z) = 0 \text{ at } z = h. \tag{2b}$$

This last boundary condition assumes that the ocean and its bottom are eventually underlaid with a rigid halfspace at depth $z = h$. To solve this equation by the method of separation of variables we took $p(r,z) = \phi(z)\Theta(r)$ and we obtained, for the depth dependent part, an ordinary differential equation

$$\phi_{zz} + k^2\phi = \ell\phi, \tag{3}$$

where ℓ is a constant, independent of r and z. For this equation the boundary conditions (2a) and (2b) take the form

$$\phi(0) = 0 \tag{4}$$

and

$$\phi_z(h) = 0. \tag{5}$$

It is known [Ref. 1, pg. 23] that there are countably many solutions of (3) satisfying (4) and (5) namely,

$$\phi_n(z) = \sqrt{\frac{2}{h}} \sin\left[\frac{(2n-1)\pi z}{2h}\right], \quad n = 1,2,3, \ldots . \tag{6}$$

Each of these requires a corresponding value of ℓ to be given by

$$\ell_n = k^2 - \left[\frac{(2n-1)\pi}{2h}\right]^2, \quad n = 1,2,3, \ldots . \tag{7}$$

We want to represent this set of solutions in vector notation. Let Φ be a column vector defined by

$$\Phi = [\phi_1(z), \phi_2(z), \ldots \quad]^t, \tag{8}$$

where the superscript t denotes the matrix transpose. Let L be the diagonal matrix

$$L = \mathrm{diag}(\ell_1, \ell_2, \ell_3, \ldots) . \tag{9}$$

Then Φ satisfies the vector equation

$$\Phi'' + k^2\Phi = L\Phi \tag{10}$$

and the orthogonality condition

$$\int_0^h \Phi\ \Phi^t\ dz = I, \tag{11}$$

where I is the identity matrix.

ALGEBRAIC FORMULATION OF THE PERTURBED PROBLEM

In place of (10) we will consider the perturbed vector equation

$$\Psi'' + k^2(q(z) + \epsilon\ s(z))\ \Psi = \Lambda\Psi \tag{12}$$

with the boundary conditions

$$\Psi(0) = 0 \tag{13}$$

and

$$\Psi_z(h) = 0. \tag{14}$$

In (12) the quantities in parentheses are scalars but Λ is a diagonal matrix. Although we will be only treating the case $q(z) \equiv 1$, the argument given below remains true for more general $q(z)$.

Now, let the matrix D be defined by the equation

$$\Psi = D\ \Phi\ , \tag{15}$$

with the condition

$$I = \int_0^h \Psi\ \Psi^t\ dz = \int_0^h D\ \Psi\ \Psi^t\ D^t\ dz \tag{16}$$

which implies that

$$D\ D^t = I\ . \tag{17}$$

Substituting (15) into (12) we obtain

$$D\Phi'' + k^2(1 + \epsilon\ s(z))\ D\Phi = \Lambda D\Phi\ . \tag{18}$$

Inserting the value of Φ'' from (10) into (18) we obtain

$$DL\Phi + k^2\ \epsilon\ s(z)\ D\Phi = \Lambda D\Phi\ . \tag{19}$$

Multiplying this equation by Φ^t on the right and integrating over the interval [0,h] and using (11,) we obtain

$$DL + \epsilon DA = \Lambda D\ , \tag{20}$$

where the matrix A is given by

$$A = \int_0^h k^2 \; s(z) \; \Phi \; \Phi^t \; dz \; . \tag{21}$$

Taking the transpose of (20), we have

$$(L + \epsilon A)D^t = D^t \Lambda \; . \tag{22}$$

This completes our derivation of the algebraic eigenproblem for D, the matrix of Fourier coefficients from the Fourier series expansions of the perturbed eigenfunctions ψ_n. Notice that this actually is an algebraic eigenproblem because Λ is a diagonal matrix and all the indicated matrices are arrays of constants with nothing depending on the variable z.

COMPARISON TO EXPANSIONS IN POWER SERIES IN EPSILON

Recall that the parameter epsilon multiplies the entire perturbation that we have taken. One of the standard methods [Ref. 4] for treating perturbation problems of this type is to expand everything in power series in the (presumably small) parameter epsilon. The expansions of the solutions of the perturbed depth dependent equation are expressed by

$$\psi_m(z) = \phi_m(z) + \sum_{i=1}^{\infty} \epsilon^i \; \psi_m^{(i)}(z) \; , \tag{23}$$

where each of the functions $\psi^{(i)}$ is expressed as

$$\psi_m^{(i)}(z) = \sum_{p=1}^{\infty} \alpha_p \; \phi_p \quad , \tag{24}$$

which is an infinite Fourier series expansion in the eigenfunctions of the idealized equation.

Obviously, both series must be truncated for practical applications. This implies that the accuracy of the conventional method is dependent upon the rate of convergence of both the power series in epsilon and the Fourier series.

Compare this to the algebraic formulation where no expansion in epsilon is made. We consider this an important advantage of the algebraic method, which expresses each eigenfunction of the perturbed depth dependent equation as only a Fourier series expansion of the unperturbed eigenfunctions. The power series method requires that the perturbation is uniformly small. It can be useful that our approach permits the perturbation to be unrestricted in size. The effect of truncating the Fourier series must, of course, still be estimated [Ref. 5] .

Equation (21) defines A as a matrix with entries of the form

$$a_{mn} = \int_0^h k^2 \; s(z) \; \phi_m(z) \; \phi_n(z) \; dz \; . \tag{25}$$

Using the identity, $\sin\alpha \sin\beta = \cos(\alpha-\beta) - \cos(\alpha+\beta)$, allows (25) to be expressed as the sum of two integrals I of the form

$$I = \int_0^h k^2 \; s(z) \cos \frac{P\pi z}{h} \; dz, \tag{26}$$

where P is an integer. The matrix A may be represented as the sum of a Toeplitz matrix (where m-n is constant) and a Hankel matrix (where m+n is constant). Exploiting this special structure would yield another advantage: the set of algebraic equations is solvable more quickly than usual [Refs. 6 - 10].

REFERENCES

1. D. J. Ahluwalia and J. B. Keller, Exact and asymptotic representations of the sound field in a stratified ocean, in Wave Propagation and Underwater Acoustics, J. B. Keller and J. S. Papadakis, eds. Lecture Notes in Physics, vol. 70, Springer Verlag, New York, 1977.
2. M. D. Duston, G. R. Verma and D. H. Wood, Bottom interaction effects on normal modes: an algebraic approach, Proceedings of the SACLANT Ocean and Seismo-Acoustic Conference, Plenum Press, New York, 1986 (to appear).
3. M. D. Duston, G. R. Verma and D.H. Wood, Changes in eigenvalues due to bottom interaction using perturbation theory, Proceedings of the 11th IMACS World Congress, B. Wahlstrom, et al., eds., Norwegian Society of Automatic Control, Oslo, Norway, 1985, vol 5, pp 389-392.
4. E. C. Titchmarsh, Eigenfunction Expansions Associated with Second-Order Differential Equations, Part II, Oxford University Press, Oxford, England, 1958.
5. D. Gottlieb and S. A. Orsag, Numerical Analysis of Spectral Methods: Theory and Applications, CBMS-NSF Regional Conference Series in Applied Mathematics, vol. 26, Society for Industrial and Applied Mathematics, Philadelphia, 1977.
6. G. A. Merchant and T. W. Parks, Effecient solution of a Toeplitz-plus-Hankel coefficient matrix system of equations, IEEE Trans. Acoustics, Speech, and Signal Processing, Vol. ASSP-30, No. 1, 1982, pp 40-44.
7. H. Lev-Ari, Cholesky factorization of structured matrices, Second SIAM Conference on Applied Linear Algebra, Raliegh, North Carolina, April 29 - May 2, 1985.
8. P. J. Davis, Circulent Matrices, John Wiley and Sons, New York, 1979.
9. U. Grenander and G. Szego, Toeplitz Forms and Their Applications, Chelsea Publishing Company, New York, 1984.
10. D. Savio and D. H. Wood, A fast algorithm for perturbation of normal modes, extended abstract for this Congress.

FAST COMPUTATION OF PERTURBED NORMAL MODES

Dominic Y. Savio* and David H. Wood*+

*Department of Mathematics, The University of Rhode Island
Kingston, Rhode Island 02881

+Code 3332, New London Laboratory, Naval Underwater Systems Ctr.
New London, Conn. 06320

ABSTRACT

This paper presents an operations count and a convergence theorem for modeling sound fields in the ocean using a special algorithm for computing perturbed normal modes. We convert the perturbed normal mode problem to an infinite algebraic eigenproblem, truncate this to finite size, and estimate its operations count to be of the order of $n^2\log(n)$.

INTRODUCTION

We expand the perturbed normal modes as infinite Fourier series in the normal modes of the unperturbed problem and obtain an infinite eigenproblem for the matrix of the coefficients of the Fourier series as in Ref. 1. This procedure is generally referred to as the Galerkin method (Ref. 2) for approximating the normal modes. This is to be contrasted with the classical perturbation technique (Ref. 6) where the Fourier series coefficients are themselves further expanded in power series in a small parameter and the first few, lowest order, terms are found.

The Galerkin procedure is not often used because of two drawbacks: 1) each element of the matrix must be generated by performing a numerical integration and 2) the resulting matrix is not a banded matrix (for which efficient methods are known).

For our problem we overcome these two drawbacks as follows. Because the unperturbed normal modes are known (Ref. 5) to be sine functions, it turns out that the necessary numerical integrations needed to generate the matrix can be done with a single Fast Fourier Transform (FFT) that yields 2n+1 Fourier coefficients, where n is the number of terms to be used in the Fourier series expansion of the normal modes. In Ref. 3, this observation has been used in a similar way. We make the additional observation that while the matrix of interest is not banded, it does have a special structure: it is the sum of a Hankel matrix and a Toeplitz matrix. We exploit this structure in an algorithm that finds the eigenvalues and eigenvectors efficiently provided that the perturbation of the index of refraction from its average value is not too large.

ALGEBRAIC FORMULATION FOR THE NORMAL MODES

We now sketch the derivation of the infinite algebraic eigenproblem that generates the perturbed normal modes. More details can be found in Refs. 1 and 4. We consider equations that govern the perturbed normal modes in vector form (primes denote differentiation with respect to z)

$$\Psi''(z) + (k^2 + \epsilon\, k^2\, s(z))\Psi(z) = \Lambda\Psi(z) \tag{1}$$

with boundary conditions

$$\Psi(0) = 0 \tag{2}$$

and

$$\Psi'(h) = 0. \tag{3}$$

In (1), the quantities within parentheses are scalars, but Ψ is a vertical vector of perturbed normal modes and Λ is a diagonal matrix of the corresponding eigenvalues. When $\epsilon=0$, the unperturbed case for which exact results are known in terms of sine functions (Ref. 5), we denote the vector of normal modes by Φ and the diagonal matrix of eigenvalues by L.

Now, let the matrix D be defined by the equation

$$\Psi(z) = D\, \Phi(z), \tag{4}$$

that is to say, the elements of the pth row of the matrix D are the (constant) coefficients of the Fourier series expansion of the pth perturbed normal mode. We also require that the perturbed normal modes be L_2 normalized, that is,

$$\int_0^h \Psi(z)\, \Psi^t(z)\, dz = I, \tag{5}$$

where I denotes the identity matrix. This equation implies that

$$D\, D^t = I, \tag{6}$$

because we assume that the unperturbed normal modes have already been normalized.

Substituting (4) into (1) we obtain

$$D\, \Phi''(z) + (k^2+\epsilon\, k^2 s(z))\, D\, \Phi(z) = \Lambda\, D\, \Phi(z)\ . \tag{7}$$

Setting $\epsilon=0$ in (1), we obtain the differential equation governing the vector Φ of unperturbed normal modes. Solving this equation for $\Phi''(z)$ and substituting it into (7), we obtain

$$D\, L\, \Phi(z) + k^2\, \epsilon\, s(z)\, D\, \Phi(z) = \Lambda\, D\, \Phi(z)\ . \tag{8}$$

Multiplying this equation by $\Phi^t(z)$ from the right and integrating with respect to z on the interval (0,h), we obtain

$$D\, L + \epsilon\, D\, A = \Lambda\, D\ , \tag{9}$$

where the matrix A is given by

$$A = \int_0^h k^2\, s(z)\, \Phi(z)\, \Phi^t(z)\, dz\ . \tag{10}$$

Taking the transpose of (9), we have the following algebraic eigenproblem:

$$(L + \epsilon A)\, D^t = D^t \Lambda \ . \tag{11}$$

THE ALGEBRAIC EIGENPROBLEM FOR COEFFICIENTS

In the algebraic eigenproblem (11), the columns of D^t are the eigenvectors of $L + \epsilon A$ and Λ is the diagonal matrix of perturbed eigenvalues. All of the indicated matrices are arrays of constants because nothing depends on the variable z. Let us truncate D^t to an nxn matrix and rename it B. This corresponds to calculating only n normal modes each as a Fourier series truncated to n terms.

To find the desired coefficients of the truncated Fourier series, we wish to solve the eigenproblem

$$(\epsilon A + L) B = B \Lambda \ , \quad \text{with} \quad B^t B = I. \tag{12}$$

Here L is the known diagonal matrix of eigenvalues of the unperturbed problem and the matrix A was defined in (10) in terms of known quantities. Note that in (12) all matrices have been truncated to size nxn. This problem is to be solved for Λ, the diagonal matrix of eigenvalues and B, the matrix of l_2 normalized eigenvectors.

A FAST ALGORITHM TO SOLVE THE EIGENPROBLEM

We now present an algorithm that converges for sufficiently small values of the parameter ϵ and, most importantly, can be executed in $O(n^2\log(n))$ operations because our matrix A has a special structure.

<u>Algorithm</u>

0. Initialize: let B(0) = I.

Iterate the following steps for k=0,1, ...

1. let Λ(k+1) = {diagpartof[B(k)]}**(-1)*diagpartof[ϵ A B(k)] + L ;

2. solve for the elements off of the main diagonal of the new matrix B(k+1): L B(k+1) - B(k+1) L = ϵ A B(k) + B(k) [Λ(k+1) - L] ;

3. define the diagonal elements so that diagpartof[B^t(k+1) B(k+1)] = I ;

4. increment until B(k) and Λ(k) converge .

ANALYSIS OF THE PROPOSED FAST ALGORITHM

The matrix A can easily be shown (Refs. 1 and 4) to be the sum of a Toeplitz matrix (constant on diagonals) and a Hankel matrix (constant on counterdiagonals). This fact makes it possible to execute our algorithm in $n^2\log(n)$ operations. In fact, we prove a somewhat more general result.

<u>Theorem 1</u>

If L is an nxn diagonal matrix and A is an nxn matrix that is a linear combination of a diagonal matrix, a Hankel matrix, a Toeplitz matrix and a circulant matrix, the above algorithm can be executed in $O(n^2 \ln n)$ operations.

The proof of Theorem 1 is based on two observations. One is tha matrices with the specified structure can multiply an arbitrary matrix (o the same size), as required in step 1 of the algorithm, in $O(n^2\log(n)$ operations (Ref. 7). Secondly, when we examine the general element of th equation in step 2, we observe that this step requires only an additiona n^2 operations.

For the estimated operations count to be of interest the algorith must converge.

Theorem 2

Let A be a normal matrix and L be a diagonal matrix with distinc eigenvalues. The eigenproblem (12)

$$(\epsilon A + L) B = B \Lambda \quad , \quad \text{with } B^t B = I,$$

has a unique solution and the matrices B(k) and Λ(k) generated by the abov algorithm converge to the solution of this eigen-problem for sufficientl small ε.

The proof of Theorem 2 demonstrates that the algorithm reduces th error by a multiple of ε at each iteration. Estimates of the errors can b given in terms of the elements of the matrices as follows. Let

$$|\delta_{ij}(k)| = |b_{ij} - b_{ij}(k)|$$

be the absolute value of the error in the general element of the matri B(k). Then

$$|\delta_{ij}(k+1)| \leq \epsilon F_1 |\delta_{ij}(k)| \quad , \quad \text{when } i \neq j,$$

$$|\delta_{ii}(k+1)| \leq \epsilon F_2 |\delta_{ii}(k)| \quad ,$$

$$|\Lambda_{ii}(k+1) - \Lambda_{ii}| \leq \epsilon G |\Lambda_{ii}(k) - \Lambda_{ii}| \quad ,$$

and (using a matrix norm)

$$\| B^t(k+1)B(k+1) - I\| \leq \epsilon H \| B^t(k)B(k) - I\| \quad ,$$

where F_1 , F_2 , G and H are constants.

REFERENCES

1. D. H. Wood, M.D Duston and G.R. Verma, Bottom interaction effects on normal modes: an algebraic approach, Proceedings of the SACLANT Ocean Seismo-Acoustic Conference, Plenum Press, New York, 1986 (to appear).
2. G. Birkhoff and R.E. Lynch, Numerical Solution of Elliptic Problems, SIAM, Philadelphia, 1984.
3. L. M. Delves, A Fast method for the solution of Fredholm integral equations, J. Inst. Maths. Applics. (1977) 20, 173-182.
4. M. D. Duston, G. R. Verma and D. H. Wood, Perturbation modeling for ocean sound propagation, extended abstract, 12th ICA.
5. D. J. Ahluwalia and J. B. Keller, Exact and asymptotic representation of the sound field in a stratified ocean, in Wave Propagation an Underwater Acoustics, J.B. Keller and J.S. Papadakis, eds., Lecture Notes in Physics, vol. 70, Springer-Verlag, New York, 1977.

6. E. C. Titchmarsh, Eigenfunction Expansions Associated with Second-Order Differential Equations, Part II, Oxford University Press, Oxford, England, 1958.
7. A. V. Aho, J. E. Hocroft and J. D. Ullman, The Design and Analysis of Computer Algorithms, Addison Wesley, 1974.

OMNIDIRECTIONAL AMBIENT NOISE MEASUREMENTS IN THE SOUTHERN BALTIC SEA DURING SUMMER AND WINTER

R.A. Wagstaff and J. Newcomb

Naval Ocean Research and Development Activity
Code 245, NSTL, MS 39529

ABSTRACT

Omnidirectional ambient noise at frequencies from 20 Hz to 2 kHz, shipping surveillance, and water temperature-depth profile measurements were conducted in five areas of the southern Baltic Sea during both summer and winter from a maritime patrol aircraft. Hydrophone depth settings were 18m and 121m. The deeper setting put the hydrophone on the bottom at four of the sites.

INTRODUCTION

Ambient noise and associated environmental and noise source data were measured in five geographic areas of the southern Baltic Sea (see Fig. 1) during both summer and winter. Generally, measurements in each area were conducted on two different days of each season. During each measurement six calibrated sonobuoys were deployed from an aircraft at each site in a triangular pattern with 10 nmi sides. At each apex one sonobuoy was set for 18m depth and another was set for 121m depth. At most locations this latter sonobuoy rested on the bottom. In addition, a temperature profile was obtained before and after the acoustic measurements by AXBTs and the sea state was estimated. Finally, prior to and after completion of the acoustic measurements, radar surveillances of all the shipping in the entire southern Baltic Sea (below 57° 30' N) were conducted.

DATA ACQUISITION/PROCESSING

The acoustic data were monitored during acquisition by listening to an aural output on a speaker or a headset and by visually observing the audio signals on a bank of oscilloscopes. Periods of abnormal signal behavior were logged on the voice channel of the analog tape recorder. The information on the voice channel was later used during the data processing as an aid for the selection of the data sets to be processed. The acoustic data from each of the six channels recorded at a given site were run through a strip-chart recorder to display time histories in a broad spectral band. From these time histories it was easy to determine when the quality of the data was good or was

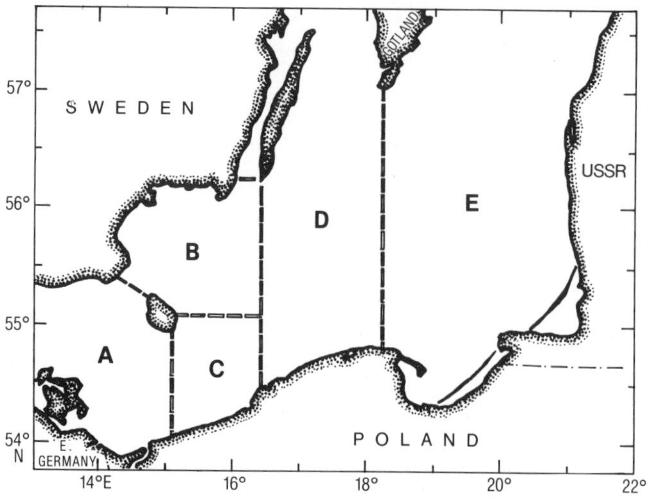

FIG. 1. Areas for ambient noise measurements in the Baltic Sea

adversely affected by abnormal circumstances. Time periods of good data were selected from the strip-charts for further processing. Those analog data tapes of acceptable quality were digitized and the six acoustic channels were simultaneously sampled at a rate of 5120 Hz. The first 512 data points of each sample interval were written to 9-track magnetic tape and then processed one channel at a time. Spectral analysis was accomplished by performing a Hann shaded Fast Fourier Transform (FFT) on each set of 512 data points and by averaging the FFT outputs over the specified number of samples. The resulting spectrum was then adjusted to an absolute sound pressure level.

The calibrated spectra were sampled in both frequency and time and then various statistics were calculated to aid in the assessment of data quality and to characterize the ambient noise field. Only the data that passed the data quality tests were used in the analysis of the ambient noise. Most of the analyses were based on spectra obtained during four minute time periods. The time periods were chosen as being representative of the data at a given site, hydrophone depth, and season. The ambient noise levels, depth dependence, and temporal and spatial variability are best illustrated by comparing these spectra. However, other statistics such as percentile values, skew, and kurtosis are more easily presented and analyzed in tabular form.

The signals from the AXBTs were recorded on tape and plotted in real-time. Salinity-depth profiles obtained near each site were later used with the temperature-depth profiles to obtain sound speed-depth profiles. The sound speed-depth profiles were then used to calculate the acoustic propagation from near-surface sources to the hydrophones at each site, depth, and season. The results were used in the interpretation of the measured ambient noise data.

From one of the surveillance flights for each noise measurement day, the probability of at least one ship being in each "square" (0.5 deg latitude x 1.0 deg longitude) was calculated. This probability was then entered above the line of each square. The median number of ships, observed in a given square, given that at least one ship was observed, was recorded below the line.

RESULTS AND DISCUSSION

Acoustic/environmental conditions

The nominal depth in each area was 30m, 80m, 85m, 37m, and 120m for areas A, B, C, D, and E respectively. The surface temperatures in the summer were all about 17°C while in the winter they ranged from about 1.5° to 3°C. The summer sound speed profiles indicated that acoustic propagation is strongly downward refracting and bottom interacting. For operational reasons, it was also not possible to obtain a temperature profile in area C-summer. The winter sound speed profiles indicated that the propagation is upward refracting, directing acoustic energy toward the sea surface and reducing bottom effects.

The observations of sea state from the aircraft during the low altitude (about 100-200m) passes to drop the sonobuoy patterns were recorded and correlated with the measured spectra. The results were typical of shallow water areas.

Shipping

Figure 2 presents the summer and winter results of the shipping surveillance. The probability of the square being occupied (above the line) and the most likely number (median) of ships to be expected when occupied (below the line) are included for each grid box. An approximation to the density can be obtained by multiplying these two numbers together.

A comparison of the summer and winter median numbers for particular grid boxes indicates that shipping during the winter was about 40 percent greater than during the summer. The main reason for this was believed to be the increased commercial fishing activity during the winter. It was not uncommon during the winter to either modify the location of the buoy pattern slightly (as much as 10 nmi) to avoid high level noise contamination caused by a fishing fleet or to select an alternate site which didn't have the same problem that day. Fortunately, the fishing fleets usually didn't occupy the same areas for more than one or two days.

Data Quality

There were numerous causes of data quality degradation. However, the quantity of data collected was usually sufficient that time periods could be chosen in which the data were of high quality.

The winter survey presented an unexpected problem. The substantial increase in shipping and fishing activity coupled with the very strong positive sound speed gradient (upward refracting) created a condition in which clipping was rarely avoidable at the 18m hydrophone depth. This made shallow data during the winter unusable.

Ambient noise depth dependence

Most of the southern Baltic Sea is less than 121m deep; therefore, the deep hydrophones were on the bottom when set for 121m

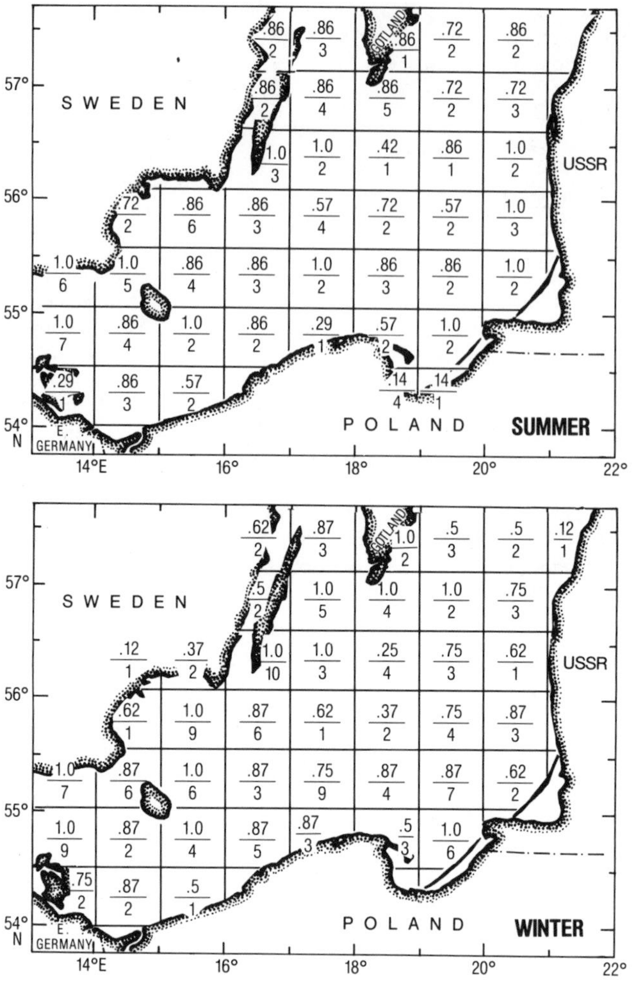

FIG. 2. Probability of an area being occupied by ships (above line) and the median number of ships observed (below line) for occupied areas during summer and winter.

except in area E. This restricts the investigation of depth dependence to comparisons between hydrophones at 18m and the bottom depth, except in area E.

There were two significant depth effects observed in the data. The first was observed during data acquisition. The noise in the winter at the 18m hydrophone depth was sufficiently high to cause clipping in the sonobuoy electronics. This happened at all times and at all sites. This extremely high noise level near the surface was believed to be due to the high density winter shipping (mostly commercial fishing vessels) combined with the strongly upward refracting winter acoustic propagation conditions which tended to propagate sound away from the bottom and toward the surface.

The second depth effect was observed in the summer data at

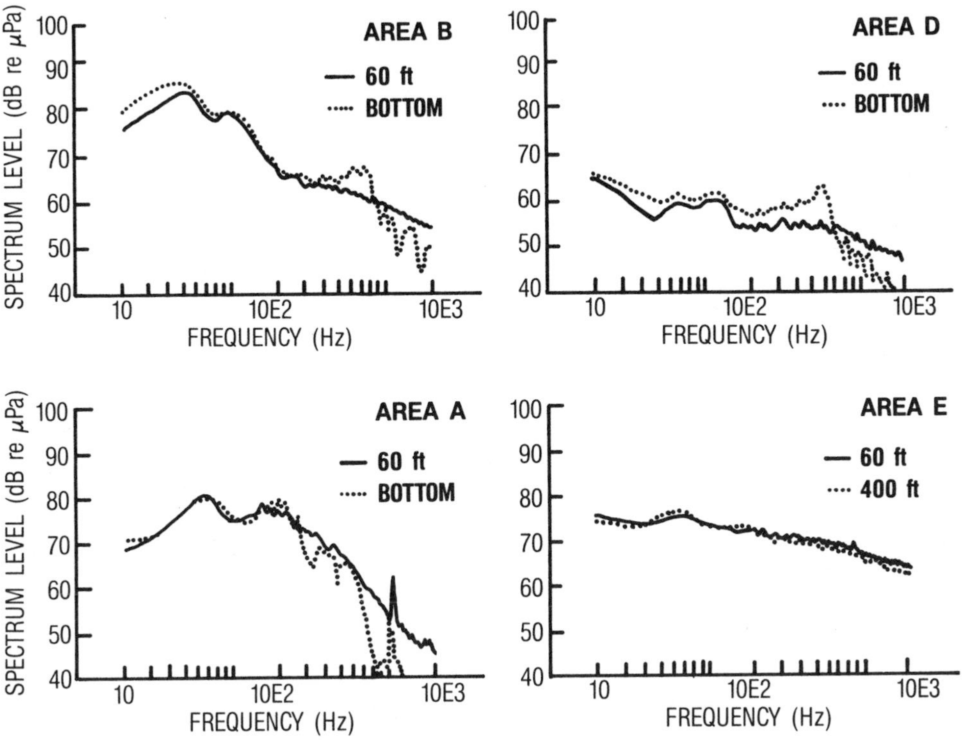

FIG. 3. Ambient noise spectra for hydrophones at 60 ft (18m, solid curves) and on the bottom (except in area E, it is at 400 ft, 121m) during the summer in areas A, B, D, and E.

frequencies above 200 Hz. In all cases, the spectra for the hydrophones, which were on the bottom, exhibited an unusual character. This is illustrated by examples in Fig. 3. The solid curves are spectra for hydrophone depths of 18m in areas A, B, D, and E. The dotted curves are corresponding spectra for hydrophones on the bottom in areas A, B and D and at 121m (just off the bottom) in area E. The character of the high frequency (above 200 Hz) noise for the bottomed hydrophones differs considerably from that for the shallow hydrophones. The erratic structure versus frequency is indicative of constructive and destructive interference. The noise around 500 Hz on the bottomed hydrophones in areas B and D is about five to eight decibels greater than the corresponding noise on the shallow hydrophones. At higher frequencies it is one to eight decibels less. The same noise on the deep hydrophone in area A is greater at about 200 Hz and less at higher frequencies.

The results of the summer measurements in area E, where the deep hydrophones were not on the bottom, are quite different. The deep and shallow spectra are nearly identical. This agreement suggests that the unusual deviations of the bottomed hydrophone spectra at the other sites are a result of the hydrophone being in contact with or buried within the bottom causing interference effects. These effects are evident in the noise from the wind and seas (i.e., sea state noise) which arrives at relatively high vertical angles as well as the noise

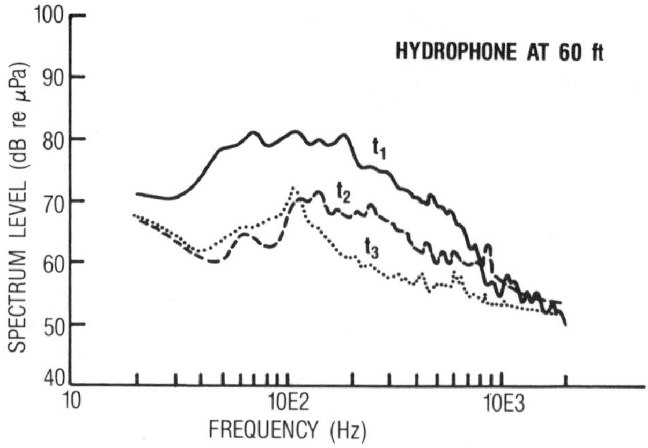

FIG. 4. Temporal variation of the ambient noise in area A for three times within one half-hour.

due to shipping which would arrive at more horizontal angles. This conclusion is supported by the 1000 Hz "line" component in the area A results, which must be from shipping. This component is affected the same as the sea state noise. Its signal-to-noise ratio is nearly the same in both spectra, but its level is about 10 dB less in the bottomed hydrophone spectrum. The source of this interference was determined by testing in the laboratory to be mechanical interference between the hydrophone and the sediment.

Ambient Noise Short Time Dependence

The maximum variation in the ambient noise with time (see Fig. 4) ranges from about 15 to 20 dB. Not only do the levels change significantly but the shapes of the spectra change as well. Range dependent attenuation is a possible cause for the differences in the spectra. The spectra from the shallow hydrophone had few characteristics in common, which suggests that the causes for the variations, in this case, are not as simple as just range dependent attenuation. Something more complicated is at work here.

One source of range dependent attenuation, which is not easily predicted, is the combined effects of surface decoupling loss (Lloyd's Mirror Effect or surface image interference) and bottom loss. The decoupling loss decreases with increased ray angle (or mode number) while the bottom loss usually increases. As the range from the hydrophone to the ship increases, the number of bottom reflections increases, selectively attenuating the energy at higher ray angles greater than at the more horizontal angles. The surface decoupling, however, decreases with increasing vertical angle. Furthermore, the bottom loss usually increases with frequency. The net effect is that the attenuation of sound with range will increase in a rather complicated way, but it is predictable provided the attenuation characteristics of the bottom are accurately known. A more complete discussion of this phenomenon is given by Bannister and Pederson[1]. This phenomenon could account for some of the differences in the spectra in Fig. 4. Aspect of the ship is another possibility that should not be ignored. The ship will present different aspects to the measurement site, as it passes, which can account for changes in the spectral content of the noise received from the ship.

Spatial Variability

The ambient noise varied considerably from site to site. The highest levels were measured at the sites surrounded by the greatest number of ships. The shipping density was highest to the northeast near the coast of Sweden and just south of Gotland. This appeared to be the main shipping lane for ships entering and leaving the Baltic Sea. The density of ships was high in other areas not near obvious shipping lanes but, in general, it was due to commercial fishing activity. Hence, the most important parameter in the prediction of the ambient noise in the southern Baltic Sea is the location and density of the shipping. Predicting the location and density of the merchant fleet could probably be done with reasonably high confidence; however, prediction of the sizes and locations of the fishing fleets most likely could not be done with confidence.

Seasonal Variability

The ambient noise changed considerably from summer to winter. The most dramatic seasonal effect observed was the relatively high noise level at low frequencies at the shallow depth (18m) at all sites during the winter. The ambient noise was too high for a valid measurement by the sonobuoys. It caused clipping in the sonobuoy electronics which was impossible to overcome during the processing of the data. The cause of this seasonal variability in the ambient noise was believed to be due to the increased shipping in the winter and the differences in the acoustic propagation conditions.

The increased winter shipping (including commercial fishing activity) that also contributed to the higher levels of noise in the winter is illustrated by Fig. 2. The top plot is for summer and the bottom one is for winter. In some places the shipping increases by as much as a factor of three during the winter. For example, off Sweden the median number of ships observed in one grid block was three in the summer and ten in the winter. The differences in the summer and winter ambient noise in the nearest area, area D, ranged from 6 dB (48 Hz) to about 18 dB (145 Hz).

The summer noise levels were on the average about 12 dB less than those during the winter except in area E. At this site the shipping and commercial fishing activity were light and didn't change appreciatively with time (i.e., hour, day, or season).

CONCLUSIONS

The number of ships in the southern Baltic Sea was about 40 percent greater during winter than during summer. This is partly due to the increased commercial fishing activity during the winter.

The ambient noise spectra from sonobuoy hydrophones resting on the bottom show interference effects at frequencies above 200 Hz. Increased noise by about five to eight decibels around 500 Hz was observed at some sites. Above 800 hz the noise on the bottomed hydrophones was less at all sites (the deep hydrophone was not on the bottom in area E) and had an erratic character due to mechanical interference between the hydrophone and the sea bottom.

The ambient noise in area E, where the 121m hydrophone was not on the bottom, showed negligible depth dependence during the summer. However, there was a significant depth dependence in the ambient noise at all sites during the winter, when the acoustic propagation conditions were strongly upward refracting and the shipping density was relatively high.

The ambient noise measured on bottomed hydrophones at all frequencies during the winter was about 12 dB greater than during the summer. The differences are attributed to increased shipping and commercial fishing in the winter and less winter acoustic propagation loss resulting from strongly upward refracting conditions.

Temporal and spatial variations in the ambient noise in the southern Baltic Sea of the order of 20 dB were observed and believed to result from differences in the densities of nearby shipping and the fact that the shallow water acoustic propagation conditions effectively discriminate against noise arriving from all but nearby ships.

REFERENCES

1. R.W. Bannister and M.A. Pedersen, Low-frequency surface interference effects in long-range sound propagation. J.Acoust.Soc.Am. 69:76 (1981).

AMBIENT NOISE OVER THICKLY SEDIMENTED CONTINENTAL SLOPES

D. R. Del Balzo, M. J. Authement and D. A. Murphy

Naval Ocean Research and Development Activity
NSTL, MS 39529-5004, U.S.A.

ABSTRACT

Calculations of ocean ambient noise over a continental slope are presented at 30 and 100 Hz for both summer and winter conditions. The implicit finite-difference formulation of the parabolic equation method, with high-angle (40 deg) capability and a full geoacoustic bottom description is utilized. The generic slope of interest is typical of the North Atlantic Ocean, has a "thick," 500 m sediment and a vertical angle of 2.8 deg. The noise field is generated by summing contributions from surface ships/storms both in the deep-water basin and over the shallow-water shelf. A "noise notch" is predicted over the slope with greatest magnitude (27 dB) at shallow depths and which decays rapidly with depth. The range extent of the "notch" is 50 km near the ocean surface and 30 km at a depth of 300 m.

INTRODUCTION

One aspect of underwater acoustic propagation that has received only limited attention is the interaction of sound with a sloping sea floor, especially for signals propagating in the upslope direction. Now that advanced computational methods are available, accurate studies of the effect of slopes on acoustic energy are possible. This paper addresses the combined acoustic fields generated by surface noise originating both in the deep basin and over the continental shelf.

Early acoustic propagation models utilizing ray theory were developed for deep water situations with little or no bottom interaction, especially at frequencies above 50 Hz. Eventually, researchers developed the normal mode approach for shallow water, range-independent environments. However, the development of the parabolic equation (PE) method made a significant contribution to efficient full-field computations at all frequencies and water depths in range-dependent environments. Recent modifications in the PE approach allow for steep angles and discontinuous interfaces in the geoacoustic description, which are essential for the sloping environment of interest.

A precursor to this environmental-acoustic study was an effort to characterize all of the world's large-scale, linear bathymetric slopes into generic classes and to assign representative geoacoustic descriptions. A large-scale, linear slope is defined as a feature with at least 2 km of vertical relief, with a length many times its width, and an average slope in excess of 1 deg. By this definition most continental margins and island arc systems, as well as numerous submarine ridges and some fracture zones, are included. The identification and statistical characterization (such as average slope and height) were made from unpublished bathymetric maps produced by the U.S. Naval Oceanographic Office. The details of this analysis, including maps showing the global distribution of slopes, are presented in Green and Matthews.[1] This work produced five generic classes, and the most common of these (Passive Margin) was used for the analysis represented herein.

This paper begins with a discussion of the geoacoustic, oceanographic, and acoustic field models. The results of slope interaction on surface noise generated both in the deep-water basin and over the shallow-water shelf follows.

GEOACOUSTIC, OCEANOGRAPHIC, AND ACOUSTIC MODELS

Geoacoustic

The passive-margin class of slope was chosen for the following analysis because it represents over one-third of all large-scale, linear ocean slopes.[2] This class is typical of much of the North Atlantic Ocean and is geoacoustically simple in that less than half of this class has outcropping rock or shallow acoustic basement. As shown in Figure 1, it consists of a 0.5 deg clayey-silt sediment fan in the basin from 150 into 75 km, a 2.81 deg silty continental slope from 75 into 22 km, and a 0.26 deg continental shelf. The shelf is terminated at a reference range where the water depth is 100 m. The shelf/slope break occurs at 200 m depth and the base of the slope is at 2800 m depth. In each region, the sediment is 500 m thick with depth dependent geoacoustic properties, overlaying a constant property half-space.

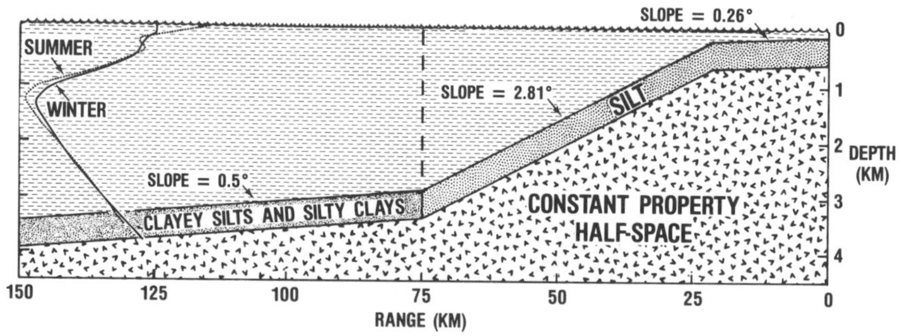

Figure 1. Geologic model for passive-margin with summer and winter Sargasso Sea sound speed profiles.

Oceanographic

Since the passive-margin class is representative of many sites along the east coast of the United States, a Sargasso Sea sound speed profile (SSP) was selected for the computer calculations. In order not to confuse oceanographic effects with bathymetric effects, a single range independent SSP was used. Calculations were made for the summer and winter SSPs shown in Figure 1.

Acoustic Field

The results were obtained using the implicit finite-difference (IFD) formulation of the parabolic equation (PE) method with high-angle capability. The high-angle capability extends the angular range of applicability to about 40 deg above or below the horizontal. This study deals with environments in which multiple reflections of sound between the sea surface and the sea bottom lead to acoustic ray paths greater than 30 deg above and below the horizontal, which makes the high-angle capability essential.

ACOUSTIC FIELD CALCULATIONS

A thorough treatment of noise is beyond the scope of this study and would require many detailed calculations for the appropriate horizontal distribution of noise sources and detailed geoacoustic and oceanographic descriptions in all directions. A very simple, but somewhat complete, noise model would involve noise coming from only three directions--the deep-water basin, along the continental slope/shelf, and directly landward of the receiver. In those situations where shipping over the shelf is extensive, one must include the coastal component of noise. However, those ships along the continental coast (but not directly landward) will probably be highly attenuated at a receiver positioned over the slope for two reasons. First, the entire propagation path will be in relatively shallow water so that the received level will be low due to bottom interaction. Second, this sound will propagate in a wedge and thus be refracted out to sea and away from the receiver after several bounces. Therefore, only noise sources in the deep-water basin and sporadic noise sources directly landward of the receiver, need be considered.

A suite of PE calculations was performed at 30 and 100 Hz for both summer and winter conditions. A source depth of 7 m was chosen to simulate a large surface ship. Surface decoupling losses play a major role in explaining the calculated results, so the source directivity patterns are given in Figure 2. The major axis of the 100 Hz energy is 33 deg with 3 dB down points at 16 and 55 deg. The 30 Hz energy is much steeper (3 dB down at 41 deg) and will consequently involve more bottom interaction and higher losses.

Acoustic fields were calculated from several ranges in the basin (upslope) and from several ranges over the shelf (downslope). The general character of all the upslope runs was the same and similarly, the downslope runs were qualitatively the same. Thus, for simplicity, a single position in the basin (150 km) and a single position over the shelf (0 km) were chosen to describe ambient noise over a continental slope. Obviously, this simplistic approach will not give absolute levels, but it does shed insight on the particular feature of interest - the noise notch.

Figure 3 gives the results for propagation up and down a thickly sedimented slope for a winter SSP and at a receiver depth of 20 m for both 30 and 100 Hz. The upper curves correspond to a noise source in the basin at a range of 150 km from the reference position over the shelf. The curves are displaced by approximately 8 dB due to the surface decoupling loss discussed previously, i.e., the 30 Hz energy is so steep that severe bottom loss is encountered at short range, even in deep water. At a range of 75 km, the slopes of the propagation loss curves change as energy begins to interact with the base of the continental slope. Beginning at 50 km, there is a dramatic change in the character of the acoustic field, due to increasing penetration of sound into the bottom.

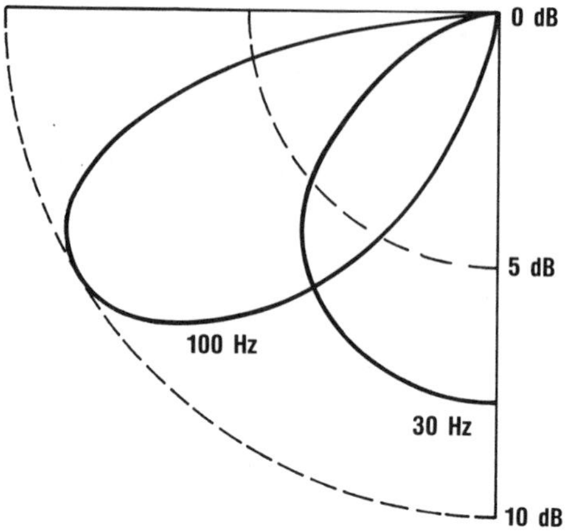

Figure 2. Source directivity patterns for 7-m source at 30 and 100 Hz.

The middle curves of Figure 3 represent the propagation loss for a noise source over the shelf at a range of 0 km. Over the shelf, the curves are displaced due to surface decoupling at an ever increasing rate due to the large amount of bottom interaction associated with the steep energy at 30 Hz. At a range of 22 km (shelf/slope break), the energy at the shallow 20 m depth decreases dramatically for both frequencies because the downward refracting SSP directs the sound into the sloping bottom where the vertical distribution of energy is decreased by twice the slope angle (2 x 2.81 deg) at each encounter. In effect, shelf noise "falls" down the slope and is trapped in the deep sound channel, thereby contributing little to the total ambient noise field at shallow depths over the slope.

The lower curves on Figure 3 represent the sum of the basin and shelf noise, assuming equal source level contributions from each. These summed results indicate a notch in the noise field over the slope with significant magnitude and range extent.

Figure 4 is a similar plot for a summer SSP. As before, the upper curves are displaced due to surface decoupling and nearly parallel out to a range of 75 km. At this point, the 100 Hz curve follows a similar behavior to that shown in Figure 3 for both frequencies. However, the 30 Hz propagation loss curve shows an extreme case of slope enhancement, which may be due to energy at the base of the slope grazing the boundary and thereby suffering little loss and no increased vertical distribution of energy. This behavior is real, but somewhat anamolous because calculations at other source locations generally show little to no enhancement. This case is included as an extreme to help bound the results on noise notch characteristics.

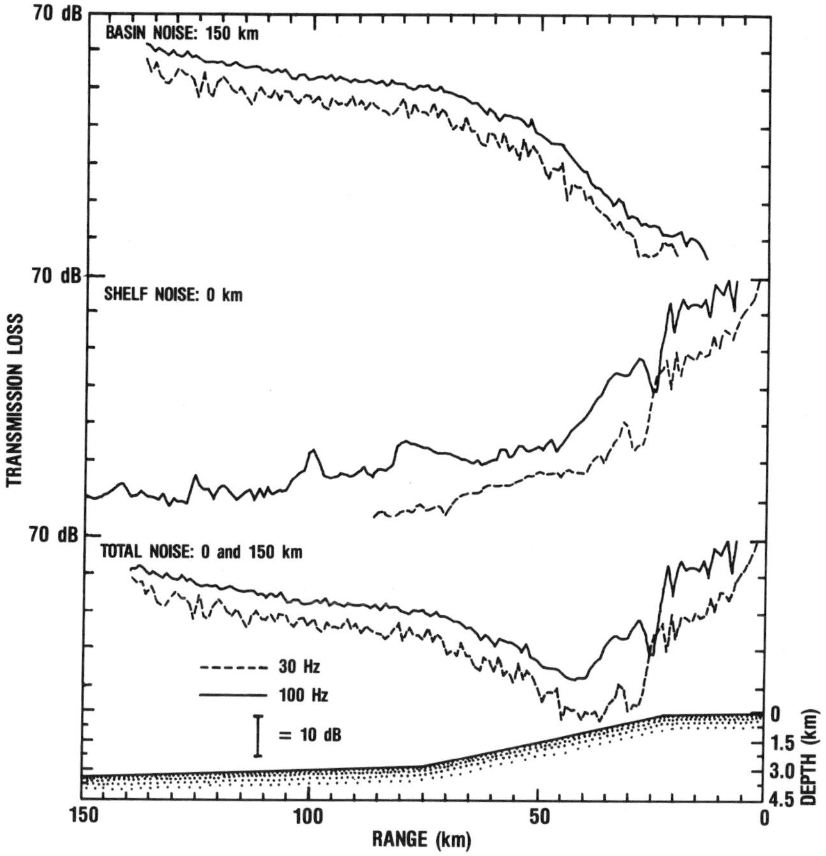

Figure 3. Propagation loss curves in the winter at 30 and 100 Hz for a 20 m receiver.

The middle curves on Figure 4 show the propagation loss values for downslope conditions. Again, surface decoupling reduces the 30 Hz energy levels to even a greater extent than in the winter because of the severe downward refraction caused by the summer SSP. At the shelf/slope break, the 100 Hz propagation loss increases dramatically when compared to the 30 Hz values. This can be explained in terms of the source directivity pattern, where the 100 Hz energy begins in relatively shallow angles (as low as 16 deg) and progressively becomes more shallow out to the shelf/slope break. At this range much of the energy is so shallow that it propagates down the slope with little additional interaction with the surface, especially for the summer SSP. In contrast, the 30 Hz energy originates at angles so steep (greater than 41 deg) that by the time it interacts with the slope it is still steep enough to have significant energy at shallow depths.

The lower curves in Figure 4 represent the sum of the basin and shelf noise, assuming equal source level contributions. As in winter, a noise notch is present at approximately the same range. Even for the extreme case of upslope noise enhancement at 30 Hz, a noise notch is observed.

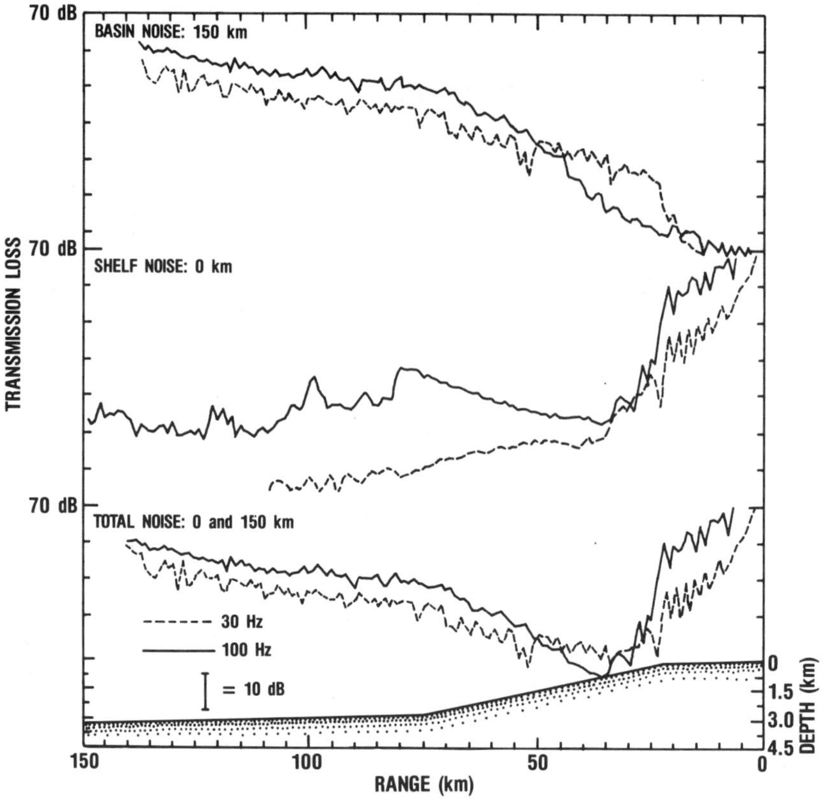

Figure 4. Propagation loss curves in the summer at 30 and 100 Hz for a 20 m receiver.

In order to quantify these results, straight line eyeball curve fits were made to the propagation loss curves at three receiver depths. These are shown in Figure 5 as transmission loss versus range relative to the deepest point of the notch. The main conclusion is that the notch is the most significant at the shallowest receiver depths. This is mainly controlled by shelf generated noise which is low near the ocean surface and high near the bottom.

Looking at this same data in another way, the magnitude and range extent (referenced to the propagation loss at the base of the slope) of the notch are plotted against receiver depth in Figure 6. The most significant notch has a magnitude of 27 dB and occurs in the summer at 100 Hz at a 20 m receiver depth. The summer, 30 Hz values are pessimistically low due to the extreme upslope noise enhancement. The winter results at the 20 m receiver depth show a 17 to 20 dB magnitude of the notch which is frequency independent within the accuracy of the calculations. The range extent of the noise notch is also independent of frequency and covers a set of ranges between 30 and 50 km, again with the greatest values at the shallowest receiver depths.

Two components of noise which were deferred from this study into ongoing work are (a) distant generated SOFAR noise and (b) local wind-generated noise. Some comments about these components are appropriate. The SOFAR noise occurs as ships and storms move over distant, sloping bathymetric features, or in cold, high-latitude waters, and steep energy is thereby converted into shallow angles which can propagate for long distances. The energy loss both at the bathymetric features (due to bottom interaction) and along the propagation path (due to absorption) is highly dependent on frequency. It is expected that in many locations, this component will be significant at frequencies below 100 Hz. The local wind-generated noise occurs at steep angles and is also highly frequency dependent as well as sea-state dependent. Previous measurements indicate that in many locations for frequencies around 100 Hz this component will be insignificant for wind speeds less than 5 kt. Thus, for certain realistic situations (e.g., for large ocean basins with boundaries, seamounts and cold water areas at large distances from the receiver, for sufficiently high frequencies to attenuate distant noise, for low local wind speeds, etc.) the predicted noise notch discussed in this paper should exist.

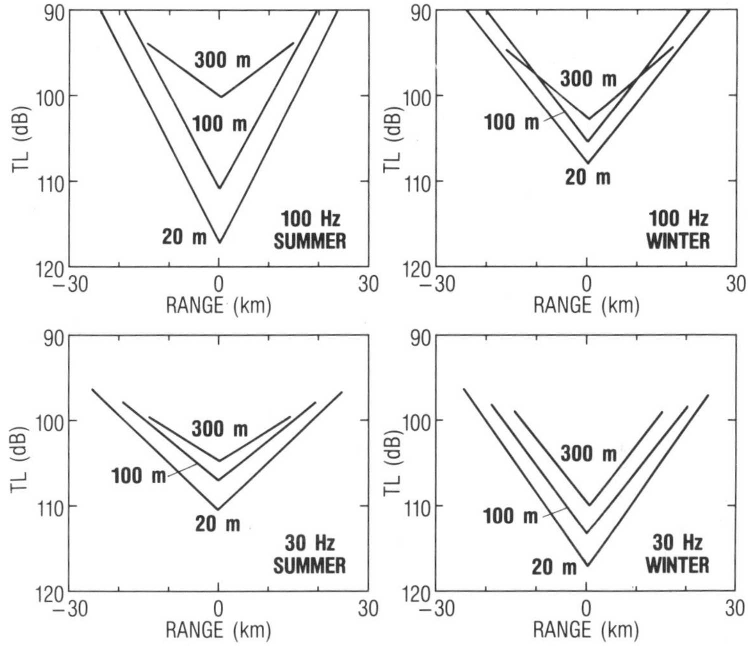

Figure 5. Frequency, seasonal, and receiver depth dependence of the noise notch.

CONCLUSIONS AND RECOMMENDATIONS

A high-angle version of a parabolic equation model (IFD formulation) with a full geoacoustic description of the bottom was used to calculate acoustic noise fields over thickly sedimented continental slopes. A combination of fields from noise sources in the deep water basin and from over the continental shelf indicated the presence of a noise notch with a magnitude of 20 dB and greater at shallow receiver depths. The notch extends over a large range interval (30 to 50 km) which is independent of season and frequency.

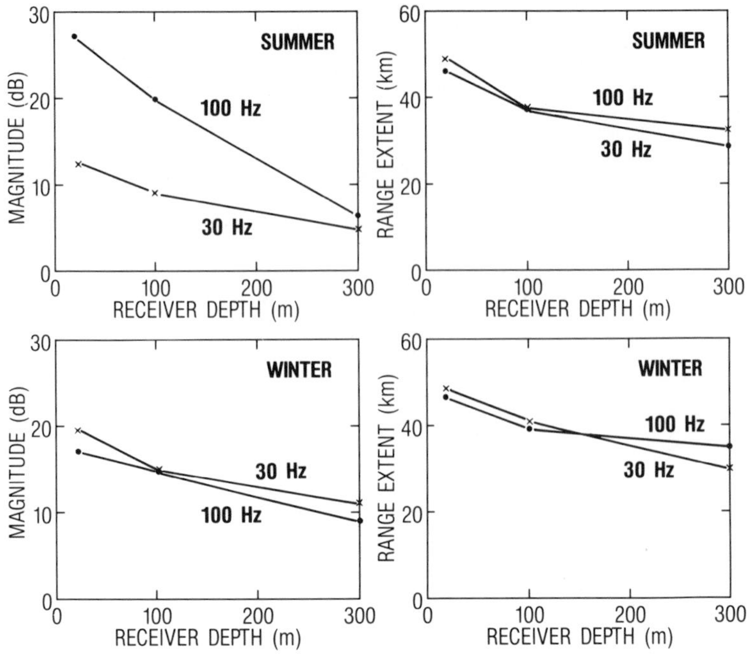

Figure 6. Magnitude and range extent of noise notch.

Additional work is needed in other types of environments to check the validity for various slope angles, subbottoms, and SSPs. The frequency dependence needs investigation and a more complete noise field description is desired (especially to include locally generated ship and wind noise and shallow angle noise generated at high latitudes and/or over distant continental boundaries). Finally, measurements should be taken using a high spatial sampling density in search of decreased ambient noise over slopes, to confirm the conclusions of this paper.

1. J. A. Green and J. E. Matthews, "Global Analysis of the Shallow Geology of Large-Scale Ocean Slopes," NORDA Tech Note 197 (1983).
2. D. R. Del Balzo, J. E. Matthews, J. V. Soileau, and C. Feuillade, "Acoustic Propagation over Large-Scale Linear Ocean Slopes," in: OCEAN SEISMO-ACOUSTICS, T. Akal and J. M. Berkson, ed., Plenum Press, New York (1985).

ACKNOWLEDGEMENT

This work was supported by the Office of Naval Technology, Bottom Interaction Program.

SOME NEW CHALLENGES IN SHALLOW WATER ACOUSTICS

E. C. Shang[1]

Institute of Acoustics, Academia Sinica
P. O. Box 2712
Beijing, China

ABSTRACT

Nowadays, we are still facing challenges in shallow water acoustics. Three topics of shallow water acoustics are discussed in this paper: (1) long-range reverberation modeling in shallow water; (2) source location in shallow water; and (3) new features of the transmission loss in shallow water.

INTRODUCTION

In recent years, shallow water acoustics has been receiving more attention. It is well known that the sound field in shallow water is much more complex than in deep water because of strong bottom interaction. The strong waveguide effect leads to some things that are not so readily understandable. One example is the long-range reverberation in shallow water. The present status of reverberation prediction and modeling in shallow water is still poorly developed. The experimental data on sound scattering by the bottom at low frequencies and low grazing angles are very sparse. The reason for this is the lack of an adequate reverberation model which can be used for extracting data representing the characteristics of the real world. Recently, in the Institute of Acoustics, Academia Sinica, two theoretical models of long-range reverberation in shallow water have been developed. In part (1) of this paper, two models – the angular power spectrum model and the normal mode model based on the Bass perturbation method – are briefly introduced. It is shown that the results of the two models are consistent. The second example of the challenging problems that we are facing in shallow water acoustics is source localization in shallow water. Owing to the complex modal interference field structure, the conventional beamforming technique is no longer effective. New methods of passive source ranging and source depth estimation based on mode decomposition are discussed in part (2). Even for transmission loss (TL) modeling which has been well developed, there are still a few factors needing further investigation. Some new features of TL in shallow water, such as "optimum" frequency and "trap frequency" of TL, are reported in part (3) of the present paper.

(1). LONG-RANGE REVERBERATION MODELING IN SHALLOW WATER

Since the pioneer work developed by Mackenzie (1962) and Urick (1970), little progress has been made on this difficult task. The basic theoretical form of reverbera-

[1]For present address, see end of paper.

tion in deep water has been established quite well for a long time. The backscattering strength of the sea bed has also been reported by a number of investigators (Mackenzie, 1961; Patterson, 1963; Burstein and Keane, 1964; Merklinger, 1968; Smailes, 1978; Bunchuk and Zhitkovskii, 1980). In deep water the paths from source to bottom and back to receiver are simple and the grazing angle of the sound on the bottom is readily determined from the ray diagram. In shallow water, however, the matter is not so simple. The problem is that multipath transmission in the shallow water waveguide and multi-angle scattering at the bottom must be taken into account. Recently, a phenomenological model of long-range reverberation in shallow water was developed based on the "angular power spectrum" (average flux) method of the sound field (Zhou et al., 1982). In addition to the phenomenological model, a normal mode theory of long-range reverberation in shallow water caused by the roughness of the bottom has been developed using the Bass perturbation method (Gao, 1986; Tang et al., 1986).

Angular Power Spectrum Model

The sound field excited by a point source averaged in a certain space interval can be expressed as (Brekhovskikh, 1965; Weston, 1980; Smith, 1974; Zhou, 1980):

$$I(r) = \int_0^{\pi/2} F(\theta, r)\, d\theta \tag{1}$$

where $F(\theta, r)$ is the angular spectrum of the field at horizontal range r and θ is the grazing angle of the interaction with the sea bottom. For an isovelocity layer and small grazing angles we have:

$$F(\theta, r) = \frac{2}{Hr} \exp\left[-\frac{Qr}{H}\theta^2\right] \tag{2}$$

where H is the water depth and Q is the bottom loss parameter, defined as:

$$-\ln |V_b(\theta)| = Q\theta \ , \tag{3}$$

where $V_b(\theta)$ is the plane wave reflection coefficient at the interface between the water and the sea bed.

By introducing a bi-angular scattering strength $M(\theta, \phi)$ describing the scattering characteristics phenomenologically, the long-range reverberation can be expressed as:

$$R(r) = \frac{\pi c \tau}{H^2 r} \left\{ \int e^{-\frac{Qr}{H}\theta^2} \int e^{-\frac{Qr}{H}\phi^2} M(\theta, \phi)\, d\phi\, d\theta \right\} \tag{4}$$

where c is the sound speed in the water, τ is the pulse duration of the emitted signal, and $r = ct/2$ is the distance from the source to the scattering area.

The process of "reverberation derived" backscattering strength can be established by assuming $M(\theta, \phi)$ as follows:

$$M(\theta, \phi) = \left[\sum_i^N \sqrt{\mu_i}\Delta_i(\theta)\theta^{n_i}\right] \times \left[\sum_i^N \sqrt{\mu_i}\Delta_i(\phi)\phi^{n_i}\right] \tag{5}$$

where:

$$\Delta_i(x) \equiv \begin{cases} 1 & x \in [x_{i-1}, x_i], \\ 0 & \text{otherwise.} \end{cases} \tag{6}$$

The backscattering strength is:

$$S_b(\theta) = 10 \log \left[M(\theta, \phi)|_{\phi=\theta}\right] = 10 \log \left[\sum_i^N \Delta_i(\theta)\mu_i\theta^{2n_i}\right] \tag{7}$$

Substituting Eq.(6) into Eq.(5), we get:

$$R(r) = \sum_i^N R_i \Delta_i(r) \tag{8}$$

$$R_i \approx \frac{\mu_i(\pi c\tau)}{(n_i+1)^2}\left[H^{(1-n_i)}Q^{(1+n_i)}r^{(2+n_i)}\right]^{-1} \tag{9}$$

$$r_i \equiv \frac{H}{Q}\theta_i^2. \tag{10}$$

From Eq.(9), the scattering indices μ_i and n_i can be derived from two data points on the experimental curve of reverberation level near r_i. Substituting θ_i, μ, and n_i into Eq.(7), the "reverberation derived" backscattering strength S_b corresponding to grazing angle θ_i can be obtained. So, a transformation between the angle dependence of bottom backscattering strength and the range dependence of reverberation level can be established through Eqs.(8–10). By using this model the "reverberation derived" backscattering strength for the frequency band 0.8–4.0 kHz and grazing angles of 2°–10°are shown in Fig. 1. For comparison, the dependence of S_b on θ obtained by some observers from deep water measurements are also included.

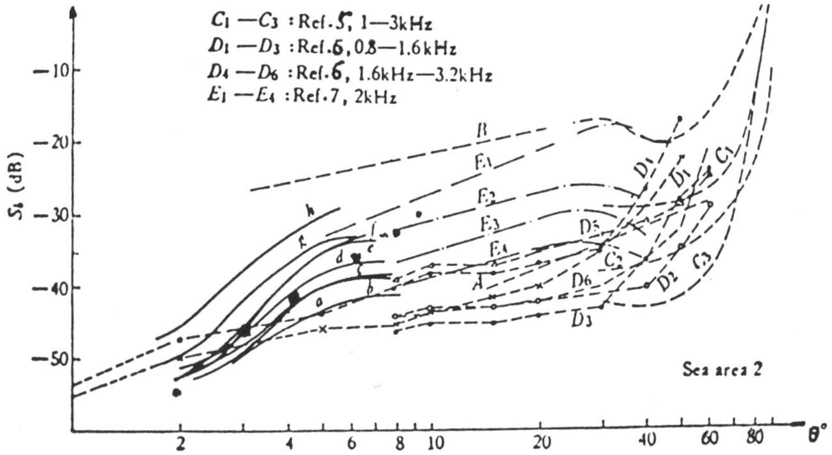

Fig. 1. $S_b(\theta)$ derived from reverberation data. References 5, 6, and 7 refer to Burstein and Keane (1964), Merklinger (1968), and Smailes (1978), respectively.

Normal Mode Theory

Let us consider a shallow water waveguide with a rough bottom (see Fig. 2). The fluctuating height of the rough bottom is ς, and $<\varsigma>=0$. By using the Bass perturbation theory, the continuity of the field at the rough interface is replaced by the inhomogeneous boundary condition at the smooth interface $z = H$, as follows:

$$\frac{\partial}{\partial z}u_1 - \frac{\partial}{\partial z}u_2 = v(\vec{r})\cdot G \tag{11}$$

$$\rho_1 u_1 - \rho_2 u_2 = p(\vec{r})\cdot G \tag{12}$$

where

$$v(\vec{r}) = \left(1-\frac{1}{m}\right)[\nabla_\perp(\varsigma\nabla_\perp)] + \left(k_1^2 - \frac{1}{m}k_2^2\right)\varsigma \tag{13}$$

$$p(\vec{r}) = (\rho_2-\rho_1)\varsigma\frac{\partial}{\partial z} \tag{14}$$

where $m = \rho_2/\rho_1$, and u_1, u_2 are the scattered field in the water and bottom respectively.

According to the Green formula, the scattered field of a unit point source in water is given by:

$$U_1(\vec{R}, \vec{R_0}) = \iint \left[G(R, r_s) v(\vec{r_s}) G(r_s, R_0) - \frac{1}{\rho_1} \frac{\partial}{\partial z} G(R, r_s) p(\vec{r_s}) G(r_s, R_0) \right] dr_s \quad (15)$$

where $G(R, r)$ is the Green function of the waveguide with the smooth bottom.

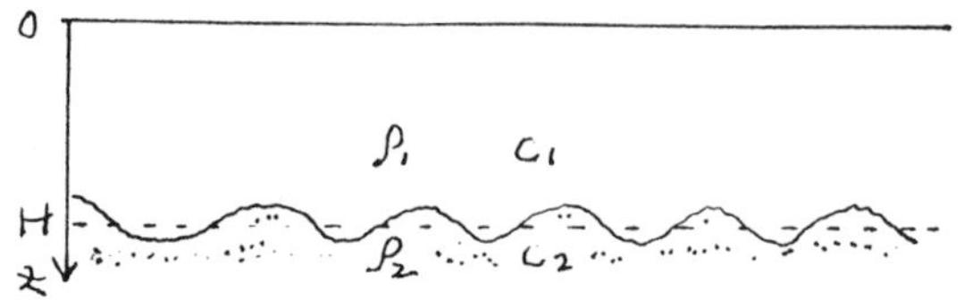

Fig. 2. Shallow water waveguide with rough bottom.

For an isovelocity water layer, we have:

$$G = \sum_l P_l \cos \left[\xi_l H + \frac{\ln V_b(\theta_l)}{2i} \right] e^{-ik_l(r_0 - r)} \quad (16)$$

where

$$P_l = \frac{-4\sqrt{\pi} e^{i\pi/4}}{H\sqrt{2k_l r_0}} F_l \cos \left[\xi_l z_0 + \frac{\ln V_b(\theta_l)}{2i} \right] e^{ik_l r_0} \quad (17)$$

where r_0 is the horizontal distance from the source to the scattering center, and

$$k_l^2 + \xi_l^2 = k_1^2 \quad (18)$$

$$\xi_l H = l\pi + \frac{i}{2} \left[\ln V_b(\theta_l) + \ln V_s(\theta_l) \right] \quad (19)$$

$$V_b(\theta_l) = \frac{m\xi_l - i\sqrt{k_l^2 - k_2^2}}{m\xi_l + i\sqrt{k_l^2 - k_2^2}} \quad (20)$$

$$V_s = -1 \quad (21)$$

$$F_l = \left[1 + \frac{1}{2iH} \left(\frac{1}{V_b} \frac{\partial V_b}{\partial \xi_l} + \frac{1}{V_s} \frac{\partial V_s}{\partial \xi_l} \right) \right]^{-1} \approx 1 . \quad (22)$$

Substituting Eq.(13), Eq.(14) and the Green function into Eq.(15), we get:

$$U_1(R_0, R_0) = \sum_l \sum_j P_l P_j \frac{f_{lj}}{4\sqrt{V_b(\theta_l) V_b(\theta_j)}} \iint \varsigma(\vec{r_s}) e^{(k_l + k_j)(r - r_0)} dr_s \quad (23)$$

where:

$$\begin{aligned} f_{lj} &= (-1)^{l+j} k_1^2 \left\{ \left[\left(1 - \frac{n^2}{m} \right) + (1 - \frac{1}{m}) \cos\theta_l \cos\theta_j \right] [1 + V_b(\theta_l)][1 + V_b(\theta_j)] \right. \\ &\quad \left. + (m-1) \sin\theta_l \sin\theta_j [1 - V_b(\theta_l)][1 - V_b(\theta_j)] \right\} . \end{aligned} \quad (24)$$

Table 1. Calculated scattering strengths.

grazing angle $\theta°$	$\mathcal{M}_{lj}^{S}$ dB (sand)	$\mathcal{M}_{lj}^{S}$ dB (silt)
2	-67	-63
3	-60	-56
4	-55	-51
5	-52	-48
6	-49	-45
7	-47	-44
8	-45	-41
9	-43	-40
10	-42	-39

The reverberation intensity is:

$$I_R = \overline{|U_1|^2} = \sum_l \sum_j + \sum_l \sum_j \sum_{l' \neq l} \sum_{j' \neq j} . \tag{25}$$

The second term in Eq.(25) might be neglected:

$$\begin{aligned} I_R &= \sum_l \sum_j |P_l|^2 |P_j|^2 \left| \frac{f_{lj}}{4\sqrt{V_b(\theta_l)V_b(\theta_j)}} \right|^2 \\ &\times \iint_S \iint_{S'} \overline{\varsigma(r_s)\varsigma(r_{s'})} \exp\left[i(k_l + k_j)(\vec{r}_s + \vec{r}_{s'})\right] d\vec{r}_s \, d\vec{r}_{s'} \end{aligned} \tag{26}$$

or

$$I_R = S \sum_l \sum_j |P_l|^2 |P_j|^2 \mathcal{M}_{lj}^S \tag{27}$$

where S is the scattering area, and:

$$\mathcal{M}_{lj}^S = (2\pi)^2 \sigma^2 \left| \frac{f_{lj}}{4\sqrt{V_b(\theta_l)V_b(\theta_j)}} \right|^2 \widetilde{W}(k_l + k_j). \tag{28}$$

Eq.(27) can be explained as follows: the reverberation intensity I_R is the sum of the scattering intensity of the j-th mode excited by the incident intensity of the l-th mode on the interface specified by the scattering coefficient $\mathcal{M}_{lj}^S$ (mode conversion matrix).

The $\mathcal{M}_{lj}^S$ defined by Eq.(28) has been calculated for four different types of bottom roughness correlation function given by (Brekhovskikh, 1974):

$$\begin{aligned} &(a) \quad w_1 = \exp[-(\rho/\rho_0)^2] \\ &(b) \quad w_2 = J_n(\rho/\rho_0)/(\rho/2\rho_0)^n \qquad n = 1,2,3,\ldots \\ &(c) \quad w_3 = \exp[-|\rho/\rho_0|] \\ &(d) \quad w_4 = (\rho/\rho_0)^n K_n(\rho/\rho_0)/2^{n-1} \qquad n = 1,2,3,\ldots \end{aligned} \tag{29}$$

It was shown that types (a) and (b) can never fit experimental data with reasonable parameters. The numerical results for type (d) are shown in Table 1, with ρ_0=7 m and σ=0.1 m for a frequency of 1000 Hz (for sand m=2.03, n=0.8; for silt m=1.74, n=0.94). In the frequency range of 0.5–4.0 kHz there is almost no frequency dependence of $\mathcal{M}_{lj}^S$.

It is interesting to compare the normal mode result given by Eq.(27) with the "angular spectrum" result given in Eq.(4). For this purpose, the "smoothing" procedure along depth z has to reduce to $|P_l|^2$ and $|P_j|^2$ in Eq.(27). We get:

$$\widehat{|P_l|}^2 \widehat{|P_j|}^2 \doteq \frac{(4\pi)^2}{H^4 k_0^2 r^2} \exp\left[-Q \frac{\pi^2 r}{k_0^2 H^3}(l^2 + j^2)\right]. \tag{30}$$

Then, by using the dispersion equation, the relation between θ_l and l is approximately given by:

$$\theta_l \approx \left(\frac{l\pi}{k_0 H}\right), \qquad d\theta_l \approx \left(\frac{\pi}{k_0 H} dl\right). \tag{31}$$

Transforming the summation of l,j in Eq.(27) into integration of θ_l and θ_j gives:

$$I_R = \frac{4^2 S}{H^2 r^2} \int e^{-\frac{Qr}{H}\theta_l^2} d\theta_l \int e^{-\frac{Qr}{H}\theta_j^2} \mathcal{M}_{lj}^S \, d\theta_j. \tag{32}$$

If $S = (\pi c \tau) r$ is considered, then:

$$I_R = \frac{4^2(\pi c \tau)}{H^2 r} \int e^{-\frac{Qr}{H}\theta_l^2} \int e^{-\frac{Qr}{H}\theta_j^2} \mathcal{M}_{lj}^S \, d\theta_l \, d\theta_j. \tag{33}$$

Comparing Eq.(33) with Eq.(4), we find that the normal mode theory and the "angular spectrum" theory give the same result, provided:

$$16 \mathcal{M}_{lj}^S = M(\theta, \phi). \tag{34}$$

(2). SOURCE LOCATION IN SHALLOW WATER

It was pointed out by C. S. Clay (1966) that owing to the complex multimode interference structure of the field in shallow water, the conventional beamforming approach is no longer effective for source location. For example, let us consider the bearing of a source by using a long horizontal array at depth z. The receiving field at the n-th hydrophone, corresponding to a source located at (r_0, z_0) and with a bearing angle θ_0, is given by:

$$P_n = \pi i \sum_m^M U_m(z) U_m(z_0) \sqrt{\frac{2}{\pi k_m r_0}} e^{-k_m (r_0 + nd \sin\theta_0) - \beta_m r_0}. \tag{35}$$

Then, the conventional beamformer output of the array is:

$$\begin{aligned} D(\theta, \theta_0) &= \sum_n^N P_n e^{-ik_0 nd \sin\theta} \\ &= \sum_n^N |P_n| e^{i[\phi_n(r_0, z_0, z, \theta_0) - ik_0 nd \sin\theta]}. \end{aligned} \tag{36}$$

Because of the multipath effect, in general (except for θ_0=0) we have:

$$\phi_{n+1} - \phi_n \neq \text{const} = (k_0 d \sin\theta_0).$$

Then the performance of the beamformer will be degraded especially for the endfire case. A numerical example of the degradation of the array in an isovelocity layer of shallow water with a sand bottom is shown in Fig. 3. the phase differences on the array are shown in Table 2. For comparison, the $D(\theta, \theta_0)$ of the same array in free space is shown in Fig. 3 by the dotted line.

However, the modal interference as a sort of holograph can be used for source location provided a proper information extraction method is considered. New passive ranging and source depth estimation approaches have been developed based on the mode filtering technique (Shang et al.; Shang, 1985). By using a mode filter, the field sample data for a vertical array are transformed to a data set as:

$$S_m \sim U_m(z_0) e^{ik_m r_0}, \qquad m = 1, 2, \ldots \tag{37}$$

Table 2. Phase difference on the array

θ_0=90°	H=30 m
f=220 Hz	r=2000 m
d=6 m	z=3 m
n=20	z_0=3 m

n	$\Delta_{n,n+1}$°	n	$\Delta_{n,n+1}$°
1	40	11	67
2	60	12	67
3	73	13	66
4	77	14	66
5	74	15	67
6	70	16	60
7	66	17	34
8	65	18	24
9	65	19	51
10	66	20	61

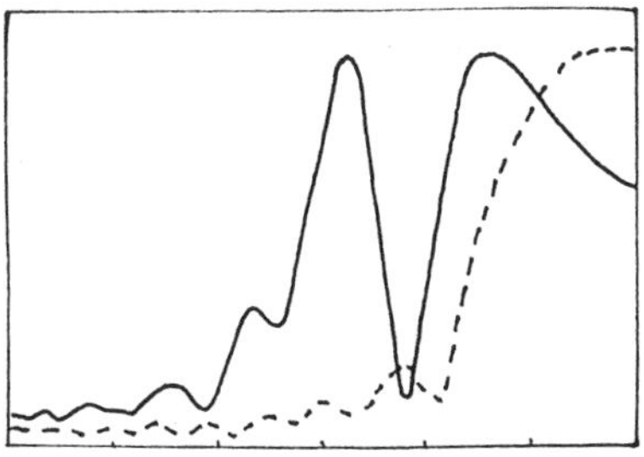

Fig. 3. Bearing performance of array.

Both the source range and depth information are included in this data set. The source range information can be extracted by measurement of three individual mode phases. The source range was expressed in terms of the "mode interference distance" D_{ij} as follows:

$$r_0 = D_{ij}[L_{ij} + (\delta\phi_{ij}/2\pi)] \tag{38}$$

where D_{ij}=$2\pi(k_i - k_j)$, $\delta\phi_{ij}$ is the phase difference of the i-th mode and j-th mode, k_i is the wave number of the i-th mode given by a numerical code, and L_{ij} is a certain integer. The value of L_{ij} can be estimated by means of a comparison of the phase of the j-th mode with another mode, say the m-th mode, and then L_{ij} would be estimated by solving the following equation:

$$(\delta\phi_{ij}/2\pi) = \text{Fractional part}\left\{\left(\frac{D_{ij}}{D_{jm}}\right)[L_{ij} + (\delta\phi_{ij}/2\pi)]\right\} \tag{39}$$

The phase related to r_0 in the data set S_m can be compensated provided the source range information is extracted. After phase compensation, we get a new data set:

$$S_m \sim U_m(z_0), \qquad m = 1, 2, \ldots \tag{40}$$

A correlation operator is used for source depth estimation:

$$F(z) = \sum_{m}^{M} U_m(z_0)U_m(z). \tag{41}$$

The resolution of this estimator is about H/M, where M is the effective mode number. The effective mode number M is dependent on the bottom reflection loss, if a linear loss is taken for small grazing angle, as:

$$\ln |V_b(\theta)| = -Q\theta. \tag{42}$$

Then for an isovelocity water layer, M is given by (Wang and Shang, 1981):

$$M \doteq \sqrt{\frac{0.7H}{Qr}} \left(\frac{k_0H}{\pi}\right). \tag{43}$$

At least three modes are needed for range estimation, so the limiting range in which the source information can be maintained is:

$$r_{lim} = \frac{0.7H}{9Q} \left(\frac{k_0H}{\pi}\right)^2. \tag{44}$$

Recently, a high resolution source depth estimation method by using Pisarenco's orthogonal decomposition has been proposed (Shang and Wang, 1986). The performance of this depth estimator is shown in Fig. 4.

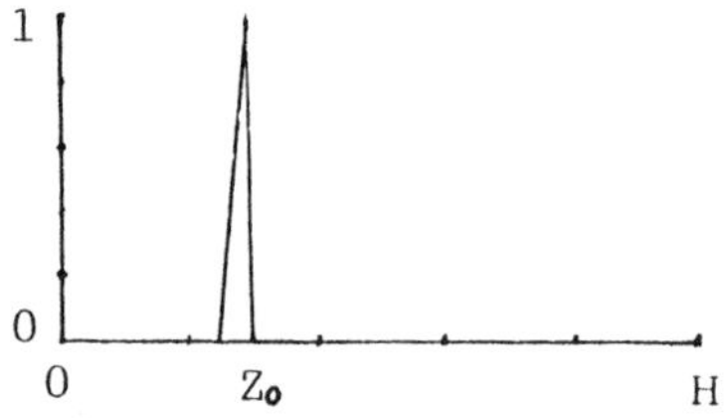

Fig. 4. Source depth: 0.15; estimated depth: 0.15.

(3). NEW FEATURES OF THE TL IN SHALLOW WATER

The frequency dependence of the TL in shallow water has been investigated for many years. It is found that an optimum frequency F_{opt} is often present in shallow water (Jensen and Kuperman, 1983). In fact, the frequency dependence of sediment attenuation can play a dominant role on the existence of F_{opt} (Zhang et al., 1985; Zhou and Rogers, 1986). So, we are facing a new task to collect the information on the frequency dependence of sediment attenuation at lower frequencies. In the present paper, a new feature of the TL in shallow water is reported. In some of our sound propagation experiments conducted in shallow water, a "trap" frequency can often be found in most of the summer-condition TL data. An explosive signal was used in the experiments, the spectrum at near distances is shown in Fig. 5A, and the spectrum at r=27.7 km is shown in Fig. 5B. Around the "trap" frequency f=630 Hz the TL can be 20 dB lower than the adjacent frequency components. The signal at this "trap" frequency suffered a heavy spreading in the time domain. In Fig. 6, the received signals at r=37.2 km are shown (Fig. 6A – f=400 Hz; Fig. 6B – f=630 Hz; Fig. 6C – f=800 Hz). The "trap" frequency is variable for different experiments in the range of 500–2000 Hz. The mechanism causing this phenomenon is not very clear yet.

A

B

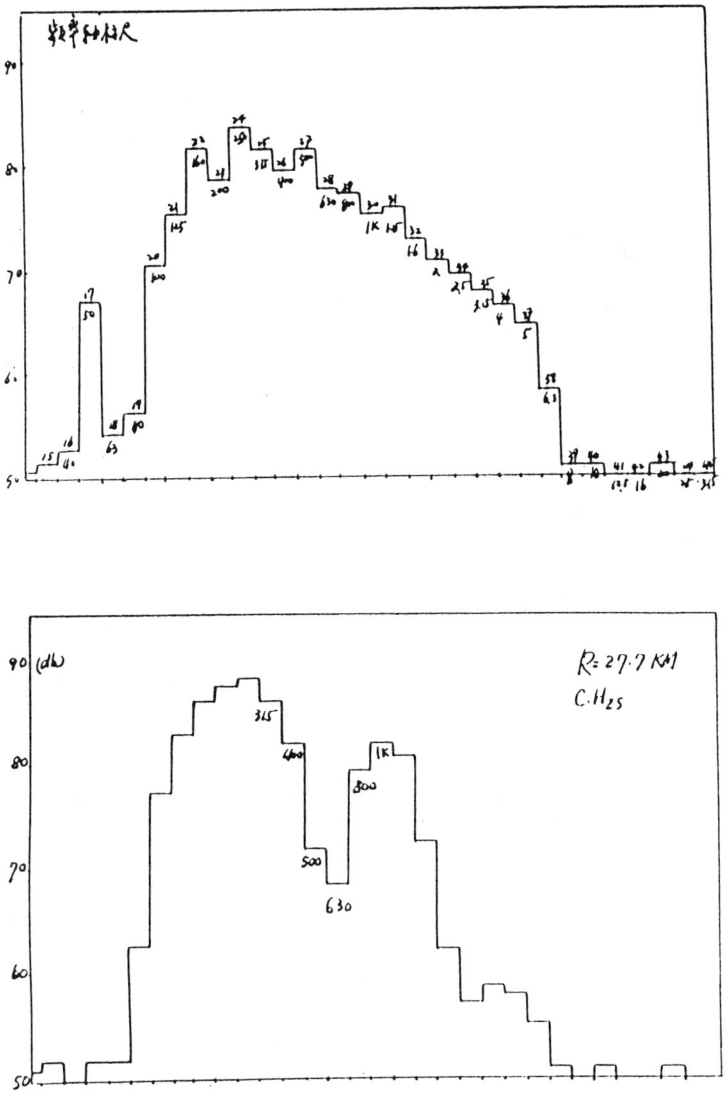

Fig. 5. Signal spectrum: (A) at short range; (B) at $r = 27.7$ km.

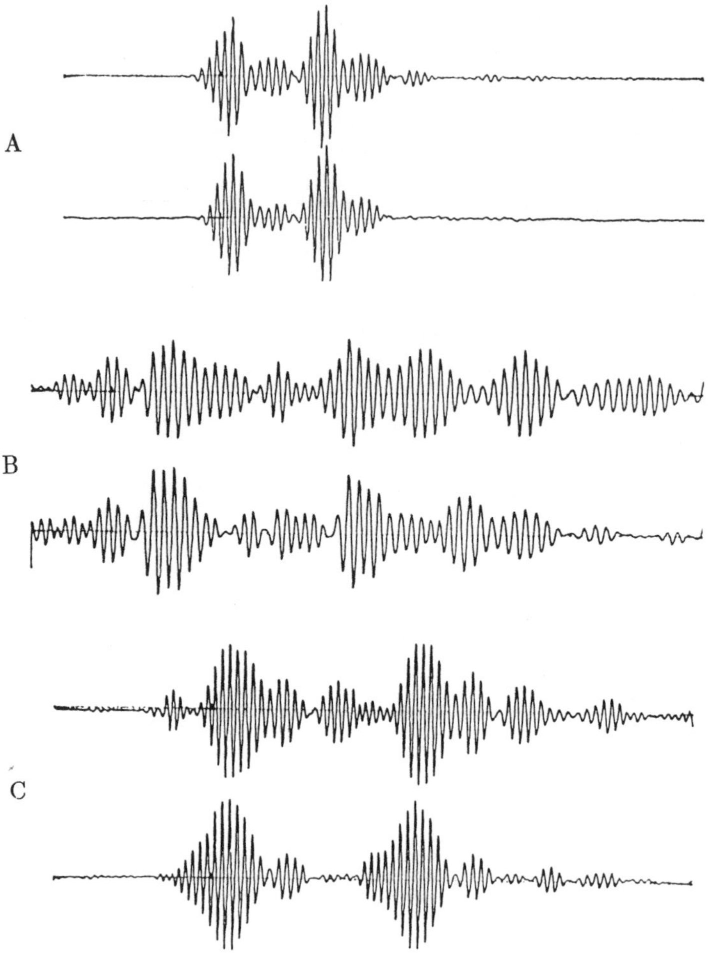

Fig. 6. Signal at $r = 37.2$ km: (A) $f = 400$ Hz; (B) $f = 630$ Hz; (C) $f = 800$ Hz.

REFERENCES

Bass, F. G., and Fuks, I. M., 1979, "Wave Scattering From Statistically Rough Surfaces," Pergamon, New York.

Brekhovskikh, L. M., 1965, The average field in an underwater sound channel, Sov. Phys. Acoust., 11:126.

Brekhovskikh, L. M., 1974, "Ocean Acoustics," Ch. 4.

Bunchuk, A.V., and Zhitkovskii, Yu. Yu., 1980, Sound scattering by the ocean bottom in shallow-water regions (review), Sov. Phys. Acoust., 26:383.

Burstein, A. W., and Keane, J. J., 1964, Backscattering of explosive sound from ocean bottoms, J. Acoust. Soc. Am., 36:1596(L).

Clay, C. S., 1966, Waveguides, arrays, and filters, Geophysics, 31:501.

Gao, T. F., 1986, A relation between waveguide and nonwaveguide sound scattering from rough interfaces, (to be published).

Jensen, F. B., and Kuperman, W. A., 1983, Optimum frequency of propagation in shallow water environments, J. Acoust. Soc. Am., 73:813.

Mackenzie, K. V., 1961, Bottom reverberation for 530- and 1030-cps sound in deep water, J. Acoust. Soc. Am., 33:1498.

Mackenzie, K. V., 1962, Long-range shallow-water bottom reverberation, J. Acoust. Soc. Am., 34:62.

Merklinger, H. M., 1968, Bottom reverberation measured with explosive charges fired deep in the ocean, J. Acoust. Soc. Am., 44:508.

Patterson, R. B., 1963, Backscatter of sound from a rough boundary, J. Acoust. Soc. Am., 35:2010.

Shang, E. C., 1985, Source depth estimation in waveguides, J. Acoust. Soc. Am., 77:1412.

Shang, E. C., Clay, C. S., and Wang, Y. Y., 1985, Passive harmonic source ranging in waveguides by using mode filter, J. Acoust. Soc. Am., 78:172.

Shang, E. C., and Wang, Y. Y., 1986, Mode decomposition approach for source depth estimation in shallow water, 12 ICA, Toronto, Canada.

Smailes, I. C., 1978, Bottom reverberation measurements at low grazing angles in the NE Atlantic and Mediterranean Sea, J. Acoust. Soc. Am., 64:1482.

Smith, P. W., 1974, Average sound transmission in in range-dependent channels, J. Acoust. Soc. Am., 55:1197.

Tang, D. J., Shang, E. C., and Gao, T. F., 1986, Theoretical analysis of shallow water reverberation, (to be published).

Urick, R. J., 1970, Reverberation-derived scattering strength of the shallow sea bed, J. Acoust. Soc. Am., 68:287.

Wang, T. C., and Shang, E. C., 1981, "Underwater Acoustics" (in Chinese), Science Press, Beijing, China.

Weston, D. E., 1980, Average flux methods for oceanic guided waves, J. Acoust. Soc. Am., 68:287.

Zhang, X. Z., Zhou, J. X., and Shang, E. C., Oct. 1985, Computer simulation of TL in shallow water with layered structure bottom, The 4-th Acoustics Conference in China.

Zhou, J. X., 1980, Acta Acustica, May:86, (in Chinese).

Zhou, J. X., Guan, D.H., Shang, E. C., and Luo, E. S., 1982, Long-range reverberation and bottom scattering strength in shallow water, Chinese Journal of Acoustics 1:54 (in English).

Zhou, J. X., and Rogers, P., 1986, Effect of nonlinear frequency dependence of seabottom sound attenuation on low-frequency acoustic response in shallow water, J. Acoust. Soc. Am. Suppl. 79:S40.

[1]Current address:

E. C. Shang
Department of Computer Science
Yale University
P.O. Box 2158 Yale Station
New Haven, CT
USA 06520-2158

PROPAGATION OF SEISMIC AND ACOUSTIC WAVES IN HORIZONTALLY STRATIFIED MEDIA WITH STOCHASTICALLY ROUGH INTERFACES

Henrik Schmidt

SACLANT ASW Research Centre
Viale San Bartolomeo 400
19026 La Spezia, Italy

ABSTRACT

A boundary perturbation method has recently been extended to treat scattering at a randomly rough interface separating elastic media, and the general seismic/acoustic propagation model SAFARI has been modified to include this feature. This has yielded the possibility of treating propagation more realistically in ocean environments where rough surface scattering is an important attenuation mechanism. The effect of rough surface scattering on propagation in different ocean environments is analysed, with special emphasis on low-frequency seismic interface waves along a randomly rough sea bed.

INTRODUCTION

With the objective of obtaining exact solutions to the wave equation, horizontal stratification is a common assumption in underwater and atmospheric acoustics as well as in crustal seismology. To extend the applicability of the numerical models to cases where the interfaces are globally horizontal, but characterized by a small stochastic roughness, a series of approximate theories have been presented, based on a perturbational approach. These theories have earlier been applicable to rough interfaces between fluid media only, but recently the perturbational approach was extended by Kuperman and Schmidt[1] to cover also the elastic case and the theory has been implemented in the general seismic-acoustic SAFARI code[2,3]. In underwater acoustics this has yielded the possibility of modelling sound propagation more realistically in environments where both roughness and shear properties are important.

Here the theory behind the roughness modification of the SAFARI code is first briefly outlined. Then the numerical model is used to illustrate the influence of interface roughness on seismic and acoustic propagation in both deep and shallow water environments, with special emphasis on the propagation of seismic interface waves along rough sea beds.

THE NUMERICAL MODEL SAFARI

The SAFARI model is a general application code for seismic/acoustic

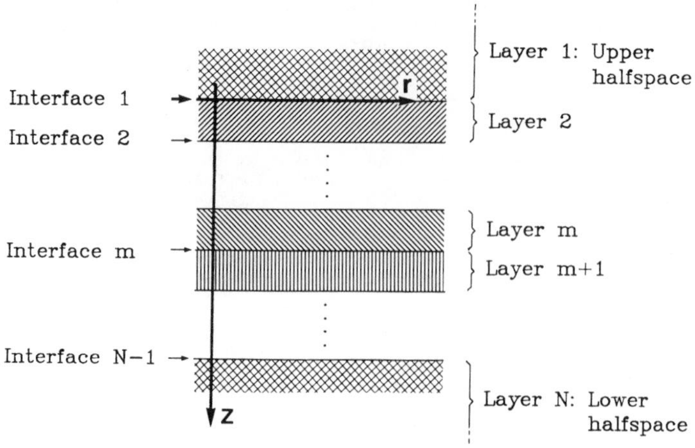

Fig. 1 Horizontally stratified medium.

propagation in horizontally stratified fluid/solid environments. The model and its mathematical background has been described in Refs. 2,3, and here only the details pertinent to the rough interface perturbation shall be outlined.

The medium is assumed to be a horizontally stratified stack of solid or fluid layers, Fig. 1, and a source of sound can be present in any layer. The exact integral representations for the field in such an environment can be obtained by integral transform techniques, and a whole class of numerical models have been developed based on this technique, differing mainly in the numerical approach taken. Traditionally, hardware restrictions have promoted the use of recursive solution techniques like the Thomson-Haskell propagator approach or invariant embedding. In SAFARI an unconditionally stable global matrix approach has been applied [2,3], which greatly simplifies implementation of the rough surface scattering as shown below.

In a cylindrical coordinate system with the axis passing through the source, the radial and vertical displacements in a general solid layer are expressed in terms of displacement potentials as

$$u_n = \frac{\partial \phi_n}{\partial r} - \frac{\partial \psi_n}{\partial z} \tag{1}$$

$$w_n = \frac{\partial \phi_n}{\partial z} + \frac{\partial \psi_n}{\partial r} \tag{2}$$

where the potentials satisfy homogeneous wave equations, with the solutions

$$\phi_n(r,z) = \int_0^\infty [A_n^-(\eta)e^{-iz\eta_{z,n}(h)} + A_n^+(\eta)e^{iz\eta_{z,n}(h)}]\eta J_0(r\eta)d\eta, \tag{3}$$

$$\psi_n(r,z) = \int_0^\infty [B_n^-(\eta)e^{-iz\eta_{z,n}(k)} + B_n^+(\eta)e^{iz\eta_{z,n}(k)}]\eta J_0(r\eta)d\eta, \tag{4}$$

where A_n^-, A_n^+, B_n^-, and B_n^+ are arbitrary functions of the horizontal wavenumber, representing the amplitudes of up and down going conical waves. $\eta_{z.n}(h)$ and $\eta_{z.n}(k)$ are the corresponding vertical wavenumbers of the compressional and shear waves, respectively. In a fluid layer only the compressional potential, (3), exists. In the source layer, integral representations similar to (3) and (4) for the source field has to be added, and he total field at any interface n, separating layers n and n+1, now has to atisfy the boundary conditions of displacement and stress continuity. The ntegral representations for the displacements are obtained by inserting 3) and (4) in (1) and (2), and those for the stresses follow by Hooke's aw. It is obvious, that the integral representations for these physical arameters are of the same form as (3) and (4); hence the boundary conitions have to be satisfied by the kernels, i.e. for each horizontal avenumber. For the general interface between two solid media, this leads o 4 equations in the 8 unknown wavefield amplitudes in the two layers. or convenience, these equations are expressed in the operator form

$$B(\tilde{\chi}_j; \tilde{\chi}_{j+1}) = 0 \tag{5}$$

here $\tilde{\chi}_j$ represents the kernels in both (3) and (4) for a solid layer and he kernel of (3) for a fluid layer.

In the global matrix approach, the coefficients in all local systems f equations (5) are first calculated. Then these equations are collected n a global system of equations, the solution of which yields the field in ll layers simultaneously. Compared to the traditional recursive pproaches, the global technique has a number of computational advantages, he most important being the efficient computation of the field at many eceiver depths[4].

In the present context, however, it is important that the numerical nplementation closely follow the original boundary value problem forulation. Hence, the wavefield amplitudes are retained as the basic nknowns, facilitating implementation of more complex media models. It is ell known that for a fluid medium with $1/c(z)^2$ linear, the exponential epth behaviour in (3) is simply replaced by Airy functions. This type of luid layers is therefore easily implemented in the global solution techique. Also more complicated solid media can be treated. Traditionally, ne solid layers are assumed to be homogeneous and isotropic. The last onstraint can, however, be slightly softened, as only transverse isotropy s required in the present cylindrically symmetric formulation. The only nange required is the replacement of the simple geometric determination of ne vertical wavenumber by a calculation of a more complicated function of ne horizontal wavenumber[5]. All other parts of the solution technique are naffected, including the formal structure of the boundary condition operaor (5). Finally, it turns out[1], that the effect af a small stochastic oughness of any interface in the layer stack can be accounted for by a elatively simple modification of the associated boundary condition operaor (5). This important feature is briefly outlined in the following secion.

OUGH INTERFACE PERTURBATION

The boundary perturbation method developed by Kuperman[6] to treat cattering of sound at a randomly rough interface has been implemented in everal numerical models in underwater acoustics. Due to the complexity of ne algebra involved, only the perturbations for the simplest interface ypes, involving solely fluid media, has been developed and implemented. ecently, however, it was shown by Kuperman and Schmidt[1], that the perurbed boundary conditions can be formulated in terms of a set of effective potentials", to which the standard boundary condition operator 5) should be applied. The "effective potentials" are themselves related the mean field potentials through the same operator (5). As a result ne solution of the global system of equations resulting from the perturbed boundary condition operators will yield the coherent component of the tota field. Due to the fact that the boundary operator (5) is one of the basi building blocks in the global solution technique, the implementation in th SAFARI code was relatively straighforward, even in the general solid case as most of the algebra presented in Ref. 6 can be taken care of by the com puter. As demonstrated in Ref. 1 the perturbed boundary conditions becom especially simple in the Kirchhoff approximation, valid for roughness smal compared to the vertical wavenumber. The initial implementation has there fore been based on this approximation. Here the main theoretical result from Ref. 1 are outlined, and in the following section applications to pro pagation problems in different ocean environments will be presented.

Let a rough interface separating two isovelocity layers j and j+1 b defined by z = $\gamma(r)$, with mean zero, $\langle \gamma(r) \rangle = 0$. The total field i then expressed as a sum of the mean field and a scattered field as

$$\chi = <\chi> + s; \begin{cases} \phi = <\phi> + p, \\ \psi = <\psi> + q, \end{cases} \tag{6}$$

where the scattered field s originates at the interface and thus contain only components travelling away from the interface. For a solid medium th mean and scattered terms in the formal representation (6) represent bot the compressional and shear waves. After some algebra it can be shown tha the scattering is taken into account by replacing the local continuit equations (5) by

$$B(\tilde{\chi}_j^*; \tilde{\chi}_{j+1}^*) = 0 \tag{7}$$

where $\tilde{\chi}_j^*$ is the "effective potential"

$$\tilde{\chi}_j^* = <\tilde{\chi}_j> \left(1 - \frac{<\gamma_j^2>}{2}\eta_{z,j}^2\right) + <\gamma_j \widetilde{\frac{\partial s_j}{\partial z}}> . \tag{8}$$

The last term in general involves scattering integrals, but in th Kirchhoff approximation the expressions simplify significantly, and th scattering terms are related to the mean field as

$$B\left(<\gamma_j \widetilde{\frac{\partial s_j}{\partial z}}>; <\gamma_j \widetilde{\frac{\partial s_{j+1}}{\partial z}}>\right) =$$

$$- <\gamma_j^2> B\left(i\eta_{z,j}\frac{\partial <\tilde{\chi}_j>}{\partial z}; -i\eta_{z,j+1}\frac{\partial <\tilde{\chi}_{j+1}>}{\partial z}\right). \tag{9}$$

Thus all terms in the effective potentials (8) are expressed in terms c the mean fields, and the new local system of equations (7) can therefore b considered as equations in these new unknown functions. When assembled i the global system of equations, the standard solution technique for th non-perturbed case can be used to obtain the mean field in all layer simultaneously.

Due to the local nature of the scattering formulation, multiple scat tering is not treated, but on the other hand there are no restrictions c the number of rough interfaces that can be treated by the current tech nique. The modified boundary conditions (7) can be applied at any inter face, and any type of interface, whether involving solid or liquid medi is treated in a consistent manner.

In the actual implementation the operations involved in (8) and (9) are taken care of by simple matrix operations on already existing matrices, i.e. the coefficient matrices in the unperturbed boundary condition operator (5). This again stresses the convenience of the boundary-value-problem formulation of the global-matrix-solution technique.

NUMERICAL EXAMPLES

The scattering of sound by rough surfaces can be an important attenuation mechanism for the coherent component of sound propagating in the ocean. Up to now the modelling capabilities have been limited to rough interfaces between fluid media. The models have therefore mainly been applicable to scattering by waves on the ocean surface. It is well known, however, that the shear properties have significant influence in several areas of ocean acoustics, but a detailed analysis of scattering from rough interfaces between media supporting shear waves has so far not been possible. With the implementation of the generalized boundary perturbation algorithm in the SAFARI code, such an analysis is now feasible for roughness within the limits of the Kirchhoff approximation. In the following some examples will be given on propagation in environments where the combined effect of shear and roughness is important. The model has been verified by comparison with available theoretical results for limiting cases. In the complicated solid cases a rigorous comparison with experimental results have to be performed, thus the applications presented here are initial results.

Reflection from rough ice cover

Recently, experimental transmission loss data from the Arctic was analysed by DiNapoli and Mellen[7], using available numerical models including different liquid-medium scattering approximations, and it was found, that none of these could account for the relatively high transmission losses observed at low frequencies. This data set therefore formed a challenging initial application of the modified SAFARI code, and it was shown in Ref. 1

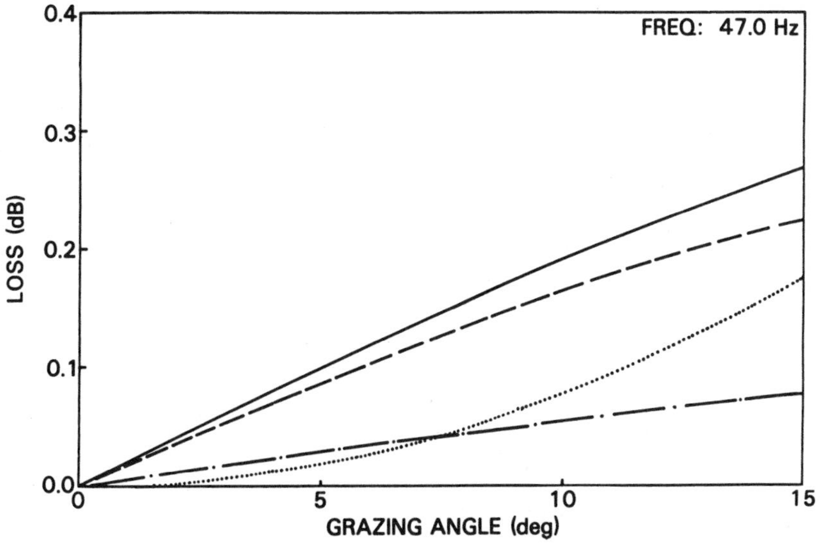

Fig. 2 Reflection loss of ice cover. Solid curve: full result. Dashed: no free surface roughness. Dashed-dotted: no roughness. Dotted: rough pressure release surface.

that the combined effect of shear properties and roughness may be able to account for the high losses observed.

As an example, Fig. 2 shows the calculated plane wave reflection coefficients for 47 Hz for an ice cover of 3.9 m thickness. Compressional speeds of 3000 m/s and 1440 m/s were assumed for ice and water, respectively, together with a shear speed of 1300 m/s for the ice. The associated shear attenuation was estimated to 2.5 dB/λ. The water/ice interface has an RMS roughness of 1.9 m, and the ice/air interface a roughness of 0.6 m. The solid curve shows the full elastic solution. The dashed curve indicates the result obtained if the ice/air interface is smooth, whereas the dashed/dotted curve shows the reflection loss when both interfaces are smooth. For comparison the Kirchhoff approximation for the 1.9 m rough free water surface is shown as a dotted curve.

By comparing the different curves it is clear, that the full elastic solution gives higher losses than the sum of the free surface scattering result and the volume attenuation contribution in ice. Thus the scattering into shear waves is an important loss mechanism at low grazing angles. It should be stated that the ice roughness structure is probably at the limit of applicability of the Kirchhoff approximation, and the present results are therefore mainly qualitative. A non-Kirchhoff theory currently under development, however, is expected to yield even higher values of the reflection loss.

Shallow water propagation

It is well established that the bottom shear properties are important for shallow water propagation. To demonstrate that also scattering into shear waves can be a significant attenuation mechanism in such an environment, the model has been used to analyse the propagation in the environment shown in Fig. 3. A soft sedimentary layer of 10 m thickness is overlying a basalt subbottom, and the water/sediment interface is assumed to have an RMS roughness of 0.5 m. Fig. 4 shows the calculated transmission losses

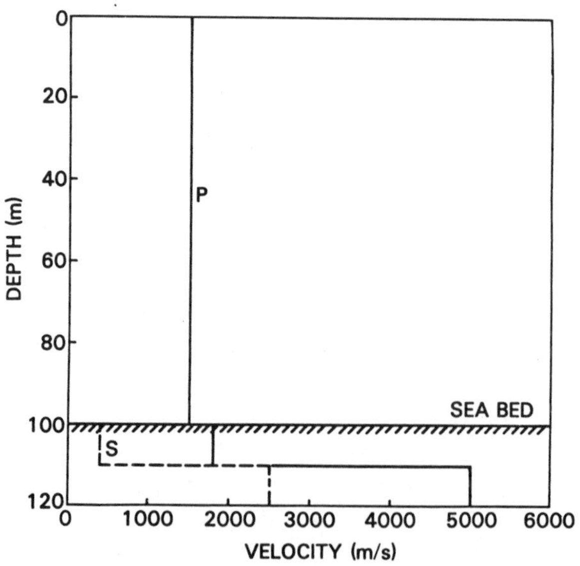

Fig. 3 Wave speed profile for shallow water environment.

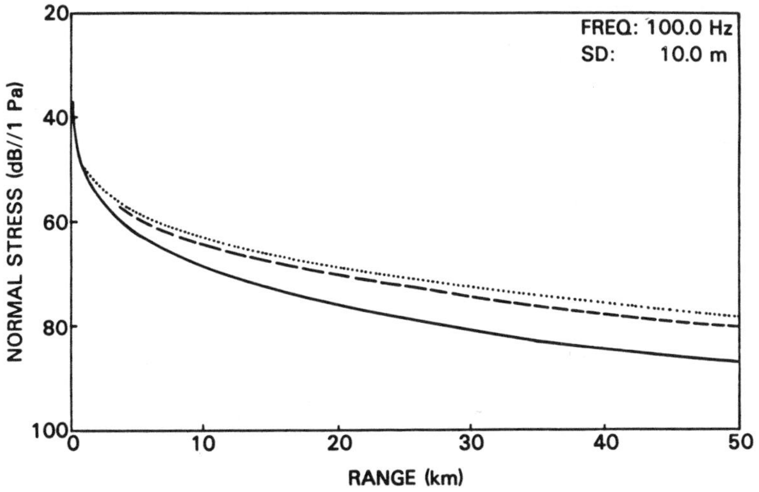

Fig. 4 Depth-averaged transmission loss for shallow water environment. Solid curve: full elastic scattering. Dashed: compressional scattering only. Dotted: no scattering.

averaged over the water depth. The solid curve represents the full result, whereas the dashed curve shows the losses obtained if the scattering into shear waves is ignored and only scattering into compressional waves is considered. The dotted curve shows the result obtained for a smooth sea bed, and the influence of the shear wave scattering is again significant.

Seismic interface waves

Due to the fact that the seismic interface waves are almost entirely controlled by the sediment shear properties, they yield a convenient basis for shear property inversion. During the last decade significant effort has been put into the investigation of the propagation characteristics of these waves. A review is given in Ref. 8, where also a special data set was analyzed in order to determine not only the shear speed profile, but also the associated shear attenuations. It was concluded that a volume attenuation being quadratic in frequency had to be assumed in order to model the data. No available theory predicts this behaviour at the very low frequencies considered, and it is therefore interesting to look for another explanation for this frequency dependence. One possibility is that the natural roughness of the sea bed gives rise to scattering of the interface wave energy into incoherent compressional and shear waves. Due to the limitations of the available theories the modelling of this effect has not been feasible, but with the development of the scattering model described above, this is now possible.

To highlight the scattering effect a very simple environmental model is chosen. An isovelocity water column of 100 m depth is overlying a solid isovelocity bottom with the compressional and shear velocities being 1800 m/s and 600 m/s, respectively. A 30 Hz point source is assumed to be placed 5 m above the sea bed and a hydrophone is placed just above the bottom. In Fig. 5, the amplitude of the depth-dependent Green's function (the kernel of (3)) is shown as a function of the horizontal wavenumber for 0 m (a), 1.0 m (b) and 2.0 m (c) sea bed roughness. As can be observed, the two waterborne normal modes,'1' and '2' are somewhat attenuated, whereas the interface mode, '0' clearly decreases in amplitude with increasing roughness. This result supports the assumption, that interface roughness

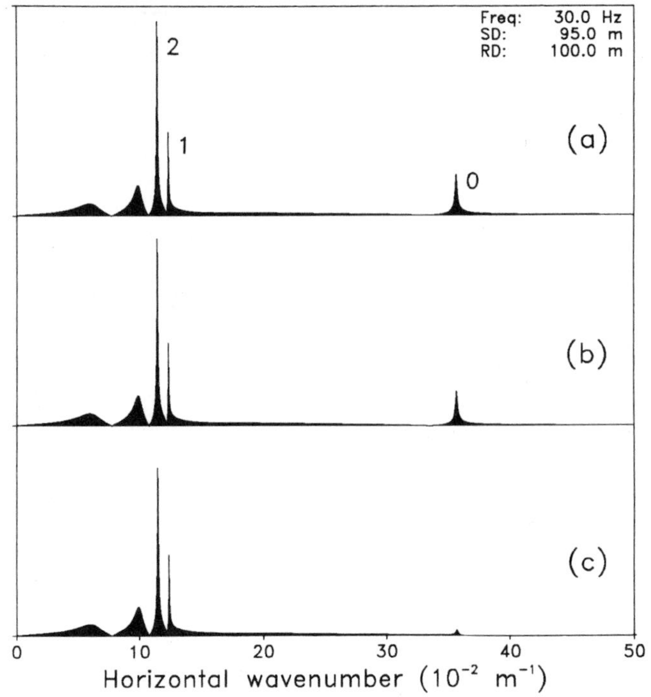

Fig. 5 Wavenumber spectrum at rough sea bed. (a) no roughness. (b) 1 m roughness. (c) 2m roughness.

may be responsible for some part of the attenuation experimentally observed.

Although roughness information is not available for the environment treated in Ref. 8, we will use that dataset to evaluate this hypothesis. The inversion performed in Ref. 8 yielded the shear speed profile shown in Fig. 6. The shear attenuation profile shows the values in dB/λ at a frequency of 3 Hz. We will here assume that this attenuation profile is independent of frequency. As also done in Ref. 8, the comparison of experimental and synthetic results will be performed through the so-called

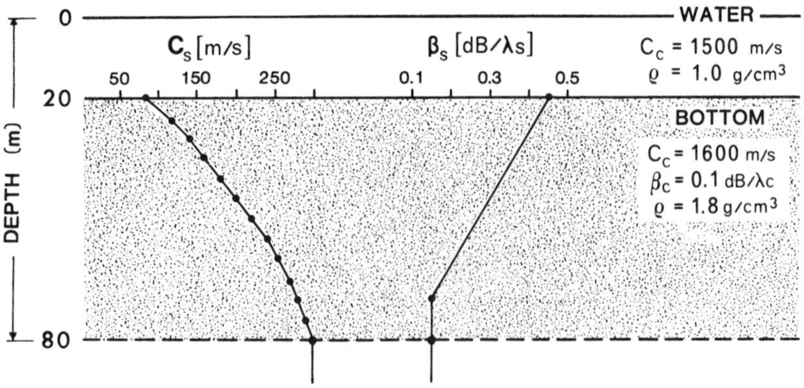

Fig. 6 Shear speed and attenuation profiles obtained by inversion of seismic interface wave data[8].

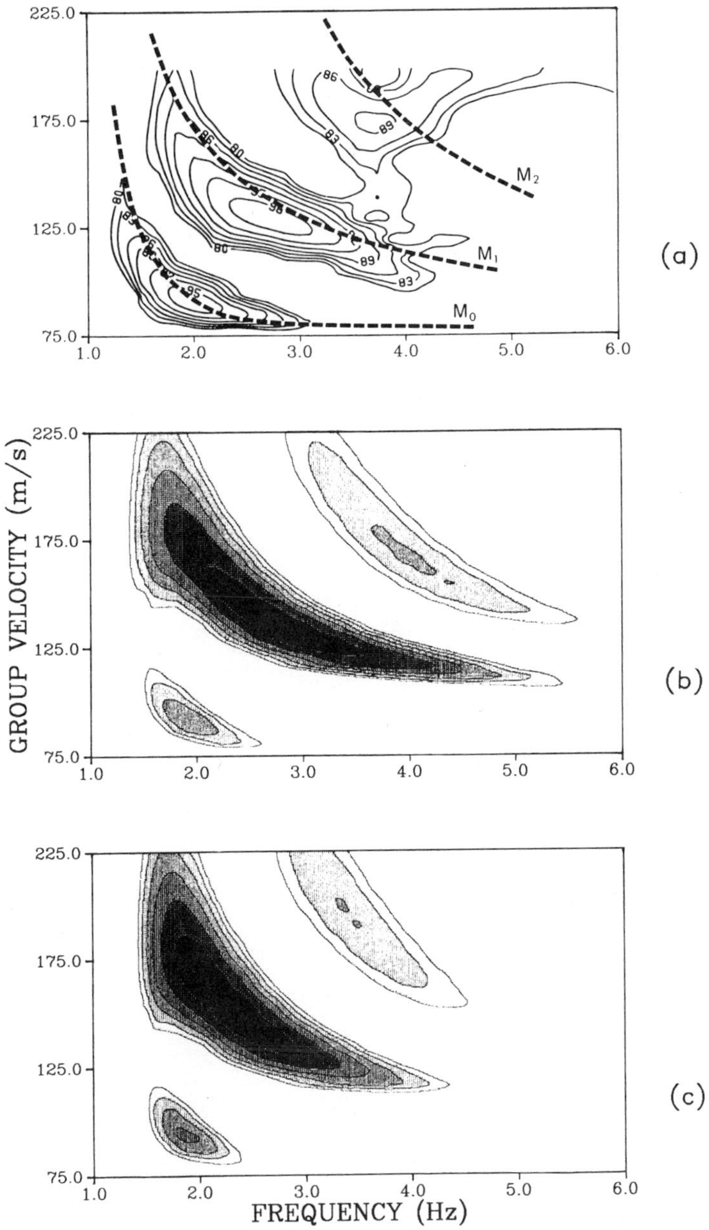

Fig. 7 Gabor matrix for seismic interface waves. (a) Experimental results. (b) Synthetic result for smooth sea bed. (c) Synthetic result for 0.75 m roughness.

Gabor matrix, obtained by a multiple filtering technique[9]. Fig. 7a shows the Gabor matrix for the measured radial particle velocity at range 1700 m[9], with a contour interval of 3 dB. The corresponding synthetic result for a smooth interface is shown in Fig. 7b. It is obvious, that too much high-frequency energy is present in the first mode (M_1) in the synthetic result, and this behaviour lead to the assumption of a quadratic frequency dependence of the attenuation in Ref. 8. Here, instead, a 75 cm RMS roughness of the sea bed was introduced, and the Gabor matrix in Fig. 7c was obtained. The resultant reduction of the high-frequency energy is clearly observed.

It should be stated, however, that the overall agreement between the Gabor matrices in Fig. 7a and 7c is not as good as was the case in Ref. 8. In particular, the relative energy content of the fundamental mode (M_0) and the first mode (M_1) is different. This disagreement may however be removed by slight changes in the other parameters, but the scope here is not to perform a full inversion of the dataset, but just to demonstrate qualitatively that realistic values for the bottom roughness can account for part of the attenuation observed in experimental data. A new dataset is, however, presently being collected[10], including measurements of the bottom roughness. This is expected to yield the possibility of a more rigorous quantitative evaluation of the roughness effect on the propagation characteristics of seismic interface waves. Further, the non-Kirchhoff theory is expected to be necessary to accurately model the scattering in this case.

CONCLUSIONS

A new numerical model for treating wave propagation in horizontally stratified fluid/solid environments has recently been extended to include the effect of interface roughness on the coherent component of the field. Since this model is not restricted to roughness between fluid media, propagation can be modelled for environments where scattering into incoherent shear waves is important. Here a few real and canonical examples have been given, demonstrating the importance of including this additional loss mechanism in the prediction of sound propagation in the ocean. On the background of these initial investigations it is expected, that this new modelling capability will yield the possibility of explaining some of the unclarified propagation effects in underwater acoustics. The application of the present model is restricted by the Kirchhoff approximation, but an extension to a non-Kirchhoff approximation is relatively straightforward in the present formulation, and this feature is currently being developed. In particular, this improvement is expected to be necessary to quantitatively explain the propagation effects observed in the Arctic and for ocean bottom seismic interface waves.

REFERENCES

1. W.A. Kuperman and H. Schmidt, Rough surface elastic wave scattering in a horizontally stratified ocean. J. Acoust. Soc. Am. (in press).
2. H. Schmidt, Modelling of pulse propagation in layered media using a new fast-field program. In: "Hybrid Formulation of Wave Propagation and Scattering" L.B. Felsen ed., Martinus Nijhoff Publishers, The Hague (1984): 337-356.
3. H. Schmidt and F.B. Jensen, A full wave solution for propagation in multilayered viscoelastic media with application to Gaussian beam reflection at fluid-solid interfaces. J. Acoust. Soc. Am. 77: 813-825 (1985).
4. H. Schmidt and G. Tango, Efficient global matrix approach to the computation of synthetic seismograms. Geophys. J.R. Astr. Soc. 84: 331-359, (1986).
5. M. Schoenberg, Reflection of elastic waves from periodically stratified media with interfacial slip. Geophysical Prospecting 31: 265-292 (1983).

6. W.A. Kuperman, Coherent component of specular reflection and transmission at a randomly rough two-fluid interface. J. Acoust. Soc. Am. 58: 365-370 (1975).
7. F. DiNapoli and R.H. Mellen, Low frequency attenuation in the Arctic Ocean. In: "Ocean Seismo-Acoustics", T. Akal and J. Berkson eds., Plenum Press, N.Y. (1986).
8. F.B. Jensen and H. Schmidt, Shear properties of ocean sediments determined from numerical modelling of Scholte wave data. In: "Ocean Seismo-Acoustics", T. Akal and J. Berkson eds., Plenum Press, N.Y. (1986).
9. B. Schmalfeldt and D. Rauch, Explosion-generated seismic interface waves in shallow water: Experimental results, SACLANTCEN Report SR-71, (1983).
10. M. Snoek, SACLANTCEN, private communication.

THE EFFECT OF VARIABLE ROUGHNESS OF A GRANITE SEABED ON LOW-FREQUENCY SHALLOW-WATER ACOUSTIC PROPAGATION

Philip R. Staal, David M.F. Chapman, and Pierre Zakarauskas

Defence Research Establishment Atlantic
P.O. Box 1012, Dartmouth, Nova Scotia
Canada B2Y 3Z7

ABSTRACT

Previously (Staal and Chapman, 1985), we reported unexplained high loss in acoustic propagation in a shallow water area with a rough granite seabed, in the frequency range 10 Hz - 100 Hz. We had a chart of seabed roughness for this area prepared for us. The chart divides the area into four roughness provinces: 0 m to 2 m, 1 m to 6 m, 4 m to 8 m, and 8 m to 20 m. We performed a second propagation experiment, placing explosive sources at 65 m depth along a circle of 13 km radius centred on a receiving location near a boundary between rough and smooth provinces. In this way, the radial propagation paths from different sources to the same receiver covered areas on the seabed which varied in roughness from path to path. The propagation data collected show correlation between the roughness and the propagation loss. However, the main high-loss feature was not removed by choosing a propagation path with a smooth bottom. We conclude that either bottom roughness alone is not responsible for this main feature, or that our measure of bottom roughness is not appropriate.

INTRODUCTION

Many researchers, including ourselves, have successfully modelled shallow-water acoustic propagation over *thickly sedimented* seabeds with all-fluid models. However, we have noticed that over seabeds with little or no sediment cover, all-fluid models do not explain the experimentally-observed high acoustic propagation loss.

Experimental measurements over *chalk* seabeds (Staal, 1983) show high propagation loss, especially below 100 Hz. Most of this loss can be explained by shear-wave loss into the seabed, using a flat-seabed model (Akal and Jensen, 1983). This model predicts very good acoustic propagation over a *flat granite* seabed. We observed experimentally just the opposite: high propagation-loss for a *rough granite* seabed (Staal and Chapman, 1985).

To verify that the sound speed parameters of our geoacoustic model were correct, we compared theoretical curves of normal-mode group speed with experimental dispersion data. This comparison showed that the seabed was indeed hard granite and not some softer, more absorbing material.

Since the theoretical propagation loss was for a *smooth* seabed and the experiment was over a *rough* seabed, we took two approaches to try to determine whether or not seabed roughness caused the extra propagation loss. First, we started to include seabed roughness in our models. Second, we performed an experiment over smooth and rough granite seabeds. This experiment is described here, along with the modelling of the normal-mode dispersion.

THE EXPERIMENT

In June 1985, DREA conducted an acoustic propagation experiment in a shallow water area of the eastern Canadian continental shelf for which the seabed was rough granite with little or no sediment cover. We deployed our receiving array of hydrophone sensors, called Hydra (Staal et al., 1981), with a horizontal leg on the sea bottom and a vertical leg in the water, as shown schematically in Fig. 1. Prior to the experiment, we had a bottom roughness chart of the area prepared from echo-sounder surveys. The chart divides the area into provinces having different roughness scales collected into four groups: 0 m to 2 m, 1 m to 6 m, 4 m to 8 m, and 8 m to 20 m. Part of this chart is reproduced in the centre of Fig. 2, which also shows that we located the Hydra array on a smooth part of the seabed adjacent to a very rough area. In this way, propagation tracks along different radial directions would pass over areas on the seabed having quite different roughness profiles, so we could investigate the possible correlation of propagation loss and seabed roughness.

Our quiet research ship, CFAV Quest, stood by to receive acoustic data telemetered from the Hydra array and to control the array by telemetering commands to the microprocessor controller. A second vessel travelled on a circular course (marked as the source track in Fig. 2) 13 km from the array, dropping 0.45 kg explosive charges set to explode at 65 m depth at roughly 3 km intervals. Although Hydra consisted of several hydrophones in the vertical and the horizontal, we will only report here the results of a single hydrophone on the bottom, at a depth of 146 m. A bottomed hydrophone is more sensitive to interface wave signals and is less susceptible to potential self-noise problems.

The bathymetry and sound speed profiles along the source track are shown in Fig. 3, along with the charge depth. It may be seen that a slight sound channel existed. The sea-bed sloped very gently downward from the lower right to the upper left of the roughness map in Fig. 2.

The seabed in the experimental area is one large body of granitic and volcanic rock of Hadrynian age, with very little sediment cover. The seabed has several metres of relief, sometimes as much as 10 or 20 metres. The spatial periods of the relief displayed in Fig. 2 are in the range of 25 to 1000 m.

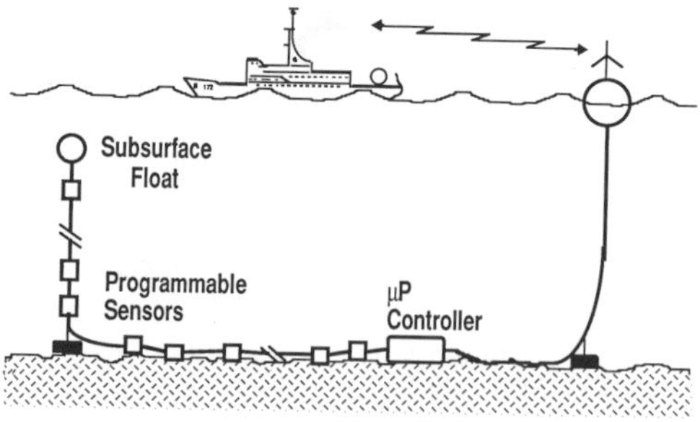

Fig. 1. Typical deployment of the Hydra array on the seabed.

RESULTS

By relating the pressure-time signatures in Fig. 2 to the roughness map, it is clear that there is a correlation between high frequency loss and bottom roughness: more high-frequency loss is evident over the rougher areas. A low-frequency dispersed interface wave can be seen on most of the signatures.

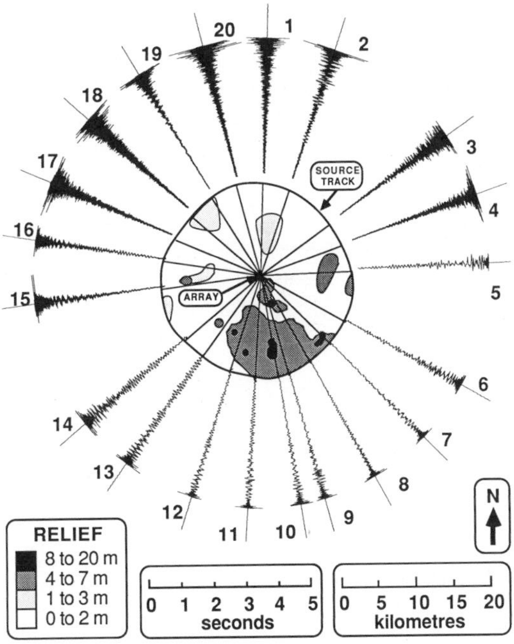

Fig. 2. Bottom roughness map and shot pressure-time signatures for the experiment. The radial lines indicate acoustic propagation paths from individual shots at 65 m depth along the source track, to a bottom-mounted hydrophone in the middle at a depth of 146 m. Part of the acoustic pressure versus time signature for each shot is aligned with its propagation path. Time goes radially inwards. True north is marked.

In order to examine more closely the correlation between the bottom roughness and the frequency content of the received shots, we calculated 1/3 octave band propagation loss from the shots. As shown in Fig. 4, there is generally better acoustic propagation at low frequencies (< 30 Hz) over the rough bottom, and at high frequencies (> 30 Hz) over the smoother bottom. However, the high loss around 30 Hz is evident for all bottom types in this experiment.

THEORY

The Modelling of Modal Dispersion

Fig. 5 contains a time-frequency-intensity plot of a signal received at Hydra from a single explosion; it clearly shows that the energy arrives in a set of discrete modes, each displaying a unique travel time vs. frequency characteristic. We modelled this dispersion using a simple two-layer propagation model (Ellis and Chapman,1985) that includes shear wave propagation in the seabed. The environment we chose consisted of a water layer of depth $H = 150$ m and sound speed $c_w = 1460$ m/s overlying a semi-infinite granite layer of compressional speed $c_p = 5500$ m/s, and shear speed $c_s = 3300$ m/s; the ratio of densities water:granite was $r = 1{:}2.6 = 0.385$.

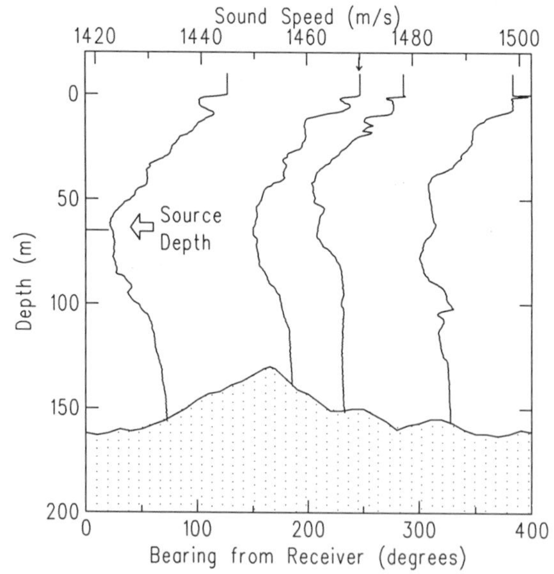

Fig. 3. Bathymetry and sound speed profiles for the experiment. The sound speed scale is at the top; vertical ticks at the sea surface are reference values of 1470 m/s. Profiles join the seabed at the appropriate bearing (in degrees relative to true north) from the receiving array.

The propagation model is a normal mode model wherein the modal phase speeds c_n at a given frequency f are the real roots of the characteristic equation:

$$4\sqrt{1-\left(\frac{c}{c_s}\right)^2}\sqrt{1-\left(\frac{c}{c_p}\right)^2}-\left[2-\left(\frac{c}{c_s}\right)^2\right]^2=r\left(\frac{c}{c_s}\right)^4\sqrt{\frac{1-\left(\frac{c}{c_p}\right)^2}{1-\left(\frac{c}{c_w}\right)^2}}\tanh\left[\frac{2\pi fH}{c}\sqrt{1-\left(\frac{c}{c_w}\right)^2}\right] \quad (1)$$

This equation was introduced by Essen (1980) to describe the dispersion of the interface wave (mode 0 in our case) in such a two-layer environment, but the equation also gives the acoustic modes. Note that Eq. (1) is well-behaved at $c=c_w$ and that solutions for $c>c_w$ are obtained by replacing $(1-(c/c_w)^2)^{1/2}$ by $((c/c_w)^2-1)^{1/2}$ and by replacing tanh[...] by tan[...]. At a specified frequency f there are $N+1$ real roots such that:

$$c_0 < c_1 < c_2 < \ldots < c_N . \quad (2)$$

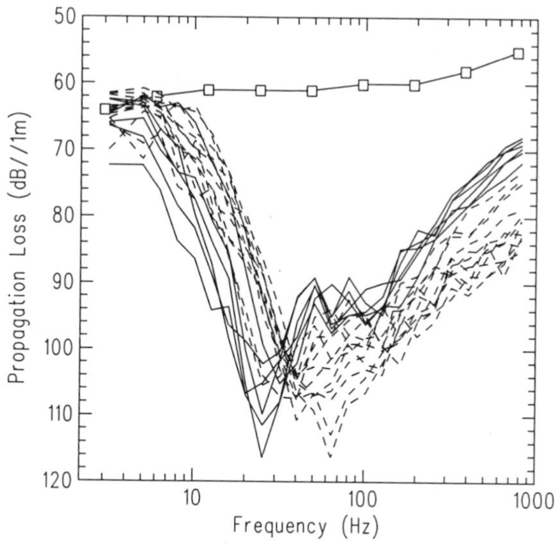

Fig. 4. Measured propagation loss versus frequency at several bearings for a bottom-mounted hydrophone, at a depth of 146 m. Source depth was 65 m, and the ranges were about 13 km. Losses for shots 5 to 16, which correspond to paths over rougher bottoms, are marked by dashed lines. Modelled propagation loss is shown by open square symbols.

The lowest order mode n=0 is the interface mode; the modes n=1,2,3...N are the acoustic modes. The acoustic modes each have a cut-off frequency f_n below which they do not propagate; the interface mode has no such cut-off frequency. The spectrum of cut-off frequencies has the property:

$$0 < f_1 < f_2 < \ldots < f_N < \infty. \tag{3}$$

As frequency increases from zero, the phase speed of the interface wave mode decreases monotonically from the speed of the Rayleigh wave in the granite to the speed of the Scholte wave at the water/granite boundary (as $f \rightarrow \infty$). For each acoustic mode, as frequency is increased from the cut-off frequency f_n, the phase speed decreases monotonically from c_s to c_w (as $f \rightarrow \infty$).

The speed of energy transport is not given by the phase speed but by the group speed, which is less than or equal to the phase speed in this case. The modal group speeds v_n are given by:

$$v_n = c_n / (1 - (f/c_n)(dc_n /df)). \tag{4}$$

Using our simple two-layer propagation model, we calculated the modal phase speeds numerically from Eq. (1), then we calculated the modal group speeds from Eq. (4).

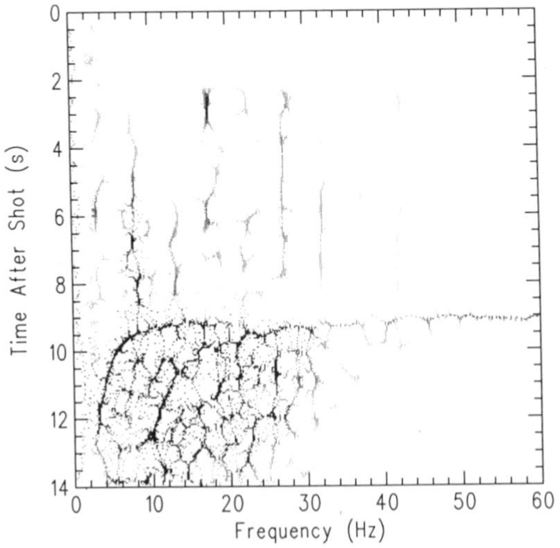

Fig. 5 *Spectrum level (plotted as intensity) as a function of time after explosive shot number 13 (at a range of 13 km), and as a function of frequency.*

We converted the time-frequency-intensity plots for five explosions into group-speed-frequency intensity plots using the known source-receiver ranges, and then we averaged these together to form an average dispersion curve representing the modal dispersion for the experiment. This curve is shown in Fig. 6 along with the modelled dispersion curves for modes $n = 0, 1, 2$. With this simple model, the remarkable agreement of experiment and theory shows that our description of the environment is essentially correct. Of course, the modelled curves do not include any information about absorption or scattering in the medium. There seems to be little energy in mode 2 in the experimental data: we have already shown that propagation is poor at these frequencies; also, mode 2 is likely to have a null very close to the source depth of 65 m. Although they do not appear in Fig. 6 due to low amplitude, the frequencies of the high speed arrivals in the data (i.e. the short-time arrivals of Fig. 5) are the same as those suggested by the modelled dispersion curves.

The Modelling of Propagation Loss

In reporting a previous, similar experiment (Staal and Chapman, 1985), we had difficulty accounting for the high propagation loss in the band 10 Hz to 100 Hz. Although our dispersion analysis (similar to that above) indicated that our bottom model was essentially correct, the modelled propagation loss using the same bottom model did not reproduce the notch observed in the experimental propagation loss data. At the time, we proposed that scattering at the rough seabed may have an effect but the results reported above show that scattering loss, although significant, is a second-order effect and does not explain the notch in propagation loss.

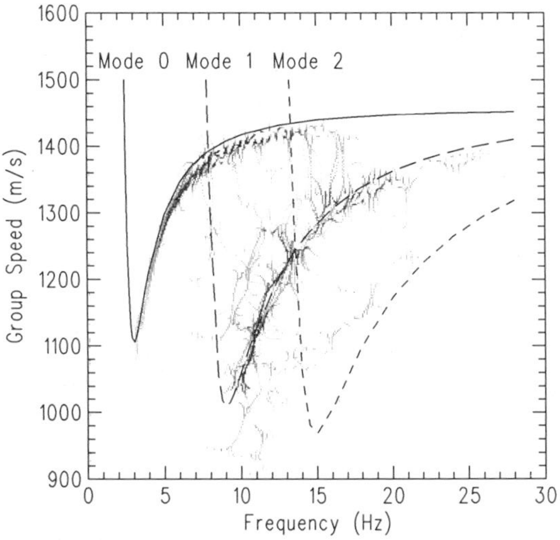

Fig. 6 *Spectrum level (plotted as intensity) as a function of group speed (assuming straight-line propagation) and as a function of frequency, for the average of shots 6, 13, 14, 15, and 16. Modelled dispersion curves for the Scholte mode and the next two higher modes are superimposed.*

Using the SAFARI model of Schmidt (1983), which handles shear waves in the seabed, we calculated propagation loss over a smooth granite seabed having acoustic characteristics: $c_p = 5500$ m/s, $c_s = 3300$ m/s, $r = 0.385$, compressional attenuation 0.55 dB/wavelength, and shear attenuation 1.65 dB/wavelength. The modelled propagation loss is shown in Fig. 4 along with the experimental data; it is clear that we have a large discrepancy between experiment and model that needs explaining. Since the dispersion modelling was so successful, it would not seem reasonable to introduce seabed layers having significantly different speed parameters in order to improve the propagation modelling. This leaves the absorption parameters open to question: perhaps the upper granite layers have different absorption characteristics than the deeper layers; perhaps they depend upon frequency in a more complicated manner; or perhaps scattering *within* the upper layers may be a loss mechanism. Initial modelling attempts using a shear-wave absorption coefficient that depends upon the square of the frequency have given promising results.

CONCLUSIONS

Our most recent propagation loss experiment over a rough granite seabed has shown that the high propagation loss observed in the band 10 Hz to 100 Hz is only secondarily related to the seabed roughness. Although there is a significant difference in the propagation loss measured over smooth and rough seabeds, the presence of a deep notch in propagation loss seems independent of the seabed roughness. We have successfully modelled the dispersion of the normal modes using a simple two-layer geoacoustic model; our attempts at modelling propagation loss have proven unsuccessful. Either our measure of the seabed roughness is inappropriate, or seabed roughness alone is not responsible for the observed losses. In either case, we need more information about the geoacoustic environment and improved propagation models that can handle seabed roughness and shear wave effects.

REFERENCES

Akal, T., and Jensen, F. B., 1983, Effects of the sea-bed on acoustic propagation, in: "Acoustics and the Sea-Bed," N.G. Pace, ed., Bath University Press, Bath, UK, p. 225.

Ellis, D. D., and Chapman, D.M.F., 1985, A simple shallow water propagation model including shear wave effects, J. Acoust. Soc. Am., **78**: 2087-2095

Essen, H. -H., 1980, Model computations for low-velocity surface waves on marine sediments, in: "Bottom-Interacting Ocean Acoustics," W. A. Kuperman and F. B. Jensen, eds, Plenum Press, NewYork, p. 299.

Schmidt, H., 1983, Excitation and propagation of interface waves in a stratified sea-bed, in: "Acoustics and the Sea-Bed," N.G. Pace, ed., Bath University Press, Bath, UK, p. 327.

Staal, P. R., Hughes, R. C., and Olsen, J. H., 1981, Modular digital hydrophone array, in: "Proceedings of IEEE Oceans '81", Boston, p. 518.

Staal, P. R., 1983, Acoustic propagation measurements with a bottom mounted array, in: "Acoustics and the Sea-Bed," N.G. Pace, ed., Bath University Press, Bath, UK, p. 289.

Staal, P. R., and Chapman, D. M. F., 1985, Observations of interface waves and low-frequency acoustic propagation over a rough granite seabed, in: "Proceedings of Ocean Seismo-Acoustics", a conference held in La Spezia, Italy, 10-14 June 1985, Plenum Press, New York.

OCEAN SEISMO-ACOUSTIC PROPAGATION

Tuncay Akal and Finn B. Jensen

SACLANT ASW Research Centre
Viale San Bartolomeo 400
19026 La Spezia, Italy

ABSTRACT

When acoustic energy propagates through the ocean medium, the energy often remains confined within a duct defined by the sound speed structure, the sea-bottom and/or the sea-surface. For long-range propagation, acoustic energy at low frequencies suffers little volume attenuation but interacts increasingly with the sea-floor as the acoustic wavelength increases. The increase of propagation loss as the frequency decreases has been demonstrated down to the cut-off frequency of the ocean waveguide. Below the cut-off frequency where waterborne propagation is extremely poor, seismic waves become important propagation mechanisms.

INTRODUCTION

The ocean and its boundaries form a medium for the propagation of acoustic energy. Within the ocean the sound-speed profile and water depth control the influence of the boundaries. Waterborne acoustic energy may propagate into and out of the sea-bed, as well as be scattered at the sea-surface and the sea-floor.

Propagation of acoustic energy in a range-dependent ocean environment has received considerable attention within the ocean acoustics research community. Various approaches have been attempted for a theoretical and experimental description of complex range-dependent acoustic propagation.

ACOUSTIC INTERACTION WITH THE SEA BED

Figure 1 shows results of an experiment designed to measure frequency-dependent transmission loss in a range-dependent ocean[1]. The receiver (60 m) is situated in deep water, while the source (50 m) moves onto the continental shelf. In deep water, propagation is characterized by decreasing loss with decreasing frequency. At a range of 70 km, at the edge of the continental shelf, the propagation characteristics change drastically. Low-frequency sound (<100 Hz) is highly attenuated.

The thickness of the propagation channel, defined by the surface and the bottom, controls the number of interactions of sound rays with the sea-bottom. When the water becomes shallow, acoustic energy interacts more with the bottom and hence more energy moves into the bottom. There is another important phenomenon controlled by the thickness of the propagation channel. Wave theory predicts that for any kind of ducted propagation there is a cut-off frequency below which the duct ceases to act as a waveguide. Below the cut-off frequency, waterborne propagation is extremely poor while seismic waves become important propagation paths.

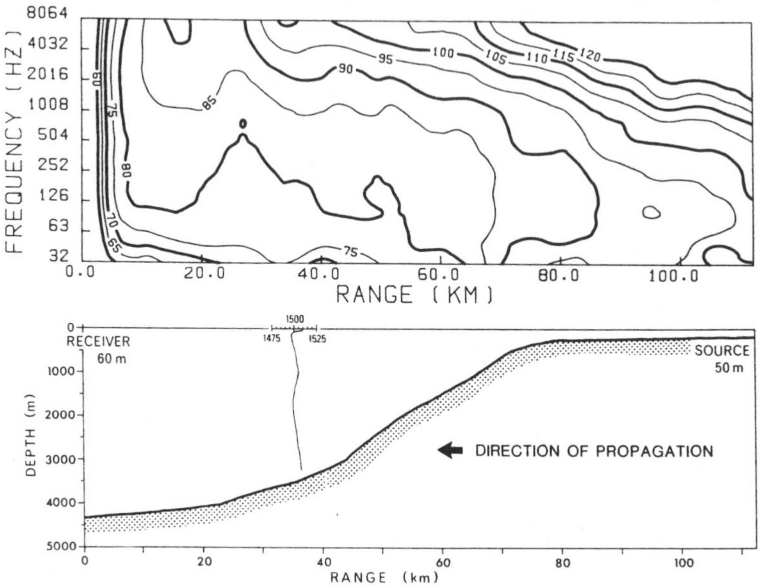

Fig. 1 Measured transmission loss over an acoustic path changing from deep to shallow water.

For a theoretical description of this cut-off process the Parabolic Equation (PE) method[2], which includes both diffraction and mode-coupling effects was applied to up-slope propagation in an isovelocity, wedge-shaped ocean with a penetrable, isovelocity bottom. The environment consists of an initial 5 km flat region, with a water depth of 200 m, followed by a bottom slope of 1.5^{o}. Figure 2 depicts contoured propagation loss (for 25 Hz) versus depth and range. As schematically indicated in the ray picture (corresponding to a given mode), the grazing angle increases as the ray propagates upslope, and at a certain point the angle exceeds the critical angle at the bottom (high reflection loss), where energy starts propagating in the bottom. The point where this happens corresponds to the cut-off depth of the equivalent mode.

We next consider propagation under the simplified environmental condition indicated in the upper-left corner of Fig. 3, i.e. 100 m of isovelocity water (1500 m/s) overlying an homogeneous bottom. The source is placed at mid-depth (50 m) and the receiver placed on the bottom at 10 km distance from the source. Propagation loss calculations between 0.1 and 100 Hz were performed by a fast field program (FFP)[3]. The frequency-dependent propagation characteristics of the environment are illustrated in Fig. 3 for two homogeneous bottom types of silt (compressional wave velocity of 1600 m/s and shear wave velocity of 200 m/s) and sand (compressional wave velocity of 1800 m/s and shear wave velocity of 600 m/s). Starting from 100 Hz, propagation loss does not vary much down to 5-10 Hz, where there is an apparent cut-off. Data from experimental results obtained under similar conditions (water depth of 80-110 m and layered sea-bed) are shown in the same figure. Note that the data

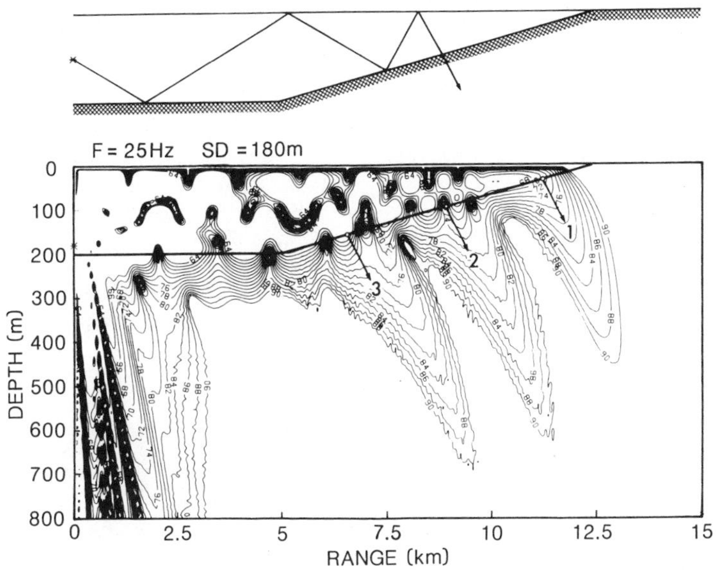

Fig. 2 Computed modal cut-off during upslope propagation.

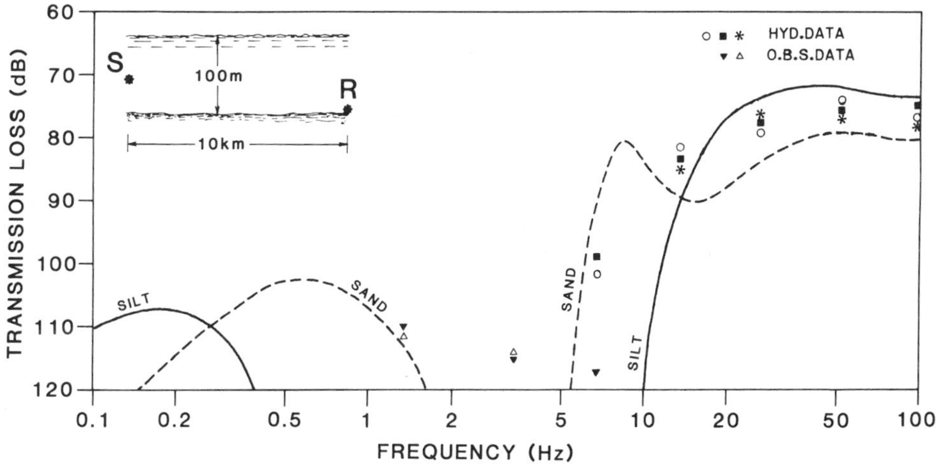

Fig. 3 Measured and calculated transmission losses of seismic and water-borne propagation modes.

demonstrate similar cut-off characteristics for water-borne acoustic energy as predicted by the model. At much lower frequencies (0.1-2.0 Hz) seismic propagation becomes important with acoustic energy generated in the water column now propagating as seismic waves through the sea-floor. Figure 3 shows the measured and calculated propagation losses of seismic interface and water-borne waves for the simple environmental conditions considered.

SEISMIC WAVES IN THE VICINITY OF THE WATER/SEDIMENT INTERFACE

When acoustic energy interacts with the sea-bed the energy creates two basic types of deformation, translational and rotational. Solution of the equation of the wave motion shows that each of these types of deformation travels outwards from the source (where acoustic energy interacts with the sea-bed) with its own velocity. These two types of deformations (compressional and shear) belong to a group of waves (body waves) which propagate in an unbounded homogeneous medium. However, when the medium is layered and/or possesses a free surface, another group of waves (surface waves) may develop. These basic types of waves and their characteristics are illustrated in Fig. 4.

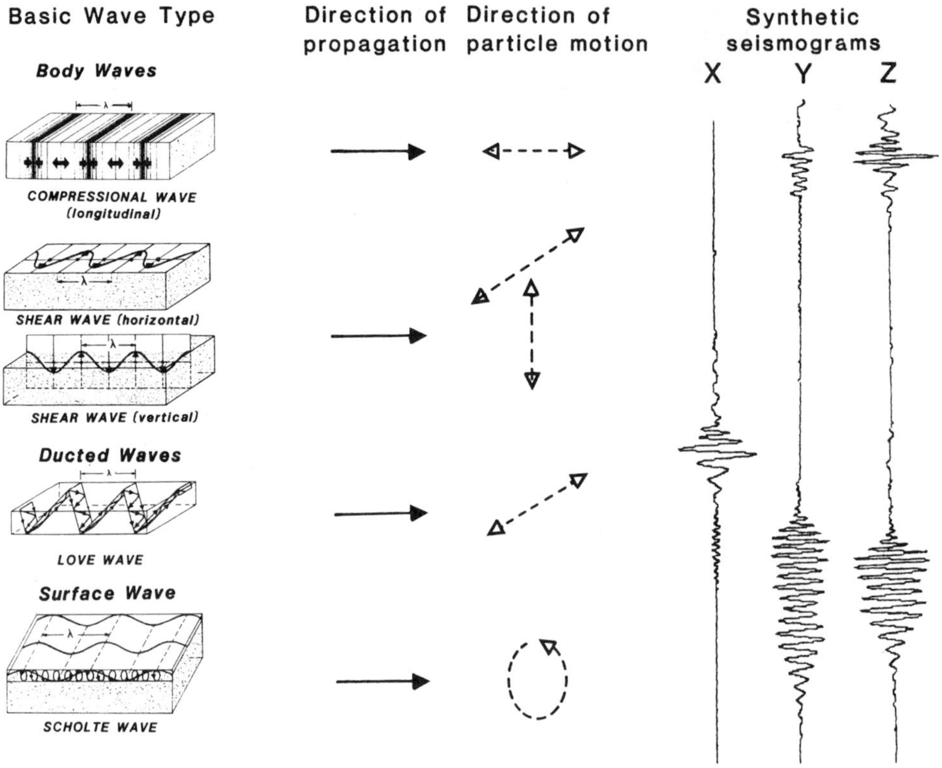

Fig. 4 Basic seismic wave types in the vicinity of a water/sediment interface and their characteristics.

Under realistic conditions, i.e. for an inhomogeneous, bounded and anisotropic sea-floor, some of these waves convert from one to another. The different wave types may travel with different speeds or together, and they generally have different attenuations. Figure 5 displays signals from an explosive source (at 1.5 km distance) received by a hydrophone and three orthogonal geophones placed on the sea-floor. It is evident that the signal generated by an explosive source (500 g TNT) dispersed over nearly 18 s allows for time separation of the various modes of propagation. These features are indicated in the figure for four different sensors. They can be identified in order of arrival time as:

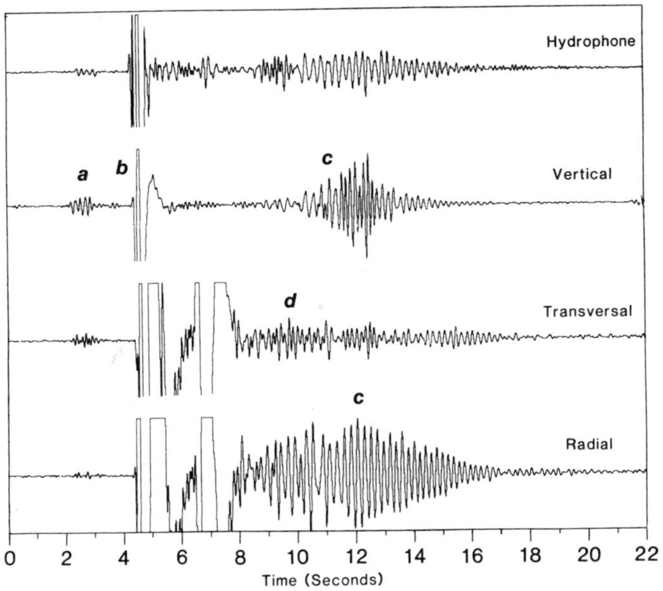

Fig. 5 Signals from an explosive source (at 1.5 km distance) received by a hydrophone and three orthogonal geophones.

a. Head wave : A wave generated at critical incidence and propagating horizontally in the bottom with energy leakage back into the water.

b. Water arrival (compressional wave): A high-frequency arrival traveling with a speed of 1500 m/s.

c. Interface wave: This type of wave develops at water/sediment or sediment/sediment interfaces and the wave amplitude decays rapidly with increasing distance from the guiding interface. The propagation deformation of the wave is a combination of both dilatation and vertical shear strains. The particle motion takes place in a vertical plane parallel to the direction of propagation and has an elliptical orbit (Fig. 5a).

d. Love wave: This wave is another type of surface wave which develops in low-velocity (unconsolidated) upper sedimentary layers. [Horizontally-polarized shear waves (SH)] (Fig. 5b). This wave always exhibits dispersion. As is the case with interface waves, at high frequencies the Love wave velocity approaches the horizontal shear velocity in the upper layer, while for low frequencies it approaches the shear velocity in the deeper layers.

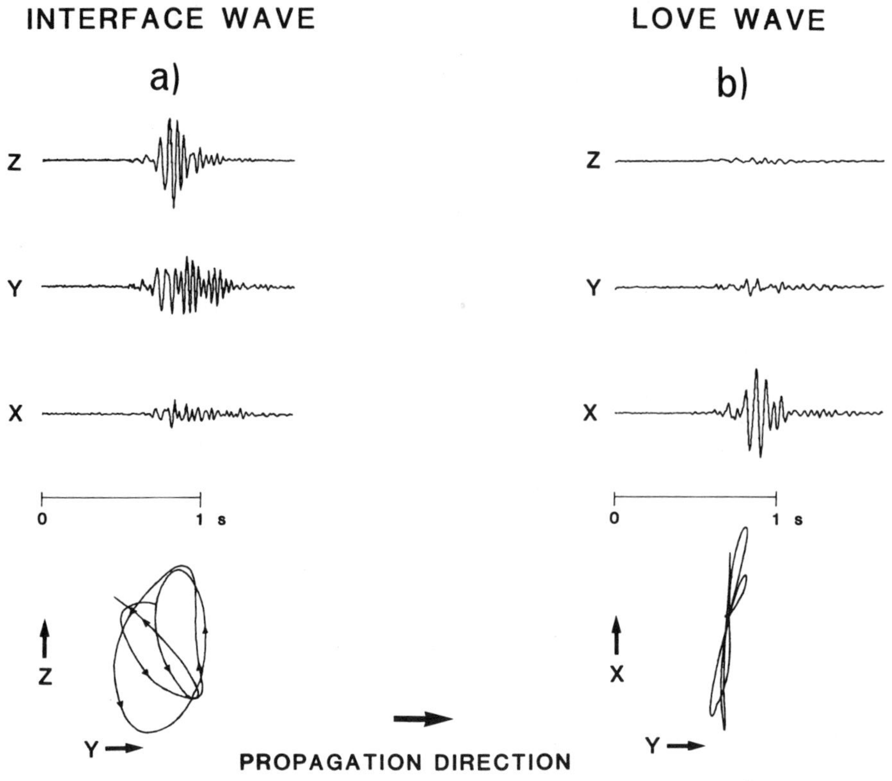

Fig. 5a Interface wave and its particle motion. Fig. 5b Love wave and its particle motion.

AMBIENT NOISE COUPLING INTO THE SEISMO-ACOUSTIC FIELD

Ambient noise in the ocean can be related to various known sources[4]. The dependence of ambient noise spectra on propagation characteristics as well as on source distribution is well understood[5]. The role of propagation on ambient noise becomes more important for low frequencies for which distant sources are often the dominant contributors. Figure 6 identifies several well-known sources that introduce energy into the water column. The energy propagates through the ocean until it reaches the continental shelf edge where it starts interacting with the sea-bed. Ambient noise measurements made with seismic and acoustic sensors demonstrate the presence of different types of seismic waves in the noise field[5].

Figure 7 shows ambient noise spectra measured on the continental shelf with a tri-axial geophone and a hydrophone on the bottom. Even though the hydrophone and the geophone respond to different physical parameters (pressure and particle motion, respectively), we note the similarity in spectral shapes above 25 Hz. In the transition region between low and high frequencies (5 to 25 Hz) there is a significant level difference (low levels on the hydrophone signal) which is probably due to the cut-off of waterborne noise. For the water depth and the sediment type found in the test area, the waterborne cut-off frequency is approximately 10-15 Hz. Apparently, the noise source is distant shipping in deep water with energy propagating initially within the water column, but being coupled into seismic waves on the continental shelf in connection with the waterborne energy cut-off. Hence energy leaving the water column couples into the seismic waves which are detected by the geophones. The seismic energy appears as high levels within the 5 to 25 Hz frequency band on the geophone spectra.

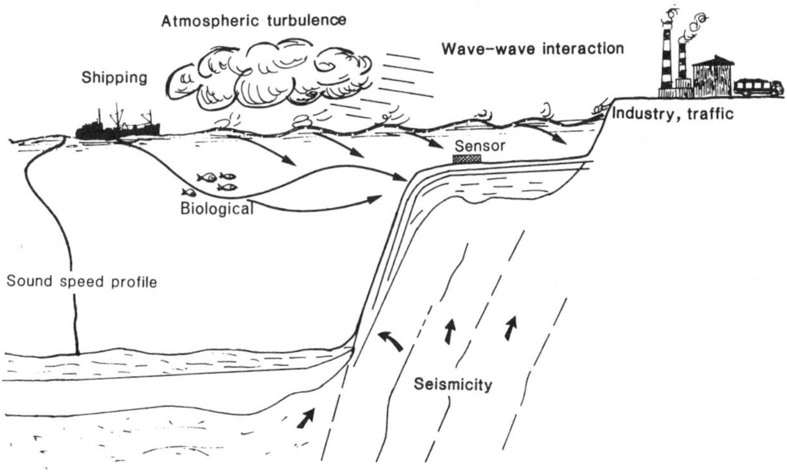

Fig. 6 Basic physical sources which transfer low frequency energy into the underwater ambient noise field.

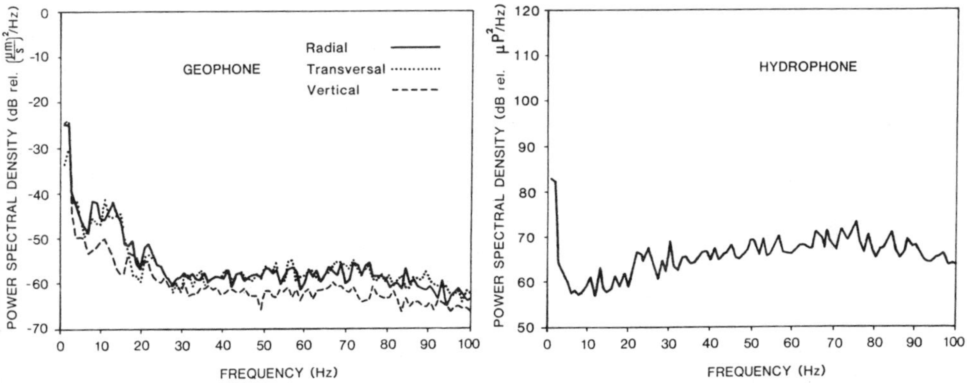

Fig. 7 Seismo-acoustic ambient noise at sea state zero.

CONCLUSIONS

When waterborne acoustic energy is cut off on the continental shelf it starts propagating in the bottom as seismic waves. There is experimental evidence that most of the energy is converted into some form of guided seismic wave (Scholte wave or Love wave). However, the distribution of energy on the different seismic wave types is not easily determined. Experimental data on seismo-acoustic coupling is extremely sparse, but with the increasing interest in very-low-frequency acoustics, the experimental effort in this area is expected to grow.

REFERENCES

1. T. Akal and F.B. Jensen, Effects of the sea-bed on acoustic propagation. In: "Acoustics and the Sea-Bed," N.G. Pace, ed., Bath University Press, Bath (1983): 225-232.
2. F.B. Jensen and W.A. Kuperman, Sound propagation in a wedge-shaped ocean with a penetrable bottom, J. Acoust. Soc. Amer. 67: 1564-1566 (1980).
3. H. Schmidt and F.B. Jensen, A full wave solution for propagation in multilayered viscoelastic media with application to Gaussian beam reflection at fluid-solid interfaces, J. Acoust Soc. Amer. 77: 813-825 (1985).
4. D. Ross, Role of propagation in ambient noise. In: "Underwater Ambient Noise," R.A. Wagstaff and O.Z. Bluy, eds., Rep. CP-32. SACLANT ASW Research Centre, La Spezia, Italy (1982): 1-1 to 1-18.
5. T. Akal, A. Barbagelata, G. Guidi and M. Snoek, Time dependence of infrasonic ambient seafloor noise on a continental shelf. In: "Ocean Seismo-Acoustics," T. Akal and J. Berkson, eds., Plenum, New York (in press).

MEASUREMENTS OF WIND DEPENDENT ACOUSTIC TRANSMISSION LOSS IN SHALLOW WATER UNDER BREAKING WAVE CONDITIONS

P. Wille, D. Geyer, L. Ginzkey, and E. Schunk

Forschungsanstalt der Bundeswehr für Wasserschall- und Geophysik
Klausdorfer Weg 2-24, D-2300 Kiel 14, Germany

ABSTRACT

The enhancement of transmission loss induced by breaking waves in shallow water of 30 m depth, measured at a fixed 10 km-range under isothermal conditions rather depends on wind speed than on wave height. The attenuation coefficient, growing weakly at low wind speeds increases after a thresholdlike transition by the third to the fifth power of the wind speed, probably due to air bubble suspension. The transition wind speed around 10 m/s decreases with increasing frequency. The time lag between changing wind speed and attenuation is of the order of minutes. The loss differs at increasing and decreasing fetch. The close relation between wave height and attenuation at low sea states and predominating rough boundary scattering disappears at high sea states.

INTRODUCTION, MEASURING SETUP

The fixed acoustic range at the research platform NORDSEE /1/ in the German Bight (10.5 km range length, 30 m water depth, sand bottom) permits controlled propagation loss measurements up to high sea states where the assumption of pure boundary scattering appears doubtful. Since air bubble screening of surface induced noise has been observed at high wind speeds /1/ a corresponding effect is expected for horizontal propagation and frequent surface bounces.

The paper refers to continuous automatic measurements during 35 days under isothermal conditions, wind speeds of nearly calm to 23 m/s and wave heights of 0.3 to 4.5 m, covering 98 % of a three years cumulative distribution of wind speeds of that area. Third octave signals, centered at 1, 3.2 and 8 kHz, continuously transmitted at 20 m depth by horizontally oriented directional sources (24 , 12 and 5 beamwidth between the -3 dB points) and received at 28 m depth are presented, together with wind speed and wave height as 10 min averages. The geometrical spreading loss has been derived from separate shipborne propagation measurements such as described in /5/. The environmental quantities are recorded at the platform (anemometer height 46 m, wave rider buoy some 100 m distant). Apart from meteorological fronts, both wind speed and wave height have been measured to be nearly stationary along the propagation range.

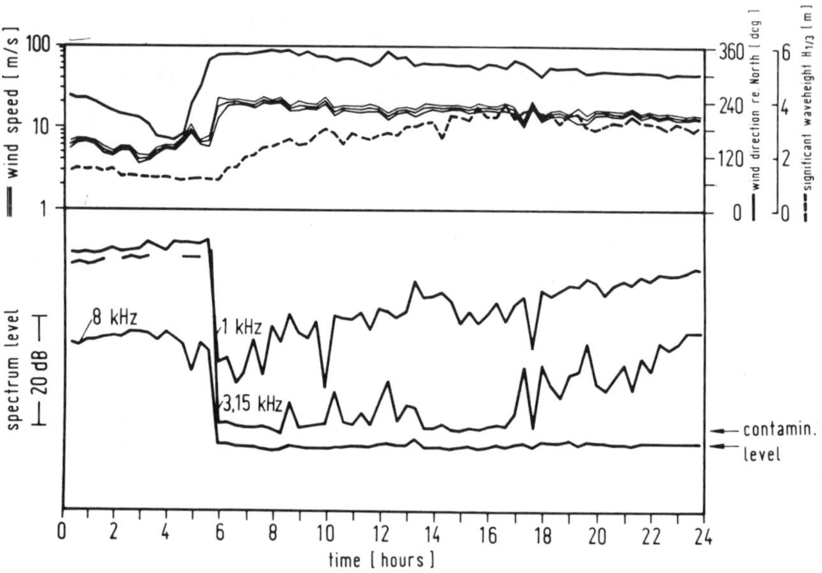

Fig. 1: Spectrum level of received sound signals and environmental parameters (10 min averages)
Top: wind speed $\pm$ σ (3 thin lines), wind direction (strong line) and wave height (broken line)
Bottom: Third octave spectrum level at 1, 3.2 and 8 kHz

LOSS DEPENDENCE ON WIND SPEED AND WAVE HEIGHT

Fig. 1 demonstrates that the wind speed rather than the wave height determines the increase of the transmission loss: a wind speed jump from 6 to 20 m/s within minutes is immediately followed by a drastic decrease of the receiving level which recovers again, together with a slow wind speed decrease during the next hours, despite an increasing wave height. According to the wind direction and the bearing of the transmission range of 320 , the wind jump had passed the hydrophone location less than 9 min before the arrival at the anemometer at the source location. Small superimposed wind maxima (e.g. 10.00 or 17.40 h) originate instantaneous sound level minima and vice versa (e.g. 12.20, 17.20 h).

A close relation between transmission loss and wave height, as observed at low sea states, where rough boundary scattering should predominate (Fig. 2a, 2b, left) disappears at higher sea states. The dominating wind speed dependence (Fig. 2a, 2b, right) is obvious also regarding the non-isotropic distribution of a given spectrum level in the wave height-wind speed domain of fig. 3. The thresholdlike loss increase near 10 m/s where breaking waves become frequent supports a supposition of bubble induced absorption since the bubble density is closely related to the wind speed /2/, and the absorption cross section of bubbles at resonance exceed their geometrical size by orders of magnitude /3/.

LOSS DEPENDENCE ON WIND HISTORY

Though the wind dependent bubble production has been observed to depend also on properties of the surface roughness /4/, the parametrization of transmission loss enhancement by the wave height appears similarly inappropriate as it has been found for sea surface noise production which requires breaking waves as well /1/. However, a reduction of the transmission loss spread is achieved when the wind-wave height history is considered. The loss beyond the critical wind speed appears lower at increasing than at decreasing wind; the shift amounting to 2.5 m/s (fig. 4), hardly depending on the slope of the wind speed increase. Since offshore and onshore wind directions (fig. 5) correspond to increasing and decreasing wind speeds respectively in relation to the resulting average transmission loss, the sign of the fetch change appears to decide on lower or higher loss. Since the rising time of resonant bubbles lasts at most minutes, even for the 8 kHz-size, the wind history effect cannot simply result from remaining bubbles. The simplest explanation would be that the larger waveheights at decreasing fetch enhance the bubble attenuation by scattering loss. However, a greater convection depth of the bubbles at a given wind speed and decreasing fetch may contribute as well.

The attenuation coefficients for the three frequencies (fig. 6) are derived from the transmission loss by correction with the geometrical spreading loss, determined separately by shipborne propagation measurements. The attenuation coefficient is connected with the wind speed by a power law between 3.5 and 4.5 within the high wind speed regime. A similar exponent has been found for the number density of bubbles of a given size at a given depth as a function of the wind speed /2/.

ATTENUATION DEPENDENCE ON FREQUENCY

Fig. 6 indicates a shift of the critical wind speed by signal frequency. Based on the three third octave bands, the frequency dependence of the attenuation is depicted in fig. 7. The weak slope at low wind speeds corresponds to earlier shipborne measurements at low sea states /5/. The drastic increase of the slope by the wind speed may be attributed to the combined effect of the steep rise of the bubble production by the wind

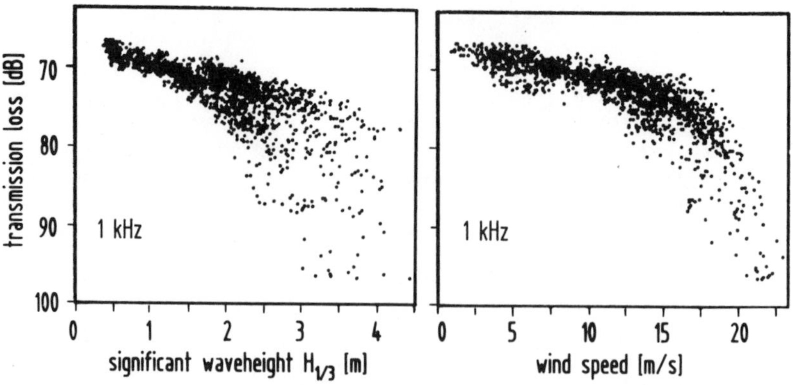

Fig. 2a: Transmission loss at 1 kHz versus wave height (left) and wind speed (right). Each point represents 10 min average

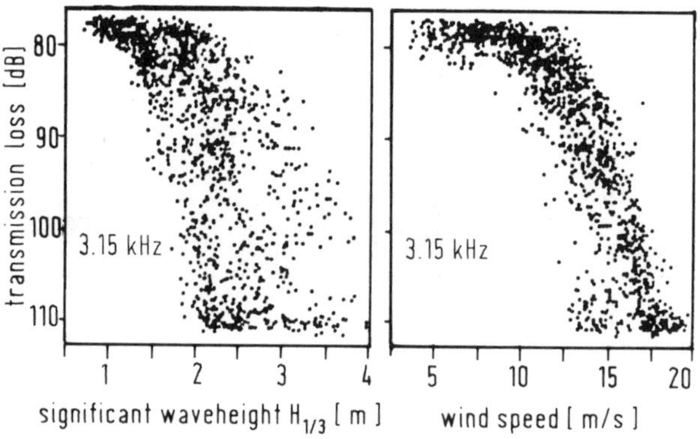

Fig. 2b: Transmission loss at 3.2 kHz versus wave height (left) and wind speed (right). Each point represents 10 min average

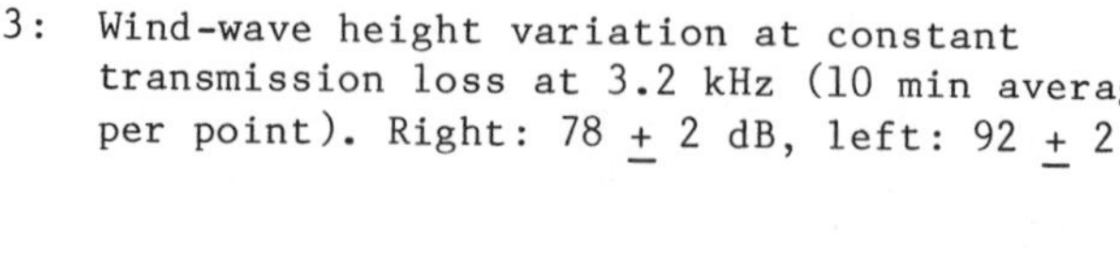

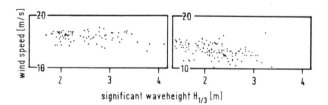

Fig.3: Wind-wave height variation at constant transmission loss at 3.2 kHz (10 min average per point). Right: 78 ± 2 dB, left: 92 ± 2 dB

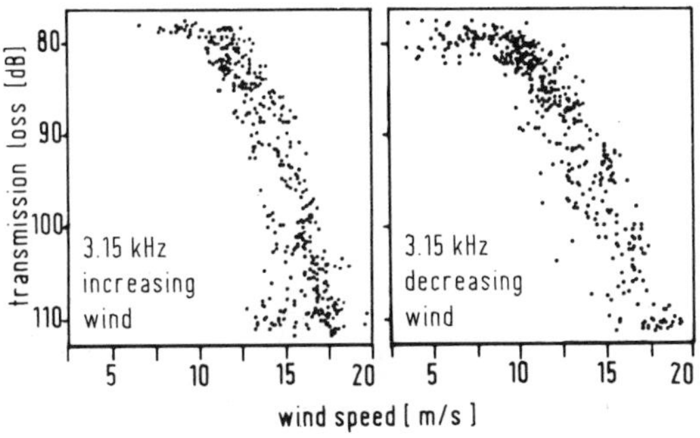

Fig. 4: Influence of wind history on transmission loss

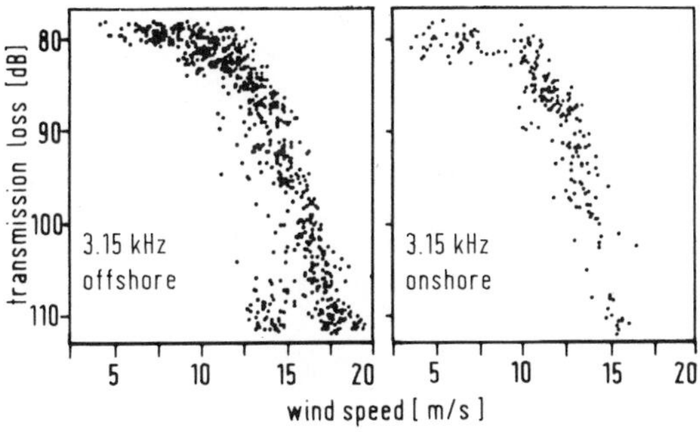

Fig. 5: Influence of spatial fetch on transmission loss

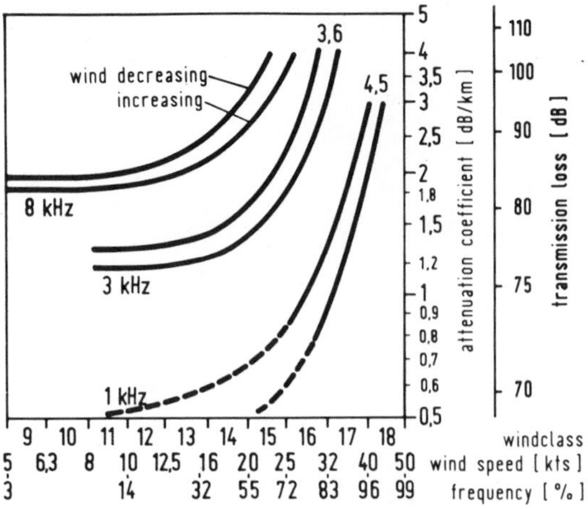

Fig. 6: Attenuation coefficient α versus wind speed (log-log-scale). The slope at high wind speeds u represents the exponent n of $\alpha \sim u^n$.
Lowest scale: cumulative probability of wind speeds

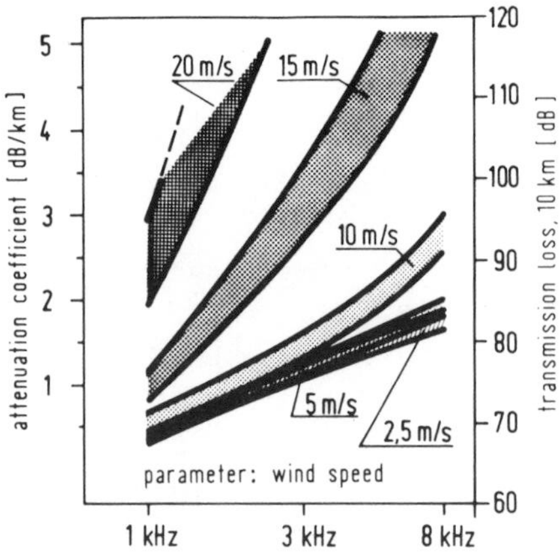

Fig. 7: Dependence of transmission loss on frequency

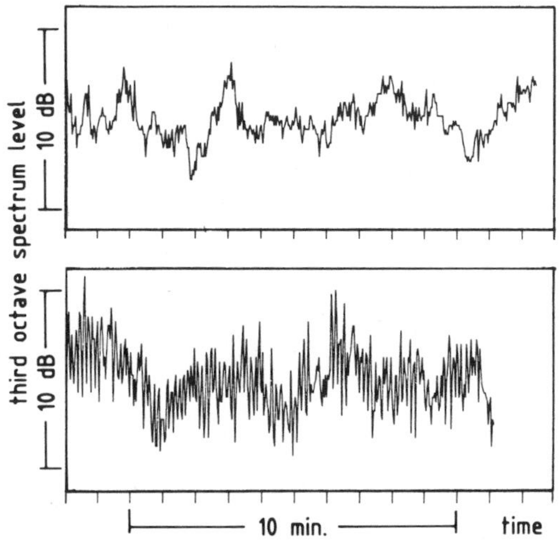

Fig. 8: Short time fluctuations of spectrum level at 1 kHz
Top: 2 m/s wind speed, 0.5 m significant waveheight
Bottom: 15 m/s wind speed, 3 m significant waveheight

speed /2/, the strong decrease of the bubble concentration and the bubble size probability with depth.

LOSS FLUCTUATIONS

Despite consideration of the wind history, the remaining spread of the 10 km-transmission loss at a given wind speed in the breaking wave region is hardly less than 10 dB (fig. 4,5). How far a further reduction of the spread can be achieved - perhaps by additional parametrisation with the steepness of the short wave portion of the sea waves, which tend to break most frequent /7/, needs further investigation. However, measurements of short time (2 s averages) third octave level fluctuations /8/ which cover a level spread comparable to the 10 min average levels indicate the limits of transmission loss prediction accuracy. The fast fluctuations of the time function of the received third octave level, such as depicted in fig. 8, bottom, are typical for high wind speed situations. They correspond to periods of 8 s, close to the main sea wave period of 3 m characteristic height. However, there are superimposed variations in the time scale of minutes which are present as well at low sea states (fig. 8, top) and which do not directly correspond to a motion process in the same time scale.

CONCLUSION

The assumption of bubble induced attenuation enhancement is supported by a model computation, applying a uniform layer of resonant bubbles beneath the surface /6/. The short time lags between increase and decrease of wind speed and attenuation indicate that short sea waves of a few meters wave lengths should be primarily responsible for the production of bubbles /7/ which disappear rapidly again due to the short rising time of bubbles of resonance size.

Whereas the attenuation enhancement by rough boundary scattering is indirectly effective through redistribution of sound energy towards the bottom , the bubble induced attenuation works by direct absorption near the surface and should rather depend on properties of the sea surface than on regional bottom features. This indicates the possibility of sound attenuation monitoring by satellite remote sensing under the conditions considered.

REFERENCES

1. Wille, P.C. and Geyer, D. "Measurements on the Origin of the Wind-dependent Ambient Noise Variability in Shallow Water", J. Acoust. Soc. Am. 75(1), 173-185 (1984).
2. Wu, J. "Bubble Populations and Spectra in Near-Surface Ocean: Summary and Review of Field Measurements", J. Geophys. Res. 86, 457-463 (198
3. Clay, C.S. and Medwin, H. Acoustical Oceanography, John Wiley & Sons, New York (1977).
4. Thorpe, S.A. "On the Determination of KV in the Near-Surface Ocean from Acoustic Measurements of Bubbles", J. Phys. Oceanogr. 14(5), 841-854 (1984) or "A Model of Turbulent Diffusion of Bubbles below the Sea Surface", J. Phys. Oceanogr. 14(5), 841-854 (1984) - (I don t know which, - Ed.)
5. Schellstede, G. and Wille, P.C. "Measurements of Sound Attenuation in Standard Areas of the North Sea and Baltic" SACLANTCEN Conference Proceedings No. 14, Vol II (Conference on Sound Propagation in Shallow Water 23-27 Sept 1974) 52-68 (1974).
6. Schneider, H.G. "Modelling Wind Dependent Acoustic Transmission Loss Du to Bubbles in Shallow Water" presented at 12 ICA Associated Symposiu on Underwater Acoustics, Halifax 16-18 July 1986. Published in this book, next paper.
7. Stolte, S. Fwg Internal Report 1985-4.
8. Ginzkey, L., Kroll, W., Scholz, B. and Wille, P. "Temporal and Spatial Variability of Transmission Loss Measured in Shallow Water and Break ing Wave Conditions" (in progress).

MODELLING WIND DEPENDENT ACOUSTIC TRANSMISSION LOSS DUE TO BUBBLES IN SHALLOW WATER

Hans G. Schneider

Forschungsanstalt der Bundeswehr fur Wasserschall und
Geophysik, Klausdorferweg 2-24, D-2300 Kiel 14
Federal Republic of Germany

ABSTRACT

The acoustic transmission loss data from 1 to 8 kHz reported by Wille et al. (preceding paper) over a wide range of windspeeds in shallow water of 30 m depth exhibits a significant loss component which is attributed to bubbles. Indeed, the standard loss mechanisms, losses due to spreading, medium attenuation, bottom absorption and rough surface scattering are not sufficient to explain the data. Hence an absorbing layer of bubbles beneath the sea surface is additionally included into a stochastic ray-tracing routine to model the transmission loss. From the fitted attenuation coefficients a bubble density is estimated with a slope of -3.5 versus bubble radius which is in agreement with literature. In an intermediate range of windspeeds and waveheights the attenuation due to rough sea surface scattering is of equal importance to the transmission loss as the attenuation by bubbles, while at sufficiently high windspeeds the transmission loss is rather insensitive to the significant waveheight which is also observed in the measured data.

An extrapolation to 120 m waterdepth yields basically the same results, except for the insensitivity of the TL to waveheight at high windspeeds.

INTRODUCTION

The high transmission loss reported in the preceding paper by Wille, Geyer, Ginzkey and Schunk (1986, preceding paper) cannot be explained by using the standard loss mechanisms as geometrical spreading, medium attenuation, bottom absorption and rough surface scattering. It is the aim of this paper to link this extremely high loss to extinction by resonant bubbles near the sea surface.

The experimental setup consists of a fixed acoustic range of 10.5 km length in the North Sea, which is continuously and automatically monitored from the manned research platform NORDSEE. The waterdepth is a constant 30 m, the bottom is up to 1 m mainly sand with an even harder subbottom. The source and receiver depths were 20 and 28 m respectively.

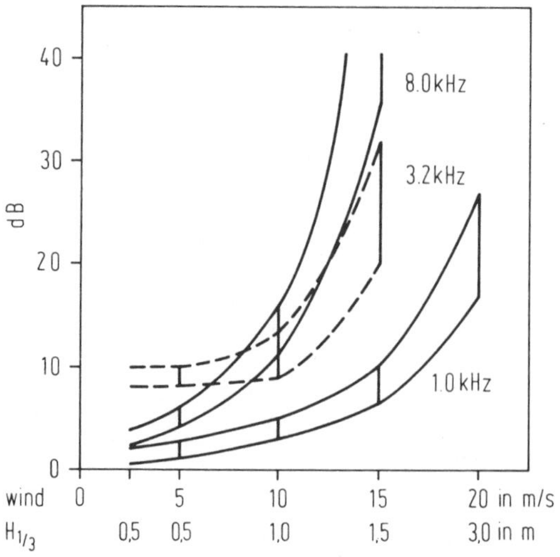

Figure 2: Difference of measured transmission loss to computed loss (without bubbles) for three frequencies. The data are parametrized according to windspeed, while this computation uses only waveheight; the relation between windspeed and waveheight follows roughly the upper boundary of the domain in Fig. 1 and is listed at the bottom of Fig. 2.

by a rough sea surface according to the theory of Lynch and Wagner which has been modified for swell. The scattering is energy conserving and generally broadens the angular energy distribution thus giving rise to an increased loss via the reflection coefficient at the bottom interface.

For the following computations the relation between windspeed and waveheight, as given by the upper boundary of the domain in Fig. 1 and is listed at the bottom of Fig. 2 is used. Fig. 2 depicts the difference of the measured transmission loss to the computed one including scattering at the sea surface. This considerable loss component at only 10.5 km range is to be explained. For windspeeds of less than 10 m/s the curves for 3.2 kHz are unreliable because of signal clipping at low loss situations and will be disregarded further on.

LOSS INCLUDING BUBBLES

Because of the shallow water depth with a relatively good reflecting bottom and the existing windspeed/waveheight conditions, the formulation of the loss due to bubbles in an isothermal subsurface layer as given by Novarini and Bruno (1982) for deep water is not applicable here. Even without scattering, the loss would be too high for the higher frequencies and it would not be possible to relate different waveheights to one windspeed or vice versa.

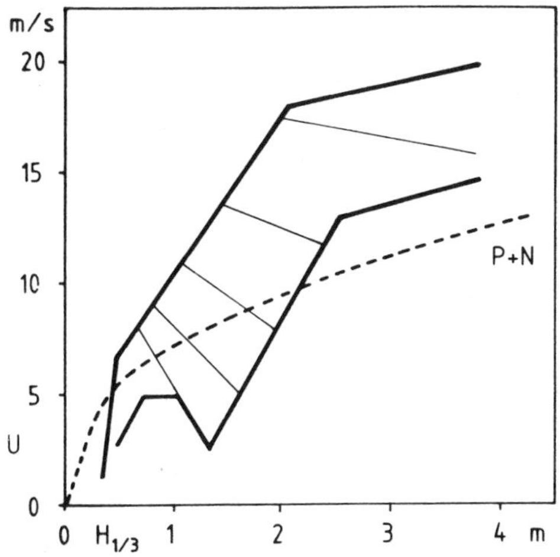

Figure 1: Windspeed/waveheight domain; thick lines: boundary of the domain for the acoustic measurements; thin lines: lines of constant transmission loss; dashed line: waveheight as a function of windspeed according to Pierson and Neuman.

Three broadband signals with center frequencies of 1, 3.2 and 8 kHz were transmitted and recorded in 1/3 octave bands as 10 min samples. The water was isothermal during the measurements.

Fig. 1 summarizes the windspeed/waveheight conditions during the acoustic recordings. Only a few events are outside the broad lines. The thin inner lines indicate the trend of values of constant transmission loss. Two items are noteworthy. First the windspeed/waveheight dependence differs significantly from deep water relations as given by, e.g. Pierson and Neuman, which is indicated by the dotted line. This is partially due to the shallowness of only 30 m water depth and also to not fully developed seastate conditions because of fetch limitations, as the wide spread of waveheight values for one windspeed implies. Second, the iso-loss lines are steeper for low windspeeds indicating a stronger dependence on waveheight than on windspeed, while for high windspeeds the iso-loss lines are more horizontal and thus are more sensitive to windspeed than to waveheight. This observation holds for all three frequencies.

LOSS WITHOUT BUBBLES

First a model computation of the transmission loss was done without bubbles to separate the known loss. The model used is based on a stochastic ray-tracing scheme (Schneider, 1976) which includes scattering

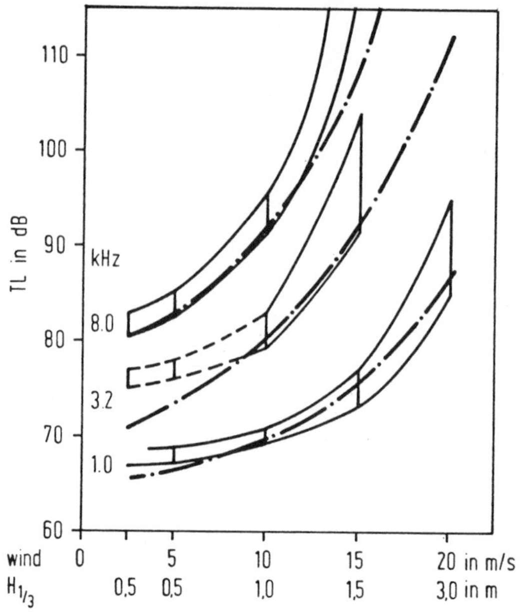

Figure 3: Measured and computed transmission loss for three frequencies; continuous lines: measured transmission loss; dashed lines: computed transmission loss including attenuation by bubbles. Windspeed/waveheight relations used are listed at the bottom of the figure.

The experimental frequencies used correspond to bubble radii at resonance of a = 3.4 mm for 1 kHz, a = 1.1 mm for 3.2 kHz and a = 0.42 mm for 8 kHz. Unfortunately, investigations of wind generated bubbles at sea are available only for smaller radii and the knowledge of bubbles in excess of 1 mm is rather limited because of the inherent measuring difficulties with large bubbles which stay relatively close to the surface, tend to be deformed, form clusters, etc. Thus the extinction of sound waves by bubbles had to be fitted via an assumed loss per length.

According to Wu (1981) the maximum depth of these large bubbles is less than 2 m for a = 0.4 mm and much smaller for the larger bubbles. Hence a layer with an uniform bubble density and a vertical extent of 1 m was assumed below the sea surface, causing an acoustic attenuation β in dB/m.

Fig. 3 shows the computed loss with β fitted to match the lower values of the transmission loss data. The agreement is sufficiently close for the purpose considered here.

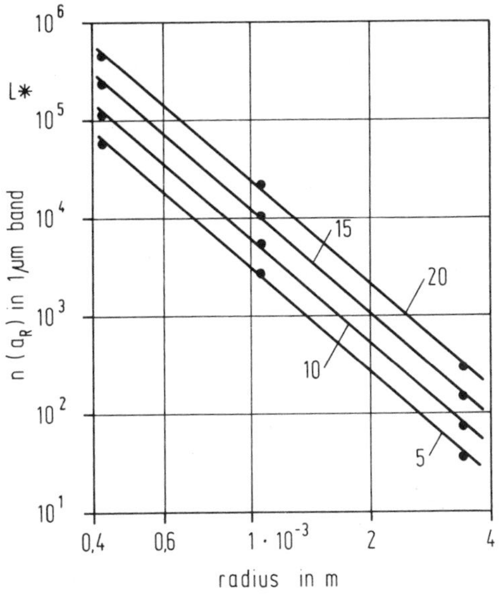

Figure 4: Number of resonant bubbles in 1μm band versus bubble radius for four windspeeds (5, 10, 15, 20 m/s) as deduced from the attenuation fitted to match the acoustic data (Fig. 4). The value indicated by L on the left is an extrapolation of the data from Lovik (1980) from 8 m depth to the surface for a frequency of 12 kHz and 10 m/s windspeed.

Assuming (Clay and Medwin, 1977) that the main contribution to the attenuation results from the resonance peak of the bubble extinction cross section, the bubble density n(a) can be estimated from β. This density for which it is assumed that all resonant bubbles are uniformly distributed in a layer of 1 m thickness is depicted in Fig. 4 for four windspeeds. The slope of this density is -3.5 which is well within the range of other investigations. Wu (1981) reports data from Kolovayev and Johnson and Cook with exponents varying between -3.5 and -5. Clay and Medwin (1977) give a value of -4, while Lovik (1980) measured -4.2 for a <0.1 mm and -2.6 for a>0.1 mm defining an average slope of -3.5.

To relate the magnitude of the deduced bubble density to other data, the measurements of Lovik for 12 kHz and 20 knots windspeed were extrapolated to the sea surface. The corresponding value is noted by L in Fig. 4.

So far the parametrization used followed the upper boundary of the windspeed/waveheight domain in Fig. 1. In the following values corresponding to the lower boundary of the domain are considered and lines of equal loss are given for one sample frequency of 3.2 kHz in Fig. 5. The values computed are annotated with dB levels for transmission

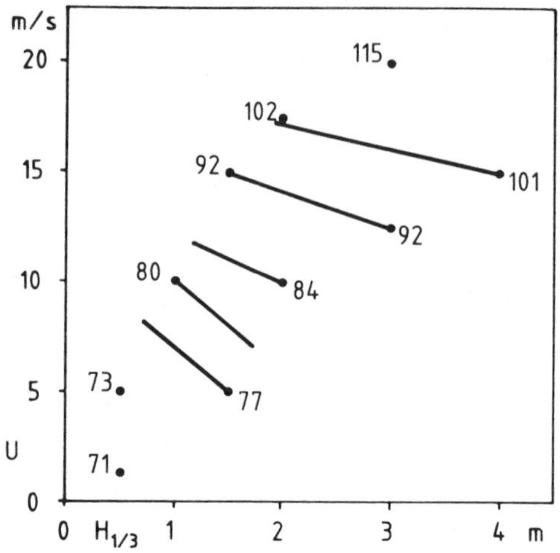

Figure 5: Lines of constant transmission loss (computation including bubbles) in the waveheight/windspeed domain; waterdepth: 30 m frequency: 3.2 kHz range: 10.5 km isothermal conditions compare to data in Fig. 1.

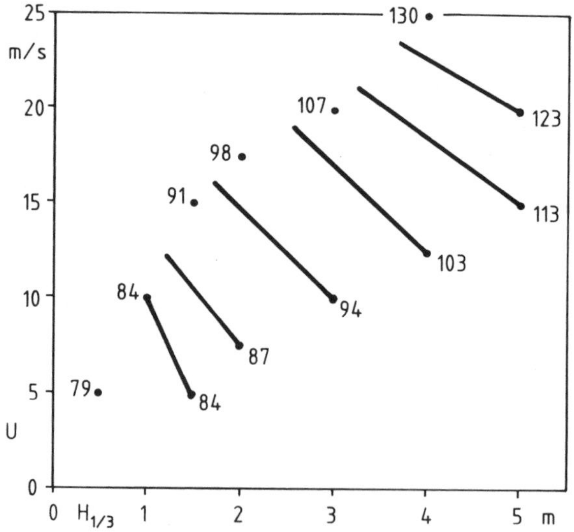

Figure 6: As Fig. 5 except for depth: 120 m; range: 20 km

loss. As in the measurement of the slope the iso-loss lines decreases with increasing loss. In the windspeed/waveheight domain from 5 to 12 m/s and 1 to 2 m the scattering of sound at the rough boundary and the attenuation due to bubbles seem to be of equal importance. Windspeed and significant waveheight are independent quantities under these circumstances. At sufficiently high windspeeds however, the transmission loss is rather insensitive to the exact waveheight, which would justify a parametrization of this loss component by windspeed only.

The assumption of almost independent windspeed and waveheight which is characteristic for a not fully developed sea allows for a considerable variation of the transmission loss due to changes in windspeed only.

Since this investigation is for very shallow water of only 30 m depth it is thought worthwhile to transfer this parametrization to deeper but still shallow water. Fig. 6 displays the computed transmission loss for a frequency of 3.2 kHz at 20 km range in isothermal water of 120 m depth. Comparison with Fig. 5 reveals that at high windspeeds the transmission loss is more sensitive to the waveheight than in the shallower water. This would then require a loss parametrization with both windspeed and waveheight.

CONCLUSION

The transmission loss measured under breaking wave conditions in shallow water can successfully be modeled only if attenuation due to resonant bubbles is assumed in addition to scattering at the rough sea surface. The order of magnitude of the bubble density as well as the dependence on frequency is in agreement with data from literature, as can be deduced from the attenuation used to match the measured transmission loss.

In the intermediate windspeed/waveheight domain both scattering and attenuation due to bubbles are important for the transmission loss. The same must then be expected for the backscattering. At sufficiently high windspeeds the transmission loss is rather insensitive to the exact waveheights which agrees well with the data. In the case of variable windspeeds a considerable variability of the transmission loss due to resonant bubbles must be expected.

A similar transmission loss dependence is found in a model simulation for a greater waterdepth of 120 m. In this case the transmission loss is also sensitive to the waveheight at high windspeeds.

ACKNOWLEDGEMENT

It is a pleasure to thank Prof. P. Wille for initiating this study and the many encouraging discussions. Further I thank my colleagues for the data acquisition and processing.

REFERENCES

Clay, C.S. and H. Medwin, 1977, "Acoustical Oceanography" J. Wiley & Sons, New York.

Lovik, A., 1980, Acoustic Measurements of the Gas Bubble Spectrum in Water, in "Cavitation and Inhomogeneties in Underwater Acoustics", W. Lauterborn, ed., Springer-Verlag, 1980.

Novarini, J.C., and D.R. Bruno, 1982, Effects of the sub-surface bubble layer on sound propagation, J. Acoust. Soc. Am. 72(2), Aug 1982, p.510.
Schneider, H.G., 1976, Rough Boundary Scattering in Ray-Tracing Computations, ACUSTICA Vol. 35(1), 1976, p. 18-25.
Wille, P.C., D. Geyer, L. Ginzkey, E. Schunk, 1986, Measurement of Wind Dependent Acoustic Transmission Loss in Shallow Water Under Breaking Waveconditions, presented at 12 ICA Associated Symposium on Underwater Acoustics (this book, preceding paper).
Wu, J., 1981, Bubble Population and Spectra in Near-Surface-Ocean: Summary and Review of Field Measurements, J. Geophys. Res. Vol. 86, No. Cl, p.457.

FOURIER AND FRESNEL FOCI AND LEVEL DISTRIBUTIONS IN INTERFERENCE FIELDS

D.E. Weston

Admiralty Research Establishment, Portland, Dorset
DT5 2JS England

ABSTRACT

The distribution of levels in the deterministic wave field in a simple environment may have a stochastic look, but it has the advantage that in principle it can be calculated exactly. Of course any movement will convert the spatial variability into temporal fluctuation. An important example is the field due to many modes propagating in a simple waveguide; e.g. underwater acoustic waves in shallow water. The modes combine to form a whole series of high-intensity regions sometimes known as Fourier and Fresnel foci. A preliminary calculation is made for the typical distribution in level. The presence of the foci leads at both very high and very low levels to a probability exceeding that for the Rayleigh distribution equivalent.

INTRODUCTION

This paper concerns fluctuations, especially those in underwater acoustic propagation. Our first objective is to point out that one can start with a fully determined simple medium and yet arrive at a calculated wave field which is very complicated indeed, and has, at first sight, the appearance of a stochastic nature. It is not really stochastic and we can take advantage of this to calculate the distribution of levels, i.e. the probability density function.

As our second objective, we can illustrate this for a simple homogeneous waveguide with plane parallel boundaries. The wave field complications include an unexpected set of high-intensity focal regions, descriptions of which have appeared in connection with underwater acoustics and with fibre optics[1-6]. Sample wavefields calculated with the parabolic equation appear in Figure 1. These were prepared at ARE with D.G. Gleaves after similar calculations by F.B. Jensen and W.A. Kuperman. Figure 1 is further complicated because the source is at midwater in one case and near the bottom in the other.

WAVEGUIDE INTERFERENCE FIELD

The waveguide may be for any medium, but to concentrate the mind we can think of acoustic waves in homogeneous water of depth H. The boundaries are assumed plane, parallel and perfectly reflecting for the modes under consideration and assumed to show a sharp cut at a relatively large mode number N. Our formulae will assume phase change zero at both boundaries, or π at both, but with other reasonable assumptions there is unlikely to be any substantial change in the character of the results. We make the paraxial assumption so that the grazing angles associated with the modes are all small and for mode n given by

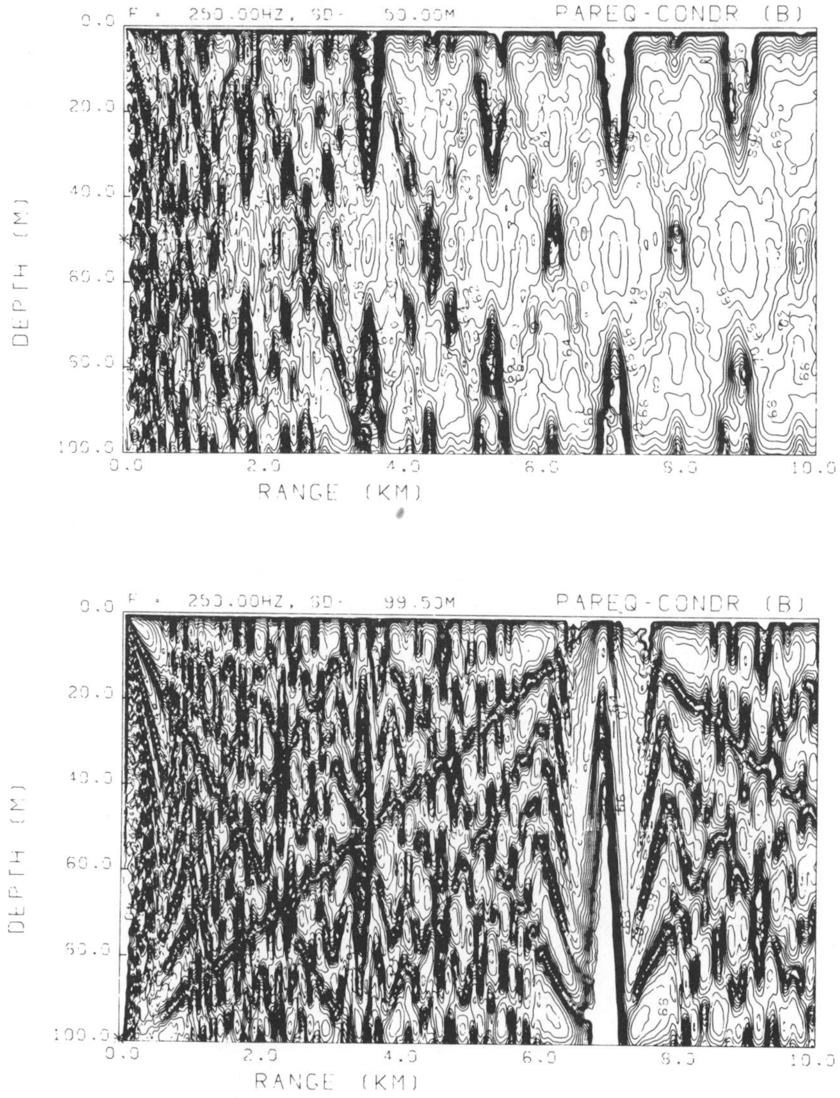

Figure 1: Waveguide Interference.

$$\phi_n = \pm\, n\lambda/2H \tag{1}$$

where λ is wavelength.

From consideration of phase velocity, the references cited[1-6] show that all the modes will be in phase at distances from the source which are multiples of $8H^2/\lambda$, and so we have a series of focal points (the Fourier images or foci) at the source depth. There are additional focal points at intermediate ranges of multiples of $4H^4/\lambda$, occurring at the complementary depth to the source (see Figure 1). In fact the whole field is specified by what happens over the basic range interval $4H^2/\lambda$, and we concentrate on this interval.

All the propagating energy passes through the above foci but there are, in addition, auxiliary or lesser or Fresnel foci where the energy at a given range is shared among p foci. They occur at fractions q/p of the basic range interval, where p and q are integers having no common factor. Thus not all values of q are allowed, and the proportion is found to be a fairly bumpy function of p. The average eventually settles down and is given by the product,

$$\prod_{2}^{\infty} (1-y^{-2}), \tag{2}$$

involving the prime numbers y. This may be evaluated using exponential integrals or simply identified as the Riemann zeta function $[\zeta(2)]^{-1}$, equal to $6/\pi^2$ or 0.6079. It is also admitted that behaviour is influenced by source depth: we will be dealing with a depth that is "typical", without wishing at this stage to define precisely what this means.

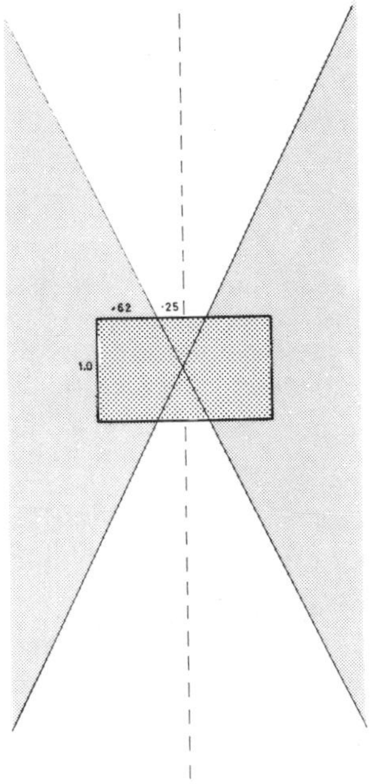

Fig 2. Focus with Butterfly Wings.

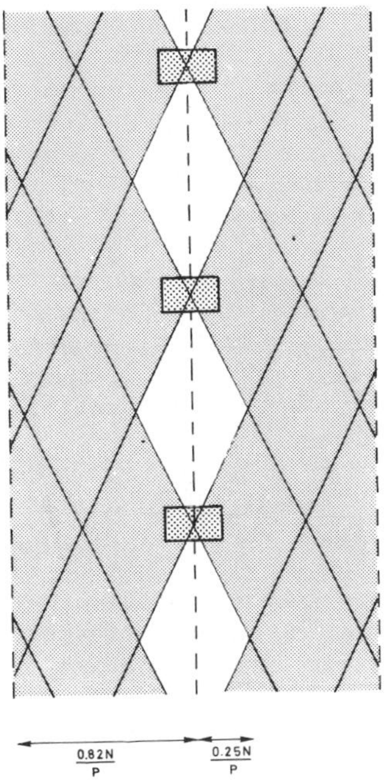

Fig 3. Focal Domain and Quiet Diamonds.

The shape in the focal region will be the same as that for the usual optical focus[7], though strictly we must work in two dimensions or specify a line optical source. This is shown schematically in Figure 2. For mode n the effective or phase wavelength in the vertical is λ cosec ϕ_n, and we can integrate over the modes by treating n as continuous up to its maximum value N. We find an intensity proportional to

$$\left(\frac{\sin k\phi_N \Delta h}{k\,\phi_N \Delta h}\right)^2 \qquad (3)$$

where k is wavenumber and Δh is vertical displacement. This equals $(2/\pi)^2$ or 0.4 or -4dB at $\Delta h = \pm H/2N$. We can take $2\Delta h$ or H/N as the focal height or one vertical pixel; and note that there are N vertical pixels in the depth interval. This sets the upper limit of p as N.

The phase wavelength in the horizontal is λ sec ϕ_n and a similar integration, followed by a maximization with respect to phase, gives an intensity variation with horizontal displacement Δr as

$$\frac{C^2(x) + S^2(x)}{x^2} \quad \text{where } x = \frac{N}{H}\sqrt{\frac{\lambda \Delta r}{2}}, \qquad (4)$$

C and S are Fresnel integrals. We find the half-power or 3 dB downpoint occurs at $x^2 = 1.738$, implying a focal width $2\Delta r$ of $4 \times 1.738\ H^2/N^2\lambda$. It is convenient to define one horizontal pixel as $4H^2/N^2\lambda$ so there are N^2 horizontal pixels in our basic range interval, and focal width is 1.738 horizontal pixels.

Note that at large distances the fall-off in the vertical is fast, with average intensity proportional to $(\Delta h)^{-2}$. In the horizontal it only varies as $(\Delta r)^{-1}$, and the level can be greater at other values of ϕ. It tends to spread out reasonably uniformly within the angles $\pm\ \phi_N$, to give a butterfly wing effect. In pixel space these angle or slope limits become $\pm$ 2; Figure 2 is drawn and marked in pixel units.

Our basic range interval may now be described as N pixels deep by N^2 pixels long, a total of N^3 pixel areas.

It remains to comment on the distribution and the separation of those ranges at which focussing occurs, and on what happens in between. For a type p focus the mean focal separation in the vertical is roughly N/p pixels, and so neighbouring butterfly wings will touch when Δr is about N/4p. The wings enclose a set of blank or quiet areas which are roughly diamond-shaped, illustrated in Figure 3. For small p these quiet diamonds can be very impressive features since they are much larger than the foci themselves. For large p they can shrink to become smaller than the foci, and are not at all impressive.

A related calculation extends the above ideas by assuming that we can associate the area around a given focus with that focus, and that the focus domain extends beyond N/4p to a horizontal pixel range a/p. The coefficient a may be found by treating p as continuous, integrating over all foci, and equating with the basic range interval of N^2 pixels.

$$\int_0^N \frac{2a}{p} . 0.6079\ p\ .\ dp = 1.216\ aN = N^2\ , \qquad (5)$$

Whence a is 0.8225N, and the domain extends to more than three times the wings-touching range (Figure 3).

The horizontal separation of two foci will tend to equal the sum of their two domain radii. It must be stressed that this is only a statistical description, but numerical investigations show that it works surprisingly well both qualitatively and quantitatively.

BASIC PROBABILITY DENSITY FUNCTION FOR FOCI

If we sample all points in our depth and basic range interval we can construct or calculate the pdf (probability density function) for the values of intensity of level. Since we are seeking a "typical" distribution, we can, within narrow limits, allow convenience to dictate the form. We start with the foci.

Let I be the intensity normalized to a mean value of unity, and assume a large range to obviate geometrical spreading complications. For a focus of type p we have

$$I = N/p, \qquad (6)$$

i.e. for the major focus with p = 1 the intensity is N times its mean value. This point lies behind the whole paper: we expect to find a distribution with probability still quite high for high levels. Treating p as continuous, Eq. (6) leads to

$$p = \frac{N}{I}, \left| \frac{dp}{dI} \right| = \frac{N}{I^2} . \qquad (7)$$

For a given p (i.e. dp = 1) there are before censoring p^2 foci, a fraction p^2/N^3 of the total pixel areas. Allowing for the proportionality factor and the focal area the contribution to the pdf is

$$Z = 0.6079 \times 1.738 \times \frac{p^2}{N^3} . \left| \frac{dp}{dI} \right| . \qquad (8)$$

Substituting from Eq. (7)

$$Z = 1.057\ I^{-4} \simeq I^{-4} . \qquad (9)$$

Notice that N has cancelled out, and that the two numerical coefficients almost balance each other out. The formula (9) only holds for I between 1 and N. It is, of course, a statistical approximation and is likely to be particularly inaccurate near these limits. Among other matters, there is no censoring at p = 1 and the proportionality factor should be 1.

The simple equation (9) is the chief result of the paper: by itself it confirms that for high levels the probability will be higher than the e^{-I} figure predicted for a Rayleigh distribution.

PROBABILITY DENSITY FUNCTION FOR THE WHOLE FIELD

We will now massage or slightly improve the calculation for the pdf of the foci, add the contributions from the quiet areas and from the rest. In doing this, we must ensure that the separate probabilities S sum to unity, and also that the separate fractions E of the total energy also add to unity.

The above calculation for the foci treats them as square-topped. In reality they are rounded at their vertical edges and slowly decaying in the butterfly wings. Instead of a proper modelling of this, we will merely convolve the Eq. (9) law with a convenient shape for the area distribution of intensity for a single focus. We finally choose a two stage process, which retains some of the discontinuity. Three quarters of the square top is preserved, with pdf

$$Z \simeq (3/4) I^{-4} \ . \tag{10}$$

The other quarter top is altered to fall linearly to zero over a doubled area, with pdf

$$\left.\begin{aligned} Z &\simeq 1/8 &&, \quad I < 1 \ , \\ &\simeq (1/8) I^{-4} &&, \quad I > 1 \ . \end{aligned}\right\} \tag{11}$$

The fractional S and E values for the foci are respectively

$$S \simeq \frac{5}{4} \int_1^{\infty} I^{-4} dI = \frac{5}{12}, \quad E \simeq \int_1^{\infty} I^{-3} dI = \frac{1}{2} \ . \tag{12}$$

The quiet area calculation sums area or probability over all the diamond-shaped quiet areas and then subtracts the overlap with the focal areas

$$S = N^{-3} \left\{ \int_0^N \frac{0.6079 \ N^2}{4} dp - 0.5x0.5x0.25x4x0.6079 \int_0^N p^2 \ dp \right\} \tag{13}$$

$= 0.1013$, arbitrarily rounded down to 1/12.

Again, arbitrarily, it is assumed that the delta function is really of the exponential form in intensity, appropriate to a Rayleigh function, and at the 1/e point for Z that the intensity I is still only -15dB.

$$Z \simeq 2.64 \ e^{-31.6 I} \ , \tag{14}$$

giving the above S and a negligible E value.

The rest consists of the remaining area in between the focal columns and from Eqs. (12) and (13) we see we must take $S = 1/2$. The natural choice is a Rayleigh distribution in amplitude specified here by pdf

$$Z = \frac{1}{2} e^{-I} \ . \tag{15}$$

The energy fraction for the rest then becomes

$$E = \int_0^{\infty} \frac{1}{2} I \ e^{-I} = \frac{1}{2} \ . \tag{16}$$

We can now confirm that the three different sets of contributions add up properly.

$$\left.\begin{aligned} \Sigma S &= \frac{5}{12} + \frac{1}{12} + \frac{1}{2} = 1 \quad , \\ \Sigma E &= \frac{1}{2} + 0 + \frac{1}{2} = 1 \quad . \end{aligned}\right\} \tag{17}$$

Figure 4 shows the combined pdf, compared to an e^{-I} law as reference. Of course the discontinuity around I = 1 is not meaningful. The pdf equations do not contain N, nevertheless the curve should not be used for I>N and in practice N will decrease with range[8,9]. The important point is that for large I the pdf exceeds that for the Rayleigh distribution, and this it has in common with some other theoretical and experimental distributions. It is perhaps even more interesting and unusual that for small I the pdf again exceeds that for Rayleigh. Thus the dB standard deviation should exceed that for Rayleigh; we would expect it to be more than 6dB.

Further work is needed to check the accuracy of the paraxial approximation, but the greatest need is to move on from "typical" conditions to tightly defined conditions. We should investigate different values of H/λ, N, source depth etc., all with different boundaries and different slices of the receiving field.

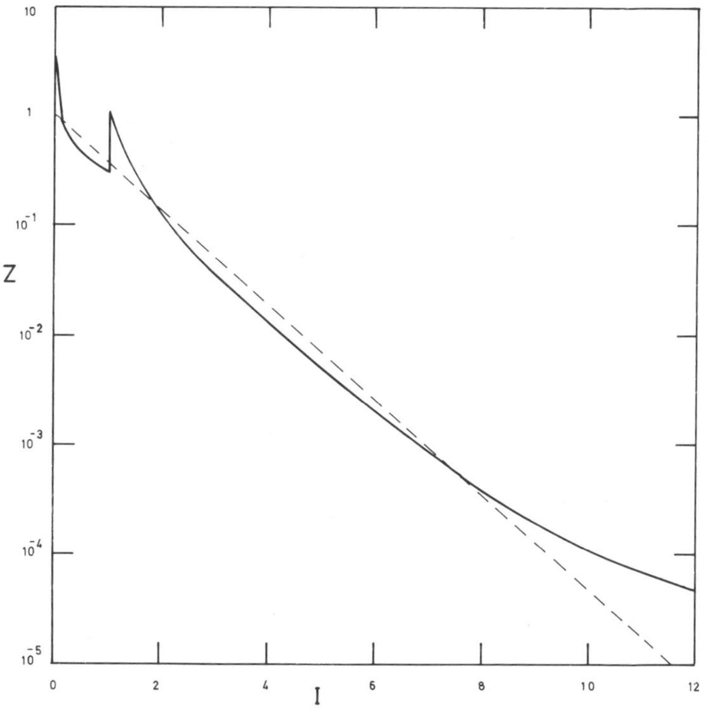

Figure 4. Preliminary Calculation of Probability Density Function (Solid) Compared to Rayleigh (Dashed).

CONCLUSIONS

The preliminary result in Figure 4 shows a pdf including a contribution of I^{-4} form due to the foci. Even with a fully determined field the pdf is similar to Rayleigh, but with excesses at both high and low levels.

REFERENCES

D.E. Weston, A moiré fringe analog of sound propagation in shallow water, J. Acous. Soc. Am. 32: 647-654 (1960).

D.E. Weston, Sound focusing and beaming in the interference field due to several shallow-water modes, J. Acoust. Soc. Am. 44: 1706-1712 (1968).

L.A. Rivlin and V.S. Shul'dyaev, Multimode waveguides for coherent light, Radiophys. Quantum Electron. 11: 318-321 (1968).

E.E. Grigor'eva and A.T. Semenov, Waveguide image transmission in coherent light (review), Sov. J. Quantum Electron. 8: 1063-1073 (1978).

D.C. Chang and E.F. Kuester, A hybrid method for paraxial beam propagation in multimode optical waveguides, IEEE Trans Microwave Theory and Tech MTT-29: 923-933 (1981).

D.E. Weston, "Rays, modes and flux", in: Hybrid Formulation of Wave Propagation and Scattering, L.B. Felsen, ed. pp 46-60, Martinus Nijhoff, Dordrecht (1984).

M. Born and E. Wolf, Principles of Optics (Ch 8), 5th ed, Pergamon, Oxford (1975).

D.E. Weston, Intensity-range relations in oceanographic acoustics, J. Sound. Vib. 18: 271-287 (1971).

G.A. Grachev and G.N. Kuznetsov, Attenuation of the interference maxima of a sound field in shallow water, Sov. Phys. Acoustics 31: 408-410 (1985).

A MODEL OF SOUND PROPAGATION IN AN OPEN DOCK

P.J.T. Filippi and D. Habault

C.N.R.S., Laboratoire de Mécanique et d'Acoustique
31, chemin J.Aiguier, BP 71, 13402 Marseille Cedex 9. France

ABSTRACT

A theoretical and experimental study has been carried out to establish a model of sound progagation in a approximately 50x20x12 m^3 floating dock. The theoretical study was twofold : to determine the acoustical behaviour of the bottom and to establish a model of the sound propagation in the layer of water. In correspondence with these two parts, two types of sound levels measurements have been made : in the first series of experiments, the signal emitted was a short periodic signal, the results were used to determine the acoustical properties of the bottom ; in the second series, continuous large bandwidth signals were emitted, the results were used to develop a model of propagation in the dock and check its validity.

INTRODUCTION

The propagation domain is a sea-water dock which will be used for measurements of sound levels and vibrations of structures. Because of the small dimensions of the dock, compared to the wavelengths studied (frequencies lower than 3000 Hz) the echoes from the boundaries must be taken into account, this makes a model of sound propagation in the layer of water necessary.

To carry this study out, the methods used are based upon the results obtained in outdoor sound propagation above absorbing plane grounds. Indeed, since 20 years, a lot of papers have been published and the simple problem of propagation above a homogeneous plane ground is now solved. Several representations of the sound pressure emitted above a plane ground by a harmonic, point source are known /1/; they have been obtained for several types of ground models, mainly when the ground surface is characterized by a specific normal impedance, that is the "locally reacting" model. Because the determination of the impedance by using a standing wave tube leads to some well known problems, numerical methods have been developed to obtain the impedance values from a few sound level measurements above the ground /2/. These methods are based on a least-square minimization and the impedance value is computed by an iterative process to minimize the difference between experimental and theoretical sound levels. Generally, the sound levels are measured for one position of the source and N positions of a receiver microphone ($5 \leqslant N \leqslant 10$, for example) above the ground ; the easiest way (for both experimental and theoretical reasons) is to put the source and the microphone on the ground surface. The signal emitted can be periodic signal (then the impedance value is obtained for one frequency) or a large band-width signal (then the impedance values are obtained for a lar frequency band if an impedance model is used).

The first part of our study has then consisted in modelling the bottom of the dock as a " locally reacting surface " and evaluating the corresponding impedance values from sound level measurements. In the second part, a model of sound propagation in the layer of water has been developed ; it is based on an image method which is improved by using an exact representation of the diffracted pressure by an absorbing plane ground. To determine the behaviour of the bottom and to develop the sound propagation models, two types of experiments have been carried out.

EXPERIMENTAL STUDY

The floating dock in which the experiment was conducted has a trapezoidal shape, the largest surface is the sea-air interface. The side boundaries are inclined. The interface sea-sediments is quite irregular but can be a priori considered as smooth, compared to the frequencies studied (200-2600 Hz). Furthermore, the composition of the bottom is not well-known : the surface is covered with a 0.70m - deep layer of mud. This mud seems to be liquid enough to be acoustically ignored, at low frequency.

The source and seven microphones were brought down, at 1.2m above the rigid bottom (0.5m above the interface mud-water). Each of them were held by a rope from the sea surface (see figure 1). They were on a horizontal axis situated as far as possible from the side boundaries. Unfortunately, because the dock is sea open, the movements of the sea can make their positions vary, but these variations do not have a big influence, at low frequency. The distances between the source and the microphones were: 5.2m,10.2m,15.2m, 20.2m and 30.2m. One microphone was fixed at 0.2m from the source and used as a reference (on this microphone, the amplitude of the signal directly emitted from the source is quite larger than the amplitude of the echoes). The directivity pattern of the source can be considered as spherical for the frequencies studied.

Two types of signals were emitted :

- short periodic signals : the length of the signal had to be short enough such that it was possible to isolate the direct signal and the echo coming from the bottom from the echoes coming from the top and the walls. Then the number of periods of each signal depends on its frequency ;
- continuous large band-width signals : in this case, the sound levels received at every microphone were recorded, they include all the echoes coming from every direction.

All these measurements have been averaged on 40 or 50 samples ; furthermore, experiments have shown that the sound levels can vary from 0 to 3 dB, one day from the other. Finally, because of the difficult experimenta conditions, the accuracy of the sound levels is estimated at around 3 or 4 dB.

EVALUATION OF THE ACOUSTICAL IMPEDANCE OF THE BOTTOM

Because of the inaccuracy of the measurements, an impedance model versus frequency has been chosen instead of determining an impedance value for every frequency studied (that is 200, 400, 600, 800, 1000, 1300, 1600 and 2600 Hz).

The Delany-Bazley model /3/ has been chosen. It was empirically established for industrial porous materials (like rockwool,...). The reduced normal specific impedance ς is described by the formula :

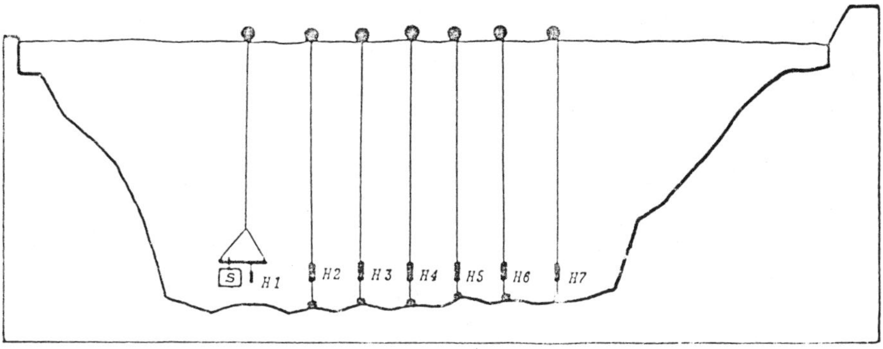

Figure 1

S represents the source

Hj represent the microphones

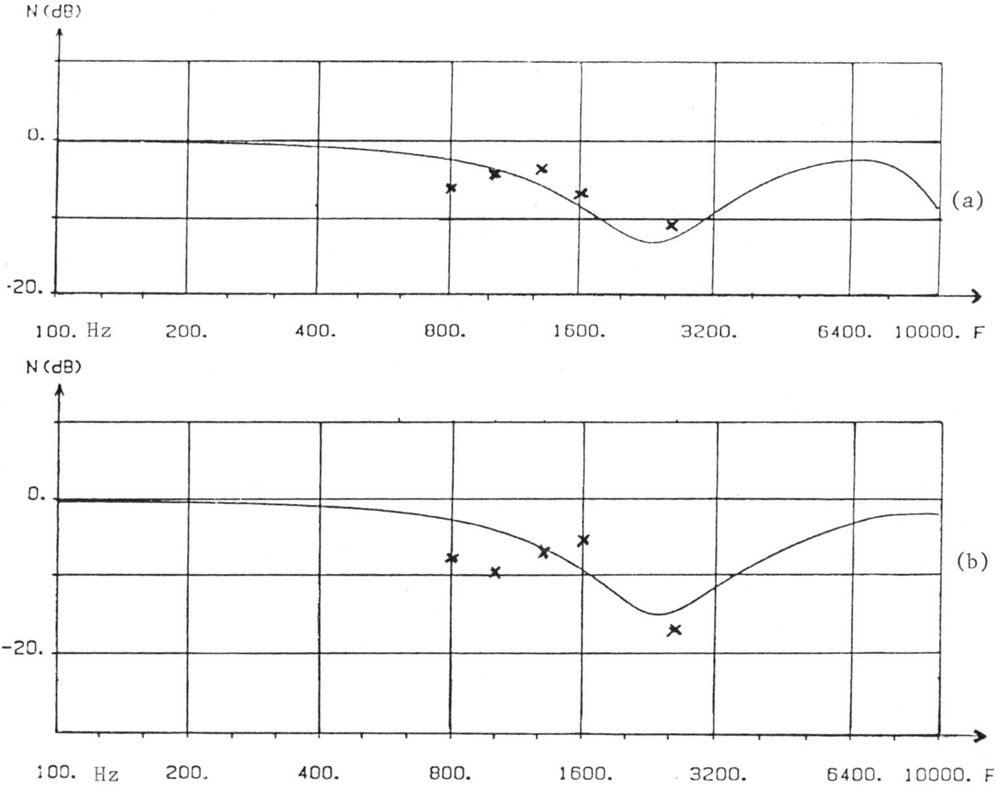

Figures 2

Sound levels compared to those that would be obtained above a perfectly reflecting plane.

× × × measured sound levels

——————— computed sound levels . σ = 880.

Source and microphone height = 1.2m

Distance source-microphone = (a) 20.2m
(b) 30.2m

$$\xi = 1. + 9.08\left(\frac{\sigma}{f}\right)^{0.75} + i\ 11.9\left(\frac{\sigma}{f}\right)^{0.73} \qquad (1)$$

where f is the frequency and σ the flow resistance (σ is a real parameter which does not depend on the frequency).

This model has been extensively used for outdoor sound propagation and give satisfying results for several types of absorbing grounds and the most commonly studied frequency ranges (1000-5000 Hz, for example). That is the reason why, although there is no rigorous justification, this model has been chosen to describe the bottom of the dock.

Then the parameter σ has been determined so that it minimizes the function :

$$F(\sigma) = \sum_{i=1}^{N} \left(P^m(\sigma, X, f_i) - P^c(\sigma, X, f_i) \right)^2 \qquad (2)$$

where P^m and P^c represent the measured and computed sound levels for the frequency f_i, at a point X. For an impedance model, only one measurement point is needed.

A value of σ was then obtained for every microphone. The fact that σ does not vary "too much" at one microphone from the other is a good test of the validity of the model. The table 1 presents an example of the results obtained.

Table 1
Values of
Source and receiver heights:1.2m

Distance source-microphone	10.2m	15.2m	20.2m	30.2m
σ (C.G.S. units)	1340	740	880	870

These values have been deduced from the sound levels measured at 800, 1000, 1300, 1600 and 2600 Hz. They are quite close one from each other, at least for 15.2, 20.2 and 30.2m, and it must be noticed that at 10.2m using σ = 880 instead of σ = 1340 does not lead to an error larger than 2 dB. So that it can be considered that the Delany-Bazley model can give satisfying results with σ = 880, compared to the 3-4 dB error on the measurements. Figures 2-a and b show a comparison between the theoretical curve corresponding to σ = 880 and the experimental curve, for distances 20.2 and 30.2m.

A SIMPLE MODEL FOR THE SOUND PROPAGATION IN THE LAYER OF WATER

In a first step, it was decided to ignore the echoes from the walls of the dock. This can be done because the slope of the walls is quite large, so that the main echoes which can arrive in the middle of the dock correspond to the reflections on the top and on the bottom of the dock. That is the reason why the following simple model has been developed . If E is the interface sea-air and Σ the interface sea-bottom, E is characterized by a homogeneous Dirichlet condition and Σ by a "locally reacting surface" condition. Because of the Dirichlet condition, the images of S (S') and of Σ (Σ') are introduced.

Figure 3

Then, the pressure emitted by the point source S at a point M was approximated by :

$$p(S,M) \simeq -\frac{e^{ikR(S,M)}}{4\pi R(S,M)} + \frac{e^{ikR(S',M)}}{4\pi R(S',M)}$$

$$+ p_d(S,\Sigma) - p_d(S',\Sigma') + p_d(S,\Sigma') - p_d(S',\Sigma) \qquad (3)$$

where k is the wavenumber and the notation $p_d(X,\Gamma)$ means the diffracted pressure emitted by the source X, above a plane Γ .

The expression of $p_d(X,\Gamma)$ is well known /1/ when Γ is a locally reacting surface.

This kind of approximation should be better than a 2-term expression obtained by a classical method of images because the diffraction by the absorbing plane is exactly taken into account.

From a numerical point of view, the approximation (3) is quite convenient, since every term p_d includes a surface wave term and a Laplace type integral which lead to very fast computations.

The figures 4-a and b show a comparison between the experimental curves and the theoretical curves corresponding to the approximation (3) and σ = 880. On figure 4-a the distance is 10.2m and the frequency range is 940-2500 Hz. The description of the sound field is satisfying ; the gaps which appear on the computed curve and not on the experimental curve probably correspond to the fact that the measured sound levels were obtained through a third-octave filter while the computed levels were obtained for a very small ratio $\Delta f/f$ ($\simeq 10^{-2}$). On figure 4-b, the distance is 20.2m and the frequency range 464-1000 Hz. The theoretical levels are 10 dB above the experimental ones but both curves describe the same interference pattern; this probably means that the value of σ for these low frequencies is not good enough but that the model of sound propagation in the layer of water is correct.

CONCLUSION

For this study of sound propagation in water, we have used the numerical methods developed for the sound propagation in air above absorbing grounds. Although there is no theoretical obstacle to use these methods in water, some difficulties arise :

- the first one is concerned with the experimental conditions. Of course, sound level measurements in the floating dock are easier to carry out than in open sea but they are more difficult than in the air. Also, it is always better to let the source and the microphones lie on the ground surface so that there is no interference system and the sound level curves are smooth ; it is then easier to evaluate the accuracy of the measurements. In the present study, this was not possible because of the mud and because the bottom of the dock was too irregular ;
-another problem is the choice of a model to describe the acoustical behaviour of the bottom ; although a lot of experiments and articles showed that a local reaction model is quite satisfying for an absorbing ground,

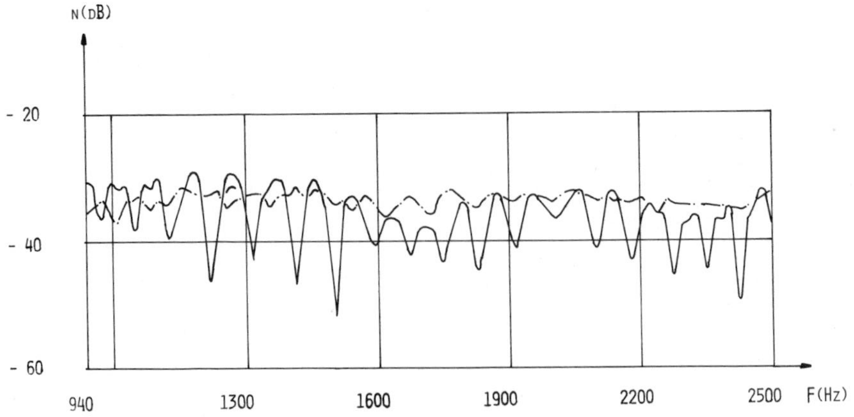

Figure 4 - a

Sound levels obtained in the layer of water

– – – – – measured sound levels
——— computed sound levels. σ = 880. (the echoes from the top and from the bottom of the layer are taken into account)
Source and microphone height : 1.2m
Distance source-microphone : 10.2m

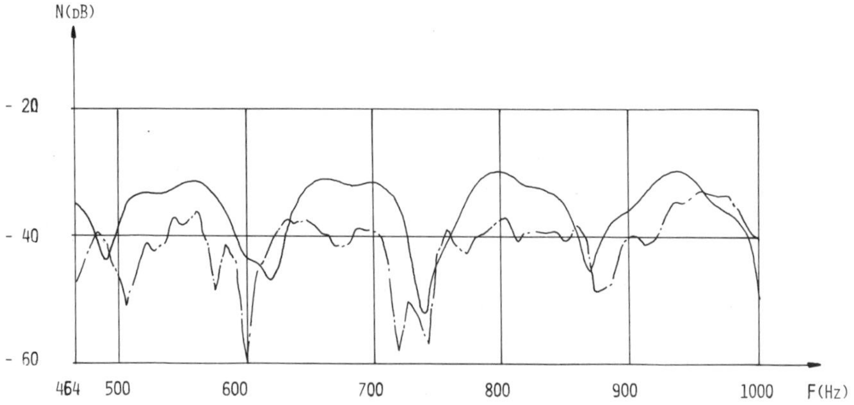

Figure 4 - b

Sound levels obtained in the layer of water

– – – – – measured sound levels
——— computed sound levels. σ = 880. (the echoes from the top and from the bottom of the layer are taken into account).
Source and microphone height : 1.2m
Distance source-microphone : 20.2m

this assumption has not been checked by enough experimental results in water.

From the comparison made between experimental and theoretical results in this study, it seems that the simple model of sound propagation in the layer of water can provide a correct prediction of the sound levels. But the one-parameter impedance model cannot be used for a large frequency range ; results could certainly be improved by computing a value of σ for smaller intervals, like a third octave, for example. Of course, because of the inaccuracy of the measurements, it is not useful to develop a too complicated model. The next step of this study will be then to increase the number of measurements on the same frequency range.

REFERENCES

1. P.J.T. Filippi, "Extended source radiation and Laplace type integral representations. . .", J. of Sound Vib., Vol. 91(1), p.65-84, 1983.

2. D. Habault and G. Corsain, "Identification of the acoustical properties of a ground", J. of Sound Vib., Vol. 100(2), p.169-180, 1985.

3. M. Delany and E. Bazley, "Acoustical properties of fibrous absorbent materials", Applied Acoustics, Vol. 3(2), p. 105-116, 1970.

SOUND PROPAGATION IN A RANGE-DEPENDENT SHALLOW OCEAN WITH A BOTTOM CONTAINING VERTICAL SOUND SPEED GRADIENTS

John F. Miller†, Anton Nagl*, and Herbert Überall*

†The Singer Company, Link Simulation Systems Division
11800 Tech Road, Silver Spring, MD 20904, USA

*Department of Physics, The Catholic University of America
Washington, DC 20064, USA

ABSTRACT

A coupled normal-mode model of acoustic propagation for a gradually range-dependent ocean environment is developed to accommodate an arbitrary layering of the ocean and sediment with piecewise-linear range segmentation. This paper specifically focuses on the detailed solution of the range-separated equations. Theoretically, no limit is imposed on the number of layers and range segments, or on the number of modes that are to be calculated and summed. Practically, however, the model is best applied to low-frequency propagation in a quasi-stratified shallow ocean. Furthermore, such applications best exploit the model's capacity to accurately treat bottom penetration in combination with sound speed gradient layering. Such gradients are here shown to influence decisively the character of bottom-limited upslope propagation over a continental shelf. To wit, previous studies showed for isospeed profiles that upon mode cutoff in the water, sound penetrates steeply downward into the bottom and is lost. We demonstrate that upward-refracting water columns weaken this effect, although it is still present. More importantly, we show that for upward-refracting bottom gradients, as are often present in sediments, sound will continue propagating shorewards beyond cutoff underneath the ocean bottom, as a potentially receivable signal.

BACKGROUND

Normal mode theory arises from consideration of the Helmholtz equation for the acoustic velocity potential $\varphi(z,\rho)$ that is due to a unit point source of harmonic frequency ω located on the depth axis at $(z_0,\rho_0 \equiv 0)$. The coordinate system (along with the general cylindrically symmetric environmental model) is illustrated in Fig. 1. The Helmholtz equation (Officer, 1958; Skudrzyk, 1971):

$$(\nabla^2+k^2)\varphi = -\frac{1}{2\pi\rho}\,\delta(\rho)\delta(z-z_0), \tag{1}$$

with wavenumber $k \equiv \omega/c$, was first solved via normal-mode theory for a single ocean layer of constant sound speed, c (Pekeris, 1948). Numerous layered-ocean analyses followed, each modeling a range-independent sound speed profile (SSP), $c=c(z)$, with flat boundaries (Bucker, 1970).

The work was subsequently extended to incorporate gradually sloping boundaries and a SSP having the range as a secondary parameter, $c=c(z,\rho)$ (Pierce, 1965; Milder, 1969). The method employs a quasi-separation of variables,

$$\varphi(z,\rho) = \sum_{n=1}^{N} \Psi_n(\rho) u_n(z,\rho), \tag{2}$$

wherein the N eigenfunctions (or normal modes) $u_n(z,\rho)$ solve a depth-separated eigenvalue problem locally at each range site ρ,

$$\frac{\partial^2 u_n}{\partial z^2} + \left[k^2(z,\rho) - k_n^2(\rho)\right] u_n(z,\rho) = 0, \tag{3}$$

and where the N local eigenvalues $k_n(\rho)$ replace their constant "Pekeris" counterparts. The eigenfunctions form a complete, locally orthonormal set over the composite waveguide/basement of thickness z_F (Tolstoy and Clay, 1966), with the depth-dependent density $\eta(z)$ as a weighting function:

$$\int_0^{z_F} \eta(z)\, u_n(z,\rho)\, u_m(z,\rho)\, dz = \delta_{nm}. \tag{4}$$

When Eq.(2) is substituted into Eq.(1), a set of N range equations ensues (Graves et al., 1975; Nagl et al., 1978; Chwieroth et al., 1978):

$$\left[\frac{d^2}{d\rho^2} + \frac{1}{4\rho^2} + k_n^2(\rho)\right] f_n(\rho) = -\frac{\delta(\rho)}{2\pi\rho^{1/2}}\, \eta(z_o) u_n(z_o, 0)$$

$$- 2 \sum_m \left(\frac{df_m}{d\rho} - \frac{f_m}{2\rho}\right) M'_{nm}(\rho) - \sum_m f_m M''_{nm}(\rho), \tag{5}$$

where each range function,

$$f_n(\rho) \equiv \rho^{1/2}\Psi_n(\rho), \qquad (n=1,\ldots,N) \tag{6}$$

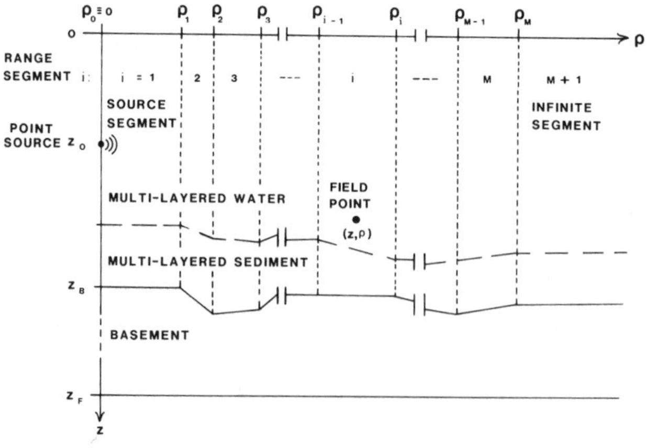

Fig. 1. Coordinate system and the range-segmented environment.

is coupled in all the depth functions, with first-order coupling coefficients given by

$$M'_{nm}(\rho) \equiv \int_0^{z_F} \eta(z) u_n \frac{\partial u_m}{\partial \rho} \quad dz, \tag{7}$$

and second-order coefficients given by

$$M''_{nm}(\rho) \equiv \frac{1}{\rho} M'_{nm}(\rho) + V_{nm}(\rho), \tag{8}$$

with

$$V_{nm}(\rho) \equiv \int_0^{z_F} \eta(z) u_n \frac{\partial^2 u_m}{\partial \rho^2} \quad dz. \tag{9}$$

For the calculations discussed below, our solution of the eigenvalue Eq.(3) at each fixed range site ρ_i (see Fig. 1) is based on a previous, layer-matched Airy-function approach (Nagl et al., 1978), but was extended for the present work as follows: (a) sediment layers with different densities and variable SSPs have been included in the code, in order to accommodate generalized bottom penetration; and (b) the continuous eigenvalue-spectrum modes (Ewing et al., 1957) have been included [albeit in a discretized form by introducing a mathematical pressure-release boundary deep in the basement (McDaniel, 1977; Evans, 1983)], in order to accommodate coupled-mode transfers of energy, specifically at waterborne-mode cutoff.

The set of normal modes thus obtained at each fixed ρ_i is used as one of the elements in a piecewise-linear range segmentation of the environment. In the next section, we concentrate on decoupling and solving the range-separated Eqs.(5). The model is then applied to upslope propagation in a wedge-shaped ocean.

SOLUTION OF THE RANGE EQUATIONS

The quasi-stratified environmental model (already presented in Fig. 1) assumes M finite range segments bounded in the far-field by an infinite segment. Introducing a superscript i into the Eq.(5) that is applicable to the $i^{\underline{th}}$ segment ($\rho_{i-1} \le \rho \le \rho_i$), we note that the solution $f_n^i(\rho)$ will have two unique coefficients, say α_n^i and β_n^i, which are determined by matching the segment solutions across the vertical boundaries using the physical conditions of continuity of pressure (thus f_n^i) and particle velocity (thus $f_n^{i\prime}$, the radial derivative of f_n^i).

In the first and last range segments, the coupling terms drop out of Eqs.(5) and the solutions are trivial:

$$f_n^1(\rho) = \rho^{1/2} \left\{ \alpha_n^1 H_0^{(1)}[k_n(\rho_o)\rho] + \beta_n^1 H_0^{(2)}[k_n(\rho_o)\rho] \right\} \tag{10}$$

$$f_n^{M+1}(\rho) = \rho^{1/2} \alpha_n^{M+1} H_0^{(1)}[k_n(\rho_M)\rho], \qquad (n=1,\ldots,N)$$

where an outward radiation condition has been applied to eliminate β_n^{M+1}. In each intermediate segment, a modal decoupling can be performed by first combining Eqs.(5), for all modes, as a single matrix equation:

$$\left[\frac{d^2}{d\rho^2} + \underline{k}^2\right] \vec{f}^i(\rho) + 2\underline{M}'\vec{f}^{i\prime}(\rho) + \underline{V}\vec{f}^i(\rho) = 0, \qquad (i=2,\ldots,M) \tag{11}$$

where we note the following: at the ranges of applicability, the source term on the RHS of Eq.(5) drops out and the term $1/4\rho^2$ on the LHS is negligible; $\vec{f}^i(\rho)$ is a column vector composed of the N unknown range functions f_n^i (n=1,...,N); $\underline{k}^2$ is an N x N diagonal matrix whose elements are the eigenvalues squared, assumed constant and evaluated at the segment midpoint $\overline{\rho_i}$, $k_n^2(\overline{\rho_i})$; $\underline{M}'$ and $\underline{V}$ (similarly assumed constant) are N x N coupling matrices (thus non-diagonal) whose nmth elements are $M'_{nm}(\overline{\rho_i})$ and $V_{nm}(\overline{\rho_i})$; and the second-order $M''_{nm}(\rho)$ has been redistributed using Eq.(8). The assumption of segmentwise constant $\underline{k}^2$, $\underline{M}'$, and $\underline{V}$ (evaluated at $\overline{\rho_i}$) is reasonable for a realistically fine range segmentation.

A series of transformations is next performed in order to repeatedly diagonalize matrices and regroup terms in Eq.(11) until the modes are decoupled. Dropping the segment index i (as was already done in writing $\underline{k}^2$, $\underline{M}'$, and $\underline{V}$), we first find the matrix $\underline{S}_1$ that diagonalizes $\underline{M}'$ to have elements λ_n:

$$\underline{\lambda} = \underline{S}_1^{-1}\underline{M}'\underline{S}_1. \tag{12}$$

Another diagonal matrix $\underline{\sigma}(\rho)$ is then formed having elements $\exp(-\lambda_n\rho)$, followed by a regrouping of terms into the matrix $\underline{T}$ [where, as before, we assume $\underline{\sigma} \equiv \underline{\sigma}(\overline{\rho_i})$]:

$$\underline{T} \equiv \underline{\sigma}^{-1}\underline{S}_1^{-1}(\underline{k}^2 + \underline{V})\underline{S}_1\underline{\sigma} - \underline{\lambda}^2. \tag{13}$$

In this manner, Eq.(11) is transformed into

$$\left(\frac{d^2}{d\rho^2} + \underline{T}\right) \vec{\ell}(\rho) = 0, \tag{14}$$

where

$$\vec{\ell}(\rho) \equiv \underline{\sigma}^{-1}\underline{S}_1^{-1}\vec{f}(\rho). \tag{15}$$

Although $\underline{T}$ is constant and the first derivatives are gone, the range equations are still coupled because $\underline{T}$ is not diagonal. They are finally decoupled by finding the matrix $\underline{S}_2$ that diagonalizes $\underline{T}$ to have elements Λ_n:

$$\underline{\Lambda} = \underline{S}_2^{-1}\underline{T}\,\underline{S}_2. \tag{16}$$

By forming the overall transformation matrix with elements U_{nm}:

$$\underline{U} \equiv \underline{S}_1\underline{\sigma}\,\underline{S}_2, \tag{17}$$

and reinstating the segment superscript i, the solutions are finally found to be

$$f_n^i(\rho) = \sum_{m=1}^{N} U_{nm}^i(\alpha_m^i \cos q_m^i\rho + \beta_m^i \sin q_m^i\rho), \tag{18}$$

$$(n=1,\ldots,N;\ i=2,\ldots,M)$$

where q_n^i are the diagonal elements $\Lambda_n^{1/2}$ in the ith segment.

THE UPSLOPE WEDGE PROBLEM

Although the model developed above will treat a general depth and (gradual) range dependence in the ocean environment, we consider here the rather specific problem of a single-angle bottom slope with several simplified ocean and sediment SSPs.

Isospeed Case

Notwithstanding its considerable history of research effort, the topic of upslope sound propagation in an ocean wedge with penetrable bottom has previously received attention solely in the case of an isospeed wedge and bottom. This problem was first tackled using the parabolic equation model (Jensen and Kuperman, 1980). Following this were several notable analytic studies (Pierce, 1982; Kamel and Felsen, 1983) and the use of the "slab model", in which a succession of constant-depth regions forms a staircaselike bottom (Evans, 1983). The results of this latter investigation showed that a coupled-mode approach is applicable, with the wedge-trapped discrete modes cutting off in upslope succession. Each cutting-off mode transfers a considerable amount of energy into the continuous-spectrum modes that are not trapped, but propagate into the bottom and that may be called "radiating modes." The assumption of an isospeed bottom makes this appear like a discrete "ray-bundle refraction" of sound energy into the bottom (although this refraction does not correspond to that of a typical ray picture, but occurs separately for the individual modes). According to all the cited works, the energy then propagates steeply downward into the bottom and gets lost in the depths. Thus it does not appear possible that signals originating out in the ocean and traveling shorewards up the continental slope, could be received any closer to shore than the cutoff point of the last mode in the water wedge. We have found, however, that when the basement includes a surficial sediment layer having a realistic, upward-refracting SSP, the resulting acoustic field is quite different.

General Cases

Our calculations were carried out for the bathymetry shown in Fig. 2 and the SSPs shown in Fig. 3. Case 1 is our reference example. From the 25-Hz source (at 112m depth) shorewards, a homogeneous ocean with a sound speed c=1,500 m/sec, a density η=1, and a depth of 200m, reaches over a 5,000m range, then slopes linearly upward to a depth of 60m at a range of 10,173m (i.e., with a slope of 1.55°), remaining flat from there onwards. An isospeed basement of c=1,704.5 m/sec and η=1.15 underlies this water wedge. We take the deep pressure-release boundary to be at z_F=1,200m, together

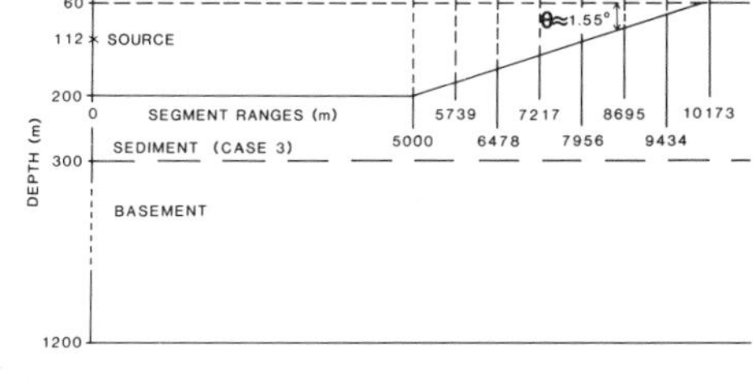

Fig. 2. Assumed geometry of ocean wedge, sediment, and basement.

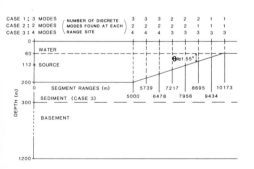

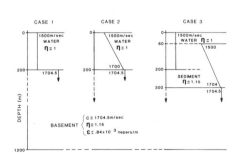

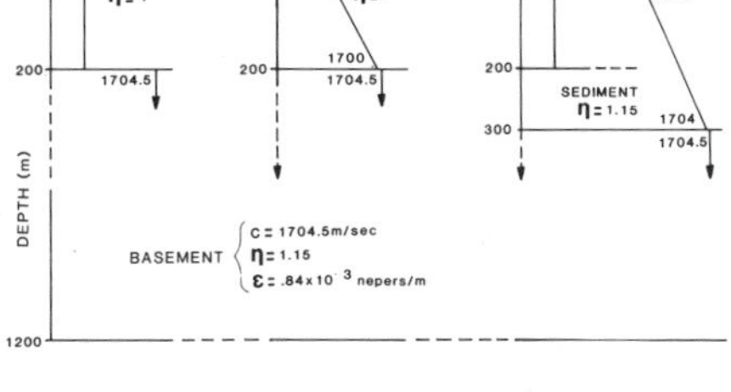

Fig. 3. Environmental parameters for the three cases.

with a basement attenuation constant of $\varepsilon = 0.84 \times 10^{-3}$ nepers/m in order to eliminate spurious reflections from this boundary. Our case 2 uses basically the same two layers, except the water column assumes a constant, upward-refracting sound speed gradient of 1 sec^{-1}. For our case 3, we assume the two isospeed layers of case 1, but for which the basement reaches from a top-depth of 300m downward. A sediment layer, sandwiched between the water wedge and the basement, is taken to have a constant upward-refracting gradient of 0.85 sec^{-1}, and constant values of η=1.15, and of c=1,704 m/sec at its interface with the basement.

Our case 1 (isospeed water/basement) corresponds exactly to that considered previously (Evans, 1983). Three trapped modes are present in the source segment as indicated at the top of Fig. 2: one cuts off at a range around 7,000m; another cuts off near 9,000m but the source depth has been chosen to coincide with a null of this mode so that it does not contribute to our acoustic field. The last mode does not cut off, but propagates into the flat region at the top of the wedge.

Results

Our results for this case are shown in Fig. 4. A total of 36 modes exist (3 trapped and 33 radiating ones at the source); the acoustic field of Fig. 4 was calculated using 22 modes, although the results were seen to stabilize with only 15 or so modes. At the cutoff point of the mentioned mode (near 7,000m), the sound is coupled into the basement, and owing to the isospeed profile, is radiated steeply downward, only to be lost in the depths. The results of Fig. 4 correspond closely to those of Evans (1983).

For case 2, there are two modes initially trapped in the positive-gradient water column. The higher-order mode cuts off near a range of 8,000m. Again using 22 modes, the results are shown in Fig. 5, where a basement-radiated beam is still present, but shifted out in range as compared to case 1 (corresponding to the increased mode cutoff range). In contrast to case 1, the narrower beam, accompanied by higher losses at shallower basement depths, supports the conclusion that the upward-refracting water concentrates more energy near the sea surface, with less energy available for coupling to the basement.

The situation is completely changed for the realistic, upward-refracting sediment-layer case 3. Four trapped modes are present in the source segment. One of these cuts off near a range of 7,000m while the three remaining ones continue propagating into the flat region at the top of the wedge. The acoustic field is shown in Fig. 6, using 22 modes as before. This time, near the cutoff point of the mentioned mode (which was strongly excited at the assumed source depth), no downward-propagating radiation "tongue" appears. Instead, the radiation spreads into the sediment in two nearly horizontal layers, the top layer dipping just below the water-sediment interface, the bottom layer delving somewhat deeper, but never reaching as far down as the sediment-basement interface. We conclude that the cutting-off of a mode propagating upslope in the wedge leads to a transfer of its acoustic energy into the bottom, but in such a manner that the sediment-borne sound continues traveling upslope beyond the cutoff point, rather than downward into the basement where it would be lost. This indicates that appropriately bottom-mounted receivers can be expected to register signals originating out in the ocean, after traveling up the continental shelf beyond final mode cutoff, at locations closer to shore than previously predicted.

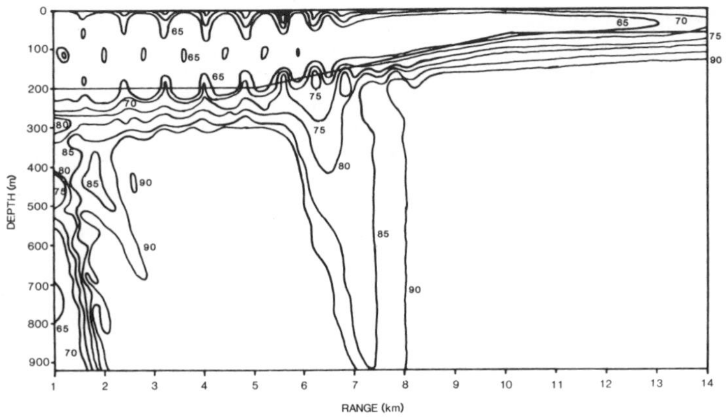

Fig. 4. Contour plot (in dB) of acoustic field for case 1 (two isospeed layers).

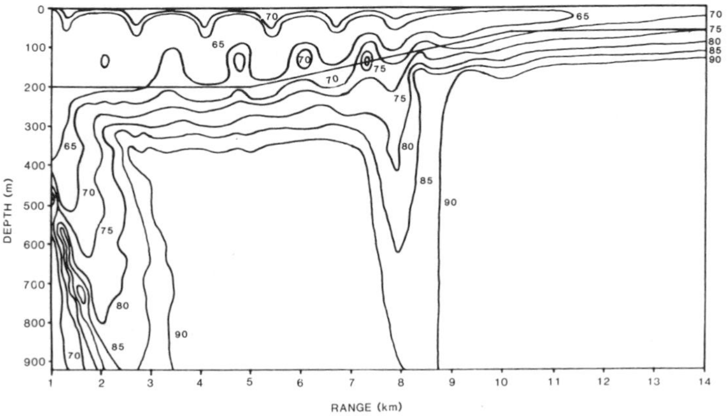

Fig. 5. Contour plot (in dB) of acoustic field for case 2 (upward-refracting wedge, isospeed basement).

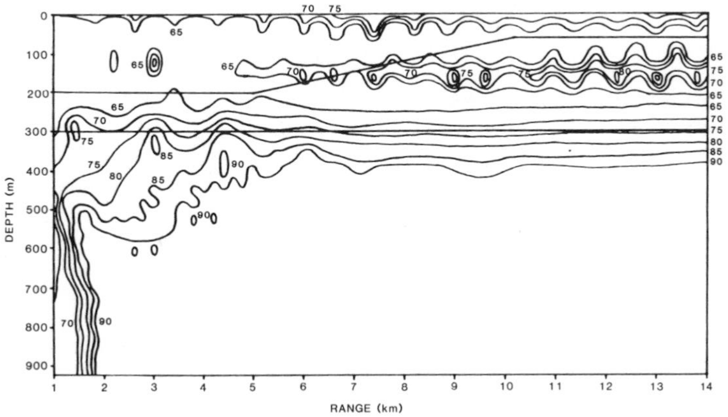

Fig. 6. Contour plot (in dB) of acoustic field for case 3 (isospeed wedge, upward-refracting sediment, isospeed basement).

ACKNOWLEDGMENTS

This work was supported by The Singer Company, Link Simulation Systems Division, and by the Office of Naval Research. Calculations were supported by the Catholic Univeristy Computer Center. Special thanks are due to Deborah Wilson, who typed the manuscript with impressive attention to detail.

REFERENCES

Bucker, H. P., 1970, Sound propagation in a channel with lossy boundaries, J. Acoust. Soc. Am., 48:1187.

Chwieroth, F. S., Nagl, A., Überall, H., Graves, R. D., and Zarur, G. L., 1978, Mode coupling in a sound channel with range-dependent parabolic velocity profile, J. Acoust. Soc. Am., 64:1105.

Evans, R. B., 1983, A coupled mode solution for acoustic propagation in a waveguide with stepwise depth variations of a penetrable bottom, J. Acoust. Soc. Am., 74:188.

Ewing, W. M., Jardetzky, W. S., and Press, F., 1957, "Elastic Waves in Layered Media," McGraw-Hill Book Co., Inc., New York.

Graves, R. D., Nagl, A., Überall, H., and Zarur, G. L., 1975, Range-dependent normal modes in underwater sound propagation: application to the wedge-shaped ocean, J. Acoust. Soc. Am., 58:1171.

Jensen, F. B. and Kuperman, W. A., 1980, Sound propagation in a wedge-shaped ocean with a penetrable bottom, J. Acoust. Soc. Am., 67:1564.

Kamel, A. and Felsen, L. B., 1983, Spectral theory of sound propagation in an ocean channel with weakly sloping bottom, J. Acoust. Soc. Am., 73:1120.

McDaniel, S. T., 1977, Mode conversion in shallow-water sound propagation, J. Acoust. Soc. Am., 62:320.

Milder, D. M., 1969, Ray and wave invariants for SOFAR channel propagation, J. Acoust. Soc. Am., 46:1259.

Nagl, A., Überall, H., Haug, A. J., and Zarur, G. L., 1978, Adiabatic mode theory of underwater sound propagation in a range-dependent environment, J. Acoust. Soc. Am., 63:739.

Officer, C. B., 1958, "Introduction to the Theory of Sound Transmission," McGraw-Hill Book Company, New York.

Pekeris, C. L., 1948, Theory of propagation of explosive sound in shallow water, Geol. Soc. Am. Mem., 27.

Pierce, A. D., 1965, Extension of the method of normal modes to sound propagation in an almost-stratified medium, J. Acoust. Soc. Am., 37:19.

Pierce, A. D., 1982, Guided mode disappearance during upslope propagation in variable depth shallow water overlying a fluid bottom, J. Acoust. Soc. Am., 72:523.

Pierce, A. D., 1982, Augmented adiabatic mode theory for upslope propagation from a point source in variable-depth shallow water overlying a fluid bottom, J. Acoust. Soc. Am., 74:1837.

Skudrzyk, E., 1971, "The Foundations of Acoustics," Springer-Verlag, New York.

Tolstoy, I. and Clay, C. S., 1966, "Ocean Acoustics: Theory and Experiment in Underwater Sound," McGraw-Hill Book Company, New York.

CANONICAL PROPAGATION PROBLEMS FOR A WEDGE SHAPED OCEAN: I. LAYERED FLUID-SOLID BOTTOM; II. BOTTOM WITH LINEAR SURFACE IMPEDANCE VARIATION

I. T. Lu and L. B. Felsen

Dept. of Electrical Engineering & Computer Science
Weber Research Institute, Polytechnic University
Farmingdale, New York 11735

ABSTRACT

The wedge shaped ocean has emerged as one of the most important test models for range-dependent bottom-dominated underwater acoustic propagation. This paper deals with spectral synthesis of exact prototype field solutions for the following two classes of bottom profiles: a) planar fluid and (or) solid stratification, and b) linear surface impedance variation. Category a) is inherently nonseparable, while category b) is separable in a polar coordinate frame centered at the apex. In both categories, one may construct mode-like spectral objects, the intrinsic modes, which propagate without coupling to other intrinsic modes. The exact construction in full generality is trivial for category b) but formidable for category a); however, the latter becomes well tractable for small bottom slopes. The spectral underpinnings of both types of intrinsic modes, and their relation to adiabatic modes, are emphasized. To illustrate the effects of fluid-solid layering, intrinsic mode forms are calculated numerically for a homogeneous fluid sediment layer between the homogeneous water wedge and a homogeneous solid basement. Special attention is given to the upslope transition through cutoff.

I. INTRODUCTION

The wedge shaped ocean has emerged as probably the most important test configuration for studies of range dependent propagation in a shallow water environment. For homogeneous water and fluid bottom media, several recent analytical and numerical investigations [1-3] have sought to clarify the existence, or not, of mode-like field distributions which propagate without coupling to other "modes" in this intrinsically nonseparable propagation channel, with particular emphasis on the upslope transition through cutoff. In these explorations, the conventional adiabatic modes, which are well-trapped in the water column sufficiently far downslope from cutoff, have played an essential interpretative and initiating role. The adiabatic modes are defined on planar local cross sections of the equivalent range independent waveguide. They fail in the cutoff transition, and coupling between them is weak or even negligible for very small bottom slopes but can become appreciable for steep bottom inclinations.

The *intrinsic* modes [4], which represent exact source-free field solutions in the ocean wedge configuration, overcome the adiabatic mode coupling and transition problem. They account compactly for forward and backward coupling between adiabatic modes, for arbitrary bottom slopes. They also appear to be decoupled from one another although no orthogonality or completeness theorem has as yet been found to substantiate this assertion. However, a *rigorous* theory exists that incorporates them within the Green's function for the wedge domain [5]. Their proper computation [6] continues to be the subject of present activity, as is their relation to "wedge modes" [7] observed in recent experiments, wherein the adiabatic mode profile impressed on a *circular* apex-centered cross section generates a waveform that propagates downslope without distortion even for relatively large bottom slopes.

An important feature of intrinsic mode theory is its capability of accommodating a bottom structured of homogeneous fluid and (or) elastic layers parallel to the water-bottom interface. These generalized test configurations can therefore clarify the role of compression and shear waves, and of bottom profiling, on sound propagation in the water column as well as in the bottom itself. In a previous investigation, we have calculated typical two-dimensional intrinsic modes when the bottom is a homogeneous elastic solid [8]. These studies have now been extended to the case where a fluid sediment layer is inserted between the water and the solid basement. Typical mode shapes are presented to illustrate the influence of the layer on the propagation characteristics of the intrinsic modes, especially in regard to the transition from the well trapped downslope to the radiative (leaky) upslope regime. These calculations are performed for small enough bottom slopes to warrant ignoring certain remainder terms that complicate the numerical evaluation.

An interesting question is how to incorporate interface waves within the intrinsic mode format. Such waves decay away from the guiding boundary on either side. Although the recently developed rigorous intrinsic mode theory [5] can account for these phenomena, which arise from a pole singularity in the spectral boundary reflection coefficient, it is instructive to consider a special coordinate separable case that includes this effect: a homogeneous fluid wedge with a bottom described by a linearly varying surface impedance [9]. If the impedance (which may be lossy) has the proper reactive component, a boundary wave can exist. The simple exact modes in this example, valid for arbitrary bottom slopes, include not only the boundary wave but also allow a prototype study of (linear) range dependence of the bottom, with all forward *and* backward wave phenomena accounted for. The trivially obtained spectral integral form for these coordinate separable impedance wedge modes sheds further light on the spectral synthesis of the intrinsic modes.

II. INTRINSIC MODES FOR THE STRATIFIED BOTTOM

Since the intrinsic mode problem for the wedge-shaped ocean with plane layered fluid and(or) elastic bottom has been formulated previously [8], we cite here only the basic equations and the actual forms of the intrinsic mode spectral integrals used in the numerical work. The general configuration, depicted in Fig. 1, is described in a two dimensional (x,z) coordinate frame, augmented in the water wedge by a cylindrical polar (r,χ) frame when convenient. The time-harmonic $[\exp(-i\omega t)]$ acoustic pressure field $U_i(x,z)$ in the fluid wedge or bottom fluid layers satisfies the scalar wave equation

$$\left(\frac{\partial^2}{\partial x^2} + \frac{\partial^2}{\partial x^2} + k_i^2\right) U_i(x,z) = 0 \qquad (1)$$

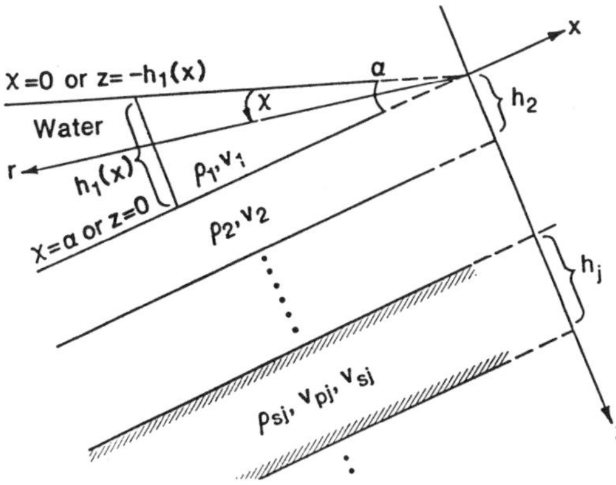

Figure 1 - Idealized fluid wedge with homogeneous slope α, on top of a z-stratified half-space comprising homogeneous fluid layers with density ρ_i, sound speed v_i and layer thickness h_i as well as homogeneous solid layers with density ρ_{sj}, pressure wave speed v_{pj}, shear wave speed v_{sj} and thickness h_j. The indexes i, j designate the layers.

with $U_1 = 0$ at the water surface $z = -h_1(x)$, whereas across an interface between adjacent fluid layers, $\rho_i U_i$ and $\partial U_i/\partial z$ are required to be continuous. Here, $k_i = \omega/v_i$ is the wavenumber, v_i is the sound propagation speed, and ρ_i is the density in the i-th fluid medium (see Fig. 1). In any elastic layer of the bottom, compressional (P) and vertical shear (SV) waves exist simultaneously. Displacement and stress for the j-th layer can be derived from the corresponding scalar potentials ϕ_j and ψ_j which satisfy

$$\left(\frac{\partial^2}{\partial x^2} + \frac{\partial^2}{\partial z^2} + k_{pj}^2\right) \phi_j = 0, \quad \left(\frac{\partial^2}{\partial x^2} + \frac{\partial^2}{\partial x^2} + k_{sj}^2\right) \psi_j = 0 \tag{2}$$

with parameters

$$k_{pj} = \frac{\omega}{v_{pj}}, \quad k_{sj} = \frac{\omega}{v_{sj}}; \quad v_{pj} = \sqrt{\frac{\lambda_j + 2\mu_j}{\rho_{sj}}}, \quad v_{sj} = \sqrt{\frac{\mu_j}{\rho_{sj}}} \tag{2a}$$

Here, v_{pj} and v_{sj} are the compressional and shear wave speeds, λ_j and μ_j are Lamé constants, and ρ_{sj} is the density. At any interface, one must invoke the appropriate conditions requiring continuity of stress and displacement. There exist systematic procedures a) for generating at the water-bottom interface the pressure plane wave reflection coefficient for arbitrary bottom layering [10-12], and b) the plane wave synthesis of fields observed in any layer in the bottom [10-12]. These plane wave building blocks, which appear in the spectral synthesis of the intrinsic modes, allow construction of the intrinsic mode form anywhere in the water or bottom region.

We list explicitly the reflection and transmission coefficients (valid away from the neighborhood of the apex) required for the special case when a single fluid sedimentary layer separates the water from the solid bottom. The <u>composite</u> reflection and transmission coefficients for the fluid <u>layer</u> between $z = 0$ to $z = h_2$ are (see Fig. 2),

$$R_D = r_d + t_u r_{pp} t_d f(1-g)^{-1}, \quad T_{DP} = t_d t_{pp} f^{1/2}(1-g)^{-1}$$

$$T_{DS} = t_d t_{ps} f^{1/2}(1-g)^{-1}, \quad f = e^{i2k_2\cos\theta_2 h_2}, \quad g = r_u r_{pp} f \tag{3}$$

The interface reflection and transmission coefficients for the fluid-fluid and fluid-solid interfaces are (see Fig. 2),

$$r_d = (Z_2 - Z_1)\Delta_1^{-1}, \quad r_u = -r_d, \quad t_d = 2Z_2\Delta_1^{-1}, \quad t_u = 2Z_1\Delta_1^{-1}, \quad \Delta_1 = Z_2 + Z_1 \tag{4}$$

and

$$r_{pp} = [Z_p \cos^2 2\theta_s + Z_s \sin^2 2\theta_s - Z_2]\Delta_2^{-1}, \quad t_{pp} = -2Z_p \cos 2\theta_s \cdot (\rho_2/\rho_s)\Delta_2^{-1}$$
$$t_{ps} = 2Z_s \sin 2\theta_s \cdot (\rho_2/\rho_s)\Delta_2^{-1}, \quad \Delta_2^{-1} = Z_p \cos^2 2\theta_s + Z_s \sin^2 2\theta_s + Z_2 \tag{5}$$

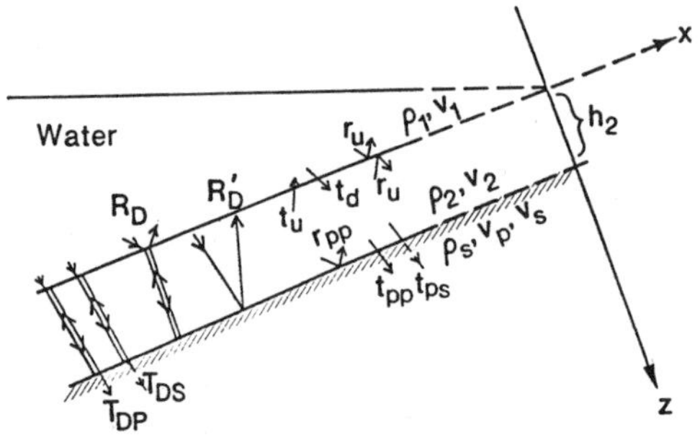

Figure 2 - Schematic of various reflection and transmission coefficients for the special case where a fluid sedimentary layer separates the water from the solid bottom. Apex region is excluded. Heavy lines signify collective effects that include all multiple reverberations between the boundaries.

Here, Z_1, Z_2, Z_p and Z_s denote, respectively, the impedance of sound waves in the wedge and in the fluid sedimentary layer, and of longitudinal and transverse waves in the solid,

$$Z_1 = \frac{\rho_1 v_1}{\sin\theta_1}, \quad Z_2 = \frac{\rho_2 v_2}{\sin\theta_2}, \quad Z_p = \frac{\rho_s v_p}{\sin\theta_p}, \quad Z_s = \frac{\rho_s v_s}{\sin\theta_s} \tag{6}$$

with θ_1, θ_2, θ_p and θ_s denoting the corresponding propagation angles which are related by Snell's Law,

$$k_1 \cos\theta_1 = k_2 \cos\theta_2 = k_p \cos\theta_p = k_s \cos\theta_s \tag{7}$$

The subscript "3" for the elastic bottom has been omitted.

The m-th intrinsic mode pressure field U_{1m} in the water wedge can be represented by a sum of upgoing (U_{1m}^+) and downgoing (U_{1m}^-) wave components [8],

$$U_{1m} = U_{1m}^+ + U_{1m}^- \tag{8}$$

where

$$U_{1m}^{\pm} = \int_C \exp[(ik_1(x\cos\theta_1 \pm z\sin\theta_1)]\{{R_D \atop 1}\}A_1(\theta_1)d\theta_1, x<0, -h_1(x)\leq z<0 \tag{8a}$$

The integration proceeds along an infinite contour C in the complex θ_1-plane along which the integrand converges, but the important portion which suffices for numerical evaluation, is a segment along the real axis between $\theta_1 = 0$ to $\pi/2$. The spectral amplitude $A(\theta_1)$, which incorporates self-consistency between the plane waves $U_{1m}^{\pm}$ after successive reflections at the wedge boundaries, is for $\alpha << 1$,

$$A_1(\theta_1) = \exp i\,[\Phi_D(\theta_1) + \Phi_U(\theta_1 + \alpha) - \frac{m\theta_1\pi}{\alpha} + \frac{1}{2\alpha}\int_{\theta_0}^{\theta_1} \Phi_D(\bar\theta) + \Phi_U(\theta+\alpha)\, d\bar\theta\,] \quad (9)$$

where θ_0 is an arbitrary reference angle while Φ_D and Φ_U are the phases of the reflection coefficients at the upper and lower boundaries, respectively, of the wedge. In our particular case, $\Phi_U = \pi$, $\Phi_D = -i\,\ln R_D$. In the sediment layer, $U_{2m} = U_{2m}^{+} + U_{2m}^{-}$, with

$$U_{2m}^{\pm} = \int_C \exp[i\,k_2(x\cos\Theta_2 \pm z\,\sin\Theta_2)]\{{R_D' \atop 1}\}\,A_2(\theta_1)d\theta_1, \quad x < 0,\ 0 \le z < h_2 \quad (10)$$

where

$$A_2(\theta_1) = A_1(\theta_1)\,t_d(1-g)^{-1}, \quad R_D' = r_{pp}f \quad (10a)$$

Here, R_D' is the downgoing reflection coefficient seen from $z = 0+$ (see Fig. 2). In the solid bottom, the intrinsic mode forms of the potentials ϕ and ψ are

$$\left\{{\phi \atop \psi}\right\} = \int_C \exp\,[ik_{p,s}(x\,\cos\theta_{p,s} + z\,\sin\theta_{p,s})]\,T_{DP,DS}A_1(\theta_1)d\theta_1, \quad x < 0,\ z > h_2 \quad (11)$$

Note that θ_1, θ_2 in (10) and (10a), and θ_1, θ_p, θ_s in (11) are related by (7). The horizontal and vertical displacements are then obtained via

$$U_x = \frac{\partial\phi}{\partial x} - \frac{\partial\psi}{\partial z}, \quad U_z = \frac{\partial\phi}{\partial z} + \frac{\partial\psi}{\partial x} \quad (12)$$

We have performed extensive calculations by direct numerical implementation of the intrinsic mode integrals for parameters that describe various wave trapping conditions in the wedge or in the layer. A typical example for the z-displacement of the $m = 2$ mode is shown in Fig. 3. In Fig. 3(a), the low velocity wedge can support an initially trapped field whereas in Fig. 3(b), the strongest trapping occurs in the low velocity layer. The plots reveal how the mode energy is redistributed during upslope propagation. The m=3 mode is shown in Fig. 4, together with its three-dimensional profile. Note that the index m is associated with the transverse periodicity in the water, regardless of what happens in the bottom. A more comprehensive treatment will be given elsewhere.

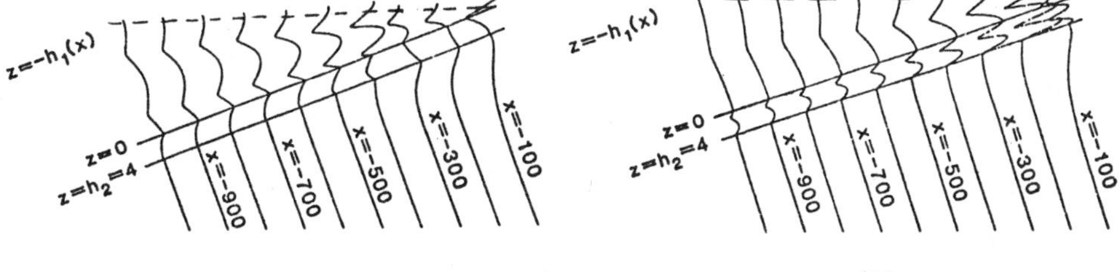

(a) (b)

Fig. 3 - z-displacement for $m = 2$ mode. Parameters: $\alpha = 0.02$, $\omega = 1$, $h_2 = 4$, $v_1 = 1$, $\rho_1 = 1$, $v_s = 2$, $v_p = 2\sqrt{3}$, $\rho_s = 2$.
(a) $v_2 = 1.5$, $\rho_2 = 1.1$ (b) $v_2 = 0.7$, $\rho_2 = 1$.

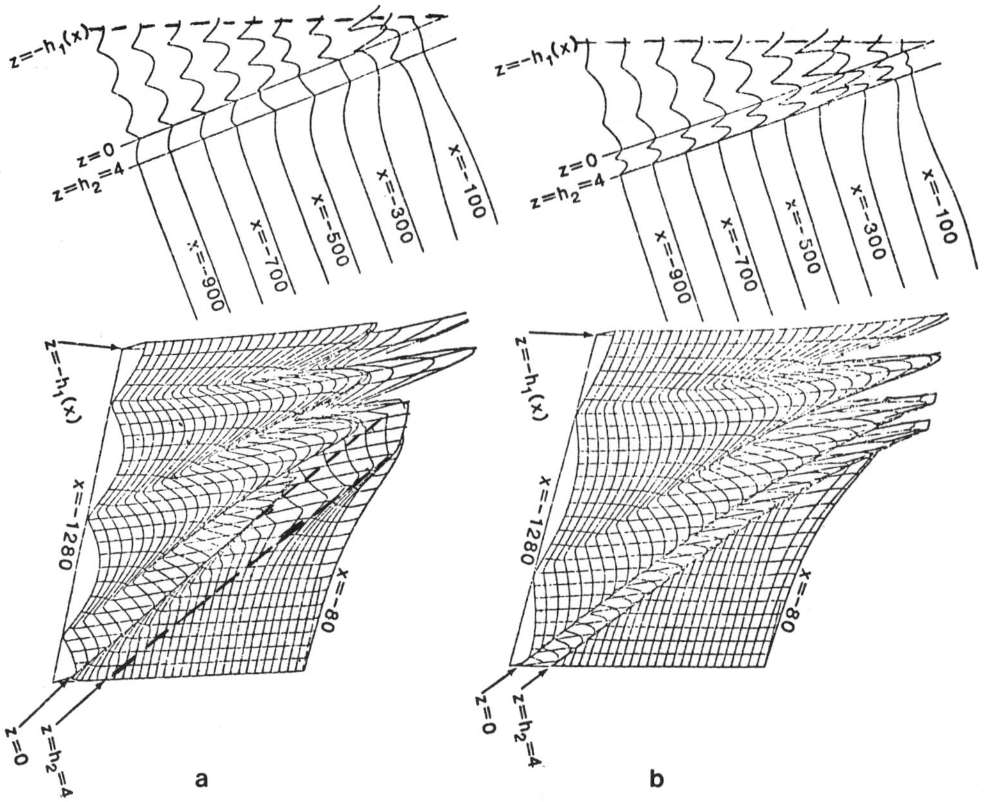

Fig. 4 Two and three-dimensional profiles of z-displacement for m = 3 mode. Parameters are the same as in Fig. 3.

III. INTRINSIC MODES FOR THE LINEAR IMPEDANCE BOTTOM

When the ocean bottom is modeled as a plane boundary with a linearly varying surface impedance Z_b, inclined at an angle α, the wave equation in (1), subject to $U = 0$ at $z = -h_1(x)(\chi = \alpha)$, and to

$$\frac{\partial U}{\partial z} = \frac{1}{r}\frac{\partial U}{\partial \chi} = \frac{1}{Z_b} U, \quad Z_b(r) = \beta r, \quad \chi = 0, \tag{13}$$

becomes separable in a polar (r, χ) coordinate system centered at the apex. The imaginary part $\mathrm{Im}\beta \leq 0$ of the complex constant β models dissipation while the real part models reactive energy which, for $\mathrm{Re}\,\beta < 0$, allows a surface wave to propagate along the bottom. The exact solution for the m-th mode everywhere inside the impedance wedge is given for arbitrary apex angle α by [9]

$$U_m = (\tilde{U}^+_{1m} + \tilde{U}^-_{1m}) + (\tilde{U}^+_{2m} + \tilde{U}^-_{2m}) \tag{14}$$

where

$$\tilde{U}^{\pm}_{\delta m} = \exp[\pm i\nu_m(\alpha-\chi)]H^{(\delta)}_{\nu_m}(kr)\left\{{R_D \atop 1}\right\}, \quad \delta=1,2 \tag{14a}$$

$H^{(\delta)}_{\nu m}$ is the Hankel function, and the angular propagation coefficient ν_m is a solution of the modal resonance equation

$$2\nu_m \alpha = 2m\pi - (\Phi_U + \Phi_D) \quad \text{or} \quad \cot\nu_m\alpha = \beta\nu_m \tag{15}$$

with

$$\Phi_U = \pi, \quad \Phi_D = \frac{1}{i}\,\ell n R_D, \quad R_D = \frac{i\nu_m\beta - 1}{i\nu_m\beta + 1} \tag{15a}$$

Φ_U and Φ_D are the phases of the upper and lower boundary reflection coefficients defined for wave spectra propagating along the angular (χ) coordinate [9]. Each of the wavefunctions $\tilde{U}^{\pm}_{\delta m}$ defines an angular wave congruence going up or down (+ or −), upslope (δ=2) or downslope (δ=1). When $\mathrm{Im}\beta < 0$, then $\mathrm{Im}\nu_m > 0$, and the resulting surface mode decays into the water region.

The impedance wedge modes are synthesized naturally in terms of the angularly propagating wave spectra. Using the integral representation of the Hankel function [9],

$$H^{(\delta)}_{\nu}(kr) = \frac{1}{\pi}\int_{C_\delta} e^{ikr\cos\theta + i\nu(\theta-\pi/2)}\,d\theta \tag{16}$$

with $C_{1,2}$ chosen as Sommerfeld contours [9], we can write after a few trivial changes of variable and use of (15),

$$\tilde{U}^{\pm}_{\delta m} = \frac{1}{\pi}\int_{C^{\pm}_{\delta}} \exp[-ikr\cos(\theta \pm (\alpha-\chi)) + \frac{i}{2\alpha}(\theta - \frac{\pi}{2})(\Phi_U + \Phi_D - 2m\pi)]\left\{{R_D \atop 1}\right\}d\theta \tag{17}$$

Noting that Φ_U and Φ_D are independent of θ and $-r\cos(\theta \pm (\alpha-\chi)) = x\cos\theta \mp z\sin\theta$, one can write (17) as

$$\tilde{U}^{\pm}_{\delta m} = \frac{e^{im\pi^2}}{\pi}\int_{C^{\pm}_{\delta}} \exp[ik(x\cos\theta \mp z\sin\theta)]\left\{{R_D \atop 1}\right\}$$

$$\cdot \exp i\,[\frac{1}{2\alpha}\int_{\theta_0}^{\theta}(\Phi_U + \Phi_D)d\bar{\theta} - \frac{1}{2}(\Phi_U + \Phi_D) - \frac{m\,\theta\pi}{\alpha}]d\theta, \quad \theta_0 = \frac{\pi}{2} - \alpha \tag{18}$$

Comparing the intrinsic mode wavefunction $U_{2m} = \tilde{U}^{+}_{2m} + \tilde{U}^{-}_{2m}$ for the variable impedance wedge with the corresponding intrinsic mode form in (8), (8a) and (9) for the plane stratified penetrable bottom, one observes substantial spectral simplifications in (18). This is due to the fact that for the coordinate separable surface impedance, the angular spectral reflection coefficients are constant after successive reflections; the conventional plane wave reflection coefficients are not and therefore require the phase integral in (9), which is generated by the modal closure condition (see [4]). Note also that separability (i.e., absence of mode coupling) for the impedance wedge is tied to the polar coordinate system, which gives rise in the adiabatic mode regime to local mode profiles defined on circular cross sections centered at the apex. Using the customary adiabatic modes defined on a plane cross section would introduce coupling between these modes. This circumstance lends further support to the preveously mentioned experimental observations of Hoaek, et al. [7].

IV. SUMMARY

We have examined inherently exact intrinsic mode solutions for two vastly different bottom configurations in a wedge shaped ocean: a) planar fluid-solid-stratification, and b) linear surface impedance variation. The synthesizing spectra of the intrinsic mode wavefunctions have been examined and related to the resulting observed wavefields, which propagate without coupling to other intrinsic modes in their respective environments. Mode shapes have been computed for a small angle wedge with a bottom composed of a fluid sediment layer above a solid basement to illustrate the effects of stratification as well as compression-shear wave coupling, especially during the cutoff transition encountered during upslope propagation. The exact theories for these special configurations can be generalized. When the water and bottom environments deviate weakly from the canonical wedge prototypes, local intrinsic modes, in two and three dimensions, form the basis for constructing an approximate propagation theory of broad scope [13,14]. Green's functions can be synthesized--exactly for the canonical prototypes [5] and approximately under more general conditions. Adiabatic transforms can be defined to relate wavefields to spectra in these weakly range-dependent layered geometries [14], with implications for profile inversion. These aspects, as well as the more precise numerical evaluation of the intrinsic modes, are presently under investigation.

ACKNOWLEDGEMENT

This work has been supported by the U.S. Office of Naval Research under Contract No. N-00014-79-C-0013 and by the National Science Foundation under Grant No. EAR-8416052.

REFERENCES

1. A.D. Pierce, "Extension of the Method of Normal Modes to Sound Propagation in an Almost-Stratified Medium," J. Acoust.Soc. Am. 37, 19-27, (1965).
2. F.B. Jensen and W.A. Kuperman, "Sound Propagation in a Wedge-Shaped Ocean with Penetrable Bottom," J.Acoust.Soc.Am. 67, 1564-1566,(1980).
3. A. Kamel and L.B. Felsen, "Spectral Theory of Sound Propagation in an Ocean Channel with Weakly Sloping Bottom," J.Acoust.Soc.Am. 73, 1120-1130, (1983).
4. J.M. Arnold and L.B. Felsen, "Rays and Local Modes in a Wedge-Shaped Ocean," J. Acoust. Soc. Am. 73, 1105-1119, (1983).
5. J.M. Arnold and L.B. Felsen, "Theory of Wave Propagation in a Wedge Shaped Layer," submitted to Wave Motion.
6. E. Topuz and L.B. Felsen, "Intrinsic Modes: Numerical Implementation in a Wedge-Shaped Ocean," J. Acoust. Soc. Am. 78, 1735-1745, (1985).
7. H. Hoaek, C.T. Tindle, and T.G. Muir, "Downslope Propagation of Normal Modes in a Shallow Water Wedge," J.Acoust.Soc.Am. 78, S70, (1985).
8. I.T. Lu and L.B. Felsen, "Intrinsic Modes in a Wedge-Shaped Ocean With Stratified Elastic Bottom," to appear in the Proceedings of the Conf. on Ocean Seismo-Acoustics, LaSpezia, Italy, June 1986.
9. L.B. Felsen and N. Marcuvitz, "Radiation and Scattering of Waves," Prentice-Hall, Englewood Cliffs, N.J. 1973, Sec. 6.6.
10. W.W. Ewing, W.S. Jardetsky and F. Press, "Elastic Waves in Layered Media," McGraw-Hill, N.Y. (1957).
11. B.L.N. Kennett, "Seismic Wave Propagation in Stratified Media," Cambridge University Press, N.Y. (1983).
12. I.T. Lu and L.B. Felsen, "Matrix Green's Functions for Array Type Sources and Receivers in Multiwave Layered Media," Geophy.J.R. Astr. Soc., 84, 31-48, (1986).
13. J.M. Arnold and L.B. Felsen, "Local Intrinsic Modes: Layer with Nonplanar Interface," Wave Motion, 8, 1-14, (1986).
14. I.T. Lu and L.B. Felsen, "Adiabatic Transforms for Spectral Analysis and Synthesis of Weakly Range Dependent Shallow Ocean Green's Functions," submitted to J. Acoust. Soc. Am.

RIGOROUS THEORY OF HORIZONTAL REFRACTION IN A WEDGE SHAPED OCEAN

J.M. Arnold[+] and L.B. Felson*

[+]Department Electronics and Electrical Engineering
The University Glasgow, G12 8QQ, Scotland

*Department Electrical Engineering and Computer Science
Polytechnic Institute New York, Farmingdale, N.Y. 11735

ABSTRACT

A rigorous method for the analysis of acoustic propagation in a 3-dimensional wedge is described, with boundary conditions which are either perfectly reflecting or penetrable. In particular, we show how the intrinsic mode concept can be generalized to 3-dimensional structures, with explicit reference to horizontal refraction phenomena exhited by the solutions.

INTRODUCTION

The problem of the proper theoretical description of acoustic wave propagation in nonuniform waveguides has been subjected to intense scrutiny recently, following a model numerical computation by Jensen and Kuperman (1980) of the upslope propagation behaviour of a wedge-shaped shallow ocean with a fluid-fluid interface to represent the bottom. This computation, carried out with the parabolic equation (PE) algorithm clearly demonstrated the emergence of discrete radiation beams into the bottom medium, each emergent beam appearing to originate from a point on the water-bottom interface whose depth corresponds with the cut-off depth of a local normal mode. While it may be intuitively clear that an initially guided mode will begin to radiate after propagating upslope beyond its cut off section, at the time of publication (1980) these results could not be quantitatively explained by any means other than the PE calculation, which is itself a purely numerical procedure for solving parabolic partial differential equations. Subsequently, Pierce (1982, 1984), Arnold and Felsen (1983), Kamel and Felsen (1983) and Evans (1983) all found other methods by which the phenomena first predicted by Jensen and Kuperman could be reproduced. The theory used by Evans (1983) is essentially conventional coupled mode theory (Pierce, 1965; Chwieroth et al, 1978), with careful attention paid to those parameters to which coupled-mode theory is known to be highly sensitive, such as discretisation and truncation of the continuous spectrum of radiation modes. Pierce (1982, 1984) modified the lowest order solution of coupled mode theory, the 'adiabatic' approximation, by the use of matched asymptotic expansions. Kamel and Felsen (1983) constructed an integral whose residue contributions are adiabatic modes similar to the Fourier representation for translation invariant waveguides but with modifications to accommodate range dependence Arnold and Felsen (1983) obtained an explicit representation for the plane wave spectrum of the acoustic field excited by a line source in a wedge with penetrable bottom, and obtained asymptotic approximations for the acoustic field inside the wedge which reduce to the adiabatic mode solution This method was shown subsequently to lead naturally to a completely new concept, an 'intrinsic mode', which substantially explains the phenomena observed by Jensen and Kuperman (Arnold and Felsen, 1984a, 1986c).

In a wedge with penetrable bottom, an intrinsic mode is a self-consistent plane wave spectrum consisting of two components which reflect into each other at top and bottom boundaries, satisfying each boundary condition as they do so. There is one condition which must be imposed on these spectra to ensure consistency, which is that a certain invariant should equal an integer multiple of π ; the invariant is essentially the geometrical invariant of Harrison (1977) and Weston (1959). The particular integer q appearing in this condition indexes intrinsic modes, which therefore form an infinite discrete spectrum. Each intrinsic mode looks locally like an adiabatic mode where the latter is guided, and like a 'leaky' adiabatic mode beyond its cut off; each intrinsic mode corresponds with exactly one adiabatic mode, and vice-versa. Independent numerical calculations conducted by Topuz and Felsen (1985) and Arnold et al (1985) show that each intrinsic mode radiates a single beam into the bottom which originates at the cross section where the corresponding adiabatic mode reaches cut off, and good agreement between the two codes has been established. It has further been shown (Arnold and Felsen, 1984) that the acoustic field far from a line source located in the wedge can be represented, for small wedge angle α, as a superposition of intrinsic modes with various indices q. Thus, each beam radiating into the bottom medium in the Jensen-Kuperman model is associated with a discrete intrinsic mode, which is in turn associated with a specific adiabatic mode.

Recent theoretical investigations of the properties of intrinsic modes have begun to address the important problem of field calculations in full 3-dimensional geometries with arbitrary but smooth variations of depth (Arnold and Felsen 1986a). The adiabatic mode theory for these configurations was worked out for underwater acoustics by Weinberg and Burridge (1974) using the concept of horizontal rays. Roughly speaking, when viewed from above the plane forming the water surface, adiabatic modes propagate along paths lying in that plane called horizontal rays; when the water depth is variable, due to a nonuniform water-bottom interface, these paths are curved, giving rise to the phenomenon of horizontal refraction (Weston, 1961). These paths can be traced by an exact analogue of Snell's law, using a concept of 'effective' refractive index; the effective index n_e is the local phase propagation coefficient β of the adiabatic (local normal) mode, normalized by some suitable position independent scale length.

The approximate horizontal ray theory described above encounters difficulties due to asymptotic nonuniformities; these occur particularly at caustics of the horizontal rays and in the vicinity of local cut off of the adiabatic modes which propagate along horizontal rays. The intrinsic mode concept, however, can be generalized to include these cases (Arnold and Felsen, 1986) even for arbitrary variations of depth; in all cases the intrinsic mode is uniform where the adiabatic mode is not.

In this paper we present a simple construction of the intrinsic modes of a wedge-shaped ocean with a linear dependence of depth on position in the 2-dimensional range space, and a penetrable water-bottom boundary. This problem has an exact solution when the wedge is bounded by perfectly reflecting planes, and this solution has been extensibely studied by Buckingham (1983). When, as in our case, one boundary is penetrable the intrinsic mode we describe is also in principle an exact solution, although computational considerations reveal a very complex structure except in the asymptotic limit of small wedge angle α.

INTRINSIC MODES IN THE INTERIOR OF A WEDGE

Consider the wedge-shaped geometry of Figure 1. Suppose a plane wave is propagating in the wedge interior in the plane of symmetry normal to the direction of translation invariance at an angle θ with respect to the top boundary B_2, directed towards the top boundary. Multiple reflection of this plane wave in both boundaries generates two discrete sets of plane waves whose directions of propagation have an angular separation of 2α, where α is the wedge angle (Figure 2). If the reflection coefficient is -1 at each boundary (perfectly reflecting case), and the initial plane wave is assigned unit amplitude, then all waves in the first spectrum (odd number of reflections) have amplitude -1, all those in the second spectrum (even number of reflections) have amplitude $+1$. In each discrete set, plane wave amplitudes are properly connected by the laws of reflection. These plane wave directions and amplitudes are discrete samples of a continuous function which assigns amplitudes to all plane wave directions, and suitable continuous functions can easily be found:

$$\tilde{u}^{\pm}(\theta) = \pm e^{-iq\pi\theta/\alpha} \tag{1}$$

where θ is a general angle of plane wave propagation and q is an arbitrary integer. For $\theta \equiv \theta_o \bmod 2\alpha$, then

$$\tilde{u}^{\pm}(\theta) = \tilde{u}^{\pm}(\theta_o) \tag{2}$$

as required by the discrete samples constructed above. These values may now be taken as plane wave amplitudes in a <u>continuous</u> spectrum, to generate wavefunctions

$$u^{\pm}(\underline{x}) = \pm\int_C e^{-\frac{iq\pi\theta}{\alpha}}\, e^{-ikr\cos(\theta \mp \chi)}\, d\theta \tag{3}$$

where (r, χ) are polar coordinates for an observation point $\underline{x}$. The superposition

$$u(\underline{x}) = u^{+}(\underline{x}) + u^{-}(\underline{x}) \tag{4}$$

may be expected to satisfy the boundary conditions $u = 0$ on the two boundaries B_1 and B_2 ($\chi = \alpha$ and $\chi = 0$, $r > 0$) because all the spectral plane waves which constitute u are properly connected through the laws of reflection at each boundary. In fact it can be shown (Arnold and Felsen,

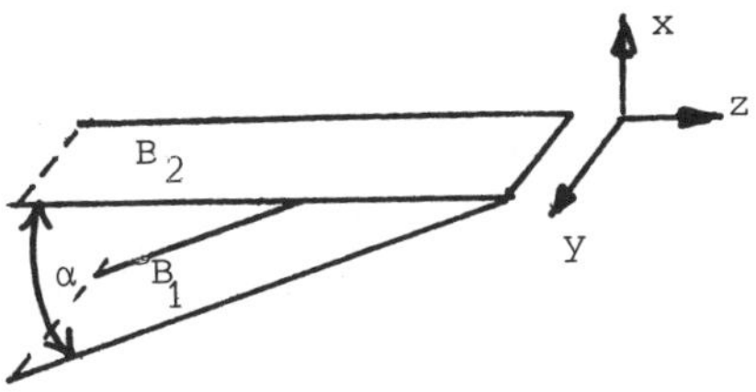

Figure 1: Wedge Geometry

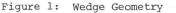

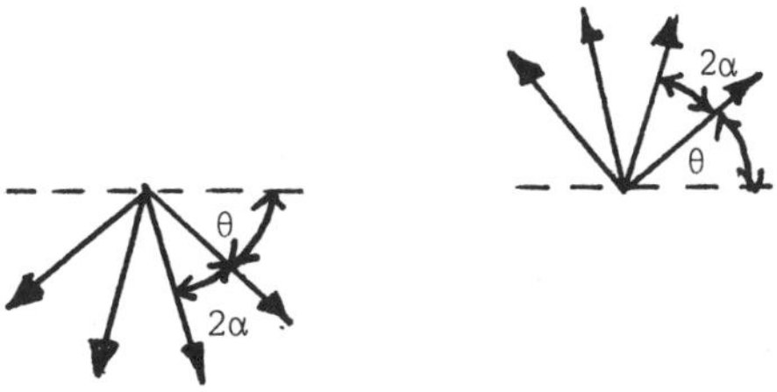

Figure 2: Discrete Plane Waves

1986c) that (3) and (4) are integral representations for exact solutions which can be expressed as Bessel functions. (The fact that $\exp(-iq\pi\theta/\alpha)$ is many-valued in $\theta\varepsilon(0,2\pi)$ for $q = 0$ when π is not an integer multiple of α is of no consequence if the integral is taken over a contour C in the complex θ-plane). The method of stationary phase can be applied to the integrals in (3). The phase of the integrand is stationary when

$$\frac{\partial}{\partial\theta}\left\{\frac{q\pi\theta}{\alpha} + kr\cos(\theta \mp \chi)\right\} = 0$$

which reduces, when those quantities are neglected which vanish as $\alpha \to 0$, to

$$q\pi = kh\sin\theta \qquad (5)$$

where $h = \mathrm{Lim}(r\alpha)$ (= $O(1)$ if r is assumed to be $O(\alpha^{-1})$ in the limiting procedure); h is then the local depth of the wedge at the observation point $\underline{x}$. Eq (5) gives the direction θ of the plane waves from which the dominant contribution to the integral arises, and is identical to the modal eigenvalue equation for a parallel waveguide of depth h and perfectly reflecting boundaries. For this reason, the leading order stationary phase contibution to the field $u(\underline{x})$ is precisely the adiabatic mode of lowest order coupled mode theory. From equation (5) the local phase propagation coefficient k_z can be determined as

$$k_z = k\cos\theta = (k^2 - q^2\pi^2/h^2)^{\frac{1}{2}} \qquad (6)$$

Equations (5) and (6) actually determine two possible stationary phase points (θ and $\pi - \theta$ or $\pm k_z$), which may both contribute to the wavefield $u(\underline{x})$, in the case of the guided mode region ($h > q\pi/k$), due to reflection of an upslope adiabatic mode at the cut off range where $h = q\pi/k$. (This point will be clarified subsequently).

The transition to a fully 3-dimensional theory can be made quite easily in the case of the wedge, because there exists a direction of translation invariance and a corresponding plane of symmetry normal to this direction. In the previous analysis we assumed that, in generating the discrete planewave sets in Figure 2, the initial plane wave's direction of propagation lay in this plane of symmetry, and therefore so do all its multiple reflections in Figure 2. We could equally well have assumed that the initial plane wave had an oblique inclination with respect to the plane of symmetry, specified by a non-vanishing component k_y of wave-vector parallel to the direction of translation invariance. In that case, this component of wave-vector is conserved in all reflections, since it is parallel to both boundaries B_1 and B_2; hence, the discrete set of plane wave wavevectors generated by multiple reflection lies, not in a plane as in Figure 2, but distributed around the surface of a cone (Figure 3), such that the axis of the cone lies parallel to the direction of translation invariance. When boundaries B_1 and B_2 are perfectly reflecting, the problem of interpolating between these discrete plane waves to generate

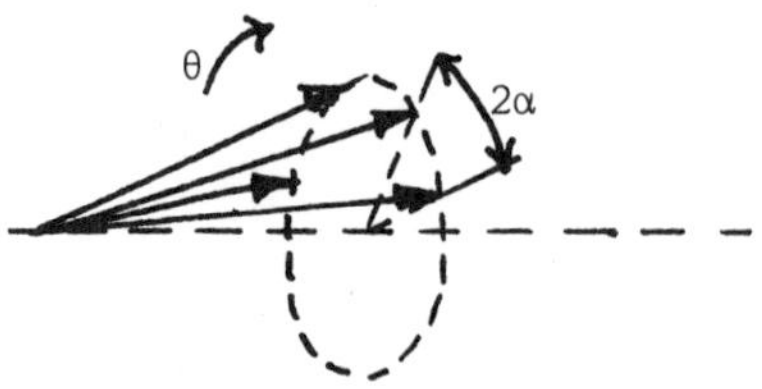

Figure 3: Cone of Multiply Reflected Wavevectors

a continuous spectrum is no more difficult than previously (when $k_y = 0$). Equation (3) holds for this more general case if we make the replacement $k \rightarrow k\cos\psi$, where $k_y = k\sin\psi$, and premultiply by a factor of $\exp(ik_y y)$. This replacement also transforms the stationary point in (5) and (6) according to $k_z \rightarrow \pm (k^2\cos^2\psi - q^2\pi^2/h^2)^{\frac{1}{2}}$. The two numbers (k_y, k_z) are then the resolved components of the local wave-vector of the wavefield in the (horizontal) y - z plane, and indicate the direction of propagation of the local adiabatic mode. They can be exhibited on the y - z plane as a vector field (Figure 4), with two wavevectors at each point. This vector field is translation invariant and possesses integral curves (shown as broken lines in Figure 4); these curves are the trajectories along which local adiabatic modes propagate, and can be derived from Snell's law in 2-dimensions using the 'effective refractive index' $n_e = \beta/k$ with $\beta^2 = k_y^2 + k_z^2$ (Arnold and Felsen, 1986a). These trajectories turn out to be identical to the 'horizontal rays' of Weinberg and Burridge (1974).

An interesting phenomenon arises when $k_z = 0$, which requires $\beta = k_y$. At such points the z-direction of the horizontal ray is reversed, at a caustic. This is the analogue in 3-dimensions of the 2-dimensional reflection of upslope adiabatic modes referred to earlier.

PENETRABLE BOTTOM BOUNDARY

When the bottom boundary is penetrable the discrete samples of wavevectors due to multiple reflection in the wedge boundaries are exactly the same as illustrated already in Figure 3, equally spaced around a cone. However, in this case the amplitudes on each plane wave are not equal, because the reflection coefficient at each boundary depends on the angle of incidence. Thus the interpolation problem (to obtain a continuous spectrum from these discrete samples) is more difficult. If $\underline{k}_n$ is the wavevector of the n'th plane wave in the cone of Figure 3, and $\tilde{u}(k_n)$ is the amplitude, then

$$\tilde{u}(\underline{k}_{n+1}) = R(\underline{k}_n)\tilde{u}(\underline{k}_n) \qquad (7)$$

where $R(\underline{k}_n)$ is the combined reflection coefficient after reflection in top and bottom boundaries, for an initial incident wavevector $\underline{k}_n$. Writing

$$\tilde{u}(\underline{k}) = e^{iS(\underline{k})} \quad , \quad R(\underline{k}) = e^{i\phi(\underline{k})} \quad , \qquad (8)$$

(7) reduces to

$$S(\underline{k}_{n+1}) - S(\underline{k}_n) = \phi(\underline{k}_n) \quad , \qquad (9)$$

which can be solved by the Euler-Maclaurin formula, exactly as for the 2-dimensional case described by Arnold and Felsen (1984, 1986c). The result of this is a continuous function $S(\underline{k})$ satisfying the difference

Figure 4: Field of Wavevectors

equation (9) for all values of $\underline{k}$. Denoting angle around the axis of the cone in Figure 3 by θ, and contracting $S(\underline{k}(\theta))$ to $S(\theta)$ (with $S(\underline{k})$ given by (9)), the wavefunctions resulting from this construction are

$$u^{\pm}(\underline{x}) = e^{iky\sin\psi}\int_C e^{iS(\theta)}e^{-ikr\cos(\theta\mp\chi)\cos\psi}d\theta \qquad (10)$$

where $\pi/2 - \psi$ is the half-vertex angle of the cone. The intrinsic mode is then given by (4) with (10). Actually, (10) represents a whole family of wavefunctions because $-2q\pi$ can be added to the right hand side of (9) without affecting its validity. Hence we write

$$u^{\pm}(\underline{x}) = W_q^{\pm}(\underline{x}) = e^{iky\sin\psi}\int_C e^{iS(\theta)}e^{-\frac{iq\pi\theta}{\alpha}}e^{-ikr\cos(\theta\mp\chi)\cos\psi}d\theta \qquad (11)$$

where the symbol $W_q^{\pm}$ has been introduced to denote the wavefunctions of the qth intrinsic mode, and the factor $\exp(-iq\pi\theta/\alpha)$ arises from interpolation, over intervals of width 2α, of the $-2q\pi$ term which may be added to (9). The stationary phase point of (11) occurs when

$$\frac{\partial S}{\partial\theta} = \frac{q\pi}{\alpha} - kr\sin(\theta\mp\chi)\cos\psi \qquad (12)$$

The derivative $\partial S/\partial\theta$ is a complicated function of θ, but in certain circumstances it can be approximated by $(2\alpha)^{-1}\Delta S$, where ΔS is the difference in values of S at two wavevectors spaced by an angle 2α around the cone of Figure 3. ΔS is given by (9), so we can approximate (12) for small α by

$$kh\sin\theta\cos\psi + \tfrac{1}{2}\phi(\theta) = q\pi , \qquad (13)$$

where $\phi(\theta)$ is a contraction of $\phi(\underline{k}(\theta))$, and the limit $\alpha \to 0$ has been taken, as before. Equation (13) is again the eigenvalue equation for a parallel guide of depth h, so again the adiabatic mode is the dominant approximation to the intrinsic mode as $\alpha \to 0$.

The same method as in section 2 leads to the horizontal ray construction. An additional complication here is that complex rays may have to be considered when the trajectories travel into a region where the local adiabatic mode is cut off. However, the wavefunctions $W_q^{\pm}(\underline{x})$ are uniform in this case, which is not true for the (approximate) adiabatic modes.

SUMMARY AND CONCLUSIONS

The intrinsic mode theory of the wedge-shaped waveguide naturally extends to 3-dimensions, wherein it explicitly exhibits the phenomenon of horizontal refraction, including complete reflection at caustics on horizontal rays. The constructed wavefunction is uniform at these caustics, and at transitions from bound to leaky behaviour of the local adiabatic modes which propagate along horizontal rays, because the wavefunction for an intrinsic mode is a plane wave spectral integral.

For simplicity of presentation we have only considered field variations of the form $\exp(ik_y y)$ along the direction of translation invariance; arbitrary variations can be constructed by Fourier synthesis of these elementary distributions.

The problem of coupling intrinsic modes to sources is under very active investigation at the present time. In the case of the wedge-shaped ocean waveguide (with flat bottom interface) the problem of a uniform line source has been completely solved (Arnold and Felsen, 1983, 1986c), with the result that the field far from the source is approximately a superposition of intrinsic modes. The extension of this result to a line source with $\exp(ik_y y)$ variation is relatively trivial using arguments of the type we have deployed in this paper; final extension to a point source can be achieved by Fourier synthesis. In the case of a point source, the intrinsic mode fields possess horizontal ray trajectories which all pass through a single point, in contrast to the translation invariance we have used earlier. It is worth noting that the effective index from which these trajectories are calculated is different for each intrinsic mode, indicating that the concept of effective index is more complex than might appear at first sight. Nevertheless, the insight provided by the intrinsic mode theory is of undoubted value in understanding propagation processes in complex waveguide environments.

ACKNOWLEDGEMENTS

This work was supported in part by the office of Naval Research under contract No N-00014-79-C-00/3, and in part by the Science and Engineering Research Council, UK.

REFERENCES

Arnold, J M and Felsen, L B, 1983, Rays and local modes in a wedge shaped ocean, J Acous Soc Am, 73:1105-1119.

Arnold, J M and Felsen, L B, 1984, Intrinsic modes in a nonseparable ocean waveguide, J Acous Soc Am, 76:850-860.

Arnold, J M and Felsen, L B, 1986a, Local intrinsic modes: layer with non planar boundary, Wave Motion, 8:1-14.

Arnold, J M and Felsen, L B, 1986b, Coupled mode theory of intrinsic modes in a wedge, J Acous Soc Am, 29:31-40.

Arnold, J M and Felsen, L B, 1986c, Theory of wave propagation in a wedge shaped layer, submitted to Wave Motion.

Arnold, J M, Belghoraf, A and Dendane, A, 1985, Intrinsic mode theory of tapered waveguides in integrated optics, IEE Proc (J), 132:314-318.

Buckingham, M J , 1983, Acoustic propagation in a wedge-shaped ocean with perfectly reflecting boundaries, in, "Hybrid formulation of wave propagation and scattering", L B Felsen ed , Nato ASI Series E, 86, M Nijhoff, Dordrecht.

Chwieroth , F S, Nagl, A, Uberall, R, Graves, R D and Zarur, G L, 1978, Mode coupling in sound channel with range dependent velocity profile, J Acous Soc Am, 64:1105-1112.

Evans, R B, 1983, A coupled mode solution for acoustic propagation in waveguide with stepwise depth variation of penetrable bottom, J Acous Soc Am, 74:188-195.

Harrison, C H, 1977, Three dimensional ray paths in basins, troughs and near seamounts by use of ray invariants, J Acous Soc Am, 62:1382-1388.

Jensen, F B and Kuperman, W A, 1980, Sound propagation in a wedge shaped ocean with penetrable bottom, J Acous Soc Am, 67:1564-1566.

Kamel, A and Felsen, L B, 1983, Spectral theory of sound propagation in an ocean channel with weakly sloping bottom, J Acous Soc Am, 73:1120-1130.

Pierce, A D, 1965, Extension of the method of normal modes to sound propagation in an almost stratified medium, J Acous Soc Am, 37:19-27.

Pierce, A D, 1982, Guided mode disappearance during upslope propagation in variable depth shallow water overlying a fluid bottom, J Acous Soc Am, 72:523-531.

Pierce, A D, 1984, Augmented adiabatic mode theory for upslope propagation from a point source in variable depth shallow water overlying a fluid bottom, J Acous Soc Am, 74:1837-1847.

Topuz, E and Felsen, L B, 1985, Intrinsic modes: numerical implementation in a wedge shaped ocean, J Acous Soc Am, 78:1746-1756.

Weinberg, H and Burridge, R, 1974, Horizontal ray theory for ocean acoustics, J Acous Soc Am, 55:63-79

Weston, D E, 1959, Guided propagation in a slowly varying medium, Proc Phys Soc, 73:365-384.

Weston, D E, 1961, Horizontal refraction in a three-dimensional medium of variable stratification, Proc Phys Soc, 78:46-52.

SPECTRAL DECOMPOSITION OF PE FIELDS IN A WEDGE-SHAPED OCEAN

Finn B. Jensen and Henrik Schmidt

SACLANT ASW Research Centre
Viale San Bartolomeo 400
19026 La Spezia, Italy

ABSTRACT

The parabolic-equation technique is a widely used approach for solving range-dependent problems in underwater acoustics. However, in order to compare total-field PE solutions with alternative techniques (coupled modes, intrinsic modes), it is desirable to decompose the PE field at a given range into its spectral components and analyse the propagation situation in terms of its modal structure. The spectral decomposition is accomplished by using the PE field as a source field in the SAFARI FFP code (H. Schmidt and F.B. Jensen, JASA 77, 813-825, 1985), which, in turn, constructs the "local" Green's function corresponding to the source field excitation. The "local" Green's function is the spectral decomposition in horizontal wavenumbers allowing us to identify modes, and hence to study mode coupling in a wedge-shaped ocean. Numerical results are given for both adiabatic and coupled-mode situations, for up- and down-slope propagation.

INTRODUCTION

Range-dependent ocean acoustic problems can be solved by a variety of modelling techniques comprising ray tracing, adiabatic modes, coupled modes, intrinsic modes, and the parabolic equation. The latter is probably the most widely used technique for solving low-frequency propagation problems. The PE technique, however, does not explicitly provide information on the propagation of various spectral components (eigenrays, modes), and hence does not provide the desired physical insight needed to understand and explain complex propagation situations. To overcome this limitation we have devised a scheme for decomposing the PE field at a given range into its spectral components versus horizontal wavenumber. This spectral decomposition not only permits a detailed modal analysis of propagating PE fields, but, as shown by McDaniel[1], it also facilitates a comparison with alternative ray or mode-based modelling results.

The major range-dependent feature of many ocean environments is the variation of water depth with range. In its simplest form of propagation over a constant slope, this acoustic problem has recently received much attention. Thus propagation upslope or downslope in a wedge-shaped ocean has been solved by using the parabolic equation[2], adiabatic modes[3], coupled modes[1,4,5], intrinsic modes[6], and rays with beam displacement[7]. Here we shall present PE solutions for selected acoustic problems in sloping bottom environments, and analyse the propagation situation in terms of energy conversion between individual modes, as well as between discrete and continuous spectral components.

PLANE WAVE DECOMPOSITION

Assume that the PE has been used in plane geometry to propagate the acoustic field to some range r_0, where the resulting field (velocity potential) as a function of depth is $F(r_0,z)$. The scope is now to decompose this field into a "local" set of plane waves. Due to the structure of the ocean, a horizontal baseline is the natural choice; hence we have to determine the kernel in the integral representation

$$F(r_o, z) = \int_{-\infty}^{\infty} f(k, z) e^{-ikr_o} dk \tag{1}$$

where k is the horizontal wavenumber. In this formulation, all parts of the spectrum are included, i.e. not only the normal modes, but also the continuous and evanescent spectra.

Due to the complexity of the field, it is inconvenient to invert Eq. (1) directly. Instead we construct a range-independent model environment with the sound speed and density profile identical to the local properties at range r_0. It is well known that the field in such an environment is of the form

$$G(r, z, z_o) = \int_{-\infty}^{\infty} g(k, z, z_o) e^{-ikr} dk \tag{2}$$

where z_0 is the source depth, i.e. a form very similar to Eq. (1). As the environment has been changed, the boundary conditions requiring continuity of pressure and horizontal particle velocity cannot both be satisfied at r_0. Thus we have to select one of those boundary conditions to be fulfilled. Here it is convenient to choose the horizontal particle velocity $u(r_0,z)$, i.e. the range derivative of Eq. (1). The field in the model environment is then[8]

$$F(r, z) = \int_{0}^{\infty} u(r_o, z_o) G(r, z, z_o) dz_o, \quad r > r_o \tag{3}$$

By reversing the order of integration in Eqs. (2) and (3) we obtain the following approximate expression for $f(k,z)$ in Eq. (1),

$$f(k, z) = \int_{0}^{\infty} u(r_o, z_o) g(k, z, z_o) dz_o \tag{4}$$

By discretizing Eq. (4) the problem is reduced to finding the depth dependence of the field in a range-independent environment with multiple, arbitrarily phased sources in a vertical array. The SAFARI fast-field code is directly set up to provide this for any number of sources[9], and has therefore been straightforwardly interfaced to the PE by using the range derivative of the PE field at range r_0 as the source weighting function.

Since the wave amplitude function, Eq. (4), is dependent on the receiver depth z, it is necessary to average over depth in order to obtain a complete spectral amplitude distribution. This averaging is a standard option in the SAFARI code.

NUMERICAL EXAMPLES

We have constructed prototype problems expected to show all the fundamental features of up- and down-slope propagation including the spectral redistribution of energy during mode conversion and mode cutoff.

Weak Coupling

The environment illustrated in Fig. 1a consists of a gentle upslope situation (0.7^{o}) on the initial 10 km with water depth decreasing from 200 to 80 m, followed by a 5 km flat section, and terminated by a steeper downslope section (1.4^{o}) of 5 km length, where the water depth increases from 80 to 200 m. The sound speed profile and the bottom properties are given in Table 1. Note that we are considering an inhomogeneous water column overlying a homogeneous bottom. In doing the field calculations we have removed geometrical spreading (plane geometry) and considered both water and bottom to be lossless.

The computed total-field solution with the parabolic equation[10] is shown as a contoured loss on Fig. 1a, while the spectral decomposition every 1 km in range is given in Fig. 1b. Let us first analyse the propagation situation as it appears from the field contours on Fig. 1a. For a frequency of 50 Hz we can have a maximum of four modes at the source, and only two modes in the shallow section of 80 m depth. With the source being placed at the null of the 2nd mode and with mode 4 not being computed, the starting field consists of modes 1 and 3 only. We see a regular two-mode interference pattern on the initial 7 km, at which point mode 3 cuts off and leaks into the bottom. If this were an adiabatic situation, we would expect only mode 1 to be present in the shallow part (mode 3 is cut off and mode 2 not excited at the source), but there is a weak interference structure indicating that some energy has been transferred to mode 2. The downslope part is clearly dominated by the fundamental mode.

The above qualitative analysis of the propagation situation is the only information that can be extracted from the total-field contours of Fig. 1a. We shall now turn to Fig. 1b and demonstrate that important propagation aspects can be clarified and quantified by analysing the spectral decomposition of the PE field versus horizontal wavenumber. Note that the spectrum is divided into two parts: the discrete spectrum corresponding to energy (modes) trapped in the water column, and the continuous spectrum corresponding to energy propagating primarily in the bottom. The local Green's function has in this case been averaged over the upper 300 m of the environment. This means that we do not have an exact representation of the

Table 1 Environmental input

Sound speed profile (m)	(m/s)	Bottom profile (km)	(m)
0	1530	0	200
10	1530	10	80
30	1510	15	80
100	1511	20	200
200	1513		

Bottom properties		
speed	:	1600 m/s
density	:	1.0 g/cm^3
attenuation	:	0.0 dB/λ

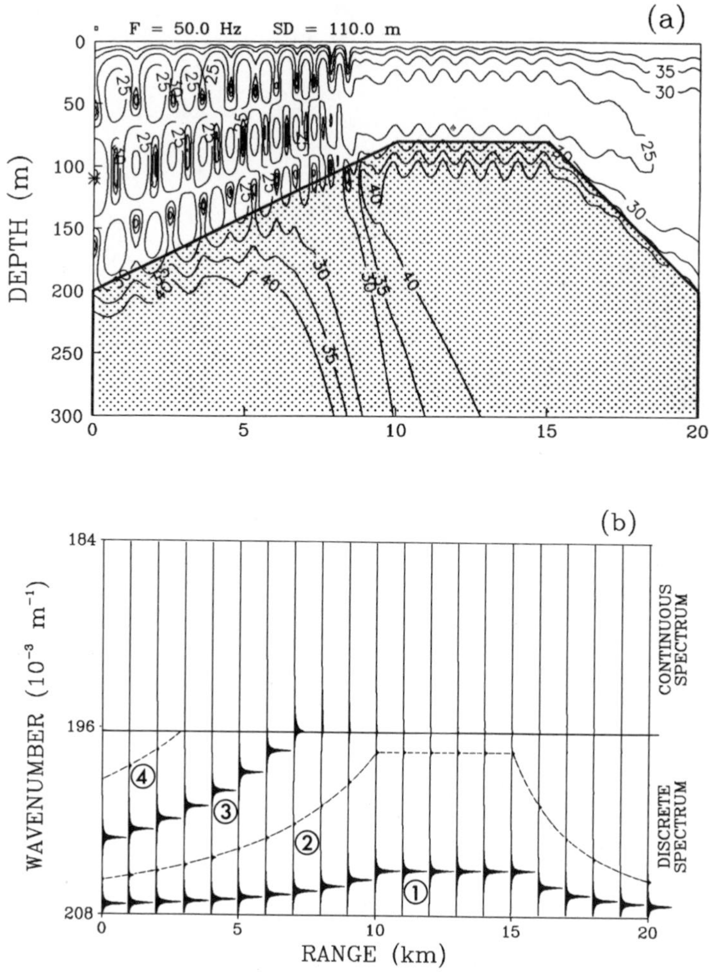

Fig. 1 Modal decomposition of PE field for an environment with gentle bottom slopes. a) Computed loss contours for a modal starting field with only modes 1 and 3 being excited. Note the cutoff of mode 3 at a range of 7 km. b) Spectral decomposition of the acoustic field at 1 km intervals indicating negligible coupling between modes (adiabatic propagation).

continuous-spectrum energy, since part of it is present below 300 m depth in the bottom. We note the presence of two spectral peaks at range zero corresponding to modes 1 and 3. During upslope propagation (0–10 km) energy couples weakly into modes 2 and 4, of which mode 4 cuts off at range 3 km. The energy on mode 3 is seen to remain constant until cutoff of the mode at 7 km. Here the energy moves into the continuous spectrum with negligible coupling to mode 2. In the flat region (10-15 km) and during downslope (15-20 km) the field is totally dominated by mode 1. Considering that the coupling of energy into mode 2 is very small (< 3%), we are clearly dealing here with an adiabatic situation.

Strong Coupling

To obtain pronounced mode coupling for the environment used in the above example, we just need to increase bottom slopes or, alternatively, to increase the frequency. We have chosen to vary the bottom slope and for simplicity to consider only the upslope part of the problem. Hence water and bottom properties are as given in Table 1, and the initial water depth is 200 m. The starting field again consists of modes 1 and 3 only, which are propagated upslope and into a flat section of 80 m depth.

The computed field solutions with the parabolic equation[10] are shown in Fig. 2 for four different bottom slopes. As in the former example we have removed the geometrical spreading loss in order to check energy conservation. A qualitative analysis of the propagation situation is easily obtained from the loss contours of Fig. 2. The initial field consists of modes 1 and 3 only, of which mode 3 cuts off during upslope propagation. On the shelf (80 m water depth) two trapped modes can exist, but since only the fundamental mode is excited at the source, the presence of energy in mode 2 is a measure of the importance of mode coupling. For an initial bottom slope of 5° the lack of interference pattern on the shelf indicates that the energy is entirely contained in the fundamental mode (adiabatic propagation). With increasing bottom slope a pronounced two-mode interference structure develops on the shelf, showing that more and more energy is coupled into mode 2.

By doing a spectral decomposition of the PE fields on the shelf we can analyse the coupling process in considerable detail and determine the total energy arriving on the shelf as a function of bottom slope as well as the relative energy distribution on the two guided modes. The result of this analysis is given in Fig. 3, with the modal energy normalized to the energy present in the first mode at the source. Since the PE solution was expected to be inaccurate for slopes as steep as 20°, we generated a coupled-mode reference solution using the code of Evans[5]. However there is no evidence of the PE result being inaccurate, since the computed energy distributions on the shelf are in agreement to within 5% (0.2 dB). We see from Fig. 3 that the total energy (mode 1 + mode 2) arriving on the shelf is constant to within about 10%, but with energy being shifted to mode 2 for slopes above 10°. It is interesting to notice the sharp transition from an adiabatic behavior to strong mode coupling at around 10°. This result implies that adiabatic mode theory is indeed applicable to a wide class of range-dependent acoustic problems, even though it is difficult to formulate generally valid criteria for the adiabaticity.

We can try to answer one more question concerning the coupling of energy into mode 2. Does the energy come primarily from mode 1 or mode 3? The answer can be given by inspection of Fig. 4, which displays the decomposed PE fields on a 20° bottom slope. Figure 4a is for the standard source field (modes 1 and 3), while in Figs. 4b and 4c the two modes are propagated individually. Thus starting with just mode 1 (Fig. 4b), there is some coupling into both modes 2 and 3, and similarly starting with just mode 3 (Fig. 4c) there is coupling into modes 1 and 2 as well as into mode 4. However the contribution of energy to mode 2 is seen to be primarily from mode 3.

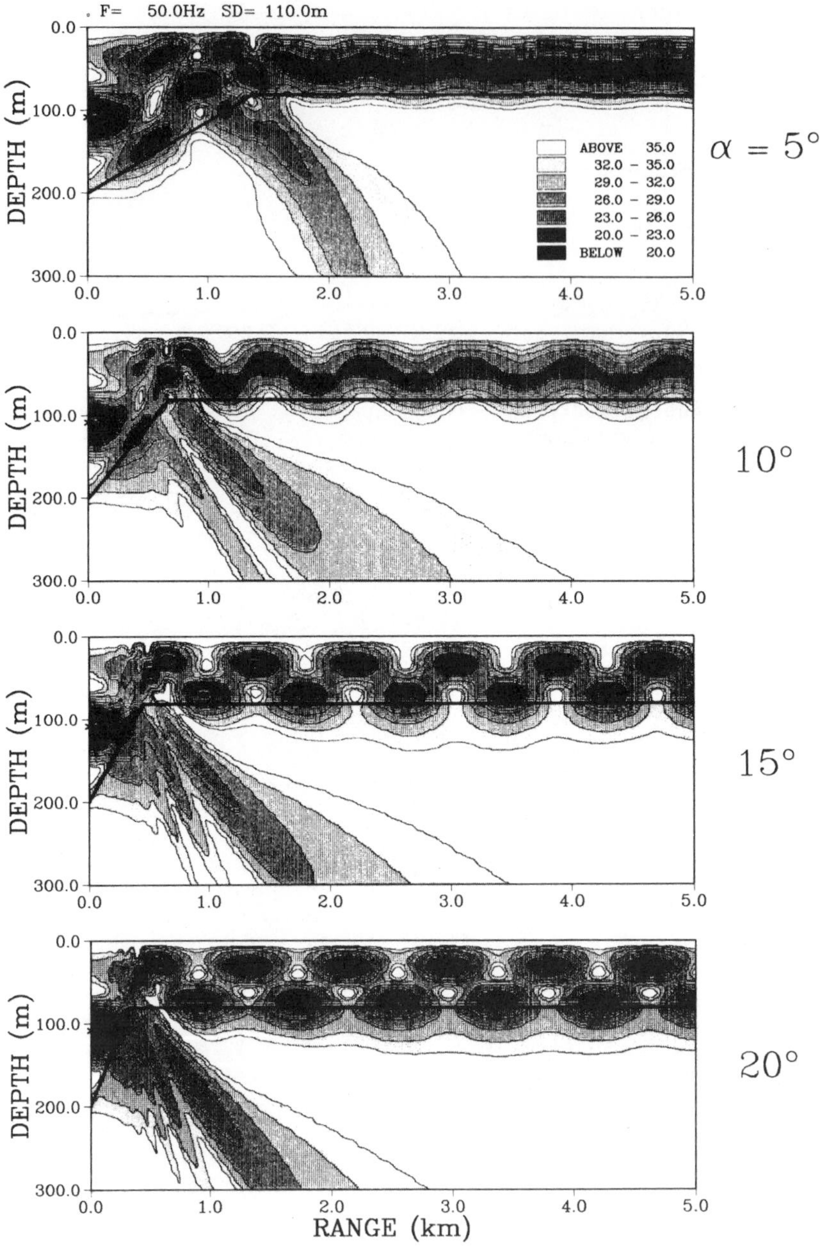

Fig. 2 Computed loss contours for environments with steep bottom slopes. The starting fields are identical in all four cases with only modes 1 and 3 being excited. During upslope propagation mode 3 cuts off, and the energy arriving on the shelf can propagate only in modes 1 and 2. Note that the energy on the shelf is entirely contained within mode 1 for a slope of 5°, while there is an increasing transfer of energy to mode 2 with increasing slope.

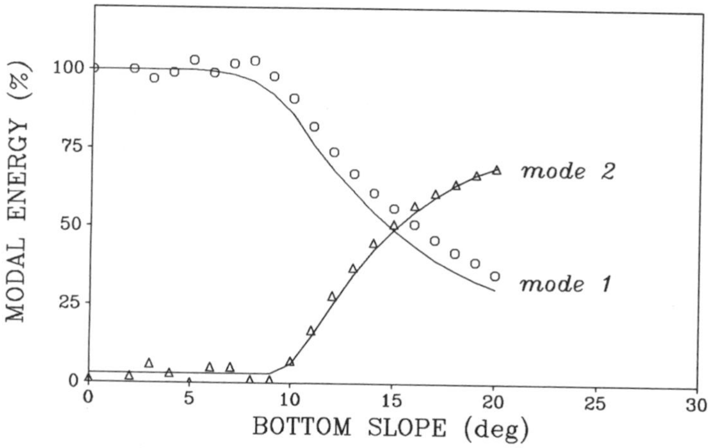

Fig. 3 Energy distribution among the two modes on the shelf as a function of bottom slope. The full lines are coupled-mode reference solutions, while the symbols are decomposed PE results at 1° intervals.

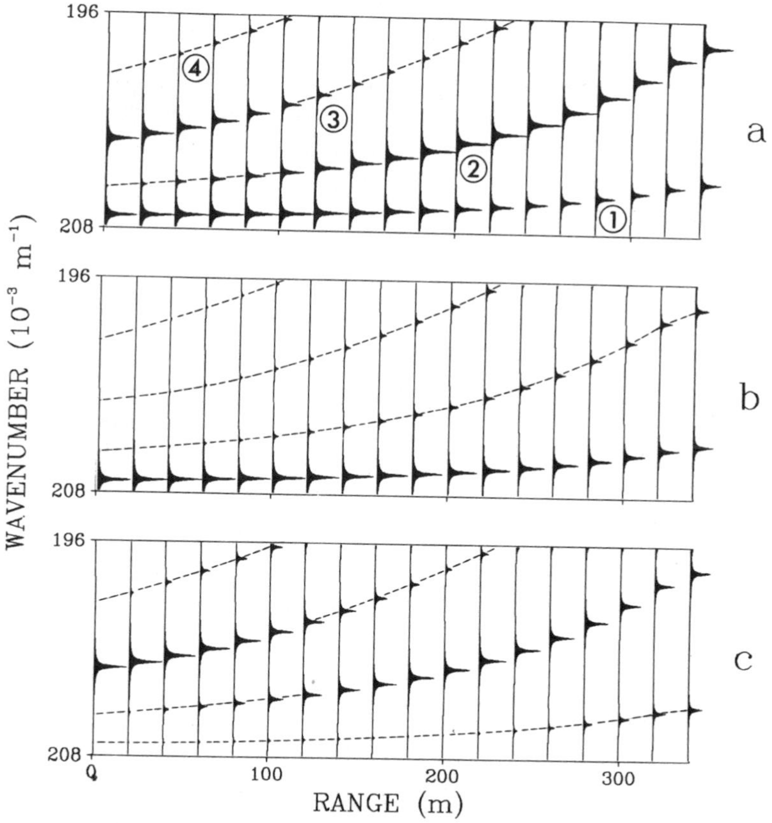

Fig. 4 Spectral decomposition of PE fields for a 20° bottom slope using different source fields: a) modes 1 and 3 only, b) mode 1 only, c) mode 3 only. The transfer of energy into mode 2 is seen to start immediately with the highest contribution coming from mode 3.

CONCLUSIONS

The parabolic equation technique is a widely used approach for solving range-dependent propagation problems in underwater acoustics. By adding to the standard PE total-field output the spectral energy distribution as a function of range, we have obtained a wave-theory analysis package that allows for an in-depth study of complex range-dependent propagation situations.

REFERENCES

1. S.T. McDaniel, Mode coupling due to interaction with the seabed, J. Acoust. Soc. Amer. 72: 916-923 (1982).
2. F.B. Jensen and W.A. Kuperman, Sound propagation in a wedge-shaped ocean with a penetrable bottom, J. Acoust. Soc. Amer. 67: 1564-1566 (1980).
3. A.D. Pierce, Augmented adiabatic mode theory for upslope propagation from a point source in variable-depth shallow water overlying a fluid bottom, J. Acoust. Soc. Amer. 74: 1837-1847 (1983).
4. I.J. Thompson, Mixing of normal modes in a range-dependent model ocean, J. Acoust. Soc. Amer. 69: 1280-1289 (1981).
5. R.B. Evans, A coupled mode solution for acoustic propagation in a waveguide with stepwise depth variations of a penetrable bottom J. Acoust. Soc. Amer. 74: 188-195 (1983).
6. E. Topuz and L.B. Felsen, Intrinsic modes: Numerical implementation in a wedge-shaped ocean, J. Acoust. Soc. Amer. 78: 1735-1745 (1985).
7. C.T. Tindle and G.B. Deane, Sound propagation over a sloping bottom using rays with beam displacement, J. Acoust. Soc. Amer. 78: 1366-1374 (1985).
8. P.M. Morse and K.U. Ingard, "Theoretical Acoustics," McGraw-Hill, New York (1968): 376.
9. H. Schmidt and F.B. Jensen, A full wave solution for propagation in multilayered viscoelastic media with application to Gaussian beam reflection at fluid-solid interfaces, J. Acoust. Soc. Amer. 77: 813-825 (1985).
10. D.J. Thomson and N.R. Chapman, A wide-angle split-step algorithm for the parabolic equation, J. Acoust. Soc. Amer. 74: 1848-1854 (1983).

ANALYTICAL SOLUTION OF THE PARABOLIC EQUATION FOR GUIDED MODE DISAPPEARANCE AT A CRITICAL DEPTH

Allan D. Pierce

School of Mechanical Engineering
Georgia Institute of Technology
Atlanta, Georgia 30332, USA

ABSTRACT

The fundamental problem of guided mode disappearance at a critical depth during upslope propagation in a shallow water waveguide with variable depth overlying a fluid halfspace is reconsidered. Jensen and Kuperman's numerical solutions based on a parabolic equation (PE) have stimulated a number of analytical and numerical studies based on models in which PE approximations were not invoked. One such analytical study carried out by the present author developed a characteristic transition solution for the region near a critical depth and matched it asymptotically to the adiabatic mode solution that applies in regions somewhat removed from critical depths. Although numerical agreement between the latter augmented adiabatic mode solution and the PE calculations are very good, the reasons for such good agreement require further clarification. The present paper extends a recently derived analytical result that the adiabatic mode theory and a suitably refined PE model (in which the characteristic wavenumber k_o varies with range) must give the same results to leading order at ranges without critical depths. If this refined PE model is taken instead of the Helmholtz equation as a starting point for the development of the augmented adiabatic mode theory for guided modes near critical depth, virtually the same analytical results emerge as previously. Thus the augmented adiabatic mode model is, in reality, a type of parabolic equation model, so its agreement with numerical solutions of a PE equation becomes better understood.

INTRODUCTION

The topic of guided mode disappearance at a critical depth has previously been examined by Jensen and Kuperman [1] with direct numerical integration of an appropriate parabolic equation. Other theoretical treatments not explicitly invoking the parabolic approximation have been given by the present author [2,3], by Coppens, Humphries, and Sanders [4] by Evans [5], and by Topuz and Felsen [6]. All of these theories are approximate and their interrelationships are not obvious, but the general numerical agreement of the separate theories with parabolic equation (PE) computations tends to give greater credibility to PE methods in ocean acoustics. The present paper contributes to the further understanding of what physical situations are adequately modelled by parabolic equation methods by showing that a particular PE formulation yields analytical results which, for environments that vary sufficiently slowly with range, are virtually equivalent to the author's augmented adiabatic mode theory of guided mode disappearance, even though the latter was originally developed without *conscious* invoking of a PE approximation at the outset.

Carrying through the derivation for the PE formulation also allows the author an opportunity to clarify some of the mathematical steps of the earlier paper. In particular, the manner in which the transition length scales L_x and L_y are identified is described more explicitly here. Also, the identification of the appropriate expansion parameter ϵ is described in a more explicit manner. The derivation of the simplified parabolic equation for the bottom fluid in the transition region is also given in a more direct and simpler manner.

PARABOLIC EQUATION FORMULATION

The complex pressure amplitude at range r and depth z can be written

$$p(z,r) = r^{-1/2}\, e^{i\chi(r)}\, F(r,z) \tag{1}$$

where

$$\chi(r) = \int_o^r k_o\, dr \tag{2}$$

with customary PE approximations. There are a number of variations [7] of the parabolic partial differential equation for F; that adopted here is the "narrow-angle" "range-dependent-environment" PE model [8] previously derived (using various approximations motivated by basic physical principles) by the author,

$$\left(\frac{k_o}{\rho}F\right)_r + \frac{k_o}{\rho}F_r = i\left[\left(\frac{1}{\rho}F_z\right)_z + \frac{k_o^2}{\rho}(n^2-1)F\right] \tag{3}$$

Here $n = \omega/ck_o$ is an apparent index of refraction.

In addition, the general model requires that the reference wavenumber k_o must be related to F by the quotient

$$k_o^2 = \frac{\int(\omega/c)^2\rho^{-1}|F|^2\,dz - \int\rho^{-1}|F_z|^2\,dz}{\int\rho^{-1}|F|^2\,dz} \tag{4}$$

This results from requiring that the disturbance be consistent with Rayleigh's observation that kinetic and potential energies should be equal in a progressive wave. In general, this leads to the implication that k_o must be range-dependent.

ADIABATIC MODE SOLUTION OF PARABOLIC EQUATION

In another previous paper [9], it is shown that the solution (at ranges somewhat removed from those at which a guided mode encounters a critical depth) of Eqs. (3) and (4) is equivalent to that predicted by the adiabatic mode theory. If only mode n is present and if k_o is set to k_n at any initial range, then at other ranges,

$$k_o(r) \approx k_n(r) \tag{5}$$

$$F(r,z) \approx \frac{A}{k_n^{1/2}}\,\frac{Z_n(z|r)}{\left(\int\rho^{-1}Z_n^2\,dz\right)^{1/2}} \tag{6}$$

where A is a constant; k_n^2 and $Z_n(z|r)$ are the n-th mode's local eigenvalue and eigenfunction of the Sturm-Liouville problem

$$\rho\left(\frac{1}{\rho} Z_{n,z}\right)_z + \left[(\omega/c)^2 - k_n^2\right] Z_n = 0 \tag{7}$$

with Z_n being 0 at the free surface.

In accord with the purpose stated in the introductory remarks above, the present paper seeks to determine an approximate solution of Eqs. (3) and (4) that applies near the critical range $r_c(n)$ of mode n, beyond which mode n does not exist.

SHALLOW WATER MODEL

The environmental model adopted is that of a Pekeris wave guide: a shallow water layer (Fig. 1) of variable depth and constant sound speed c_1 and density ρ_1 overlies a fluid bottom (c_2 and ρ_2) having a higher sound speed and idealized as a half-space. The depth $H(r)$ decreases slowly with range. Rather than use the coordinates r and z in the analysis, it is convenient to introduce an orthogonal curvilinear coordinate system (x, y) with origin at the bottom interface;the x axis is along the water- bottom interface; the y axis points obliquely downwards. The curvature of lines of constant x and constant y is neglected in accordance with the assumption that $|H''H| \ll 1$. The free surface is therefore at $y = -H(x)$, and the interface is at $y = 0$. For $-H < y < 0$, the sound speed c is c_1 and the density ρ is ρ_1, while these are c_2 and ρ_2 for $y > 0$. The origin for x is chosen such that x is positive upslope of the critical range and negative downslope.

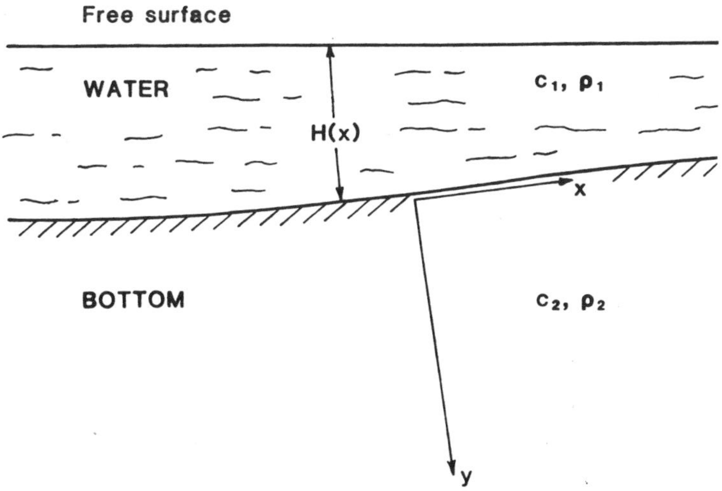

Figure 1. Pekeris shallow water waveguide of variable depth.

For moderately large negative values of x, the adiabatic solution, (5) and (6), is valid, with the function Z_n taken as

$$Z = \sin[q(y + H)], \qquad -H \leq y \leq 0 \tag{8a}$$

$$= \sin(qH)e^{-\gamma y} \qquad y \geq 0 \tag{8b}$$

Here the subscript n is dropped for simplicity; the various parameters that appear here are as defined in a previous paper [2] by the author. One defines μ as the solution lying between 0 and $\pi/2$ of the transcendental equation

$$k_2 H = \frac{(n+1)\pi - \tan^{-1}(b^{-1}\tan\mu)}{\beta \sin\mu} \tag{9}$$

where β abbreviates $[(c_2/c_1)^2 - 1]^{1/2}$ and b abbreviates ρ_1/ρ_2. Then the parameters q and γ that appear in Eq. (9) are given by

$$q = k_2\beta\sin\mu; \qquad \gamma = k_2\beta\cos\mu \tag{10}$$

The corresponding modal wave number k_n, which we abbreviate in what follows by κ, is given by

$$\kappa = k_2[1 + \beta^2\cos^2\mu]^{1/2} \tag{11}$$

where k_2 is ω/c_2. One should note also [8] that the k_o as would be computed from Eq. (4) is the same as the κ of Eq. (11) in this adiabatic mode approximation to the solutions of Eqs. (3) and (4).

OUTER BOUNDARY CONDITIONS

For upslope propagation, the adiabatic mode approximation breaks down as the depth $H(x)$ approaches the critical depth

$$H_c = \frac{[(n+\frac{1}{2})\pi]}{k_2\beta} \tag{12}$$

at which $\mu = \pi/2$, $\kappa = k_2$, $q = k_2\beta$, and $\gamma = 0$. The manner in which these limits are approached determines the outer boundary condition on the transition region solution as well as the approximations that are appropriate for the transition region. Those aspects of the solution [given jointly by Eqs. (1), (2), (5), (6), and (8)] which vary most rapidly with range as the critical depth is approached, given that H(r) is slowly varying with r, are the exponential factors $\exp(i\chi)$ and $\exp(-\gamma y)$. Examination of the two exponents here in the neighborhood of $x = 0$ allows one to write

$$\chi \simeq k_2 x + (1/3)(k_2^3\beta^4/2b^2)(H')^2 x^3 \tag{13a}$$

$$\gamma y \simeq -(k_2^2\beta^2/b)H' xy \tag{13b}$$

and these expressions in turn allow (or *force*) one to identify unique (apart from a multiplicative constant of the order of unity) length scales L_x and L_y for horizontal and depth distances in the transition region. The identification of these scales is a crucial step in the application of the method of matched asymptotic expansions. The first term in Eq. (13a) will be present regardless of whether or not the depth is varying so it is ignored in the identification process; the second term must accordingly be of the order of x^3/L_x^3. Similarly the right side of Eq. (13b) must be of the order xy/L_xL_y. The guiding principle here is that the appropriate deviations of the exponents are of order unity when x and y are of order L_x and L_y, respectively.

Following the presciption just descibed and with a convenient (from hindsight, the choice being clarified further below) definition of a parameter ϵ as being such that

$$\epsilon^3 = b^{-1}\beta^2|H'| \tag{14}$$

appropriate identifications of the two length scales are

$$L_x = 2/(k_2\epsilon^2); \qquad L_y = 1/(k_2\epsilon) \tag{15}$$

The natural coordinates in the transition region $(-L_x < x < L_x)$ are $\xi = x/L_x$ and $\eta = y/L_y$.

The identification of the appropriate expansion parameter (which turns out to be the ϵ defined above) emerges after one rewrites the governing equations and adiabatic mode solution results in terms of ξ and η. The depth H is regarded as a function of ξ rather than of x, and all derivatives of $L_x^{-1}H$ with respect to ξ are regarded as being of order unity. One approximates the adiabatic mode solution in the vicinity of $\xi \simeq 0$ consistent with the assumption that $|\xi| \ll 1$, keeps only the dominant terms, and then examines the order of magnitude of the neglected terms. One finds, for example, that

$$k_o L_x \approx 2/\epsilon^2 + 4\xi^2 + \epsilon\Psi(\xi) \tag{17}$$

where the generic function $\Psi(\xi)$ is of the order of unity when ξ is of the order of unity. Hence the order of magnitude of what one is neglecting, by keeping the leading terms only, is that of ϵ, so ϵ is an appropriate expansion parameter.

One identifies the outer boundary condition on the lowest order transition region solution of Eq. (3) to be that F approach, at large negative values of ξ, the simplest (lowest order in ϵ) nonzero approximation to Eq. (6), when the latter is expressed in terms of ξ and η. This approximation turns out to be

$$F \approx (-\xi)^{1/2} K \sin[k_2\beta(y+H)] \quad -H < y < 0 \tag{18a}$$

$$F \approx (-\xi)^{1/2} K \sin(k_2\beta H) e^{2\xi\eta} \quad y > 0 \tag{18b}$$

Here the constant K is easily related to the A in Eq. (6), the result being

$$K = 2(-1)^n \rho_2\, \epsilon^{1/2} A \tag{19}$$

TRANSITION REGION SOLUTION

Within the transition region (i.e., near where the depth attains its critical value for the mode under consideration), it can be demonstrated that the dominant contributions to the integrals in Eq. (4) come from the bottom layer, $y > 0$, where $c = c_2$. Furthermore, $|F_z|^2$ in this region is substantially smaller than $(\omega/c_2)^2|F|^2$. Consequently, it is sufficient to approximate k_o by k_2 in Eq. (3). Aiso, even though Eq. (3) was derived for rectangular Cartesian coordinates, it is still permissible to use it when the propagation is described in terms of the curvilinear coordinates x and y. In such a coordinate system the bottom interface formally appears as flat, but the water-air interface formally appears to be sloped.

There is, however, one small subtlety that emerges when one approximates k_o by k_2 within the transition region, because doing so requires that one make a corresponding approximation to the phase χ that appears in Eqs. (1) and (13a). If F_{TR} is the transition region solution of the approximate version of Eq. (3), with k_0 approximated by k_2, then the matching criterion has to be changed to that the $F\,\exp(i\chi)$

from the adiabatic mode theory solution match $F_{TR}\exp(ik_2x)$ when ξ is of the order of unity. In the vicinity of the critical range, χ is (apart from a constant) given by Eq. (13a), which can equivalently be expressed $k_2x + (4/3)\xi^3$. Consequently, the *matching criterion is not* that F_{TR} match the F of Eqs. (18), but that *instead it match* $\exp(i4\xi^3/3)$ *times* the F of Eqs. (18) when ξ is negative and of the order of unity. In what follows the subscript TR is omitted for brevity on F_{TR}, so what we are referring to as F is in reality F_{TR}.

For the Pekeris waveguide, and with the terminology used above, Eq. (3) in the transition region can be written as

$$\epsilon^2\frac{\partial F}{\partial \xi} + \left[\frac{1}{k_2^2}\frac{\partial^2 F}{\partial y^2} + \beta^2 F\right] = 0 \tag{20a}$$

for the water layer, where y is between $-H$ and 0. Similarly, one has to lowest order in ϵ

$$\frac{\partial F}{\partial \xi} + i\frac{\partial^2 F}{\partial \eta^2} = 0 \tag{20b}$$

for the field in the bottom fluid.

Examination of Eq. (20a) and of Eq. (18a), which hold for the water layer portion, suggests that in this layer the term $k_2^{-2}F_{yy}$ should be regarded as of zeroth order in ϵ in the context of the ordering scheme described in the preceding section. The x-derivative term is of a higher order, so one has the approximate equation

$$F_{yy} + k_2^2\beta^2 F = 0 \tag{21}$$

which is only an ordinary differential equation with constant coefficients. Given the boundary condition that $F = 0$ at $y = -H$, one identifies

$$F = G(\xi)\sin[k_2\beta(y + H)] \tag{22}$$

for $-H < y < 0$. Here $G(\xi)$ is a function of ξ (or of x) which remains to be determined.

The most important aspect of Eq. (13) insofar as the overall method of solution is concerned is that it furnishes an impedance boundary condition for solution of the partial differential equation (20b) for the bottom layer ($y > 0$). The above equation yields

$$\frac{F_y}{F} = k_2\beta\cot(k_2\beta H) \quad \text{for } y = 0^- \tag{23a}$$

which in turn, because F and $\rho^{-1}F_y$ are continuous at a interface, yields

$$\frac{F_y}{F} = (\rho_2/\rho_1)k_2\beta\cot(k_2\beta H) \quad \text{for } y = 0^+ \tag{23b}$$

In both of these equations, the ratio is x-dependent because the water depth H varies with range. At the critical depth H_c, where $x = 0$ (or $\xi = 0$), one has

$$\cos(k_2\beta H_c) = 0 \tag{24}$$

so the right side of Eq. (23b) can be approximated, to leading order in ϵ by $b^{-1}|H'|k_2^2\beta^2 x$, which is just a constant times ξ. One consequently obtains the approximate boundary condition

$$\frac{\partial F}{\partial \eta} = 2\xi F \quad \text{at } \eta = 0^+ \tag{25}$$

which is the same as Eq. (35a) in reference [2].

The partial differential equation (20*b*), derived here for the bottom fluid in the transition region is just the same partial differential equation (Eq. (35*b*) in reference [2]) derived in the author's earlier paper. The previous derivation, however, used the artifice of a variational principle and was not as direct. Here, as before, the resulting boundary value problem can be solved exactly in terms of Airy functions of complex argument; conscious use of the parabolic equation approximation at the outset does not change the result (provided, of course, that the reference wavenumber is understood to be calculated according to Eq. (4)).

ACKNOWLEDGMENTS

The author thanks Leopold Felsen, Finn Jensen, and Ding Lee for helpful discussions during the course of this research.

REFERENCES

[1] F. B. Jensen and W. A. Kuperman, Sound propagation in a wedge-shaped ocean with a penetrable bottom, *J. Acoust. Soc. Am.* **67**(5), 1564–1566 (May 1980).

[2] A. D. Pierce, Guided mode disappearance during upslope propagation in variable depth shallow water overlying a fluid bottom, *J. Acoust. Soc. Am.* **72**(2), 523–531 (August 1982).

[3] A. D. Pierce, Augmented adiabatic mode theory for upslope propagation from a point source in variable- depth shallow water overlying a fluid bottom, *J. Acoust. Soc. Am.* **74**(6), 1837–1847 (December 1983).

[4] A. B. Coppens, M. Humphries, and J. V. Sanders, Propagation of sound out of a fluid wedge into an underlying fluid substrate of greater sound speed, *J. Acoust. Soc. Am.* **76**(5), 1456–1465 (November 1984).

[5] R. B. Evans, A coupled mode solution for acoustic propagation in a waveguide with stepwise depth variations of a penetrable bottom, *J. Acoust. Soc. Am.* **74** (1), 188–195 (July 1983).

[6] E. Topuz and L. B. Felsen, Intrinsic modes: Numerical implementation in a wedge-shaped ocean, *J. Acoust. Soc. Am.* **78**(5), 1735–1735 (November 1985).

[7] D. Lee,*The state-of-the-art parabolic equation approximation as applied to underwater acoustic propagation, with discussions on intensive computations, NUSC Technical Document 7247* (Naval Underwater Systems Center, New London, Connecticut, 1 October 1984).

[8] A. D. Pierce, The natural reference wavenumber for parabolic approximations in ocean Acoustics, *Computers and Math. with Appl.* **11**(7/8), 831–841 (1985).

[9] A. D. Pierce, The relation of the parabolic equation method to the adiabatic mode approximation, to appear in *11th IMACS World Congress, Oslo, Norway, 5-9 August 1985; Vol. 1—Numerical Analysis and Applications*, edited by R. Vichnevetsky and J. Vignes (North-Holland, Amsterdam, 1986).

VERTICAL DIRECTIONALITY OF ACOUSTIC SIGNALS PROPAGATING DOWNSLOPE TO A DEEP OCEAN RECEIVER

N.R. Chapman, J.M. Syck* and G.R. Carlow

Defence Research Establishment Pacific
FMO Victoria, BC
VOS 1BO

*Naval Underwater Systems Centre
New London, CT

ABSTRACT

The vertical directionality of signals from shallow explosive charges has been measured in an experiment to study downslope propagation to a deep ocean receiver off the west coast of Canada. Large angle bottom reflected arrivals were observed for charges deployed in deep water at the beginning of the track where the ocean bottom was flat, however, the signals from charges deployed near the edge of the continental shelf were received at very shallow angles. For these signals ray theory predicts that the steep angle propagation paths are converted by interaction with the sloping bottom into low loss shallow angle paths which are ducted in the sound channel. Measurements of the propagation loss are consistent with the ray model predictions, with a downslope enhancement observed for the charges deployed near the edge of the continental slope. These results provide an experimental verification of the slope interaction model which has been proposed to account for the vertical directionality of ambient noise in the ocean.

INTRODUCTION

Measurements of the vertical directionality of ambient ocean noise (Anderson, 1979; Wagstaff, 1980; Wales and Diachok, 1981) indicate that the noise intensity at low frequencies is strongest at small angles to the horizontal. Since low frequency noise is generally attributed to shipping, the question arises as to how the noise generated near the sea surface is coupled into shallow angle propagation paths in the sound channel. Initially, Wagstaff (1980) and Wales and Diachok (1981) proposed that their measurements could be explained by conversion of bottom interacting energy into low loss ducted paths as the sound propagated down continental slopes. Dashen and Munk (1984) have completed a detailed study of three coupling mechanisms, including scattering by internal waves, coupling with the near surface sound channel at high latitudes, and conversion of shipping noise generated in shallow water by interaction with a sloping bottom. Their results showed that only the coupling via slope interaction could account for the experimental data.

Downslope conversion of shipping noise is capable of describing the data obtained in experiments carried out in the oceans of the northern hemisphere, but Bannister (1986) has pointed out that this mechanism is not appropriate for modelling vertical noise directionality in the southern hemisphere. Although the measured directionality spectrum is similar, the coastal shipping density in the southern oceans is not large enough to account for the measured intensity in the horizontal. Bannister (1986) has pointed out that the measurements could also be explained by considering the wind noise from high latitudes where the sound channel is near the sea surface. It is likely that both mechanisms contribute to the observed vertical directionality, and that the geographical location will determine which process is dominant.

This paper presents measurements of the vertical directionality of sound propagating downslope to a deep ocean receiver from explosive charges deployed over the continental slope and onto the continental shelf. The results provide an experimental verification of the slope interaction model. In the remainder of the paper the experiment and data analysis procedure are described, and a ray model is presented to provide a physical picture of downslope propagation. Following this the measurements of vertical signal directionality and propagation loss are presented and discussed. The results are summarized in the final section.

DOWNSLOPE PROPAGATION EXPERIMENT

Downslope propagation of sound from small explosive charges was studied in an experiment carried out off the Canadian west coast. The data were obtained using a 200 m vertical line array deployed in deep water at the axis of the sound channel about 130 km off shore. The charges (0.8 kg SUS) were deployed at 1.5 km intervals at an average depth of 23 m as the source ship opened range along a track over the continental slope and onto the shelf. The ocean environment is shown in Figure 1; the continental slope is generally monotonic, with two ledges at depths of 2000 m and 900 m, and rises to about 200 m at the edge of the continental shelf at a range of 90 km. The steep near surface sound speed gradient ensures strong coupling with the ocean bottom for the shallow shots. The depth excess is about 700 m for the initial deep water portion of the course but the sources are bottom limited for ranges greater than 60 km, i.e. beyond the 2000 m ledge.

The vertical arrival angles at the array were estimated using a cross correlation algorithm to determine the time differences of the signals at hydrophones separated by the array aperture of 200m. Both the broadband signal (10-400 Hz) and a low-pass filtered component (10-60 Hz) were considered in the analysis. The propagation loss, H, was determined in 1/3 octave bands from 10-400 Hz for the dominant arrivals at each range according to the relationship (in dB)

$$H(f) = SL(f) - RL(f) \qquad (1)$$

Here SL and RL are the source level (Chapman, 1986) and the received multipath energy levels, respectively.

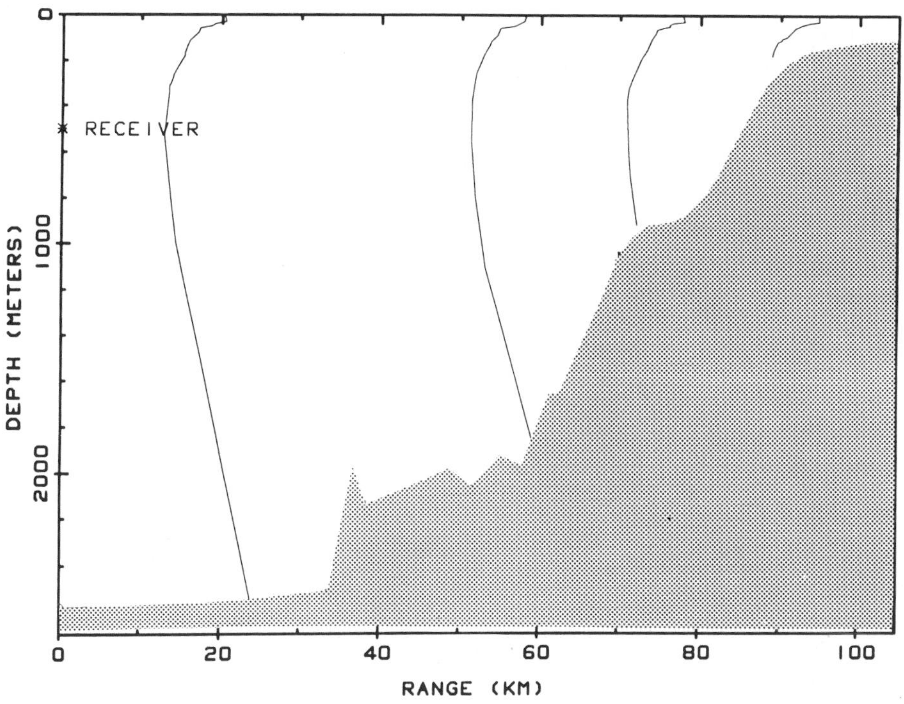

Figure 1. Bathymetry and sound speed profiles measured during the experiment. The surface sound speed was 1490 m/s.

RAY MODELLING

The downslope enhancement of sound propagating over a sloping bottom was first reported by Northrop et al (1968) who measured the propagation loss at a deep water site from explosive charges deployed on the continental shelf. The enhancement is easily understood in terms of ray theory since the ray parameter of bottom reflecting propagation paths is increased with each interaction on the slope. Consequently, the steep angle bottom interacting paths are converted into shallow angle paths which propagate with low loss in the sound channel. This conversion process accounts for the decrease in propagation loss observed in downslope propagation experiments.

Numerical calculations with a ray model (Sen and Fraser, 1985) using the ocean environment measured during the experiment indicated that the conversion to ducted propagation paths occurred after two bottom reflections for sources near the edge of the continental shelf. This behaviour is illustrated in Figure 2 which presents a comparison between the bottom bounce propagation paths for sources at 60 km in the deep water portion of the track, and the ducted propagation for sources at 90 km near the edge of the shelf.

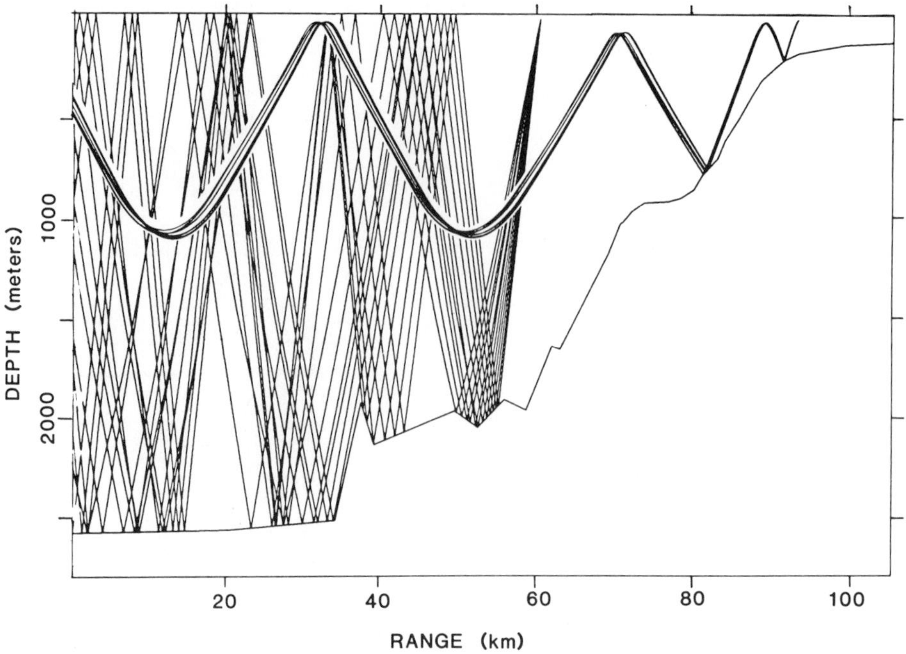

Figure 2. Ray tracing for sources at 60 km and 90 km.

The magnitude of the downslope enhancement depends on several conditions including the bottom loss, the ocean depth, the source depth and the sound speed profile. In this experiment there was relatively strong coupling with the sound channel since the conversion process took place at depths approximately equal to the depth of the sound channel axis. However this effect could be offset by the losses due to the interaction with the ocean bottom, or by the surface decoupling loss (Bannister and Pedersen, 1981) for the shallow source depths, and it is difficult to estimate the enhancement strength.

VERTICAL DIRECTIONALITY

The measurements of the vertical arrival angles for the broadband signals are presented in Figure 3. Similar results were obtained for the low-pass filtered data. The numerals indicate the order of the bottom arrival as determined by ray modelling of the paths between the source and receiver. Examination of Figures 1 and 3 shows that the bathymetry has a strong effect on the vertical directionality of the signal. Multipath arrivals were observed for the initial deep water portion of the track with bottom reflections from first to fourth order out to a range of 60 km. The 2000 m ledge does not have a significant effect because it is deeper than the reciprocal depth of the sources. At the onset of the continental slope the multipath structure is still evident, but there is a shift of the arrival angle for each order to smaller values. The angles continue to decrease for shots deployed over the 900 m ledge. The significant feature of these measurements is the absence of the multipath structure and the near horizontal angles of propagation for the shots deployed at about 90 km near the edge of the continental shelf. This result was predicted by the ray model and is clearly evident in the ray diagram shown in Figure 2.

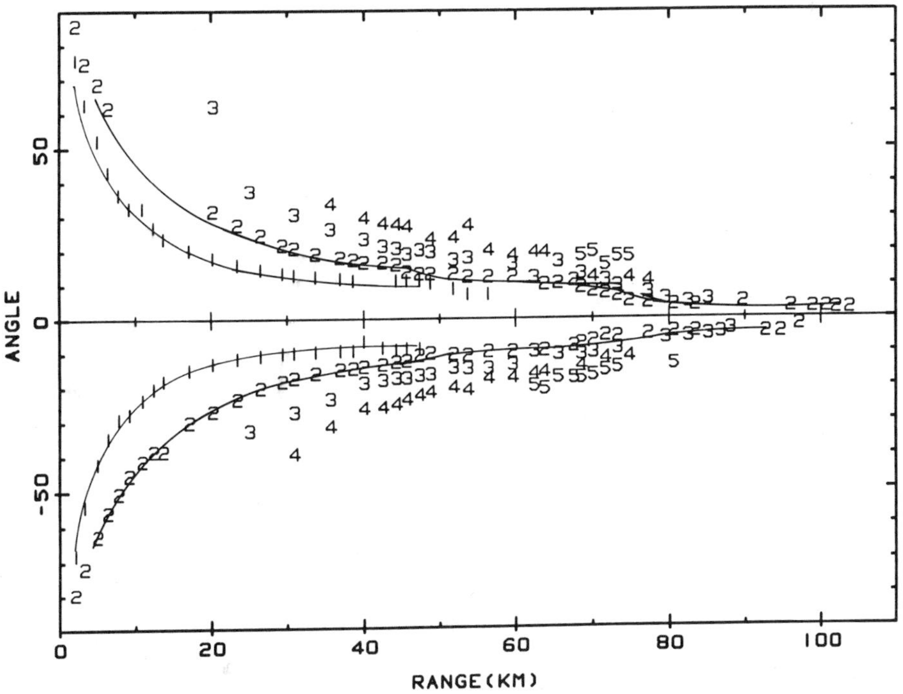

Figure 3. Measured vertical arrival angles at the array. The solid curves represent predicted arrival angles based on ray modelling.

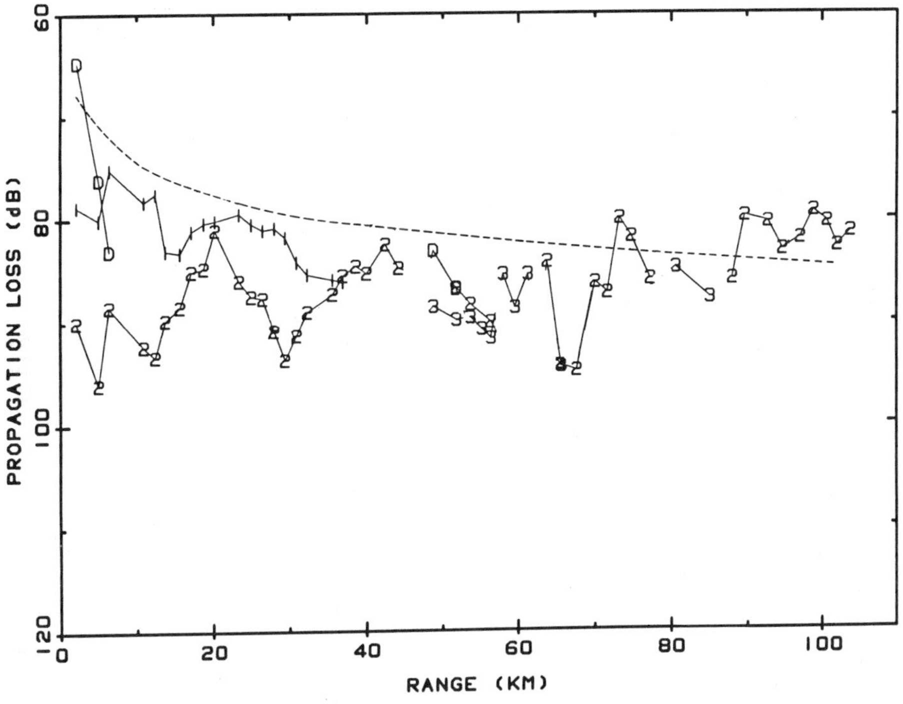

Figure 4. Propagation loss for the dominant arrival at each range. The solid curve represents cylindrical spreading loss.

The ray theory predictions for the vertical angles at the receiver are indicated by the solid lines in Figure 3. The numerical calculations are in excellent agreement with the data. The angles at the source were also calculated for the propagation paths shown in Figure 3 for the shots deployed near the edge of the shelf. Owing to the steep sound speed gradient near the sea surface, very small angles within 3° of the horizontal were predicted. Consequently, it is to be expected that the surface decoupling loss at low frequencies (< 40 Hz) will be relatively high for signals from these shots.

PROPAGATION LOSS MEASUREMENTS

Downslope enhancements were observed (Syck and Chapman, 1985) over the entire frequency band at ranges of 70 and 90 km corresponding to the onset of the 900 m ledge and the edge of the continental shelf. The propagation loss for the dominant arrival at each range is shown in Figure 4, for the 100 Hz band; the numerals indicate the order of bottom reflection. Comparison with Figure 3 indicates that the enhancements are observed at ranges where the vertical arrival angles become nearly horizontal. The relatively low loss is to be expected for the shallow angle ducted propagation paths for the signals from these shots. The enhancement was greatest for the shots at the edge of the shelf where the propagation loss decreased by about 6-8 dB from that expected for cylindrical spreading in the deep water range independent environment. This result is due to the stronger coupling with the sound channel for the shots at 90 km than for those at 70 km where the water depth is much greater than the sound channel axis depth.

The magnitude of the downslope enhancement at 90 km was constant over the frequency band between 40-200 Hz. The loss increased at lower frequencies due to the large surface decoupling loss associated with the shallow source depths for these shots. The bottom interaction does not limit the magnitude of the enhancement since the grazing angles on the slope are less than 10° and the bottom loss is uniformly small at frequencies in this band in this region of the continental slope. However, at higher frequencies the enhancement decreases due to the increased bottom loss above 200 Hz.

SUMMARY

The measurements presented here demonstrate that sound from shallow sources on the continental shelf is received at a deep water site at very shallow angles near the horizontal. Also, in contrast to the characteristic multipath expected for deep water propagation, there is very little travel time dispersion. Downslope enhancements were observed for ranges where the signal vertical arrival angles were nearly horizontal. These observations are consistent with the slope interaction model for conversion of sound generated in shallow water into propagation paths ducted in the deep sound channel.

REFERENCES

Anderson, V.C., 1978, Variation of vertical directionality of noise with depth in the North Pacific, J. Acoust. Soc. Am. 66, 1446-1452.

Bannister, R.W., 1986, Deep sound channel noise from high latitude winds, J. Acoust. Soc. Am. 79, 41-48.

Bannister, R.W., and Pedersen, M.A., 1981, Low frequency surface interference effects in long range sound propagation, J. Acous. Soc. Am. 69, 76-83.

Chapman, N.R., 1986, Measurements of the source levels of shallow explosive charges, J. Acoust. Soc. Am. 79, S55.

Dashen, R. and Munk, W., 1984, Three models of global ocean noise, J. Acoust. Soc. Am. 76, 540-554.

Northrop, J., Loubridge, M.S., and Werner, E.W., 1968, Effects of near source bottom conditions on long range sound propagation in the ocean, J. Geo. Res. 73, 3905-3908.

Sen, M.K. and Frazer, L.N., 1986, (private communication).

Syck, J. and Chapman, N.R., 1985, Propagation loss measurements in a region of complex bathymetry over the continental slope, J. Acous. Soc. Am. 77, S14.

Wagstaff, R.A., 1980, Low frequency ambient noise in the deep sound channel - the missing component, J. Acoust. Soc. Am. 69, 1009-1014.

Wales, S.C. and Diachok, O.I., 1981, Ambient noise vertical directionality in the northeast Atlantic, J. Acoust. Soc. Am. 70, 577-582.

A THEORETICAL MODEL OF ACOUSTIC PROPAGATION AROUND A CONICAL SEAMOUNT

Michael J. Buckingham

Radio & Navigation Department
Royal Aircraft Establishment
Farnborough, Hampshire GU14 6TD
England

and

Institute of Sound and Vibration Research
The University
Southampton SO9 5NH
England

ABSTRACT

The ocean around a conical seamount is an acoustic waveguide whose depth increases linearly with range measured out from the apex. As a result of horizontal refraction, the field created by an harmonic point source in such a channel is three-dimensional in character. A full solution for this 3-D field, derived from the inhomogeneous Helmholtz equation, is discussed in this paper. The solution, giving both amplitude and phase of the field, consists of a sum of uncoupled normal modes, which are functions of the angular depth about the apex. The range and azimuthal dependence of the field is contained in the mode coefficients. These coefficients exhibit three features which are characteristic of 3-D fields in general: acoustic shadowing in the horizontal; strong spatial variations within each mode (intra-mode interference); and apparent bearing shifts in the source position, due to ray curvature in the horizontal.

INTRODUCTION

It has been estimated (Menard and Ladd, 1963) that throughout the world's oceans there are $2x10^4$ seamounts, about half of which lie in the Pacific. Some of these seamounts, such as Dickens Seamount in the Northeast Pacific Ocean, are elongated, ridge-like structures submerged well below the surface. Acoustic shadowing by Dickens has been investigated by Chapman and Ebbeson (1983) and Medwin et al. (1984), the latter adopting a wedge geometry in a hybrid analysis of energy propagating over the peak and leaking into the shadow region behind the seamount.

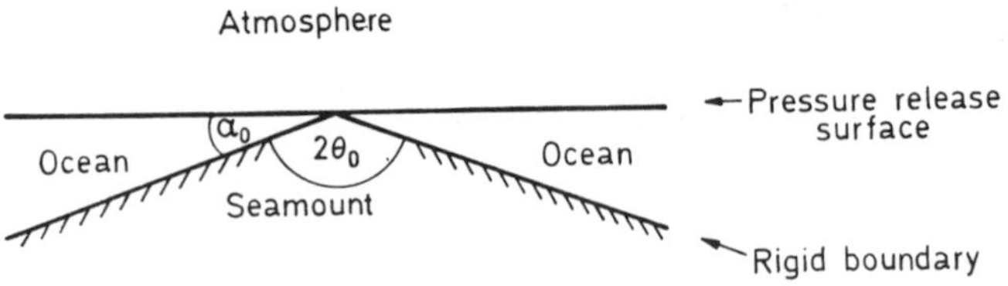

Fig. 1. Vertical section through the apex showing the geometry and boundary conditions used in the conical seamount analysis.

Many seamounts are less elongated than Dickens, with their peak much closer to the surface. Indeed, in some cases the apex penetrates the surface to form an island. These cone-like seamounts are basaltic structures, formed by volcanic action, with shallow slopes of less than 22°. On the younger conical seamounts, sedimentary deposits form only a very thin layer (Herzer, 1971) which has only a minor effect on acoustic propagation at frequencies below about 100 Hz. In this frequency range the seamount approximates to an acoustically rigid reflector.

Acoustic energy from a point source in the ocean close to a shallow, conical seamount propagates principally around the structure, rather than over the top of it as in the case of a deep seamount such as Dickens. The idealized geometry shown in Fig.1 provides a reasonable basis for analysing the field in the ocean around a conical seamount. As we discuss below, the field is intrinsically three-dimensional in character, showing well-defined acoustic shadows behind the seamount and intra-mode interference effects in the ensonified regions.

THE WAVE-THEORETIC MODEL

Spherical polar coordinates are the natural choice for the geometry illustrated in Fig.1. By placing the origin at the apex with the polar axis vertical, the coordinate surfaces can be matched to the boundaries of the channel. With perfectly reflecting boundaries (which is our case) the Helmholtz equation is then separable and, in principle at least, can be solved exactly for the acoustic field. A source term, which for a point source is a delta function, must be included on the right of this equation.

A systematic approach to the problem of solving the inhomogeneous Helmholtz equation is to apply a series of three integral transforms to both sides. These transforms must be chosen carefully to ensure that the boundary conditions are satisfied and that the eigenfunctions are orthogonal over the angular interval of the channel. The result of these operations is a triply transformed expression for the field (with harmonic time dependence suppressed). The field itself is then obtained by applying the corresponding sequence of inversion integrals. Although this procedure sounds straightforward, a variety of difficulties are encountered along the way. A full account of the theory is given elsewhere (Buckingham, 1986a). The essential points in the analysis and the conclusions are outlined below.

The three integral transforms to be applied to the inhomogeneous Helmholtz equation are a finite Fourier cosine transform over azimuthal angle, a generalized Legendre transform over polar angle (i.e. angular depth through the ocean channel) and a generalized Hankel transform over range. The inverse of the Fourier transform is a Fourier cosine series which is very slow to converge. The kernel of the Legendre transform is a linear combination of associated Legendre functions of the first and second kind. A family of such functions represents the normal modes for the problem. The member functions of the family are orthogonal over the channel angle, and each satisfies the boundary conditions. The range dependence of the field is obtained through the Hankel transform, whose kernel is a Bessel function of the first kind of order very much greater than unity. The corresponding inversion integral is a known form, involving a product of a Bessel function and a Hankel function, both having the same (high) order, which varies with the mode number.

Although the normal modes are given exactly by the family of associated Legendre functions, this is a computationally difficult formulation. A small-channel-angle approximation clarifies the solution, as well as leading to a substantial simplification in which the associated Legendre eigenfunctions reduce to familiar trigonometric forms. The algebra is lengthy, yielding eventually the following expression for the velocity potential (excluding the time dependence) of the field:

$$\psi \simeq (2/\pi\alpha_o)[\cos(\alpha)\cos(\alpha')]^{-1/2}\sum_{m=1}^{\infty} A_m \sin[(m-1/2)\pi\alpha/\alpha_o]\sin[(m-1/2)\pi\alpha'/\alpha_o] \ , \qquad (1)$$

where $A_m \equiv A_m(r,r',\phi)$ is the (complex) coefficient of the mth mode. The mode coefficients are themselves given by an infinite series, viz. the Fourier cosine series over azimuth mentioned earlier. The symbols (r,α,ϕ) used here represent range, angular depth and azimuth of the receiver, the primes indicate the corresponding coordinates for the source (whose azimuthal coordinate is $\phi'=0$), and α_o is the channel angle.

Eq.(1) represents the full solution for the field in the water column around the seamount. The eigenfunctions containing the angular depth dependence are the same as those encountered in a wedge with similar boundary conditions (Buckingham, 1984). The range/azimuth dependence of the field is contained entirely in the mode coefficients A_m, which are therefore of particular interest.

The Fourier series for the mode coefficients is

$$A_m(r,r',\phi) = [j\pi/(4rr')^{1/2}] \sum_{n=-\infty}^{\infty} J_\mu(kr)H_\mu^{(1)}(kr')\cos(n\phi) \qquad (2)$$

when $r<r'$, with a similar formulation applying when $r>r'$ but with r and r' interchanged. The order of the Bessel and Hankel functions in Eq.(2) is

$$\mu = \left\{[(m-1/2)\pi/\alpha_o]^2 + (n^2-1/4)\right\}^{1/2} \ . \qquad (3)$$

Because the series in Eq.(2) converges so slowly, the computation of A_m using this formulation is prohibitively lengthy. The difficulty can be overcome by reformulating the expression for the mode coefficients using Poisson's sum formula. The result is the following expression containing a single finite integral:

$$A_m = [j\pi/(4rr')^{1/2}]\left\{H_o^{(1)}(kR_\phi) - kr_{cm}\int_{|\phi|}^{\pi} w(w^2-\phi^2)^{-1/2}H_o^{(1)}(kR_w) \times J_1[kr_{cm}(w^2-\phi^2)^{1/2}]dw\right\}, \qquad (4)$$

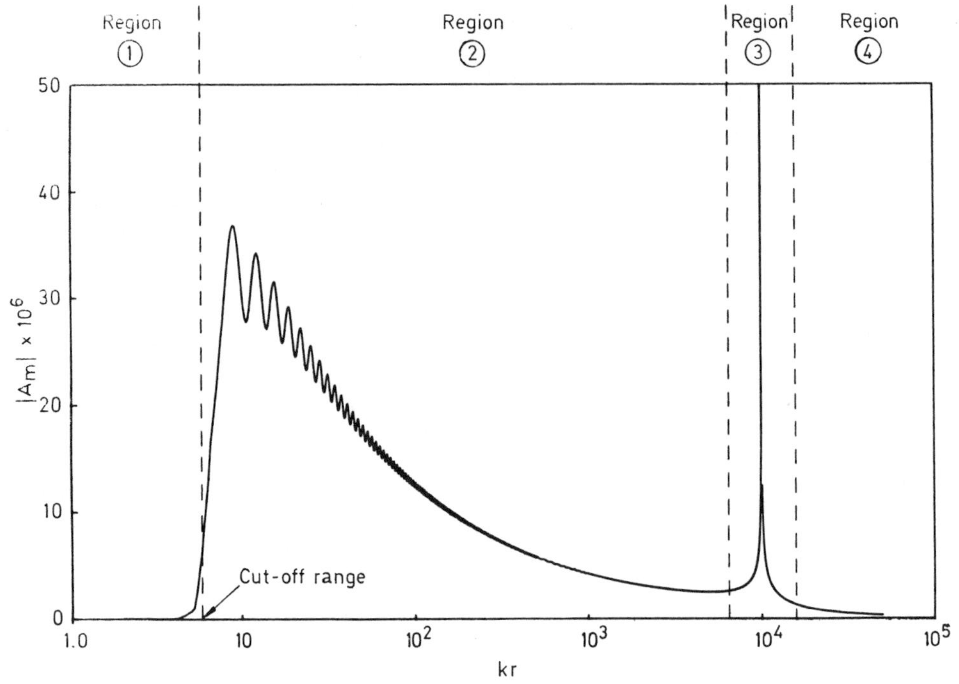

Fig. 2. Range dependence of the first mode coefficient in the $\phi=0$ plane. ($kr'=10^4$, $\alpha_o=15°$).

where

$$R_x = [r^2 + r'^2 - 2rr'\cos(x)]^{1/2} \quad , \qquad (5a)$$

$$kr_{cm} \simeq (m-½)\pi/\alpha_o \qquad (5b)$$

and k is the wavenumber. Note that the Hankel and Bessel functions in Eq.(4) are of order zero and unity, respectively (cf. the order $\mu \gg 1$ in Eq.(2)).

Computationally, Eq.(4) is a much more manageable expression than Eq.(2), and it can be simplified still further by using the method of stationary phase to evaluate the integral (Buckingham et al., 1986). However, this is beyond the scope of the present discussion.

RESULTS

Fig.2 illustrates the range dependence of the first mode in the vertical plane $\phi=0$, calculated from Eq.(4). The curve may be conveniently divided into four range regions, each with its own distinctive features. In region (1), at ranges less than the cut-off range, the field intensity is extremely low because the depth of the channel is not sufficient to support the mode. On passing through the cut-off range into region (2) the field rises very steeply and then shows rapid spatial oscillations (intra-mode interference). On moving out to greater ranges, into region (3) spanning the source, the field diverges to infinity at the source position. Region (3) is thus the near field region, where the boundaries have little effect and the source appears to be radiating into an infinite medium. Beyond the source, in region (4), the field decays to zero as the receiver range extends to infinity.

The azimuthal behaviour of the first three modes is illustrated in Fig.3, showing the mode coefficients calculated from Eq.(4) with r and r' fixed. It is clear that the field in each mode is a highly oscillatory function of azimuthal angle ϕ, with a well defined shadow at the rear of the seamount. The angular width, $(2\beta_m)$, of the shadow in the mth mode increases with increasing mode number, and is given approximately by the expression

$$2\beta_m \simeq 3\pi[(m-\tfrac{1}{2})(r^2+r'^2)^{1/2}/(krr'\alpha_o)]^{2/3} \tag{6}$$

This result is derived from the wave-theoretic expression in Eq.(4), and is in full agreement with the shadow width obtained from ray-theoretic arguments (Harrison, 1979). Since the shadow broadens as the mode number rises, the shadow width in the total field is determined by the shadow in the lowest order mode.

The rapid spatial oscillations in the modes in Fig.3 indicate the presence of strong intra-mode interference. This phenomenon is peculiar to three-dimensional fields, where horizontal refraction ensures that at any point in the channel the acoustic energy in a given mode is partitioned into two components propagating in different directions. These two components of the mode interfere, to produce spatial patterns like those shown in the figure.

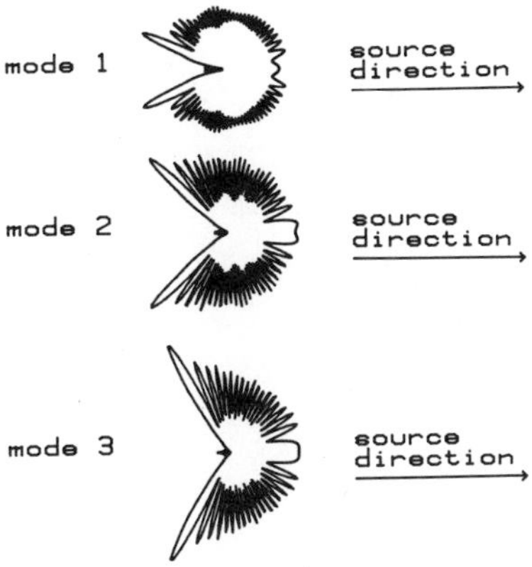

Fig. 3. Polar plots showing the azimuthal dependence of the first three mode coefficients, $|A_m|$, computed from Eq.(4). ($kr'=10^4$, $kr=10^2$, $\alpha_o=15°$).

RAY-THEORETIC INTERPRETATION

Each mode in the channel is associated with a family of rays. The members of the family all have the same grazing angle when launched from the source, but set off on different headings. An expression for the modal ray paths in the (r,ϕ) plane has been derived by Harrison (1979) using ray invariants. A modal-ray family calculated from his result, in this case for the first mode, is illustrated in Fig.4. The curvature of the ray paths in the horizontal, due to the mechanism of horizontal refraction, gives rise to an ensonified region spreading around the seamount, which cuts off sharply to form a shadow at the rear of the structure. Thus, the shadow is just the region where the modal rays cannot penetrate because of the constraints imposed by the geometry of the channel.

The prominent lobes adjacent to the shadow zones in Fig.3 can be understood on the basis of the ray diagram in Fig.4. These lobes represent "geometrical" caustics, attributable directly to horizontal refraction. In the ray picture, the density of rays is high along the edges of the shadows, consistent with the wave-theoretic result.

Within the ensonified region in Fig.4 multiple crossings of ray paths occur. This gives rise to mutual interference, from which it may be inferred that the modal field in this region will show highly oscillatory spatial variations. Such behaviour is qualitatively in agreement with the intra-mode interference observed in the modes derived from the wave-theoretic analysis. Similarly, both ray and wave theories show modal shadows, with quantitative agreement on the position of the shadow edge. Thus, the two approaches are consistent in predicting the major features of the 3-D field around the seamount. It is difficult to make more detailed comparisons because the ray theory does not readily yield amplitude and phase information.

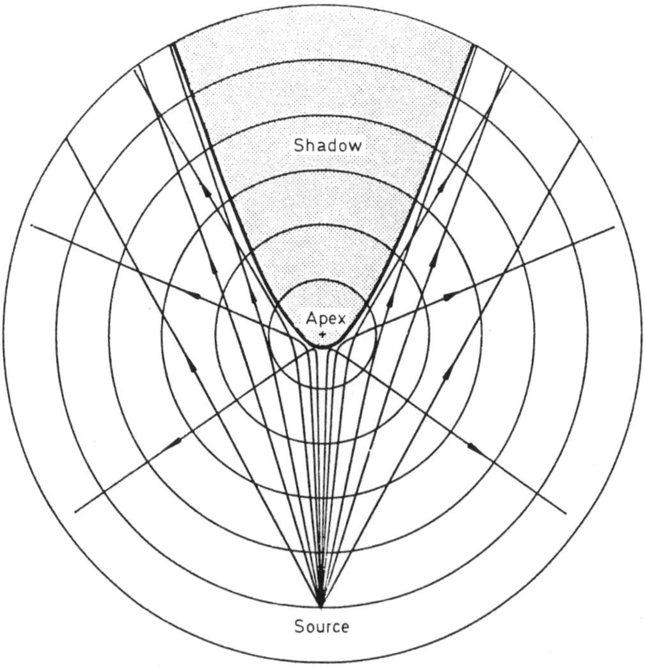

Fig. 4. Ray paths in the (r,ϕ) plane for mode 1. The circular depth contours are separated by units of kr=20. (α_o=15°).

CONCLUDING REMARKS

The theoretical model presented here of acoustic propagation in the ocean around a conical seamount gives a full description of the field in three dimensions. Thus, features of the field appearing in the horizontal, particularly modal shadows towards the rear of the seamount and intra-mode interference in the ensonified regions, emerge naturally from the analysis.

A perfectly reflecting bottom boundary is assumed in the theory, which is probably a reasonable approximation for low frequencies, below about 100 Hz, where the wavelength is considerably greater than the depth of the sediment overlying the basaltic bedrock of the seamount. At higher frequencies a better approximation may be a penetrable bottom. This would introduce substantial difficulties into the analysis of the field, but it is conceivable that an approach similar to that adopted in the case of the penetrable wedge (Buckingham, 1986b) could be applied with some success.

In a more general model of the seamount the apex would lie below (or above, in the case of an island) the sea surface. This would also increase the complexity of the problem, since the boundary surfaces of the ocean channel could then no longer be matched to the surfaces of one of the separable coordinate systems. It may be possible to address this type of problem using a hybrid approach based on wave/image theory.

REFERENCES

Buckingham, M.J., 1984, Acoustic propagation in a wedge-shaped ocean with perfectly reflecting boundaries, in: "Hybrid Formulation of Wave Propagation and Scattering", L.B.Felsen, ed., Martinus Nijhoff, Dordrecht.

Buckingham, M.J., 1986a, Theory of acoustic propagation around a conical seamount, to be published in J.Acoust.Soc.Am.

Buckingham, M.J., 1986b, Theory of three-dimensional acoustic propagation in a wedge-like ocean with a penetrable bottom, to be submitted for publication in J.Acoust.Soc.Am.

Buckingham, M.J., Jones, S.A.S., and Harriman, P.N., 1986, Stationary phase evaluation of the acoustic field around a conical seamount, to be published in J.Acoust.Soc.Am.

Chapman, N.R. and Ebbeson, G.R., 1983, Acoustic shadowing by an isolated seamount, J.Acoust.Soc.Am. 73, 1979-1984.

Harrison, C.H., 1979, Acoustic shadow zones in the horizontal plane, J.Acoust.Soc.Am. 65, 56-61.

Herzer, R.H., 1971, Bowie Seamount, a recently active flat-topped seamount in the Northeast Pacific, Can. J. Earth Sci. 8, 676.

Medwin, H., Childs, E., Jordan, E.A., and Spaulding, Jr. R.A., 1984, Sound scatter and shadowing at a seamount: Hybrid physical solutions in two and three dimensions, J.Acoust.Soc.Am. 75, 1478-1490.

Menard, H.W., and Ladd, H.S., 1963, Oceanic islands, seamounts, guyots and atolls, in: "The Sea", vol.3, pp.365-387, M.N.Hill, ed., Interscience Publishers, New York.

EXPERIMENTAL BEARING SHIFTS AS EVIDENCE OF HORIZONTAL REFRACTION IN A WEDGE-LIKE OCEAN

R. Doolittle *, A. Tolstoy *, and M. Buckingham **

* Acoustics Division, Code 5120, Naval Research Lab, Wash. DC USA 20375-5000

** Royal Aircraft Establishment, Farnborough, Hampshire, GU146TD, UK

ABSTRACT

Analysis of experimental data obtained in the region of the East Australian Continental Slope has been found to be consistent with theoretical predictions of energetic horizontal refraction due to sound-slope interaction where the observed azimuthal characteristics are in accordance with both normal mode and ray theoretical solutions. The experiment, conducted with two ships (one towing a source and the other an array), started at a 400 m depth contour with an initial separation of 33 km and proceeded to deep water on a diverging but constant line of bearing course. Beamformed data bearing shifts showing changes of tens of degrees over a few hours demonstrated the dominance of horizontally refracted arrivals for this geometry. A combination of source-array positions and critical angle effects served to limit both the azimuthal extent of energy received and the total number of modes arriving at any one position.

INTRODUCTION

Energy from an acoustic, electromagnetic, or seismic point source located in a wedge-like environment (flat upper surface and sloping bottom surface) propagating obliquely toward the wedge apex (not directly upslope) can repeatedly reflect from the upper and lower boundaries with a consequent change in direction away from the wedge apex with each bottom interaction. The energy when viewed from above appears to horizontally refract or curve. As a result, determination of source location (bearing, distance, and depth) can be affected.

The possibility of horizontal refraction in the ocean has been recognized for many years but has never been experimentally confirmed with CW sources and directional receivers. Horizontal refraction requires repeated bottom reflections, and detractors argued that in shallow water such an effect could not be measured because the energy would be absorbed by the bottom. However, significant bottom absorption requires steep vertical angles of incidence for the acoustic energy, and some horizontally refracted paths do not achieve such angles. Theoretically this issue has been difficult to tackle since an easy to calculate analytical solution to this problem for a time harmonic source for even the perfectly reflecting bottom has only recently been devised [1]. Such a solution for an absorbing bottom does not yet exist.

Experimental data obtained in late 1984 by NRL (in conjunction with scientists from Australia and New Zealand) in the region of the East Australian Continental Slope has confirmed the existence of energetic horizontal refraction due to sound-slope interaction in a shallow water environment.

EXPERIMENTAL RESULTS:

The experiment consisted of two ships initially 33 km apart, 30 km from shore, and in water 400 m deep over 1.2 degree slope with one ship towing a 152 Hz CW sound source and the other a horizontal array of sensors. Over most of the acoustic path between source and receiver the bottom consisted of a smooth, sandy sediment about 200 m thick overlying a rock basement (Figure 1a). The bottom topography was smooth early in the run but became irregular later (Figure 1b). Measured bottom slopes varied from 0.43 degrees (near shore) to 17.1 degrees (50 km from shore). Each ship proceeded out to deeper water on diverging paths but on courses chosen so that both ships remained equidistant from shore, and source ship direction and array direction remained nearly parallel (see Figure 2). The ship towing the array had a resultant heading nearly due east at about 1.5 knots, and there was a local current nearly due south at 2 knots. Hence, the array had a final heading of 38 degrees where depth and heading sensors confirmed the linear and horizontal attitude of the array. Navigation was kept by both SATNAV systems and by triangulation with known points on shore.

By combining the signals received at each acoustic sensor (beamforming), the angle of arrival for the energy received from the source could be determined. Both conventional and maximum-likelihood algorithms were used where the latter has the advantage of suppressing sidelobe interference and, when spatially whitened, suppresses broadband noise including towship interference. Without horizontal refraction the energy would be expected to arrive at the array at a nearly constant 38 degree angle. Figure 3 (produced with conventional processing where each horizontal trace represents a two minute average) shows the arrival angles as a function of time (the top of the trace corresponds to the beginning of the run). Early in the run energy arrives at approximately 40 degrees while at later times we see vivid evidence of horizontal refraction with some arrivals nearly broadside to the array (90 degrees). The white dots show the arrival angle of energy expected in the absence of horizontal refraction.

THEORETICAL RESULTS:

Aspects of the experimental behavior have been qualitatively predicted by combining the highly idealized wedge solutions [1] with ray theoretical predictions [2] based on more reasonable idealizations of this environment. First, wave theoretical simulations of the acoustic field at the array were generated using the ideal wedge solution with geometrical parameters matching as closely as possible those at the beginning of the experiment where bottom contours were fairly smooth (see Table 1). This complex field was then beamformed with results similar to the experimental data seen at the top of Figure 3 but with additional energy incident at several higher discrete arrival angles. It was felt that this additional energy would disappear if consideration of bottom absorption could be included. Consequently, the role of acoustic absorption by the bottom was studied next by means of ray theoretical calculations where this property was represented by a critical angle of 30 degrees. Since the ray model was not used to compute field intensity but rather to compute the three-dimensional arrival angle of the energy at the array, we could treat ray-bottom interactions simply. In particular, the rays hitting the bottom at grazing angles less than 30 degrees continued to propagate (reflected at the angle appropriate according to Snell's law) while at angles greater than 30 degrees the rays were stopped/absorbed. For this study we selected rays corresponding to the strong modes as predicted by the wave theoretical model.

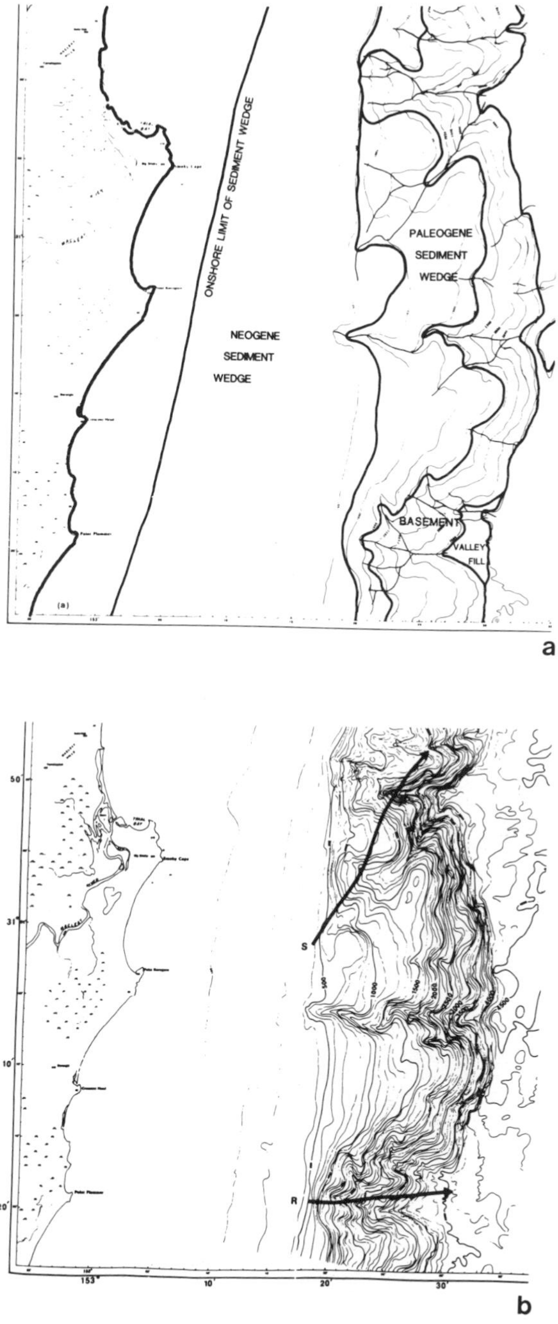

Fig. 1. Bottom geology (a) and bathymetry (b) in the experimental area.

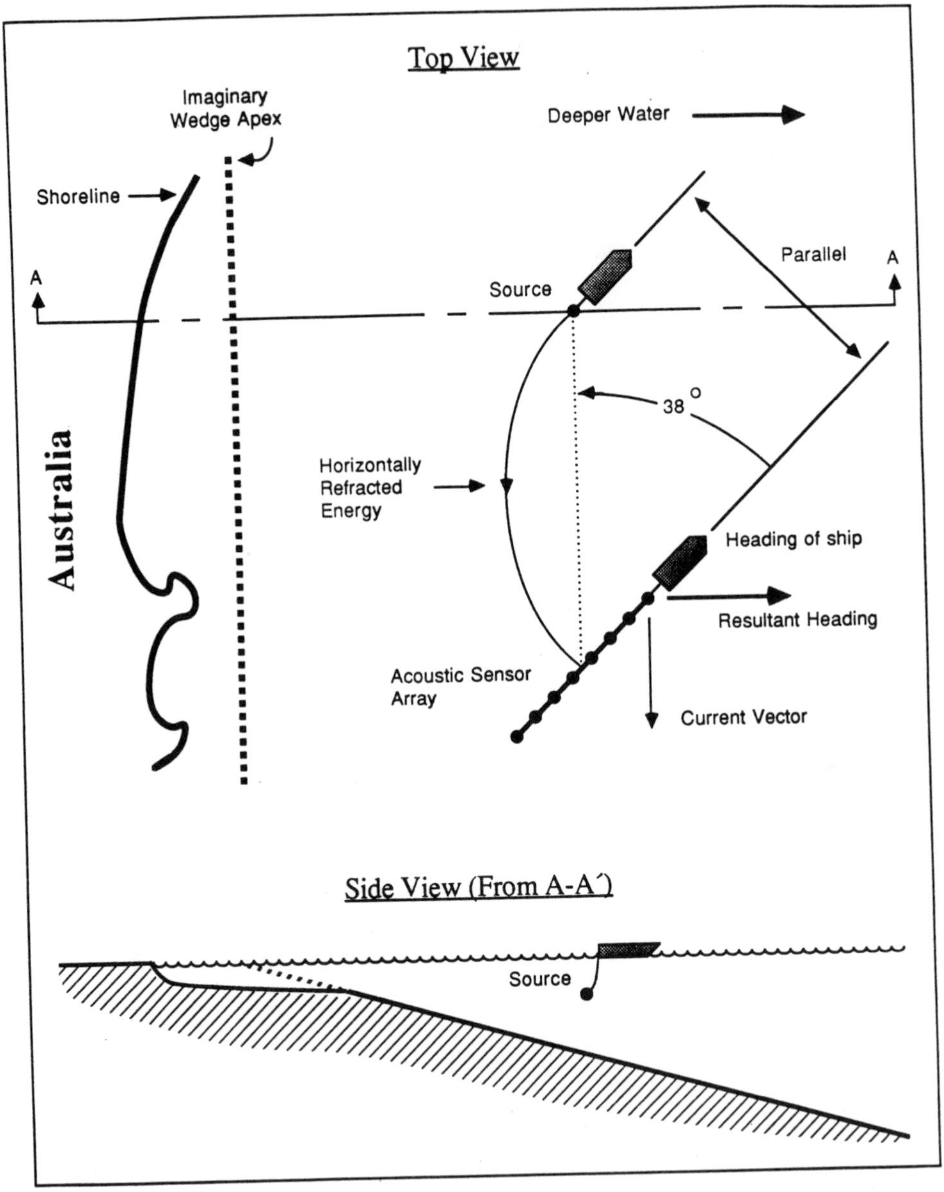

Fig. 2. Geometry of the experiment (top and side views) at the beginning of the run.

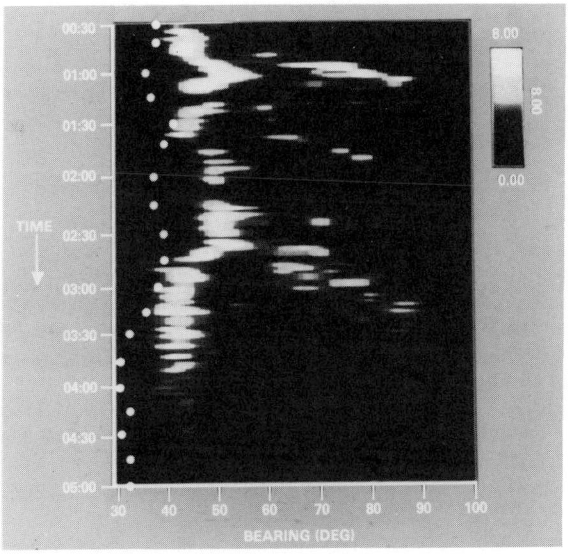

Fig. 3. Bearing-time trace for the acoustic signal at the array of acoustic sensors. This data represents 4.5 hrs of data beginning at the top where the ships are least separated and nearest shore. Highest energy levels are shown in white with scale shown in the upper right (8 dB from white to black). The white dots show the arrival angle of energy expected in the absence of horizontal refraction.

Table 1. Summary of Experimental Details

Source depth = 80 m
Receiver array depth = 80 m
Bottom depth at source and receiver = 400 m
Separation distance between source and receiver = 33 km
Bottom slope = 1.2 deg
Frequency = 152 Hz (wavelength = 9.87 m, wavenumber = 0.64/m)
Array heading = 38 deg True

Another advantage of the ray model was that we were able to consider a faceted bottom consisting of a number of wedges rather than just the one required by the wave theoretic model. Thus, we could approximate the bottom features a bit more realistically. In the end these efforts lead to predictions which accounted for the lack of high angle, highly refracted energy paths at the beginning of the experiment (top of Figure 3), while also accounting for their presence at somewhat later times.

CONCLUSIONS:

The measurements shown here, conducted in a wedge-like shallow ocean, confirm theoretical predictions of energetic horizontal refraction. Two idealized theoretical models, used to interpret the results obtained in the experiment, provide an informative description of the acoustic field observed at the array. There is, nevertheless, a need for a more general modelling capability, which would allow the three-dimensional field in an ocean with a complicated environment (non-planar boundaries, non-uniform refractive index) to be computed with relative ease. This is particularly true now that energetic horizontal refraction in the ocean has been confirmed.

REFERENCES

1. M. J. Buckingham, Acoustic propagation in a wedge-shaped ocean with perfectly reflecting boundaries, Proceedings of the NATO Advanced Research Workshop on Hybrid Formulation of Wave Propagation and Scattering, IAFE, Castle Gandolpho (Rome), Italy, 77-105 (August 30 to September 3, 1983).

2. D. E. Weston, Horizontal refraction in a three-dimensional medium of variable stratification, Proc. Phys. Soc. 78:46 (1961).

UNDERWATER SOUNDBEAMS CREATED BY AIRBORNE LASER SYSTEMS

Allan D. Pierce and H.-A. Hsieh

School of Mechanical Engineering
Georgia Institute of Technology
Atlanta, Georgia 30332, USA

ABSTRACT

The configuration and sequencing of laser-induced heating of the ocean surface with an airborne laser can be controlled with judiciously designed optical systems. The present analysis relates the surface heating to the acoustic field at intermediate and large ranges and shows that the spatial and temporal pattern of energy deposition can be selected such that a transient narrow band acoustic pulse of very narrow beamwidth propagates obliquely downward into the water. Existing technology makes it feasible to create easily detectable acoustic signals at distances up to the order of 10 km.

INTRODUCTION

There is an inherent attractiveness in air-to-sea communication systems where laser beams are launched from air-borne platforms (Fig. 1) onto the water surface and subsequently generate sound within the water. No hardware would be in the water, so the system would be highly mobile.

The present authors' studies on laser-generated sound have led to the tentative conclusion that the effective signal detection range could be markedly increased if the laser energy were deposited into the water in a judiciously preselected spatial and temporal pattern. The present paper gives an overall summary of the mathematical model and analysis that supports this conclusion.

A principal tenet is that present day optical technology allows the possibility of one's moving the impinging beam over the surface at virtually any speed and of controlling the contour of the trajectory. This tenet is supported by recent experiments reported by Berthelot and Busch-Vishniac [1] in which a laser beam was caused to move over a water surface at supersonic speeds by using the relatively simple device of first reflecting a stationary beam from a rotating prism with reflecting surfaces. Relatively clear and reproducible acoustic pulses were detected in the large water tank over the surface of which the laser beam moved. Another tenet is that the optical system can be developed such that the beam is split and broadened and such that, at any instant, the pattern of the light impinging on the surface has the form of a sequence of parallel slabs of light (Fig. 1). That one can indeed develop such an optical system seems beyond question; the actual development is nevertheless anticipated to be a challenging problem in state-of-the-art optical design. Here we proceed to examine the benefits that might result from such a design.

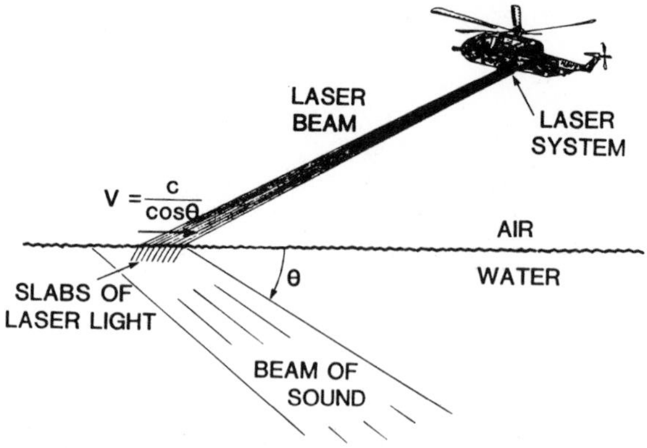

Figure 1 Concepts associated with the creation of an underwater acoustic beam using an airborne laser.

ANALYTICAL APPROACH

Most principal features of the generated sound field can be quantitively examined with the aid of the idealized model sketched in Fig. 2. A spatially periodic and amplitude modulated pattern of light with spatial period $2\pi/k_o$ moves over the water in the x-direction with a speed V, which is taken to be greater than the sound speed c. An angle θ whose cosine is c/V is subsequently identified as the angle with which the generated beam of sound makes with the horizontal direction. The trace velocity matching principle leads to the conclusion that this sound beam will have a wavelength λ of $(2\pi/k_o)\cos\theta$ and a frequency c/λ.

Because the laser power is finite, the area over which this pattern extends at any instant is limited; this is taken into account with an envelope enclosing the periodic pattern which is gaussian shaped in both horizontal coordinates x and y. This double envelope, moving with the speed V, is described at the water surface by the function

$$E_M\,(x - Vt, y) = e^{-(x-Vt)^2/L_x^2}\, e^{-y^2/L_y^2} \tag{1}$$

where L_x and L_y characterize the regional extent of the envelope; the subscript M on E_M is mnemonic for 'moving'. Also, because the total energy in a laser pulse is limited, an additional envelope factor $E_T\,(t) = \exp(-t^2/T^2)$ is introduced, with T characterizing the duration of the overall irradiation.

For most efficient production of sound, it is apparent that the laser light, after refraction at the air-water interface, should enter the water at an angle θ with the vertical. The laser intensity dies out exponentially as $\exp(-\mu s)$ with propagation distance s through the water, so one deduces that the laser energy q deposited in the water per unit time and per unit volume must vary with the spatial coordinates and time t as

$$q = F_P\,(k_o N)\, E_M\,(N, y)\, E_T\,(t)\, e^{-\mu_e z} \tag{2}$$

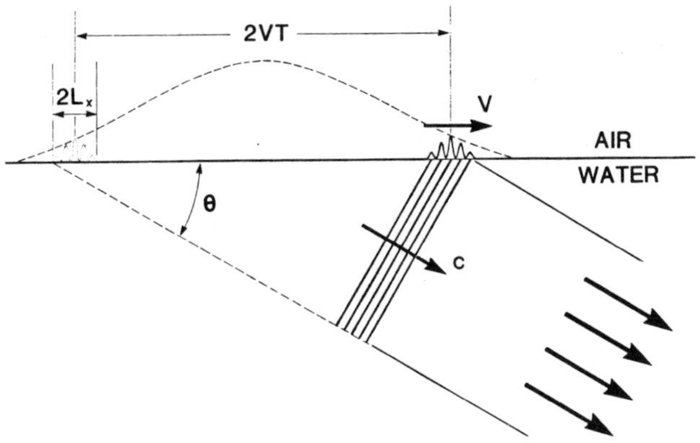

Figure 2 Principal features of analytical model, indicating the considered spatial and temporal configuration of the heat deposition near the water surface and the resulting soundbeam.

where μ_e is $\mu/\cos\theta$ and N abbreviates the combination

$$N = x + z\tan\theta - Vt \tag{3}$$

with z denoting depth below the water surface. The function F_P is periodic and can, for simplicity, be taken as

$$F_P(k_o N) = K(1 + \cos k_o N) \tag{4}$$

such that it oscillates back and forth between 0 and $2K$. The constant K can be related to the total energy E deposited in the water with the requirement that the total integral of q over volume and time must be E, so

$$K = \frac{\mu_e E}{\pi^{3/2} L_x L_y T} \tag{5}$$

The model assumes that the total power entering the water at time t varies with t as $E_T(t) = \exp(-t^2/T^2)$.

Determination of the acoustic disturbance in the water requires solution of the inhomogeneous wave equation [2-4] that governs the generation of sound when thermal energy is added to a fluid,

$$\nabla^2 p - \frac{1}{c^2}\frac{\partial^2 p}{\partial t^2} = -\frac{\beta}{c_p}\frac{\partial q}{\partial t} \tag{6}$$

where β is the coefficient of thermal expansion, c_p is the specific heat. The heat q added to the water by the laser beam per unit time per unit volume is taken as given by Eq. (2) above. In addition one has the boundary condition that $p = 0$ at $z = 0$, along with the causality condition, imposed on the solution.

The general solution approach expresses p as a triple inverse Fourier transform of a function of "k-space variables" ω, k_x, and k_y, along with the depth coordinate z, such that one writes

$$p(x,y,z,t) = \int_{-\infty}^{\infty}\int_{-\infty}^{\infty}\int_{-\infty}^{\infty} \psi(k_x,k_y,z,\omega)e^{-i\omega t}e^{ik_x x}e^{ik_y y}\,d\omega dk_x dk_y \tag{7}$$

where the integrand function ψ is readily found to satisfy an inhomogeneous ordinary differential equation in the depth coordinate z, this being

$$\frac{d^2\psi}{dz^2} + l^2\psi = \frac{i\omega\beta}{c_p}Q \tag{8}$$

where

$$l^2 = (\omega/c)^2 - k_x^2 - k_y^2 \tag{9}$$

and

$$Q = \frac{1}{8\pi^3}\int_{-\infty}^{\infty}\int_{-\infty}^{\infty}\int_{-\infty}^{\infty} q(x,y,z,t)e^{-ik_x x}e^{-ik_y y}e^{i\omega t}\,dxdydt \tag{10}$$

The latter triple integral for Q, given the expression (2) for $q(x,y,z,t)$, readily integrates to

$$Q = Q_o + Q_+ + Q_- \tag{11}$$

where

$$Q_+ = G_+(k_x,k_y,\omega)e^{(-\mu_e + ik_x\tan\theta)z} \tag{12}$$

$$G_+ = \frac{KL_xL_yT}{16\pi^{3/2}}e^{-\frac{1}{4}T^2(\omega-k_xV)^2}e^{-\frac{1}{4}L_x^2(k_x-k_o)^2}e^{-\frac{1}{4}L_y^2k_y^2} \tag{13}$$

Analogous definitions apply for Q_-, G_-, Q_o, and G_o; the three terms in Eq. (11) correspond, respectively, to the terms 1, $\frac{1}{2}\exp(ik_oN)$, and $\frac{1}{2}\exp(-ik_oN)$ that sum to the factor $1 + \cos k_oN$ in the expression (4) for $F_P(k_oN)$.

The ordinary differential equation (8) is solved by Green's function techniques. With the pressure release boundary condition taken into account, the portion ψ_+ of ψ that arises from the term Q_+ in (11) is readily found to be

$$\psi_+(k_x,k_y,z,\omega) = \frac{i\omega\beta}{c_p}G_+\int_{-\infty}^{\infty} \text{sign}(z_o)e^{(-\mu_e+ik_x\tan\theta)|z_o|}\frac{1}{2il}e^{il|z-z_o|}dz_o \tag{14}$$

Here it should be understood that the phase of l is either the same as that of ω or is $\pi/2$ for real k_x, k_y, and ω.

The integral in Eq. (14) is readily evaluated and approximates, when the condition $\exp(-\mu_e z) \ll 1$ holds, to

$$\psi_+(k_x,k_y,z,\omega) \simeq -\frac{i\omega\beta}{c_p}\frac{G_+e^{ilz}}{(-\mu_e + ik_x\tan\theta)^2 + l^2} \tag{15}$$

Subsequent substitution of this expression into Eq. (7) yields, for the p_+ term that contributes to the acoustic pressure p,

$$p_+ \simeq \frac{i\beta K L_x L_y T}{16 c_p \pi^{3/2}} \int_{-\infty}^{\infty} \int_{-\infty}^{\infty} \int_{-\infty}^{\infty} F e^{\Phi} \, d\omega dk_x dk_y \tag{16}$$

where

$$\Phi = ik_x x + k_y y + ilz - i\omega t - \frac{1}{4}T^2(\omega - k_x V)^2 - \frac{1}{4}L_x^2(k_x - k_o)^2 - \frac{1}{4}L_y^2 k_y^2 \tag{17a}$$

$$F = \frac{\omega}{(-\mu_e + ik_x \tan\theta)^2 + l^2} \tag{17b}$$

The principal challenge in completing this solution is the analytical reduction of the three-fold integral in Eq. (16) (or of its counterpart for p_-) to a simpler albeit approximate form that is amenable to calculations and useful for identification of trends. It is evident that for cases of practical interest one desires that the solution describe a beam of nearly constant frequency propagating obliquely downward into the water and that such will only be achieved if each of the quantities $k_o cT$, $k_o L_x$, and $k_o L_y$ are all much larger than unity. Also, we are interested in propagation distances from the heat deposition region that are somewhat larger than cT, L_x, and L_y, although not necessarily much larger than $k_o L_x^2$, etc. The latter implies that an ultimate asymptotic limit yielding an expression corresponding to the Fraunhaufer limit may be crude; instead one desires a Fresnel-type expression. All this suggests that the triple integral in Eq. (16) be evaluated by using some techniques that are analogous to the saddle point technique for integration over one complex variable.

Formal implementation of such a technique taking into account all of the terms in the exponent factor Φ, and taking the saddle point for any one integration as being where the corresponding derivative of Φ vanishes, leads to a rather intricate expression which is unnecessarily complicated. One recognizes, however, that one usually does not desire a good quantitative estimate of the sound pressure at all points in the far field and at all times, but only at those points and times where the amplitude is significant. Such points are those within the dominant beam of sound and the appropriate times are those when the dominant sound pulse is arriving. These considerations suggest (after some detailed supporting analysis) that one should expand Φ about the point in $\{k_x, k_y, \omega\}$ space where the portion

$$-\frac{1}{4}T^2(\omega - k_x V)^2 - \frac{1}{4}L_x^2(k_x - k_o)^2 - \frac{1}{4}L_y^2 k_y^2$$

is stationary and that one need only keep up to second order terms. The factor F may be approximated by its value at this stationary point.

Details of how the derivation just outlined is carried through are too lengthy to include with the present paper, so we give only the result, this being

$$p_+ = (\text{Const.})\ (\text{PWF})\ (\text{PE})\ (\text{BE})\ (\text{SF}) \tag{18}$$

where the five factors are identified further below as being a multiplicative constant (Const.), a plane wave factor (PWF), a pulse envelope (PE), a beam envelope (BE), and a spreading factor (SF). The constant is given by

$$\text{Const.} = \frac{E}{2\pi^{3/2} L_x L_y cT} \, \frac{\beta c^2}{c_p} \, \frac{i\cos\theta}{[(\mu/k_B) - i\sin 2\theta]} \tag{19}$$

where $k_B = k_o / \cos\theta$ is the wavenumber of the sound in the beam (which is propagating obliquely downward at an angle of θ with the horizontal).

The plane wave factor in Eq. (19) corresponds to a plane wave of angular frequency $\omega_B = ck_B$ propagating in the direction of the center line of the beam and is given by

$$\mathrm{PWF} = e^{i(k_B s - \omega_B t)} \tag{20}$$

where $s = x\cos\theta + z\sin\theta$ is distance along the direction of the sound beam from the origin. This plane wave is amplitude modulated in time by a pulse envelope factor

$$\mathrm{PE} = e^{-(s-ct)^2/L_s^2} \tag{21}$$

where $L_s = L_x \cos\theta$ is the characteristic half-length of the pulse along the beam axis; $c^{-1}L_s$ is the characteristic half-duration of the pulse. The beam is gaussian in both the y direction and the w direction, where $w = x\sin\theta - z\cos\theta$ is the transverse coordinate that is perpendicular to both the y direction and the axis of the beam. This gaussian property causes the plane wave to be encased by a beam envelope factor given by

$$\mathrm{BE} = e^{-w^2/\xi_w^2} e^{-y^2/\xi_y^2} \tag{22}$$

Here, in accord with the well-known properties of gaussian beams, the lengths ξ_w and ξ_y, whose squares appear in the denominators of the corresponding exponents, are actually complex numbers whose squares have an imaginary part that increases linearly with propagation distance; the appropriate expressions are

$$\xi_w^2 = W^2 + i(2s/k_B); \qquad \xi_y^2 = L_y^2 + i(2s/k_B) \tag{23}$$

where $W = VT\sin\theta$ is a characteristic w-dimension half-width of the beam in the early stages of propagation. The emergence of a spreading factor in the solution, given by

$$\mathrm{SF} = \frac{WL_y}{\xi_w\,\xi_y} \tag{24}$$

is a natural byproduct of the requirement that the gaussian beam envelope times the plane wave factor times the spreading factor should be a solution of the scalar Helmholtz equation.

Regarding the other two terms, p_o and p_-, that arise from Q_o and Q_- in Eq. (10), the term p_o is essentially a quasi-static term, so its contribution to the integral (7) is concentrated near $\omega = 0$. The factor ψ, however, has a multiplicative factor of ω, as is evident from the right side of Eq. (8). Consequently, p_o is negligibly small compared to p_+ when $T \gg 1/\omega_B$. The remaining term p_- is just the complex conjugate of p_+. Hence the overall solution for the acoustic pressure, with the approximations as stated above, is simply

$$p(x, y, z, t) = 2\mathrm{Re}(p_+) \tag{25}$$

DISCUSSION

The overall theory outlined above is extendable such that one can estimate required peak laser power, pulse duration, and pulse energy for an "easily detectable" acoustic signal at range r in direction θ. Such an estimate must take into account the attenuation of sound in sea water. The procedure is to determine what peak acoustic pressure is needed at the receiver location to be detected in the presence of background noise and then work backwards, given the general functional form of the

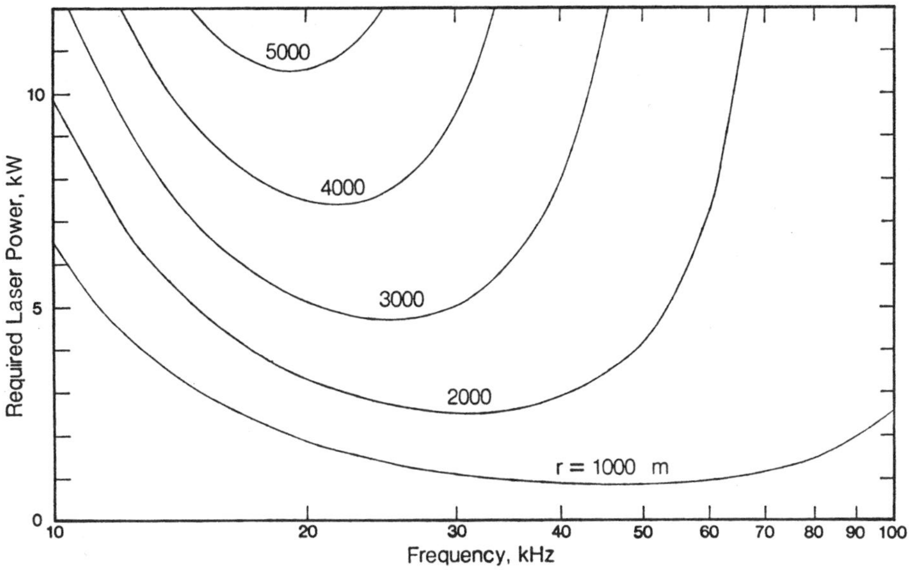

Figure 3 Plots of the required laser power versus frequency, for various acoustic propagation ranges *r*.

function q as described above, to determine what possible combinations of peak laser power and laser pulse duration would be sufficient to generate such a signal. Such an analysis suggests (Fig. 3) that with commercially available lasers it should be possible to configure the heating deposition pattern such that pulse signals having dominant frequencies of the order of 30 kHz can be detected at ranges from 5 to 10 km.

A principal feature of the acoustic disturbance generated by the heating configurations of the type considered in the present study is that the sound beam is highly collimated—the region in which the heat is deposited acts essentially as a highly directional array. Moreover, the optical to acoustic energy conversion efficiency is increased, because successive slabs of light tend to add heat in just the right places within the evolving sound wave— such that the amplitude is being pumped upward. The beam collimation tends to postpone spherical spreading; moreover, at ranges where spherical spreading begins to become evident, the angular beamwidth is extremely narrow. What this implies is that, at points within the center of the acoustic beam, the apparent source strength can appear much higher (Fig. 4) than one would ordinarily expect [5] from calculations based on a model where the acoustic energy spreads without any exceptional directionality.

ACKNOWLEDGMENTS

The authors thank Yves Berthelot, Ilene Busch-Vishniac, Peter Rogers, and Raymond Fitzgerald for helpful discussions during the course of this research. The work reported here was supported by ONR, Code 425-UA.

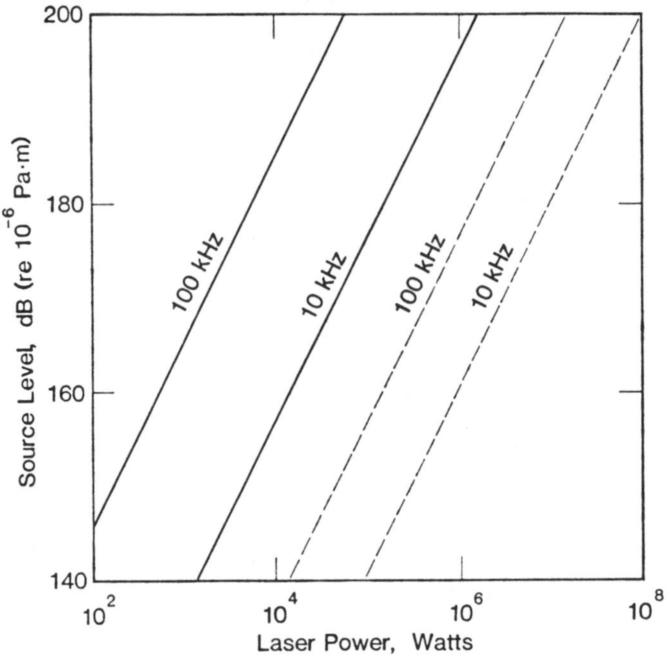

Figure 4 Predicted apparent maximum source levels for underwater acoustic signal received at a range of 2 km from a surface region where laser-generated heat is added. Dashed lines are predictions of Muir, Culbertson, and Clynch [5] for when laser beam is stationary and modulated sinusoidally in time; solid lines are predictions of the present paper's model.

REFERENCES

[1] Y. H. Berthelot and I. J. Busch-Vishniac, Thermoacoustic radiation of underwater sound by a moving high-power laser pulse, *J. Acoust. Soc. Am. Suppl. 1*, **77**, S103 (1985).

[2] B.-T. Chu, Pressure waves generated by addition of heat in a gaseous medium, *NACA Technical Note 3411* (National Advisory Committee for Aeronautics, Washington, June 1955).

[3] P. J. Westervelt and R.S. Larson, Laser-excited broadside array, *J. Acoust. Soc. Am.* **54**, 121–122 (1973).

[4] F. V. Bunkin and V. M. Komissarov, Optical excitation of sound waves, *Akust. Zh.* **19**, 305–320 (May–June l973); *Sov. Phys—Acoustics* **19**(3), 203–211 (Nov.–Dec. 1973)

[5] T. G. Muir, C. R. Culbertson, and J. R. Clynch, Experiments on thermoacoustic arrays with laser excitation, *J. Acoust. Soc. Am.* **59**, 735–743 (1976).

OPTICAL GENERATION OF SOUND: EXPERIMENTS WITH A MOVING THERMOACOUSTIC SOURCE. THE PROBLEM OF OBLIQUE INCIDENCE OF THE LASER BEAM

Yves H. Berthelot† and Ilene J. Busch-Vishniac

Applied Research Laboratories, The University of Texas at Austin
P.O. Box 8029, Austin, Texas 78713-8029

† Present address: School of Mechanical Engineering
Georgia Institute of Technology, Atlanta, Georgia 30332

ABSTRACT

An intensity modulated laser beam was reflected by a rotating mirror over a body of water to produce a moving thermoacoustic source. Pressure waveforms were recorded at several locations in the field and for different values of the the source velocity. A strong increase in sound pressure was observed when the source was moving at a velocity close to that of sound in water. The effect of oblique incidence of the laser beam on the water surface is then studied analytically for farfield radiation and numerically for the nearfield case. It is shown that in general the effect of oblique incidence is to accentuate the broadside characteristic of the acoustic radiation.

INTRODUCTION

The generation of underwater sound from an airborne laser source has been extensively studied during the last two decades. (See for instance the review papers by Lyamshev and Sedov,[1] and by Lyamshev and Naugol'nykh[2]). The main advantage of this optoacoustic technique over other conventional methods is that the laser beam trajectory can be accurately controlled by optical components (rotating mirrors, beam splitters, lenses, etc...). Consequently, a virtually unlimited number of source configurations are achievable without any hardware in the water.

The principal mechanism of interest in the production of sound by a laser is the thermo-acoustic effect in which optical energy is converted into an acoustic wave by thermal expansion of the medium. Chu[3] showed that the production of sound in a medium containing a heat source is proportional to the time derivative of the heat added into the medium per unit time and unit volume. Thus, in the case of a laser beam illuminating a body of water, the acoustic pressure is proportional to the time and spatial derivatives of the laser intensity. It means that sound can be produced if the laser intensity is modulated[4] or if the laser beam is moving.[5] The first part of this paper contains experimental data that were obtained with a moving thermoacoustic source, while the second part addresses the problem of the oblique incidence of the laser beam on the water surface and its effect on the directional properties of the sound field.

I- EXPERIMENTS WITH A MOVING THERMOACOUSTIC SOURCE

An experiment was designed to produce a moving thermoacoustic source (TS) in fresh water, by means of a rotating mirror installed about 4.1 m above the water surface. The rotational speed was adjustable between 150 and 4500 rpm, so that source Mach number on the water surface could be varied between $M = 0.1$ and $M = 2.6$ where the Mach number M is defined as the ratio of the source velocity v to the sound speed in water c. The laser pulse, which was deflected from ground level to the rotating mirror by a total internal reflection prism, provided up to 5 joules over approximately 1 ms during which the intensity could be modulated monochromatically between 5 and 80 kHz. The modulation was achieved by an externally driven Pockels cell. The experimental results were obtained either with a neodymium-glass laser (optical wavelength 1.06 μm) or with a ruby laser (optical wavelength 0.6943 μm) so that the penetration depth of the source could be either fairly short ($\sim$ 0.1 m with neodymium-glass) or fairly long ($\sim$ 1 m with ruby light). A more detailed discussion of the experiment is given in reference 6, but here we restrict the discussion to pressure waveforms, sound pressure level (SPL) dependence on source Mach number, and propagation curves.

Pressure waveform

Figure 1 shows the normalized pressure waveform radiated by a TS moving at Mach 1.6. The lasing element was a neodymium glass rod (optical absorption in water α = 13.7 Np/m). The laser intensity was modulated at a frequency f_o = 35 kHz and the laser pulse duration τ_o was about 0.8 ms. The receiving transducer was a standard H-56 hydrophone (sensitivity of -171 dB re: 1V/μPa at 35 kHz) placed 3.94 m ahead of the source and about 0.70 m below the water surface, so that the distance of observation was 4 m and the angle between the vertical axis and the source-receiver axis was θ =80°. Figure 1 shows that the exponential decay of the laser pulse is received before its peak value. This is just an effect of the time reversal which occurs with supersonically moving acoustic signals. Note also that both the received frequency f_d and the received pulse duration τ_d are affected by the Doppler factor $D = |1 - M \sin\theta| \approx 0.575$. As expected, one finds that $f_d \approx f_o/D$ and $\tau_d \approx D\tau_o$. In general, experimental results agree within reason with theoretical predictions that can be made from a time domain approach.[6]

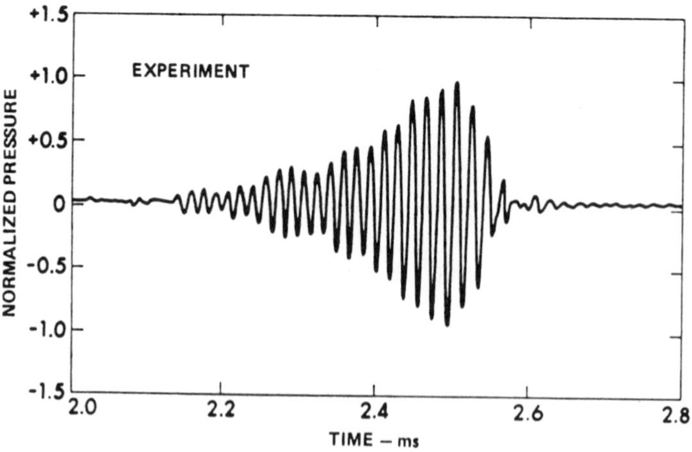

Fig. 1. Pressure waveform produced by a supersonic thermoacoustic source ($M = 1.6$, $\alpha = 13.7$ Np/m , $f_o = 35$ KHz)

Sound pressure level dependence on source velocity

As pointed out in reference 7, the efficiency of the thermoacoustic mechanism of sound generation with a laser is extremely low. However, by moving the source at the velocity of sound, one can considerably increase the received acoustic pressure[8] because wavelets emitted by the laser heating "pile-up" on each other in front of the beam and move with the beam at the sound speed. Figure 2 shows the sound pressure level dependence on source velocity. The modulation frequency of the laser intensity was f_o =25 kHz, the lasing frequency was 1.06 μm, and the distance source-receiver was about 4.2 m. The angle between the vertical axis and the source-receiver direction was $\theta = 80^o$. The receiving transducer was an H-56 hydrophone. In Fig. 2, the acoustic levels are normalized with respect to an equivalent stationary source. The dotted line represents the theoretical predictions of a linear model discussed in reference 6, and the "windows" show the type of normalized pressure signal detected at the hydrophone, for subsonic, transonic, and supersonic source velocities. It is interesting to note that the theory agrees reasonably well with the experimental data in

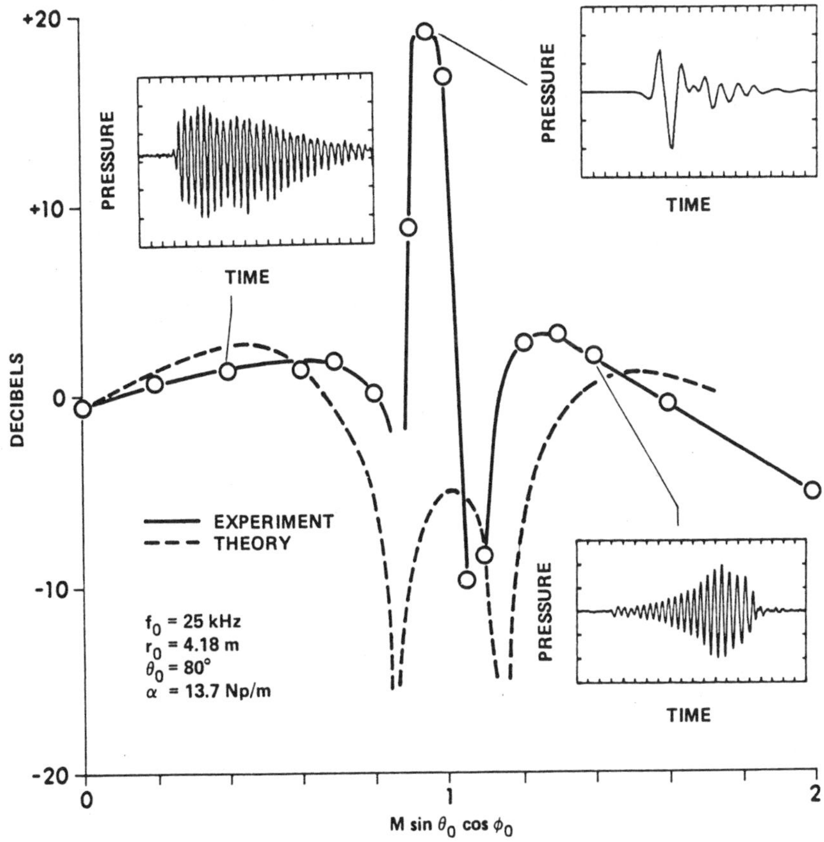

Fig. 2. Sound pressure level dependence on source velocity

the subsonic and supersonic cases. However, it seems to underestimate the acoustic pressure for transonic source velocities. The model does predict however the symmetric dips around Mach one that were observed experimentally. These dips are the result of diffraction effects induced by the finite width of the laser beam. When the beam diameter is much smaller than the acoustic wavelength, diffraction effects are unimportant. However, when the source is moving, the acoustic wavelength is Doppler shifted and, for source velocities sufficiently close to Mach one, it becomes comparable to the laser beam diameter. In such a case, diffraction effects are important and tend to reduce significantly the acoustic pressure. In other words, we may estimate that dips occur when $M \sin\theta \approx 1 \pm (a/\lambda_o)$, where a is the laser beam radius, and $\lambda_o = c/f_o$ is the acoustic wavelength. Applying this estimate to the data relevant to Fig. 2, one finds that dips should occur at Mach .8 and 1.2 . This is confirmed experimentally. Also, for a higher modulation frequency, diffraction effects are expected to spread over a wider region. This was verified experimentally for a modulation frequency of 35 kHz.

Propagation curves

The sound pressure level dependence on distance of observation was determined for a TS moving at Mach 1.5 towards the receiver. In this experiment, the lasing element was a ruby rod and consequently the coefficient of optical absorption was

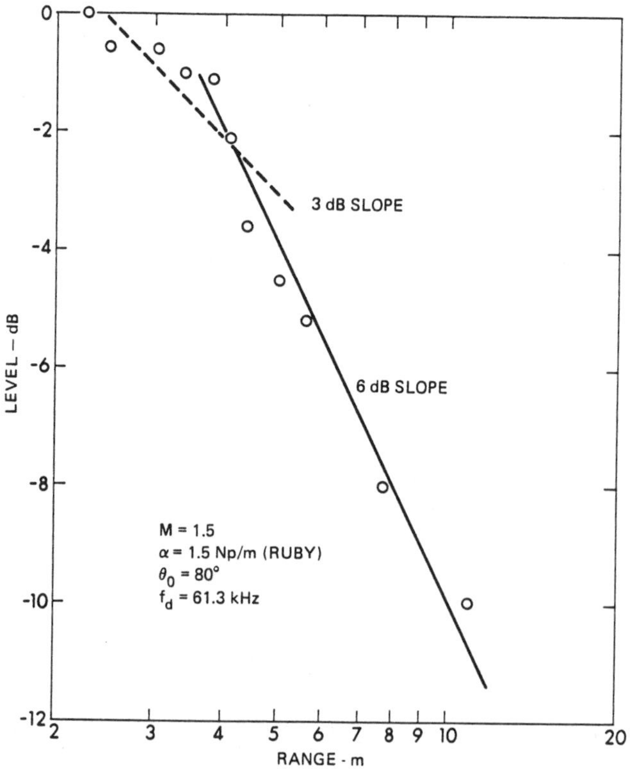

Fig. 3. Sound pressure level dependence on distance of observation

fairly small ($\alpha = 1.5$ Np/m). The acoustic pressure signal was detected by an H-56 hydrophone at distances ranging between 2.25 m and 10.83 m. The angle between the vertical axis and the source-receiver axis was kept constant ($\theta = 80^o$) so that the Doppler shifted frequency would remain also constant (f_d =61.3 kHz). During this experiment the Pockels cell modulator failed and it was replaced by a mechanical modulator which consisted of a series of wooden sticks placed evenly over the water surface, along the laser beam path, so that the laser light would penetrate only intermittently. The thickness of the sticks was equal to the spacing between the sticks, and, in order to simulate a sinusoidal modulation, this spacing was made equal to the laser beam diameter on the water surface. This scheme turned out to be quite effective. Experimental results are reported in Fig. 3. Sound pressure levels are normalized so that the first data point (distance of 2.25 m) is 0 dB. Far enough from the source, the spreading appears to be the usual 6 dB per doubling of distance, whereas fairly close to the source, a cylindrical spreading (3 dB slope) seems to dominate. It should be noted that, in general, pressure waveforms produced with the ruby laser were not as clean and as repeatable as the waveforms produced with the neodymium glass laser.

II- OBLIQUE INCIDENCE OF THE LASER BEAM

In most experiments involving moving thermoacoustic sources, the laser beam does not strike the water surface at normal incidence, but rather at an oblique angle. A very promising design for instance, which has been suggested by Pierce and Hsieh,[9] uses non-perpendicular incidence of a laser beam moving at a slightly supersonic velocity. The objective of this section is to provide some understanding of the effect of oblique incidence on the directional properties of the resulting sound field. We restrict the discussion to the case of a stationary source, and we analyze successively farfield and nearfield source directivities.

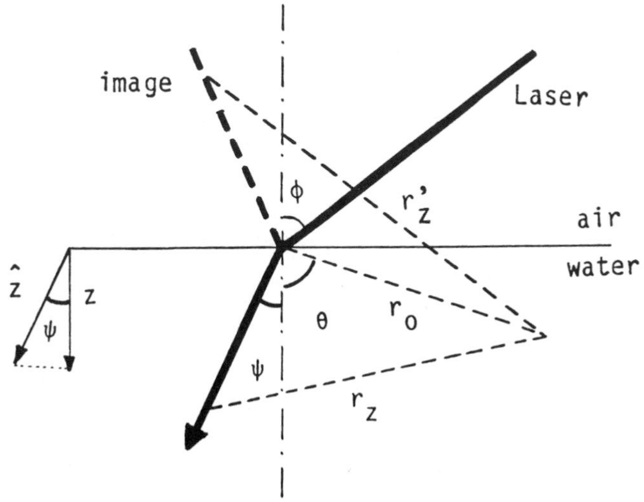

Fig. 4. Geometry of the problem for oblique incidence

Farfield analysis

The geometry of the problem is shown in Fig. 4. A laser beam impinges upon a water surface at an angle ϕ with respect to the vertical axis. Snell's law of refraction requires that the laser beam be refracted at an angle ψ such that $\sin\psi = (c_w/c_a)\sin\phi$ where c_w and c_a are the speeds of light in water and in air, respectively. For a lasing frequency in the visible (or in the near infrared), $c_w/c_a \approx 0.75$, so that refraction tends to minimize the effects of oblique incidence. For a CW laser whose intensity is monochromatically modulated, the acoustic pressure can be expressed in the form[4]

$$p = K\int_0^\infty e^{-\alpha\hat{z}}\left(\frac{e^{ikr_z}}{r_z}\right)d\hat{z} \;-\; \text{mirror image} \qquad (1)$$

where K is a constant, α is the coefficient of absorption of light in water, and k is the acoustic wavenumber defined as the ratio of 2π times the modulation frequency f_o over the speed of sound in water. The pressure release characteristic of the water-air interface indicates that the mirror image contribution has to be subtracted from the integral in Eq. (1). The depth coordinate z is related to the depth coordinate along the tilted axis of the beam, $\hat{z}$, by the relation $\hat{z} = z/\cos\psi$. It is assumed in Eq. (1) that the laser beam radius is very small compared with an acoustic wavelength $\lambda = 2\pi/k$. The quantity r_z in Eq. (1) denotes the distance between a source point along the tilted beam and the receiver. It can be shown that, in the farfield, $r_z \approx r_o - \hat{z}\cos(\theta+\psi)$, and that, similarly for the mirror image, $r_z' \approx r_o + \hat{z}\cos(\theta-\psi)$. As indicated in Fig. 5, r_o is the distance between the receiver and the point where the laser beam impinges on the surface; the angle θ is the angle between the vertical axis and the source-receiver axis. With these notations, the directionality of the source can be shown to be:

$$D(\theta) = \frac{|\; 2\sigma\; \cos\theta\; \cos\psi \;|}{\left[\left(1+\frac{\sigma^2}{2}(\cos 2\theta + \cos 2\psi)\right)^2 + 4\sin^2\theta\sin^2\psi\right]^{1/2}} \qquad (2)$$

where $\sigma = k/\alpha$ is a measure of the length of the source in terms of the acoustic wavelength. It can be verified that Eq. (2) satisfies the pressure release boundary condition at the water-air interface, and that for $\psi = 0$, it reduces to the normal incidence directivity of a thermoacoustic source.[1] Figure 5 shows the resulting farfield beam patterns for three different values of the angle of incidence ($\psi = 0^o, 15^o$, and 45°) for both a long ($\sigma = 10$) and a short ($\sigma = 3$) laser penetration depth in terms of an acoustic wavelength. As expected for an array excited virtually simultaneously (at the speed of light) along its axis, its farfield directivity is essentially a broadside type directivity. It can be seen from Fig. 5 and from an analysis of Eq. (2) that the angle $\hat{\theta} = \theta + \psi$ between the source and the axis of the broadside acoustic lobe tends asymptotically towards 90° as the angle of incidence ψ increases. It reaches faster its asymptotic value for large values of σ. In other words, the presence of the pressure release interface is not felt as strongly for large values of the angle of incidence, and for large values of σ, so that the broadside characteristic ($\hat{\theta} \to \pi/2$) of the acoustic radiation is emphasized. Also, in the farfield, the half-power beamwidth of the acoustic lobe seems fairly independent of the angle of incidence of the laser beam.

Nearfield analysis

The nearfield directivity of a TS can be found by taking the Fourier transform of the impulse response of the optoacoustic system. Such an analysis is given in reference 10 for the case of a laser beam striking a water surface at normal incidence. It shows that, in the nearfield, one can expect some sidelobes in the directivity pattern. A

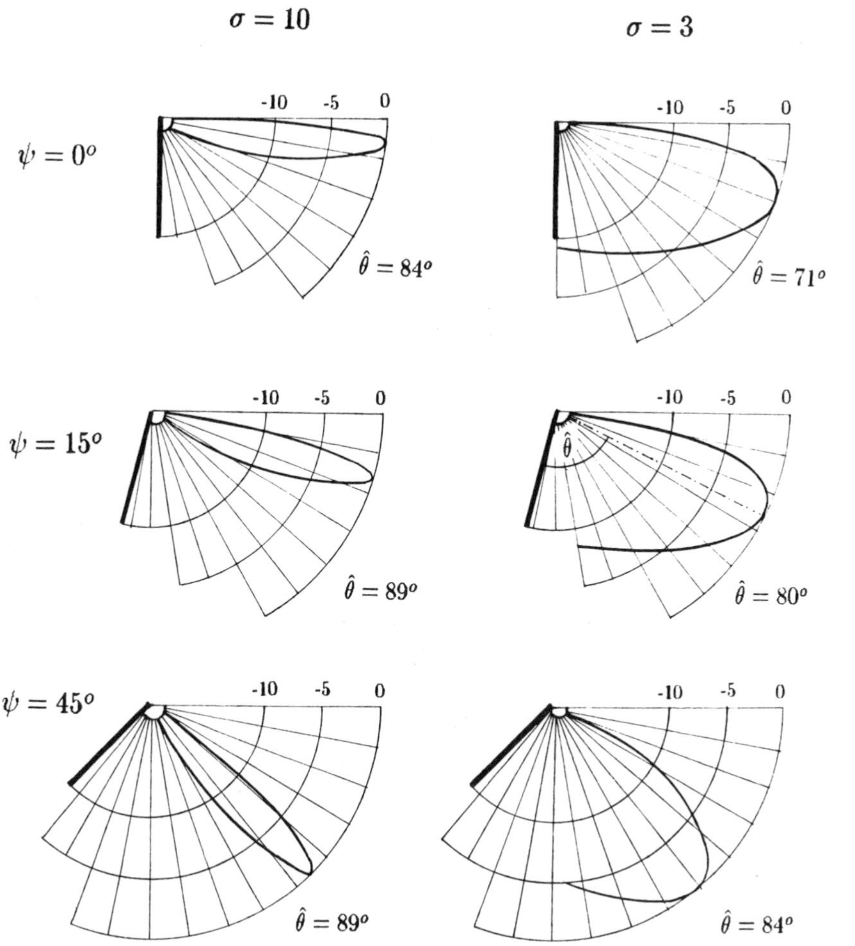

Fig. 5. Farfield directivity patterns for $\psi = 0^o$, 15° , 45° and $\sigma = 3$, 10

similar analysis can easily be performed for the case of oblique incidence of the laser beam, and one would find that, in the nearfield, the directivity of the source is given by

$$D(\theta) = 20 \log \left\| \frac{H_\omega(\theta)}{H_{max}} \right\| \tag{3}$$

where the function H_ω , at a given modulation frequency, is the Fourier transform of the impulse response of a tilted thermoacoustic source, and can be shown to be:

$$\begin{aligned} H_\omega(\theta) = K_1 \int_0^{u_o} & \cosh(\gamma \sinh u) \ \exp(-i\eta \cosh u) \ \mathrm{S}(\frac{\pi}{2} - \hat{\theta}) \ \mathrm{du} \\ + K_2 \int_{u_o}^{\infty} & \exp(-\gamma \sinh u) \ \exp(-i\eta \cosh u) \ du \\ - K_3 \int_{u'_o}^{\infty} & \exp(-\gamma' \sinh u) \ \exp(-i\eta' \cosh u) \ du \end{aligned} \tag{4}$$

where $\gamma = \alpha r_o \sin\hat{\theta}$, $\gamma' = \alpha r_o |\sin\beta|$, $\eta = k r_o \sin\hat{\theta}$, $\eta' = k r_o |\sin\beta|$, $u_o = \cosh^{-1}(1/\sin\hat{\theta})$, and $u'_o = \cosh^{-1}(1/|\sin\beta|)$, with $k = 2\pi f_o/c$, $\hat{\theta} = \theta + \psi$, and $\beta = \pi - \theta + \psi$. $S(\lambda)$ in Eq. (4) is a step function which is zero for $\lambda \leq 0$, and unity otherwise. K_1, K_2, and K_3 are some constants defined by $K_1 = [A\beta\alpha\exp(-\alpha r_o \cos\hat{\theta})]/2\pi c_p$, $K_2 = K_1/2$, and $K_3 = K_2 \exp(+2\alpha r_o \cos\theta \cos\psi)$, where A is the optical transmissivity between air and water (which depends on the angle of incidence of the laser beam), β is the coefficient of thermal expansion of water, and c_p is the specific heat of water measured at constant pressure.

The nearfield directivity is therefore easily evaluated numerically and it shows that, for $\sigma \gg 1$, the directivity pattern of a thermoacoustic source tilted by an angle ψ is nearly identical to that of a non-tilted source, after a rotation of the whole beam pattern by an angle ψ. Also, as the distance r_o between the source and the receiver is increased, the side lobe structure vanishes and the resulting directivity pattern is that predicted by Eq. (2).

ACKNOWLEDGEMENTS: This work was supported by the the U.S. Office of Naval Research, Code 0425 UA.

REFERENCES

1. L. M. Lyamshev and L. V. Sedov, Sov. Phys. Acoust. 27 (1), 4-18 (1981)
2. L. M. Lyamshev and K. A. Naugol'nykh, Sov. Phys. Acoust. 27 (5), 357-371 (1981)
3. B. T. Chu, NACA Technical Note 3411 (1955)
4. P. J. Westervelt and R. S. Larson, J. Acoust. Soc. Am. 54, 121-122 (1973)
5. A. I. Bozhkov et al., Sov. J. Quantum Electron. 7 (4), 536-537 (1977)
6. Y. H. Berthelot, Technical Report ARL-TR-85-21, The University of Texas at Austin (1985)
7. T. G. Muir et al., J. Acoust. Soc. Am., 59, 735-743 (1975)
8. F. V. Bunkin et al., Sov. J. Quantum Electron. 8(2), 270-271 (1978)
9. A. D. Pierce and H.-A. Hsieh, J. Acoust. Soc. Am., Suppl. 1, 77, S104 (1985)
10. Y. H. Berthelot and I. J. Busch-Vishniac, J. Acoust. Soc. Am., 78 (6), 2074-2082 (1985)

ACOUSTIC MEASUREMENTS AND APPLICATIONS OF KINETIC IMPACTS ON ICE

Philippe de Heering

Staff Scientist, Advanced Systems Group
Canadian Astronautics Limited
Ottawa, Ontario
Canada

ABSTRACT

This paper summarizes the results of an investigation in the properties and applications of the acoustic impulses generated in ice-covered water as a result of a kinetic impact on ice. The impulses generated in this manner are characterized by high axial peak pressure levels (235 dB Re 1 micro Pa and more), good repeatability and significant spectral level from a few hundred Hertz to a few kiloHertz. A phenomenological model is presented, which accounts for the observed pressure levels, and two remote sensing applications are discussed, together with examples from field data.

INTRODUCTION

As part of a contract with the Canadian Hydrographic Service, Department of Fisheries and Oceans, Canadian Astronautics Limited (CAL) has developed an acoustic source, which can generate, as a result of a kinetic impact on the ice surface, a high amplitude, wide-band impulse in the water underneath the ice. Although this source was developed for the specific purpose of non-contact through-the-ice sounding, it also has more general applications in the field of acoustic remote sensing, in the Arctic in particular.

In the following sections, we describe and characterize the acoustic source and we give examples of its application to under-ice depth and ice properties measurements.

SOURCE DESCRIPTION

When a high energy projectile impacts on a solid or liquid surface, part of the kinetic energy of the decelerating projectile is converted into an acoustic pulse, which propagates in the solid or liquid. This method of excitation, which is not unknown in geophysical exploration [1-3], had not, to the best of the author's knowledge, been applied before to acoustic remote sensing of ice or water.

Table 1. Projectile Characteristics

Calibre mm(in)	Bullet Mass g(grains)	Kinetic Energy kJ
5.55 (.224)	3.2-3.6 (50-55)	1.71-3.30
7.62 (.308)	6.5-11.8 (100-180)	3.10-5.95
11.35 (.458)	19.6-39.3 (300-600	6.77-11.4

For the work described here, the projectiles used were .22, .30 and .4 calibre solid brass or copper-jacketed spire point bullets, fired fro standard and specially adapted rifles. Most of the loads and bullets use were specially developed for the purpose of this work. Table 1 summarize the range of characteristics of the projectiles used in these experiments.

SOURCE PROPERTIES

A number of experiments were carried out to measure the sound pressur levels, as well as the spatial and spectral characteristics of the nois impulses associated with the impacts onto the ice. The results presente below were obtained during experiments carried out on the ice of the Ottaw River near Ottawa in February 1984 and 1985, over (fresh water) ice 75 c thick. The bullets were shot at the ice either from an elevated platfor and, later, from the ice level, after it was determined that the muzzl blast did not contaminate the acoustic measurement. Acoustic pressur levels were measured in the water by (B&K 8103) hydrophones suspended in th water directly below the impact point, and corrected for spherical spreadin to refer them to one meter distance from the point of impact. Thi correction, which does not take into account the refraction between ice an water, is nonetheless approximately valid when the depth of the hydrophon is several times the ice thickness, as was the case in these experiments.

Figure 1 shows a typical pressure waveform (low pass filtered a 10 kHz) recorded as the result of the impact of a specially designed 7.6 calibre, 9.1 g bullet on ice at a nominal velocity of 1144 m/s. The puls generated is high level, clean and short, and has a significant energ content up to some 5 kHz. It was also observed that the acoustic pulse generated by consecutive impacts are repeatable within one to two decibels.

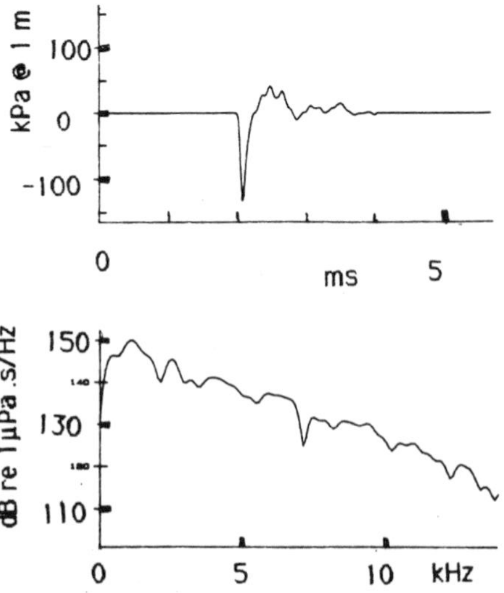

Figure 1 Sample Pressure Impulse and Energy Spectral Level.

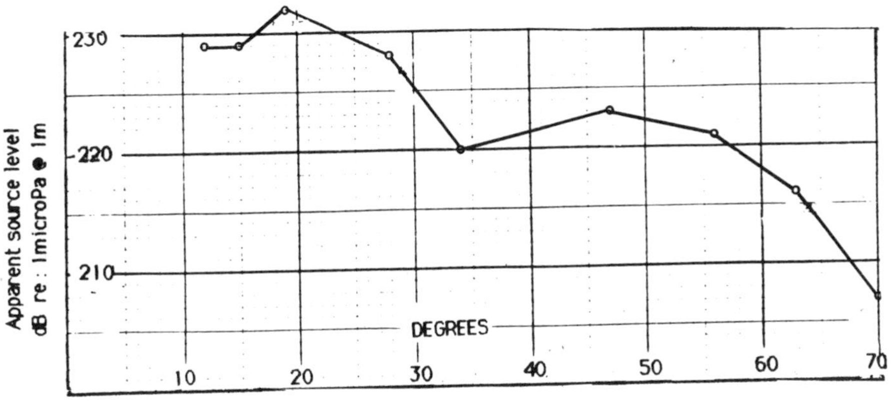

Figure 2 Sample Beam Pattern (.30 calibre bullet).

Bullet impacts on water were observed to result in peak should pressure source levels several decibels higher than the corresponding impacts on fresh water (river) ice, whereas the corresponding impacts onto sea ice resulted in peak sound pressure source levels that were in general lower by several decibels.

Figure 2 illustrates a beam pattern effect on the peak pressure level: this effect is easily explained as a consequence of the refraction at the ice water interface and results in most of the acoustic energy being radiated in the water within 30° of the vertical.

The acoustical impulse generation process is a complicated one, as the deceleration of the bullet in the ice is accompanied by ice fracturing into pieces of various sizes, as well as by ice melting. Experiments were therefore carried out to derive a phenomenological model of the sound generation.

Figure 3 illustrates the dependence of the peak sound pressure (source) level (PSPL, in Pa) on the kinetic energy (KE, in J) divided by the calibre (CAL, in m). For comparison, the inhomogeneous relation

$$PSPL = (.9)\ KE/CAL \tag{1}$$

is also plotted, and the agreement is seen to be fair. Dependence on the same independent variable is also observed[4] for the axial energy flux density.

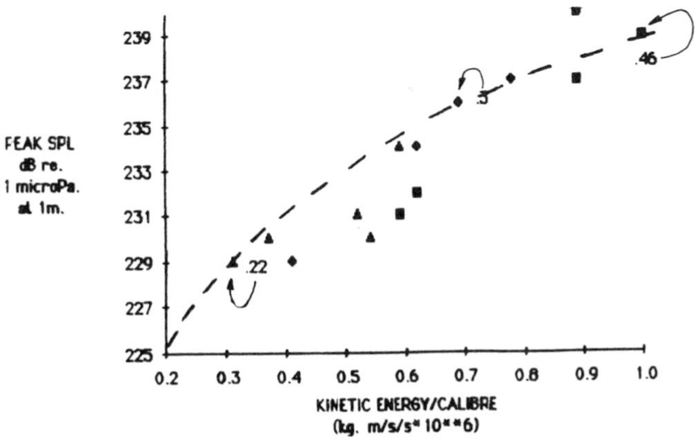

Figure 3 Peak Axial Level vs. Kinetic Energy/Calibre.

The normalization of the kinetic energy by the calibre in the rig hand side of Eq. 1 can be understood in a qualitative manner as follows:

(i) For equal kinetic energy, small calibres penetrate deeper in the ic the coupling of the bullet energy into the ice is thus improved as the to free surface of the ice is farther removed from the place where the acoust energy is generated.

(ii) For equal kinetic energy, smaller calibres fracture the ice less, an therefore, a larger portion of their energy is available for conversion in sound.

It is pertinent to note that a similar effect has been noted geophysicists using ballistic impact sources, who report [1,3] that shooti in a hole a few feet deep rather than at the topsoil surface, significant improves the level of the generated sound. In the latter case, contributing factor is the greater cohesion of the soil at depth wi respect to the surface.

NON-CONTACT THROUGH-THE-ICE SOUNDING

In order to chart the water depth in the Canadian Arctic, through-th ice sounding methods have often to be used, as many of these waters a covered by ice a large part of the year. Present methods generally invol physical contact with the ice and thus feature a relatively slow mappi rate. The method most often used consists in applying a transducer (mount on a helicopter) against the ice surface. The transducer is then used as conventional echo-sounder.

Canadian Astronautics proposed (see Figure 4) to use a kinetic impa source 12 mounted on a helicopter to produce an impulse 17 in the ice. Th impulse travels through the ice and then through the water, where reflects from the sea floor; after travelling through the water and the ic a second time, it is transmitted into the air where it is received by microphonic device 30 suspended from the helicopter. The water depth can calculated on the basis of the delay between transmission and reception a other relevant parameters.

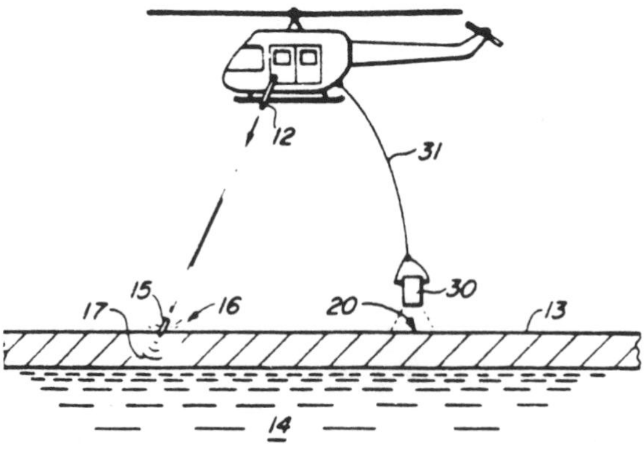

Figure 4 Proposed Sounding Method.

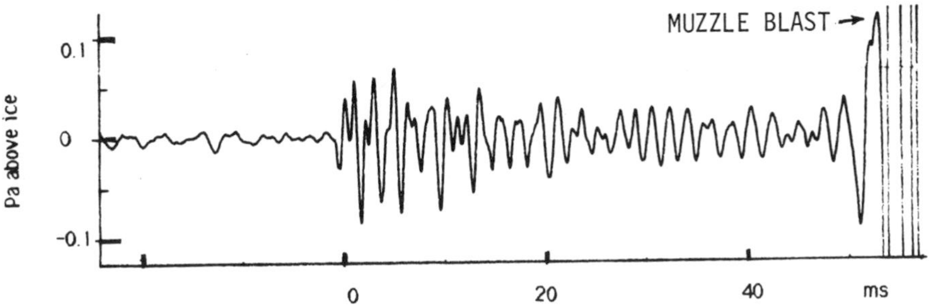

Figure 5 Airborne Echo.

Development of a prototype sounding system was carried out in 1984 and 1985 under contract to the Canadian Hydrographic Service. Figure 5 is an example of an echo obtained in 27 m depth water covered by 2 m of sea ice. The echo was received by a microphone suspended from a scaffolding at 30 cm from the ice. The impulse results from the impact of a .303 caliber British Ackley 173 grains bullet shot from the ice level at a horizontal distance from the hydrophone such that the airborne muzzle blast is received after the echo.

Figure 6 is an example of results obtained in 1985 over 2m of sea ice in a depth of 20 m of water from a helicopter in flight. The signal is received by two microphones, one noise reference (a) directed toward the helicopter; the other (b) directed toward the ice surface. These are shown high-pass filtered at 500 Hz. (c) shows (b) after adaptive noise cancelling with a 256 point filter[5]. Note the suppression of the noise impulses originating from the helicopter. In the case presented on Figure 6, because of uncertainties in the geometery, some doubt remains as to whether the received impulse is, in fact, a bottom echo, or whether it is produced by the water-coupled impact sound.

REMOTE SENSING OF ICE PROPERTIES

The kinetic impact source is a reasonably well calibrated source useful for the measure of mechanical properties of ice and other objects.

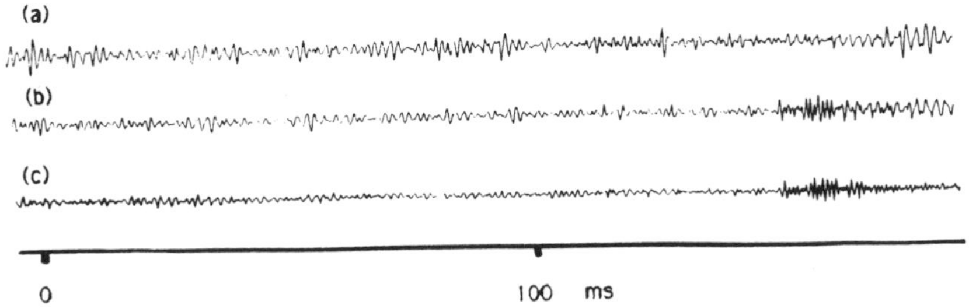

Figure 6 Airborne Returns, received by Helicopter System, and Processing.

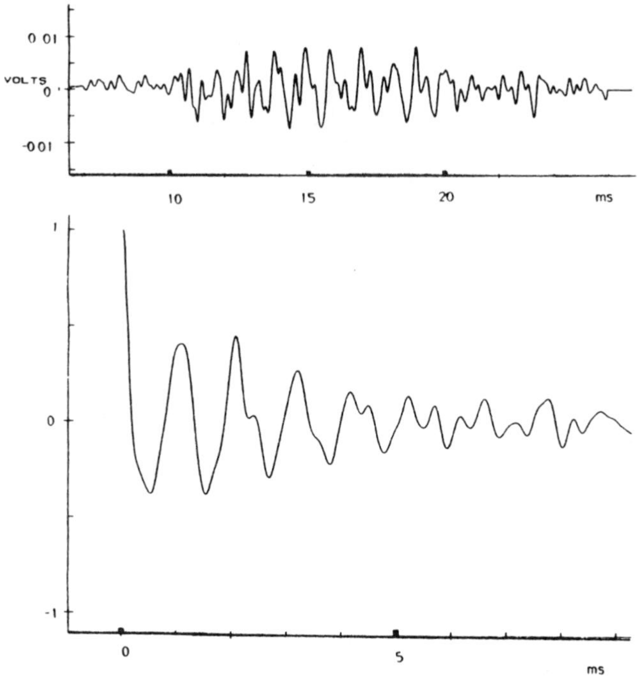

Figure 7 Ice Thickness Sensing (a) Signal (b) Autocorrelation.

Figure 7 is an example of results obtained with air microphone reception of a kinetic impact excitation of an ice sheet at 10 m range. The sea ice was approximately 1.90 m thick. The autocorrelation of the received impulse exhibits a peak at approximately 1 ms, in agreement with the ice thickness and the average sound speed in ice of some 3700 m/s.

CONCLUSION

Several aspects of kinetic impact sources have been discussed, with particular emphasis on their interaction with ice. Promising applications include remote sensing of water depth and ice properties. Other applications, involving other media, are currently being studied.

ACKNOWLEDGEMENTS

The support of Bryan White and George McDonald, the scientific authorities from the Canadian Hydrographic Service for part of the work, is gratefully acknowledged. Polar Continental Shelf Project (Energy, Mines and Resources, Canada), provided the indispensable logistic support during the Arctic field work. Peter Sutcliffe and Mike DesParois, both of Canadian Astronautics Limited, did most of the instrumentation, signal processing and ballistics work.

REFERENCES AND NOTES

1. D.W. Steeples, "High-resolution seismic reflections at 200 Hz", Oil and Gas, 86-92, Dec. 3, 1984.

2. There is at least one commercially available source: the Betsy (TM) Seisgun Source, available from MAPCO, Tulsa, Oklahoma, U.S.A. This is an 8-gauge gun manufactured by Remington and specifically adapted to be used as a seismic source.

3. J.A. Hunter et al, "Shallow seismic reflection mapping of the overburden - bedrock interface with the engineering seismograph - some simple techniques", Geophysics, 49, 1381-1385 (1984).

4. P. de Heering, P. Sutcliffe, "Continuous through-the-ice sounding; prototype design, integration and test", Canadian Astronautics Limited report Ref. 290 for the Canadian Hydrographic Service, August 1985.

5. B. Widrow, S.D. Stearns, "Adaptive Signal Processing", Prentice Hall Inc., Eaglewood Cliffs, N.J., U.S.A., 1985.

AXIAL FOCUSING BY PHASED CONCENTRIC ANNULI

H.D. Mair and D.A. Hutchins

Department of Physics, Queen's University
Kingston, Ontario, Canada K7L 3N6

ABSTRACT

Focusing transducers such as spherical bowls, annular arrays and planar transducers fitted with lenses have many uses in underwater acoustics, and various other areas involving acoustical imaging. In this paper, the characteristics of the first two types of transducer will be compared theoretically, with a comparison to experiment for the case of the spherical bowl transducer.

INTRODUCTION

The radiated fields of focused radiators were investigated the early work of O'Neil[1], who solved the surface integral at a single frequency assuming an infinite baffle (Huygen's boundary condition). The result was an approximate solution along the axis and in the focal plane, assuming a slightly curved radiator with $ka >> 1$, where k is the wave-number and a the transducer radius. An interesting result, noted in the present investigation, was that the point of maximum pressure did not always occur at the geometric focus.

Other authors have also considered continuous wave fields from focusing transducers. Lucas and Muir[2] found a single integral solution for any point in the field of a focused source by applying a Hankel transform to the surface integral. They used this integral to make a comparison to the experimental fields from two focused sources. Archer-Hall and Ali Bashter[3] studied how a surface integral method differed from classical theory while Swindel et al[4] suggested that by simple geometric considerations, the field from a focused source could be calculated as a single sum.

With the growth of interest in acoustical imaging, the fields from pulsed radiators became of increasing interest. An efficient technique for the calculation of transient fields is to use an impulse response method, where the waveform at a particular point in the field is calculated by performing a convolution between the impulse response of the radiator at that point and the waveform with which the transducer is driven. An analytical solution for the impulse response of a focused radiator was first derived by Pentinen and Luukkala[5], who used it to predict the fields of focused radiators under continuous wave excitation. Several other workers have used this impulse response to find fields from pulsed focused radiators. For example, Arditi et al[6] characterised concave annular arrays and presented a different form for the impulse response of a focused transducer. Goodsit et al[7] theoretically and experimentally studied pulse shapes from focusing radiators in attenuating and non-attenuating media. In addition, Cobb[8] used a frequency domain impulse response for an attenuating media, to overcome sampling problems in discrete Fourier transforms that occur when high frequency components are present in the time domain impulse response. The field pattern calculated by this method was then compared to conventional diagnostic transducers. Weyns[9] theoretically studied the pulsed fields of disc and ring concave transducers, noting how the fields varied with pulse length. Finally, using spatial Fourier transform methods, Guyomar and Powers[10] were able to calculate the impulse response from a focusing transducer with arbitrary velocity distribution across its face.

It would appear that to date there has been no study that compares the experimental and theoretical fields from focusing transducers throughout the nearfield and the farfield. It would also appear that in previous work, the range of bowl-shaped transducers studied have been only slightly curved, because, to date, any practical theoretical solutions involve this approximation.

In this paper, the axial and full fields from two types of focusing transducers - concentric annuli and spherical bowls - will be examined over a range of frequencies. The experimental fields from three bowl transducer geometries will also be compared to theoretical predictions using a surface integral technique. Also included is a theoretical comparison between the pressure fields of bowl transducers and phased annular arrays.

THEORY

The axial pressure variations of a concave spherical transducer, expressed in terms of a scalar velocity potential $\phi(z)$, may be written as

$$\phi(z) \propto \frac{1}{p} \left[e^{-jks_2} - e^{-jks_1} \right] \qquad (1)$$

where $s_1 = R_c - p$ and $s_2 = (R_c^2 + p^2 - 2p\sqrt{R_c^2 - a^2})^{1/2}$.

Here, R_c is the radius of curvature, a is the aperture radius, k the wavenumber and p the axial distance from the geometric focus. A comparable expression for a single annulus, with outer and inner radii a_o and a_i respectively, is

$$\phi(z) \propto (e^{-jks_2} - e^{-jks_1}) \qquad (2)$$

where $s_1 = (a_i^2 + z^2)^{1/2}$ and $s_2 = (a_o^2 + z^2)^{1/2}$. Using Eq. (2), the axial field from a series of phased, concentric annuli may be found by summing the contributions from each annulus, multiplying by the appropriate complex number to obtain constructive interference at the desired point.

The above analysis may be extended to off-axis positions (ρ, z) for both transducer types by making use of a single integral expression for a plane piston, of the form[11]

$$\phi(\rho,z) = \begin{bmatrix} 0 \\ 1/2 \\ 1 \end{bmatrix} e^{-jkz} + \frac{1}{\pi} \int_0^{\pi} e^{jks} \frac{(a\rho\cos\psi - a^2)\,d\psi}{(a^2 + \rho^2 - 2a\rho\cos\psi)} \qquad (3)$$

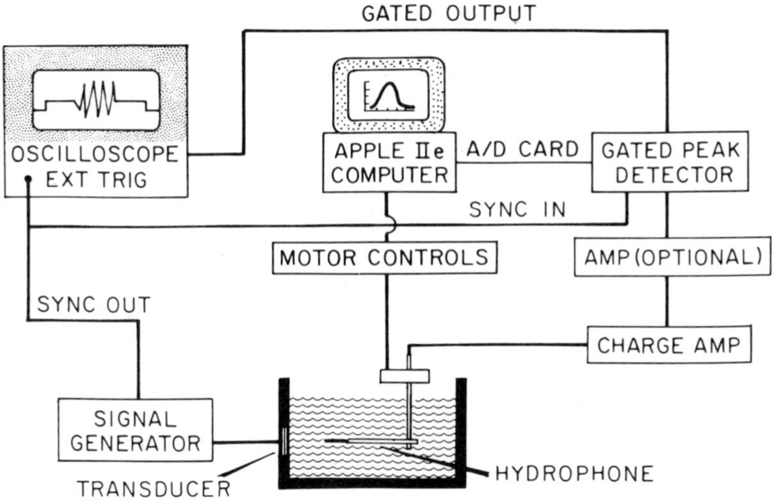

Fig. 1: Schematic diagram of apparatus.

where ρ is a radial coordinate, and s is given by

$$s = (\rho^2 + a^2 - 2a\rho\cos\psi + z^2)^{1/2} \quad . \tag{4}$$

Note that the first term in (3) is multiplied by 1 when $\rho < a$, 1/2 when $\rho = a$ and zero when $\rho > a$. The field from an annulus can then be found by subtracting the field of one plane piston from that of another of the same vibrational amplitude but larger radius. The fields of the bowls, cones and other slightly curved non-planar transducers can be calculated by adding the fields of concentric annuli displaced appropriately in the z plane (the "thin annulus approximation"). In addition, the fields of planar array transducers, phased to focus on a specific point, can be found by multiplying the fields from each annuli by the appropriate complex number, as described above for the axial case. The above approaches will be used to compare the fields of spherical bowls and phased concentric annuli transducers, with the former also being investigated experimentally, as will now be described.

APPARATUS

A schematic diagram of the apparatus, used to investigate the pressure fields of bowl transducers experimentally, is presented in Fig. 1. The transducers were constructed from 90 μm thick PVDF film, bonded to a bowl-shaped aluminum backing. To avoid wrinkling, the PVDF polymer was

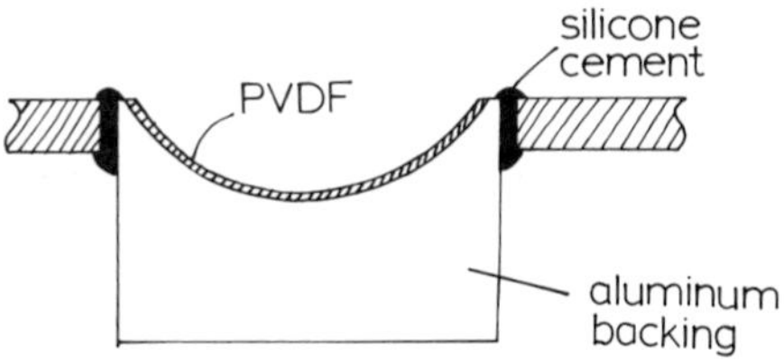

Fig. 2: Mounting arrangement of PVDF bowl transducers to simulate infinite baffle conditions.

cut into sectors, and applied to the spherical surface under pressure using a custom mould for each radius of curvature. The transducers were then mounted through one wall of the water tank, resulting in the configuration of Fig. 2. PVDF was chosen as the piezoelectric material to avoid spurious resonances, and to result in uniform vibrational amplitudes across the active area.

The pressure field was sampled using a 1 mm diameter PZT hydrophone. In these model experiments, the scans were undertaken at frequencies ≦ 300 kHz, leading to a resolution of ≦ $\lambda/5$ in the measurement in water. Note also that these frequencies were well below the resonance of the hydrophone or the PVDF films. The hydrophone was scanned under computer control, and the data collected using the same computer with a 12 bit A/D converter to sample the d.c. output of a gated peak detector. To simulate C.W. behaviour, a gated sine wave was used to drive the transducers. The gate of the gated peak detector was adjusted so as to sample the signal after C.W. behaviour had been established, yet before any reflections had

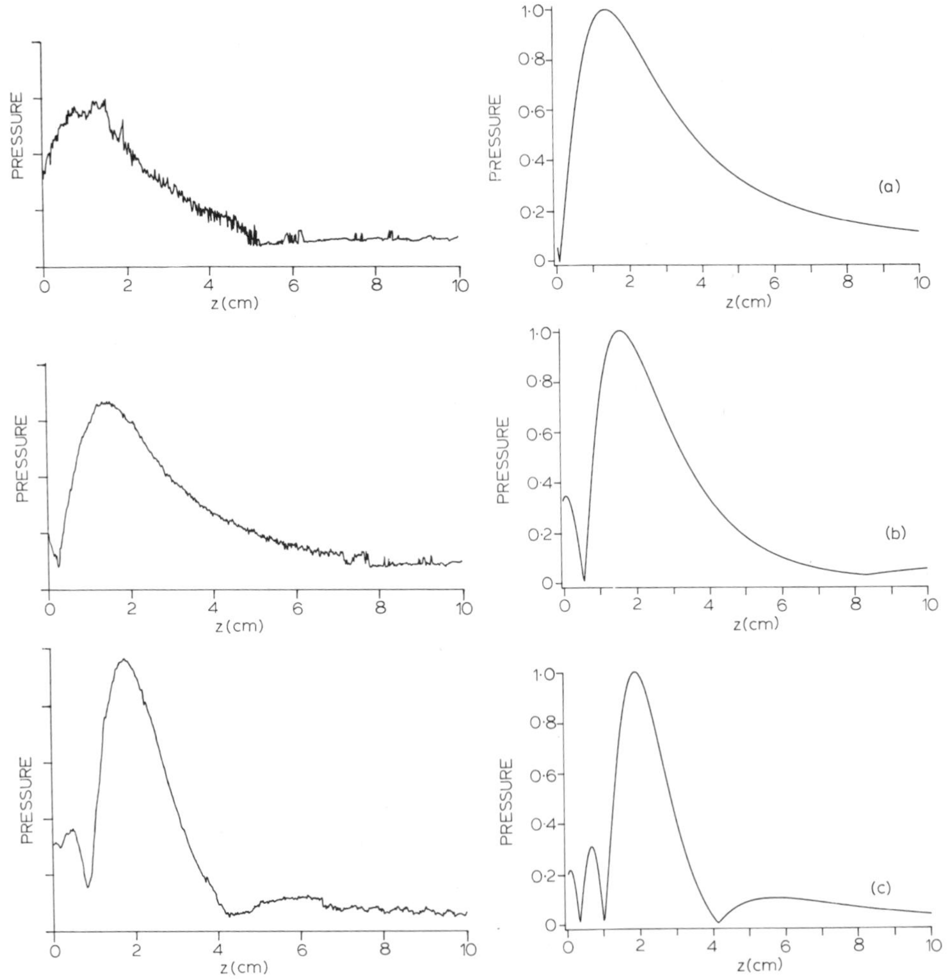

Fig. 3: Experimental (left) and theoretical (right) axial pressure fields for a spherical bowl, of aperture 5 cm and radius of curvature 3.2 cm, excited at (a) 120 kHz, (b) 180 kHz and (c) 300 kHz.

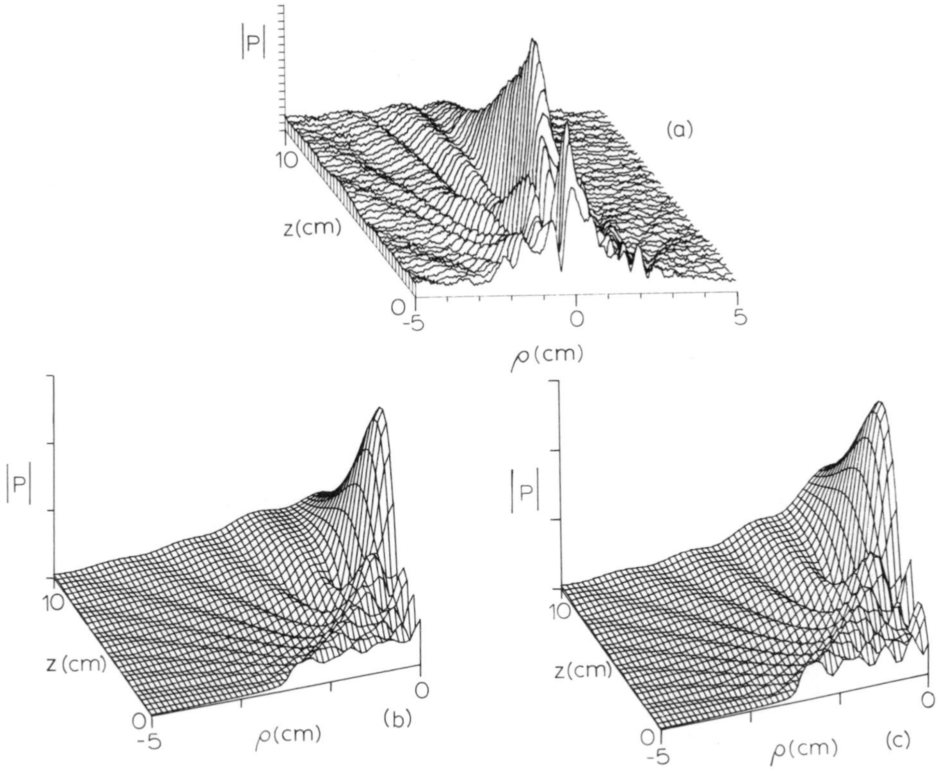

Fig. 4: 3D pressure fields for focusing transducers, assuming a 5 cm aperture and excitation at 300 kHz. (a) Experiment and (b) theory for a bowl with a 5 cm radius of curvature; (c) theory for a 20 element annular array, phased to focus on z = 5 cm.

returned from the back or sides of the tank.

RESULTS

(a) Comparison of Theory and Experiment for Spherical Bowl Transducers

Using the apparatus described above, axial pressure fields were first recorded for a PVDF bowl transducer with a 3.2 cm radius of curvature, excited at 120 kHz, 180 kHz and 300 kHz. The results are presented in Figs. 3(a) - (c) respectively, with experimental fields on the left and theory (Eq. (1)) to the right. It is evident that theory is reasonably consistent with experiment over the frequency range examined. Note that for the configurations shown, a 5 cm diameter at the bowl aperture corresponds to apertures of 4, 6 and 10 wavelengths respectively. It might be expected that for a bowl with a reasonable curvature, deviations from theory would result at low frequencies due to self diffraction effects. However, the above results demonstrate that the theories used were adequate, without the need for a further diffraction correction, even for relatively marked bowl curvatures.

The comparison may be extended to three dimensions, using a raster scan experimentally and surface integral techniques for theoretical predictions. The result is shown in Figs. 4(a) and (b) and again reasonable agreement is evident for a 5 cm radius of curvature at 300 kHz.

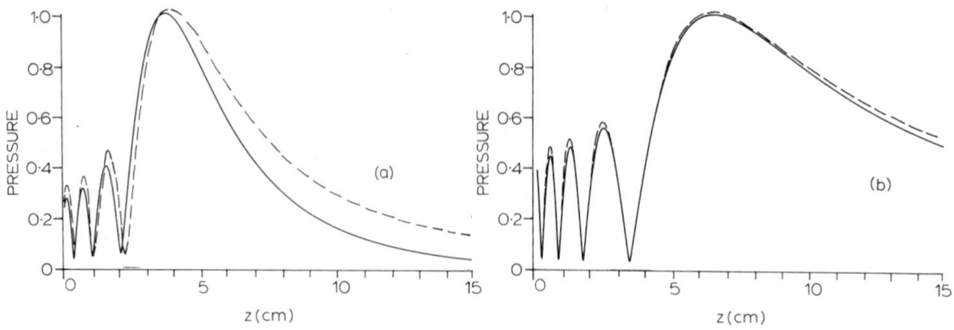

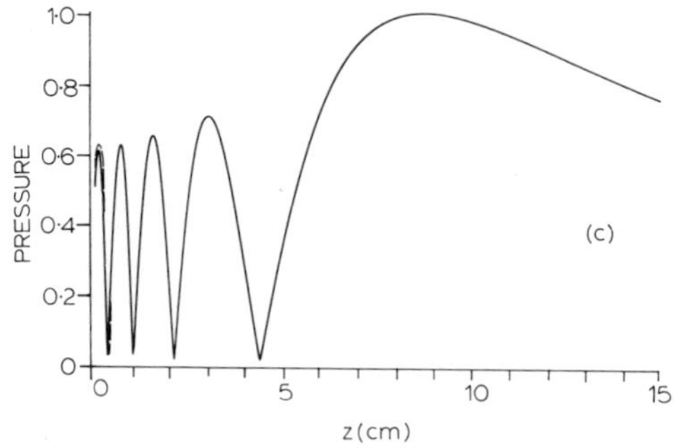

Fig. 5: Comparison of the theoretical axial fields at 300 kHz of bowls and annular arrays, focused on (a) 2a, (b) 4a and (c) 8a, where a = 5 cm. ——— bowls, ----- phased annuli.

(b) Theoretical Comparison of Bowls and Annular Arrays

The theoretical approach outlined above has been used to predict the 3D pressure field from a 5 cm diameter planar transducer, assumed to be fabricated from 20 concentric annuli with phases chosen for a 5 cm focus at 300 kHz. The predicted theoretical field is shown in Fig. 4(c), and comparison to theory and experiment for an equivalent bowl transducer (Figs. 4(a) and (b)) indicates that the pressure fields are very similar. A more detailed comparison may be made along the axis, for focusing at 2a, 4a and 8a axial (z) distances respectively, and the results are presented in Fig. 5. As can be seen, there are only minor differences between the two transducer types, with the axial maxima being slightly larger for the phased annular transducer. Reference to Fig. 4 will also indicate that side lobe levels are also somewhat higher.

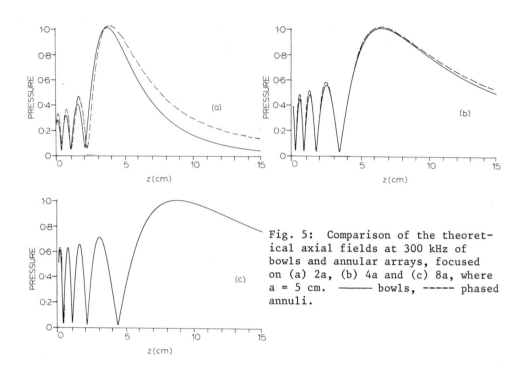

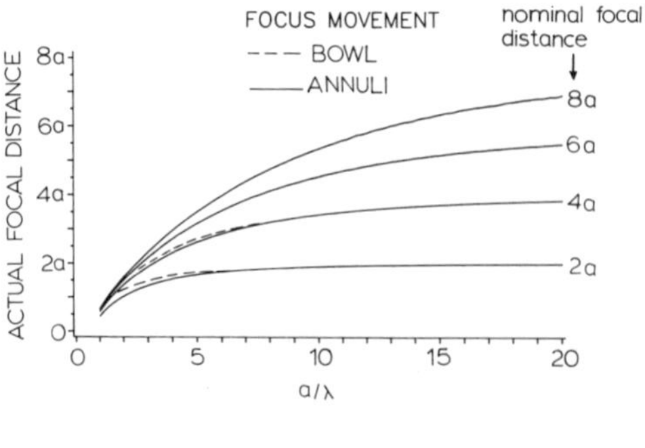

Fig. 6: Curves of actual focal length against a/λ, for a range of nominal focal distances.

A point of interest in Fig. 5 is that the focal region (the last axial maximum) is not at the radius of curvature of the bowl transducer, and also not at the expected position assuming the phases of the annular transducer. Indeed, the focus is always closer to the transducer than the nominal value. This, in fact, is an expected result. Consider the axial field of a plane piston, of 5 cm radius excited at 300 kHz. The last axial maximum would occur at $z \sim a^2/\lambda$, or at 12.5 cm. This is the extreme case, with an infinite radius of curvature. Hence, making the transducer more curved, or introducing appropriate phases to an annular transducer, will bring the focus closer to $z = 0$. This trend is plotted in Fig. 6, which shows the actual axial focal distance as a function of a/λ, for a series of nominal transducer foci. It will be seen that for all cases, the focus moves closer to the transducer as a/λ decreases (i.e. as the frequency or the transducer radius decreases). This also implies that the focus is closer to its intended value at smaller distances from the transducer.

A final criterion is the number of annuli in the phased array design. This has been investigated theoretically, with the results shown in Fig. 7. For a transducer with 2 annuli, Fig. 7(a), the width of the annuli is in excess of two wavelengths and little focusing is achieved. However, with 5 annuli (Fig. 7(b)), where the annular width is one wavelength, a reasonable focus is achieved. The quality of the focus increases with the number of annuli, and in fact little further improvement is achieved beyond 10. As technical considerations limit the number of phased annuli in any practical implementation, it is important that this factor be considered for each frequency and transducer radius of interest.

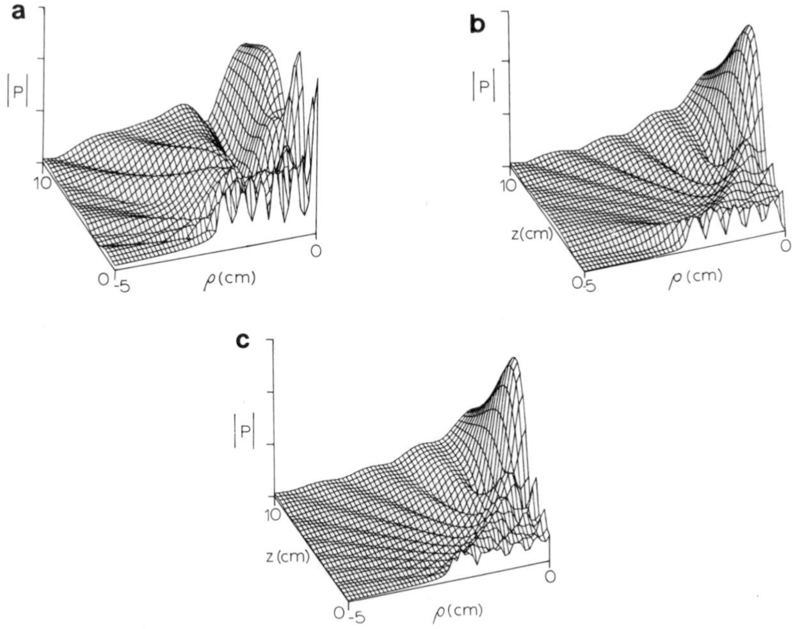

Fig. 7: Pressure variations at 300 kHz of phased annular transducers, focused on 2a with a = 2.5 cm. Transducer contained (a) 2, (b) 5, (c) 20 annuli.

CONCLUSIONS

A comparison has been presented between the fields radiated by spherical bowl and annular array transducers. It has been shown that for an annular array transducer with ten or more elements, the radiated pressure field is very similar to an equivalent bowl transducer. Both types have a focal position which deviates from the nominal design value, this deviation being marked as the nearfield/farfield boundary is approached. It has also been demonstration that the theoretical pressure fields of bowl transducers show a good correlation with experiment at a single frequency.

ACKNOWLEDGEMENTS

This work was funded by NSERC Canada.

REFERENCES

1. H.T. O'Neil, "Theory of focusing radiators", J. Acoust. Soc. Am. 21: 516 (1949).
2. B.G. Lucas and T.G. Muir, "The field of a focusing source", J. Acoust. Soc. Am. 72: 1289 (1982).
3. J.A. Archer-Hall and A.I. Ali Bashter, "The diffraction pattern of large aperature bowl transducers", NDT International April 1980: 51 (1980).
4. W. Swindel, R.B. Roemer and S.T. Clegg, IEEE Ultrasonics Symposium: 750 (1982).
5. A. Penttinen and M. Luukkala, "The impulse response and pressure nearfield of a curved ultrasonic radiator", J. Phys. D. 9: 1547 (1976).
6. M. Arditi, F.S. Foster and J.W. Hunt, "Transient fields of concave annular arrays", Ultrasonic Imaging 3: 37 (1981).
7. M.M. Goodsitt, E.L. Madsen and J.A. Zagzebski, "Field patterns of pulsed, focused, ultrasonic radiators in attenuating and non-attenuating media", J. Acoust. Soc. Am. 71: 318 (1982).
8. W.N. Cobb, "Frequency domain method for the prediction of the ultrasonic field patterns of pulsed, focused radiators", J. Acoust. Soc. Am. 75: 72 (1984).
9. A. Weyns, "Radiated field calculations of pulsed ultrasonic transducers: Part 2 spherical disc- and ring-shaped transducers", Ultrasonics 18: 219 (1980).
10. D. Guyomar and J. Powers, "Transient fields radiated by curved surfaces - Application to focusing", J. Acoust. Soc. Am. 76: 1564 (1984).
11. D.A. Hutchins, H.D. Mair, P.A. Puhach and A.J. Osei, "Continuous-wave pressure fields of ultrasonic transducers", J. Acoust. Soc. Am., accepted for publication.

COMPUTER SIMULATION OF BEAM PATTERNS FOR A SONAR

PHASED CYLINDRICAL ARRAY

A. Stepnowski, J. Szczucka* and L. Pankiewicz

Institute of Telecommunications
Technical University of Gdansk
80-952 Gdansk, Poland

* Institute of Oceanology PAS
81-712 Sopot, Poland

ABSTRACT

This paper presents the results of a computer simulation of beam patterns for horizontally phased cylindrical arrays which are utilized in a high-resolution beamforming sonar. The first part outlines the theory of cylindrical arrays. The successive parts give the numerical simulations of horizontal beam pattern, for various combinations of the array parameters: radius of curvature, angular aperture, transparency, element spacing, weighting and directivity. The sensitivity of beam pattern behaviour for changing the array parameters was identified, and thus practical indications for array design are obtained.

INTRODUCTION

The cylindrical arrays are widely used in current sonar systems for directional transmission and reception of acoustical signals. The conventional application of the cylindrical arrays in search sonars was aimed for constructing the projecting arrays, which have a wide active angular aperture and are driven in phase. By these means a high acoustic source level is achieved, over the broad, fan-shaped, horizontal beam, providing the power density in the ceramic does not exceed the acceptable level. The recent development in multibeam sonars, particularly employing efficient serial phase-shift beam-forming techniques, has extended the application of cylindrical arrays for synthesis of the narrow multiple beams. These arrays utilize horizontal phasing to compensate for geometrical phase-delays of incident or transmitted waveform, which result in a narrow beam formation. Additionally, the circular symmetry of the array provides the beam rotation around the array (scanning), if sequential sampling of its successive elements is performed[1].

The beam pattern of widebeam cylindrical arrays, driven in phase, can be simply calculated analytically, by summing the significant lower-order terms of its expansion into a series of Bessel functions[2]. On the contrary, for phased arrays this approach is not effective, and as a consequence the designing of these arrays is rather difficult, when based on analytical calculation of their beam pattern[3]. The presented numerical approach gets round this inconvenience, and allows rapid monitoring of beam pattern behaviour, under an arbitrary choice of array parameters.

BEAM PATTERNS OF CYLINDRICAL ARRAY

Consider a cylindrical array composed of M circular sub-arrays each consisting of N equispaced elements, positioned in (x,y,z) coordinate system, as shown in Fig. 1.

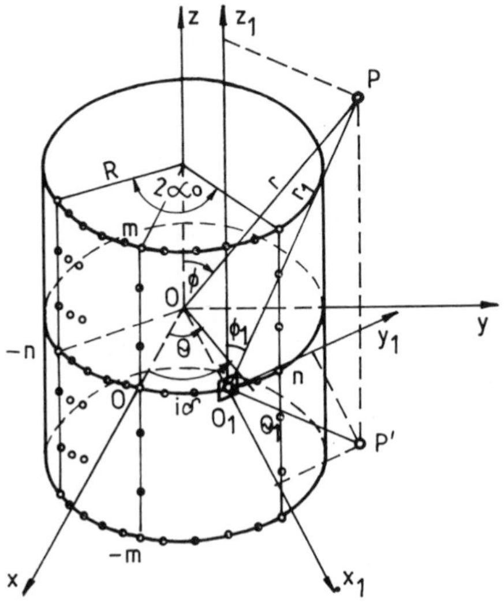

Fig. 1. Geometry for a cylindrical array.

Let the horizontal "rows" and vertical columns contain odd numbers of elements, i.e., M = 2m + 1 and N = 2n + 1. The horizontal element spacing is d, the vertical spacing is d_v. The angular spacing is $\delta = d/R = \alpha_0/n$, where R is the radius of curvature, and $2\alpha_0$ is the angular aperture of the array. Assume a single element as a uniformly vibrating small rectangular aperture in an infinite rigid cylindrical baffle. The farfield pressure due to such a source at observation point P, assuming unit particle velocity, is[3]

$$P_{il}(\Theta,\phi) = \frac{2\rho c b}{\pi}\,\frac{e^{-jk(r-ld_r\cos\phi)}}{r\sin\phi}\,\frac{\sin(kb/2\cos\phi)}{kb/2\cos\phi}\times \left(\frac{\alpha_1}{H_q^{(1)'}(kR\sin\phi)}+\sum_{q=1}^{\infty}\frac{\sin(q\alpha_1)\cos q(\Theta-i\delta)}{qH_q^{(1)'}(kR\sin\phi)}\,e^{-jq\pi/2}\right) \tag{1}$$

where c is the acoustic wave impendance, $k = 2\pi/\lambda$ is the wave number, b is the vertical dimension of element, $2\alpha_1$ is the angular width of element, $H_q^{(1)'}(\cdot)$ is the derivative of Hankel function of the first kind, and r, Θ, ϕ are spherical coordinates of point P. The beam pattern of the entire array can be written in terms of these elementary pressure distributions a:

$$B(\Theta,\phi) = \frac{p(\Theta,\phi)}{p(\Theta_o,\phi_o)} = \frac{\sum_{i=-n}^{n}\sum_{l=-m}^{m} \tilde{w}_{il}\, p_{il}(\Theta,\phi)}{\sum_{i=-n}^{n}\sum_{l=-m}^{m} \tilde{w}_{il}\, p_{il}(\Theta_o,\phi_o)} \tag{2}$$

where $\tilde{w}_{il} = w_{il}e^{j\arg \tilde{w}_{il}}$ is the complex amplitude (weight) of "il"-th element and (θ_o, ϕ_o) is the reference direction, MRA.

If the individual elements have the same beam patterns b_{il} (θ_l, ϕ_l), in their own coordinate system (x_l, y_l, z_l) all elementary pressure distributions are given as: p_{il} $(\theta_l, \phi_l) = b_{il}$ $(\theta_l, \phi_l) \times p(0, 0)$, and resultant array beam pattern can be obtained by summing p_{il}, transformed to the common coordinate system. Relating the angles θ_l, ϕ_l with the axes x_l, y_l, z_l (see Fig. 1) we have:

$$\begin{aligned} x_l &= \sin \phi_l \cos \theta_l \\ y_l &= \sin \phi_l \sin \theta_l \\ z_l &= \cos \phi_l \end{aligned} \tag{3}$$

and by the parallel translation of coordinate system (x_l, y_l, z_l) to the common origin 0 we obtain:

$$\begin{aligned} x_l &= x \cos i\delta + y \sin i\delta \\ y_l &= y \cos i\delta - x \sin i\delta \\ z_l &= \cos \phi \ . \end{aligned} \tag{4}$$

The comparison of the relations (3) and (4) gives the required transformation of variables:

$$\begin{aligned} \theta_l &\Rightarrow \theta - i\delta \\ \phi_l &\Rightarrow \phi \end{aligned} \tag{5}$$

what allows to write formula (2) in simpler form:

$$B(\Theta,\phi) = \frac{\sum_{i=-n}^{n}\sum_{l=-m}^{m} \tilde{w}_{il}\, b_{il}(\Theta - i\delta, \phi)\, e^{-jkR\sin\phi\cos(\Theta-i\delta) - jkld_r\cos\phi}}{\sum_{i=-n}^{n}\sum_{l=-m}^{m} \tilde{w}\ b\ (\Theta_o - i\delta, \phi_o)\, e^{-jkR\sin\phi_o\cos(\Theta_o-i\delta) - jkld_r\cos\phi_o}} \tag{6}$$

Phased arrays

The array curvature introduces the geometrical phase delays $\Delta\Psi i$ for successive elements ($i\delta$ - directions). To compensate these for in the reference direction (θ_o=0, ϕ_o = 90°), the "steering" phase-delay is required

$$-\Delta\psi_i = k\Delta x_i = kR\cos i\delta \tag{7}$$

which is formally equivalent to put the arg $\tilde{w}_{il}$ equal to the value of exponent in the denominator of formula (6), viz.

$$\arg \tilde{w}_{il} = kR\sin\phi_o\cos(\Theta_o - i\delta) - kld_r\cos\phi_o = kR\cos i\delta \quad \begin{pmatrix} \Theta_o = 0 \\ \phi_o = 90^\circ \end{pmatrix}$$

The resultant beam pattern of the cylindrical array, utilizing horizontal phasing, is then:

$$B(\Theta,\phi)=\frac{\sum_{i=-n}^{n}\sum_{l=-m}^{m} w_{il}\, b_{il}(\Theta-i\delta,\phi)\, e^{-jkR[\sin\phi\cos(\Theta-i\delta)-\cos i\delta]\; jkld_r\cos\phi}}{\sum_{i=-n}^{n}\sum_{l=-m}^{m} w_{il}\, b_{il}(-i\delta,\pi/2)} \tag{8}$$

The horizontal beam pattern of phased array is easily obtained by substitution of angle $\phi = 90^o$ to formula (8)

$$B(\Theta)=\frac{\sum_{i=-n}^{n} w_i\, b(\Theta-i\delta)\, e^{jkR[\cos(\Theta-i\delta)-\cos i\delta]}}{\sum_{i=-n}^{n} w_i\, b(-i\delta)} \tag{9}$$

The derived formulae neglect the diffraction effects and are valid for transparent arrays only (i.e. if array surface does not disturb free-field conditions). For nontransparent arrays (e.g. elements in rigid baffle) the additional term $G_i(\theta)$ must be included, accounting for number of elements reduction ("shadowing"). And finally, the beam pattern formula, adequate as well for transparent, as for nontransparent arrays is:

$$B(\Theta)=\frac{\sum_{i=-n}^{n} w_i\, G_i(\Theta)\, b(\Theta-i\delta)\, e^{-jkR[\cos(\Theta-i\delta)-\cos i\delta]}}{\sum_{i=-n}^{n} w_i\, G_i(\Theta)\, b(-i\delta)} \tag{10}$$

where

$$G_i(\Theta)=\begin{cases}1 & \text{for } i \leqslant n\,\dfrac{90^\circ-|\Theta|}{\alpha_o}\\[2ex] 0 & \text{for } i > n\,\dfrac{90^\circ-|\Theta|}{\alpha_o}\end{cases}$$

Phased cylindrical array versus linear array

Consider the horizontal beam pattern of phased array (9), applying the change of variables $y_i = R\sin i\delta$, shown in Fig. 2.

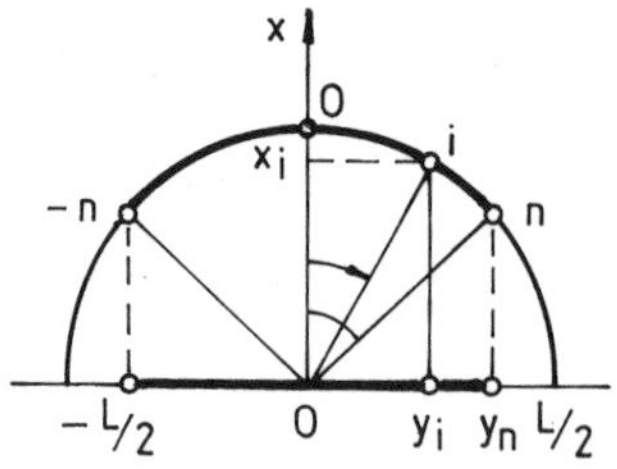

$$x_i = R\cos i\delta$$
$$y_n = L/2 = r\sin\alpha_o$$

Fig. 2. Transformation of phased circular array to linear aperiodic array

This allows us to rewrite the formula (9) to the following form:

$$B(\theta) = \frac{\sum_{i=-n}^{n} \tilde{w}(y_i)\, e^{-jky_i \sin\theta - j\alpha(y_i)}}{\sum_{i=-n}^{n} \tilde{w}(y_i)\Big|_{(\theta=0^\circ)}} \tag{11}$$

which represents the beam pattern of a discrete linear array of length $L = 2R \sin \alpha_0$, having the aperture distribution, complex weighting, $\tilde{w}(y_i) = w(y_i) \exp(-j\alpha(y_i))$

where $w(y_i) = \dfrac{w(i\delta)\, b(\theta - i\delta)}{\cos i\delta}$ is the amplitude distribution,

$\alpha(y_i) = k(1-\cos\theta)(R^2 - y_i^2)^{1/2}$ is the phase distribution.

This equivalent linear array has nonuniform element spacing and for this reason does not demonstrate the real grating lobes, but rather quasi-grating lobes, similarly, as it does the phased cylindrical array.

COMPUTER CALCULATION

The numerical simulation of the horizontal beam pattern for a phased cylindrical array was carried out. The amplitude beam pattern, given as a modulus $|B(\theta)|$ of the complex beam pattern (9), has been calculated for various choices of the array parameters[4]. Three values of the angular aperture (60°, 90°, 120°) and two values of the radius of curvature (5λ, 12λ) combined with six values of the element spacing (0.3, 0.5, 0.6, 0.7, 0.8, 1.0) were chosen as adequate examples for these computations. Directivity of the array elements was modeled by the beam patterns of three types of the elementary sources and one line source, as shown in Fig. 3, for transparent and nontransparent arrays. Two kinds of the amplitude weighting were chosen $w_i(\theta) = 1$ (unweighted array) and $w_i(\theta) = \cos\theta$. The beam patterns have been calculated using 90 points along the θ axis, and were normalized to 0 dB level for MRA ($\theta = 0°$) direction.

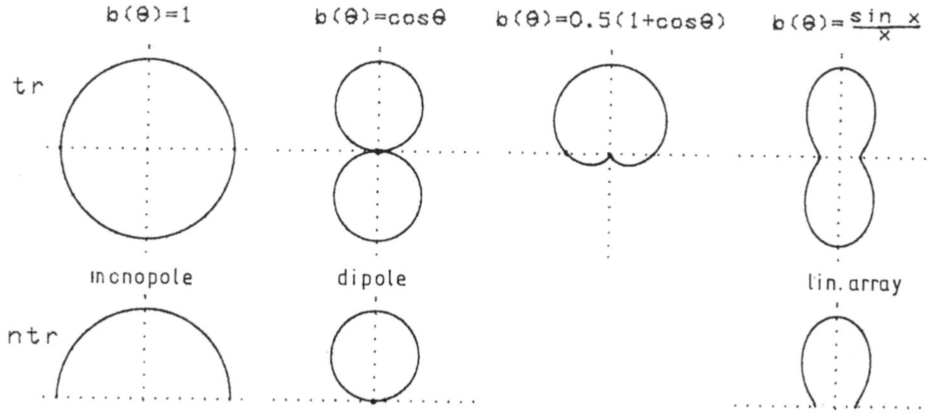

Fig. 3. Directivity patterns for array elements used in simulation (tr-transparent, ntr-nontransparent array).

Figures 4 and 5 show two sets of the beam patterns, calculated for an acoustically transparent array of nondirectional elements, using two chosen pairs of the array radius and its angular aperture, with element spacing as a parameter. All plots demonstrate two regions: the first, covering the main lobe and the side-lobes, whereas the second - the quasi-grating lobe. As may have been expected the beamwidth and the side-lobe number and position depend on the effective array aperture, however, there is no apparent difference in side-lobes level for various aperture and element spacings. The most interesting features are observed in the second region. As spacing increases, i.e. the spatial sampling rate decreases, the quasi-grating lobe tends to arise for lower spatial frequencies (small angles) and exhibits the multi-lobe structure, despite an array aperture. These results in an increased quasi-grating lobe level for visible angles ($|\theta| < 90^o$) when greater spacing is used.

Figure 6 gives the plots of the beam pattern of transparent array, calculated for a different elemental directivity pattern - as specified in Figure 3. In general, there is no remarkable influence of this parameter on the overall beam pattern performance, however, some minor effects are observed. If one considers the side-lobe level, the best ratio can be found using an elemental beam pattern of the type $b(\Theta) = \cos(\Theta)$, which corresponds to a dipole source. There is also no apparent difference in the quasi-grating lobe region, except of the second plot (i.e. cardioidal elemental directivity pattern) which does not demonstrate the multi-lobe form.

Figures 7 and 8 show the similar plots as in Figure 6, while additionally the different element spacing was introduced. Both previously mentioned effects are observed in a more clear form. The extra reduction of the quasi-grating lobe level, for cardioidal elemental directivity pattern is found, particularly when small element spacings were used.

The comparison of the array beam pattern for transparent and nontransparent array is shown in Figure 9. The improvement of the beam pattern performance dure to quasi-grating lobe level for the latter case is evident, especially for a smaller element spacing.

Figure 10 shows the sample of the array beam pattern, calculated for the weighted and unweighted arrays, compared to a beam pattern of an equivalent linear array. The similarity between the beam pattern of a cosine wighted cylindrical phased array and that of linear array is demonstrated, for a lower range of spatial frequencies - covering the main lobe and the several first side-lobes.

CONCLUSION

The use of the presented simple numerical simulation approach seems to be of considerable value in evaluating the beam patterns for cylindrical phased sonar arrays. It enables rapid calculation of beam patterns and monitoring of its performance for a quite great choice of array parameters. Several interesting features of the considered beam patterns, which have not been reported in the literature were observed. The first, and of the major interest, is one related to the element spacing. As seen from Figures 4 and 5, if spacing does not exceed the value of 0.7λ, the quasi-grating lobe level for a visible region can be maintained at a tolerable level of the first side-lobe. This was checked and confirmed for a great choice of array parameters having of practical interest. Thus, some practical analog of sampling theorem in spatial frequency domain, for the phased cylindrical arrays, can be derived, which requires spatial sampling interval $d/\lambda = 0.7$.

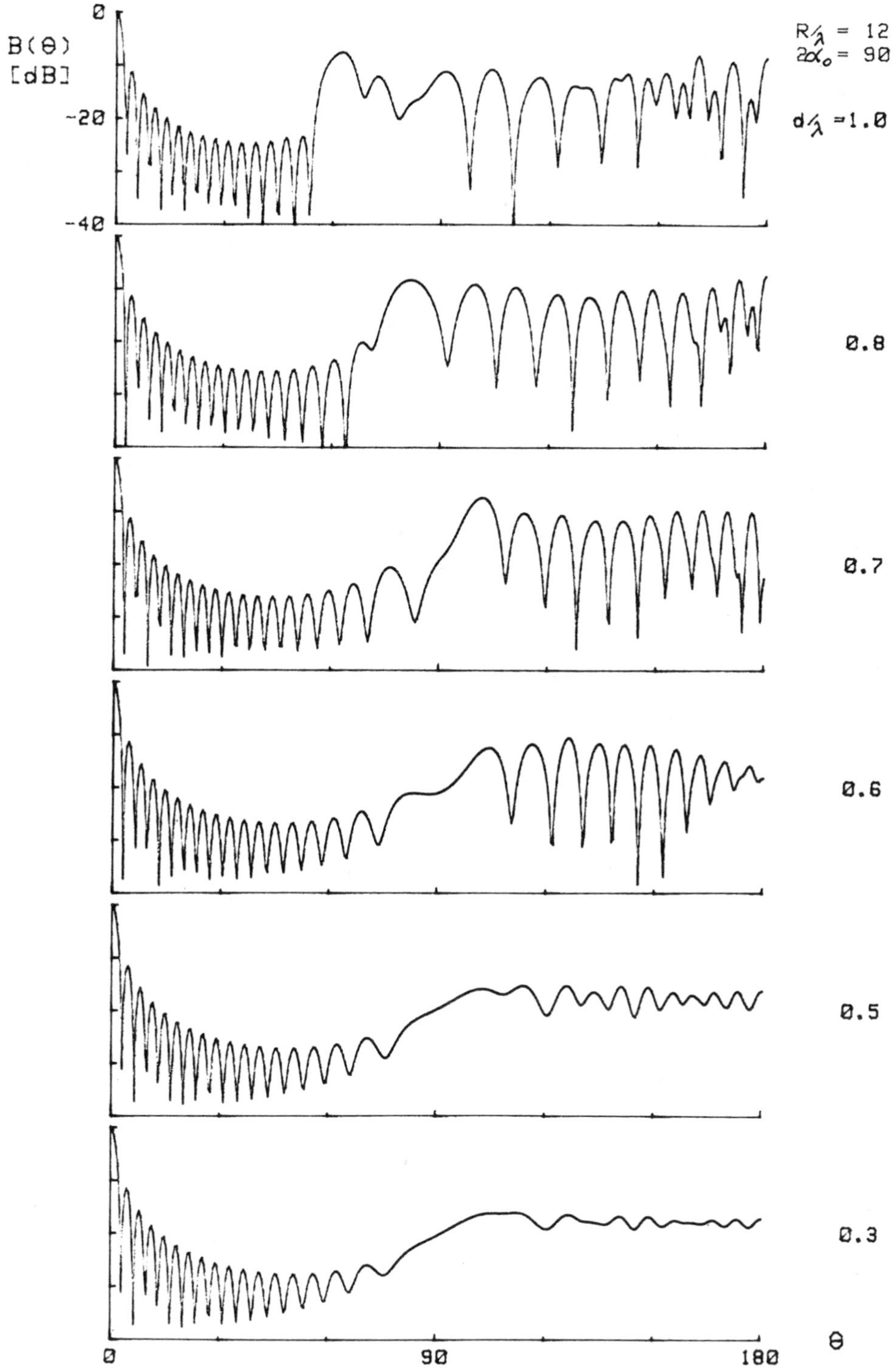

Figure 4. Horizontal beam patterns for phased cylindrical array with $R/\lambda = 12$, $2\alpha_0 = 90^\circ$, $b(\Theta) = 1$, using different element spacing $d/\lambda = 0.3, 0.5, 0.6, 0.7, 0.8, 1.0$.

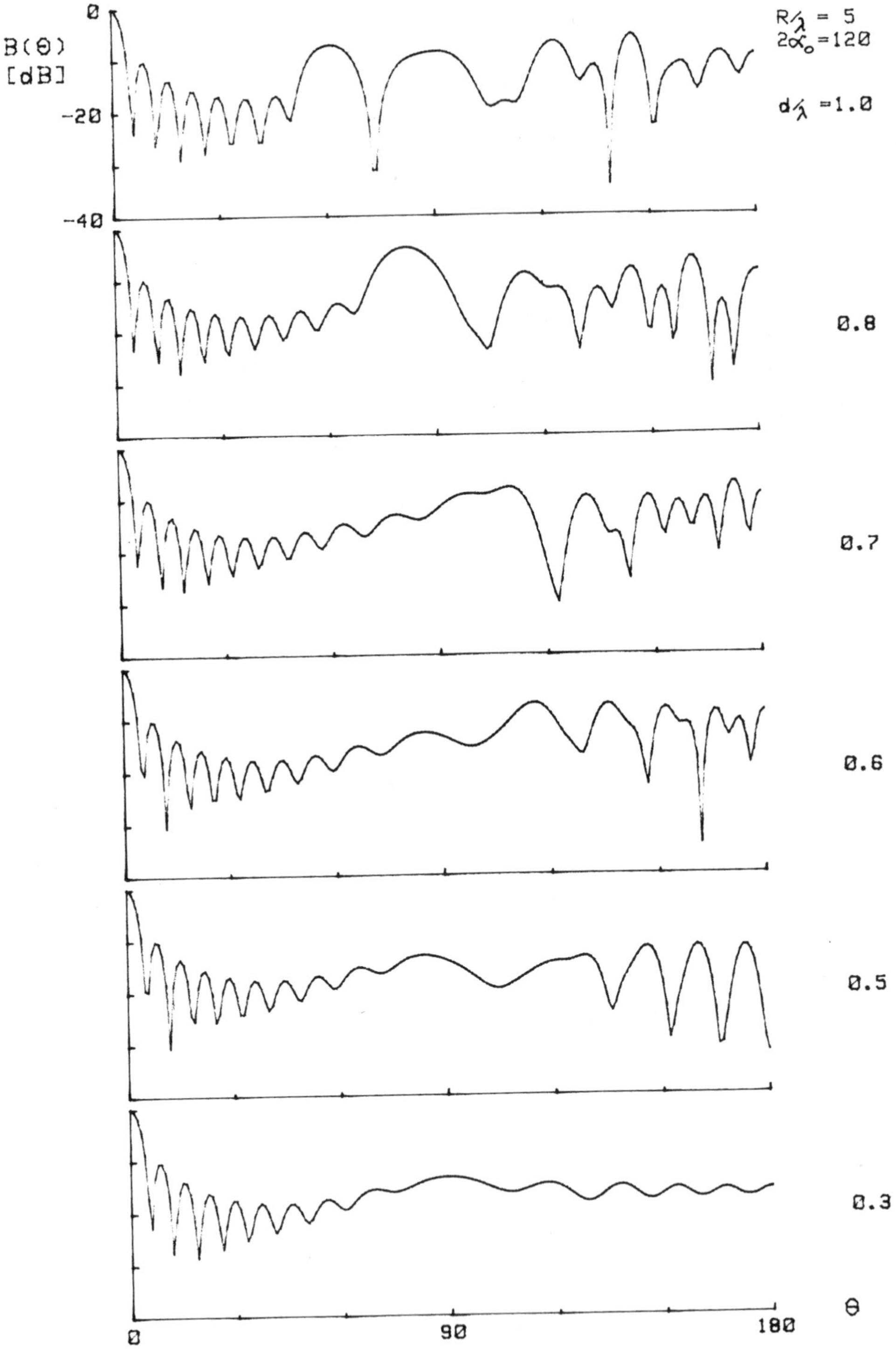

Figure 5. As Figure 4 but for $R/\lambda = 5$, $2\alpha_0 = 120^\circ$.

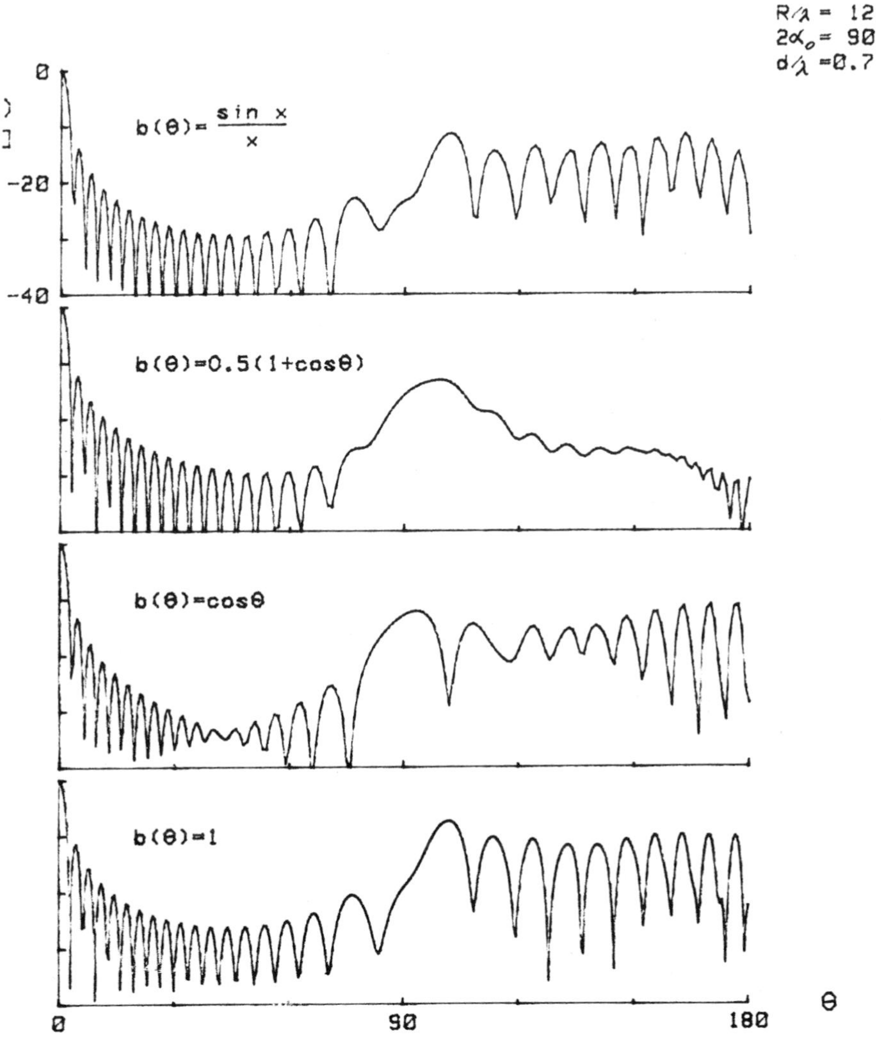

Figure 6. As Figure 4 but for $^d/\lambda$ = 0.7 and for different elemental directivity pattern b(Θ).

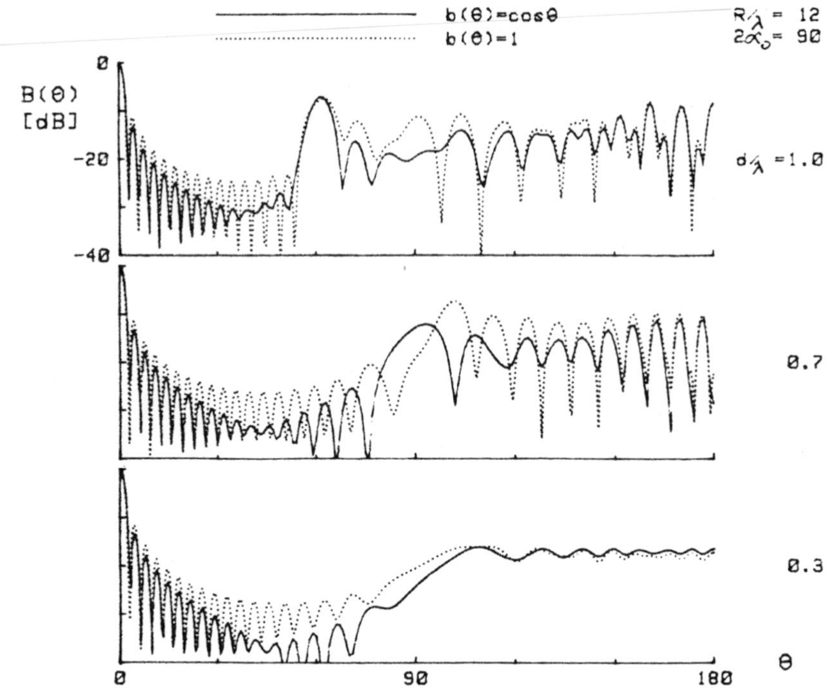

Figure 7. As Figure 4 but for b(Θ) = cos Θ and b(Θ) = 1 using $^{d}/\lambda$ = 0.3, 0.7, 1.0.

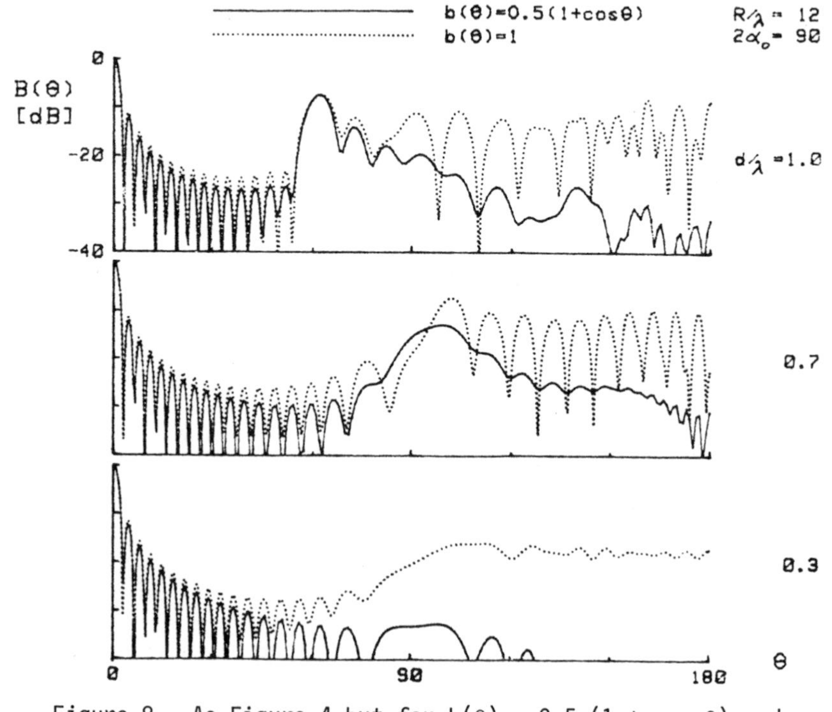

Figure 8. As Figure 4 but for b(Θ) = 0.5 (1 + cos Θ) and b(Θ) = 1, using $^{d}/\lambda$ = 0.3, 0.7, 1.0.

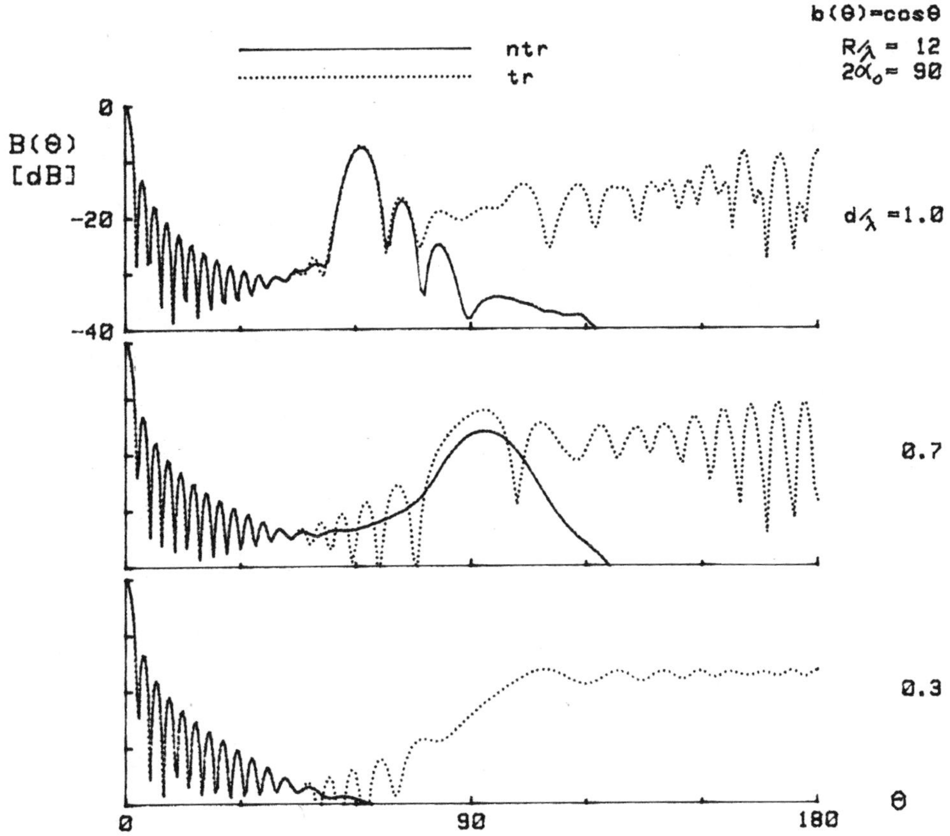

Figure 9. As Figure 4 but for transparent (tr) and non-transparent (ntr) array, with $b(\Theta) = \cos\Theta$, using $d/\lambda = 0.3, 0.7, 1.0$.

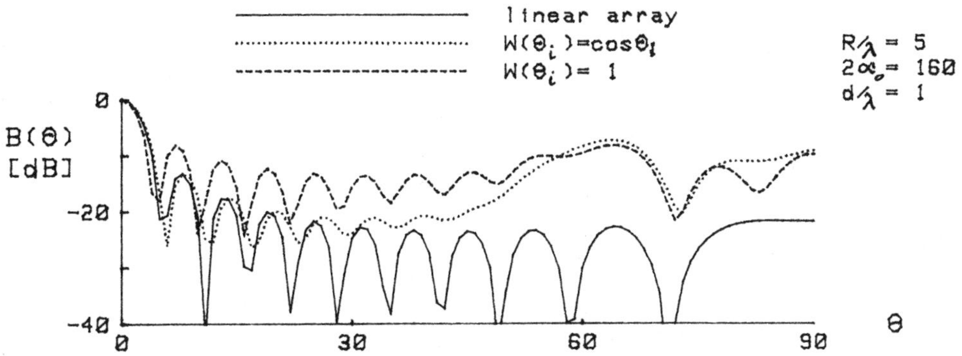

Figure 10. As Figure 4 but for $R/\lambda = 5$, $2\alpha_0 = 160°$, $d/\lambda = 1$, and for weighting $W(\Theta_i) = \cos\Theta_i$, $W(\Theta_i) = 1$, in comparison with beam pattern of linear array.

The second observation derived from Figures 6 to 8 allows us to conclude a rather weak influence of the elemental directivity on the overall beam pattern of the array in the first region, whereas this influence becomes quite remarkable in the quasi-grating lobe region. However, it should be noted that not all simulated elemental directivity patterns have practical interest, due to their dependence on the actual boundary condition on the array surface. Thus, the possibilities of the beam pattern formation by means of the elementary aperture directivity are rather confined. One the other hand, most sonar arrays can be considered as a rigid nontransparent aperture, which allows us to expect reduction of quasi-grating lobes as seen from Figure 9.

The aperture weighting, although not examined extensively, shows the expected reduction of the side-lobes (but not of the quasi-grating lobe) at the cost of the widening of the main lobe. The chosen weighting allows the cylindrical array beam pattern to approach that of a linear array, due to the relation of its weighting function to the former one (see Formula 11). However, this similarity holds only for lower spatial frequencies and becomes apparent for wider angular aperture.

REFERENCES

1. A. Stepnowski and R. Salamon, High resolution sampled phase-delay beamformer, Acoustical Imaging 14, Plenum Press, New York (1985)

2. E. Skudrzyk, "Foundation of Acoustics", Springer-Verlag, Vien (1972)

3. M.D. Smaryshev, "Napravelennost gydroakusticheskyh antenn", Sudostroyenye, Leningrad (1973)

4. A. Stepnowski, "Modelling of the beam patterns for cylindrical arrays" (in Polish), Proc. of XXXII Open Seminar on Acoustics, 2:229, Cracov (1985)

SOME ASPECTS OF TRANSDUCER DESIGN BY FINITE ELEMENT TECHNIQUES

J. R. Dunn

Dept. of Electronic & Electrical Engineering
University of Birmingham
P.O. Box 363, Birmingham B15 2TT, England

ABSTRACT

Finite element techniques have been applied to the design of piston transducers for the range 20 to 50 kHz. The head should be light and, ideally, rigid for maximum bandwidth, and the bending of a real head is minimised if it is driven by the piezo-electric stack over an annulus between one half and two thirds of the diameter of the head. The loss in electromechanical coupling coefficient can also be demonstrated by considering the behaviour if imaginary materials with very high stiffness are substituted for the real head and tail materials, and this loss can be significant. Improved versions of 30 and 50 kHz transducers have been designed and tested satisfactorily, showing improved bandwidth and coupling coefficient.

INTRODUCTION

This paper is concerned with the design of low frequency transducers of the type generally known as piston or 'tonpilz' elements, in which the basic vibrating structure consists essentially of two masses and an intervening spring. The range of resonant frequencies for this type of element is around 2 to 60 kHz, or with some difficulties in the practical design up to 100 kHz. One mass is the head which is in contact with the water and which is usually about one half to three quarters of a wavelength in water across, and the other mass acts as a countermass and it is usually at least as heavy as the head, frequently much heavier for wider bandwidth. The spring is a piezo-electric stack which provides the essential coupling between the mechanically and electrically stored energies; alternatively it could be of piezo-magnetic material, for which the analysis would be very similar to that described here for the piezo-electric material. The stack usually has a smaller cross-section area that either the head or the tail, with the aim of maximising the bandwidth by minimising the mass in front of the nodal plane. One more, usually essential, item in the design is a prestressing bolt through the complete stack, whereby under conditions of high power electrical drive the piezo-electric material remains in compression, but for the purposes of the present analysis the bolt has been ignored, which is a fair approximation if it is of minimum weight and stiffness and decoupled from the tail by a compliant washer. The design criteria which are particularly addressed in this paper are for a wide bandwidth and a high coupling coefficient. The former may be dictated by the needs of the sonar system for which the transducer is being designed, but it is also advantageous in permitting a wider tolerance on the resonant frequency, and the high coupling is desirable from the point of view of the electrical terminating conditions. Traditional design methods, as discussed in the next section, are in many cases adequate for producing reasonably successful designs, but these can be refined by the application of finite element methods; in other cases with more demanding specifications, such methods become virtually essential for fine-tuning the theoretical design.

THE CLASSICAL DESIGN METHOD

The "Lumped Element" Approach

The basis of this is the assumption that the transducer can be regarded simply as two masses separated by a spring, in which the masses are rigid and the spring has no mass, with one of the masses coupled to the radiation load. The individual components are designed most readily by considering the equivalent circuit in terms of electrical parameters, wherein masses are modelled as inductances, springs (compliances) as capacitances and the radiation load as a resistance with an inductive component due to the dimensions of the head being small in wavelengths. From this the actual masses and the compliance can be calculated using electrical network theory, including filter design theory (Morris 1971), and a minor adjustment can be made to account for the mass of the piezo-electric stack. A refinement to the equivalent circuit permits the inclusion of the prestressing bolt, which may have a significant effect on the longitudinal compliance. Further refinements are possible by treating each section as a transmission line instead of a lumped mass or compliance and using the same theory that is used in the analysis of inter-connected electrical transmission lines; this can be more readily applied to the tail, which usually has a constant cross-section area and may be as long as one eighth of a wave-length. A fundamental assumption is made that the cross-section is small and that the behaviour is essentially one dimensional, i.e. that there is no bending in the transverse plane. Purely radial strain is brought into account automatically by using the appropriate value of Young's modulus in the calculations of longitudinal compliance. It is primarily in order to take properly into account the bending of the head that the finite element method is being applied in the design process, but it also greatly simplifies the inclusion of the longitudinal strain in the head, which usually has a non-uniform cross-section. However this simple design method would still be used for the initial design.

Limitations of the Simple Approach

The principal limitation of the "lumped element" design procedure is that it does not take into proper account the compliance of the head, particularly in bending. The usual recommendation is that the lowest bending mode resonance of the head should be sufficiently removed upwards from the working frequency, but the formulae used in estimating this resonance do not properly take into account the way in which the head is actually driven by the piezo-electric stack, nor do they indicate directly how much the head bends at the resonant frequency of the whole transducer. It is more difficult to take into account the effective compliance of the head, which affects the resonant frequency, but it probably causes significant uncertainties in the design only for the higher frequencies (above 40 kHz), for which the stack is likely to be very short and stiff, and for head diameters around three quarters of a wavelength in water. The corresponding difficulties at the tail end are not likely to be as important, since the tail is usually more massive than the head and the departures from the ideal (or simplified) behaviour are likely to have relatively unimportant effects on the parameters of the complete transducer as compared with the approximations made for the behaviour of the head; thus the tail can be analysed to a sufficient degree of accuracy on a transmission line basis. An additional factor is that the abrupt changes in cross section area, associated with designing for wide bandwidth, could lead to undesirable concentrations of stress and it would be useful to be able to predict these.

THE FINITE ELEMENT METHOD

General remarks

The essence of the finite element method is the division of an object into a large number of elementary units, the behaviour of each of which can be described in a simple way for the static case, by the relationships between the displacements of certain points (the nodes) and the internal and external forces at these points, together with the assembly of these individual relationships into a global matrix of equations describing the behaviour of the complex body. The extension to dynamic situations with sinusoidal drive is made in a straightforward way by including terms representing the inertia at each node, these having the dimensions of mass times frequency squared, and the dissipation of energy can be included by the addition of terms proportional to frequency and having a 90 degree phase shift. The solution in the latter case is more complicated, since the equations now include real and imaginary terms.

Broadly speaking there are two classes of situation, the two dimensional and the three dimensional; axisymmetric bodies can be analysed by an extension to the two dimensional case, and it is these which are considered in detail in this paper. The elements are defined geometrically by their shape and the number of nodes in each; for the present analysis triangular elements with three nodes were used, implying a constant mechanical strain throughout each element, but computer programs have been developed for six noded triangles, in which the strain may have a linear variation and in this case the real stress/strain distribution can be modelled more accurately. The programs used for this analysis included neither the piezoelectric coupling nor the radiation of power into the water; therefore the analysis is applicable only for the prediction of the axisymetric modes of vibration and of the resonant frequencies. Because of the relatively coarse mesh of elements used, only the lowest mode of vibration has been investigated, but it could be a worthwhile exercise in some cases to investigate the next higher resonance in case it occurs within the range of frequencies over which the transducer may be used. The modelling of the electromechanical coupling is circumvented by using the piezo-electric parameters appropriate to the "constant electric field" condition, corresponding to voltage excitation and hence equivalent to the short-circuited condition, and by applying balanced axial forces on the two planes defining the end of the stack. This is admittedly an approximation, but it has permitted an investigation into the critical area of head resonances and it gives fairly accurate predictions of unloaded resonant frequency, given that there are in any case uncertainties in allowing for the compliance of the bonds in the stack.

Applications to Piston Transducers

The first investigation (Dunn 1984) was into the flapping resonances of isolated heads of a typical tapered shape on which the drive was concentrated near the centre, with a varying ratio of diameter to thickness, for comparison with an analytical formula. The excitation was modelled by an axial force round the outer circumference, and the inner face where the

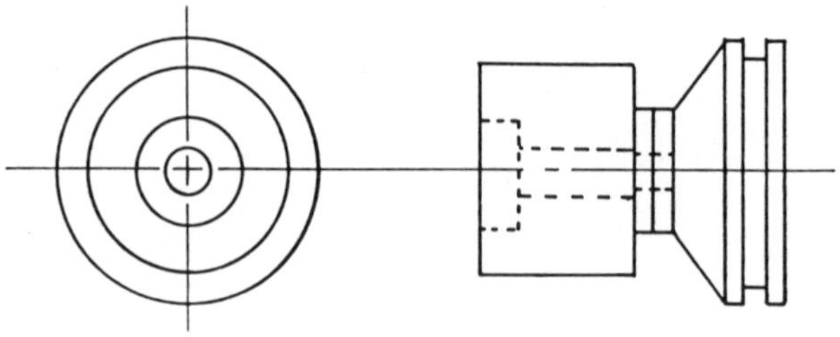

Fig. 1 Original design for 50 kHz

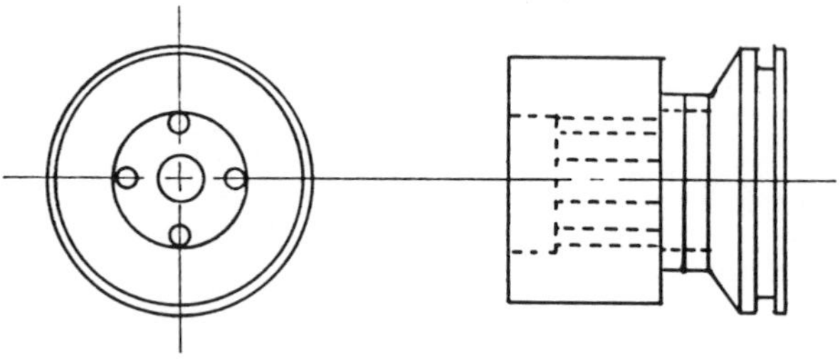

Fig. 2 Improved design for 50 kHz

driving stack would normally be located was either blocked or free, these two conditions giving slightly different resonances, which coincided for a particular value of diameter/thickness; however neither condition properly represents the conditions under which the head actually operates. For the range of head diameters commonly used, up to 0.90 wavelengths, the finite element method gave for either terminating condition higher resonant frequencies for the same shape than the analytical method.

The more significant investigation was into means of optimising the design so as to reduce the amplitude of the flapping mode at the resonant frequency of the transducer without regard to the frequency of the flapping resonance while at the same time minimising the weight of the head. The importance of this lies in the fact that with a very heavy tail mass the bandwidth is more or less inversely proportional to the mass of the head, and in the original design which had the drive near the centre a reduction in the thickness of the head would have lowered the flapping resonant frequency, as calculated by the previous approach, too far. In order to reproduce the proper driving conditions on the head, the complete transducer was modelled, i.e. including the stack and the tail, but not however the pre-stressing bolt. The variable parameter was the mean radius of the annulus over which the head was driven, the area of this annulus remaining constant, so that the stiffness of the stack, and hence the resonant frequency of the transducer, remained nearly constant, apart from variations due to changes in the apparent stiffness of the head. It was found that there was an optimum mean radius for which the ratio of the maximum and minimum axial deflections of the outer face of the head (i.e. the normal radiating face) was nearest to unity. Additional data which could be extracted from the calculations are the deformations of the head and tail, from which it would be possible to calculate the elastically stored energy for comparison with that in the driving stack, and from this the loss in coupling coefficient below the ideal could be derived. The coupling coefficient may be critical for very wide bandwidths, and any reduction below the figure for the piezo-electric material alone is to be avoided; thus it

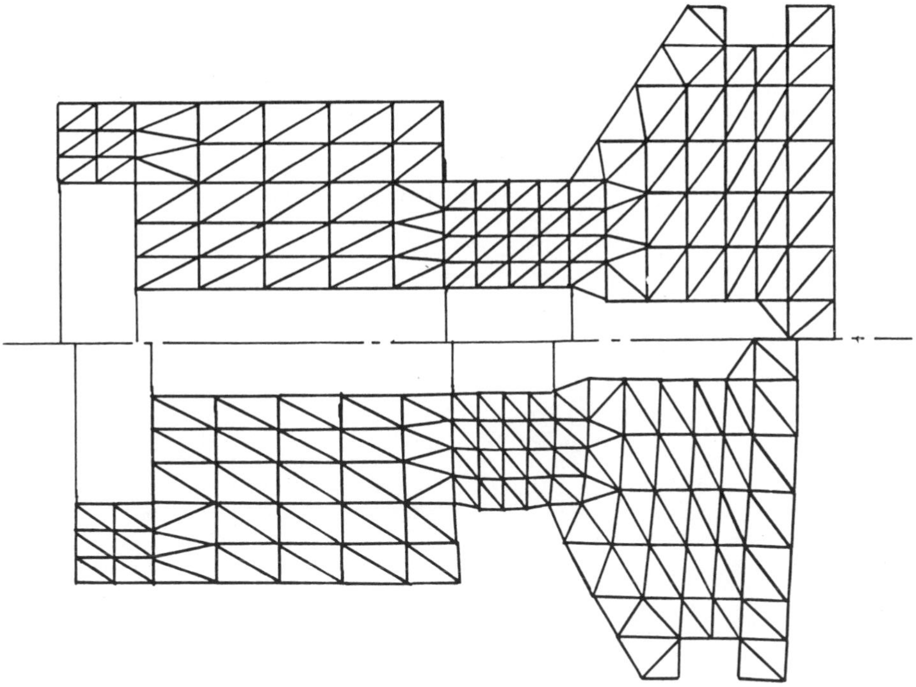

Fig. 3 Displaced finite element mesh for the original design

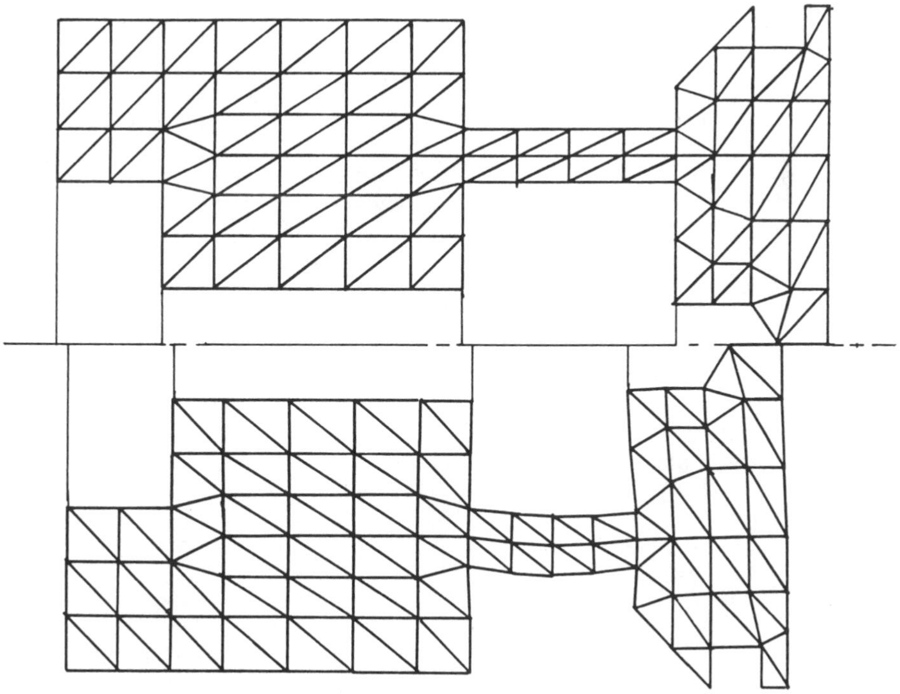

Fig. 4 Displaced finite element mesh for the improved design

is highly desirable that the highest proportion of elastically stored energy should be in the piezo-electrically active stack and not in the inactive materials.

THEORETICAL INVESTIGATIONS

One important feature for an efficient radiator is that the radiating face should behave as nearly as possible as a rigid piston, and so in the analysis of unloaded transducers close attention was paid to the variation in the amplitude of the vibration across the face at frequencies close to the resonance of the transducer as a whole. Precisely at resonance and in the absence of losses, with a constant sinusoidal driving force the displacements should go to infinity, in which case arithmetical errors could arise in the computation; therefore the final calculations were made at frequencies close to, but not actually at, resonance. For the original design (fig.1) with an aluminium head the ratio of the maximum displacement (at the outside edge) to the minimum (close to the centre line) was 1.45 close to the resonant frequency for the finite element model at 51.91 kHz. For the modified version of fig. 2 the ratio of displacements was 1.16 close to 49.36 kHz, and for a similar element with a titanium head of reduced thickness it was 1.15 at 51.67 kHz, the maximum displacement still being at the outside edge. The patterns of the displacements are shown to an exaggerated degree in fig. 3 and fig. 4, in which the distorted triangular finite elements are mirrored against the undistorted outlines of the two designs. In this way the relative amplitude of the bending distortion has been reduced by a factor of three by moving the drive from the piezo-electric stack away from the centre line in spite of the head being thinner and lighter.

The significance of another factor, the amount of mechanical strain in the head and tail, can be assessed by considering the change in resonant frequency if the stiffness of the materials of the head and tail are increased substantially. If the effective compliance of these parts contribute significantly to the overall compliance, then the resonant frequency would change appreciably. With two orders of magnitude increase in the stiffnesses, the computed resonant frequency of the old standard element changed from 51.91 kHz to 74.04 kHz, the overall compliance therefore changing by a factor of 0.487, assuming no change in the effective mass. The modified version with a titanium head changed from 51.67 kHz to 59.18 kHz, the corresponding factor being 0.762, and therefore the head in the latter case is behaving in a much stiffer manner; the aluminium version changed from 49.36 kHz to 55.85 kHz, the factor being 0.781. Hence the new aluminium head appears to be slightly stiffer than the titanium one.

PRACTICAL INVESTIGATIONS

The principal motive behind these developments was the need to modify an existing, reasonably successful, design of transducer resonant at 50 kHz and having a head diameter of 25 mm. (0.85 wavelength) by replacing the aluminium head by one of titanium. The difficulty in the redesign lies in the titanium being about two thirds as heavy again as aluminium whereas its stiffness moduli are only about one third greater. The original design was done initially by the "lumped element" method, combined with experimental modifications (one early version did in fact suffer from problems due to the head flapping because it was too thin), and it was expected therefore that simply redesigning the head with the new material to have the same weight would have made it too thin and compliant. New elements for 50 kHz, with aluminium or titanium heads of reduced thickness, have been designed using the techniques described above and constructed. Preliminary results show that they have a lower value of Q than the first design, 3.2 (instead of 5.0) for aluminium and 4.0 for titanium. The effective coupling coefficient, a function of the ratio of the motional and clamped capacitances in the electrical equivalent circuit, has been increased from an average of 0.33 for the original design to at least 0.41 in the new design, and this increase is consistent with the reduced theoretical contribution of the head to the overall compliance of the element, as indicated in the previous section. The value of 0.41 was shown by an element with temporary bonds, and a higher value should be shown by a fully bonded structure. At the time of writing the experimental work is still proceeding, and it is expected that small changes to the design will be necessary to achieve the desired resonant frequency of 50 kHz in water, one problem being that because of the reactive part of the radiation load being equivalent to an added mass, the wider the relative bandwidth then the greater is the change in resonant frequency from loading by air to loading by water.

A design for 31 kHz has also been investigated, using parts with basically the same dimensions as those used in the 50 kHz design, but with twice the length of piezo-electric stack. In order to adjust the resonant frequency within small limits while using the same piezo-electric rings, the tail has been designed to have a short stub with the same cross section as the rings, so that the effective compliance may be increased slightly. In this case there are of course reduced difficulties with the head flapping because of the lower resonant frequency of the element for the same diameter of the head, but there is a greater difference between the resonant frequencies in air and in water, the head diameter in wave-lengths being smaller and the reactive loading relatively greater.

REFERENCES

Dunn, J.R., 1984, An elementary introduction to finite element analysis. Proc. Inst. Acoustics vol. 6 pt. 3:79.

Morris, J.C., 1971, Transducer design based on filter theory, Brit. Acoust. Soc. Spring Meeting, paper No. 71SC7.

A 10-kW RING-SHELL PROJECTOR

G. W. McMahon and B. A. Armstrong*

Defence Research Establishment Atlantic
P.O. Box 1012, Dartmouth, Nova Scotia
Canada B2Y 3Z7

ABSTRACT

A ring-shell underwater sound projector is described that resonates at 600 Hz, has a power output capability of 10 kW, and weighs 220 kg. Its -3dB bandwidth is 160 Hz and its electroacoustic efficiency is over 90%. A passive pressure compensation system allows an operating depth of 300 m or more, depending on an initial charging pressure of dry gas.

INTRODUCTION

Ring-shell projectors have been under research and development for several years at the Defence Research Establishment Atlantic. The ability to generate high acoustic power at low frequency from a light, compact package makes the ring-shell design very attractive. Sparton of Canada Limited has continued the ring-shell development in recent years under DREA contract. Essentially, the transducer comprises a piezo-ceramic ring sandwiched between two shallow spherical shell segments. Radial motion of the ring drives the shells in flexure. Early units, without pressure compensation[1], were restricted to relatively shallow depths, in common with other flextensional transducers. In more recent models[2,3], an internal water bladder provides pressure compensation, increasing the operating depth to some hundreds of metres, and protecting the transducer from damage at even greater depths.

The ring-shell design is very versatile in terms of the size, resonance frequency, bandwidth, and power output that can be selected. Mathematical analysis, using finite elements, allows the design goals to be accurately modelled prior to construction. The math-modelling, construction, and acoustic performance of small (2kW) projectors have been described in Ref. 3. Here, we present the predicted and measured performance of a much larger projector, which has a resonance frequency of 600 Hz, a bandwidth of 160 Hz, and an expected source level of 212 dB re 1μPa at 1m (>10 kW).

CONSTRUCTION

The construction of the projector is basically the same as that described in Ref. 3. The driver ring is made from lead zirconate titanate ceramic plates, with steel staves interspersed between each pair of ceramic plates. The ring is given a compressive bias by

*Now at Hermes Electronics Limited, Dartmouth, N.S.

an outer wrapping of fiberglass, applied under tension and consolidated with epoxy resin. The steel shells are spherical shell segments with flat flanges, which are partially cut away to form "fingers", through which the shells are bolted to the steel staves in the ring. A toroidal-shaped water bladder is supported inside the projector, so as to be free from the shells. The bladder communicates with the sea via water inlet ports that pass through special steel staves. As the depth is increased, ingress of water compresses the internal gas, providing the necessary pressure compensation. A gas inlet valve allows the interior of the projector to be pre-pressurized with dry gas, flattening the bladder, and extending the operating depth range of the projector.

Figure 1 shows a radial cross-section of the projector at a water inlet port, as it might appear at its maximum operating depth. The bladder is full but is prevented from touching the shells by a fiberglass-plastic support structure, which is compliantly attached to the ring. Any further increase in depth will force the supports against the shells, impeding their vibration, but protecting the transducer from damage at excessive depths. The outside diameter of this projector is 88 cm and its weight is 220 kg.

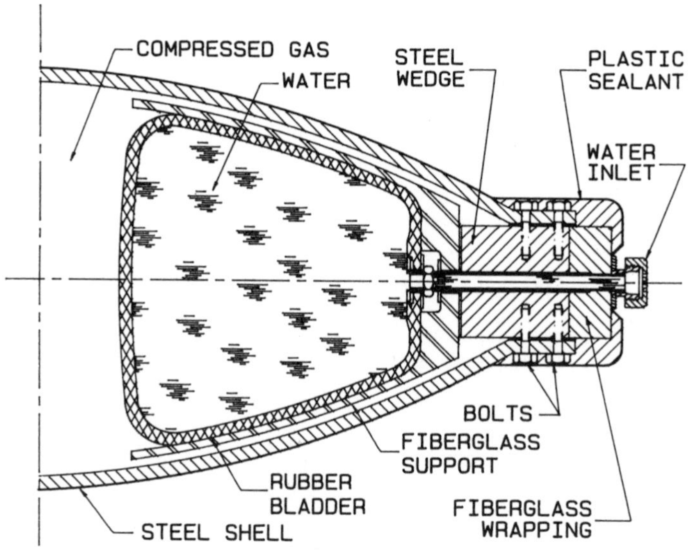

Fig. 1. Radial cross-section of the ring-shell projector as it might appear at its maximum operating depth. The diagram shows a special steel stave containing a water inlet port.

The large projector incorporates two small, but significant, design changes over the small projector: (a) The shell root diameter is greater than the inner diameter of the ring instead of being the same, and (b) A thin web of metal is left at the apex of the shell fingers rather than cutting away all of the metal between the fingers. The larger shell gives a larger radiating area, and hence a greater bandwidth than the previous design, with only a slight sacrifice of peak source level for the same ring size and resonance frequency (about 0.2 dB is predicted). A problem area on the earlier design was the reliability of the waterproof seal at the shell root, where the dynamic strain is greatest. The webbed fingers alleviate this problem and also simplify the assembly.

PERFORMANCE

The low-power performance of the projector was measured in sea water at a depth of 15 m. Its interior was pre-pressurized with dry nitrogen gas to 2.4 atm absolute, so that very little water would have entered the bladder. Figures 2 and 3 compare the predicted and measured transmitting responses in the mid-plane and axial directions, respectively. The measured electroacoustic efficiency at resonance is 93 %.

High power tests were carried out at resonance (610 Hz) at a depth of 34 m, and an internal pre-pressure of 4 atm. The maximum source level obtained in the mid-plane direction was 209.5 dB re 1μPa at 1m, limited by the available power amplifier. This represents an acoustic power output of 6300 W and an efficiency of 90%, including the matching transformer. The drive voltage was 2200 Vrms, whereas we plan to drive the projector to 3000 V, corresponding to an electric field gradient of 3 kV/cm in the ceramic. This will give a source level of 212 dB and more than 10 kW of radiated acoustic power.

Three of these projectors have been manufactured and tested. All have displayed essentially the same performance characteristics.

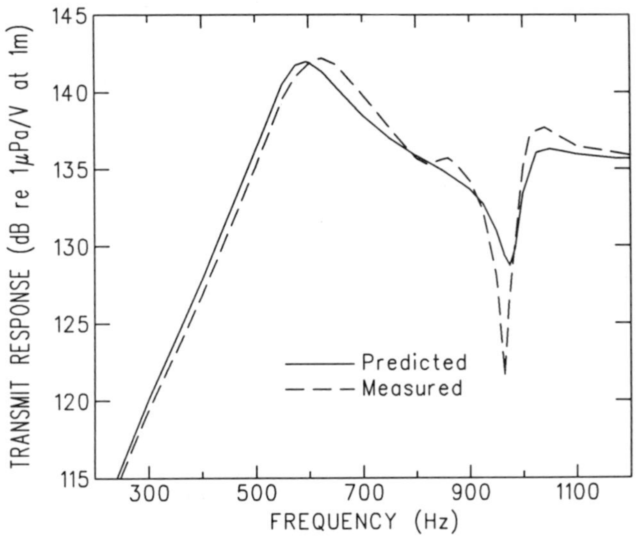

Fig. 2. Predicted and measured far-field transmitting responses on the mid-plane of the ring-shell projector.

DISCUSSION

We note that the predicted resonance frequency is about 3% lower than measured. The accuracy of a finite-element model depends strongly on the validity of the chosen material parameters. In this case, the characteristics of the piezoceramic ring were not measured before the shells were attached, so that the material properties of the composite ring were based on earlier modelling of smaller rings.

The first overtone resonance of the shells occurs at about 1000 Hz, and is closely predicted by the finite-element model. A weak resonance appears in the measured response near 800 Hz, and is likely due to the internal bladder assembly, which was not modelled in the finite-element analysis.

The finite-element model includes stress analysis and allows estimation of cavitation depth. The required depth for full power drive without cavitation is predicted to be 35 m.

The depth compensation system, with 4 atm pre-pressure, provides an operating depth range from 35 m to 300 m. This could be increased proportionately by increasing the pre-pressure up to about 10 atm. Without compensation, this projector could operate at full power to a depth of about 100 m, and would survive undriven to about 200 m. The limiting condition in this case is the tensile stress in the ceramic ring, which would exceed the compressive bias provided by the fiberglass-epoxy wrapping.

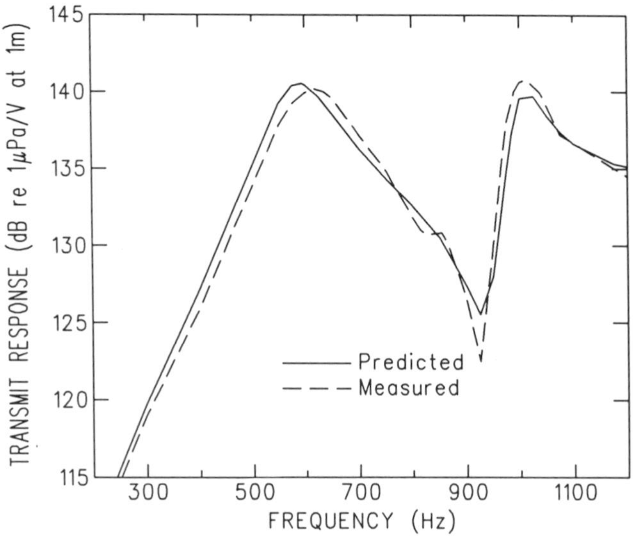

Fig. 3. Predicted and measured far-field transmitting responses on the axis of the ring-shell projector.

CONCLUSION

We have briefly described the design, construction and performance of a new high power ring-shell projector resonant at 600 Hz. While we have not yet demonstrated the full design power output of 10kW, we have achieved more than 6kW from each of the three units manufactured, and have every confidence that the full power will be obtained.

REFERENCES

[1] J. B. Lee and G. W. McMahon, "Low-frequency spherical-shell projectors", J.Acoust.Soc.Am. **61**, S57 (1977)

[2] G. W. McMahon and B. A. Armstrong, "A pressure-compensated ring-shell projector", IOA Conf. Proc., Transducers for Sonar Applications, Birmingham, U.K. Dec. 1980

[3] B. A. Armstrong and G. W. McMahon, "Discussion of the finite-element modelling and performance of ring-shell projectors", IEE Proc. **131**, 275 (1984)

BEAMFORMING WITH ACOUSTIC LENSES AND FILTER PLATES

Robert L. Sternberg

Office of Naval Research
495 Summer Street
Boston, Massachusetts 02210 U.S.A.

ABSTRACT

A survey is presented of a variety of known acoustic beamforming methods and techniques using acoustic lenses and filter plates as receivers, the former leading to receiving beams with frequency dependent or variable beamwidths and the latter to beam formation with frequency independent or constant beamwidths. In addition, a potentially useful method of forming approximately frequency independent receiving beams using twisted wavefront acoustic lenses based on a concept going back to Morris and Tucker is outlined. As transmitters it is noted that acoustic lenses are generally of little use because of power radiation limitations imposed by cavitation, but as receivers they have much in their favor.

ACOUSTIC LENS TYPES

Acoustic lenses of many types, both liquid filled and solid plastic or rubber lenses, have been used for receiving beam formation in underwater acoustics. These may be as simple as a "ball shaped", liquid filled spherical shell or a rubber lense of conventional shape with index of refraction controlled by loading the rubber with metallic powders or oxides. Because the speed of sound in typical plastics is greater than that in water, polystyrene and similar materials lead to convergent lenses, which are thinner in their centers than on their edges, and appear therefore more like divergent than convergent optical lenses. Cylindrical lenses with line foci and "barrel" shaped lenses, with or without stepping, and similar in concept to "lighthouse" lenses, are other types available for special purposes. Typical examples of several of these various types of acoustic lenses are shown in Figure 1.

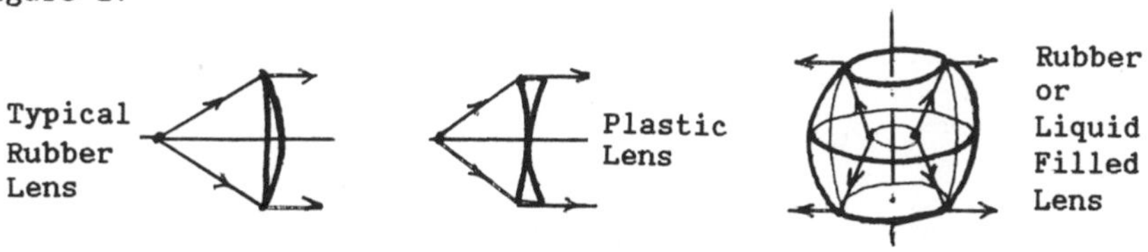

Figure 1.

Design techniques vary from use of conventional Gaussian optical methods, to the use of sophisticated aspheric design procedures originally developed for microwave radar applications based on nonlinear ordinary and partial differential equations which, together with suitable boundary and side or symmetry conditions, lead to lenses of minimal volume or weight for given aperture diameters. Included in the latter are the well known perfect focussing lenses of Cartesian oval genus going back to Huygens and Descartes.

While no full physical realizations of Luneberg lenses, or even really good approximations to such lenses, appear to have been developed for acoustic purposes, the liquid filled spherical shell lenses match both the aplanatic properties of a glass ball and, in a partial sense, the variable index of refraction lenses of Luneberg and, hence, can often be used as approximate wide angle scanners.

For all of these types of underwater acoustic lens receivers the angular width of the far field diffraction pattern, or simply the beamwidth, is largely controlled by the ratio of the aperture diameter D to the wavelength with only fractional effects attributable to variations in the shading imposed, intentionally or otherwise, on the aperture distribution of the acoustic signal. Thus one has as a rule of thumb the formula:

$$B.W. = 65\lambda/D$$

where the beamwidth B.W. is taken at the half-power points of the far field pattern and is measured in degrees. When variations in the acoustic shading are taken into account the factor 65 may vary from about 50 to 75 or 80 and, for cylindrical lenses of rectangular shape with linear receiving elements at their foci, two different beamwidths corresponding to the two different aperture dimensions of the lens result.

LENSES WITH FILTER PLATES

With the exception of the several types of spherical or "ball shaped" lenses, the beamwidths associated with all of these lens receivers can be made independent of the frequency, or constant in angular width, over variations in the frequency of the order of several octaves by the placement of one of Sternberg and Anderson's tapered metallic filter plates of high acoustic impedance immediately in front of, or behind, the acoustic lens as shown in Figure 2. Because of its radial taper in thickness from zero at its center with thickness increasing towards its periphery and the high impedance character of the metal used, the filter plate acts as an automatically variable aperture stop on the lens effectively keeping the ratio of D to λ constant, D here being the diameter of the acoustically transparent central portion of the plate over which the thickness is small compared to the wavelength in the plate material.

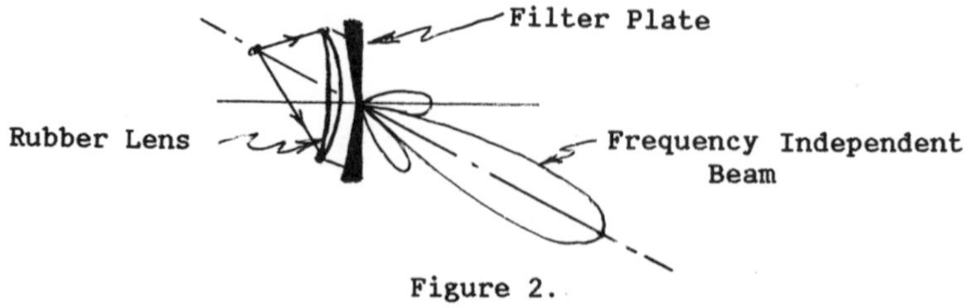

Figure 2.

SCANNING

For both acoustic lenses alone or for lens-filter plate combinations, scanning can be done either by moving the receiving element along an appropriate focal surface back of the lens, or lens filter plate combination, or by provision of a retina of fixed receiving elements on that focal surface. The scanned beams again are of either variable or constant beamwidth as a function of frequency according as the lenses are used alone or are accompanied by a filter plate.

TWISTED WAVEFRONT LENSES

In addition to the well established methods of forming receiving beams of either variable or constant beamwidth using acoustic lenses and filter plates, a new method of forming approximately constant beamwidth beams using a conceptually new type of acoustic lens alone as in Figure 3 is suggested by the twisted planar array concept, proposed some years ago by Morris as an extension of earlier ideas of Tucker and recently further extended to the more general notion of a twisted wavefront antenna via lenses by Sternberg and Anderson. Thus if a cylindrical acoustic lens has its linear receiving elements set at a slight angle to the linear rulings of the cylindrical surfaces of the lens as in the illustration, it appears that a beam of approximately constant angular width independent of the frequency will again be formed and can be scanned in the manner previously described. Similarly by altering the shape of the surfaces of a more conventionally shaped aspheric lens with rotational, or two-plane, symmetry it appears that such constant beamwidth beams can also be formed by lenses using point source elements at their receiving foci.

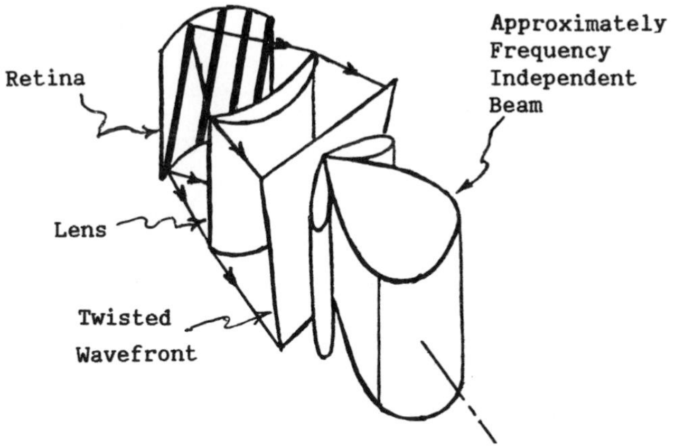

Figure 3.

While no such twisted wave front lenses have as yet been successfully developed, the problem of the design of such twisted wave front lenses again leads to nonlinear ordinary and partial differential equations of a fairly complicated nature but, which together with their accompanying boundary and side or symmetry conditions, appear to be amendable to solution by processes similar to those already developed for the minimal volume and weight aspherics previously noted.

MERITS OF FREQUENCY INDEPENDENT BEAMFORMING

The relative merits and pros and cons of frequency independent versus frequency dependent, or constant beamwidth versus variable beamwidth beam forming, for various underwater acoustic, applications are describable in a variety of ways for use in fisheries, geophysics and more general sonar problem areas. But as will be apparent, the advantages of constant beamwidth acoustic antennas can in all cases be summed up in the statement noted originally by Tucker that such beamformers eliminate the angular dependence, or modulation, of the received target spectrum as a function of the angle-off-boresight inherent in the use of conventional variable beamwidth beams whether formed by lenses, arrays or otherwise.

MISCELLANEOUS MATTERS

While no attempts appear as yet to have been made to use frequency independent or constant beamwidth beamforming methods or techniques in acoustic imaging applications, it appears that at least in principle such techniques could be applied to such matters but, many further problems would undoubtedly be introduced by such attempts and no discussion of these will be offered at present.

Similarily, while acoustic lenses and filter plates can in principle be used as sound projectors of either variable or constant beamwidth, their use for such purposes is largely impractical unless very low sound levels are to be projected, since the total sound power to be radiated must first be introduced at their principle foci by point or line sources limited by cavitation unless arrays are used to feed the system which use then would effectively eliminate the advantages of the lens or lens-filter plate combination and make their use redundant.

ACKNOWLEDGMENT

The author wishes to thank Dr. David H. Wood of the Naval Underwater Systems Center, New London, Connecticut and the University of Rhode Island, Kingston, Rhode Island for presenting the paper for the author at the Symposium on Underwater Acoustics in Halifax, Nova Scotia, Canada during the week of 16 to 18 July 1986.

REFERENCES

Tucker, D.G., "Arrays with Constant Beamwidth Over a Wide Frequency Range", NATURE, Vol. 180, 1957, pages 496-497.

Morris, J.C., "Broadband Constant Beamwidth Transducers", J. Sound Vib., Vol. 1, 1964, pages 28-40.

Sternberg, R.L., Anderson, W.A. and Stevens, G.T., "Log-Periodic Acoustic Lens-Acoustic Filter Plate Study", J. Acoust. Soc. Am., Vol. 59, 1976, pages 1104-1109.

Sternberg, R.L. and Anderson, W.A., "Frequency Independent Beamforming", NAVAL RESEARCH REVIEWS, Vol. 33, Fall/Winter 1980/81, pages 34-47.

Sternberg, R.L., Anderson, W.A., Dickson, O.P., Ilson, A.F. and Marchese, P.S., "Log-Periodic Acoustic Lens-Acoustic Filter Plate Study II, "J. Acoust. Soc. Am., Vol. 73, 1983, pages 2193-2199.

Sternberg, R.L., "A Survey of Numerical Methods for a New Class of Nonlinear Partial Differential Equations Arising in Nonspherical Geometrical Optics", PROCEEDINGS OF THE 11TH IMACS WORLD CONGRESS, Vol. 1, NUMERICAL ANALYSIS AND APPLICATIONS Edited by R. Vichnevetsky and J. Vignes, North Holland Publishers to appear circa 1986-87.

HYDROPHONE USING A FIBER FABRY-PEROT INTERFEROMETER

S. Ueha*, N.Wang*, M. Ohgaki+ and M. Okujima*

*Tokyo Institute of Technology, 4259 Nagatsuta, Midori-Ku Yokohama 227, Japan

+Oki Electric Industry Co. Ltd., 4-10-12 Shibaura Minato-Ku, Tokyo 108, Japan

ABSTRACT

This paper reports the investigation of fiber-optic hydrophone designs utilizing a Fabry-Perot interferometer. A compensator consisting of optical fibre wrapped around a piezo-electric cylinder is used to compensate for temperature fluctuations. Two fiber types (conventional single mode and polarization-maintaining single mode) are investigated experimentally. Resultant frequency response and sensitivity are reported. -Ed.

INTRODUCTION

Since the potential acoustic sensor use of fiber was pointed out,[1,2] acoustically induced phase modulation in a single mode fiber has been of interest.

As is well-known, there are two typical forms of acoustic sensors using fibers. One is an optical intensity modulation form and the other is the interferometric form. The latter contains a two-beam and a multiple-beam interferometer and the two-beam interferometer has been mainly investigated as a hydrophone.

It has been pointed out by Dr. Yoshino that the multiple-beam or Fabry-Perot interferometer can be used as a sensor.[3]

In this report, the feasibility of the optical fiber hydrophone is investigated where an optical fiber acts as a Fabry-Perot interferometer. The main problem encountered is the instability of the operating point of the hydrophone. This instability, however, has been reduced by using a stabilized laser and a servo system. It is also demonstrated that the hydrophone has a higher sensitivity than a two-beam interferometer.

PRINCIPLE AND SENSITIVITY OF HYDROPHONE

The principle of sensing is shown in Figure 1. Let us consider the case where the laser beam Iin is incident into a single mode fiber and the transmitted light Iout is detected by a photodetector.

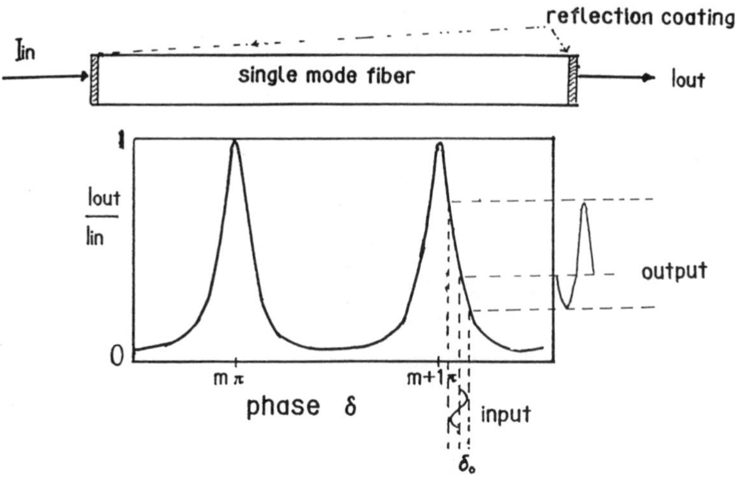

Figure 1: Fiber Fabry-Perot interferometer and its transmission characteristics

The ratio of Iout to Iin is expressed by the equation

$$\frac{Iout}{Iin} = \frac{1}{1 + F \sin^2\delta/2}\ , \quad (1)$$

where F and δ are defined by

$$F = \frac{4R}{(1 - R)^2}\ , \quad (2)$$

and

$$\delta = \frac{2\pi n L}{\lambda}\ , \quad (3)$$

respectively, R, n, L and λ are reflection coefficient, refractive index, length of fiber and wave length, respectively. Figure 1(b) shows the ratio of Iout to Iin as a function of phase δ.

The presence of strain in the fiber resulting from incident sound leads to an optical phase shift by the changes of fiber length and refractive index[4]. The phase δ is modulated in accordance with Eq. (3) if the optical fiber is exposed to sound pressure.

Then we can obtain the output signal as a change of light intensity. If the operating bias point δ_0 is adequate and the amount of modulation is small, the output signal is proportional to the input sound pressure.

The sensitivity of the hydrophone may be defined as

$$S = \frac{\Delta T}{K P L}\ , \quad (4)$$

where ΔT is defined by

$$\Delta T = \left(\frac{\partial Iout}{\partial Iin}\right) \Delta\delta = \frac{-F \sin\delta}{2\,(1 + F \sin^2\delta/2)^2}\Delta\delta \qquad (5)$$

and k denotes wave number.

If the operation bias point δ_0 is set to satisfy the following equation

$$\frac{1}{1 + F \sin^2\delta/2} = \frac{1}{2} \qquad (6)$$

the sensitivity becomes proportional to $\sqrt{F}$ as expressed by

$$S \propto \Delta T \propto \sqrt{F} * \Delta\delta \qquad (7)$$

This means that the sensitivity increases as the value of F becomes larger or the reflection coefficient R approaches to 1 as expressed by Eq. (2).

Since the sensitivity of the two beam interferometric hydrophone is given[5] by $\sqrt{F}$ = 1.27, the sensitivity is expected to be higher than that of the two beam interferometric hydrophone if $\sqrt{F}$ is larger than 1.27.

SPECIFICATIONS OF FIBER AND PRELIMINARY EXPERIMENT

Table 1: Specifications of fiber sensors

	reflection coefficient R	length (cm)	radius of fiber coil (cm)
single mode fiber (SM)	0.6	70	1.75
	0.8	70	1.5
	0.8	70	1.75
polarization-mantaining single mode fiber (PMSM)	0.6	70	1.75

As a sensor fiber, two kinds of fibers, that is, conventional (SM) and polarization-maintaining single mode (PMSM) fibers were used. Specifications of the four kinds of fiber sensors are listed in Table 1.

At first, a preliminary experiment was carried out using the simple experimental setup shown in Figure 2(a). The output signal fluctuated, as shown in Figure 2(a) and (b). Note that the two figures have different time scales. The instability of laser and the variation of the ambient temperature may be responsible for the fluctuation.

To measure the sensitivity of the hydrophone, the fluctuation had to be suppressed.

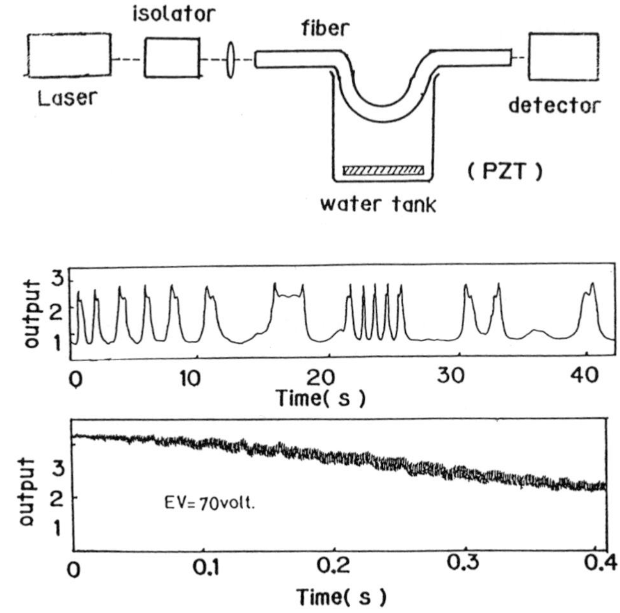

Fig.2. Preliminary experimental setup(a) and fluctuation of output signal (b),(c).

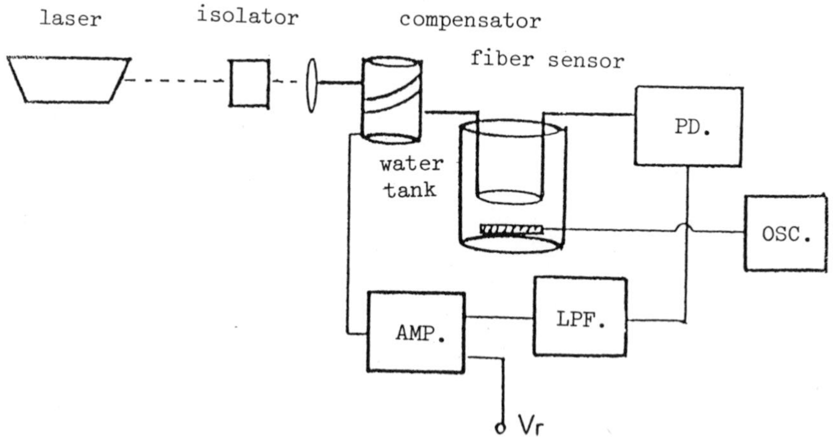

Fig.3. Experimental setup. PD, photo-detector; OSC, oscillator; AMP, comparator; LPF, low pass filter.

EXPERIMENTAL SETUP

Figure 3 shows the improved experimental setup. A stabilized transverse Zeeman gas laser (STZL-2mW He-Ne) was used. The STZL had a frequencey stability of 1.5 x 10^{-9}. The light from the laser was injected into an optical fiber by a microscope objective (20x).

Since the light coupling between the optical fiber and the resonance cavity of the laser cannot be neglected[6], an isolator was used to prevent the feedback light from coupling.

In order to compensate for the influence of temperature fluctuation, a piezoelectric optical path compensator was employed. The compensator consisted of a low pass filter, differential amplifier and a PZT element, where the fiber was wrapped around the PZT cylinder. A reference voltage V_r was applied to AMP to obtain the best operating point. The part of fiber used as a sensor was wrapped in the form of a coil and inserted into the water in the tank. As a photo-detector, a photo-diode was used.

In order to examine whether the sound pressure was equal everywhere in the tank, the measurement of sound pressure distribution was carried out using a hydrophone (B&K Type 8130). It was confirmed that the sound pressure distributes uniformly in the tank at the frequency range used.

EXPERIMENTAL RESULTS

The measurement of sensitivity-frequency characteristics were carried out changing the sensor fiber and the incident angle. The results are shown in Figure 4.

Figure 4(a) shows the effect of reflection coefficient with the sound incident angle of 90°. The measured ratio of sensitivity ranging from 2 to 3.5 is different from the theoretically predicted value of 4. It is clear, however, that the sensitivity becomes higher as the reflection coefficient increases.

Figure 4(b) shows the frequency characteristics of sensitivity measured with SM and PMSM fibers. It can be seen that the PMSM fiber sensor is more sensitive than conventional SM fiber.

Figure 4(c) also shows two characteristic curves measured at the condition shown in the corner. There is no obvious difference between the two characteristics. But it can be pointed out from the experimental results that there was a minimum value in the range of 3-5 kHz in every characteristic. But the reason for these minima has not been determined.

It is also pointed out that the minimum pressure of .01μbar was marked at f = 7 kHz.

CONCLUSION

The experimental study on the feasibility of optical fiber Fabry-Perot interferometric hydrophone has been carried out over a frequency range from 1 to 8 kHz, and the feasibility has been successfully proven.

It was also proven that this hydrophone has a higher sensitivity than a conventional two beam interferometric hydrophone and the minimum detectable pressure marked up to .01μbar at 7 kHz.

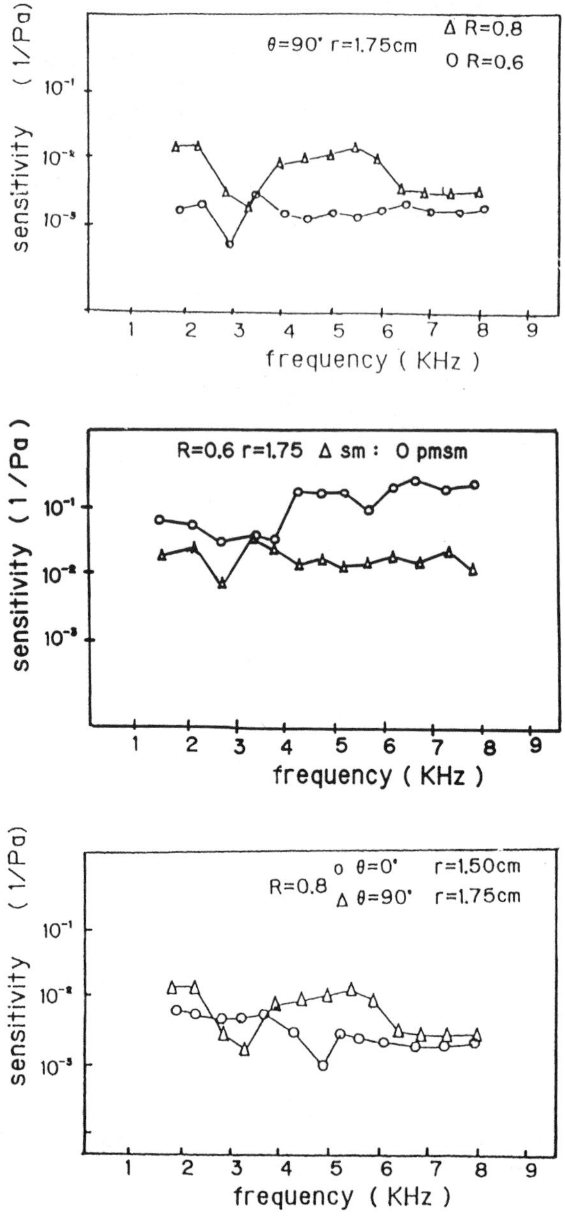

Fig.4. Frequency characteristics of sensitivity. Effect of R (a), Comparison of fiber (b) and Influence of incident angle(c).

REFERENCES

1. J.A. Bucaro, H.L. Dardy and E.F. Caromo; Appl. Opt. 16 p1761. (1977).
2. J.H. Cole, R.L. Johoson and R.G. Bhuta; J. Acoust. Soc. Am. 60, p1136. (1977).
3. T. Yoshino, K. Kurosawa, K. Itoh and T. Ose; IEEE-QE-18, No. 10, (1982) PP1624-1633.
4. J. Jarzynski, R. Hickman and J.A. Bucaro; J. Acoust. Soc. Am. 69(6), (1981).
5. M. Born and E. Wolf; Principles of Optics, Pergamon Press (Oxford, New York, 5th ed., 1975) p328.
6. T. Kaneda and K. Nawata; IEEE-QE, No. 7 (1979) p563.

AN IMPROVED METHOD FOR FLUID STRUCTURE INTERACTIONS

D.J.W. Hardie

Admiralty Research Establishment
Portland, Dorset, England DT5 2JS UK

ABSTRACT

A numerical method for solving acoustic fluid structure interactions based upon the Helmholtz integral relation is described. The method utilises the boundary element approximation comprising the surface Helmholtz equation (SHE) and its normal derivative form (NDSHE). This combination circumvents certain uniqueness problems associated with other similar methods. Difficulties due to singular kernels arising from the NDSHE are avoided for certain surface element geometries. This paper extends previous work by , consistently including an elastic description of the structure. The utility of this procedure is demonstrated by acoustic calculations on some immersed bodies.

INTRODUCTION

Much work has been conducted on the exterior acoustic problem over the past two decades. Apart from its intrinsic academic interest a widely applicable and reliable method of solution has great utility in many branches of engineering science, including the field of transducer design. Solutions in closed form can only be found for simple problems and numerical methods are often the only resort in practice . Such a scheme based upon the surface Helmholtz equation (SHE) and its normal derivative form (NDSHE) following a recommendation of Reut(1985) is proposed. This scheme incorporates both the boundary element and finite element approximations to describe the exterior fluid domain (E) and the interior elastic structure domain (I) respectively. Continuity relations at the surface interface (S) produce a consistent set of equations yielding the acoustic field quantities.

The proposed procedure avoids the need to construct an over-determined set of equations to overcome uniqueness problems. Such a requirement is necessary in the CHIEF method (Schenk 1968) and has certain disadvantages: the selection of interior collocation points is ambiguous and an expensive least-squares matrix inversion has to be performed. Our method does suffer from having to solve integral equations with singular kernels. However, this disadvantage does not arise for flat boundary element patches (Terai 1980) and provided the structure surface is well represented by a fine enough mesh a converged solution is obtained. The method has been dubbed CONDOR,

Composite Outward Normal Derivative Relation, by Reut(1985).

ACOUSTIC EQUATIONS

Consider harmonic ($e^{i\omega t}$) propogation of small amplitude acoustic wave in a fluid with density ρ incident on a structure with surface fluid interface (S). The total acoustic pressure $p(\underline{r})$ and normal velocity, $v(\underline{r})$ ($\underline{r} \in S$) are the superposition of the incident and scattered waves and satisfy both the SHE,

$$\iint \left\{ p(\underline{r}') \frac{\partial G(\underline{r}',\underline{r})}{\partial n_{r'}} - i\omega\rho v(\underline{r}') G(\underline{r}',\underline{r}) \right\} dS_{r'} = \tfrac{1}{2} p(\underline{r}) \tag{1}$$

and the NDSHE

$$\iint \left\{ p(\underline{r}') \frac{\partial^2 G(\underline{r}',\underline{r})}{\partial n_{r'} \partial n_r} - i\omega\rho v(\underline{r}') \frac{\partial G(\underline{r}',\underline{r})}{\partial n_r} \right\} dS_{r'} = \tfrac{1}{2} v(\underline{r}) \tag{2}$$

respectively. The Green function, $G(\underline{r}',\underline{r}) = e^{ikx}/4\pi x$ where $x = |\underline{r}'-\underline{r}|$, k i the acoustic wavenumber and $\underline{n}_r$ is the unit normal in the $\underline{r}$ direction.

The surface is subdivided into m boundary elements, S_j (j=1...m) and the acoustic field quantities $p(\underline{r})$ and $v(\underline{r})$ are assumed constant over an element S_j, whereby $p(\underline{r}) = \sum_j^m p_j \chi_j(\underline{r})$ and $v(\underline{r}) = \sum_j^m v_j \chi_j(\underline{r})$ and the bas functions, $\chi_j(\underline{r}) = 1$ for $\underline{r} \in S_j$ and $\chi_j(\underline{r}) = 0$ otherwise. A linear combination of the discrete forms of (1) and (2) yield the matrix equations,

$$\underset{\sim}{A}\, \underline{p} + \underset{\sim}{B}\, \underline{v} = \underline{p}_{inc} + \alpha\omega\rho\, \underline{v}_{inc} \tag{3}$$

where $\underline{p} = [p_1 \cdots\cdots p_m]^T$, $\underline{v} = [v_1 \cdots\cdots v_m]^T$ and the subscript inc refers t the incident field. The coefficient α depends on k and if Im(α) $\neq 0$, (has a unique solution for the associated (Dirichlet or Neumann) problem where ($p(\underline{r})$ or $v(\underline{r}) = 0$, $\underline{r} \in S$). Reut(1985) recommends a value of $\alpha = -$ The matrices $\underset{\sim}{A}$ and $\underset{\sim}{B}$ are given by,

$$(\underset{\sim}{A})_{ij} = \iint_{S_j} \left\{ \frac{\partial G(\underline{r}_q,\underline{r}_i)}{\partial n_q} + \alpha\, \underline{n}_i \cdot \underline{n}_q \frac{\partial^2 G(\underline{r}_q,\underline{r}_i)}{\partial n_q \partial n_i} \right\} dS_q \tag{4a}$$

and

$$(\underset{\sim}{B})_{ij} = \omega\rho \int_{S_j} \left\{ G(\underline{r}_q,\underline{r}_i) + \alpha\, \underline{n}_i \cdot \underline{n}_q \frac{\partial G(\underline{r}_q,\underline{r}_i)}{\partial n_i} \right\} dS_q \tag{4b}$$

for $i \neq j$. The evaluation of the diagonal elements (i=j) give rise to singul integrands of which the second derivative of the Green function (in 4a) is the most severe. By means of a transformation due to Terai(1980) for flat elements we obtain,

$$(\underset{\sim}{A})_{jj} = -\frac{1}{2} - \frac{\alpha}{2}\left\{ ik + \frac{1}{2\pi}\int_0^{2\pi} \frac{e^{ik\zeta_j}}{\zeta_j}\, d\theta_j \right\} \tag{4c}$$

which is readily integrable and (4b) reduces to (Reut 1985),

$$(\underset{\sim}{B})_{jj} = \omega\rho \int_{S_j} G(\underline{r}_q, \underline{r}_i)\, dS_q \tag{4d}$$

Here $\zeta_j = \zeta_j(\theta_j)$ is the polar equation of the contour C_j boardering the surface element S_j with respect to its centroid.

The scattered acoustic pressure $p_s(\underline{r})$ in the fluid ($\underline{r} \in E$) may be determined from the total acoustic field quantities evaluated at the interface ($\underline{r} \in S$). Since no uniqueness problems occur in this case only the discrete form of the SHE need be used. This is all that is required for the near field while at large r an asymptotic approximation is sufficient for the far field pressure, thus

$$p^{near}(\underline{r}) = \sum_j^m \int_{S_j} \left\{ p_j \frac{\partial G(\underline{r}_q, \underline{r})}{\partial n_q} + i\omega\rho\, v_j\, G(\underline{r}_q, \underline{r}) \right\} dS_q \tag{5a}$$

and

$$p^{far}(\underline{r}) = \frac{e^{-ikr}}{4\pi r} \sum_j^m \int_{S_j} e^{ik\cos\phi} \left\{ p_j k\cos\psi + \omega\rho\, v_j \right\} dS_q \tag{5b}$$

where $\cos\phi = \underline{r}\cdot\underline{r}_q / r r_q$ and $\cos\psi = (\underline{r}-\underline{r}_q)/|\underline{r}-\underline{r}_q|$(fig. 1).

COUPLED PROBLEM

The elastic structure which excites or is excited by the acoustic field (with frequency ω) is described by a finite element approximation giving,

$$\left\{ \underset{\sim}{K} - \omega^2 \underset{\sim}{M} \right\} \underline{q} = \underline{f} - \underset{\sim}{L}\, \underline{p} \tag{6}$$

Here $\underset{\sim}{M}$ and $\underset{\sim}{K}$ are the mass and stiffness matrices, $\underline{f}$, describes any internal applied forces and $\underline{q}$ is a generalised nodal displacement vector. The acoustic pressure field is transformed into an equivalent consistent external loading (via $\underset{\sim}{L}$). In addition, the surface normal velocities and displacements are related by continuity, where $\underset{\sim}{N}\,\underline{V} = i\omega\underline{q}$.

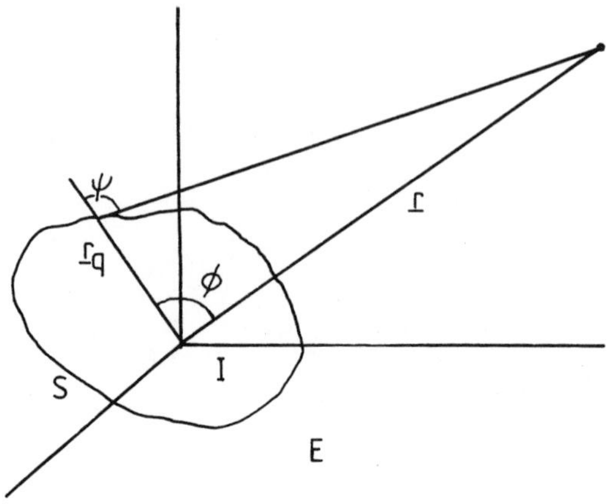

Fig. 1

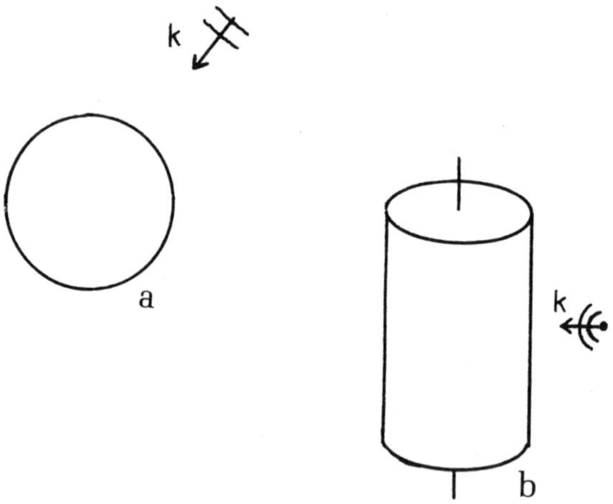

Fig. 2

Combining (3) and (6) yields the consistent set,

$$\{\underset{\sim}{A} - \underset{\sim}{Q}\} \underline{p} = \underline{p}_{inc} + \alpha \omega \rho \underline{v}_{inc} - \underset{\sim}{Q} \underline{f} \tag{7}$$

to be solved for the acoustic pressure field $p(\underline{r})$. Here $\underset{\sim}{Q} = i\omega \underset{\sim}{B} \underset{\sim}{N}^{-1} \times \{\underset{\sim}{K} - \omega^2 \underset{\sim}{M}\}^{-1}$ and is the coupling matrix. Clearly (7) may deal with radiation ($\tilde{p}_{inc} = 0$) and scattering ($\underline{f} = 0$) conditions or both.

IMPLEMENTATION

The fluid structure interface (S) is constructed by a contiguous set of triangular and quadrilateral (2D) boundary elements, S_j , each with 6 or 9 nodes respectively. Any structure surface may be composed, in principle, by a suitable combination of such elements. The acoustic field quantities are assumed constant over each element with values taken to be those interpolated at the centroid. The calculation of these values are performed using Gaussian and Trapezoidal Rule quadrature for the respective non-singular and singular integrals (4).

For our simple examples the elastic structure is assumed to be axisymmetric, homogeneous and isotropic and does not include any thin plates or struts. The structure is composed of 6-noded iso-parametric triangular axisymmetric elastic elements. A suitable transformation matrix ($\underset{\sim}{L}$) reduces the matrices $\underset{\sim}{A}$ and $\underset{\sim}{B}$ derived from the (2D) boundary elements to provide an equivalent axisymmetric loading. The set of equations (7) is then solved by numerical matrix inversion for the structural displacements, $\underline{q}$, and the external acoustic pressure field $p(\underline{r})$. The procedure can also take advantage of any geometric symmetry which the scattering or radiation situation may exhibit, to reduce computational effort.

The procedure can take account of a number of conditions: scattering by a body isonified by monochromatic plane waves or by any possible combination of single frequency finite sources (eg. dipoles) and even by broad-band noise from a particular region using repeated calculations for different ω and the principle of superposition.

APPLICATIONS

a) Spherical Shell

A scattering problem that of a plane wave (freq.= 150kHz) incident on a steel spherical shell (outer rad.= 2.0cm, inner rad.= 1.5cm) immersed in water, was considered (Fig. 2a). Comparison with an analytic solution (Wilton 1975) produced an overall agreement to within 3% using the CONDOR method for predictions of both the scattered surface pressure and normal surface displacements. Comparible results were achieved using the CHIEF method with 10% more collocation points. Agreement between the three sets of calculations for the near and far field pressures was greater than 3 sig. fig. using only 18 surface collocation points. No significant change in the CONDOR predictions was seen when varying the coupling constant α over the range, $\frac{1}{2}k < |\alpha| < 5/k$.

A radiation problem concerning a similar sphere with a first harmonic

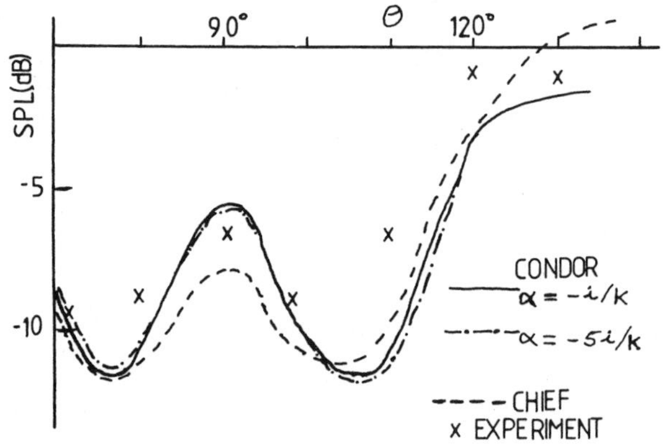

Fig. 3

surface velocity component (varying as cos θ) yielded surface pressures to within 1% of an analytic solution (Wilton 1975).

b) Cylindrical Solid

A finite steel right cylinder (len.= 305cm, diam.= 49cm) scatters acoustic radiation from a simple source at a distance of 49cm from its axis centre (Fig. 2b). The total near field sound pressure level (SPL = 20log$|P/P_{inc}|$), at the source, calculated by the CONDOR method was compared with previous work (Vrcelj 1981). The results derived from CONDOR are seen to be sensitive to varying coupling constant α , $1/2k < |\alpha| < 5/k$, at a relatively high frequency (1.86kHz) but here the mesh is coarse ($k\bar{\xi}$ > 2,see below). At a slightly lower frequency (1.0kHz), the CONDOR prediction of the SPL exhibit good accord with the previous work (within 2Db), although the mesh is still not fine enough ($k\bar{\xi} \cong 1$,see below) with the CONDOR predictions in the shadow region exhibiting some sensitivity to varying α (Fig. 3).

c) Hoop Transducer

CONDOR predictions for the far field SPL for a low frequency vibrating axisymmetric hoop transducer have been obtained and await comparison with future experimental measurements.

DISCUSSION

Since a rigorous procedure for obtaining the optimum value for the CONDOR coupling constant α is not yet available, heuristic arguments for a particular choice can only be presented. To provide comparable numerical accuracy in accounting for the singular contributions to (4), we choose, $|\alpha| k \sim 1$. Secondly we take for convenience, Re(α) = 0, although proof of uniqueness does not demand this and thirdly, dimensional arguments imply $\alpha \propto k^{-1}$. However calculations have been carried out with differing values of α to the determine the sensitivity of the results to varying α .

It is seen that provided the mesh is sufficiently fine and, by implication, the frequency is low enough, the acoustic field predictions from CONDOR are insensitive to changes in α . Conversely, however, predictions obtained with a coarser mesh, especially in shadow regions, are not so reliable and careful scrutiny of these results is recommended. An estimate of the degree of mesh fineness at a given frequency is of use here. Considering (4c) we see that the integrands in the diagonal elements of the matrix $\underset{\sim}{A}$ depend solely on $k\xi_j$, provided $\alpha \propto k^{-1}$. Denoting $\bar{\xi}$ as some mean value of $\xi_j(\theta_j)$ (averaged over θ_j and j) and assuming that a reasonable mesh requires around 5 collocation points per wavelength, then $\bar{\xi} < \lambda/8$ and therefore $k\bar{\xi} < 1$. We therefore use this criterion to determine whether a given mesh is fine ($k\bar{\xi}$ < 1) or coarse ($k\bar{\xi}$ > 1) and hence whether the results derived are reliable.

With regard to numerical implementation the CONDOR and CHIEF methods are similar; both require the construction and inversion of similar matrices. However, for a problem where the required number of surface collocation points is large the CONDOR scheme can be demonstrated to be more efficient. The matrix construction and inversion times are roughly proportional to the square and the cube of the total number of collocation points respectively. Hence for a problem with a given number of surface collocation points, the total computation time is greater using the CHIEF method due to the additional internal collocation points. Typically for a problem requiring 20% of the total number of collocation points to be internal collocation points the CHIEF method is roughly twice as slow as CONDOR for a given accuracy.

FURTHER WORK

To provide the CONDOR method with a greater applicability the procedure can be appended to a suitably general finite element code (eg. NAG finite element library). A suitable interpolation matrix ($\underset{\sim}{\ell}$) would be all that is required to interpret the acoustic field as an external loading. A combination of the acoustic boundary elements and elasto-dynamic boundary elements could be an even more efficient procedure, utilising similar numerical quadrature routines. Work on such possible schemes is continuing.

Improvements to the CONDOR method itself are possible. The relation du to Terai(1980) is proved for flat elements only and investigations into modifications of the integrals to include element curvature may prove profitable. This is likely to be only a higher order effect for sufficientl fine meshes, however any reduction in the number of surface collocation points by including element curvature can be advantageous for very complicated structures.

It is desirable to obtain a reliable scheme to optimise the coupling constant α for any mesh geometry although the present heuristic rules seem valid for most practical purposes.

Near field pressure predictions based on (5a) require careful interpretation close to the structure surface and especially in the vicinit of any surface discontinuities such as corners etc. Some workers have shown ways of improving the Helmholtz integral relations to account for such problems (Seybert et al 1985) and modifications to the CONDOR method following those recommendations are required.

CONCLUSIONS

The CONDOR method based upon a recommendation of Reut(1985) has been applied to various acoustic scattering and radiation problems involving immersed, axisymmetric bodies. Its utility has been demonstrated and the method in conjuction with a large finite element code shows promise in a wide area of application.

REFERENCES

Z.Reut (1985) Jour. Sound and Vib. 103 p267-98

H.A.Schenk (1968) Jour. Acoust. Soc. Am. 69 p71-100

A.F.Seybert,B.Soenarko,F.J.Rizzo and D.J.Shippy (1985) Jour. Acoust. Soc. Am. 77 p362-8

Z.Vrcelj (1981) M.O.D. Report No. 72196 (Unclassified)

D.T.Wilton (1975) M.O.D. Report No. 41375 (Unclassified)

ACOUSTIC PRESSURE RADIATED FROM HEAVY-FLUID LOADED INFINITE PLATE DRIVEN BY SEVERAL POINT FORCES

Jacqueline Larcher

Metravib R.D.S.
64, Chemin des Mouilles
BP 182 - 69132 Ecully Cedex- France

ABSTRACT

At circular frequency ω, we first express $G_a(\omega,x,y,z,x_o,y_o)$, the acoustic pressure at point (x,y,z) in the semi-infinite fluid, due to the radiation of an infinite thin elastic plate excited at point (x_o,y_o) by a normal unit point-force. Then we evaluate the near and far field pressures and the acoustic power radiated by the plate excited at N points by radial forces. Applied forces can be random, statistically correlated or not, or with known phase and modulus relationships between them. For various forces distributions, the acoustic radiated powers and the acoustic pressure fields are compared, respectively.

INTRODUCTION

The acoustic radiation by an infinite, heavy-fluid loaded plate driven by a time-harmonic point force has been investigated in a number of papers, for example by FEIT[1] or MAIDANIK et al[2]. In this paper we want to extend the study to include an excitation by N point forces. These forces, applied in the direction normal to the plate, may be either random or not, with known cross-correlation functions or known phase and modulus relationships between them. This problem of fluid-structure interaction is a linear one. So the resultant radiated-pressure is given by the sum of all the individual contributions due to each applied force.

First, we must derive an expression for the fluid pressure $G_a(\omega,x,y,z,x_o,y_o)$ at point (x,y,z) when the force, at circular frequency ω, acts at point (x_o,y_o). We shall see that an analytical expression for G_a may be derived in two ways, depending on whether the observation point is on the axis normal to the plate at (x_o,y_o) or not. Taking the normal derivative of G_a we obtain the corresponding expression for the fluid normal velocity.

Then, we use the expressions to evaluate the near field pressure and the acoustic power radiated by the plate driven by N forces acting at points (x_i,y_i). For the farfield pressure evaluation we use the stationary phase method and the directivity pattern is computed. Once the computations have been made, we can analyze the influence of various parameters (number of forces, relative forces positions, cross-correlation coefficients) on the acoustic radiation.

ACOUSTIC PRESSURE RADIATED BY A POINT EXCITED PLATE

The problem is set up in rectangular coordinate system with the (x,y) plane corresponding to the thin elastic plate. A heavy fluid medium is assumed to occupy the half space $z > 0$ and a vacuum the half-space $z < 0$. Attention is restricted to the frequency region below coincidence.

We must solve the simultaneous differential equations governing the plate vibrations and the pressure in the fluid together with the boundary conditions on the plate and at infinity. When we omit the time factor $e^{+j\omega t}$ which appears in all the expressions, these equations are written as :

(1) $(\partial^2 W/\partial x^2) + (\partial^2 W/\partial y^2) - (\rho_s h\omega^2/D)W = F\delta(x-x_o, y-y_o) - P(x,y,z=0)$

(2) $(\partial^2 P/\partial x^2) + (\partial^2 P/\partial y^2) + (\partial^2 P/\partial z^2) + k_o^2 P = 0$ for $z > 0$.

(3) $(\partial P/\partial z)\big|_{z=0} = + \rho_o\omega^2 W$ and the Sommerfeld radiation condition for $z \to +\infty$.

where W(x,y) is the plate normal displacement and P(x,y,z) the pressure in the fluid. D is the flexural stiffness, ρ_s the plate density, h the plate thickness, ρ_o the fluid density, k_o the acoustic wave number, δ is the Dirac distribution, F the amplitude of the applied force acting at poin (x_o, y_o) on the plate.

Taking the Fourier transform of Eq.(1) and Eq.(2) on the (x,y) variables and combining them with Eq.(3), the problem is easily solved. For a unit force F = 1, the radiated pressure labelled $G_a(\omega,x,y,z,x_o,y_o)$ is then given by the inverse Fourier transform. After some simple algebra this function G_a may be written as :

(4) $G_a = [(j\alpha e^{+j\omega t})/2\pi] \int_0^{+\infty} [(kJ_o(kr)e^{-j\sqrt{k_o^2-k^2}\,z}) / (\sqrt{k_o^2-k^2}\,(k^4-k_F^4)+j\alpha)]$

where we have used the notation :

$r = [(x-x_o)^2 + (y-y_o)^2]^{\frac{1}{2}}$ $\qquad \alpha = \rho_o\omega^2/D$ $\qquad k^2 = k_x^2 + k_y^2$

$k_F = (\rho_s h\omega^2/D)^{1/4}$, k_F is the flexural wave number of the plate <u>in vacuo</u>.

The square-root $(k_o^2-k^2)^{\frac{1}{2}}$ must be chosen in order that the pressure be bounded as z approaches infinity and remembering that the time factor is $e^{+j\omega t}$.

To evaluate the function G_a, the integration interval in Eq.(4) may be extended over the entire real axis from $-\infty$ to $+\infty$ together with $1/2\ H_o^{(2)}(kr$ instead of $J_o(kr)$. $H_o^{(2)}$ is the zero order Hankel function of second kind and J_o is the zero order Bessel function. By this way we obtain a new integral which may be evaluated by using a complex variable integration technique in the complex k plane. The integration path is shown on Figure 1 together with the locations of the allowed roots of the dispersion equation. Using Cauchy's theorem and Jordan's lemma on (C_1) and (C_2) G_a may be written as the sum of the residues at poles and of the integrals around the branch cuts.

Finally, we obtain the following expression for G_a :

(5) $$G_a = \alpha/2 \sum_{i=1}^{3} \frac{\sqrt{k_o^2-k_i^2}\ H_o^{(2)}(k_i r)\ e^{-jz\sqrt{k_o^2-k_i^2}}}{4k_o^2 k_i^2 - 5k_i^4 + k_F^4}$$

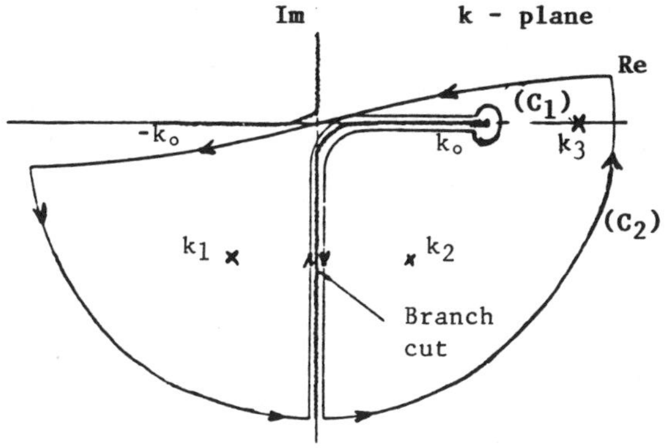

Fig.1 : Integration path for Eq.(4)

$$+ \alpha/\pi^2 \int_0^{+\infty} \frac{u K_o(ur)[\alpha \sin(z\sqrt{k_o^2+u^2}) - \sqrt{(k_o^2+u^2)}(u^4-k_F^4)\cos(z\sqrt{k_o^2+u^2})]}{(k_o^2+u^2)(u^4-k_F^4)^2+\alpha^2}\, du$$

$$- j\alpha/2\pi \int_0^{k_o} \frac{u H_o^{(2)}(ur)\,[\alpha\sin(z\sqrt{k_o^2-u^2}) - \sqrt{(k_o^2-u^2)}(u^4-k_F^4)\cos(z\sqrt{k_o^2-u^2})]}{(k_o^2-u^2)(u^4-k_F^4)^2+\alpha^2}\, du$$

where k_i are the allowed roots of the dispersion equation :

(6) $\sqrt{k_o^2-k^2}\ (k^4 - k_F^4) + j\alpha = 0.$

Integrals in Eq.(5) will be computed by gaussian quadrature techniques.

Due to the Hankel function introduced, it is not possible to evaluate G_a in this way for r = 0. Indeed, $H_o^{(2)}(kr)$ is unbounded as r approaches zero although the integral defining G_a in Eq(4) is finite for the on-axis pressure.

SPECIAL FORMULATION FOR THE ON-AXIS PRESSURE

For the on-axis pressure, in Eq.(4), $J_o(kr)$ is equal to 1 for r = 0. The integral is then split in two parts, one from 0 to k_o, the other from k_o to $+\infty$.

Substituting $u = \sqrt{k_o^2 - k^2}$ for supersonic wave numbers k and $v = \sqrt{k^2 - k_o^2}$ for subsonic wave numbers we get :

(7) $G_a(\omega, x_o, y_o, z, x_o, y_o) = (j\alpha/2\pi) \int_0^{k_o} (u e^{-juz}/Q(u))du - (\alpha/2\pi) \int_0^{+\infty} (v e^{-vz}/P(v))dv$

where Q and P are fifth degree polynomials which can be written as :

(8) $Q(-ju) = P(u) = u^5 + 2k_o^2 u^3 + (k_o^4 - k_F^4)\, u - \alpha$ or $Q(u) = \prod_{i=1}^{5}(u + j\gamma_i)$
and $P(v) = \prod_{i=1}^{5}(v - \gamma_i)$

Both integrals in Eq.(7) can now be formulated by expressing $u/Q(u)$ and $v/P(v)$ in partial fractions. It can be shown that :

$$(9) \quad dQ/du\Big|_{-j\gamma_i} = dP/dv\Big|_{\gamma_i} = P'(\gamma_i).$$

The involved integrals in Eq.(7) are expressed in terms of exponential integral sine and cosine integral functions of exponential argument[3,4]. Using the relationships between these functions defined in the complex plane cut along the half negative real axis, a very simple expression is obtained for the on-axis pressure radiated by a point excited plate :

$$(10) \quad G_a(\omega,x_0,y_0,z,x_0,y_0) = \alpha/2\pi \sum_{i=1}^{5} (-\gamma_i\ e^{-\gamma_i z}/P'(\gamma_i))\ E_1(jz(k_0 + j\gamma_i)).$$

Note that γ_i are the five complex roots of $P(v)$ as shown by Eq.(8), and P' is defined in Eq.(9). For $z \to 0$, the exponential integral function E_1 exhibits a logarithmic singularity which may be removed and it can be shown that for $z = 0$, we get :

$$(11) \quad G_a(r=0,z=0) = \alpha/2\pi \sum_{i=1}^{5} (\gamma_i/P'(\gamma_i))\ \mathrm{Log}(j(k_0 + j\gamma_i)),$$

where the correct phase determination must be chosen for the Log function argument in the complex plane cut along the half negative real axis.

INFINITE PLATE DRIVEN BY N POINT FORCES

Let us assume now that the plate is driven at N points (x_i,y_i) by forces F_i at circular frequency ω. For $i = 1,N$ F_i may be random variables. We define the cross-correlation coefficients $C_{ij} = E\{F_i.F_j^*\}$ where $E\{\ \}$ is the statistical mean.

If the forces are not random, each force may be written as $|F_i|e^{j\phi_i}$ and C_{ij} becomes $|F_i||F_j|e^{j(\phi_i-\phi_j)}$.

Clearly, equations governing the radiation problem being linear, the total radiated pressure is given by:

$$(12) \quad P(\omega,x,y,z) = \sum_{i=1}^{N} F_i\ G_a(\omega,x,y,z,x_i,y_i)$$

where G_a is given by expressions in Eq.(5) or Eq.(10) depending whether $x \neq x_i$ and $y \neq y_i$ or not.

With random excitation, assuming that $E\{F_i\} = 0$, the radiated pressure has also a null mean value. Then the only interesting term is the pressure spectral density at one point defined by :
$S_{pp}(\omega,x,y,z) = E\{P(\omega,x,y,z)P^*(\omega,x,y,z)\}$.

Near Field Pressure

The pressure field in the $z = H$ plane has been computed by using the formulation described above. Figures 2 and 3 show some results obtained for N non random in-phase forces, with the following parameters $F_i = F_j = 1/N$ for all i and j.

We can notice that, even with no force acting at $(x=y=0)$, the radiated peak pressure in the $z=H$ plane happens at $x=y=0$ (Fig.2). So a special treatment of the prussure field would be necessary to determine the location on the plate of the N points were the forces act.

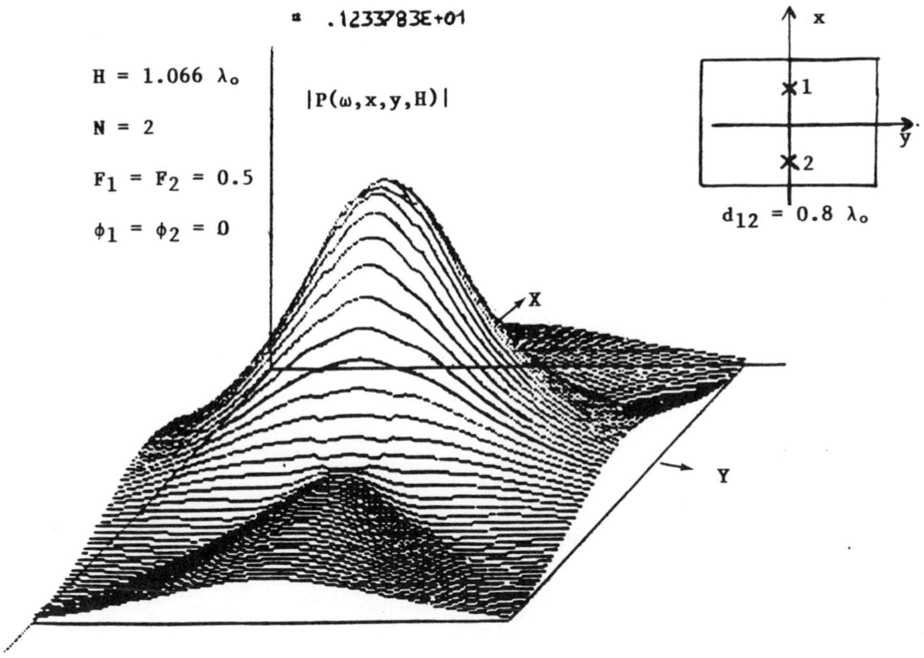

Figure 2: Amplitude of the Acoustic Pressure Radiated by the plate driven by $N = 2$ in phase, equal forces.

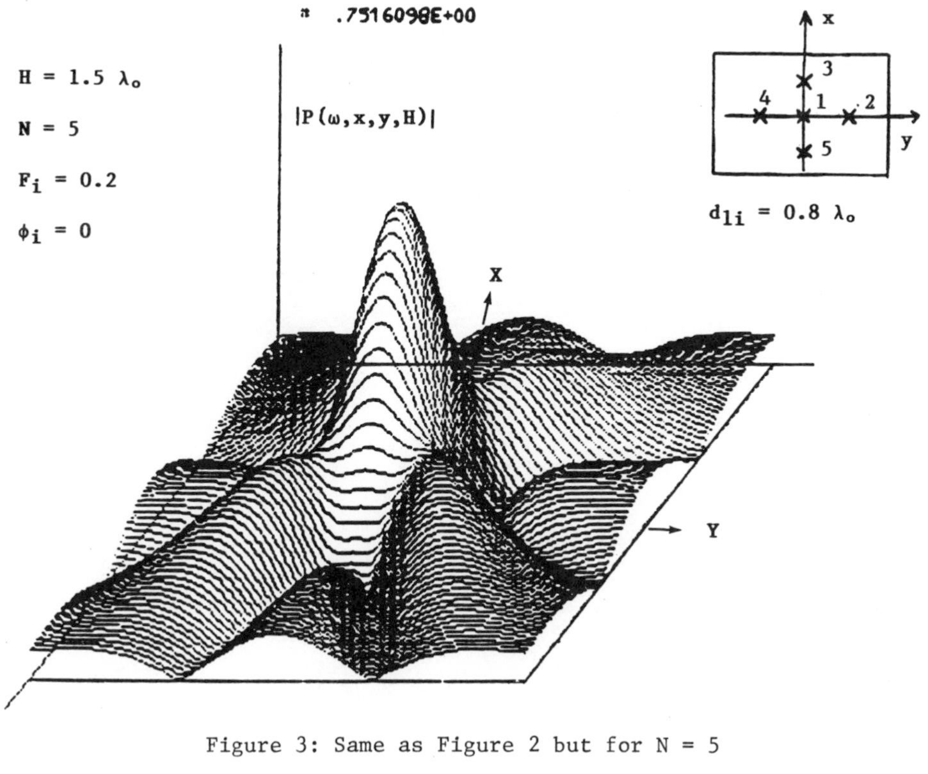

Figure 3: Same as Figure 2 but for $N = 5$

Acoustic Power

The average acoustic power radiated by a sound source can be computed by integrating the acoustic intensity on a closed surface surrounding the entire source. For the plate case, the integration surface taken is an infinite plane parallel to the z = 0 plane. Then the acoustic power for N exciting forces, can be written as :

$$(13)\quad \Pi_{ac} = \sum_{i=1}^{N}\sum_{j=1}^{N} Re(C_{ij}).(|\alpha|^2/4\pi\rho_o\omega).\int_0^{k_o} \frac{k\sqrt{k_o^2.k^2}\; J_o(kd_{ij}) \qquad dk}{|\sqrt{k_o^2-k^2}\,(k^4-k_F^4) + j\alpha|^2}$$

where $d_{ij} = (x_i - x_j)^2 + (y_i - y_j)^2$.

As we might expect, the obtained expression does not contain the z variable and involves only the supersonic components of the acoustic field. The integral in Eq.(13) has been computed by a gaussian quadrature method.

The expression in Eq.(13) leads immediately to some important results

- For random uncorrelated forces, ($C_{ij} = 0$ for $i \neq j$ and $C_{ii} \neq 0$) the total acoustic power radiated by the plate is equal to :

$$\sum_{i=1}^{N} C_{ii}.\Pi_o(\omega)$$

where $\Pi_o(\omega)$ is the acoustic power radiated by the same plate excited by a unit point force. The total acoustic power is independent of the relative locations of the application points where the uncorrelated random forces act.

- For non random case, $Re\{C_{ij}\} = F_iF_j\cos(\phi_i-\phi_j)$ and the acoustic power does depend on the relative distances between the N points where the forces act. But for the case $\phi_i = \phi_j$ and $k_od_{ij} \ll 1$ for all i and j the total acoustic power radiated is the same as if the plate were excited by one single force :

$$F_r = \sum_{i=1}^{N} F_i.$$

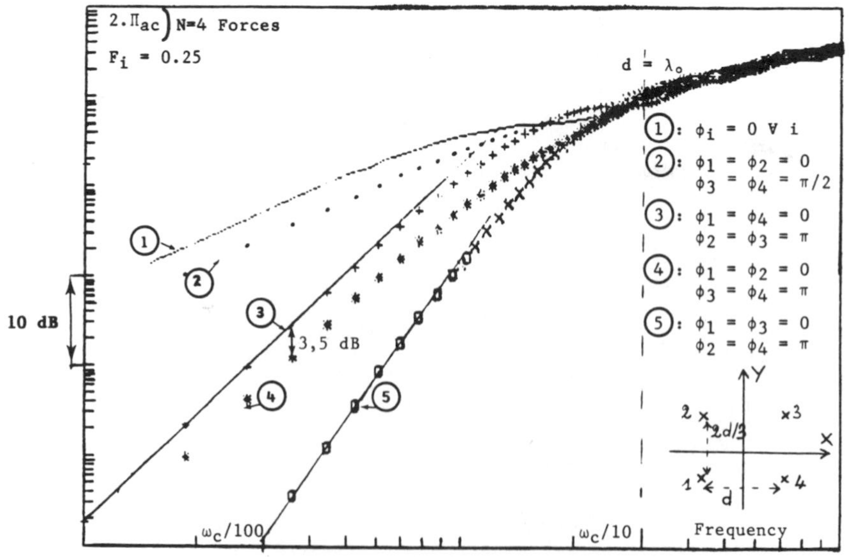

Figure 4: Acoustic power radiated by plate driven by N forces.

- Otherwise Π_{ac} depends on N, d_{ij} and $Re(C_{ij})$ and can be greatly reduced by a good choice of the phase relationships as shown in Fig. 4.

Directivity Pattern

In the farfield acoustic pressure, the directivity pattern is defined on the sphere of radius R by : $D(R,\theta,\phi,\omega)= E\{|P(R,\theta,\phi,\omega)|^2\}$. (R,θ,ϕ) are the spherical coordinate system : $x = R\cos\phi\cos\theta$ $y = R\cos\phi\sin\theta$ $z = R\sin\phi$. By noting $R_i = ((x-x_i)^2 + (y-y_i)^2 + z^2))^{\frac{1}{2}} = (r_i{}^2 + z^2)^{\frac{1}{2}}$ and $z = R_i\sin\phi_i$, R_i and ϕ_i are functions of R,θ,ϕ.

The farfield pressure radiated by a point excited plate is evaluated by the stationary phase integration method and leads to the classical expression for $G_{C.L}(R_i,\phi_i,\omega)$ for a unit point force, where R_i, ϕ_i are defined as above. Then, the directivity pattern may be computed by :

$$D(R,\theta,\phi,\omega) = \sum_{i=1}^{N} \sum_{j=1}^{N} C_{ij} \cdot G_{C.L}(R_i,\phi_i,\omega)\, G^*_{C.L}(R_j,\phi_j,\omega).$$

The directivity pattern may depend on the various parameters: N, $k_o d_{ij}$ and $C_{ij} = E\{F_i . F_j^*\}$.

Computations have been made for the same (x_i,y_i) as in Fig.4. In these cases the main results obtained are listed below :

- For uncorrelated forces, the directivity pattern is the same as if the plate were driven by a point force with amplitude :

$$F_o = \left(\sum_{i=1}^{N} C_{ii}\right)^{\frac{1}{2}} .$$

- For non random forces, various cases must be considered :

. If $\phi_i = \phi_j$ and $k_o d_{ij} \ll 1$ for all i and j the radiated pressure depends quite exclusively on the resultant exciting force :

$$F_r = \sum_{i=1}^{N} F_i .$$

. For $\phi_i \neq \phi_j$, the directivity pattern may display sidelobes, the number, angular position and width of which strongly depends on N and the relationships (phase, position) between the applied forces.

In this case too, the farfield on-axis pressure, at $\phi = \pi/2$, equals the result obtained for a plate excited by one point force F_r.

CONCLUSION

We have first derived an analytical expression of the acoustic Green's function for a thin elastic plate immersed in a heavy fluid. This function can be easily calculated at all points in the upper fluid.

Then the acoustical radiation of a plate driven by N point forces can be evaluated either for random uncorrelated forces or for forces with known phase and modulus relationships.

Several types of excitation have been tested and it has been shown that most of acoustical properties are governed by the resultant load (Directivity, On-Axis Pressure, and in some case Acoustic Power).

It has also been shown that the acoustic radiated power can be greatly reduced in the low frequency region by a good choice of the phase relationships between the applied forces.

REFERENCES

1. D. Feit, "Pressure radiated by Point Excited Elastic Plate", J. Acoust. Soc. Am. 40(6), 1489-1494 (1966).

2. G. Maidanik and E.M. Kerwin, "Influence of fluid loading on the radiation from infinite Plates below the critical Frequency", J. Acoust. Soc. Am. 40, 1034-1038 (1966).

3. I.S. Gradshteyn and I.M. Ryshik, Table of integrals, Series and Products, Academic Press (1980).

4. Abramovitz and Segun, Handbook of Mathematical Functions, Dover Publications Inc., New York, (1965).

ACKNOWLEDGEMENTS

The author gratefully acknowledges the financial Support of this work by the French Navy (STCAN).

TRANSFER PROPERTIES OF A FLUID LOADED ELASTOMER LAYER

Daniel Vaucher De La Croix

Metravib R.D.S.
64, Chemin des Mouilles
BP 182 - 69132 Ecully Cedex - France

ABSTRACT

A study was made of a rigid backed elastomer layer under heavy fluid loading. Natural modes of that physical system were determined and then response to a time harmonic point excitation was calculated. Resultant pressure levels received by a circular hydrophone embedded in the layer at varying distances from the hard surface have been obtained. Influence of the material properties on noise reduction at different frequencies was evaluated.

INTRODUCTION

The objective of this paper is to present an intermediate step in determining the effectiveness of a viscoelastic material acting as a reducer of turbulent boundary layer noise. We limit our study to the consideration of a time harmonic point force applied at the interface between the elastomer and the fluid. Section 1 reports the determination of the natural modes of the considered wave guide, while section 2 is concerned with response analysis. Effects of the elastomer layer geometry and its properties on noise reduction in this particular case are presented.

NATURAL MODES

The analytical model is a plane elastomer layer immersed in an inviscid fluid whose characteristic impedance is $\rho_o c_o$, the lower side resting on a rigid plane surface. The geometry and symbols related to the considered system are depicted in Fig. 1, where we have to omit the point driving force at the origin in our present research of the natural waves of the structure submitted to heavy fluid loading.

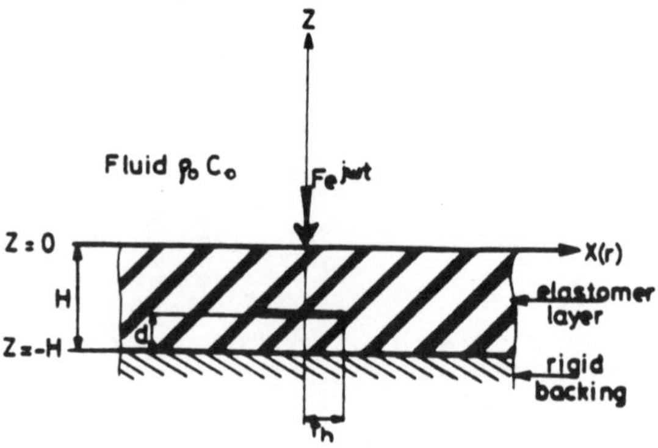

Fig.1 : Geometry the considered elastomer layer

The way we will obtain the dispersion relation associated to the studie case is now to be described. For briefness, we only retain some significant equations relevant for our purpose. A detailed description of the analytica part is presented in [1].

For the fluid medium, the basic equation is the two-dimensional wave equation for the propagation of sound, the solution of which is the pressure fluctuation.

(1.1) $\quad \Delta p_o - (1/c_o)^2 \partial^2 p_o/\partial t^2 = 0.$

where c_o is the speed of sound. If we introduce the double Fourier transform of $p_o(x,z,t)$ in the usual way, we easily obtain the solution of the transformed equation respecting Sommerfeld's radiation condition as :

(1.2) $\quad P_o(k,\omega,z) = A_o(k,\omega)\exp(-j\xi_o z), z \geqq 0$

where $\xi_o = \sqrt{k_o^2 - k^2}$, with k_o the acoustic wave number
$A_o(k,\omega)$: unknown coefficient to be determined.

Note : ξ_o must be chosen so that the pressure remains bounded as z approache infinity supposing harmonic time dependence $e^{+j\omega t}$.

We have then to describe the propagation of waves in the viscoelastic material. The elastomer layer is assumed to be isotropic, homogeneous, and quasi-incompressible. Assuming small deformations, it is possible to apply the Navier equations of the linearized theory of elasticity to the vector displacement $\underline{u}$ [2]. The resulting equation of motion writes :

(1.3) $\quad \mu\nabla^2\underline{u} + (\lambda + \mu)\underline{\nabla}\ (\underline{\nabla}\ \underline{u}) = \rho_c \partial^2\underline{u}/\partial t^2,$

where λ and μ are Lamé constants, ρ_c is the elastomer layer density. Solutio of this equation is obtained by expressing the normal and tangential components of the displacement $\underline{u}$ in terms of two scalar potentials (in the two-dimensional case). Explicitly :

(1.4) $\quad u_z = \Phi_{,z} + \psi_{,x}, \qquad u_x = \Phi_{,x} - \psi_{,z}.$

Inserting these expressions in (1.3) implies that each of the above introduced potentials obeys a two-dimensional wave equation, Φ being associated with $c_c = \sqrt{(\lambda + 2\mu)/\rho_c}$, the speed of the compressional wave and ψ with $c_s = \sqrt{\mu/\rho_c}$, the shear wave velocity. Similar to the previous treatment of the fluid pressure fluctuation, one readily obtains :

$$\left.\begin{aligned} \Phi(k,\omega,z) &= A_c(k,\omega)e^{j\xi_c z} + B_c(k,\omega)e^{-j\xi_c z} &\quad (1.6)\\ \psi(k,\omega,z) &= A_s(k,\omega)e^{j\xi_s z} + B_s(k,\omega)e^{-j\xi_s z} &\quad (1.7)\end{aligned}\right\} \quad -H \leqq z \leqq 0$$

for the double Fourier transform of the potentials $\Phi(x,z,t)$ and $\psi(x,z,t)$.

As previously, we have noted $\xi_c = \sqrt{k_{c,s}^2 - k^2}$, with $k_{c,s} = \omega/c_{c,s}$ standing for the dilatational resp. shear wave number. Determination of the five unknown coefficients A_o, A_c, B_c, A_s and B_s is achieved by taking into account boundary conditions of the interfaces $z = 0$ and $z = -H$. Rigid backing at the lower interface imposes vanishing normal and tangential components of the displacement at $z = -H$. On the other hand, the three remaining boundary conditions ar the fluid-solid interface read :

$$\left.\begin{aligned} &\partial^2 u_z/\partial t^2 = -(1/\rho_o)\partial p_o/\partial z &\quad (1.8)\\ &\tau_{zz} = -p_o &\quad (1.9)\\ &\sigma_{xz} = 0 &\quad (1.10)\end{aligned}\right\} \quad \text{at } z = 0$$

where σ_{xz} and τ_{zz} are tangential and normal stresses, respectively. The corresponding Fourier transform at these relations gives a linear system of algebraic equations in the five unknown coefficients A_o, A_c, B_c A_s, B_s, the dimension of which is readily reduced by elimination of A_o between (1.8) and (1.9). (The case $\xi_o = 0$ requires particular attention). A condensed form of the final obtained system writes :

$$[T_{ij}] < A_c, B_c, A_s, B_s >^T = 0, \quad (1.11)$$

and the desired dispersion relation results from the evident condition for non-triviality of the solution of (1.11), e.g

$$\det |T_{ij}| = 0 \quad (1.12)$$

Further introduction of the following non-dimensional parameters and variables :

$$\alpha = c_s/c_c \ , \ \alpha_o = c_s/c_o \ , \ \overline{k}^2 = (k/k_s)^2 = (c_s/c)^2$$
$$\overline{\xi}_s = (1 - \overline{k}^2)^{1/2} \ , \ \overline{\xi}_c = (1 - (\overline{k}/\alpha)^2)^{1/2} \ , \ \overline{\xi}_o = (1 - (\overline{k}/\alpha_o)^2)^{1/2}$$

allows an explicit and relatively compact expression of the above condition.

$$\begin{aligned} (1.13) \quad & 4\overline{\xi}_o\overline{\xi}_c\overline{\xi}_s(2\overline{k}^2-1) - \overline{\xi}_o\overline{\xi}_c\overline{\xi}_s(8\overline{k}^2 - 4 + \overline{k}^{-2}) \cos\xi_c H \cos\xi_s H \\ & + j(\rho_o/\alpha_o\rho_c) \left[\overline{\xi}_c \cos\xi_c H \sin\xi_s H + \alpha\overline{\xi}_c^2\overline{\xi}_s\overline{k}^{-2} \sin\xi_c H \cos\xi_s H\right] \\ & + (\overline{k}^2/\alpha)\overline{\xi}_o(8\overline{k}^2 - 4(\alpha^2 + 2) + \overline{k}^{-2}(1 + 4\alpha^2)) \sin\xi_c H \sin\xi_s H = 0. \end{aligned}$$

Such an equation presents a certain analogy with the dispersion relation associated with the free plate, as follows from earlier calculations [3]. In particular, the effect of fluid loading only appears in the emergence of additional mixed terms of the type $j\alpha_o^{-1}(\rho_o/\rho_c) \sin\xi_{c,s}H\cos\xi_{s,c}H$. This translates a.o. the fact that supersonic waves are subjected to radiation damping.

As an example, fig.2, shows dispersion curves related to the first natural modes of a theoretically undamped elastomer layer. Comparison with free modes ($\rho_o = 0$) is also given.

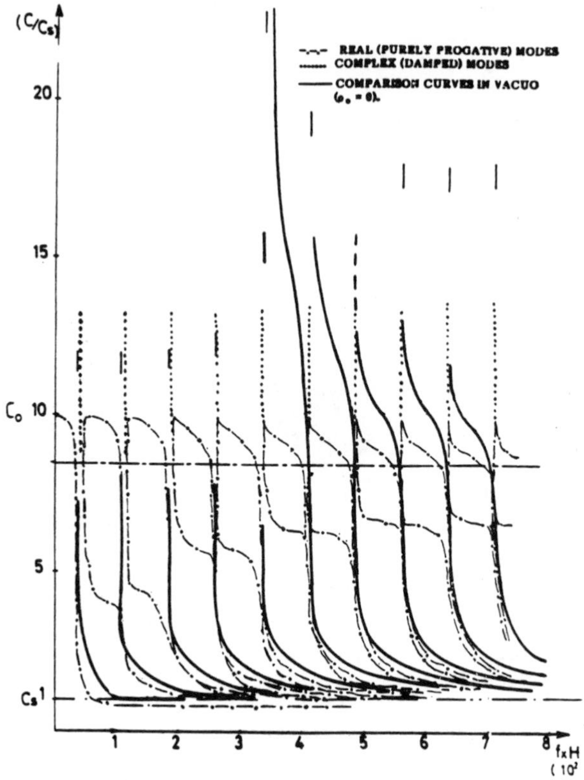

Fig.2 : Dispersion Curves of a Fluid Loaded Elastomer Layer with Rigid Backing at Z = -H

In the interesting case of fluid loading, it appears that purely progagative subsonic modes subsist. Damping of such modes will occur only as a consequence of the specific properties of the elastomer layer. Moreover, in the region $c<2-3c_s$, apart from the expected added mass effect, fluid loading leads to mode splitting around the different cut-off frequencies. Hence, the number of modes close to the asymptote c_s is increased by a factor 2, compared to the in vacuo case.

Qualitative explanations to the obtained results by means of very simple models [1] have been partially achieved. In particular, we considered the case of a fluid layer with propagation velocity c_c replacing the actual elastomer layer with identical compressional wave velocity c_c. That model successfully explains the behavior of the presented dispersion curves between c_c and c_o.

Of course, predictions about TBL noise reduction as a function of the layer properties seem quite difficult to derive only from the consideration of such dispersion curves. Nevertheless this type of preliminary study constitutes an essential reference to both validation and interpretation of the results to be obtained in the frame of the response analysis.

RESPONSE TO POINT EXCITATION

In this second part of the study, we are now considering the complete system, as described in Fig.1, e.g. the upper side of the elastomer layer exposed to point driving force at the origin. The geometrical description rests on the assumption of cylindrical symmetry. The fact that we restrain our consideration to a point unit force implies a search for the Green function of the system. The possibility of taking into account more general (realistic) types of excitation in a further study remains open, and the mathematical formalism associated to such cases won't differ from the one briefly sketched hereafter.

For the fluid medium, the equation of motion is still the Helmholtz equation (1.1) for $p_o(r,z,t)$. The propagation of waves in the elastomer layer is described by Navier's equation (1.3). In quite a similar way as in Section 1, it appears convenient to introduce two potentials Φ and $\underline{\psi}$ in order to express the components of the displacement vector $\underline{u}$. The next step consists in the definition of a so-called Galerkin vector [2], related to Φ and ψ_k in a rather complicated way. For briefness, we drop the detailed calculation [5], and concentrate on the final relations between $\underline{u}$ and $\underline{F}$, which read :

$$(2.1) \qquad 2\,\mu u_i = -\,F_{j,ji} + 2(1-\nu)\,\square_c F_i \quad (i,j = 1,2,3.)$$

where μ is one of the Lamé constants and ν is Poisson ratio. Moreover we have introduced the d'Alembert operators $\square_c$ resp. $\square_s$, usually defined as :

$$\square_{c,s} = \Delta - (1/c_{c,s}^{\;2})\,\partial^2/\partial t^2.$$

The relation (2.1) is a general representation for the u_i's which satisfies Navier's equation, provided the components of $\underline{F}$ satisfy :

$$(2.2) \qquad \square_c \square_s F_i = 0 \qquad (i = 1,2,3.)$$

Furthermore, it is noticeable that the observation of cylindrical geometry implies the vanishing of the two first components of the Galerkin vector $\underline{F}$. If we write $F_3 = Z$, we then have to solve the following system of differential equations [3]:

$$(2.3,4) \quad \square_o p_o = 0, \qquad \square_c \square_s Z = 0.$$

coupled to the boundary conditions :

$$\left.\begin{aligned} &(2.5) \quad \partial^2 u_z/\partial t^2 = -(1/\rho_o)\,\partial p_o/\partial Z \\ &(2.6) \quad \tau_{zz} = -p_o - p_{ext} \\ &(2.7) \quad \sigma_{rz} = 0 \end{aligned}\right\} \quad \text{in } z = 0$$

$$(2.8,9) \quad u_z = u_r = 0 \qquad \text{in } z = -H$$

where displacements and stresses are respectively expressed in terms of the so-called Love's strain function Z. p_{ext} stands for the applied point excitation.

The solution of the above system is obtained by taking the Hankel transform of order 0 of (2.3 -4), associated with the corresponding transformed boundary conditions (order 0 and 1 for the z resp.r components). The problem is then reduced to a linear system of algebraic equations, which finally has to be solved in the four unknown coefficients A_c, B_c, A_s and B_s, related to the transformed function $Z^\circ(k,z,\omega) = H_o[Z(r,z,\omega); r \to k]$ through :

$$(2.10) \quad Z^\circ(k,z,\omega) = A_c(k,\omega)e^{j\xi_c z} + B_c(k,\omega)e^{-j\xi_c z} + A_s(k,\omega)e^{j\xi_s z} + B_s(k,\omega)e^{-j\xi_s z}.$$

The main difference between the system resulting from the right outlined approach and (1.11) consists in the presence of the excitation term p_{ext}, making the system inhomogeneous. Once the unknown coefficients have been determined, appropriate relations [4,5], allow us to compute the transformed displacements and stresses at the excitation frequency. In the following, the spectral density response $|\overset{\circ}{\tau}_{zz}(k,z,\omega)|^2$, where z/H ranges from -1 to 0, will be of particular interest.

Validation of the present approach has been made by comparison between our results and some of those presented in [6]. The calculated frequency spectral density sensed by a circular hydrophone embedded in the elastomer layer was written as :

$$(2.11) \qquad Q(\omega) = \text{const.}\hat{P}(\omega) \int_o^\infty k \, |\overset{\circ}{\tau}_{zz}(k,\omega,z)|^2 \, S(k)dk,$$

where S(k) is the circular hydrophone function and $\hat{P}(\omega)$, the pressure spectrum resulting from a point excitation. The factor const. $\hat{P}(\omega)$ has been chosen such as to give the same value as the Corcos expression $P(k_x,k_y,\omega)$ used by Ko, in the limit of vanishing k_x and k_y : $\hat{P} \sim \omega^{-3}$. Under such an assumption, (2.11) gives good estimation of the elastomer layer efficiency in reducing turbulent boundary layer noise, provided the condition $k_s = \omega/c_s << k_{conv}$ holds, where k_{conv} is the convective wave-number associated to the considered TBL model. Comparison with [6] shows good agreement in the cases of greater values of the product (frequency)x (layer thickness).

Some calculations have been made for different test materials ; one of them appears a good compromise between high modal density $\delta = 4fH/c_s$ and low k_s/k_{conv} ratio, in the studied fH interval.

This type of elastomer has Young's modulus $E = 10^7$. Pa. Influence of loss factor ζ_s or stand-off distance d variations has been estimated and corresponding plots of Q(ω) are shown in fig.3(a)-(b). The reference level 0 dB has been arbitrarily assigned to the configuration with parameter values d = 0 and $\zeta_s = 0.05$.

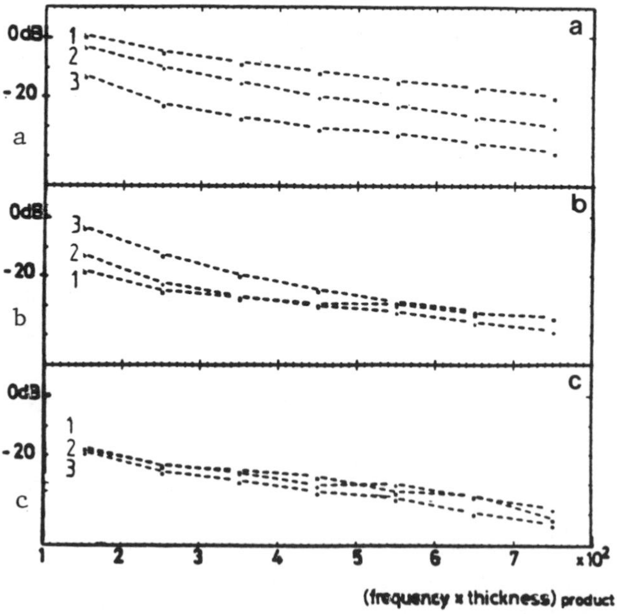

Fig. 3 Power Spectral Level $\tilde{Q}(\omega)$, Arbitrary Units

a) varying loss factor at d/H = 0,5 : ζ_s = 0.05(1),0.10(2),0.30(3)
b) varying stand-off distance for ζ_s = 0.30 : d/H = 0.(1),0.50(2),0.75(3)
c) parameter variation as in a), circular hydrophone of radius r_h = 0.75 H.

In both cases, gains of about 10 dB or more are achieved for point-drive noise reduction between extreme configurations when we consider a point hydrophone. It is noticeable that the optimal stand-off distance is not necessary d = 0 due to rigid surface conditions at z = -H. The effect of spatial averaging inherent in the introduction of a circular hydrophone tends to reduce the effect of both d and ζ_s (fig. 3c). A reflexion resting on similarity confirms the intuitively clear fact that more noise reduction is obtained as the layer thickness increases for a given hydrophone dimension. Various hydrophone sizes have also been considered.

Starting from the same physical configuration as in[6,7] but restricting our modeling under the assumption of point drive excitation, we have obtained good agreement with the results presented there.

Future work will include the consideration of turbulent boundary layer noise, based on the Corcos or on the Chase model, applied to the same type of configuration.

ACKNOWLEDGEMENTS

The present work is supported by the G.E.R.D.S.M. (Groupe d'Etude et de Recherche en Détection Sous-Marine, le Brusc, France) and this is gratefully acknowledged.

REFERENCES

1. D. Vaucher de la Croix, "Modèle du guide d'ondes - rapport intermédiaire", internal report (juillet 1985).

2. Y.C. Fung, Foundations of solids mechanics, Prentice Hall (1965).

3. I.A. Viktorov, Rayleigh and Lamb Waves, Plenum Press (1967).

4. I.N. Sneddon, The use of integrals Transforms, McGraw Hill (1972).

5. D. Vaucher de la Croix, "Calcul de réponse", internal report (1985).

6. S.H. Ko, "Theoretical prediction for the reduction of flow induced noise on a plane surface", International Union of Theoretical and Applied Mechanics, Symposium on , Lyon (1985).

7. S.H. Ko, "Turbulent boundary layer noise reduction by means of a viscoelastic material", Proc. EUROMECH Colloquium 188, Leeds (1984).

PARAMETRIC ACOUSTIC ARRAY APPLICATION USING LIQUIDS WITH LOW SOUND SPEED

A.W.D. Jongens* and P.B. Runciman†

*Central Acoustics Laboratory, University of Cape Town
†National Underwater Acoustic Centre, Simonstown
South Africa

INTRODUCTION

Humphrey[1] first demonstrated the use of a parametric array to measure the insertion loss of small, half a metre square, panels of material using a 1MHz primary source and positioning the test specimen within the interaction zone of the array at a range of 700mm from the primary source. A parametric array is capable of producing a very narrow beam over a wide frequency range with a single, relatively small, projector. Humphrey was thereby able to extend the 50kHz lower frequency limit obtained with conventional methods of measurement to approximately 10kHz. Decreasing source level and increasing diffraction to sound around the specimen with decreasing frequency limits the lowest useable frequency.

A need arose to measure the insertion loss of materials at frequencies lower than 10kHz. Previous experiments[2] had shown that a significant increase in difference frequency pressure and in directivity of the beam pattern could be obtained by permitting part of the interaction of primary to difference frequency waves to occur in a liquid with low sound speed. The liquid was contained in a thin walled tube whose length was short compared to the interaction length of the array.

The present paper discusses the feasibility of applying this method of increasing the source level and directivity to extend the lowest measuring frequency by a further decade.

MEASUREMENT OF INSERTION LOSS

The acoustic characteristics of materials immersed in water can be determined by measuring the insertion loss and echo reduction of rectangular panels of material. The insertion loss is obtained by recording a projector-to-receiver signal before and after the insertion of the panel between the two transducers.

Theoretical predictions of insertion loss and echo reduction assume plane waves to be propagating and panels of infinite extent. In practice, the insonifying beam is not plane but spherical and the panels are of finite extent.

See page 810 for Abstract.

A factor limiting the lowest useable frequency is the diffraction of th wave around the edge of the finite sized test specimen. Diffraction becomes increasingly significant as the frequency is decreased until a limit is reached when it is no longer possible to resolve the diffracted wave from th transmitted wave in time. For conventional methods the lower limiting frequency is approximately 50kHz.

Use of a directional source will provide a reduction of the edge diffracted wave. However, this is difficult to achieve at lower kilohertz frequencies when using conventional methods. A means to overcome this is to employ a parametric acoustic source.

THE PARAMETRIC ARRAY

The radiation of two finite amplitude primary waves, closely spaced in frequency, from a directional source will produce a wave at the difference o two primary frequencies due to nonlinear interaction of the primaries in the liquid. The difference frequency wave can be considered to be produced by virtual sources distributed throughout the volume of interaction of the primary beams. The amplitude and phasing of these sources form a virtual end-fire array producing, in the farfield, a narrow beam without side lobes.

The formation of the difference frequency wave is cumulative, with the axial pressure increasing rapidly within the collimated nearfield of the primary source and reaching a maximum in the near/farfield transition region Beyond the nearfield, the rate of growth is offset by spherical spreading an absorption of the primary wave. The difference frequency amplitude starts t decrease until, at a range approximately equal to the inverse sum of the primary absorption coefficients, there is insignificant further formation of difference frequency and the amplitude decreases according to the spherical spreading law.

MEASURING INSERTION LOSS USING A PARAMETRIC SOURCE

The difference frequency beam has not been fully formed at the primary near/far transition range. However, a relatively narrow beam with reduced sidelobe level has already been produced which, together with a maximum in axial pressure and minimum spot size at approximately the same range, provides an optimum range from the primary Source for the insonification of test specimens of small size.

The interaction zone needs to be truncated at the cost of a slight broadening of the beam pattern by inserting a low-pass acoustic filter. Thi is necessary so as to attenuate the primary frequencies sufficiently in orde to reduce further interaction beyond the filter to an insignificant level whilst permitting the low frequencies to pass through without significant attenuation. The filter must also provide sufficient attenuation of the primaries to avoid nonlinear distortion occurring in the measuring hydrophone. Humphrey used a 13mm thick sheet of cork filled butyl rubber which provides over 40 dB attenuation at 1MHz while producing less than 2dB attenuation below 100kHz[1].

Use of a parametric source has made it possible to extend the lower frequencies to a new limit of approximately 10kHz. Due to the relatively short length of the array at this frequency, there is no longer significant amplitude shading of the difference frequency beam at this frequency. In addition, due to a decrease of difference frequency pressure amplitude with decreasing frequency, the signal-to-noise ratio becomes very small.

An increased source level and increased directivity is required to extend measurements to lower frequencies.

USE OF LIQUIDS WITH LOW SOUND SPEED

The difference frequency pressure amplitude is proportional to,

$$\frac{f_d^2 p_1 p_2 \beta}{\rho c^4 \alpha}$$

where β is the nonlinearity parameter of the liquid, f_d is the difference frequency, p_1 and p_2 the primary pressure amplitudes, ρ the liquid density, c the sound speed and α the primary absorption coefficient.

f_d decreases by 12dB for a halving of the difference frequency. This imposes a lower frequency limit on the difference frequency depending on the signal-to-noise ratio, signal processing method used and insertion loss of the material.

The difference frequency pressure amplitude is inversely proportional to sound speed to the fourth power and directly proportional to β. Previous experiments[2] had shown that a significant increase in amplitude of the difference frequency and in directivity could be obtained by permitting part of the interaction of primary-to-difference frequencies to occur in a small quantity of Freon 113, a liquid with a sound speed half that of water.

Propagation through the interface between low-speed liquid and water will cause some of the energy to be reflected, unless the characteristic impedances are the same. If the two liquids have the same characteristic impedance, the gain in amplitude of the difference frequency is then proportional to sound speed to the third power.

A decrease in sound speed is accompanied by a proportionate decrease in wavelength. The difference frequency amplitude, measured on the transducer axis in water outside the tube, is accompanied by a narrowing of the difference frequency beam pattern due to an increase in the number of wavelengths within the nearfield length.

The results of the reported experiments suggested the possibility of extending the useable low frequency limit in insertion loss measurements by using a low-speed liquid in a tube fitted with a projector at one end and incorporating an acoustic low-pass filter at the other end of the tube.

This could provide an additional benefit. The distance between the source and truncating filter needs to be kept constant during measurements. Any relative movement, for example due to disturbance of the water, will cause the amplitude of the difference frequency beyond the filter to vary. The tube can now maintain a constant distance between source and filter. The question arises whether the size of the tube can be kept small enough for a portable system to be realised.

THE SLOW-WAVEGUIDE

The Freon in the tube immersed in water acts as a slow-waveguide acting to concentrate the sound waves within the tube by progressive refraction.

The phase speed, C_z, at which the difference frequency wave travels in the tube is not equal to the characteristic sound speed of the unbounded Freon, C_f, but lies between that of water, C_w=1480m/s, and that of Freon, C_f=715m/s; C_z is a function of frequency and tube diameter approaching C_f at high frequencies and C_w at low frequencies.

The guided sound waves tend to be increasingly confined within the tube a the frequency is increased[3,4]. Fig. 1 shows curves of the dependence of C_z as a function of frequency, f, for tube diameters of 110mm and 220mm calculated according to methods derived in reference (4). The ratio of the sound power flow within the tube to that outside the tube can be calculated[4] and is given in Fig. 2.

The high frequency primary waves, propagating at sound speed C_f are essentially collimated within a radius equal to the source transducer and do not "see" the Freon/water interface except at the very end of the tube. The virtual endfire array thus propagates at velocity C_f.

For high difference frequency waves, which are generated by the virtual sources, the C_z is essentially also that of the unbounded Freon with most of the power flow being within the tube. A decrease in sound speed relative to that of water is accompanied by a proportionate decrease in wavelength and increase in number of wavelengths within the primary nearfield. The increas in axial difference frequency amplitude, as measured in water outside the tube, is accompanied by a beam pattern with greater directivity and lower sidelobe level compared to an array in water measured at the same range.

For lower difference frequencies there is an increased difference frequency power flow outside the tube for decreasing frequency. Outside the tube the waves propagate at a higher speed, C_w, than within the tube, C_z. Waves propagating through the Freon/water interface are refracted towards th axis causing a narrowing of the beam pattern and hence an increase in directivity. This offsets somewhat the beam broadening due to the decrease in the number of wavelengths accommodated within the tube.

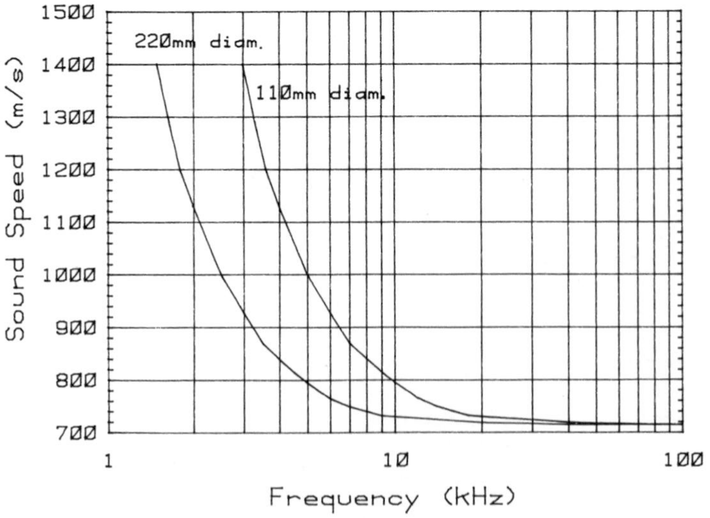

Fig. 1 Phase Speed versus Frequency

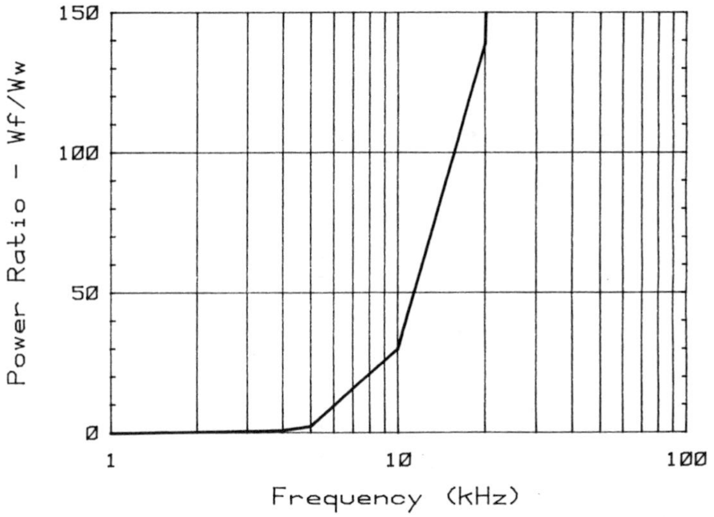

Fig. 2 Ratio of Power in Freon to that in Water for 110mm Diameter Tube.

An additional effect within the tube at lower frequencies is that the generated difference frequency waves travel at a higher speed, C_z, than the virtual sources which travel at speed C_f. Each wavelet is then not quite in phase with the previously generated wavelet. This will result in increasing sidelobe level with decreasing difference frequency.

A lower useful frequency limit for a given tube radius is reached when Cz rapidly approaches C_w as the difference frequency is decreased. Although there may be some increased source level, there will no longer be an improvement in beam pattern compared to an array in water. From Fig. 1, it is anticipated that this lower frequency limit is approximately 5kHz for a 110mm diameter tube and 2.5kHz for a 220mm diameter tube. The ratio of power flow within the tube to that without is 2.5 at these frequencies.

EXPERIMENTS

A 38mm diameter piezoceramic transducer with a Q of 11 radiated two closely spaced high frequency pulsed waves at a mean primary frequency of 1MHz. A B&K 8103 hydrophone with a flat response up to 100kHz was positioned at a range of 800mm from the projector. Cork filled butyl rubber was not available. In place thereof an acoustic low-pass filter comprising two quarter wavelength (1mm) thick brass plates 400mm square positioned at ranges of 400mm and 800mm from the transducer. This provided 46dB attenuation at 1MHz. The hydrophone was positioned at a range of 810mm.

The difference frequency was obtained by adjusting the frequency of each generator by half the diference frequency on either side of the transducer resonant frequency. The output was gated before being fed to a power amplifier. The hydrophone was then moved along a path perpendicular to the transducer axis to obtain the pressure across a plane.

The measurements were repeated with the projector mounted in one end of a 0.46mm thick PVC tube 110mm in diameter and 340mm long. The tube was filled with Freon 113, a liquid with a measured sound speed of 715m/s, density 1553kg/cubic metre and β estimated to be 6.5. The interaction length in Freon was approximately 22 metres. The length of the tube was equivalent to the Fresnel length in Freon at the mean primary frequencies. The input electrical signal was kept constant in both cases.

Fig. 3 shows 80, 40, and 10kHz difference frequency beam patterns for interaction in water and in Freon. There is a significant increase in directivity and it can be seen that the beam patterns after interaction in Freon at a particular frequency are equivalent to that in water at double that frequency.

The increase in difference frequency pressure due to interaction in Fre relative to that in water was calculated to be 24dB for the same array length and allowing for a measured difference in electro-acoustic efficiency of the transducer in the two liquids.

Table 1 records the half power beamwidth measured at 800mm for interact ion in water and Freon respectively as well as the gain in difference frequency pressure at several frequencies.

TABLE 1 Measured half-power beamwidths in water and Freon and difference frequency pressure gain due to interaction in Freon

Frequency (kHz)	Beamwidth (deg.) Water	Freon	Pressure Gain (dB)
80	12.4	6.8	17
40	19.4	10.8	17
20	25.4	18.4	14
10	46.0	26.0	14

The measured beamwidths in water were within 10% of calculated values assuming an end-fire array of virtual sources of constant amplitude. For the array in Freon the measured values were 20% narrower than that calculate on the basis of a simple line array.

An insertion loss measurement was carried out on a 450mm by 600mm by 12.4mm thick mild steel plate. The calculated and measured normal incident insertion loss are shown in Fig. 4.

The signal applied to the transducer in this experiment was a tone burs centred on the resonant frequency of the transducer and modulated by a haversine function. The pulse length was adjusted for different frequency ranges to maximize the produced signal. The received signal was amplified, and digitized by a transient recorder. A frequency spectrum was obtained by FFT techniques on 100 sample averages.

The measured insertion loss values were within 1dB of the calculated values for frequencies down to 6kHz. Below this frequency the values varied increasingly from the calculated values. This appeared to confirm the low frequency limit of the tube used in the experiment.

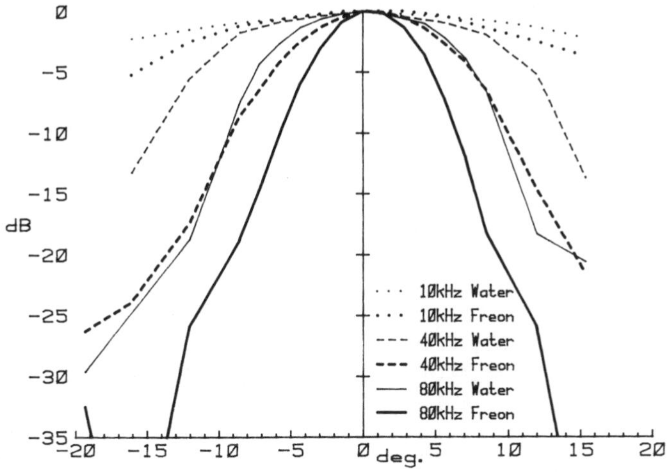

Fig. 3 Difference Frequency Beam Patterns after Interaction in Water and Freon.

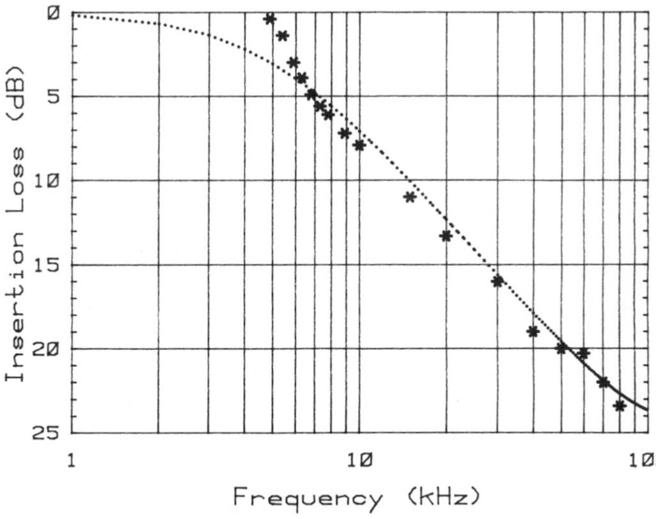

Fig. 4 Calculated and Measured Insertion Loss of 12.4mm Thick Mild Steel.

CONCLUSION

A significant reduction in beamwidth and increase in difference frequency pressure over a wide range of frequencies was obtained using a length of low-speed liquid which was short compared to the interaction lengt of the parametric array. It is anticipated that the use of a tube with twic the diameter of that used in the experiment as well as a liquid such as Freon C51-12, with a sound speed of 540m/s, will extend measurements down to 1kHz.

REFERENCES

1. Humphrey,V.F. The measurement of acoustic properties of limited size panels by use orf a parametric source. J. Sound & Vibration, 98(1), 67-81, 1985.
2. Jongens,A.W.D. Improving the conversion efficiency of the parametric acoustic array. Proc. 11 I.C.A., July 1983.
3. Jacobi,W.J. Propagation of sound waves along liquid cylinders. JASA, 21,2, 120-127, 1949.
4. Haswegawa,A. Propagation of guided sound waves in acoustic slow-waveguides. Acustica, 52, 237-245, 1983.

SIGNAL PROCESSING FOR PRECISE OCEAN MAPPING

Oivind Heier

Simrad Subsea a/s, Strandpromenaden 50
P.O. Box 111, N-3191 Horten, Norway

ABSTRACT

The paper gives a description of signal processing for wide swath mapping systems based on data obtained by delayed sampling giving inphase and quadrature components of the complex envelopes of the incoming signals. It includes a description of expected echo duration, phase information between two half beams which constitute one beam, receiving beamforming, amplitude and phase detection in each beam, and time-of-arrival estimation. Theoretical and measured values are compared. Measurements are made at 95 kHz for depths down to 500m.

INTRODUCTION

Compared with single beam echo sounding methods giving depth values along a single track directly beneath the survey ship's track, wide swath bottom mapping techniques present an attractive and efficient means of obtaining bathymetric data over large areas of the bottom. The decisive performance of such a system is the ability to make a correct bottom detection. Knowing how the characteristics of bottom return signals for a single beam echo sounder vary with system parameters, depth and bottom (type, roughness, slope . . .) such systems require an increase in system complexity and signal processing for simultaneous detection and estimation of time-of-arrival (TOA) for bottom return signals from different directions. Therefore these systems have increased numbers of transducer elements, power amplifiers, preamplifiers, and electronic circuits for beamforming, signal processing, data handling and system control.

BEAMFORMING

The construction of the transducer array is an important consideration for transmitting and receiving beamforming operation. The number of elements, interelement spacing and athwartship curvature of the transducer have to fulfill the requirements for making transmitting and receiving beams with specified beamwidths and sidelobe levels within a specified athwartship sector (Steinberg, B., 1976). The beams also have to be stabilized against roll and pitch. In the fore and aft direction there is only one beam looking in the vertical and so the number of elements, electronic circuits and cabling can be reduced by mechanical stabilization against pitch. Athwartship there are a lot of narrow receiving beams and the transducer has to consist of many elements and electronic stabilization of the beams or sectors against roll is easy to implement.

The transmitting sectors should be as closely matched as possible to the receiving sectors in order to reduce reflections from unwanted directions. This may be achieved with a multielement transmitting array and by controlling the signal output (amplitude and phase) to the different elements. At low frequencies and for narrow beams, the transducer size may be large. The same elements may be used for both transmitting and receiving in order to reduce the size of the total transducer and to minimize cabling between transducer and electronics.

If the beamformer is based on phase rotation and not on time-delay, a wavefront coming along a beam axis should reach all elements used in making the beam within a time (ΔT) small compared with the transmitted pulse length (T). ΔT less than T/10 has shown to be a good design criterion. The same is true for sampling of the elements used in making the beam. They have to be sampled within approx. T/10.

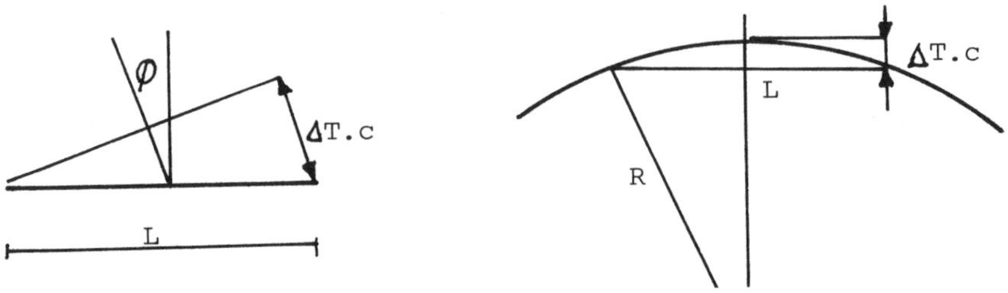

a) Plane array

b) Curved array

Figure 1

For a plane array, Figure 1a, the maximum dimension L is approx. determined by Eq. [1]:

$$L = \frac{k\ 50.6^{o}\ c}{\theta_0 f_0 \cos\phi} \tag{1}$$

where θ_0 is the minimum beamwidth (-3dB) [in degrees], c is the sound velocity [m/s], f_0 is the centre frequency [Hz], k is a constant ranging from 1 to 1.5 depending upon the weighting function determining sidelobe levels (Jenkins, J.W., 1980) and ϕ is the beamsteering angle [in degrees]. The maximum difference in time of arrival equals

$$\Delta T = L \sin\phi_{maks}/c \quad . \tag{2}$$

For a curved array with radius R, Fig. 1b, the maximum difference in time of arrival equals

$$\Delta T = R\ (1 - \cos(\text{arsin}(L/2R))/c \quad . \tag{3}$$

The transmitted pulse is usually a short pulse having a depth resolution given by Eq. [4]:

$$\Delta R = cT/2 \quad . \tag{4}$$

The received waveform is a bandpass signal with frequency content confined to $f_0 \pm \Delta f/2$, $\Delta f = 1/T$. It can be shown that all of the information in a bandpass signal is contained in the complex envelope, (Knight, W.C. et al., 1981). A bandpass signal x(t) may be described by Eq. [5]:

$$\begin{aligned} x(t) &= X_I(t)\cos(\omega_0 t) + X_Q(t)\sin(\omega_0 t) \\ &= \mathrm{Re}(X(t)e^{j\omega_0 t}) \quad , \end{aligned} \tag{5}$$

where X(t) is the complex envelope given by Eq. [6]:

$$X(t) = X_I(t) - jX_Q(t) \quad , \tag{6}$$

where $X_I(t)$ is inphase and $X_Q(t)$ is the quadrature component of the complex envelope respectively and where $\omega_0 = 2\pi f_0$.

A technique for obtaining approximate values of the inphase and quadrature components is by delayed sampling of the signals. The inphase component of the signal from element i at time t_0, $X_I(i,t_0)$ is given by the sample at time t_0 and the quadrature component $X_Q(i,t_0)$ with the sample at time $t_0+1/(4f_0)$. For a system having many elements it may be practical to multiplex signals prior to sample-and-hold (S/H) and analog-to-digital (A/D) conversion for reducing the number of electronic circuits required. The elements have to be sampled in groups and all elements used for making a beam have to be sampled within a time less than T/10.

The complex output of beam 1 at sample n, y(1,n), is given by Eq. [7] which is a sum of two half beams, left and right:

$$\begin{aligned} y(1,n) &= y_1(1,n) + y_2(1,n) \\ y_1(1,n) &= \sum_{k=1}^{N/2} x(i(k,1))w(k,1) \\ y_2(1,n) &= \sum_{k=N/2+1}^{N} x(i(k,1))w(k,1) \end{aligned} \tag{7}$$

where $x(i(k,1)) = X_I(i,n) + jX_Q(i,n)$ is the complex sample of element i at time n, w(k,1) is the complex weighting for element k and beam 1, and N is the number of elements per beam. The phase between the two half beams y_1 and y_2 can be used for the positioning of a point source within a receiving beam. The phase is given by Eq. [8]:

$$f_{i,j} = \arctan(\mathrm{Im}(Y)/\mathrm{Re}(Y)) \tag{8}$$

where $Y = y_1 \cdot y_2^*$.

The bottom is not a point source, but as long as the reflecting area is limited by pulse length and, as seen from the transducer, represents an angle less than the beamwidth, the phase gives significant information about the position of the reflecting area within the beamwidth as shown by Klepsvik (1983). This means that phase measurements are most useful on the beams having an inclination with the bottom.

REFLECTED SIGNAL CHARACTERISTICS

Expected echo duration

The duration of the bottom reflections and the signals at the output of the beamformer increase with increasing beamwidth, pulse length, depth, and angle between the receiving beam axis and the normal to the "mean bottom plane" at the intersection between the beam axis and the bottom. Due to bottom roughness and the medium, the envelopes have, superimposed, random signals making deep nulls and fluctuations in the data as shown by Morgera and Sankar (1984), and an adaptive amplitude filter has to be implemented. The expected echo duration (in number of samples), $E_{i,j}$ for beam i and ping j, is given by Eq. [9]:

$$E_{i,j} = (T + 2kR_{i,j}\tan\theta_i\tan\alpha_{tot}/c)/T_s \qquad (9)$$

where T is pulse length [s],
$R_{i,j}$ is slant range to centre of beam i, ping j [m],
θ_i is beamwidth in the plane between beam axis and normal, (n) to the mean bottom plane [in degrees],
α_{tot} is angle between beam axis and n [in degrees],
c is sound velocity [m/s],
k is constant depending on definition of beamwidth, and
T_s is sample interval [s].

$$\cos\alpha_{tot} = \cos(T_{i,j}+S_{i,j})/(\cos T_{i,j}\sqrt{1+\tan^2T_{i,j}+\tan^2L_{i,j}})$$

where $T_{i,j}$ is the athwartship bottom slope at the beam axis intersection, for ping j;
$L_{i,j}$ is the fore and aft bottom slope at the beam axis intersection, for ping j; and
$S_{i,j}$ is the angle between the beam axis and the vertical axis.

Eq.[9] requires calculation of T and L for each ping and beam before detection is made. Theoretical values of $E_{i,j}$ are shown in Figure 2 for T_s = 100μs, water depth = 200m, T = 200μs and θ_i = 2.5°. The curves show that $E_{i,j}$ increased rapidly with increasing L, especially at low T+S.

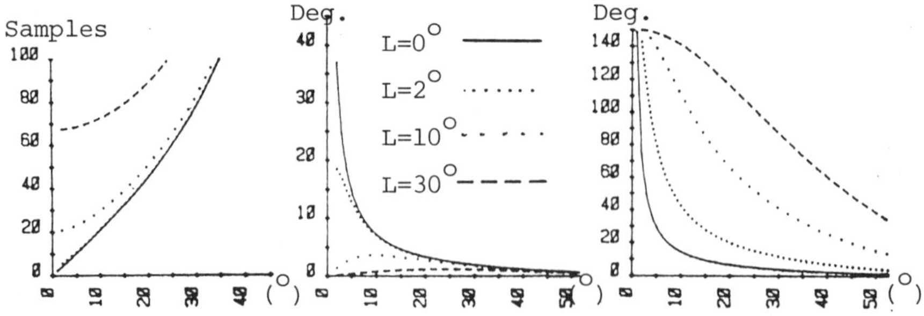

Echo duration Phase change Phase spreading

Figure 2: Theoretical echo duration, electrical phase change per sample, and phase spreading versus athwartship beam direction. θ_i = 2.5°, T = 200μs, T_s = 100μs, depth = 200 m.

Expected values of T and L have to be calculated from depth values or slant ranges in previous pings. This may be quite time consuming. If L = 0, $E_{i,j}$ can be approximated by Eq. [10]:

$$E_{i,j} = (T+k_2\Delta R_{i,j}/(2\tan\theta_s))/T_s \qquad (10)$$

where $k_2 = 2k\ \tan\theta_i/c$, θ_s = beamspacing, and $\Delta R_{i,j} = R_{i+1,j-1}-R_{i-1,j-1}$.

This is a first order approximation in the calculation of $E_{i,j}$ and increased accuracy can be attained by more interpolation between beams and pings.

<u>Expected athwartship phase change per sample</u>

The phase between the two half beams given by Eq. [8] changes with time. The expected phase change between two samples for beam i and ping j is given by Eq. [11]:

$$\Delta f_{i,j} = N\ \arctan\ (DT_s\cos(T_{i,j}+S_{i,j})/R) \qquad (11)$$

where
$D = c\ \cos\beta/(2\sin\alpha_{tot})$,
$\cos\beta = (-A\cos T+B\sin T)/\ \ 1+A^2+B^2$,
$A = (\tan L-B)/\tan T$,
$B = (\tan L+\tan T(\tan T+\tan S)/\tan L)/(1-\tan S\ \tan T)$, and
$T = T_{i,j}$, $L = L_{i,j}$, $S = S_{i,j}$

N is the ratio between electrical and mechanical angles.

Theoretical phase change per sample are shown in Fig. 2 for water depth 200 m and T_s = 100μs. If L = 0 and the bottom is supposed to be plane between beam i+1 and i-1 then α_{tot}=T+S, β=0 and $\Delta f_{i,j}$ can be given by Eq. [12]:

$$\Delta f_{i,j} = k\ \tan\theta_s T_s/\Delta R_{i,j} \qquad (12)$$

where k = 57.2 N c .

The curves in Fig. 2 show that $\Delta f_{i,j}$ decreases rapidly with increasing T+S when L is small. When L increases, $\Delta f_{i,j}$ reduces, especially at small T+S.

<u>Expected phase spreading</u>

As already mentioned, the bottom is not a point reflector and the phase given by Eq. [8] consists of contributions from different point reflectors. If the reflecting area is supposed to be limited by beam width and pulse length, the maximum difference in measured phases for two points within this area is given by Eq. [13]:

$$f_d = N\ \arctan\ \{[R_i\tan\theta_i\sin\beta +$$

$$(cT_{i,j}\cos\beta)/(2\sin\alpha_{tot})\cos(T_{i,j}+S_{i,j})]/R_i)\}. \qquad (13)$$

Theoretical values of f_d are shown in Fig. 2, showing increasing values with increasing L.

<u>Examples</u>

Fig. 3 shows theoretical echo amplitude, phase and phase spreading versus time for 2 different beam directions and depths. In both cases we have θ_i = 2.5°, T = 200μs, L = 0 and $T_{i,j}$ = 0. The phase curve goes through zero at the time when the centre of reflecting area passes the beam axis, giving the possibility of a good estimate to time arrival (TOA).

TIME-OF-ARRIVAL ESTIMATION (TOA)

Based on the depth values for previous pings, expected slant ranges and bottom slopes can be estimated and used for the determination of data acquisition windows and as input parameters for the filtering of beamformer outputs and in algorithms for determination of the time when the phase curve crosses zero.

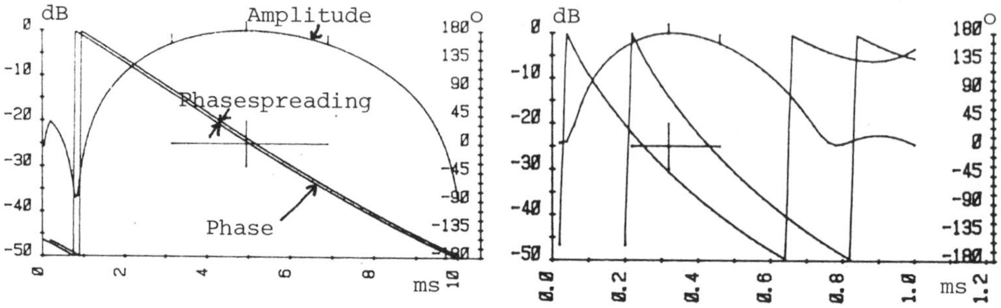

Depth=100 m, S_i=30°, T=200μs

Depth=50 m, S_i=5°, T=200μs

Figure 3: Theoretical echo envelope, phase curve and phase spreading vs. time

Amplitude detection

Due to fluctuations and differences in echo duration as given by Eq. [9], the amplitudes for the different beams have to be filtered in different filters prior to detection. The lengths of the filters are determined from the expected echo duration and the filter coefficients may be calculated to match the echo amplitude versus time. Experiences from EM100 and filter simulations have shown that an unweighted filter works quite satisfactory and is much easier to implement. When T+S is small and at short ranges it is important that the filter is not too long smoothing out the amplitude.

In an amplitude detector the definition of the bottom varies with the angle between the beam axis and the bottom. At normal incidence the start of the echo returned gives the distance to bottom. This does not coincide with the maximum amplitude and a threshold has to be used. When T+S>0, a higher threshold or max amplitude gives the best estimate of slant range at the centre of the beam.

Phase detector

The exact phase between the two half beams is given by Eq. [8] and for bottom detection we are looking for the sample of zero crossing. To avoid implementation of the arctan-function, the sign of Im(Y) and Re(Y) may be used for calculation of a phase function having maximum at the sample of zero crossing. The phase is approximated by Eq. [14]:

$$T(k) = \begin{matrix} 1, & \text{sign Im}(Y(k))=1 \\ -1, & \text{sign Im}(Y(k))=-1 \end{matrix} \qquad (14)$$

$$N(k) = \begin{matrix} 1, & \text{sign Re}(Y(k))=1 \\ -1, & \text{sign Re}(Y(k))=-1 \end{matrix} \quad .$$

These digitized values "exclusive OR" (XOR) with the two filters F_N and F_T of length M shown in Fig. 4. M is determined from expected phase change per sample and accepted phase limits, for example, $\pm 135^o$. The number of samples equals $270/\Delta f_{ij}$. The phase function F(k) is calculated according to Eq. [15]:

$$F(k) = \sum_{n=-M/2}^{M/2} (F_N(n) \quad N(k+n) + F_T(n) \quad T(k+n)) \quad . \tag{15}$$

The value $k = k_o$ where $F(k_o) = F_{maks}$ gives the sample of bottom detection. The phase detector output is accepted if $F(k_o)$ is greater than a limit. If not accepted, amplitude detection has to be used. The detector has shown good results as long as $f_{i,j}+f_d$ is less than 50°. The maximum in the detector output decreases rapidly if the estimated $\Delta f_{i,j}$ is less than the real phase change Δf_r. Curves showing F(k) when $\Delta f_{i,j}$ differs from Δf_r by P% are given in Fig. 5.

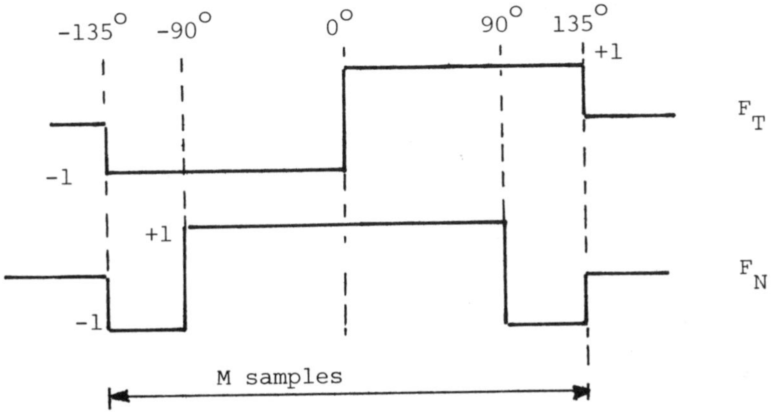

Figure 4: Filters for phase detector.

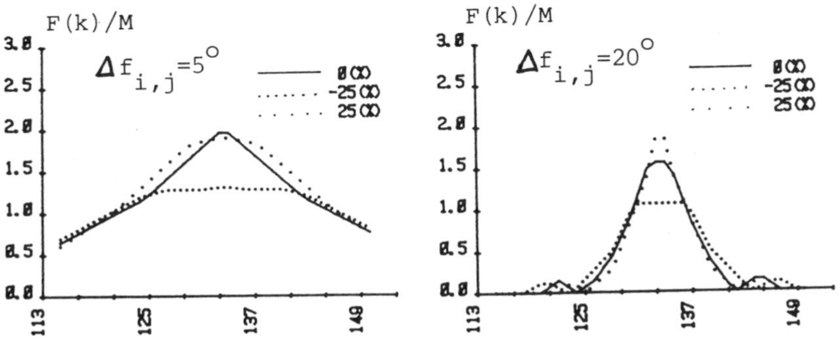

Figure 5: Phase function (F(k)) for different deviations between expected and real phase change per sample $(\Delta f_{i,j})$.

MEASURED VALUES

Measurements of the reflected signal characteristics were made with an EM100, a multibeam echo sounder working at 95 kHz, having 32 beams with beamwidths of 2.5° (athwartship) and 3° (fore and aft) and with phase measurements athwartship. The conversion factor was N = 60. Examples of measured amplitudes, phase curves and bottom detection are given in Fig. 6 and Table 1, showing good agreement with theoretical values.

Table 1: Standard deviations of measured depths

Beam no.	Beam angle degrees	Depth 45m	100m	200m
1	40	.34%	.15%	.30%
8	20	.44%	.20%	.10%
16	0	.30%	.20%	.22%

CONCLUSION

Analysis of reflected signal characteristics and measured values show that a good TOA-estimator has to combine threshold, maximum amplitude and phase detector methods. Measurements show that phase is superior to amplitude as long as sum of the phase change per sample and phase spreading is less than 50°.

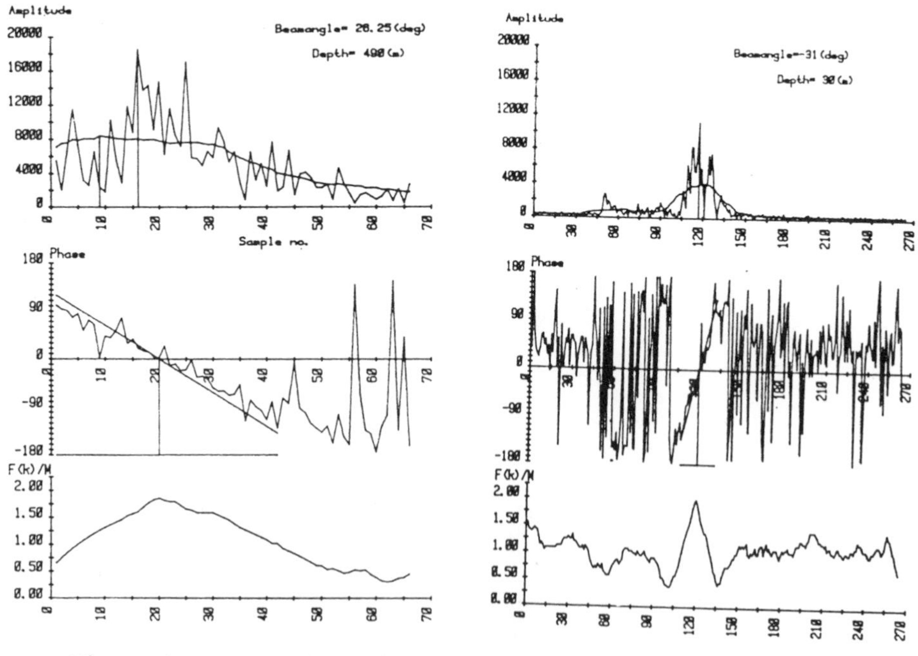

Figure 6: Measured amplitude, phase curve and phase detector output for depths of 30 m (sample interval 7.5 cm) and 490 m (sample interval 45 cm).

REFERENCES

Klepsvik, John Oddvar, 1983, Statistical characteristics of sea-bed reverberation process with application to wide swathe bathymetric mapping. Institutt for Almen Fysikk, NTH, Trondheim.

Knight, W.C., Pridham, R.G., and Kay, S.M., 1981, "Digital Signal Processing for Sonar", Proc. IEEE, 69(11) p. 1451.

Morgera, S.D. and Sankar, R., 1984, "Digital Signal Processing for Precision Wide-Swath Bathymetry", IEEE OE-9(2) p. 73.

Steinberg, B., 1976, Principles of Aperture and Array System Design, John Wiley & Son.

AN UNDERWATER ACOUSTIC POSITION FIXING SYSTEM FOR PRECISE LOW RANGE USE

Raymond Wood

Ceemaid Division
British Maritime Technology Limited
Hampshire, England

ABSTRACT

British Maritime Technology Limited has had over 70 years of experience in the field of ship hydrodynamics including the use of freely running scale models for manoeuvring and seakeeping experiments. As the position of the model is required to be measured to a high accuracy, a great deal of experience has been gained on this subject.

Ultrasonic, through-the-water, position measuring systems have been found to be accurate, reliable and cost effective. A wide range of additional applications have been possible with only minor development. Techniques developed and used under laboratory conditions have been extended for use in much harsher environmental conditions.

The introduction of microprocessor based systems has enabled the production of portable units. These can provide for very accurate measurement at a relatively low cost.

The systems and techniques developed can be adapted for a wide range of applications including R.O.V. tracking, diver tracking and underwater surveying. Current developments will lead to a range of equipment that has significant advantages over radio-based equipment and other systems which are highly complex and hence very costly.

INTRODUCTION

The use of radio-based position fixing systems to locate a vessel relative to shore references are in common use but are in general limited to metre accuracy if the vessel is underway, as for example, whilst undertaking a depth survey. For inland waters such as small rivers, canals and lakes a requirement existed in the United Kingdom for a system which although of smaller range would produce a better location accuracy, typically centimetre.

British Maritime Technology Limited has developed and used acoustic through-the-water position measuring systems for many years in conjunction with ship model manoeuvring experiments. These systems have been adapted to cover the previously mentioned requirement, in particular, to track remotely operated miniature survey vessels also developed by the company.

The resulting position measuring system uses four or more shoreline located ultrasonic receivers connected by cables to a portable control and analysis system based on microcomputer technology. The precise location of a free running ultrasonic pulse transmitter can be measured in real time. The system also provides a local measure of the speed of sound in water which can be used to enhance the accuracy of other acoustic measurement devices such as echo-sounders.

The paper describes the development of the system and its performance characteristics and also discusses a more recent extension of the capability into 3 dimensional position measurement for such activities as diver tracking around man made structures.

ACOUSTIC POSITION MEASURING SYSTEM FOR SHIP MODEL MANOEUVERING EXPERIMENTS

General Requirement for the System

British Maritime Technology in its former role as Ship Division of The National Physical Laboratory, produced its first ultrasonic position measuring system for ship models more than 20 years ago. Previous techniques such as photographic methods were very accurate but required many hours of setting up and analysis before the results from an experiment were available.

An automated system was required that could achieve a high accuracy and be set up and operated in a short time scale. The components on-board the model required to be small in both size and weight. The acoustic approach was therefore chosen as it could fulfil all the requirements.

The Original System

Fig. 1 is a block diagram of the original ultrasonic position plotting system developed for ship model manoeuvring experiments. It operated in the following way.

A pulse from the timing unit started the counters and was also sent by radio to the model where it triggered the transmitter circuitry. The resulting signal from the transmitter transducer which was suspended below the keel of the model, was a 100 microsecond burst at a frequency of 200 kHz. The radial wave travelled through the water and was received by four transducers mounted at the corners of the manoeuvring basin.

The received signals were amplified and converted into levels which turned off the counters. As the counters were fed with a reference frequency which was a multiple of the speed of sound in water, it can be seen that the counter readings corresponded to the distance of the receivers from the model transmitter.

The four counter readings were recorded at 1 second intervals to enable the model's track to be calculated using simple trigonometry.

Further circuitry was added to improve timing accuracy by synchronising model timing to that of the shore based system before the start of the experiment. Additional delays were also added to correct for transducer response times.

System calibration was carried out simply by setting the model to a known position and adjusting the counter reference frequency to compensate for any change in the speed of sound through the water.

Accuracy of the System

The accuracy of the system was a function of the frequency of the ultrasonic pulse, the accuracy of time measurement and the physical properties effecting the propagation of the pulse through the water.

Provided that the water was stable regarding temperature and impurity, which it was to a high degree, an overall accuracy of 75 mm or better was obtained. The dimensions of the basin were approximately 30m x 30m x 2.4m deep.

IMPROVED TECHNIQUES FOR MANOEUVERING BASINS

Basic Requirements

Although the original system performed very well for many years, a replacement system was required about six years ago. It was decided to further develop the system to incorporate some new facilities. Apart from improving the overall accuracy and range of the system, the elimination of the radio link was to be a fundamental requirement. Interfacing of the measurement system to a computer for real-time analysis was also necessary.

Theory of Operation

In order to eliminate the radio link between model and shore, an indirect method of transit time measurement was chosen. This had the advantage that the equipment carried on-board the model was considerably simplified.

The on-board transmitter sends out a continuous train of pulses at precise 1 second intervals. Once switched on, the transmitting system required no further attention during the experiments.

The shore based logic control system is shown in Fig. 2. The counters which measure the transit time of the ultrasonic pulse are started a finite time before the pulse leaves the model. Once detected, the circuitry can accurately determine when the next pulse is about to leave. The resulting effect is to add an unknown time to all the counter readings. This added time and hence model position were calculated using the method shown in Appendix 1. It should be noted that more complex, and hence more accurate methods could have been used, but would not have provided the speed of computation necessary for a desk top computer to analyse the results in real time.

Calibration Methods

Unlike the previous system, the improved system used a fixed reference frequency for the counters. This was not a problem as the use of an on-line computer meant that more complex calibrations could be carried out more quickly. Programmes were developed to measure the speed of sound in water and to check the position of the receiver transducers by using a portable transmitter unit. Position measurements were taken at known locations and at the receiver positions to provide a full calibration of the system.

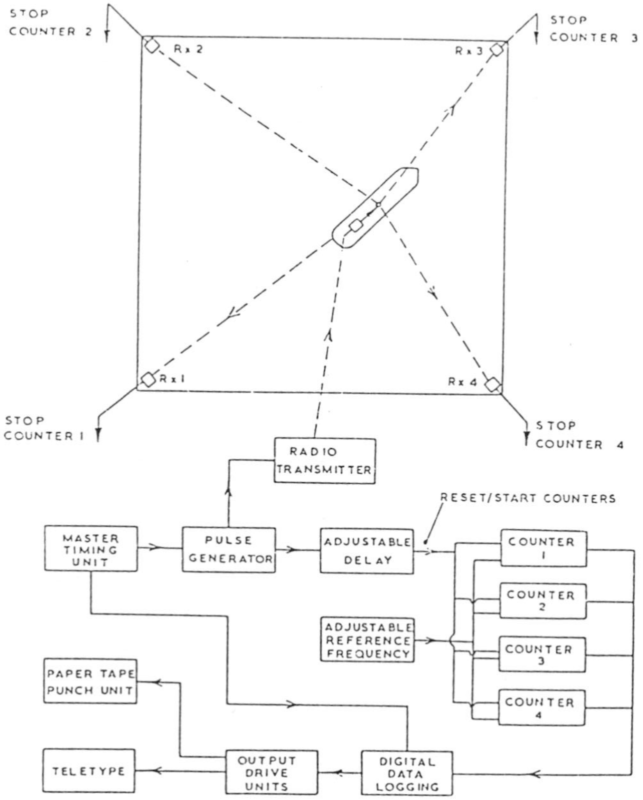

Fig. 1 The Original Sonic Position Plotting System

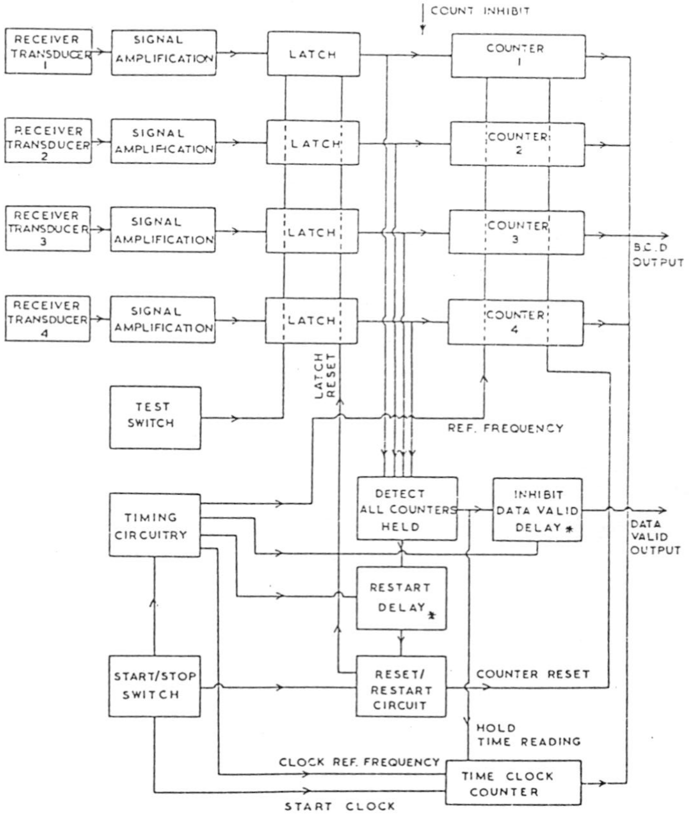

* DELAY TIME SWITCH
TO PROVIDE VARIABLE READING INTERVAL

Fig. 2 Shore Based Control Logic

Practical Experience

The improved system was checked against very accurate photographic methods and found to be accurate to 20mm on average. Over short distances a resolution of better than 10mm was obtained.

This high accuracy was maintained provided that reasonable care was taken during the experiments and calibration was checked daily. Although the temperature remained constant over a long period of time errors were produced by the water chlorination plant which produced local areas of gas filled water around the outlet pipe if left running during the experiments.

Metal structures in the water also produced errors if the transmitter came close enough for the sound to be picked up. The pulse could in this event travel faster along the metal structure and be re-transmitted to arrive at the receiver before the direct pulse through the water. These effects could be detected by the calibration process and eliminated.

Outside Manoeuvring Basin

British Maritime Technology has a large shallow water manoeuvring basin which is built in the open air and measures 60m x 60m x 1m deep. The acoustic position measuring system was installed in this basin providing a similar accuracy over the increased range.

As the water depth can be reduced for very shallow water experiments, typically 200mm or less, temperature effects are more noticeable. The largest effects can be seen during summer and autumn when the morning sun can cause localised hot spots in the cool water. This does not last for long as the temperature stabilises quickly during the early morning. Once again the calibration process identifies this effect.

DEVELOPMENTS FOR FIELD TRIALS APPLICATION

Remotely Operated Survey Vessel

Within the last two years, British Maritime Technology Limited has developed a remotely operated survey vessel. The craft, which is named HydroCAT, is intended for inshore waterways such as rivers, canals and lakes. It is a catamaran about 1.8m long and 1.2m wide which is driven by an electric outboard motor. It can be disassembled for transportation in a small van or estate car together with associated equipment.

In most cases the survey will be over small ranges requiring a high accuracy of position and water depth measurement. The ultrasonic position measuring system was therefore ideal for this application.

The System Applied to HydroCAT

In general the equipment used for the manoeuvring basin was applied directly to the survey craft. The only modifications that were required involved operating all the equipment from batteries and housing it in weatherproof enclosures. The systems also needed to be reliable and robust.

The four receiver transducers are mounted on suitable supports along the waters edge. The exact positioning is not critical, but they must be in line of sight with each other. As the range of the system is about 200m a separation of 25m between them is typical.

The received signals are sent to the combined control and analysis microcomputer via cables. The computer, which is housed in a weatherproof container, performs all time measurement functions as well as control and data analysis.

Operation of the System

As the receivers are used in a line at random positions, appropriate calibration and analysis software has been written. The operator uses a hand held transmitter to take readings at two points at a known distance apart and at the four receiver locations. The computer programme reads the time data, calculates the local speed of sound in water and the positio of the receivers relative to a base line between two of them.

Once the survey vessel has been launched its position is measured relative to the receiver baseline which can be transposed to standard grid references if required.

INVESTIGATIONS INTO THREE DIMENSIONAL MEASUREMENT SYSTEMS

The success of the 2D ultrasonic measurement system together with the advances made in the microcomputing systems has led to further recent developments.

A number of trial experiments have been carried out to determine the feasibility of developing 3D underwater measurement systems which are capable of centimetre accuracy over short ranges.

One trial involved the use of six transducers mounted in a three dimensional array around the test area. This was situated at the end of a pier in Southampton Water. A combination of salt and fresh water together with an industrial environment proved to be an ideal testing ground. Measurements were made using 180kHz and 64kHz operating frequencies over short ranges, approximately 15m x 13m x 4m deep.

The results showed that although a number of problems existed, 3D measurement was possible to a very high accuracy. Good correlation was obtained with measurements made by a tape measure and repeatibility over short timescales was as good as 5mm.

An outline design for a low cost 3D underwater measurement system has now been completed. It is expected that centimetre accuracy will be obtained over areas of 200m x 200m with a reasonable depth capability.

FUTURE APPLICATION

The 3D system under development will be suitable for a number of applications. These are being investigated at the time of writing to enabl any special requirements to be built into the system.

The use of cables to transmit the pulses from the receiver transducers to the control microcomputer may not be possible for some applications. For many others, this does not pose any problems. 3D tracking and measurement around permanent and semi-permanent man-made structures are typical examples.

As well as diver tracking for safety purposes, the diver will be able to use the system for underwater measurement. Examples of this are the accurate measurement of damage to the hulls of ships below the waterline and the surveying of wrecks for archaeological reasons. It is intended that both of these applications will be tested by British Maritime Technology in the near future.

CONCLUSIONS

British Maritime Technology Limited has used acoustic position measurement systems for many years and found them to be accurate, reliable and cost effective. It is the author's opinion that developments could lead to a wide range of future applications involving 2D and 3D position measurement. The use of advanced microcomputers together with intelligent software will lead to significant reductions in cost and systems that are highly automated.

APPENDIX 1

The Analysis of x and y co-ordinates.

Nomenclature

V_o	Velocity of sound in water
x_o	Distance between receiver transducers, x direction
y_o	Distance between receiver transducers, y direction
x	Co-ordinate of model position
y	Co-ordinate of model position
C_1, C_2, C_3 & C_4	Four counter readings
C_5	Maximum counter readings
T_1, T_2, T_3 & T_4	Four times represented by counter readings
T	Time added to all four readings
L_1, L_2, L_3 & L_4	Distance travelled by pulse in time T_1 etc
L	Effective length added to all readings
t	Transit time of pulse from model to receivers
l	Distance of travel of pulse from model to receiver
f_o	Reference frequency of counters
x') y') L') Y')	Intermediate calculations of the co-ordinates, added length and distance travelled.

The counting time is represented by $T_1 = C_1 \times \frac{1}{f_o}$ etc

The transit time of the sonic pulse:- $t_1 = T_1 - T$ etc

The distance travelled by the pulse:- $l_1 = V_o \times t_1$ etc

The equations of the x, y co-ordinates are therefore:-

$$[T_1 - T]^2 V_o^2 = x^2 + y^2 \qquad (1)$$

$$[T_2 - T]^2 V_o^2 = x^2 + (y_o - y)^2 \qquad (2)$$

$$[T_3 - T]^2 V_o^2 = (x_o - x)^2 + (y_o - y)^2 \qquad (3)$$

$$[T_4 - T]^2 V_o^2 = (x_o - x)^2 + y^2 \qquad (4)$$

$$y = \frac{[T_1 - T]^2 V_o^2 - [T_2 - T]^2 V_o^2 + y_o^2}{2y_o} \qquad (5)$$

or

$$y = \frac{[T_4 - T]^2 V_o^2 - [T_3 - T]^2 V_o^2 + y_o^2}{2y_o} \qquad (6)$$

and $$x = \frac{[T_1-T]^2 V_o^{\,2} - [T_4-T]^2 V_o^{\,2} + x_o^{\,2}}{2x_o} \tag{7}$$

or $$x = \frac{[T_2-T]^2 V_o^{\,2} - [T_3-T]^2 V_o^{\,2} + x_o^{\,2}}{2x_o} \tag{8}$$

$$T = \frac{T_1^{\,2} + T_3^{\,2} - T_2^{\,2} - T_4^{\,2}}{2[T_1 + T_3 - T_2 - T_4]} \tag{9}$$

$$L = \frac{L_1^{\,2} + L_3^{\,2} - L_2^{\,2} - L_4^{\,2}}{2[L_1 + L_3 - L_2 - L_4]} \tag{10}$$

The value of L obtained (L') using equation 10 may be inaccurate due to errors in the readings or errors produced by the calculation when close to the tank centre lines.

Although these errors may be large, L' can be used to find approximate values for the co-ordinates using equations 5 and 7 or 6 and 8 (x' and y').

A more accurate value of L can then be found using additional equations.

IMPROVEMENT OF UNDERWATER ACOUSTIC LOCALIZATION BY COHERENT PROCESSING

M. Zakharia*, P. Arzelies†, M.E. Bouhier+,
J.P. Corgiatti* and B. Fouché+

*I.C.P.I. Lyon, Laboratoire de Traitement du Signal
UA 346b, CNRS; 25, Rue du Plat, 69288, Lyon Cedex 02, France

†IFREMER, Centre de Toulon, Zone Portuaire du Brégaillon
B.P. No. 2, 83501, La Seyne-sur-Mer Cedex, France

+OCEANO Instruments, 4 AV.H. Poincaré, 92167, Antony Cedex
France

ABSTRACT

Underwater acoustic navigation systems are based on the detection of responses from immersed transponders. Long baseline systems use a set of immersed transponders to localize a surface ship or an underwater vessel, with respect to the sea bottom.

For long ranges the signal-to-noise ratio is too weak for a good detection. The noise is mainly due to the ship (or vessel) machinery and propellers.

Solving this problem should take into account all the limitations due to the natural environment (noise, multipaths, Doppler effect . . .). Using a coherent processing system can help in solving these problems. We will analyze the possibilities and the limitations of such a processing and give an example of a real-time implementation.

We will show some results obtained in sea experiments where the aim was to double the classical systems' range.

INTRODUCTION

The localization of an underwater transponder is the basis of underwater navigation systems commonly used. Detecting the transponder response to an interrogation signal will be limited by the influence of noise. The noise is mainly radiated by the ship (vessel or platform) machinery and propellers [1]. The noise level is usually reduced by a simple spatial processing on the directivity of the receiving arrays. Nevertheless, this processing is not sufficient for long ranges (more than 10 km).

The receiver detection performance is also reduced by the various limitations due to the natural environment: noise, reverberation, multipaths, Doppler effect, . . .

A coherent processing scheme can be designed to reduce the effect of these various limitations. "Pulse compression" can provide the separation between the signal duration (or energy) and the range resolution (proportional to 1/B, where B is the signal bandwidth). Widening of the signal bandwidth can increase the range resolution; it can also be very useful for maintaining Doppler tolerance (mainly by a suitable signal design [5], [6]) and for discriminating against the reverberation [7]. Increasing the signal duration increases the signal energy for a given emitter maximum level (mainly limited by the transducer linearity). Such a processing scheme can increase the signal to noise at the receiver output (by increasing the signal duration) without degrading the range resolution (given by the signal bandwidth) [8].

UNDERWATER NAVIGATION SIGNALS

The various communication signals used for underwater navigation (long baseline) are summarized in Figure 1 [2], [3].

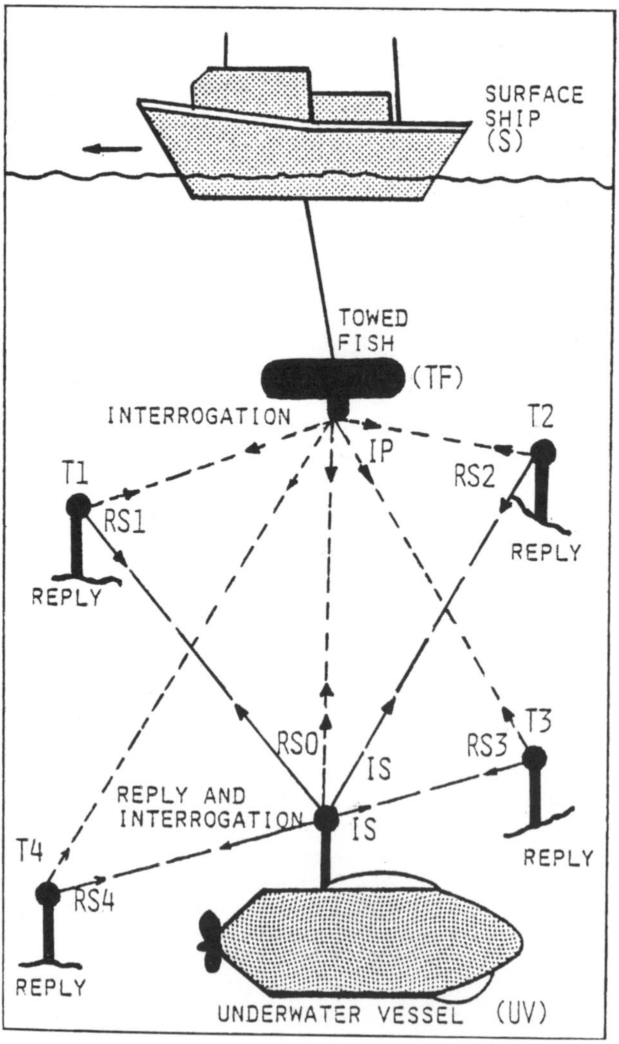

Figure 1: Underwater Navigation

A positioning sequence can be described as follows.

- The ship (Towed Fish, TF) emits an Interrogation Pulse (IP) to the Underwater Vessel (UV) and to the immersed Transponders (Tj).

- The interrogation is detected by the transponders. Each transponder Tj emits a specific Reply Signal (RSj).

- The responses are received and processed on the surface ship.

- These responses can be used for positioning the ship with respect to the "transponder set" (sea bottom).

- The Interrogation Pulse is also detected by the Underwater Vessel (UV), which re-emits an Interrogation Signal (IS) to the Transponders, and a Reply Signal to the ship. The processing of the transponder responses can be used for positioning the underwater vessel with respect to the "transponder set".

The signals commonly used are pure tone bursts. The burst duration is usually about 10 ms. Its frequency is a specific characteristic of the transponder response (or interrogation). The total available bandwidth is from 8 to 16 kHz, for "long baseline" systems.

For each channel, the processing can be briefly summarized as follows:

- band-pass filtering centered around the channel specific frequency;

- automatic gain control (and/or clipping);

- envelope detection; and

- threshold comparison.

The range and the precision of such systems depend mainly on the navigation conditions; typical values are 10 km., for the range, and a few meters, for the precision (in good propagation conditions).

NOISE ORIGINS AND PROPERTIES

The main detection problem (in underwater navigation) occurs while trying to detect the transponder Reply Signal on the surface ship (the Underwater Vessel is always closer to the Transponders and its noise is usually low compared to the surface ship noise).

The ship noise (in the signal bandwidth) is mainly due to the machinery and propellers. Its spectral and statistical characteristics depend on the navigation conditions.

We studied noise recordings from an oceanographic ship [9], [10]. The noise analysis can lead to the following general results.

- The noise level depends on the navigation conditions, nevertheless, it can be considered as "locally stationary" in the short term (a few seconds, corresponding to the signal duration).

- For the same time duration, the estimation of its probability density can be approximated by a gaussian distribution.

- In a small frequency bandwidth (1 or 2 kHz.), its estimated power spectrum can be considered as constant (white noise in the signal bandwidth).

In such conditions, the assumption of a white gaussian noise (in each signal bandwidth) seems to be a good approximation. The stationarity is verified, on the signal duration; in a practical case, the noise can be "stationnarized" by an automatic gain control (with a suitable time constant) or by a clipping.

We can reduce the ship noise influence by increasing the distance between the towed fish and the ship and by an appropriate spatial directivity design of the receiving transducers.

Even in such a case, the signal-to-noise ratio can be too low for long ranges; at a range of 9 km, we have measured a signal-to-noise ratio of about -6dB, at 13 kHz (that leads us to estimate -30dB for a range of 20 km and a frequency of 16 kHz!).

OTHER LIMITATIONS

Two main limitations (other than noises) can be encountered while operating in a natural environment: Doppler effect and multipaths.

The Doppler effect is due to the ship (or vessel) motion. The compression (or dilatation) ratio is small ($s<1.01$). Nevertheless, while using "narrow band" signals (compared to the central frequency), its influence cannot be neglected.

At the output of the matched filter, the Doppler effect can lead to a "loss" in the signal-to-noise ratio and, in some cases, to a bias. An appropriate signal design can help in solving these problems [5], [6].

Multipath problems are mainly due to the acoustic wave reflections from the sea bottom, the water surface and the ship hull. To get rid of this limitation, we need a very sharp range resolution, but we need at the same time a high signal-to-noise ratio ("high energy" signals, i.e. long time duration). This cannot be achieved by classical (envelope detector) systems.

COHERENT PROCESSING PROPERTIES

Coherent processing can help in solving this detection problem. Due to the "pulse compression", the range resolution is only related to the signal bandwidth and not to its duration. We can increase the signal-to-noise ratio at the output of the processing by increasing the signal duration, without perturbing the range resolution.

In such a case, we can find an energy-resolution compromise.

- Increasing the signal bandwidth, B, will provide better range resolution (proportional to 1/B). The first arrival can then be considered as the direct path. The signal bandwidth will be limited by the available bandwidth and by the number of transponders (channels).

- Increasing the signal duration will increase the detection performances, without affecting the range resolution. The signal duration will be limited by the power limitation of the transducers, the transponder autonomy and the real-time processing computation time.

- Choosing a suitable signal design (linear period modulation) [5], [6] will help in discrimination against the Doppler effect.

As a very rough approximation, we can assume that the processing gain, in the classical sonar equation [1] can be expressed as: 10 Log T/T_0, where T is the modulated signal duration and T_0 is the burst duration, under the assumption that both signals have the same instantaneous power (T and T_0 can be the equivalent durations in the case of amplitude modulated signals).

Performance improvements can be computed more precisely for well known receivers with a given architecture [4].

To double the range (from 10 to 20 kilometers), in that frequency range, the processing gain should be approximately 10 dB. [1]. We should use signals with an equivalent duration of about 100 ms.

DIGITAL IMPLEMENTATION

Implementing a matched filter will lead to a Finite Impulse Response (F.I.R.) digital filter (tapped delay line). The elementary operations are delay, multiplication and addition. The digital filter should operate in real time on the received signal shifted in frequency. All the needed operations should be done during the sampling interval.

Let T be the signal duration and B its bandwidth. We will need at least 2BT samples to define the impulse response of the filter matched to Z(t) and the sampling rate, F, should be at least 2B.

For our application, we have used a specialized microprocessor, the TMS 320-10, from Texas Instruments. We have added an external F.I.F.O. (First In, First Out) data memory. We can combine, in that way, the acquisition of a new sample and the sample-set shift. The total computation time is then reduced to 2BT multiplications and additions. This computation time CT should be shorter than the input sampling rate:

$$CT < 1/F .$$

This leads to the inequality:

$$CT=2BT(M+A) < 1/F < 1/2B$$

where M is the time needed for a multiplication and A is the time required for an addition.

For the TMS 320-10 processor, M+A=800 ns, including the data transfers. That leads to the inequality:

$$B^2.T < 3.125x10^5 \text{ Hz}.$$

The analytic expression for the signal used is:

$$Z(t) = \exp\left(\frac{-\log^2(t/t_0)}{2\log(g)}\right).\cos\left(\frac{2\pi b\log(t/t_0)}{\log(g)}\right) .$$

This signal has been chosen because of its interesting properties, mainly in Doppler tolerance. [5]

Figure 2 gives an example of such a signal and shows the "pulse compression" obtained by a real-time matched filter (digital implementation using TMS 320-10 processor).

The signal parameters are:

- $g = 1.15001$
- $t_0 = 56.25$ ms
- $b = -2.71253$, duration, T = 150 ms (at -37.5 dB)
- bandwidth, 100 Hz to 900 Hz (at -37.5 dB).

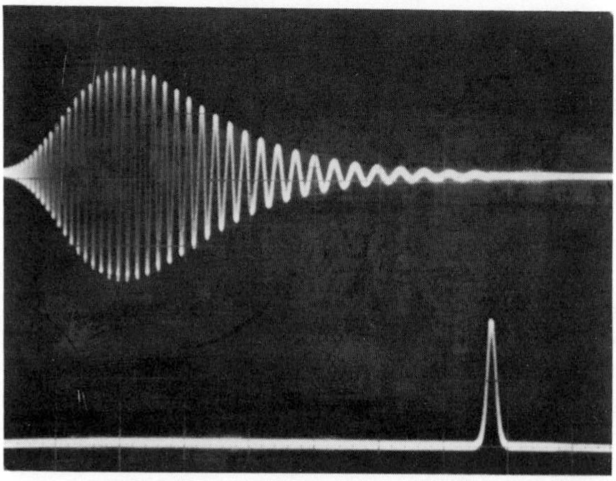

Figure 2: Coherent Processing Example (20 ms/div).

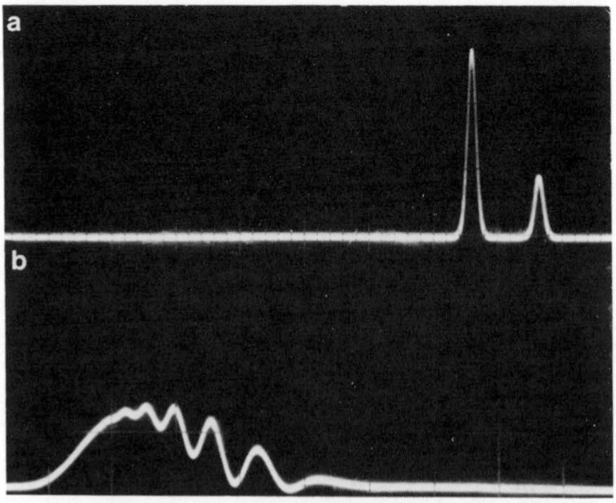

Figure 3: Range Resolution and "Pulse Compression" (20 ms/div); a) coherent processing; b) envelop detection.

LAB SIMULATION

We have simulated, in the lab, the main limitations: noise, multipath and Doppler effect, separately.

The noise simulations have been done with a noise generator and with ship noise recordings. The detection performances are very similar with natural or synthetic noise.

Figure 3 shows an example of multipath at 30 meters. We can clearly see the difference in time (range) resolution between the "coherent processing" and the "classical quadratic envelope detector". This explains our interest in "pulse compression".

The Doppler effect influence is a loss in the output signal-to-noise ratio (approximately 3 dB for a speed of 1.5 knots, similar to classical systems) and a time shift (7 ms for 1.5 knots) of the output. This time shift is theoretically equal to zero, for such signals [6]. It is due to the way the coherent processing is done: frequency shift (Doppler independent) before the correlation.

In a natural environment, there will be a combination of the perturbations due to the various effects. As a rough approximation, to try to predict the receiver performances under operational conditions, we can suppose that this perturbation will be the sum of the elementary perturbations.

SEA EXPERIMENT

For technical convenience, the emitted signal will be the sign of Z(t); the filter will be, anyhow, matched to Z(t).

For the sea application we have designed three modulated signals, SM1, SM2, SM3 with various bandwidths and time durations. Their main characteristic parameters are given in Table 1, where T is the signal duration (defined at 37.5 dB attenuation), B is the signal bandwidth (defined at the same attenuation), Te is the equivalent duration (defined as the duration of a constant amplitude signal possessing the same peak value and the same energy) and Be is the equivalent bandwidth (defined the same way).

Table 1: Signals Characteristics.

	SM1	SM2	SM3
T	100 ms	150 ms	100 ms
Te	70 ms	100 ms	70 ms
B	800 Hz	800 Hz	1200 Hz
Be	450 Hz	450 Hz	650 Hz

During the sea experiment, the same transponder can emit a tone burst or a modulated signal (at the same level). The signal choice can be made by the interrogation frequency (16 kHz or 14.5 kHz). The pure tone response signal is at 9 kHz (with 10 ms duration). The equivalent bandwidth of the classical system input filter is about 450 Hz. The modulated signals are shifted around 9 kHz.

The interrogation frequency is changed from one interrogation to another. The responses are received under the same propagation conditions. The results for the range increasing, in various conditions, are shown in Table 2.

Table 2: Comparative performances in the sea.

SIGNAL	RANGE m pure tone	RANGE m modulated signals	DEPTH m
SM1	?	9 300	3 000
SM1	5 700	10 500	3 000
SM1	?	17 000	4 000
SM1	9 600	16 550	4 000
SM2	8 900	15 700	4 000
SM2	?	14 600	4 000
SM2	?	20 900	4 000
SM2	9 600	14 900	4 000
SM2	8 500	13 800	4 000
SM3	7 800	13 200	4 000
SM3	7 050	13 400	4 000
SM3	?	15 150	4 000

CONCLUSION

The results obtained in the natural environment clearly show the performance improvements of navigation systems by using coherent processing in all the cases. In some cases the range gets close to 20 km, and even beyond.

We could increase that range further still by increasing the signal duration (the level is limited by linearity problems). Nevertheless, we cannot expect to get more important improvements easily; we will then start dealing with long range propagation and have troubles measuring the acoustic path length with sufficient precision. [1]

ACKNOWLEDGEMENTS

This work was supported by IFREMER, Centre de Toulon.

REFERENCES

1. Urick, R.J., "Principles of underwater sound", McGraw Hill, 3rd Ed., 1983.
2. Zakharia, M.E., "Coherent processing in underwater acoustic localizing", IEEE International Conference on Computers, Systems & Signal Processing, Bangalore, India, December 1984.
3. Farcy, A., "Localisation précise sous-marine de 0 à 6000m.", Association Technique Maritime et Aéronautique, 1976.
4. Bouhier, M.E., and Zakharia, M.E., "Comparaison de diverses chaines de détection en vue d'applications opérationnelles". Colloque GRETSI sur le traitement du signal et ses applications, Nice, France, May 1985.

5. Mammode, M., "Estimation optimale de la date d'arrivée d'un écho sonar perturbé par l'effet Doppler; synthèse de signaux "large bande" tolérants", Doctor of Engineering Thesis, I.N.P.G., Grenoble, France; May 81.
6. Zakharia, M.E. and Pey, J.M., "Détection et comptage de poissons par sonar haute fréquence. Possibilités de reconnaissance des espèces et d'estimation de leur taille". Colloque GRETSI sur le traitement du signal et ses applications, Nice, France, May, 1983.
7. Ol'Shevskii, V.V., "Statistical methods in sonar", Consultant Bureau, Plenum Pub., 1978.
8. Van Trees, L., Detection, estimation and modulation theory, John Wiley and Sons Inc., New York, 1968.
9. Bouhier, M.E., "Etude en vue de l'amélioration des systèmes de navigation sous-marine". Rapport de stage de D.E.A., INPG, Grenoble, France, 1984.
10. Zakharia, M.E., Bouhier, M.E., Corgiatti, J.P., Bachet, O., "Projet NACRE, étude de faisabilité du système". Report on Contract No. 83/4338, Internal report ICPI and IFREMER (CNEXO, BOM), 1984.

A BEAM STEERING PROCESS FOR SEISMIC DATA

S.P. Cheadle and D.C. Lawton

Dept. of Geology and Geophysics
University of Calgary
Calgary, Alberta, Canada, T2N 1N4

ABSTRACT

A computer program has been developed on the Cyber 205 supercomputer at the University of Calgary that performs a beam steering process on physical model seismic reflection data. The routine is applied during the synthesis of receiver array configurations used to compress raw shot records from the modeling system. Spatial and velocity filtering methods for the suppression of random and coherent noise are applied during the process. The computationally intensive approach used is facilitated by the high speed of the vector computer, generating each beam steered output trace in less than one second.

INTRODUCTION

A physical seismic modeling system has been developed at the University of Calgary (Cheadle et al, 1985). The system is currently being used to study problems encountered by reflection seismic methods in areas of the Canadian Beaufort Sea affected by subsea permafrost. Scaled models based on assumed permafrost distributions are submerged in the water-filled modeling tank and ultrasonic seismic reflection data are recorded. Experiments are being conducted to explore acquisition and processing approaches that will facilitate high resolution imaging of the shallow section in the Beaufort Sea area for engineering and exploration objectives.

The combination of strong velocity contrasts caused by variable icebonding occuring shallow in the stratigraphic sequence poses difficulties for conventional seismic methods. The streamer cables used are commonly longer than the depth to the strong reflectors associated with icebonded sediments, so much of the energy approaches the streamer at appreciable angles from the vertical. Field receiver arrays, or the hydrophone groupings summed together to generate an output trace on the seismic records, are designed to attenuate energy with nonvertical angles of incidence. As shown schematically in Figure 1, amplitude and phase distortion resulting in lost high frequency resolution may affect shallow wide angle reflections of interest. Seismic records from this and other areas are often contaminated by waterborne coherent noise due to scattering (Larner et al, 1983). The geometry of the problem also poses problems for conventional velocity analysis based on hyperbolic normal moveout as the

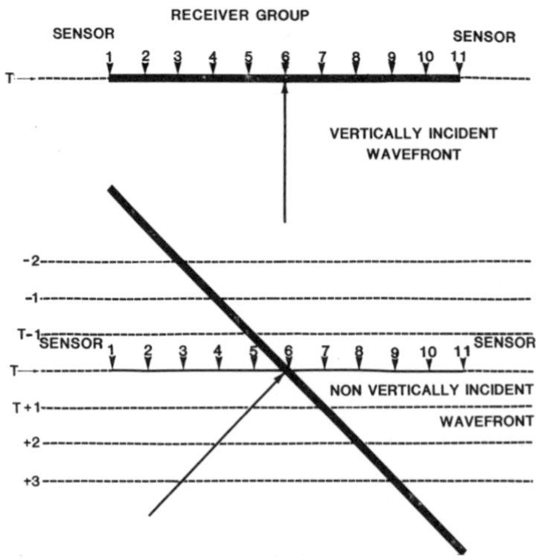

Figure 1. Schematic comparison of vertically vs. non-vertically incident wavefronts across a hydrophone receiver array.

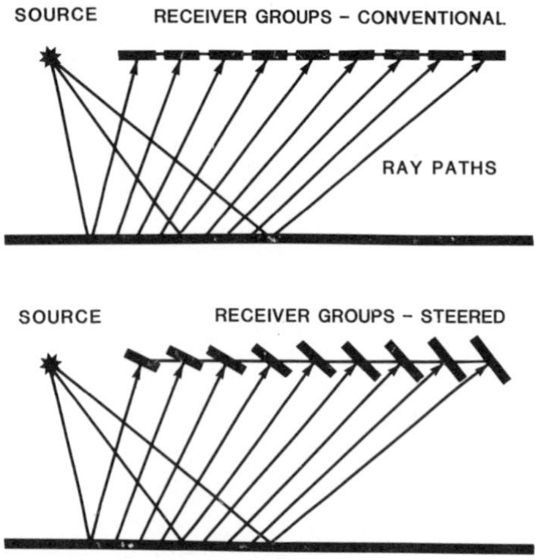

Figure 2. Schematic comparison of conventional horizontally deployed receiver arrays vs. synthetically steered receiver arrays.

offset range of data applicable to the approximation may become too restrictive.

Slant stack methods have been described by a number of workers to help remedy these problems. Biswell et al (1984) outlined a geophone beam steering method that may be used to minimize amplitude and high frequency loss during data compression by array stacking. Noponen and Keeney (1986) recently demonstrated the application of hyperbolic velocity filtering during the transformation of the time-offset shot record to the ray parameter-intercept time, or p-tau domain to attenuate coherent noise on shot records.

The p-tau transform represents a reorganization of hyperbolic reflection and linear refraction event moveouts on the time-offset record into simpler elliptic trajectories in the p-tau domain. Schultz (1982) has described a method of direct interval velocity estimation from the p-tau section that incorporates the full offset range of recorded data. Kong et al (1985) have used a slant stack approach to spatial filtering applicable to stacked CMP or common midpoint sections. As all the above methods are based on a similar slant stack operation, it is convenient to combine the beam steering and filtering processes with the generation of the p-tau section. A computer program has been developed on the Cyber 205 supercomputing facility at the University of Calgary to simulate array design and apply these techniques to the permafrost model data.

The Beam Steering Method

Data from the modeling system are recorded trace sequentially in a format similar to the raw output from a digital streamer, with each shot record including 120-240 traces with 2.5 meter - 5.0 meter spacings. Arrays are simulated by stacking or summing groups of adjacent traces to produce a compressed output shot record. Field arrays must trade off between the improved signal to noise properties of longer groups and the output trace spacing required for acceptable spatial resolution. By simulating array design in software, array lengths and output trace spacing are independent, allowing longer arrays which better match the Fresnel zone radius of illumination while permitting the desired output trace spacing.

Depending on the moveout characteristics of the model reflection events, as the array lengths increase, so does the necessity of beam steering the array to avoid phase distortion problems. Beam steering in this sense involves locally applying time shifts between the traces in the array group so that the stack is performed along linear trajectories that represent the moveout slopes of events of interest. This is analogous to time-variably pivoting a field array toward each incoming wavefront, as shown schematically in Figure 2.

The beam steering process is outlined in Figure 3 on a simple synthetic shot record. A typical array group may include from 15 to 21 traces, with the center or pivot trace representing the offset position of the output trace from the stack. Summations are taken across the array of input traces along a suite of slopes through a time sample point on the pivot trace. Values across the array are weighted with a simple triangular window. To determine the slope or moveout angle with the best local stack output, the semblance statistic is used. Semblance is a measure of the ratio of output to input energy from the stack, and reaches a maximum for the stack slope that best sums the local wavefield in phase, ie. the slope that is locally tangent to the moveout curve. The process is performed at each time point along the pivot trace, and the output is stabilized by

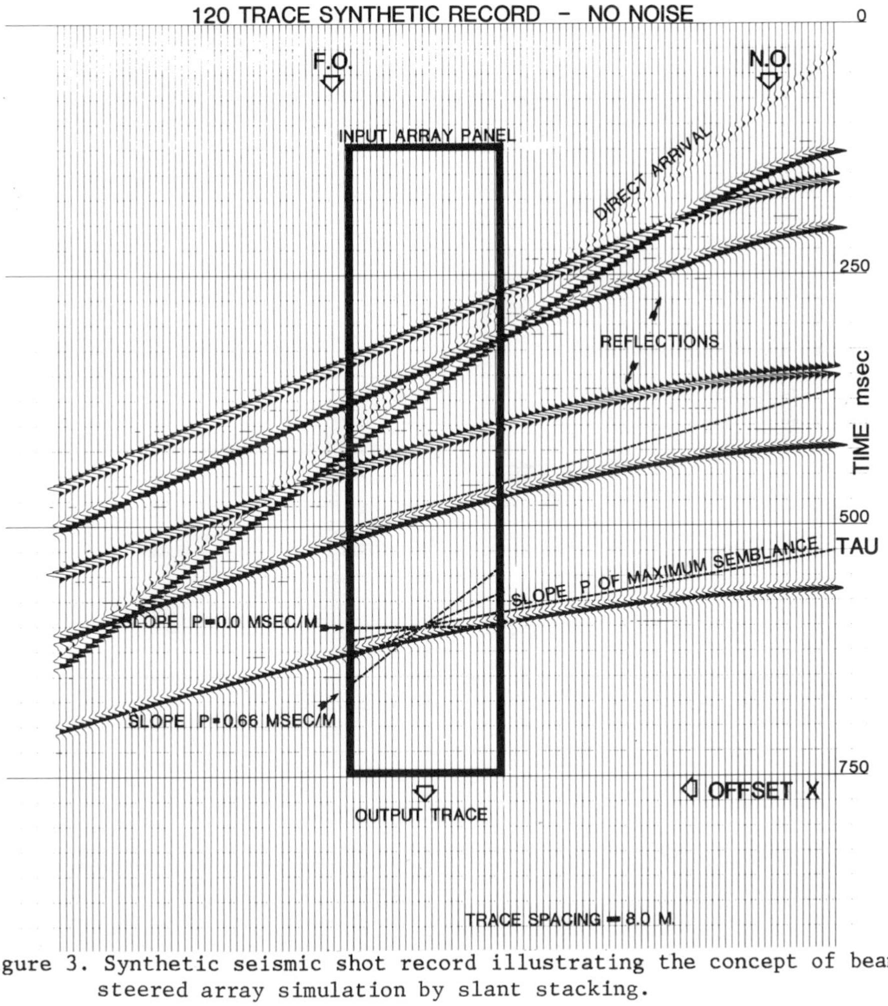

Figure 3. Synthetic seismic shot record illustrating the concept of beam steered array simulation by slant stacking.

smoothing the semblance values for each slope over a time window equal to the length of the wavelength.

Figure 4a shows the semblance filtered stack output at each of the different slope angles from an array at a near offset trace position (labelled N.O. on Figure 3). It can be seen that each event on the raw record at this offset stacks best at a particular range of angles, with the reflections at lower values and the direct arrival representative of coherent noise at steeper slopes. Figure 4b is a similar slope spread display from an output trace position at farther offset (F.O. on Figure 3), showing the general shift of events to later arrival times and steeper stack slopes. Note that the topmost reflection event is nearly asymptotic to the direct arrival at this offset range.

A beam steered output trace is generated by stacking the semblance filtered traces from the slope spread display. A filtering operation may be effected during the beam steering process by limiting the range of slope values used for the stacking and generation of the p-tau section. A fixed limit is analogous to spatially filtering arrivals depending on their angle of emergence. For example, in the marine case a limit of P=0.50 milliseconds per metre is equivalent to an angle of 45 degrees. Noponen and Keeney (1986) have described the relationship between stacking velocities and moveout slopes at a given offset-time position on the shot record based on the normal moveout approximation. Beam steering along slopes time variably limited to a range defined by minimum and maximum velocities about an assumed rms velocity vs. time function acts as a hyperbolic velocity filter which effectively attenuates coherent noise.

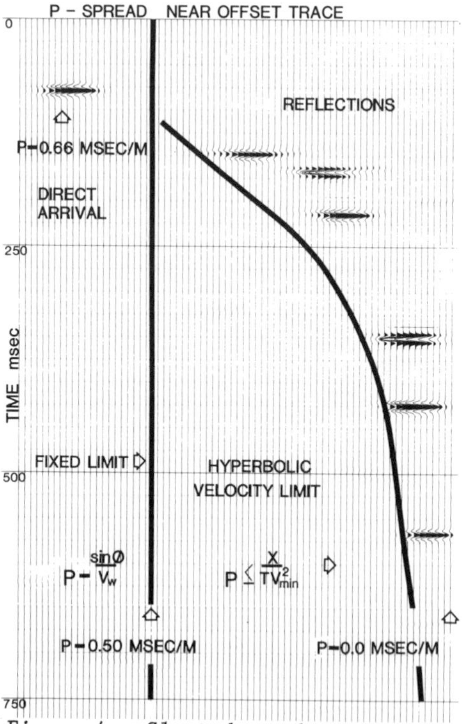

Figure 4a. Slanted stack spread from an input array of traces at a near offset range showing the spatial and velocity filtering limits that may be applied during the generation of a beam steered output trace.

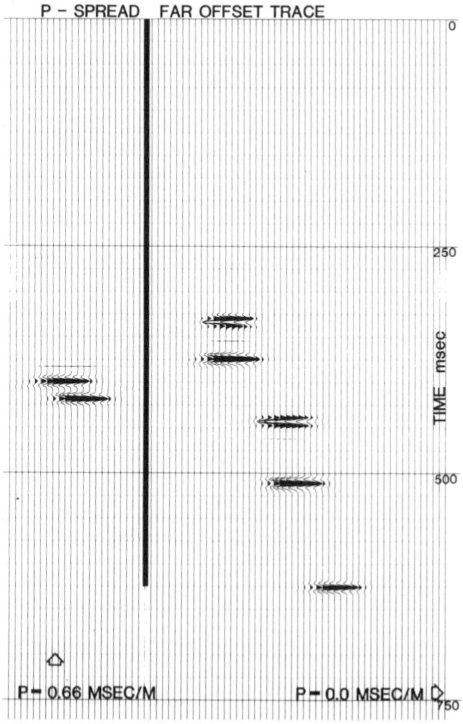

Figure 4b. Slanted stack spread from an input array at a further offset range showing the shift of semblance filtered events to steeper moveout slopes and later arrival times relative to Figure 4a.

Output values with a maximum associated semblance below a chosen threshold value are attenuated. In this manner, a filtered output trace is generated that has been stacked along moveouts optimized for each event on the trace. The array panel is then shifted across the input section to each output position. Depending on the desired trace spacing and the ratio of the compression, 24-48 trace beam steered shot records are generated. Figures 5a and 5b show a synthetic input record with no noise and the compressed beam steered output respectively. Note the elimination of the direct arrival and the asymptotic portion of the topmost reflection event. Figures 5c and 5d are similar input and output from a synthetic record with random noise added. The semblance filtering has virtually eliminated the random noise during the beam steering process. In both cases the array simulation by beam steering has preserved the original wavelet regardless of offset and local moveout.

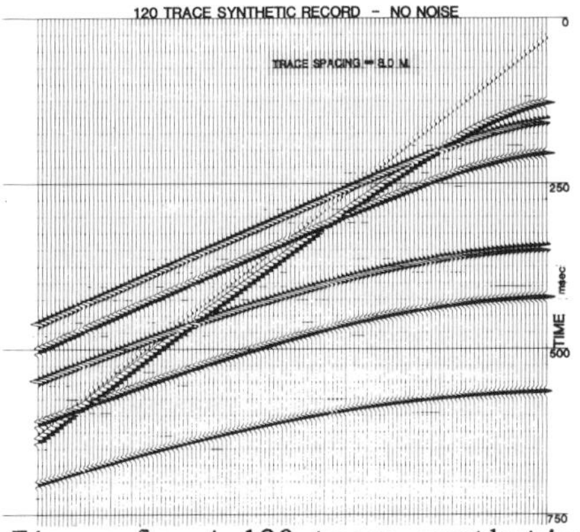

Figure 5a. A 120 trace synthetic shot record with no noise.

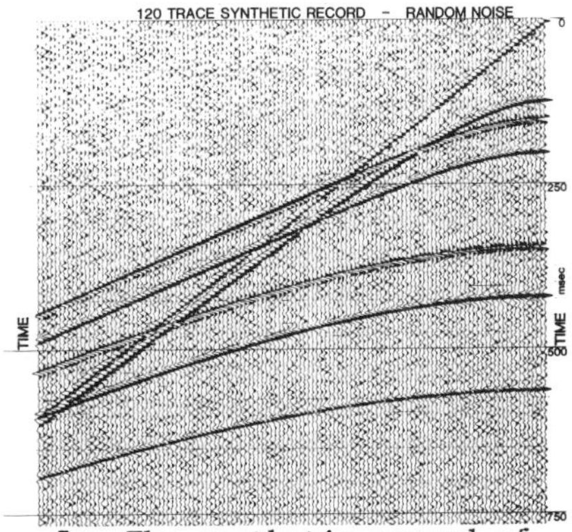

Figure 5c. The synthetic record of Figure 5a with random noise added.

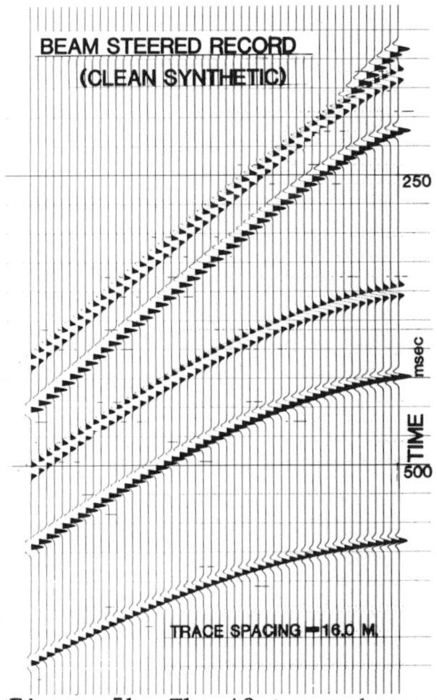

Figure 5b. The 48 trace beam steered output record compressed from the raw input section shown in Figure 5a.

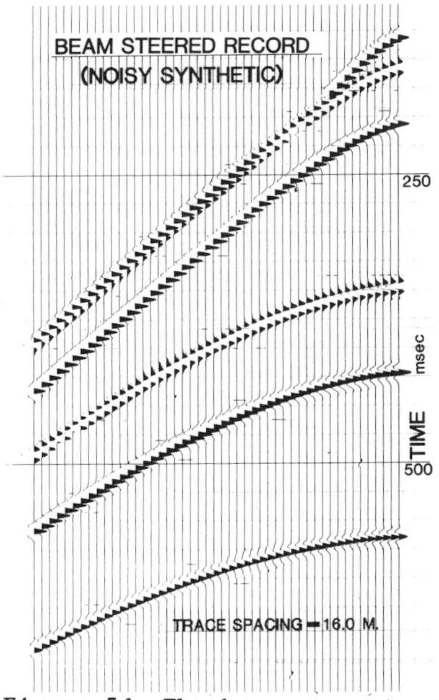

Figure 5d. The beam steered output record from the noisy synthetic section of Figure 5c showing the semblance filtering effect on the random noise component.

From the offset position and time of the pivot point the included stack slope output values can be mapped into the p-tau domain. The storage in memory of the matrix holding the beam steered output section and the p-tau transformed section may be arranged such that the offset position of contributions to the p-tau section are retained. The result of the program is an improved p-tau section, a compressed and beam steered shot record section and a map relating the two. The beam steered shot record can be subsequently treated in the normal manner with the additional benefit of some directional information. The optimum stack slope values, p, can be related to the local wavefront angle of emergence θ by $p = \sin\theta/V$, where V is the velocity of the recording medium, eg. water. Therefore the angles of approach to the recording line of each event on the beam steered section traces are uniquely determined.

The p-tau section can be used for interval velocity estimation using Schultz's method and is potentially useful for separating primary, multiple and mode-converted events. Figure 6a shows a 240 trace raw section recorded over a simple plexiglas sheet that may be used to model a permafrost affected layer in the shallow Beaufort Sea. The combination of a strong velocity contrast and wide angle geometry results in mode conversion of some of the compressional energy to shear energy. Figures 6b and 6c show the comparison of p-tau records from the clean synthetic record and the slab model record. In the absence of mode converted events, reflections typically are mapped to decreasing slope ranges with increasing arrival time. In contrast, the lower velocity mode converted events which occur at and beyond the critical angle of refraction may be distinguished by their displacement on the p-tau record toward steeper slopes.

The same algorithm may be applied to traces from a previously stacked CMP section in a manner similar to that described by Kong et al (1985). The array panel width and slope angle limits are set to pass and enhance the continuity of events within a restricted range of apparent time slopes while attenuating undesired diffraction, sideswipe or other coherent noise.

While the process is mathematically straightforward, it is computationally intensive, requiring on the order of 100-200 million operations for the generation of each output trace. The Cyber 205 vector computer can operate at speeds of up to 400 million floating point operations per second and offers a variety of features which make it highly amenable to seismic processing of the kind described. It was essential that the algorithm was structured to execute operations on values retrieved from contiguous memory locations. Seismic data is naturally arranged in a sequential format, the program was designed to perform as much of the process as possible on a tracewise basis. The large virtual memory capacity of the system enables simple data structures of very large matrices to be utilized which greatly reduces the input/output burden of smaller systems. The vector Fortran 200 syntax enables the code itself to be simplified and condensed. The initial implementation of the routine on a Perkin-Elmer 3240 system involved nearly 1000 lines of code and required 200-300 seconds to generate each output trace. A much more flexible version of the routine on the Cyber 205 system involves several hundred lines of code. A 240 trace raw record of 4096 data elements per trace can be compressed by the beam steering routine to a 48 trace record and the p-tau section generated in approximately 35 seconds on the supercomputer. Recent developments in vector and parallel computing technology should allow the beam steering approach to be extended to full three dimensional processing.

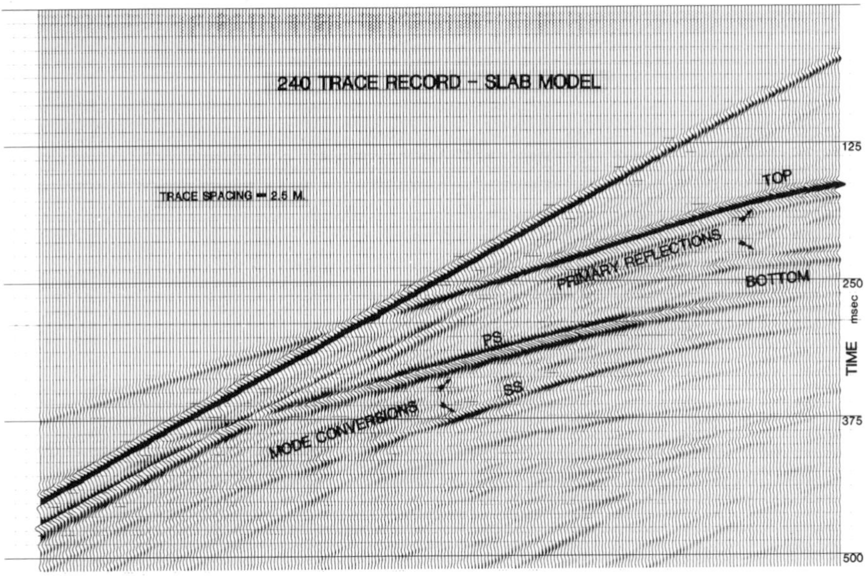

Figure 6a. A 240 trace shot record from the physical modeling system recorded over a simple plexiglas slab model, featuring primary and mode converted reflection events. PS denotes shear propagation one way through the slab while SS denotes shear both ways through the slab.

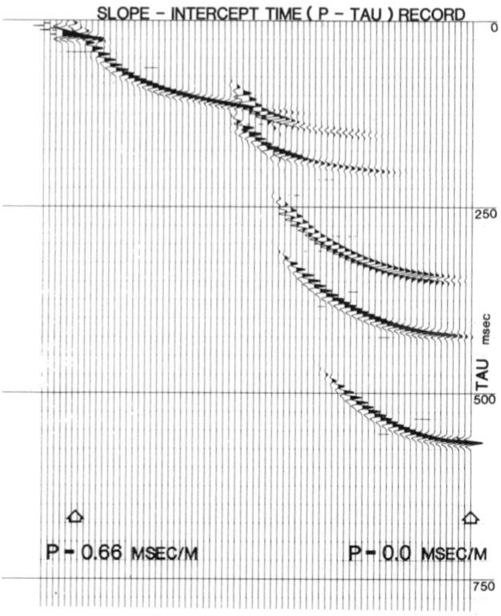

Figure 6b. The p-tau record transformed from the shot record of Figure 5a during the beam steering process.

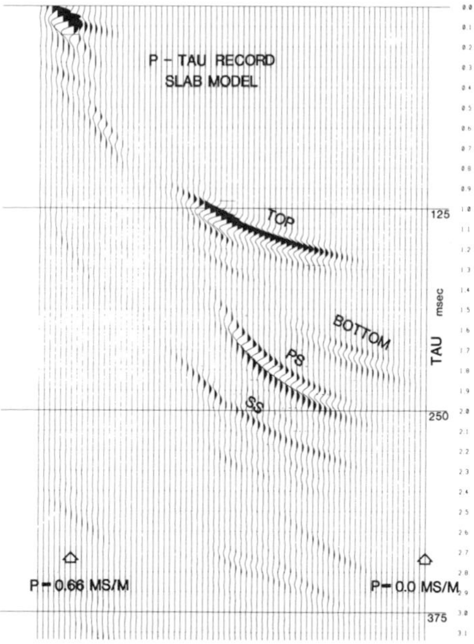

Figure 6c. The p-tau record transformed from the shot record of Figure 6a showing the displacement of the mode converted events toward steeper slopes.

ACKNOWLEDGEMENTS

The permafrost modeling project is being conducted under the supervision of Dr. Don C. Lawton, and is jointly funded by the University of Calgary and the Panel on Energy Research and Development (PERD) Task 6: Conventional Energy Systems. This program is coordinated through the Beaufort Sea project under Steve M. Blasco at the Atlantic Geoscience Centre. The development of the beam steering program on the Cyber 205 system is supported by a grant from the SuperComputer Allocation Committee, Academic Computing Services, University of Calgary.

REFERENCES

Biswell, D.E., Konty, L.F. and Liaw, A.L., 1984, A geophone subarray beam-steering process; Geophysics, vol. 49, no. 11, p. 1838-1849.

Cheadle, S.P., Bertram, M.B. and Lawton, D.C., 1985, Development of a physical seismic modeling system, University of Calgary; Current Research, Part A, Geological Survey of Canada, Paper 85-1A, p. 499-504.

Kong, S.M. Phinney, R.A. and Chowdhury, K.R., 1985, A nonlinear signal detector for enhancement of noisy seismic records; Geophysics, vol. 50, no. 4, p. 539-550.

Larner, K., Chambers, R., Yang, M., Lynn, W. and Wai, W., 1983, Coherent noise in marine seismic data; Geophysics, vol. 48, no. 4, p. 854-886.

Noponen, I. and Keeney, J., 1986, Attenuation of waterborne coherent noise by application of hyperbolic velocity filtering during the t-pau transform; Geophysics, vol. 51, no. 1, p. 20-33.

Schultz, P.S., 1982, A method for direct estimation of interval velocities; Geophysics, vol. 47, no. 12, p. 1657-1671.

FREQUENCY FREQUENCY-RATE SPECTRAL ANALYSIS

M. A. Price

Advanced Systems Development Center
Lockheed-California Company
Box 551 (Dept. 78-55, Bldg. 360, B-6)
Burbank, California, USA 91520

ABSTRACT

Target and/or sensor motion can smear a received acoustic tonal over many frequency cells when the frequency resolution is matched to the "at rest" tonal frequency width. In addition to a loss of frequency resolution there is a degradation in SNR. The FFR (Frequency Frequency-Rate spectral analysis) within limits eliminates the smearing of acoustic energy resulting in improved resolution and SNR.

FFR

Source and/or sensor motion can smear a received acoustic tonal over many frequency cells when the frequency resolution is matched to the "at rest" tonal frequency width. In addition to loss of frequency resolution there is a degradation in SNR. The FFR (Frequency Frequency-Rate spectral analysis) within limits eliminates the smearing of acoustic energy resulting in improved resolution and SNR. Figure 1 illustrates the time dependent source-sensor geometry for linear source motion.

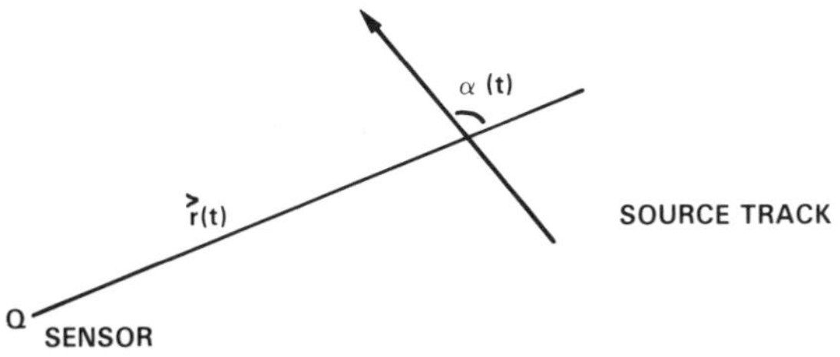

Figure 1. Time-Dependent Geometry

$\vec{r}(t)$, $\vec{v}$, f_s are the sensor-source range vector, source velocity vector and emitted tonal frequency. $\alpha(t)$ is the angle the sensor-source range vector makes with the source velocity vector. With C = speed of sound, the received frequency f(t) is given by

$$f(t) = f_s \; / \; (1 + \frac{V}{C} \cos \alpha(t)). \tag{1}$$

This time dependence of f(t) is illustrated in Figure 2.

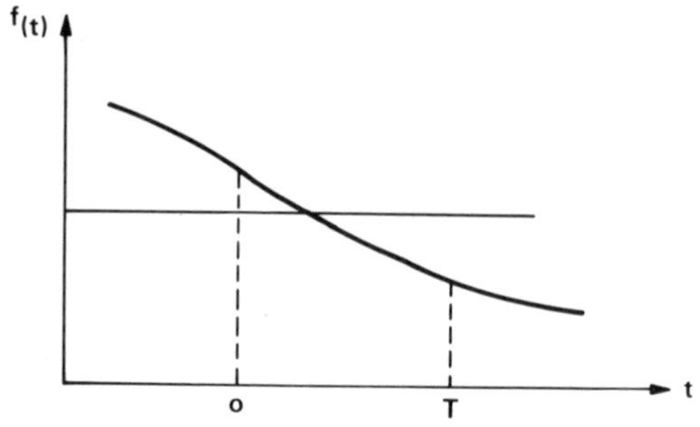

Figure 2. Received Frequency versus Time

In Fourier spectral analysis the received signal is smeared over M frequency cells where

$$M = |f(T + \Delta) - f(\Delta)| \ / \ (1/T) \tag{2}$$

and $\Delta \leq t \leq T + \Delta$ is the integration period. Δ will henceforth be zero. This spectral smearing is reduced via the FFR which is defined as

$$S(\hat{w}, \hat{\dot{w}}) = FT \left\{ s(t) \ e^{-i \ \hat{\dot{w}} \ (t - T) \ t/2} \right\} \tag{3}$$

where s(t) is the received signal, $\hat{w}$ is the spectral frequency, and $\hat{\dot{w}}$ the spectral frequency-rate.

For a tonal source

$$s(t) = A \sin \int^{t} w \ (\bar{t}) \ d\bar{t} \tag{4}$$

where $w(t) = 2\pi f(t)$.

Substituting the positive frequency component of s(t) into Eq. (3), expanding $w(\bar{t})$ in a Taylor series about T/2 and performing the integration

$$\int^{t} w(\bar{t})d\bar{t}$$

yields

$$S(\hat{w},\hat{\dot{w}}) = \int \exp i[w(T/2)t + (\dot{w}(T/2)-\hat{\dot{w}})(t-T) \ t/2$$
$$+ \text{ higher order terms}] \cdot \exp(-i\hat{w}t)dt \tag{5}$$

If $\hat{\dot{w}} = \dot{w}(T/2)$ then

$$S(\hat{w}, \hat{\dot{w}}) = \int \exp i[w(T/2)t + \text{higher order terms}] \cdot \exp(-i\hat{w}t)dt \tag{6}$$

This will result in the received signal being desmeared if

$$f(t) \approx f(T/2) + (t - T/2) \ \dot{f}(T/2) \tag{7}$$

is a good approximation for f(t). Figure 2 indicates this is a good approximation for linear source and/or sensor motion if the integration period is not excessively long. If noise is added to the received signal of Eq. (3) the noise levels added to $S(\hat{w},\hat{\dot{w}})$ are independent of $\hat{\dot{w}}$. The negative frequency component in Eq. (4) is desmeared by $\hat{\dot{w}} = -\dot{w}$ since frequency-rate is proportional to frequency.

The following are examples of FFR output. Figure 3 illustrates a 3-D FFR on simulated data. Figure 4 is the superposition of the $\dot{w}$=0 cross section and the cross section corresponding to maximum FFR amplitude of Figure 3.

Figures 5-14 are cross sections from FFRs of real data. Figures 5, 7, 9, 11, and 13 are $\dot{w}$=0 cross sections from FFRs of successive 100 sec. records of real data. The tonal is 300 Hz. Figures 6, 8, 10, 12, 14 are the frequency-rate cross sections of the same FFR set corresponding to the maximum amplitudes. Comparison of Figure 5 with 6, Figure 7 with 8, etc., indicates the FFR can improve spectral resolution and SNR over a significant time interval (in this case, 500 secs.). Figures 6, 8, 10, 12, and 14 illustrate a decreasing Doppler shifted tonal.

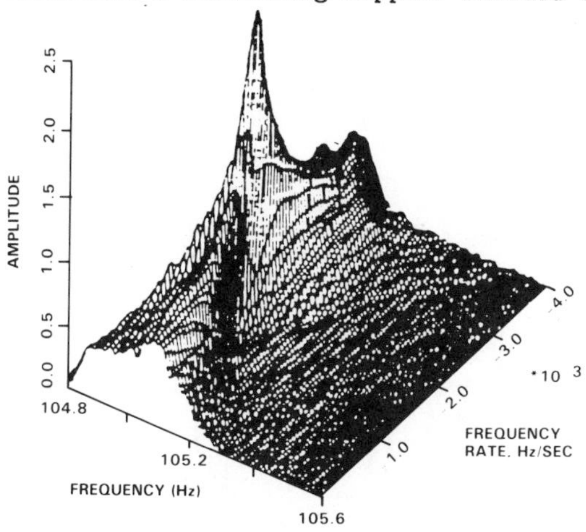

Figure 3. FFR

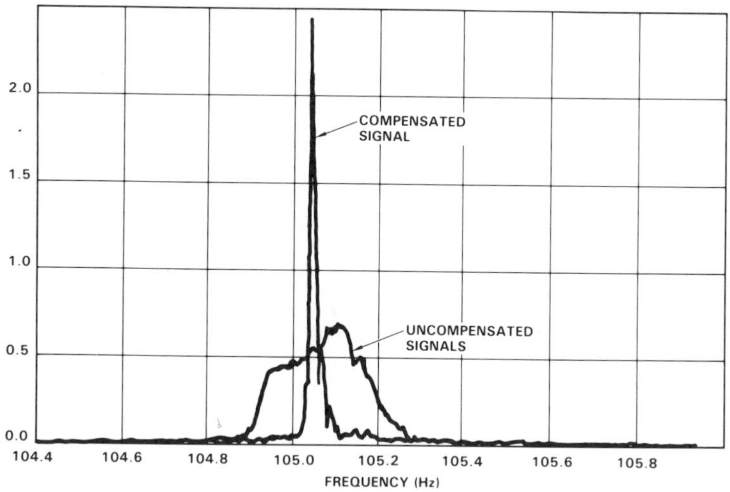

Figure 4. Superposition of Two FFR Cross Sections

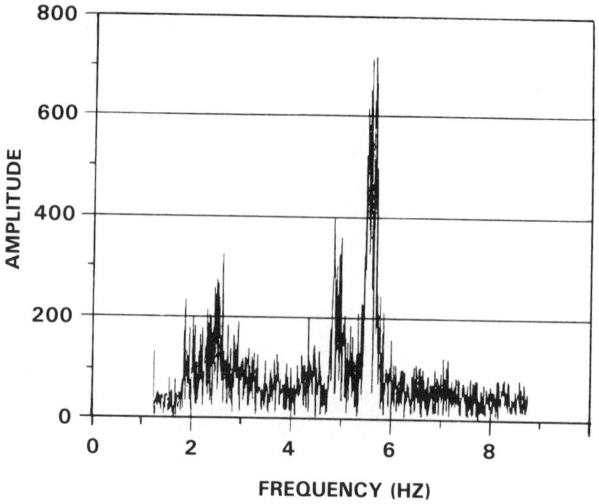

Figure 5. $\dot{w}$ = 0, 0-100 sec.

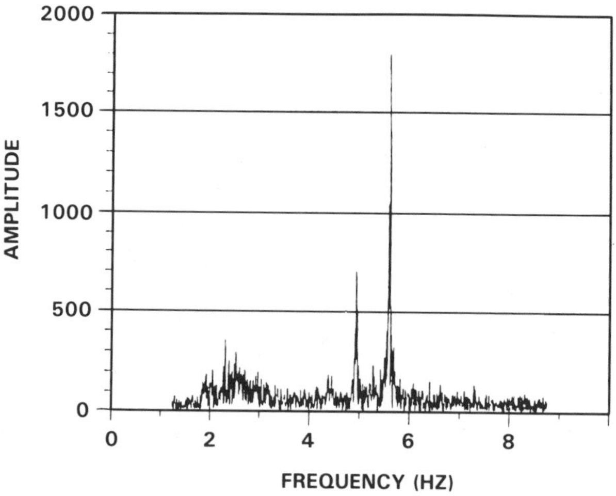

Figure 6. Maximum Amplitude, 0-100 sec.

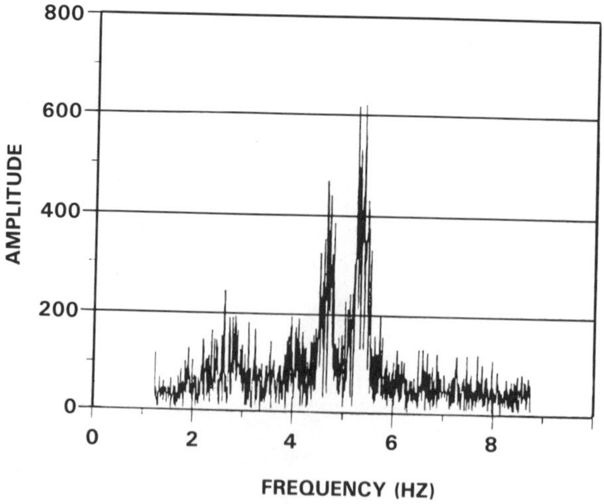

Figure 7. $\dot{w}$ = 0, 100-200 sec.

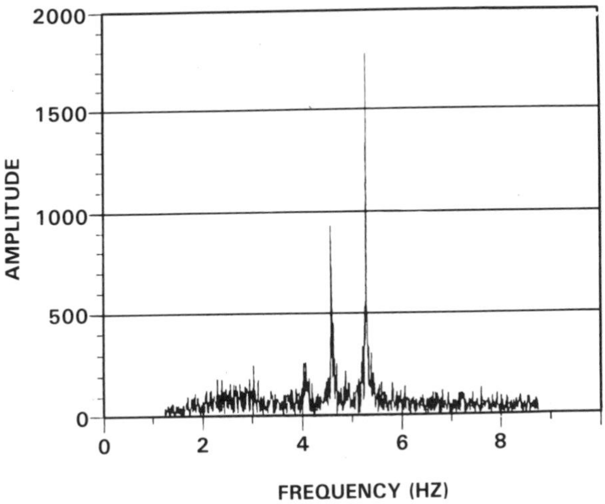

Figure 8. Maximum Amplitude, 100-200 sec.

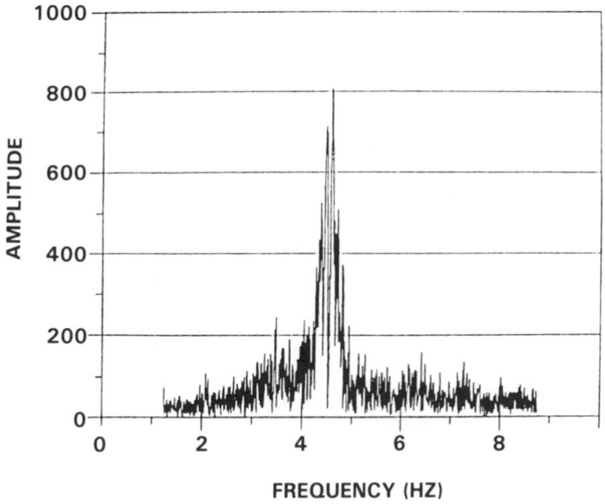

Figure 9. $\dot{w} = 0$, 200-300 sec.

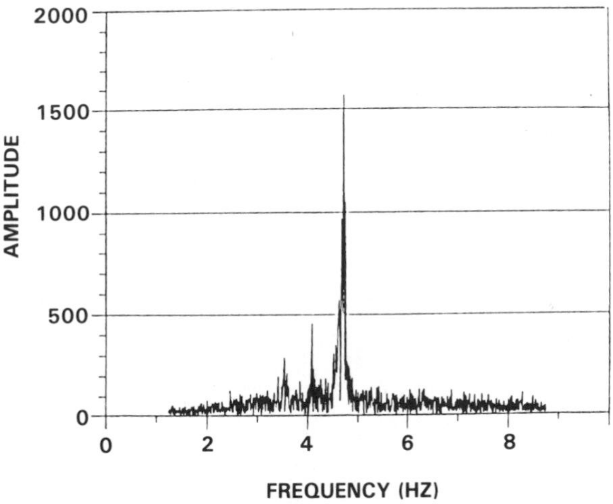

Figure 10. Maximum Amplitude, 200-300 sec.

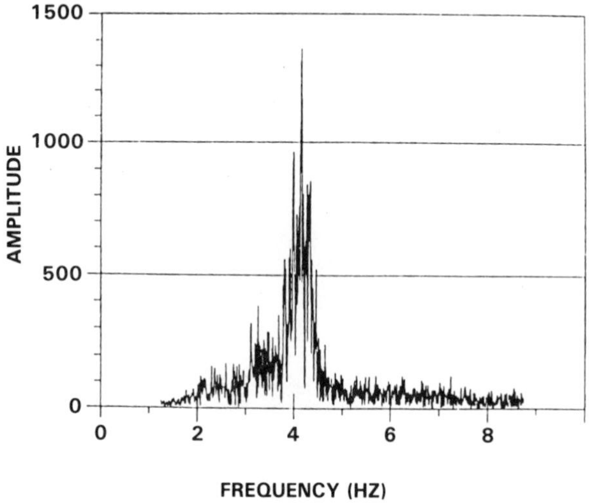

Figure 11. $\dot{w}$ = 0, 300-400 sec.

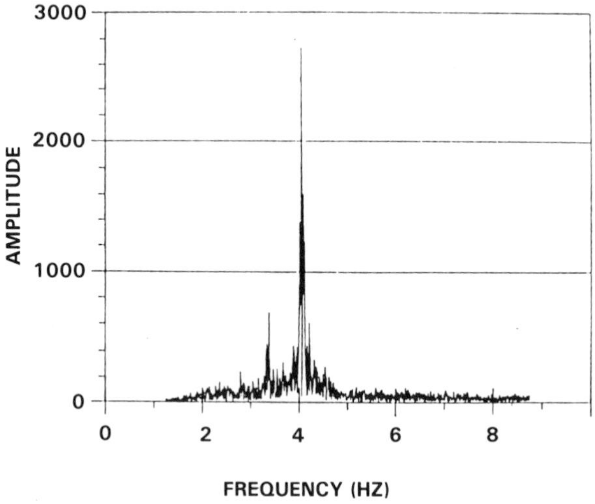

Figure 12. Maximum Amplitude, 300-400 sec.

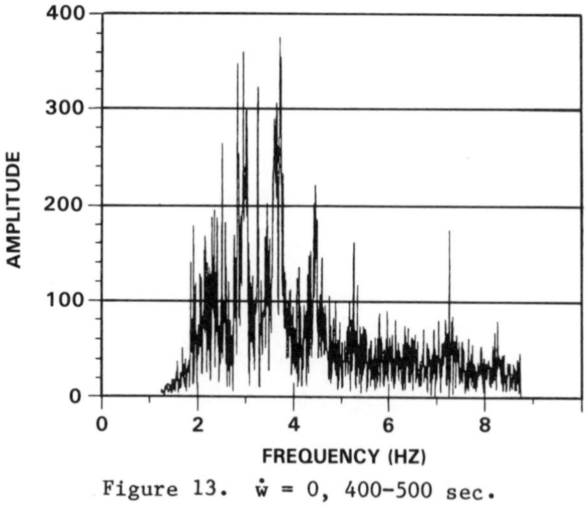

Figure 13. $\dot{w}$ = 0, 400-500 sec.

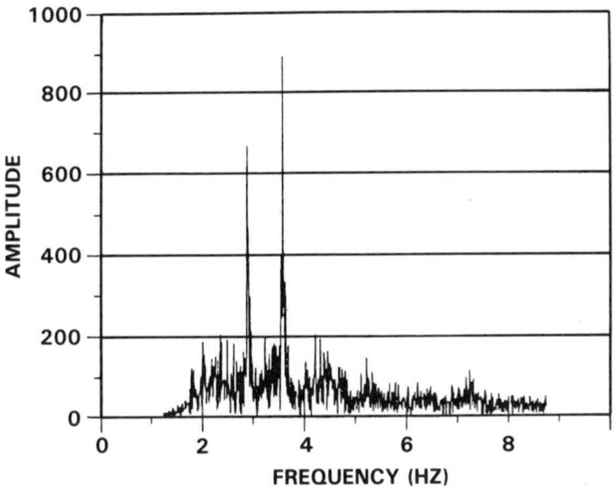

Figure 14. Maximum Amplitude, 400-500 sec.

Approximating Eq. (2) by

$$M \approx \dot{f}(T/2)T^2$$

$$= f_s \frac{V^2T^2\sin^2\alpha(T/2)}{r(T/2)} \tag{8}$$

Choose f_s = 100 Hz, V = 20 kts, C = 5000 ft/sec $\alpha(T/2)$ = 45°. The entries of Table 1 are the number of bins the tonal is smeared over as a function of source-sensor range and the integration period T.

Table 1. Spectral Smearing

r \ t	1 sec	10	50	100	300
1000'	.011	1.1	27.5	110	990
1 N.M.	.0019	.19	4.75	19	171
3 N.M.	.0062	.062	1.55	6.2	55.8
5 N.M.	.00037	.037	.93	3.7	33.3

DISCUSSION AND COMMENTS

When successive spectral records are averaged, the resulting spectral average is smeared over M_a frequency cells where

$$M_a = |f(T) - f(0)| \;/\; (1/Tr) \tag{9}$$

where T is the total time series data length and Tr is the Fourier record length. For example, assume a source whose motion is symmetric about C.P.A. with T = 200 secs., Tr = 20 secs., r_{CPA} = 5 N.M., V = 12 kts and f_s = 100 Hz. M_a = 9.6 frequency cells with the number of records averaged M = 10 or more depending on overlap.

To eliminate this smearing effect a FFR spectral analysis involving spectral record shifting is defined as

$$S_a(w,\dot{w}) = \frac{1}{M} \sum_{i=1}^{M} S_i\ (w + \delta_i(\dot{w}))$$

$$\delta_i(\dot{w}) = \text{INTEGER} \left\{ Tr^2\ \dot{w} \left(i - \frac{M+1}{2} \right) \right\}$$

$$= -\ \delta_{M+1\ -\ i} \tag{10}$$

The spectral record shifting FFR is similar in appearance to Figure 3 and the superposition of cross sections corresponding to $\dot{w} = 0$ and $\dot{w}$ that results in the greatest amplitude is similar in appearance to Figure 4.

If the frequency of the emitted tonal is a linear function of time and therefore modeled by a frequency-rate, the maximum amplitude frequency-rate of the FFR is the sum of the source and motion frequency-rates.

Eq. (7) is a good approximation if the higher order terms

$$\frac{d^n w(T/2)}{dt^n} \cdot \int^t (\bar{t} - T/2)^n d\bar{t}\ \alpha\ \frac{1}{r^n(T/2)} \int^t (\bar{t} - T/2)^n dt \quad ,\ 0 \leq t \leq T$$

of the phase expansion in Eq. (5) are negligible. However, as the source range r(T/2) decreases, the higher order terms become more significant unless the record length T is shortened.

For stable tonals, the ratio of frequency-rate to frequency ($\dot{f}/f$) is independent of frequency, dependent on source-receiver geometry and propagation path, and therefore offers to sort spectral lines into source-propagation path groups.

Frequency line trackers could be modified to predict frequency-rate which would be used to enhance SNR and frequency resolution resulting in improved tracker performance.

Estimation of frequency and frequency-rate doubles the number of measurements derived from time series data. This additional information can improve source fix-tracking estimation.

For tonals, the FFR is important when there is significant geometry change due to linear motion over the integration period. This could be a likely scenario for sources only detectable at short ranges.

SPACE-TIME PROCESSING, ENVIRONMENTAL-ACOUSTIC EFFECTS

William M. Carey

Naval Undersea Systems Center
New London, CT 06320

William B. Moseley

Naval Ocean Research
and Development Activity
NSTL, MS 39529-5004

ABSTRACT

The processing of acoustic waveforms by arrays requires an understanding of the temporal and spatial characteristics of the signal and noise fields. Temporal and spatial processing schemes are analogous transforms that can employ a variety of windows (such as Hann, Hamming, etc.). However, the ocean environment is a filter that introduces variability to a signal in both spatial and temporal domains. This randomness is superimposed on an ambient sound channel characteristic. In the case of static source and receiver combinations, the limits on horizontal broadside array resolution are due to volume scattering and surface scattering as long as the time scale is less than the signal correlation time. However, in the case of a moving source-receiver, the temporal and spatial scales are coupled through the sound channel characteristic and the fluctuation effects due to multipath or modal variations must also be considered. This paper reviews fundamental environmental effects and their influence on arrays in the deep ocean sound channel.

INTRODUCTION

Space-time processing has been extensively studied from a radar viewpoint since the 1940s, and many comprehensive reviews and texts can be found. The Winder and Loda[1] summary of acoustic-space-time processing, together with the fundamentals of statistical communication theory,[(2–5)] provides a basis for processing ocean acoustic waves. The problem of the acoustic antenna and a signal with random parameters was first treated by Bourret,[6] Berman,[7] and Bordelon[8] with additional work performed by Brown[9] and Lord.[10, 11] Shifrin,[12] with a similar approach, determined the statistical characteristics of radar antennas and described the response of antenna systems to waves with random amplitude and phase components. His results will be used as they apply to ocean acoustic arrays. Additional[(13–17)] reviews stress digital signal processing considerations and include digital Fast Fourier Transforms (FFT), shading characteristics,[(14, 15)] and high-resolution techniques.[(15–17)] These reviews mainly address spatial and temporal processing separately but have direct application to sonar processing in space-time.

With such a formidable number of existing papers, reviews, and texts, the necessity for an additional paper on this subject might be questioned. However, in underwater acoustics, the changes caused by multipath propagation and relative source-receiver motion through the severe acoustic interference field merit the discussion. This paper applies the information available in the previous reviews[(1–17)] to the characteristics of acoustic antennas in the midfrequency range. The paper addresses aspects of the problem which are unique to the acoustic antenna and results from the ocean environment and the acoustic field. It also treats the problems of spatial coherence measurements and linear array characteristics. Finally the paper discusses space-time processing when the temporal and spatial variables are weakly coupled.

NORDA Contribution 200:002:87.

BACKGROUND

This paper addresses the space-time processing of acoustic signals described by solutions that satisfy the wave equation in the absence of surfaces that scatter or diffract and are known to be continuous functions with continuous first derivatives. These solutions are also bounded by imposing a radiation condition at infinity and have the property of temporal and spatial separability. Furthermore the sound sources are harmonic and have been excited for a period sufficiently long that a steady state has been achieved. Since superposition applies, the Dirichlet conditions are satisfied and Fourier analysis can be used to describe the space-time properties.

The relationships between the space-time transforms are summarized:

(Pressure) $P(\underset{\sim}{r}, t)$ —— $F(P(\underset{\sim}{r}, t))$ ⟶ / ⟵ $F^{-1}(F_p(\omega, \underset{\sim}{k}))$ —— $F_p(\omega, \underset{\sim}{k})$ *(Wavenumber spectra)*

| |

$<P_1(\underset{\sim}{r}_1, t)\bullet P_2^{\dagger}(\underset{\sim}{r}_2, t+\tau)>$ $F_p(\omega, \underset{\sim}{k})F_p^{\dagger}(\omega, \underset{\sim}{k})$

| |

$R_p(\tau, \underset{\sim}{r})$ —— $F(R_p(\tau, \underset{\sim}{r}))$ ⟶ / ⟵ $F^{-1}(S_p(\omega, \underset{\sim}{k}))$ —— $S_p(\omega, \underset{\sim}{k})$

(Acoustic correlation function)   *(Power spectra)*

where $\boldsymbol{F}$ and $\boldsymbol{F}^{-1}$ refer to the Fourier transform and its inverse. These transformations are useful in determining properties of the pressure field $P(\underset{\sim}{r}, t)$ provided the power spectra, $S_p(\omega, \underset{\sim}{k})$ can be measured.

To describe the field $P(\underset{\sim}{r}, t)$ by the measurement of $S_p(\omega, \underset{\sim}{k})$, the response of the measurement system needs to be incorporated. If this response is described by $f(\underset{\sim}{r}, t)$, and $F_{mp}(\omega, \underset{\sim}{k})$ represents the measured transform, then

$$F_{mp}(\omega, \underset{\sim}{k}) = \boldsymbol{F}\{f(\underset{\sim}{r},t)P(\underset{\sim}{r},t)\} = \boldsymbol{F}\{f(\underset{\sim}{r},t)\}*\boldsymbol{F}\{P(\underset{\sim}{r}, t)\} = f_{lp} * F_p \, , \tag{1}$$

where * refers to convolution.

Thus the frequency-wavenumber spectra is the convolution of the transform of the system response and the transform of the field. This useful concept allows one to consider the system response and its Fourier transform independently of the acoustic field variables. Furthermore, since

$$S_{mp}(\omega, \underset{\sim}{k}) = F_{mp}(\omega, \underset{\sim}{k})F_{mp}^{\dagger}(\omega, \underset{\sim}{k}) \tag{2}$$

represents the output of the system, the conclusion that both the broadness in the system response function $f_{lp}(\omega, \underset{\sim}{k})$ and the broadness in acoustic wavenumber spectra determine the broadness of $S_{mp}(\omega, \underset{\sim}{k})$. In the wavenumber domain, the broadness of $f_{lp}(\omega, \underset{\sim}{k})$ with respect to $\underset{\sim}{k}$ is determined by the shading characteristic and length of the system. Matched processing would require that the spread in $\underset{\sim}{k}$ for both $f_{lp}(\omega, \underset{\sim}{k})$ and $F_p(\omega, \underset{\sim}{k})$ would be equal. Recall that the angular spread in $F_p(\omega, \underset{\sim}{k})$ is caused by the environment.

System Response

The description of the system response can easily be developed by the direct analogy between sampling in time and sampling in space. The particular case of a one-dimensional linear sample, a line array, is described by the illumination function, shading, or window function. Upon neglecting the time-frequency transform, one has for a rectangular window

$$F_{lp}(\underset{\sim}{k}) = \int_{-L/2}^{L/2} f(x)exp(i\underset{\sim}{k}x)dx = Lsin(L\underset{\sim}{k}/2)/L\underset{\sim}{k}/2 \tag{3}$$

$$= Lsin(\pi Lsin\,\theta/\lambda)/(\pi Lsin\,\theta/\lambda) \, .$$

The result is the same whether the sampling is in space or in time. For the case of discrete elements within the array we have

$$F_{lp}(\underset{\sim}{k}) = \int\sum_{n=1}^{N}\delta(x-nd)exp(i\,\underset{\sim}{k}\,x)dx = \frac{sin(Nk_o d\,sin\,\theta)}{N sin(k_o d\,sin\,\theta)} \, . \tag{4}$$

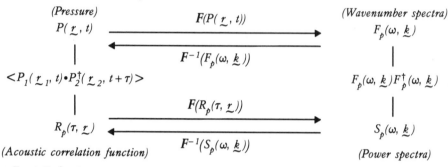

This result agrees closely with the continuous line array when the spacing d is one-half of the wavelength, λ.[1] The characteristics of different windows can be found in Harris's[14] concise review of discrete Fourier transforms and the use of windows.

Resolution

The reciprocal relationship between aperture and beamwidth can easily be seen by examining the results from the rectangular window, equation 3. The radiation pattern (here we use the principle of reciprocity) is simply $F_{lp}(\underset{\sim}{k},\omega_o)\, F^{\dagger}_{lp}(\underset{\sim}{k},\omega_o)$ and represents the Farfield radiation pattern, sometimes referred to as the Fraunhofer region as opposed to Fresnel and near-field regions. The result is as expected: for a perfectly coherent incident field, an increase in antenna length results in additional power output and a reduction in beamwidth. These effects are illustrated in fig. 1 for both space L and time T.

The resolution relationship can be explored by a variety of criteria. Rayleigh's[18] was the resolvability of two sources when the maximum of the pattern for one coincided with the first minimum of the other. This Rayleigh resolution, R, corresponds to one-half the beamwidth between first nulls (BWFN). For the rectangular window, the first null occurs when $(L\sin\theta/\lambda)$ is equal to unity. The beamwidth between the first nulls would then be

$$\Delta\theta_{FN} = 2\lambda/L\ . \tag{5}$$

The alternative description, also shown in fig. 1, is to choose the half-power (hp) points, which are

$$\Delta\theta_{hp} = 0.886\lambda/L;\ R = \lambda/L = \frac{\Delta\theta_{FN}}{2} = 1.12\,\Delta\theta_{hp}\ . \tag{6}$$

This relationship holds for the broadside case of the spatial array and can be generalized for the off-broadside and end-fire cases:

$$\text{Broadside: } \Delta\theta \bullet L/\lambda = 0.886\ \sec\theta_o;\ \text{ End Fire: } \Delta\theta(L/\lambda)^{1/2} = 2(0.886)^{1/2}. \tag{7}$$

These relations are a consequence of the trigonometric relations. The general conclusion is that

$$\Delta\theta \bullet L/\lambda = C_l \sim 1,\ \Delta f\,\Delta t = C_t \sim 1 \tag{8}$$

represents the resolution properties of space-time processors. These results are based on the deterministic response of the line array. However, as mentioned, the resolution of the system is required to be narrower than or equal to the angular spread in the incident field. The measured output is the convolution of the system response and the field frequency-wavenumber transform. In an analogous fashion, one can relate the output of the measurement system as the convolution of the array response $f_{lp}(\underset{\sim}{k},\ \omega)$ and the Fourier transform of the acoustic spatial correlation function

$$S_p(k\sin\theta,\omega) = f_{lp}(k\sin\theta,\omega) * F(R_p(\underset{\sim}{r},\ \tau)) \tag{9}$$

where we have suppressed the temporal variables. The acoustic spatial correlation function can also be defined as

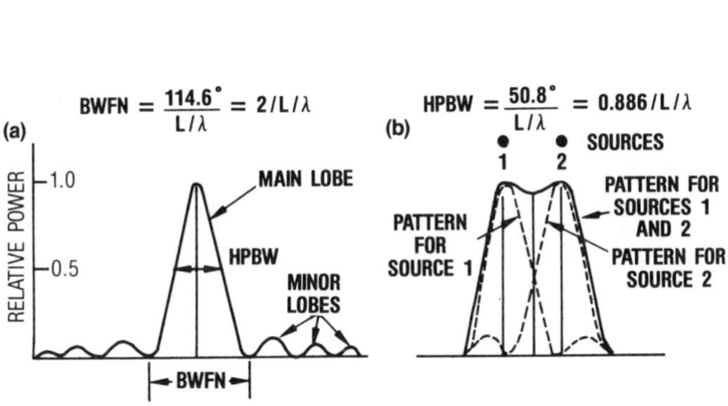

Figure 1. Farfield radiation power pattern for (a) single source and (b) for two identical sources separated by the Rayleigh resolution.

$$R_p(\underset{\sim}{r}_1, \underset{\sim}{r}_2) = \left\{ P(\underset{\sim}{r}_1)P^{\dagger}(\underset{\sim}{r}_2) \right\} \tag{10}$$

where $\underset{\sim}{r}_1$ and $\underset{\sim}{r}_2$ are the spatial location points, P is the complex acoustic pressure, † indicates conjugation, and the braces indicate ensemble averaging.

The array output as a function of horizontal angle θ is simply the convolution of the array beam pattern f_{lp} with the angular distribution of incoming acoustic signals, which is given by the spatial Fourier transform F of the acoustic correlation function.

In long-range propagation cases, it is expected that the initially smooth curve wave fronts, in the ray theory sense, will have become irregular and spatially varying descriptions of forward-scattered energy bundles. These "arrivals" are resolvable in order but, when displayed after suitable steering delays, have a phase and amplitude variability, as well as extended durations greater than the transmitted pulse. These distorted pulses decorrelate in space—a phenomenon equivalent to an angular redistribution of energy. The effects of disparities between the array properties and the incoming signal angular distribution are immediately apparent from equation 9. A short array with a beam pattern broader than the signal angular distribution is accepting unnecessary noise on the main lobe, whereas a long array with a beam pattern narrower than the signal angular spread is rejecting part of the power in the signal. Thus, the design of optimal array characteristics is facilitated by a prior knowledge of the acoustic correlation function.

The angular spread of arriving acoustic waves during a given observation period can be further broadened by the arrival of multiple rays at different angles. That is to say, off-broadside arrivals from the same azimuthal angle but with different vertical arrival angles can have conical angles such that the difference between the conical angles from multiple rays is greater than the half-power beamwidth of the array. Thus, a bifurcation or the appearance of acoustic intensity on more than one beam will be observed. The angle of arrival for a given path or group of paths depends on the relative orientation and motion between the source and receiver during the observation period. Consequently motion, both radially and with respect to relative azimuth, can have a smearing effect on the beam response.

An analogous situation exists for the frequency domain. Spread in the frequency domain can be caused by random media scatter, Doppler spread due to multipath projections of radial velocity, and accelerations. The Doppler spread due to the relative source-receiver motion is normally the dominant effect, and the remaining causes of frequency spread are small.

Statistical Characteristics[12]

The determination of the array's response to a signal with variability in phase is important to this discussion. The treatment of the response of a continuous line array to a random signal has been derived by Shifrin[12] and is included because the results do not appear to be used extensively in acoustic antenna applications despite some previous noteworthy results.[6–11]

Shifrin shows the mean power pattern, $|\overline{f(\psi)}|^2$, where $\psi = \pi L sin\theta/\lambda$, depends on the line array pattern in the absence of phase randomness and a distortion in the pattern due to the combined influence of the phase variance α and the normalized correlation length, C_m. The power in the direction of the principal maxima is also seen to be decreased by the factor $exp(-\alpha)$. The function $I(C_m, \psi)$ can be evaluated in closed form for special cases, as well as for a similar function based on the exponential form of the correlation function.

$$|\overline{f(\psi)}|^2 = exp(-\alpha) \left\{ \frac{sin^2 \psi}{\psi^2} + \tfrac{1}{4} \sum_{m=1}^{\infty} \frac{\alpha^m}{m!} I(C_m, \psi) \right\}. \tag{11}$$

Shifrin's fig. 2 shows the mean power radiation pattern $|\overline{f(\psi)}|^2$ versus ψ for various values of the phase variance α and the normalized correlation length C. His calculations show some interesting general characteristics. The randomicity in phase quantified by the variance (α) shows that the effect of an increase in α causes the principle maxima to decrease (see a, b) and a smoothing or smearing of the radiation pattern; that is, the nulls are filled or blurred. As the phase variance increases, the pattern loses detailed structure and becomes a monotonically decreasing one (see a and b for $\alpha = 3$). The effect of increasing the phase correlation length is shown in fig. 2. As C increases, the pattern approaches the ideal pattern, and the power in the direction of the principle maxima increases (see examples c and d). These numerical computations, simplified expressions, and approximate forms are discussed in his text.[12]

Further, Shifrin derives the mean directional gain with three types of behavior: the first, a region in which the gain increases proportional to $2L/\lambda$; the second, a region of gain saturation (gain independent of length); and the third, in which the gain increases by $2L/\lambda$ attenuated by $exp(-\alpha)$. (This is questionable

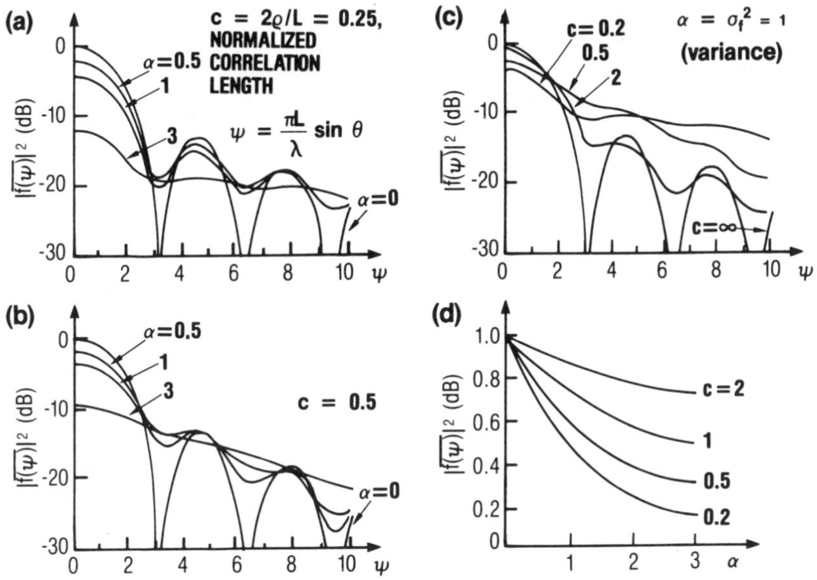

Figure 2. The mean radiation pattern $|\overline{f(\psi)}|^2$ for various values of phase variance ($\alpha = \sigma_\phi^2$) and correlation length ($C_m\phi = 2\varrho/mL$) versus $\psi = \pi L \sin\theta/\lambda$, (a) $C = 0.25$, (b) $C = 0.5$, (c) shows for a constant α, the effect of increase in phase correlation length while (d) shows for constant correlation length the effect of phase variance increase on the principal maxima.

for long-range underwater acoustic propagation.) These results were derived for a Gaussian correlation function but could also be developed for an exponential distribution. Finally the Gaussian correlation function was related to the measure of spatial coherence with a Gaussian form.

Coherence

The statistical response of an array has been developed for a signal with a random component of phase, which is Gaussian. Due to general characteristics of the Gaussian process, it can be shown that a random phase variable with a Gaussian spatial correlation function also produces a spatial coherence function, which was found to have a Gaussian form. Although not shown here, the random phase variable with an exponential correlation function produces an exponential spatial coherence function.[12]

The propagation of waves through a random medium has been treated by Chernov[19] and Tatarski.[20] The results of Chernov's work and the experimental measurements discussed by Shifrin[12] were the basis for Shifrin's choice of exponential and Gaussian phase correlation functions to determine the statistical characteristics of antennas. The Gaussian form was chosen to illustrate the general characteristics of line array when the incident wave had a normally distributed phase variable with a Gaussian correlation function. The exponential correlation function was not chosen, even though in several instances this function compares favorably with experimental data on electromagnetic propagation in the atmosphere, since an exponential correlation function represents a discontinuous physical process.

The ocean acoustics wave propagation problem is more complex than atmospheric propagation due to the spatial properties, which are anisotropic and nonhomogeneous. A thorough discussion of this subject is beyond the scope of this paper and we refer the reader to some very excellent reviews.[21–26]

The spatial coherence function can be written as

$$R_P = \langle P(\underset{\sim}{x})\, P^\dagger(\underset{\sim}{x} + \underset{\sim}{r}) \rangle = \langle P^2 \rangle \exp(-(y/L_h)^n) \quad n = 1, 1.5, 2 \tag{12}$$

where $\langle P^2 \rangle$ is the mean square pressure and is proportional to the acoustic intensity and where $n = 1$ is the exponential form, $n = 1.5$ the Beran-McCoy form, and $n = 2$ the Gaussian form. Since our goal is to describe line array performance in the linear region to the beginning of the gain saturation region, these types of coherence functions can be considered valid. Flatte[21, 22] and Dashen[23] relate the coherence to a phase structure function and in the region of interest here, the geometrical acoustics and partially saturated ($\phi\Lambda \sim 1$) regions, they give the structure function at small separations corresponding to a coherence function with $n = 2$, the Gaussian form. The exponential functional form of the coherence was employed by Cox,[27] since it enabled the development of a closed form solution to the array gain problem. Cox used the function only for small ($\lambda/2$) separations and the result is useful. However, Shifrin stated that this functional form has a finite derivative at zero separations and corresponds to a discontinuous physical process, not representative of wave propagation. The work presented in this paper is based on the $n = 1.5$ functional form after the work of Beran and McCoy.[28-30] They derived a solution in which the basic anisotropic nature of the two-point environmental statistic was retained, together with the approximate form of the power spectrum of random temperature fluctuations in the horizontal plane. One result of the Beran-McCoy formulation is given as follows

$$R_p(r, f, y, z) = I(z) exp(-E_f f^{5/2} r Y^{3/2}) = I(z) exp(-E_k(k)^{5/2} r y^{3/2}). \tag{13}$$

In this equation, the acoustic correlation function along a horizontal line transverse to the direction of propagation is an exponential function whose argument is proportional to E, an environmental parameter appropriate for the propagation path; the frequency f to the 5/2 power; the range r and the 3/2 power of y the transverse separation distance of the two correlation points. The symbol I represents the intensity at a single point receiver with the same range and depth z.

The components of the environmental parameter E for the finite source case are found by Beran and McCoy to be

$$E_f = 1.7 \left(1/C_o \bullet \frac{\partial C_o}{\partial T}\right)^2 A_T L_{\gamma m} (2\pi/C_o)^{5/2} \tag{14}$$

where C_o is the nominal speed of sound, $(\partial C_o/\partial T)$ is the partial derivative of the sound speed with respect to the temperature, A_T is the coefficient of the single term power-law spectrum representation of the random horizontal temperature variations (i.e., the nominal strength of the random temperature field), and $L_{\gamma m}$ is the correlation length of the random temperature fluctuations in the vertical direction. Thus the spatial coherence problem can be considered the experimental determination of the functional form of the coherence and the measurement of the coherence length. These specify the performance of the line array.

Spatial Coherence and the Line Array

Since the response of the array to the sound field is the convolution of the beam response of the array and the wavenumber spectra of the field, we can establish a resolution criteria by matching the angular spread in the incident field to the 3-dB width of the beam response. The spatial coherence function is a measure of this angular spread and can be evaluated as

$$S(k sin\theta) = \int_{-\infty}^{\infty} exp(-(y/L_h)^n) exp(ik sin\theta y) dy. \tag{15}$$

This integral can be evaluated in closed form for $n = 1$ and 2 and numerically for $n = 1.5$. The results of this analysis, when matched to the 3-dB width, yield the following.

	n = 1	n = 1.5	n = 2.0	f_{lp}
BW (Radian)	0.318 λ/L_h	0.457 λ/L_h	0.530λ/L_h	0.886λ/L_a
BW (Degrees)	18.2° λ/L_h	26.2° λ/L_h	30.36° λ/L_h	50.76° λ/L_a
L_a/L_h	2.72	1.89	1.64	1.00

The exponential form has an angular spread of 0.318 λ/L_h and, when matched with the 3-dB beamwidth of the line array, 0.886 λ/L_a yields an aperture length 2.72 times the horizontal coherence length. Given the same coherence length, the acoustic field angular spread for the Gaussian ($n = 2$) form is 1.67 times as large as the acoustic field angular spread for the exponential ($n = 1$) form.

In addition to matching the angular spread in the incident field to the beamwidth, the relative signal gain (RSG) can be computed, that is, the ratio of actual signal gain to ideal signal gain. It can easily be shown that

$$RSG \cong (L_b/L_a)^2 \int_{-L_a/L_b}^{+L_a/L_b} (L_a/L_b - |x|) \exp(-x^m)\, dx. \tag{16}$$

This relationship shows that the relative gain of the system (when related to an exponential form of the spatial coherence function) is a function of the coherence length. Thus the angular resolution (beamwidth) and gain of the line array are determined by the coherence length and the form of the coherence function.

Experimental Results

To compare the theoretical correlation function with the acoustic measurements the value of the environmental parameter E and the functional form of the coherence must be ascertained. Acoustic data obtained by Stickler[31] corresponded to a transmission path that intersected the receiver array at approximately a 13° elevation angle. A ray tracing program was used to describe the path of the 13° ray from the receiver by using measured profiles along a radial from the array at the azimuth of the experiment. The average sound speed along the 13° ray path was found to be 1517.5 m/s, and C_o was set equal to this value. The constant 3.6 m/s/°C was used for the partial derivative of the sound speed with respect to temperature. A_T was determined by first using the average value of the Brunt-Vaisala frequency ($N = 1.5 \times 10^{-3}$ radians/sec) along the 13° ray path. Then the measured coefficients[32] to the single term power-law spectra for the random horizontal temperature fluctuations were treated as a function of N. A linear extrapolation of the measured coefficient from the nearest measured N to the average N along the 13° ray path indicates a value of A_T of 1.5×10^{-7} °C²/m.

The normalized temperature spatial correlation function is equal to the inverse Fourier transform of the temperature spatial power spectrum normalized by the total variance, and the correlation length can be defined as the spatial length for which the value of the normalized correlation function is 0.5. Thus, the correlation length in the vertical direction of the temperature fluctuations was estimated by

$$0.5 = \int_a^b F_N(k) \cos(ks) dk \Big/ \int_a^b F_N(k) dk \tag{17}$$

where F_N is the vertical spatial power spectrum of the temperature fluctuations, k is the vertical wavenumber, s is the estimate of $L\gamma_m$, and $a = 0.209$ rad/m and $b = 20.94$ rad/m. Millard[33] has data taken southwest of Bermuda, which for the wavenumber interval under consideration, can be approximated by $F_N(k)\ n = Ck^d$, where d is between 2 and 2.5 and s is between 35 and 28 m. The theory of Garret and Munk[34] would predict d to be equal to 2.5 with s given the nominal value of 30 m. Using these values, the environmental parameter value, E, is calculated to be 4.8×10^{-17}. The environmental parameter was also calculated for sound speed profiles from the Mediterranean to be 4.88×10^{-17}, the Pacific to be 9.31×10^{-17}, and the Arctic to be 13.06×10^{-17}. The smallest E were for the Atlantic and the Mediterranean, but the largest was for the Arctic. The variation in E was 2.7 with a coherence length variation of 2. The coherence length does not change by orders of magnitude due to the averaging of the environment along the propagation path.

The work of Stickler et al.[31] using a planer receiver in combination with short (10-ms ping) acoustic signal structure provided the required path resolution for the evaluation of single path coherence. These data are shown in fig. 3 for the center frequency of 400 Hz and range of 137 km. Observe that the spread in the measured coherence values increases as the separation between receivers increases. For a fixed noise level, small values of coherence are difficult to measure, consistent with the analysis of Devilbiss et al.[35] and Carter.[36, 37] The measurement of phase in a random additive noise background requires a high signal-to-noise ratio. Even at 6 dB the uncertainty in the phase may be governed by the properties of the additive noise. Theoretical analysis of the measurement of the coherence of a fluctuating signal in an additive random noise background also shows that the coherence is a function of signal-to-noise ratio, with large confidence bounds. The conclusion drawn is that the estimation of signal phase and coherence properties in the presence of a noise field, such as the ocean ambient noise field, is difficult. This conclusion is underscored by the spread in the data obtained by Stickler. The dashed curve in fig. 3 was calculated using the environmental parameter E and the McCoy formulism. In fig. 3, the theoretical curve is repeated for all three exponential forms for the coherence function where the correlation length

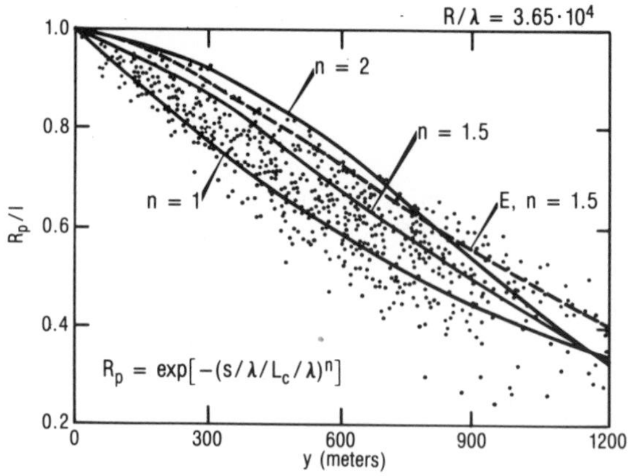

Figure 3. A comparison of $exp(-(y/L_c)^m)$ *with acoustic correlation versus transverse separation at 400 Hz.*

was determined from the estimated (*l/e*) value observed in the data. The calculation using the estimated *E* value and $n = 1.5$ fits the data, especially if one assumes that the downward spread in the coherence measurements could be due to the additive noise. Further, using the empirical estimate of coherence length, the $n = 3/2$ form also provided a match to the data bounded on the upward side by the $n = 2$ form, and the lower side, the $n = 1$ form.

We show in fig. 4 the comparison of the calculated coherence length versus range at 400 Hz based on the measurements of Stickler et al.[31] Note that the coherence lengths here are between 350 and 1647 m, compared to the value of 3.4 to 6.4 km used by Dashen.[23] Williams[38] has estimated coherence lengths at various ranges by visually smoothing phase time histories. His estimates ranged between 331 m and 1200 m at 400 Hz for ranges between 172 and 496 km as shown by the circles in fig. 4. Finally Kennedy[39] estimates the phase coherence length to be 183 m at 800 Hz and a range of 46 km. Williams' values are seen to be comparable to the Stickler et al. values. Using the functional dependence of frequency to 5/2 power and range, we estimate that the Kennedy results would correspond to a length of approximately 353 m, the lower end of our range. Nevertheless, the spread in the data stress our conclusion that the pair-wise coherence is difficult to measure and is strongly influenced by the signal-to-noise ratio.

WAVENUMBER MEASUREMENTS

The signal coherence can be inferred from the measurement of coherent gain realized by the summation of n hydrophones at a given spatial wavenumber k. An example is shown in fig. 5. One clearly sees the separation of signal energy and towship noise and in fig. 5 the beam response of the high signal level 173-Hz tone. The estimation of received coherent signal energy is easily performed at these beam signal-to-noise ratios. In these instances, the averaging interval used was too long when the source receiver motion is considered. Nevertheless, the peak signal levels provided a good measure of gain if the mean hydrophone level is known and a functional form of the coherence versus transverse distance can be hypothesized. That is to say, the difference between $20\ Log\ (N)$ and the measured gain is determined by the degree of signal phase variability and consequently the degree of coherence. The measurement uncertainties are reduced due to the large beam signal-to-noise ratios and the increased number of degrees of freedom.

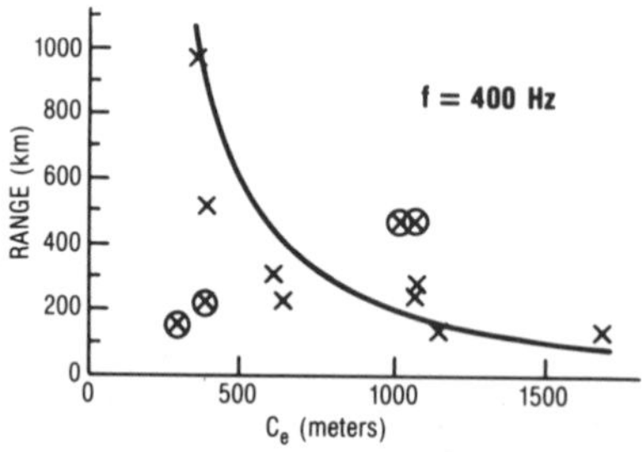

Figure 4. C_e *versus range at 400 Hz; curve represents theory.*

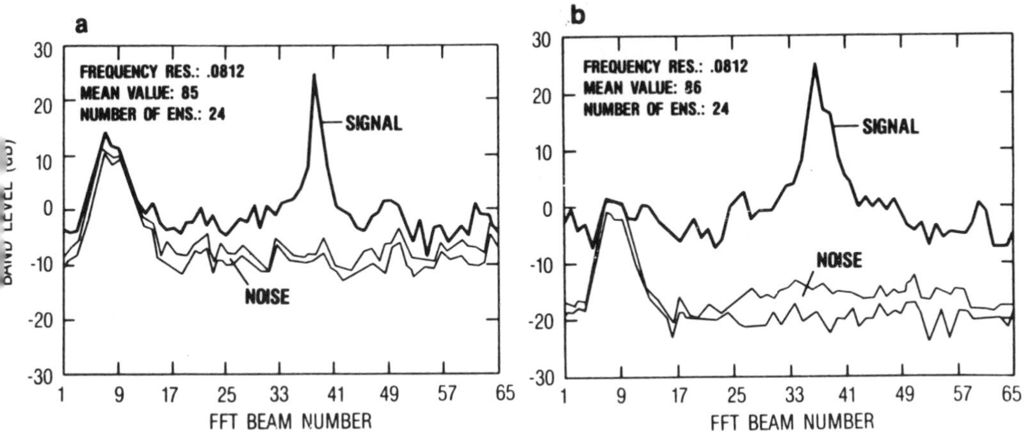

Figure 5. FFT beam response versus wavenumber (FFT beamnumber) for the signals from the 175 Hz source (a) and the 173 Hz source (b).

Angle domain measurements were made by Moseley[40,41] using as the receiver a line array of hydrophones positioned well within the sound channel in deep water. Figure 6a shows the average measured signal angular pattern (denoted by the solid curve). The reference array angular response denoted by the thin curve would result from the situation of a completely coherent plane wave impinging upon a completely straight array. As indicated in the figure, there are several major differences between the measured and reference patterns. The array signal gain (indicated by the measured value at 0°) is degraded by about 6 dB relative to the reference ideal value. The 3-dB angular width of the measured signal is about twice as wide as that of the referenced pattern. The measured angular distribution exhibits broad secondary maxima at angles different from the source direction (these will be shown to be due to multipath interference). Generally, the signal energy is distributed across angles instead of being restricted to a narrow main lobe with low sidelobe structure as in the reference case. The causes of this increased spreading in angle are threefold—acoustic multipath interference, array deformations, and acoustic scattering due to random sound speed variations in the ocean. The effects of each of these mechanisms are examined individually, and the results are compared with the measured signal angular pattern to determine their relative influence in figs. 6b, 6c, 6d. The predicted pattern combining these effects is shown in fig. 7

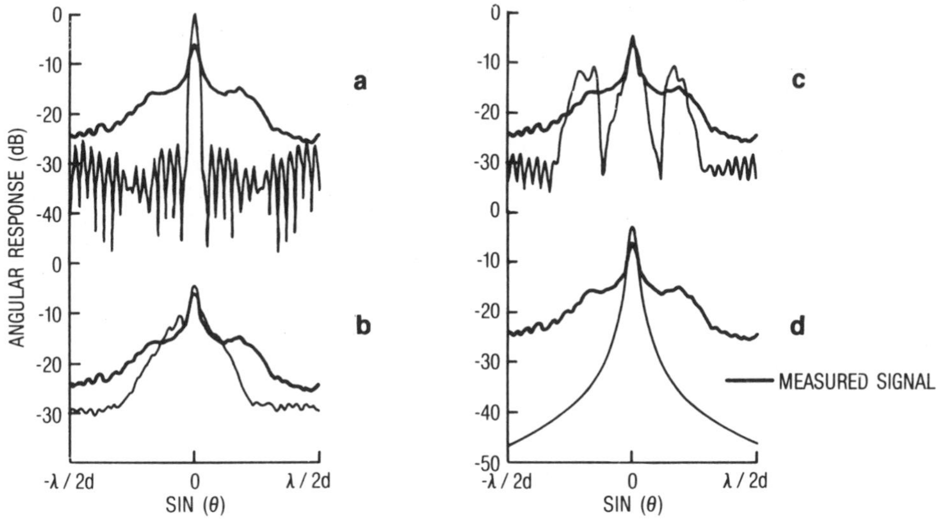

Figure 6. The average-measured beam response (dB) versus sin θ compared to computed beam response for (a) the case of an ideal array, (b) the case of a deformed array, (c) multipath interference, and (d) azimuthal scattering.

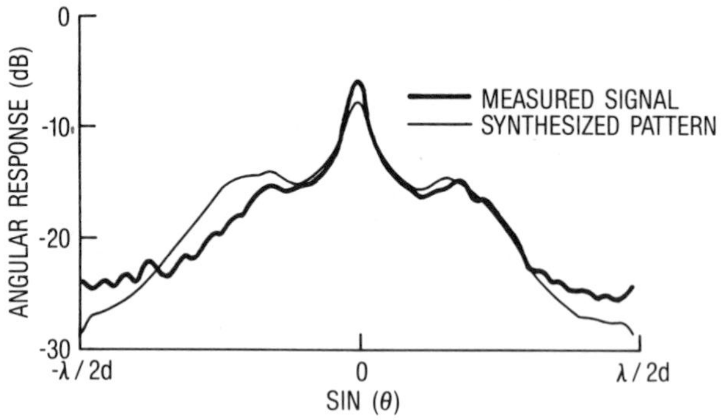

Figure 7. Comparison of measured and calculated beam angular response versus sinθ.

and was synthesized in the following fashion: with the incorporation of appropriate phase delays to simulate the measured horizontal array deformations, beamformer outputs were computed for the predominantly horizontal, but tilted, array located in the complex pressure field given for the presence of multipath interference by the corrected parabolic equation model. This angular output was then convolved with the theoretical azimuthal scattering pattern to arrive at the predicted pattern. There is now an agreement between the predicted and measured patterns to within 2.5 dB throughout almost all of the angular domain. Thus, the combination of multipath interference, array deformations, and acoustic scattering accounts for the array signal gain degradation, the change in the 3-dB angular width, the "shoulders" in the measured pattern, and the general redistribution of energy in the angle domain.

THE OBSERVATION OF MULTIPATHS

The propagation of mid- to low-frequency sound in the ocean to long range with refracted-refracted (RR) or refracted-surface-reflected (RSR) paths has been observed[42, 43] with arrays that separated the multipath arrivals. The separability of these arrivals by either a vertical or a horizontal array is determined by the length of the array, its orientation, and the coherence length of the signal. The effective coherence length of the signal is determined, in this instance, by the amount of the angular spread in the signal caused by diffused energy about a single arrival direction or by the distribution of energy in angle caused by the addition of several paths. Consider a plane wave from the θ_o, ϕ_o direction with wavenumber vector $\underset{\sim}{k}$ incident on an array along the x axis with a slight downward tilt ψ with θ_a taken with respect to the array axis and θ'_a its complement.

$$\underset{\sim}{k} \cdot \underset{\sim}{x} / |\underset{\sim}{k}||\underset{\sim}{x}| = \cos\theta_a = \cos\theta_o \cos\phi_o \cos\psi + \sin\theta_o \sin\psi = \sin\theta'_a . \tag{18}$$

We have shown from the off-broadside response of a uniform line array

$$2\Delta \sin\theta'_a = 2\Delta\cos\theta_a = 2\lambda/L \tag{19}$$

between the first nulls and for the half-power (hp) points at broadside

$$(\Delta \sin\theta'_a)_{hp} = (\Delta\cos\theta_a)_{hp} = 0.886\ \lambda/L \tag{20}$$

and at end-fire $(\Delta\theta)_{hp} = 2(0.886\ \lambda/L)^{1/2}$. For the off-broadside case, one can calculate a matched field condition where the angular spread of the signal is equal to the 3-dB beamwidth of the line array when the beam is steered in the direction of the signal.

$$\frac{L\Delta\cos\theta_a}{\lambda} = \frac{L\cos\phi_o \cos\psi}{\lambda/\Delta\cos\theta_o} + \frac{L\sin\psi}{\lambda/\Delta\sin\theta_o} . \tag{21}$$

The quantities $\lambda/\Delta\cos\theta_a = L_{eff}$, $\lambda/\Delta\cos\theta_o = L_b$, $\lambda/\Delta\sin\theta_o = L_z$ are interpreted as coherence lengths. When $L_b < L\cos\phi_o\cos\psi$, the horizontal component of the array overresolves the field as the beamwidth for a given conical angle is narrower than the field. When $L\sin\psi > \lambda/\Delta\sin\theta_o$, the vertical resolution of the array due to its vertical extent overresolves the field. When equation 21 is satisfied the field is matched to the length of the system.

The angular spreads could be due to the distribution of energy about each path resulting from scattering, the different vertical arrival angles of independent paths, or the angular spread across the array due to a change in range.

The case of vertical arrival of different paths in the deep ocean can be estimated by use of Snell's Law or Cox's ARAD[44,45] technique. Calculations with a representative Atlantic profile ($V_{os} = 1537\ m/s$, $V_a = 1526\ m/s$, $V_b = 1554\ m/s$) and Snell's Law yield for the bottom grazing and surface grazing rays $L_b = 90.9\lambda$ and $L_z = 14.4\lambda$. Using the ARAD technique with $\Delta cos\theta \sim V/V_o$, $\Delta V/D^2 = 0.17$, and $D(km)$ depth excess results in $L_b = 89\lambda/D$ and $L_z = 18\lambda/D$.

The deterministic multipath effects can be described by the methods of geometrical acoustics at the higher frequencies and by modes at the lower frequencies. The equivalence of modes and rays has been discussed by Guthrie and Tindle.[46] Here we use the ray-theoretic model, Trimain[47] to illustrate these effects.

Figure 8 shows the path arrival structure in the Atlantic as computed with the Trimain code at a frequency of 175 Hz for a source depth of 430 m and a receiver depth of 250 m. The ordinate is range in 0.2-km increments, and the abscissa is the vertical arrival angle with 0° being the horizontal. At each range increment the vertical line length represents the intensity for the eigenray at the angle corresponding to the intersection of the vertical line with the horizontal range line. Observe the angular difference and the convergence zone effects. The rate of change of arrival angle with range shows a definitive pattern with slopes between 0.44°/km to 0.16°/km. Shown in figure 8 is a similar case for the Ionian Basin of the Mediterranean. In this case the source is at a depth of 300 m, the receiver is at a depth of 250 m, and the frequency is 175 Hz. This pattern is also distinctive and has clear, separable arrivals. The response of a slightly tilted horizontal array to the field is the convolution of the beam response pattern and the angular spectrum of the incident acoustic field. At any given range one can observe multiple arrivals, the structure of which changes as the source-receiver separation is increased or decreased.

Measured data was obtained by Williams and Fisher[42] in the Pacific with a near-surface source and a receiver in the upper part of the sound channel in the midfrequency range. The arrival angles ranged between 8° and 14° in a clearly distinctive fashion, with a range rate of 0.143°/km. Thus the picture shown by data and calculation is one of multipaths at different angles, which have definitive range rate patterns. Similar results were obtained by Lawrence and Ramsdale.[43] These types of multipath patterns have also been observed to change with time as the properties of the intervening water column change.

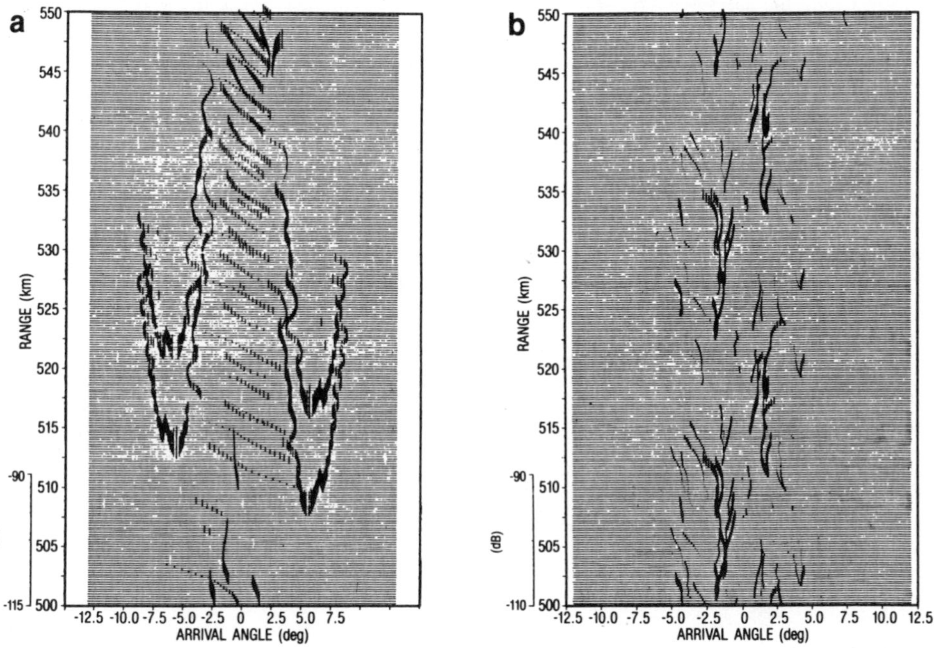

Figure 8. Computed eigenray arrival angle and intensity as a function of range at 175 Hz for (a) Atlantic and (b) Ionian

DYNAMIC EFFECTS

The space-time measurement of the ocean acoustic pressure field has been described primarily for stationary source and receivers. In these cases the limiting resolution of the line array was shown to be due to the angular spread of the incident energy about a single arrival due to scattering phenomena or due to the addition of multipaths and their variation over the aperture. Evidence obtained by Spindel and Porter,[48] as well as estimates by Dyer,[49] stresses the importance of multipath fluctuations and its dominance as relative motion between the source and receiver exceeds 1.5 m/s. In these instances the temporal and spatial resolution relationships are

$$\Delta f \cdot \Delta T \geqslant N$$

$$\Delta\theta_a \sin\theta_a (L/\lambda) = \Delta\theta'_a \cos\theta'_a (L/\lambda) \geqslant 1 \tag{22}$$

where $\sin\theta_a = \cos\theta'_a$ and $\cos\theta_a = \cos\phi_o \cos\theta_o$.

These relationships follow directly from our use of the Fourier transform and a rectangular sampling window. We observe that these relationships can become coupled by motion during the observation time.

$$\Delta\theta_{oa} = \frac{\partial\theta_{oa}}{\partial R}\Delta R + \frac{\partial\theta_{oa}}{\partial\phi}\Delta\phi = \frac{\sin\theta_o \cos\phi_o}{\sin\theta_a}\frac{\partial\theta_o}{\partial R}(L\cos\phi_o + V_R\Delta T) \tag{23}$$

$$+ \frac{\sin\phi_o \cos\theta_o}{\sin\theta_a}(V_\phi \Delta T/R)$$

This relationship can also be applied to the multipath problem as long as the angular separation between the paths is not great. We let $\Delta\theta_{oa}$ represent the angular spread due to the environmental factors and $\Delta\theta_a$ the beamwidth of the measurement system. When these two quantities are equal we have matched the measurement system to the environment. When $\Delta\theta_a < \Delta\theta_{oa}$ we have the capability to resolve multipaths, and when $\Delta\theta_a \geqslant \Delta\theta_{oa}$ the acoustic intensity is contained within a beam. A measurement problem of considerable interest is the resolution of distant shipping. In the following we will assume radial (V_R) and azimuthal velocities (V_ϕ) of 10 m/s, an azimuthal bearing of $\phi_o = 80°$, 90° being broadside, and measurement interval of 300 seconds.

By equating the aperture width with the angular spread it can be shown that for a single path with a radial motion only

$$\frac{\lambda}{L} = C_e G_o(\theta_o, \phi_o)(L\cos\phi_o + V_R\Delta T) \tag{24}$$

$$G_o(\theta_o, \phi_o) = \sin\theta_o \cos\phi_o,\ C_e = |\partial\theta_o/\partial R|$$

The constant C_e is the rate of change of single path vertical arrival angle (θ_o) and can be taken from the previously shown vertical arrival angle plots. This constant is small $0(2 \times 10^{-6})$ and near convergent zones $(0(8 \times 10^{-5}))$.

The azimuthal variation for the single path case can also be shown to be:

$$\lambda/L = \sin\phi_o \cos\theta_o V_\phi \Delta T/R \tag{25}$$

where R is the radial distance to the source, taken here to be 70 km. Similar results can be obtained for the multipath case and are given by

$$\lambda/L = \cos\phi\, \Delta\cos\theta \left\{ 1 + \left[\left(\frac{1}{\Delta\theta_o} \right) \left| \frac{\partial\Delta\theta_o}{\partial R} \right| + G_1 \left| \frac{\partial\theta_o}{\partial R} \right| \right] V_R\Delta T \right\} \tag{26}$$

$$+ \frac{\cos\theta_o \sin\phi_o}{R} V_\phi \Delta T$$

$$G_1 = \tan^2\phi_o/\tan\theta_{o1}(\tan^2\phi_o + \sin^2\theta_{o1}).$$

The multipath case shows a constant term determined by the initial spread in angle between the paths, a radial term that accounts for the change in spread with range, a radial term that accounts for the variation in central arrival angle with range and, finally, a term that accounts for the azimuthal rate of change.

These results are interesting, as they allow us to rank the relative importance of these effects on the ocean measurement problem. When motion is not important the single path refraction result yields

$$\frac{L}{\lambda} \doteq \rightarrow 8 \times 10^3/\sqrt{\lambda} \ . \qquad (27)$$

Comparison with measured results indicates that in this instance the measurement will be dominated by volume or surface scattering. When the motions are extreme we find for the refraction result

$$\frac{L}{\lambda} = \frac{1}{C_e G_o V_R \Delta T} \sim 3 \times 10^4 \qquad (28)$$

and again scattering, not refraction, limits the ability to measure the single path arrival. These results confirm our previous discussion regarding single path propagation and volume scattering.

We can continue the process and estimate, term by term, the relative importance of each effect. For the specific example given here we have tabulated these results in the following table.

A Comparison of Multipath/Motion Effects

Case		Remarks
Single path	L/λ	
Static refraction	$8\text{x}10^3/\sqrt{\lambda}$	$V_r = 0$ } Scattering is dominant.
Dynamic (radial) refraction	$4\text{x}10^3$	$V_r > 20$ }
Dynamic (azimuthal)	25	at R = 70 km, azimuthal motion is very important
Multipath		
Static (Atlantic)	525	$\phi_o = 80°$
	105	$\phi_o = 30°$
	506/D ~ 84 Cox's ARAD, D(km)	
Dynamic		
$V_r = V_\phi = 0$	302	Here we use a case with
$V_\phi = 0$ $V_r = 20$	278	$\theta_{o1} = 8.5°$, $\theta_{o2} = 16°$
($V_\phi = 0$ $V_r = 20$)	67.3 (Source near Cz)	
$V_r = 0$, $V_\phi = 20$	21.6	

The results, although given for only a specific measurement problem, clearly show the importance that multipath arrival structure and motion have on the resolution capabilities of the ocean measurement system. Measurements performed with arrays to measure horizontal noise directionality should carefully consider these results. We have used only horizontal arrays with no tilt or vertical extent. These results can easily be extrapolated to the case of a tilted array and the procedure applied to the case of a vertical array.

SUMMARY AND CONCLUSIONS

This paper focused on the problem of ocean-acoustic measurements with line arrays, the resolution of such systems, and the methods by which their response can be calculated.

The static source-receiver case was shown to be influenced by the randomness in signal phase due to scattering and, closely following the method and approach of Shifrin, we showed the results for Gaussian phase randomness and the importance of the coherence functional form. Calculations and data were employed to show the importance of multipath effects on the relative gain of line array measurement systems and the difficulties encountered for the determination of coherence lengths. Single path coherence lengths were found to be large and predictable using an environmental parameter E and the Beran-McCoy mutual coherence functional form. Nevertheless, multipath effects were, in our opinion, dominant.

The temporal fluctuation problem was briefly introduced for completeness but also to stress the finding that for relative source-receiver speeds of 1.5 m/s (3 knots) or greater, the fluctuations are dominated by the changes in the multipath arrivals. These effects dominate most measurements. We presented a simplified analytical formulism for determining the aperture length required to contain the multipath arrival during the measurement period or for determining the length of the system required to resolve two distinct arrivals.

In closing, we conclude that vertical and horizontal line arrays are merely spatial transforms analogous to temporal transforms. The spatial structure, angular spread, determines the length of the antenna required either to characterize the structure or to contain the structure within a spatial wavenumber interval.

When the incident signal is a plane wave with a Gaussian variable in phase, the net gain was shown to proceed according to $2L/\lambda$ until the spread in incident energy becomes greater than the spatial beamwidth half-power points, that is, a saturation effect. We conclude that this beginning of saturation is one limit on measurement system resolution.

When the vertical arrival angles from multiple paths project horizontal angles greater than the half-power beamwidths, beam bifurcation occurs. For vertical arrays we simply have the arrivals on different beams. Horizontal arrays have an azimuthal variation in the resolution from broadside to end fire. Our calculations and data lead to the conclusion that the combined effect of antenna distortion and multipath arrival structure are exceedingly important in the ability of the system to measure the incident field.

Motion of the source-receiver with respect to one another couples the space-time processing problem. We concluded that for static cases we must restrict our processing to temporal and spatial coherence lengths determined by either the temporal or the spatial correlation function. However, when the relative motion becomes important (>1.5 m/s), we are restricted to an effective length determined by the time required to perform the measurement and the variation of the multipath arrival structure.

The measurement of the signal gain of a spatial array requires the determination of the phase moments, the spatial coherence, or the coherent output of the line array. Whereas it is difficult to measure signal phase or coherence at low signal-to-noise ratio and large separations, it is possible to measure the coherently summed output of the array. This sum can be related to the mutual coherence function and, consequently, the coherence length estimated. Thus high-resolution, large arrays may provide a tool by which spatial coherence lengths can be estimated in the low- to midfrequency region.

In the case of the spatial coherence problem, a theory is required that combines and includes both the scattering and the multipath interference effects simultaneously. For the temporal coherence problem, a theory is needed that incorporates both the internal wave modulation of the acoustic path structure and the variation of the multipath interference pattern due to relative source-receiver motion. These are current areas of activity by several research groups. A complete theory that treats both temporal and spatial coherence simultaneously (including the several influential factors for each) is, in all likelihood, some years away from development. However, a precise predictive capability for either or both of the acoustic field coherence properties will influence both array and processor usage.

REFERENCES

1. A. A. Winder, C. J. Loda, *Introduction to Acoustical Space-Time Information Processing*, ONR Report ACR-63, Washington, DC, (1963), Republished, Peninsula Publishing, Los Angeles.
2. S. O. Rice, "Mathematical Analysis of Random Noise," Selected Papers on Noise and Stochastic Processes, *Bell Sys. Tech. Journal*, Dover Publications, New York, 23/24:133–294 (1954).
3. P. Middleton, *An Introduction to Statistical Communication Theory*, McGraw-Hill Book Co., New York (1960).
4. Y. W. Lee, *Statistical Theory of Communication*, John Wiley and Sons, New York (1960).
5. E. A. Robinson, *Statistical Communication and Detection, with Special Reference to Digital Signal Processing of Radar and Seismic Signals*, Hafner Pub. Co., New York (1967).
6. R. C. Bourret, Directivity of a Linear Array in a Random Transmission Medium, *J. Acoust. Soc. Am.*, 33(12):1793–1797 (1961).
7. H. G. Berman, A. Berman, Effects of Correlated Phase Fluctuations on Array Performance, *J. Acoust. Soc. Am.*, 34(5), pp. 555-562, 1962.
8. D. J. Bordelon, Effect of Correlated Phase Fluctuation on Array Performance, *J. Acoust. Soc. Am.*, 34(8):1147 (1962).
9. J. L. Brown, Variation of Array Performance with Respect to Statistical Phase Fluctuation, *J. Acoust. Soc. Am.*, 34(12):1927–1928 (1962).
10. G. E. Lord, S. R. Murphy, Reception of a Linear Array in a Random Transmission Media, *J. Acoust. Soc. Am.*, 36(5):850–854 (1964).
11. D. J. Bordelon, Comments on Reception Characteristics of a Linear Array in a Random Transmission Medium, *J. Acoust. Soc. Am.*, 37(2):387 (1964).
12. Y. S. Shifrin, *Statistical Antenna Theory*, (Published by Sovietskoye Radio, Moscow 1970). The Golem Press, Boulder, Colo. (1971).
13. B. D. Steinberg, *Principles of Aperture and Array System Design*, John Wiley and Sons, New York, (1976).

14. F. J. Harris, On the Use of Windows for Harmonic Analysis with Discrete Fourier Transform, *Proc. IEEE*, 66(1):51–83 (1978).
15. S. L. Marple, Jr., S. M. Kay, Spectrum Analysis, *Proc. IEEE*, 69(11):1380–1418 (1981).
16. W. C. Knight, R. Q. Pridham, S. M. Kay, Digital Signal Processing for Sonar, *Proc. IEEE*, 69(11): 451–1506 (1981).
17. R. A. Wagstaff, A. B. Baggeroer, *High-Resolution Spatial Processing in Underwater Acoustics*, NORDA, NSTL, Miss. (1983).
18. Lord Rayleigh, *Phil; Mag* 43, p. 259, 1897.
19. L. A. Chernov, *Wave Propagation in a Random Medium*, McGraw-Hill, New York (1960).
20. U. I. Tatarski, *Wave Propagation in a Turbulent Medium*, McGraw-Hill, New York (1961).
21. S. M. Flatte, R. Dashen, W. H. Munk, K. M. Watson, F. Zachariasen, *Sound Transmission Through a Fluctuating Ocean*, Cambridge Univ. Press, New York (1979).
22. S. M. Flatte, Wave Propagation Through Random Media: Contributions from Ocean Acoustics, *Proc. IEEE*, 71(11):1267–1299 (1983).
23. R. F. Dashen, S. M. Flatte, W. H. Munk, F. Zachariasen, *Limits on Coherent Process. Due to Internal Waves*, SRI-TR-JSR-76-14, Stanford Research Inst., Menlo Park, Calif.
24. Y. Desaubies, Statistical Aspects of Sound Propagation in The Ocean, in: *Adaptive Methods in Underwater Acoustics*, ed. H. Urban, D. Radel Pub. Co., Boston (1985).
25. J. McCoy, Propagation of Spatial Coherence in the Deep Ocean a Theoretical Framework, *Stochastic Modeling Workshop*, October 26–28, (1982), ARL/UT, Austin, Tex. (1983).
26. K. M. Guthrie, The Persistence of Acoustic Multipaths in the Presence of Ocean Volume Scattering, *Stochastic Modeling Workshop*, October 26–28, (1982), ARL/UT, Austin, Tex. (1983).
27. H. Cox, Line Array Performancce when the Signal Coherence is Spatially Dependent, *J. Acoust. Soc. Am.*, 54(16):1743–1746 (1973).
28. M. J. Beran, J. J. McCoy, Propagation through an Anisotropic Random Medium, *J. Math. Phys.* 15(11):1901 (1974). Also ref: M. J. Beran, *J. Acoust. Soc. Am.*, 56(6):1667 (1974).
29. M. J. Beran, J. J. McCoy, and B. B. Adams, *Effects of a Fluctuating Temperature Field on the Spatial Coherence of Acoustic Signals,* NRL Report 7809 Washington, D.C., (1975).
30. U. Frisch, *Probablistic Methods in Applied Mathematics*, Academic Press, New York, (1968).
31. D. C. Stickler, R. D. Worley, S. S. Jaskot, Bell Laboratories, Unpublished—Summarized by G. H. Robertson, Model for Spatial Variability Effects on Single-Path Reception of Underwater Sound at Long Ranges, *J. Acoust. Soc. Am.*, 69(1):112–123 (1981).
32. W. B. Moseley, D. R. DelBalzo, Oceanic Horizontal Random Temperature Structure, NRL Report 7673, Washington, DC, (1974).
33. R. Millard, Further Comments on Vertical Temperature Spectra in The Mode Region, Mode Hot Line News, No. 18, Private Communication, (1972).
34. G. V. Garrett, W. Munk, Space-Time Scales of Internal Waves, A Progress Report, *J. Acoust. Soc. Am.*, 80(3):271 (1975).
35. G. E. Devilbiss, R. L. Martin, N. Yen, Coherence of Harmonically Related CW Signals, NUSC-A11-C22-74, Available DTIC (1981).
36. G. C. Carter, C. H. Knapp, A. H. Nuttall, Statistics of the Estimate of the Magnitude-Coherence Function, *IEEE Trans. on Audio and Electroacoustics*, AU-21(8):388–389 (1973).
37. E. H. Scannell, C. Q. Carter, Confidence Bounds for Magnitude Squared Coherence Estimates, *Proc. IEEE Int. Conf. on Acoust., Speech and Signal Process.*, pp. 670–673 (1978).
38. R. E. Williams, Creating an Acoustic Synthetic Aperture in the Ocean, *J. Acoust. Soc. Am.*, 60(1): 60–72 (1976).
39. R. M. Kennedy, Phase and Amplitude Fluctuations in Propagating Through a Layered Ocean, *J. Acoust. Soc. Am.*, 46(3P2):737–745 (1969).
40. W. B. Moseley, Acoustic Coherence in Space and Time, *EASTCON 78*.
41. W. B. Moseley, Geographic Variability of Spatial Signal Correlation and Subsequent Array Performance, *Inter. Symp. on Underwater Acoustics*, Tel Aviv, Israel (1981).
42. B. Williams and F. Fisher, Private Communications.
43. M. Z. Lawrence, D. J. Ramsdale, Comparison of Vertical Line Array Data Taken at Two Depths in the Deep Ocean, *J. Acoust. Soc. Am.*, 77(51):GG6 (1985).
44. H. Cox, Approximate Ray Angle Diagram, *J. Acoust. Soc. Am.*, 61(2):353–369 (1977).

45. E. P. Jensen, H. Cox, Fluctuations in the Signal Response of Vertical Arrays Caused by Source Motion, *J. Acoust. Soc. Am.*, 50(SI), Paper 211 (1975).
46. K. M. Guthrie, C. T. Tindle, Ray Effects in the Normal Mode Approach to Underwater Acoustics, *J. Sound Vib.*, 47:403–413 (1976).
47. B. G. Roberts, Horizontal Gradient Acoustic Ray-Trace Program Trimain, NRL Report 7827, Washington, D.C. (1974).
48. R. P. Porter, R. C. Spindel, Low-Frequency Acoustic Fluctuations and Internal Gravity Waves in the Ocean, *J. Acoust. Soc. Am.*, 61:943 (1977).

AN EFFICIENT METHOD OF DOLPH-CHEBYSHEV BEAMFORMING

A. Zielinski

Dept. of Electrical Engineering
University of Victoria P.O. Box 1700
Victoria, B.C., Canada V8W 2Y2

ABSTRACT

Several methods have been developed for tapering or shading of line arrays to minimize minor lobes. A method developed by C.L. Dolph makes it possible to optimize the patterns so that (1), for any specified minor-lobe, the narrowest possible major lobe is achieved; or (2), for any specified major-lobe width, the lowest possible minor-lobe levels are achieved. The procedure to calculate the required set of shading coefficients involves algebraic manipulation of Chebyshev polynomials. This manipulation becomes increasingly tedious as an array size becomes larger.

The paper presents a much simpler, equivalent computational procedure which involves matrix inversion only and can be used effectively for large arrays. Sample results are presented.

INTRODUCTION

Let us consider a linear array of point receivers spaced uniformly by d and subjected to a plane waveform arriving from θ angle with respect to the normal array, as shown in Fig. 1.

If the responses of each receiver of sensitivity $A_i/2$ are summed up then the total response P of the array with an even number of receivers (elements) N is given by

$$P(u) = \sum_{i=1}^{N/2} A_i cos(2i-1)u \tag{1a}$$

where

$$u = \frac{\pi d}{\lambda} sin\theta \tag{1b}$$

and

$$\lambda = c/f \tag{1c}$$

In the above, λ denotes the wavelength, f is a frequency of the incident waveform and c is the propagation velocity.

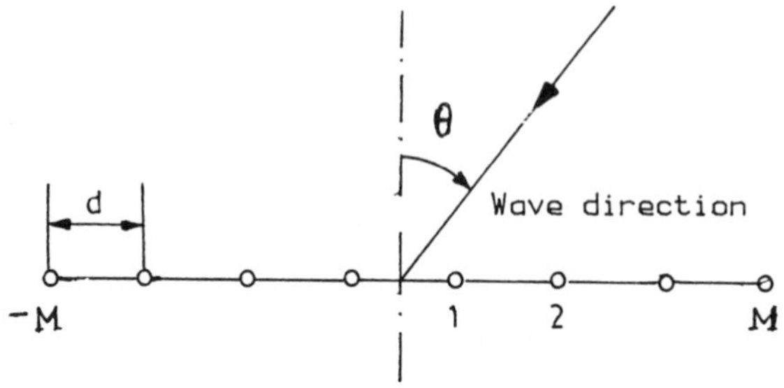

FIG. 1. THE RECEIVING ARRAY

Similarly, for an odd number of receivers

$$P(u) = \sum_{i=0}^{(N-1)/2} A_i cos2iu \tag{2}$$

The array response $P(u)$ depends through u on the incident angle θ and represents radiation pattern of the array. Often, the required radiation pattern consists of a desirable main lobe (in the $\theta = 0$ direction) and several undesirable side-lobes.

By a proper selection of coefficients A_i (or array shading), a desirable radiation pattern can be achieved as a compromise between the width of the main lobe and relative amplitudes of the side lobes.

DOLPH-CHEBYSHEV METHOD OF ARRAY SHADING

The Dolph method of array shading (Ref. 1, 3) is based on properties of Chebyshev polynomials defined as

$$T_n(x) = cosnu = \sum_{i=1}^{n} a_i x^i \tag{3a}$$

where

$$x = cosu \tag{3b}$$

and n indicates a degree of the polynomial.

It can be shown that the polynomial (3) has the following properties

$$|T_n(x)| \leq 1 \quad for \quad |x| \leq 1 \tag{4}$$

Outside the above range, the $| T_n(x) |$ is a monotonically increasing function of x. This is illustrated in Fig. 2, where the polynomial $T_n(x_0 cosu)$ is plotted as a function of u.

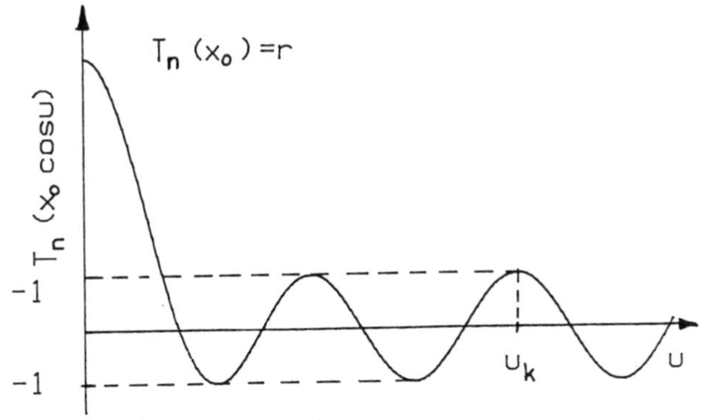

FIG. 2. THE CHEBYSHEV POLYNOMIAL.

The parameter $x_o \geq 1$ can be chosen as to achieve an arbitrary large value of $T_n(x_o) = r$ using the following relationship.

$$x_o = 1/2 \left[\left(r + \sqrt{r^2 - 1} \right)^{1/n} + \left(r + \sqrt{r^2 - 1} \right)^{-1/n} \right] \quad (5)$$

Comparing eqn. (1) or (2) with eqn. (3) we can see that the array radiation pattern can be expressed in terms of a weighted sum of Chebyshev polynomials. For example, the eqn. (1) can be written as

$$P = \sum_{i=1}^{N/2} A_i T_{2i-1}(x) = \sum_{i=1}^{N-1} B_i x^i \quad (6)$$

where the coefficients B_i are related to the coefficients A_i. By proper choice of the B_i's (and the related A_i's) one can impose the following equality

$$P(u) = \sum_{i=1}^{N-1} B_i x^i = T_{N-1}(x x_o) \quad (7)$$

An array with receivers' sensitivities $A_i/2$ such that eqn. (7) is satisfied will have a radiation pattern identical to the function shown in Fig. 2.

This method of beamforming allows for a design of arrays with arbitrary main-to-side-lobes ratios. It can also be shown (Ref. 4) that the resulting pattern yields a minimum beamwidth when the side lobe levels are fixed and a minimum side lobe level when the beamwidth is specified.

To achieve a very narrow main lobe an array with a large number of elements is required.

For longer arrays, however, the design becomes exceedingly tedious since finding the relationships between the coefficients A_i and B_i in eqn. (6) requires lengthy algebraic manipulations. For this reason methods which simplify the design have been developed (Ref. 2, 5, 6, 7).

THE MATRIX METHOD

A compact design method of Dolph-Chebyshev beamforming can be proposed by making an observation that, after applying the Dolph method of shading, the beam pattern $P(u)$ given by eqn. (1) (or (3)) has to be equal to the Chebyshev polynomial given by eqn. (7), that is

$$P(u) = \sum_{i=1}^{N/2} A_i cos(2i-1)u = T_{N-1}(x_o cosu) \tag{8}$$

The values of $T_{N-1}(x_o cosu)$ are known at several characteristic points. Specifically, the side lobes peaks -1 or $+1$ occur alternatively at

$$u_k = cos^{-1}\left[\frac{1}{x_o} cos\frac{k\pi}{N-1}\right]; k = 1, 2 \ldots \tag{9}$$

Also

$$T_{N-1}(x_o cos0) = r \tag{10}$$

The coefficients A_i can therefore be found directly from the eqn. (8) by solving a set of linear equations obtained by utilizing selected values for u as given by the eqn. (9) and (10).

In matrix formulation this can be written as

$$\begin{bmatrix} A_1 \\ A_2 \\ A_3 \\ \vdots \\ A_{N/2} \end{bmatrix} = \begin{bmatrix} 1 & \ldots & \ldots & 1 \\ cosu_1 & cos3u_1 & \ldots & cos(N-1)u_1 \\ \vdots & \vdots & \vdots & \vdots \\ \vdots & \vdots & \vdots & \vdots \\ cosu_{N/2-1} & \ldots & \ldots & cos(N-1)u_{N/2-1} \end{bmatrix}^{-1} \begin{bmatrix} r \\ -1 \\ 1 \\ \vdots \\ -1 \end{bmatrix} \tag{11}$$

A similar matrix equation can be obtained for an array with an odd number of elements. The set of equations can also be obtained utilizing zero crossings of $T_{N-1}(x_o cosu)$ or any combinations of zero crossings and side lobe peaks.

SUMMARY

The design of an array with an even number of elements involves the following steps:

. selection of the main-to-side-lobe ratio r

. calculation of the coefficient x_o as given by eqn. (5)

. formulation of the matrix equation as given by eqn. (9) and (11)

. the matrix inversion and multiplication as required by eqn. (11).

An example of a radiation pattern of an array with 40 elements and $r = 40dB$ obtained using the above approach is shown in Fig. 3. The corresponding coefficients A_i are shown in Fig. 4.

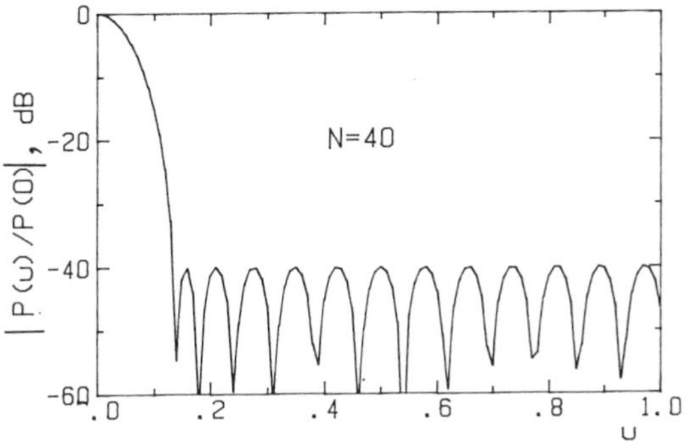

FIG. 3. THE BEAM PATTERN

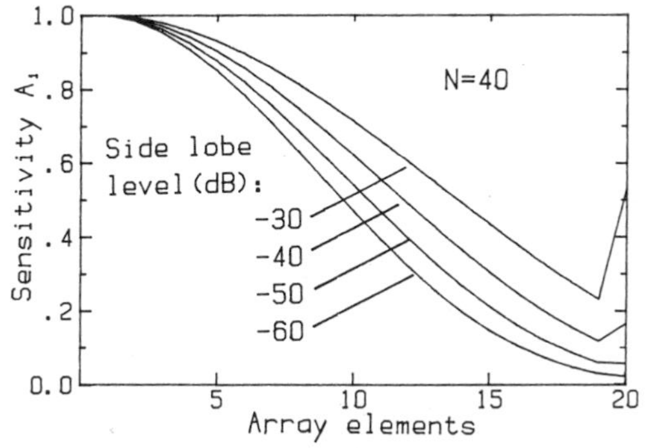

FIG. 4. THE ARRAY SHADING.

REFERENCES

1. Alberts, V.M., "Underwater Acoustics Handbook II", The Pennsylvania State University Press, 1965, pp. 188-199.

2. Barbiere, D., "A Method for Calculating the Current Distribution of Tschebyscheff Arrays", Proc. I.R.E., vol. 40, pp. 78-82, Jan. 1952.

3. Camp, L. "Underwater Acoustics", John Wiley & Sons, 1970, pp. 169-172.

4. Dolph, C.L., "A Current Distribution of Broadside Arrays which Optimizes the Relationship between Beamwidth and Side Lobe Levels", Proc. I.R.E., pp. 335-348, (June 1946); pp. 489-492, (May 1947).

5. Drane, Jr., C.J., "Dolph-Chebyshev excitation coefficient approximation,"
IEEE Trans. on Ant. and Prop., vol. AP-12, No. 6, pp. 781-782, November 1964.

6. Stegen, R.J., "Excitation Coefficients and Beamwidth of Tschebyscheff Arrays", Proc. I.R.E., pp. 1671-1674, (Nov. 1953).

7. Van der Maas, G.J., "A simplified calculation for Dolph-Tschebycheff arrays," J. Appl. Phys., vol. 25, pp. 121-124, January 1954.

MULTIPATH PROCESSING FOR IMPROVED DETECTION ON A LONG VERTICAL ARRAY

Donald A. Murphy and Donald R. Del Balzo

Naval Ocean Research and Development Activity
NSTL, MS 39529-5004, U.S.A.

ABSTRACT

Long vertical arrays provide array gain and allow separating sources due to differences in vertical arrival angle. Multipaths from one source will also be received at several arrival angles, each multipath received preferentially over a different subaperture of the vertical array. For low frequencies in the deep ocean, the optimum subaperture based upon intensity distribution will often be so large that signal multipaths will be resolved causing serious array signal gain degradation. A new processor using the sound speed profile is suggested for use on the appropriate subaperture to coherently recombine the multipaths and thus recover the lost signal gain. The arrival angle of one multipath is a function of depth, so phases are calculated to take this into account. Array signal gain (ASG) was determined by calculating the complex field at a vertical array using a model of ocean propagation. The ASG degradation (ASGD), and multipath resolution as a function of vertical aperture were calculated. For the case where several multipaths contributed substantially to the energy received, 5-7 dB of ASGD were observed.

INTRODUCTION

Several reports have described measurements of coherence as a function of aperture length made with towed and other nearly horizontal arrays. It was found that there are three factors which determine the amount of array signal gain degradation experienced (which is used as a measure of signal coherence). The least important of the three for low frequencies is spreading in arrival angle due to volume scattering. The other two are deterministic effects which, in principle, can be removed by appropriate changes to beamforming. The first is array shape, a distorted array is phased as though it is straight, thus introducing phase errors for the hydrophones into the sum in the beamformer. If the element locations are determined, these phase errors can be removed for a source in a particular direction. The second deterministic factor is that signal arrives via multipaths at different arrival angles which can be resolved if the array is not completely horizontal (tilted) or the arrivals are not at broadside (for a line array). It is this latter effect which becomes most important with a vertical array. A 120 wavelength aperture was chosen to have negligible volume scattering ASGD. When several multipaths contribute substantially to the sound field, ASGD is much worse than when there are only two multipaths with significant energy. This latter case occurs in convergence zones where the energy level is higher. Three dB of signal gain lost by resolving them can be recovered in this case.

ARRAY RESPONSE PATTERN COMPUTATION

The complex pressure field in a deep ocean area near the fifth convergence zone of a 100 meter source was calculated using the parabolic equation program PAREQ. The general intensity structure of the field is shown in Figure 1. Two range intervals were selected to sample with vertical arrays located from 300 to 1500 meters in depth. The array response was examined over 5 ranges 1 km apart to determine the exact nature of the multipath field. One of these 4 km areas was selected to include the strong signals in the convergence zone and the other was selected in an area where bottom bounce signals much lower in amplitude were dominant.

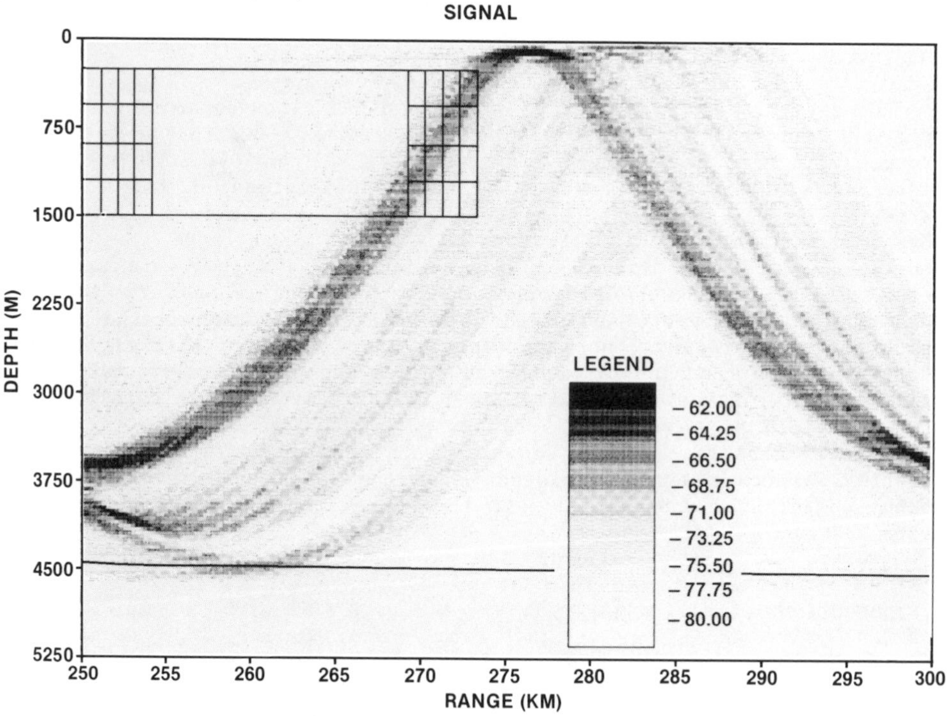

Figure 1. Shade plot of sound intensity near the fifth convergence zone showing vertical array locations.

The size of the vertical aperture was varied from 30 wavelengths to 120 wavelengths for each of the range intervals. The range interval from 250.25 km to 254.25 km, which is dominated by relatively weak bottom bounce paths will be considered first. The shortest aperture, 30 wavelengths, was examined at 4 depths. Figure 2 shows that the peak response angle varies considerably over the 5 ranges one km apart for this small aperture. This large variance was true for all four depths. It is also quite evident that this aperture does not resolve all the multipaths. The fact that they are not resolved contributes to the variance in peak arrival angle since different ratios of unresolved multipaths causes the peak to move in angle. The two 60 wavelength apertures also suffer from this same problem although their resolution is twice as great.

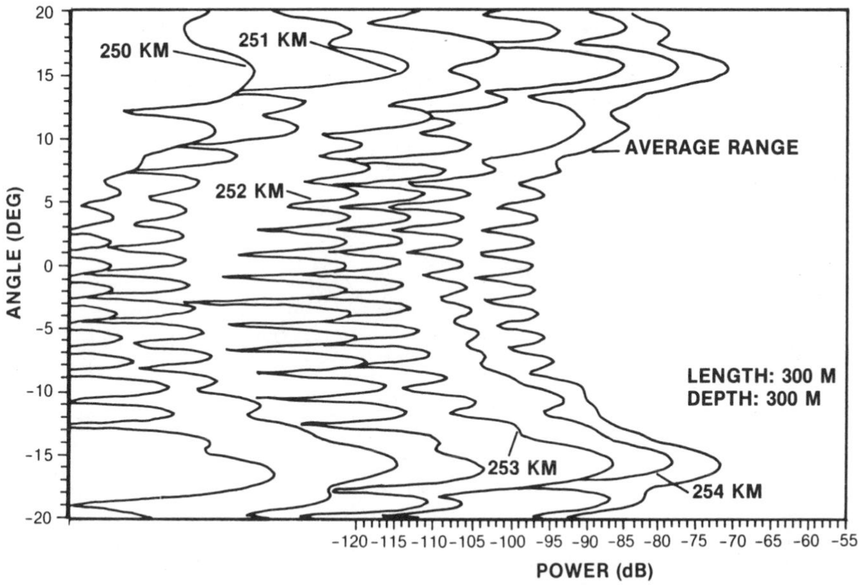

Figure 2. Array response patterns for an array extending from 300 to 600 meters as a function of range between convergence zones.

The response patterns of the 120 wavelength aperture shown in Figure 3 resolve the bottom bounce arrivals. There are many with significant energy and the peaks in the angular response patterns are relatively independent of range over the 4 km interval. The array signal gain for any one single arrival is degraded, because the many arrivals contribute to the field seen by the single hydrophone average, whereas all but one do not contribute to the energy in the beam.

The range interval between 269.25 and 273.25 km contains the refracted and RSR arrivals from the source over the depths sampled. The array response patterns over the 2 different apertures are shown in Figures 4 and 5. The arrivals are dominated by a single peak which is actually two unresolved arrivals very close together. Larger variance in angle for the shorter apertures is seen for this case also.

The 120 wavelength aperture illustrated in Figure 5 shows that the main peak is still unresolved into its two components for 4 of the 5 ranges. The fifth range where the array is mostly deeper than the main energy in the refracted arrivals shows them being resolved. The same thing occurs with a 90 wavelength aperture in spite of the shorter aperture where the array is completely below the peak energy depth. The peak at -16.5 degrees for the bottom-bounce case and the peak at 12.75 degrees for the refracted case were selected to examine for several subapertures. They remained within the "beamwidth" except for a 30 wavelength aperture above the sound channel axis.

The array signal gain for the selected intensity peak was calculated for all four 30 wavelength apertures, two 45 wavelength apertures, two 60 wavelength apertures and two 90 wavelength apertures. The results in

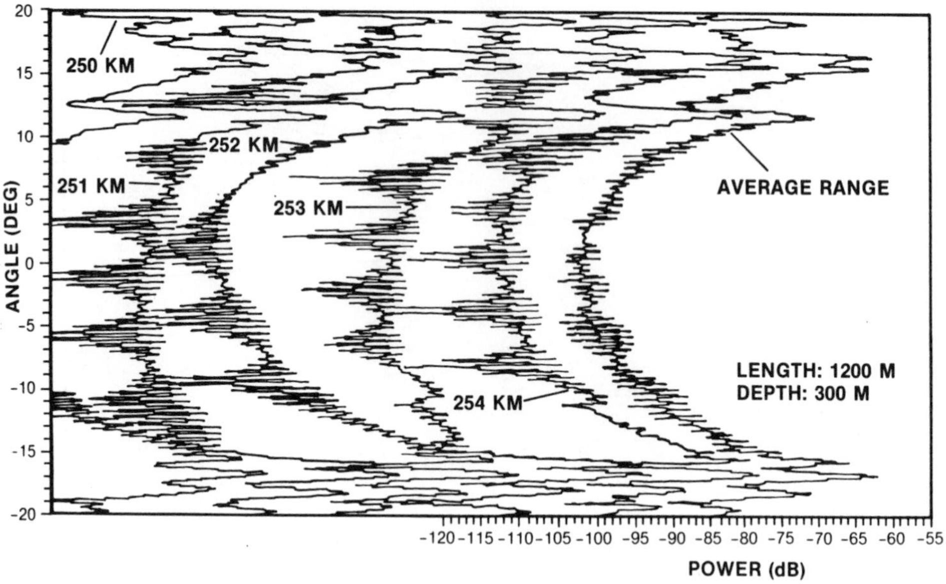

Figure 3. Array response patterns for an array extending from 300 to 1500 meters as a function of range in between convergence zones.

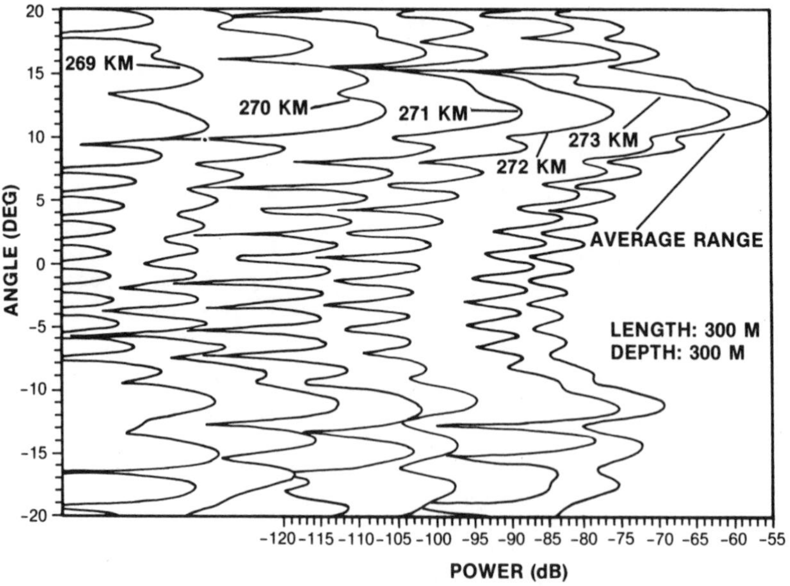

Figure 4. Array response patterns for an array extending from 300 to 600 meters as a function of range in the convergence zone.

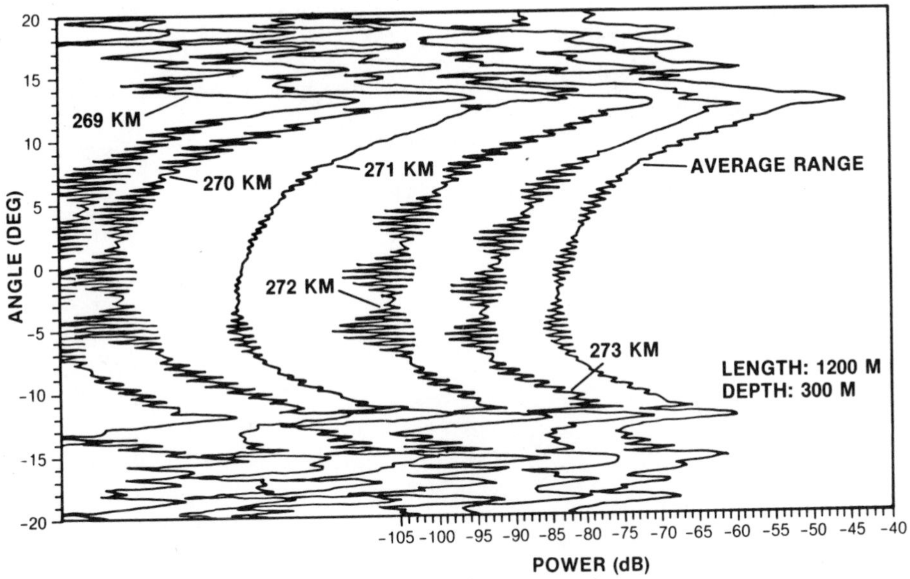

Figure 5. Array response patterns for an array extending from 300 to 1500 meters as a function of range in the convergence zone.

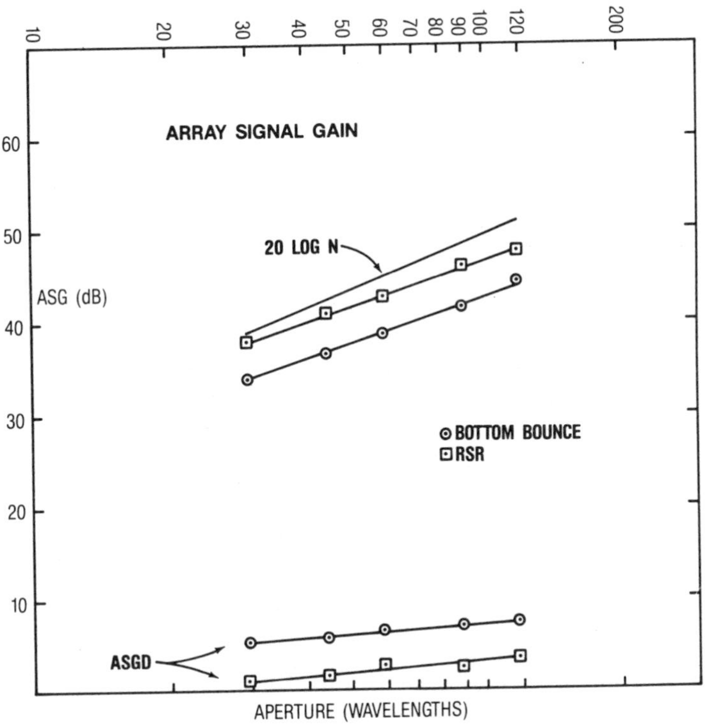

Figure 6. Array signal gain of a long vertical array and degradation due to multipaths as a function of array length.

Figure 6 show some variance about the lines for array signal gain so it is not clear whether the slope is decreasing with aperture or not. For the bottom-bounce case, array signal gain degradations from 5 to 7 dB are shown, but for the refracted case only 1 to 2.5 dB occurred.

The "beamwidths" of the array response patterns are shown in Figure 7. The bottom-bounce arrivals are close to the theoretical beamwidth of the aperture. The refracted arrival angles are spread more because of unresolved arrivals. In cases where many multipaths exist there is a significant degradation of array signal gain.

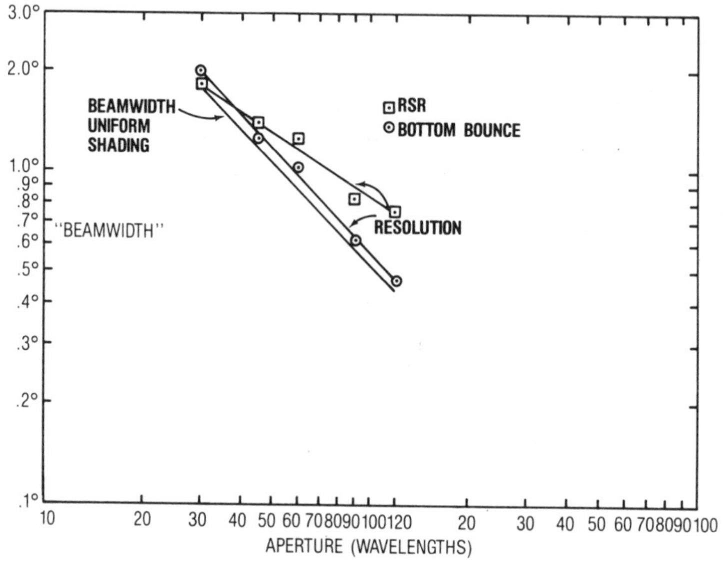

Figure 7. Array response "beamwidth" of a long vertical array as a function of array length.

WAVE FRONT SHAPE EFFECTS

The wavefronts arriving over a long vertical array are not planar even though the source may be quite distant. The sphericity is negligible in the fifth convergence zone, but the effect of refraction is not. A single multipath wavefront will have a shape described by Snell's law as shown in equation 1, where θ_1 and θ_2 are the angles of arrival of the multipath at two depths and v_1 and v_2 are the sound speeds at those depths.

$$\frac{\cos \theta_1}{v_1} = \frac{\cos \theta_2}{v_2} \qquad (1)$$

When the beamformer is modified to use these different arrival angles to calculate the phase differences for the hydrophones at different depths, the multipath energy over the entire depth interval where it exists is added in phase giving a little more array signal gain for that particular multipath.

The difference the Snell's law correction makes on the vertical arrival structure is illustrated in Figure 8. The angle of the peak is displaced because the Snell's law correction used the maximum velocity over the array aperture as a reference, and it was not referred back to the center of the array. The peaks in the pattern are a little higher, varying from .4 to as much as 2-3 dB for some cases where the vertical distribution of the particular multipath is large.

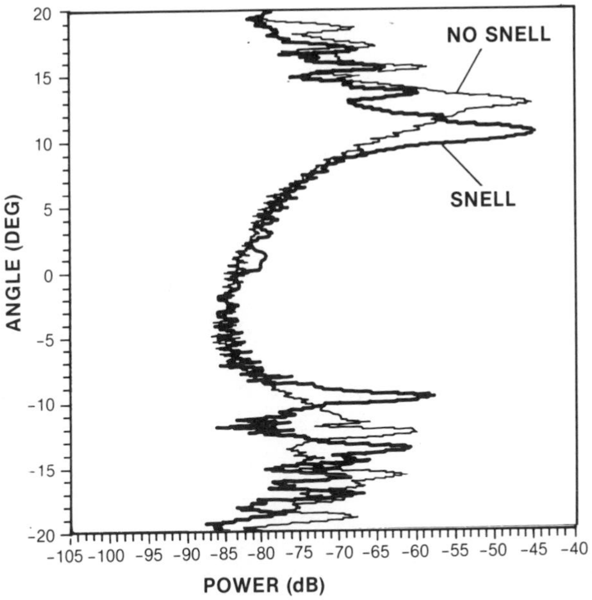

Figure 8. Array response patterns for an array extending from 300 to 1500 meters with and without Snell's law correction.

TWO EQUAL POWER MULTIPATHS

The case of two equal power multipaths is illustrated in Table 1. The values for noise and signal levels are relative to that received with a single multipath. The values shown apply to an unshaded array with N elements insonified.

TABLE 1. GAIN FOR EQUAL POWER MULTIPATHS

	PHONE NOISE	PHONE SIGNAL AVG	PHONE SIGNAL MAX	BEAM NOISE	BEAM SIGNAL AVG	BEAM SIGNAL MAX	ASG	ANG	AG
1 PATH	1	1	X	N	N^2	X	N^2	N	N
2 PATHS UNRESOLVED	1	2	4	N	$2N^2$	$4N^2$	$N^2/2N^2$	N	N/2N
2 PATHS RESOLVED	1	2	X	N	N^2	X	$N^2/2$	N	N/2
2 PATHS RECOMBINED	1	2	X	2N	$4N^2$	X	$2N^2$	2N	N

The top line illustrates the normal beamformer with one multipath. The array signal gain is N^2 and the array gain is N. The second line shows that the signal at one hydrophone can vary from 0 to 4 in the interference field of two multipaths which are not resolved. Unresolved implies that the entire array aperture spans less than one period of the interference pattern, thus there can be either twice as much pressure, no pressure at all, or anywhere in between depending on where the array is located in the interference pattern. In this case the array signal gain is N^2 and the array gain is N as before.

When the multipaths are resolved, the array spans a full interference period, and there is no difference between average and maximum levels. The array signal gain and the array gain are only $N^2/2$ and N/2 respectively for each beam containing a resolved multipath. If the multipaths are coherently recombined, the noise level is doubled, but the array signal gain is quadrupled. Thus, the array signal gain is 2N, and the array gain is N, restoring the lost array gain.

CONCLUSIONS

Multipath-caused array signal gain degradation for vertical arrays in the deep ocean up to a length of 120 wavelengths is serious between convergence zones, but is marginally significant in the fifth convergence zone.

Multipaths generally do not insonify the whole 120 wavelength aperture for either bottom bounce or convergence zone propagation and the portion of the vertical aperture insonified is different for different multipaths at a given range.

Corrections in phasing considering differing arrival angles at different depths for a single multipath can improve array signal gain from .4 to 3 dB depending on the amount of vertical aperture insonified by the multipath.

The multipath processor is able to regain the lost array signal gain caused by resolution of multipaths by the vertical array. It consists of a coherent combination of "beams" steered to the separate multipath arrival angles. A matched-field approach appears necessary to account for the less than full aperture insonification of each multipath. A ray trace approach to generating the field appears most efficient in the deep ocean, but it must be sufficiently accurate to predict the insonification level across the aperture. It can also be used for broadband signals.

ACKNOWLEDGEMENT

This work was sponsored by the Defense Advanced Research Projects Agency.

BEARING ESTIMATION OF A SINGLE MONOCHROMATIC PLANE WAVE WITH A LINEAR RECEIVING ARRAY

Didier Billon

THOMSON SINTRA Activités Sous-Marines
Route de Sainte Anne du Portzic
29601 BREST cedex - France

ABSTRACT

Three methods of estimating the arrival angle of a plane wave at linear array are compared. The three methods are Maximum Likelihood, Amplitude Comparison of adjacent beams, and phase comparison with a segmented array. - Ed.

INTRODUCTION

When a wave is impinging on a linear array of sensors, the angle between its direction of arrival and array broadside axis can be estimated from the sensor signals. In this paper it is assumed that

- there is only one wave impinging on the array,
- it is plane and monochromatic,
- its frequency is known, and
- the signals provided by sensors are corrupted by uncorrelated noise.

With these assumptions, wave angular location is an old problem that has received great attention from authors in the radar community ever since World War II (for about 30 years), especially in the case of the parabolic monopulse antenna[1,2,3]. It also occurs in the underwater acoustics field where the array antenna is obviously more convenient. But little material has been published on this problem for array processing. Emphasis is currently given on a much more difficult case: an unknown number of random sources with unknown spectral densities and unknown wave front shapes[4].

This paper aims to give a straightforward analytic treatment of the first-mentioned problem. Three bearing estimators are considered:

- the maximum likelihood estimator,
- the amplitude comparison estimator, and
- the phase comparison estimator.

This work was partly supported by the Direction des Recherches, Etudes et Techniques. Ministère de la Défense - PARIS - FRANCE.

The last two are the well known monopulse estimators. Here "monopulse" has been avoided because its meaning is related to a particular tracking radar problem.

MATHEMATICAL MODEL OF THE ARRAY SIGNALS

Let us introduce the following notation:

- θ is the angle between wave direction of arrival and array broadside axis in the plane defined by the linear array points and the wave propagation vector,
- M is the number of sensors,
- ξ_m is the abscissa of the mth sensor phase center on the array line,
- $k = 2\pi/\lambda$ is the wavenumber.

From the assumptions above, complex envelopes of array signals are the components of the random vector

$$X = sD(\theta) + N \tag{1}$$

where

$$D(\theta) = (e^{-jk\xi_1 \sin\theta}, \ldots, e^{-jk\xi_M \sin\theta})^T . \tag{2}$$

s is an unknown complex parameter that holds for the signal component induced by the wave at the output of a sensor located at abscissa $\xi=0$. N is a zero mean complex random vector, whose components are the uncorrelated complex envelopes of the noises occuring at the sensor outputs. Real and imaginary parts of each noise component are assumed to be uncorrelated and to have the same mean power $\sigma^2/2$. Therefore the signal-to-noise ratio at each sensor output is $|s|^2/\sigma^2$.

MAXIMUM LIKELIHOOD ESTIMATOR

Here the noise is assumed to be gaussian. Then the probability density function of X is

$$P_{s,\theta}(X) = \frac{1}{\pi^M \sigma^{2M}} \exp\left(- \frac{||X - sD(\theta)||^2}{\sigma^2}\right) . \tag{3}$$

X is the observed vector and we are looking for $(\hat{s}, \hat{\theta})$ such that $p_{\hat{s},\hat{\theta}}(X)$ is maximal, i.e. $||X-\hat{s}D(\hat{\theta})||$ is minimal. This occurs when

$$\hat{s} = \frac{1}{M} D(\hat{\theta})^{\dagger} X . \tag{4}$$

Then $\hat{s}D(\hat{\theta})$ is orthogonal to $X - \hat{s}D(\hat{\theta})$, hence

$$||X||^2 = ||\hat{s}D(\hat{\theta})||^2 + ||X - \hat{s}D(\hat{\theta})|| . \tag{5}$$

Therefore $||X-\hat{s}D(\hat{\theta})||$ reaches its minimal value when the magnitude of $\hat{s}$ given by Eq. (4) is maximal. $\hat{s}$ is the beam signal formed from array signals in the look direction $\hat{\theta}$. Hence $\hat{\theta}$ is the look direction of the maximal amplitude beam. In this paper it is called maximal amplitude direction (MAD).

BEAM COMPARISON ESTIMATORS

Beam comparison estimation is based on reducing the observation space dimension from M to 2 by linear beamforming.

Amplitude Comparison

Here the array is assumed to be symmetric and the abscissa origin is taken at its center:

$$\xi_{M-i+1} = -\xi_i \ . \tag{6}$$

Let y_{-1} and y_1 be the beams formed in respective look directions θ_{-1} and θ_1. From (1) they can be expressed as

$$\begin{aligned} y_{-1} &= s\ D(\theta_{-1})^{\dagger}\ D(\theta) + D(\theta_{-1})^{\dagger}\ N \ , \\ y_1 &= s\ D(\theta_1)^{\dagger}\ D(\theta) + D(\theta_1)^{\dagger}\ N \ . \end{aligned} \tag{7}$$

Because of array symmetry and abscissa origin choice $D(\theta_{-1})^{\dagger}D(\theta)$ and $D(\theta_1)^{\dagger}D(\theta)$ are real. Setting $u=\sin\theta-\sin\theta_o$ with $\sin\theta_o=(\sin\theta_{-1}+\sin\theta_1)/2$, we seek an estimate of F(u) defined by

$$F(u) = \frac{D(\theta_1)^{\dagger}D(\theta) - D(\theta_{-1})^{\dagger}D(\theta)}{D(\theta_1)^{\dagger}D(\theta) + D(\theta_{-1})^{\dagger}D(\theta)} \ . \tag{8}$$

It is recalled that from Eq. (2) that $D(\theta)$ is actually a function of $\sin\theta$. Therefore, the equation above is meaningful. F is an odd real function. It has a reciprocal if $|u|<\Delta/2$. Δ depends on array point locations and on $\delta=\sin\theta_1-\sin\theta_{-1}$. Some curves F(u) are plotted in Figure 1 for high M and for different values of the parameter δ in the case of a periodic array, i.e.

$$\xi_{m+1} - \xi_m = L/M \tag{9}$$

where L is the array length. When $\delta<\lambda/L$, Δ is approximately equal to $2\lambda/L$. It is greater if $\delta=\lambda/L$ and smaller if $\delta>\lambda/L$.

If the noise is gaussian, the maximum likelihood estimate of $v=F(u)$ is[5]

$$\hat{v} = \mathrm{Re}\left(\frac{y_1 - y_{-1}}{y_1 - y_{-1}}\right) \ . \tag{10}$$

Therefore maximum likelihood estimate of θ is

$$\hat{\theta} = \sin^{-1}(F^{-1}(\hat{v}) + \sin\theta_o) \ . \tag{11}$$

$\hat{\theta}$ is called the amplitude comparison estimate because the wave components of beams y_{-1} and y_1 have the same phase modulo π. It is clear that $|\sin\theta-\sin\theta_o|$ is always smaller than $\Delta/2$.

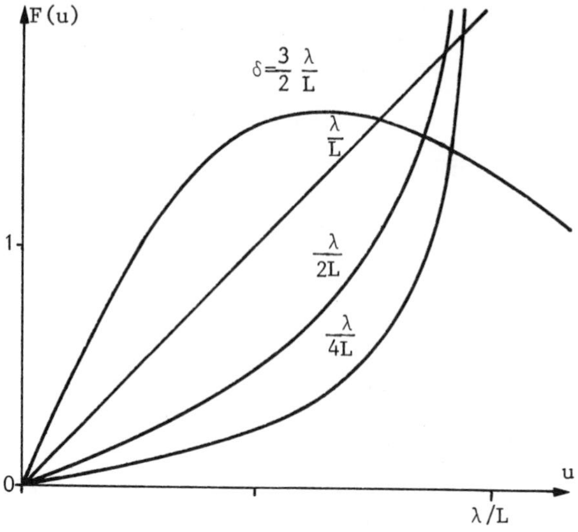

Figure 1

Phase Comparison

Here it is assumed that the last I points of the array are obtained from the first I ones by a translation of length $d \leq L/2$:

$$\xi_{M-I+i} = \xi_i + d \quad \text{for } 1 \leq i \leq I \text{ and } I \leq M/2. \tag{12}$$

Both beams z_1 and z_2 have the same look direction θ_0. Each is formed from signals provided by the I sensors, $I \leq M/2$, located at either array end. If for any vector U in $\mathbb{C}^M$, U_1 and U_2 are the vectors of $\mathbb{C}^I$ whose components are respectively the first I and the last I components of U, z_1 and z_2 can be expressed as

$$z_1 = s\, D_1(\theta_0)^\dagger D_1(\theta) + D_1(\theta_0)^\dagger N_1 \,,$$
$$z_2 = s\, D_2(\theta_0)^\dagger D_2(\theta) + D_2(\theta_0)^\dagger N_2 \,. \tag{13}$$

Signal components of z_1 and z_2 have the same amplitude and their phase difference is $kd(\sin\theta - \sin\theta_0)$. The phase comparison estimate of θ is

$$\hat{\theta} = \sin^{-1}\left(\frac{1}{kd}\arg\left(\frac{z_1}{z_2}\right) + \sin\theta_0\right) . \tag{14}$$

Because $|\arg(z)| \leq \pi$, one has $|\sin\hat{\theta} - \sin\theta_0| \leq \frac{\lambda}{2d}$.

Commentary

Amplitude comparison and phase comparison estimators are more practical than the MAD estimator. But the signal direction of arrival θ must be known approximately so that the mean direction θ_0 of the two beams is near θ. Typically $\theta - \theta_0$ must be less than the aperture angular resolution λ/L. Such *a priori* knowledge is given by a comparison of the amplitudes of the beams formed with an angular spacing less than λ/L, that are covering the antenna receiving sector. Then a coarse estimate of θ is the look direction of the maximal amplitude beam. The MAD estimate can also be obtained from these beams by interpolating between them.

ASYMPTOTIC ACCURACY

General Approach

Theorem: If f is a bounded real function on $\mathbb{R}^K$, continuous at point S, U is a random vector in $\mathbb{R}^K$ and λ is a real variable, then

$$\lim_{\lambda \to 0} E[f(S + \lambda U)] = f(S) . \qquad (15)$$

Moreover if f is differentiable at point S and U has covariance matrix $\Gamma = E[UU^T]$, then

$$E[(f(S + \lambda U) - f(S))^2] = \nabla f(S)^T \Gamma \; \nabla f(S) \; \lambda^2 + o(\lambda^2). \qquad (16)$$

∇ is the gradient operator and o(x) holds for any function such that $\lim_{x \to 0} o(x)/x = 0$. This is an obvious result if, as is usually done by several authors in monopulse literature, the first and second order moments of f(S+λU) are approximated by the first and second order moments of its first order expansion[6,7]. Actually the theorem above justifies this approximation. The exact proof is cumbersome and would have been of little interest in this paper.

Any of the three estimates considered here can be written as

$$\hat{\theta} = f(T). \qquad (17)$$

T is the observed vector in $\mathbb{C}^M$ for MAD estimation, in $\mathbb{C}^2$ for amplitude or phase comparison estimation. It can be expressed as T=AX, i.e. from Eq. (1)

$$T = s \; A \; D(\theta) + A \; N \qquad (18)$$

where A is a (M,M) or (2,M) matrix. Let us define the signal-to-noise ratio (SNR) as

$$r = \frac{M|s|^2}{\sigma^2} . \qquad (19)$$

r is the SNR of a beam pointed in direction θ. It is the highest SNR that could be achieved if θ was known by linearly combining the M array signals. Because one has f(tT)=f(T) for any vector T and any number t, it follows from Eq. (17) and Eq. (18) that

$$\hat{\theta} = f(A \; D(\theta) + \sqrt{\frac{M}{r}} \; U). \qquad (20)$$

where $U = (|s|/\sigma s) AN$ is a random vector with a covariance matrix that doesn't depend on r. It is easily seen that for each of the three estimates one has f(AD(θ))=θ. Hence applying the first part of the theorem gives

$$\lim_{r \to \infty} \hat{\theta} = \theta . \qquad (21)$$

Thus the MAD and the beam comparison estimators are asymptotically unbiased. By setting $\varepsilon^2 = E[(\theta-\theta)^2]$ it can be written from Eq. (16)

$$\varepsilon = \sqrt{M\ \nabla f(AD(\theta))\ \Gamma\ \nabla f(AD(\theta))}\ \frac{1}{\sqrt{r}} + o\left(\frac{1}{\sqrt{r}}\right)\ . \qquad (22)$$

It is worthwhile to note that in the equation above $\nabla f(T)$ is the gradient derivative of f with respect to the real vector $(Re(T)^T, Im(T)^T)^T$. In the same way Γ is the covariance matrix of the real random vector $(Re(U)^T, Im(U)^T)^T$.

MAD

The observed vector is T=X. Hence A is the identity matrix. f is defined by the equation

$$G(X, f(X)) = 0 \qquad (23)$$

with

$$G(X,\phi) = \frac{\partial}{\partial\phi}|D(\phi)^{+}X|. \qquad (24)$$

Taking the gradient derivative with respect to X of the left side in Eq. (23) gives

$$\nabla_X G(X, f(X)) + \frac{\partial G}{\partial\phi}(X, f(X))\ \nabla f(X) = 0 \qquad (25)$$

where $\nabla_X G$ and $\frac{\partial G}{\partial\phi}$ are partial derivatives of $G(X,\phi)$ with respect to $(Re(X)^T, Im(X)^T)^T$ and ϕ. By changing X in $D(\theta)$ and using the relation $f(D(\theta)) = \theta$, one finally obtains

$$\nabla f(D(\theta)) = -\left(\frac{\partial G}{\partial\phi}(D(\theta),\theta)\right)^{-1}\ \nabla_X G(D(\theta),\theta)\ . \qquad (26)$$

Γ is the identity matrix of order 2M multiplied by the factor 1/2. Computation of the right hand side of Eq. (22) gives

$$\varepsilon = \frac{M}{\sqrt{2}\ k\ \cos\theta}\ \frac{1}{\sqrt{M\Sigma\xi_m^2 - (\Sigma\xi_m)^2}}\ \frac{1}{\sqrt{r}} + o\left(\frac{1}{\sqrt{r}}\right)\ . \qquad (27)$$

For a periodic array as described by Eq. (9), Eq. (27) becomes

$$\varepsilon = \frac{1}{\pi}\sqrt{\frac{3}{2}}\ \frac{M}{\sqrt{M^2-1}}\ \frac{\lambda}{L\cos\theta}\ \frac{1}{\sqrt{r}} + o\left(\frac{1}{\sqrt{r}}\right)\ . \qquad (28)$$

Amplitude Comparison

The observed vector is $Y=(y_{-1},y_1)^T$ and A is the (2,M) matrix $[D(\theta_{-1}),D(\theta_1)]^\dagger$. From Eq. (11) the gradient derivative of f with respect to $(Re(Y)^T,Im(Y)^T)^T$ can be computed as

$$\nabla f(AD(\theta)) = \frac{2}{\cos\theta\ F'(\sin\theta)\ ((D(\theta_{-1})+D(\theta_1))^\dagger D(\theta))^2} \begin{pmatrix} -D(\theta_1)^\dagger D(\theta_0) \\ D(\theta_{-1})^\dagger D(\theta_0) \\ 0 \\ 0 \end{pmatrix}. \quad (29)$$

where F' is the derivative of F defined by Eq. (7). Then Eq. (22) gives

$$\varepsilon = \frac{M\sqrt{1 + F(\sin\theta)^2 - \rho(1 - F(\sin\theta)^2)}}{\cos\theta\ F'(\sin\theta)\,|(D(\theta_{-1})+D(\theta_1))^\dagger D(\theta)|}\frac{1}{\sqrt{r}} + o(\frac{1}{\sqrt{r}}) . \quad (30)$$

where ρ is the correlation coefficient of $D(\theta_{-1})^\dagger N$ and $D(\theta_1)^\dagger N$.

When the array is periodic one has

$$\rho = \frac{\sin(\pi L\delta/\lambda)}{M\sin(\pi L\delta/M\lambda)} . \quad (31)$$

Let us define $h(\delta)$ as the right hand side of the equation above; $\delta_{-1}=\sin\theta-\sin\theta_{-1}$ and $\delta_1=\sin\theta-\sin\theta_1$. $F(\sin\theta)$ and $F'(\sin\theta)$ are derived from Eq. (8), and Eq. (30) becomes

$$\varepsilon = \frac{1}{\pi\sqrt{2}}\left|\frac{M\sin(\pi L\delta_{-1}/M\lambda)\ \sin(\pi L\delta_1/M\lambda)}{\sin(\pi L\delta/M)\ (h(\delta_{-1})h(\delta_1) - h(\delta))}\right| \\ \times\sqrt{h(\delta_{-1})^2 + h(\delta_{-1})^2 - 2h(\delta_{-1})h(\delta_1)h(\delta)}\ \frac{\lambda}{L\cos\theta}\frac{1}{\sqrt{r}} + o(\frac{1}{\sqrt{r}}) . \quad (32)$$

Phase Comparison

$AD(\theta)$ is the vector $(D_1(\theta_0)^\dagger D_1(\theta), D_2(\theta_0)^\dagger D_2(\theta))^T$, and because of the array geometry it can be expressed as

$$AD(\theta) = D_1(\theta_0)^\dagger D_1(\theta) \begin{pmatrix} 1 \\ \exp(-jkdu) \end{pmatrix} \quad (33)$$

with $u=\sin\theta-\sin\theta_0$. For computing $\nabla f(AD(\theta))$ where f is of the form given by the right hand side of Eq. (14), that is an expression of the form f(Z), where $Z=(z_1,z_2)^T$ is the observed vector, the following expressions for the partial derivatives of arg(z) are used:

$$\frac{\partial \arg(z)}{\partial Re(z)} = -\frac{Im(z)}{|z|^2} ,$$
$$\frac{\partial \arg(z)}{\partial Im(z)} = \frac{Re(z)}{|z|^2} . \quad (34)$$

It is obtained

$$\nabla f(AD(\theta)) = \frac{1}{k\ d\ \cos\theta\ |D_1(\theta_o)^\dagger D_1(\theta)|} \begin{pmatrix} -\sin(\phi) \\ \sin(\phi-kdu) \\ \cos(\phi) \\ -\cos(\phi-kdu) \end{pmatrix} \tag{35}$$

with $\phi=\arg(D_1(\theta_o)^\dagger D_1(\theta))$. Γ is the identity matrix of order 4 multiplied by factor I/2. Then Eq. (22) gives

$$\varepsilon = \frac{\sqrt{IM}}{k\ d\ \cos\theta|D_1(\theta_o)^\dagger D_1(\theta)|} \frac{1}{\sqrt{r}} + o(\frac{1}{\sqrt{r}}) . \tag{36}$$

Both beams have the same directivity pattern and $|D_1(\theta_o)^\dagger D_1(\theta)|$ is its value in the signal direction of arrival. When the array is periodic, one has d=(M-I)L/M and Eq. (36) becomes

$$\varepsilon = \frac{1}{2\pi}\frac{L}{d}\sqrt{\frac{L}{L-d}} \left| \frac{\sin(\pi(L-d)u/I\lambda)}{\sin(\pi(L-d)u/\lambda)} \right| \frac{\lambda}{L\cos\theta} \frac{1}{\sqrt{r}} + o(\frac{1}{\sqrt{r}}) . \tag{37}$$

Commentary

For each of the three estimates the RMS error ε can be written as

$$\varepsilon = g(\sin\theta) \frac{\lambda}{L\cos\theta} \frac{1}{\sqrt{r}} + o(\frac{1}{\sqrt{r}}) . \tag{38}$$

Hence the function $g(\sin\theta)$ can be defined by the relation

$$g(\sin\theta) = \lim_{r\to\infty} \varepsilon \frac{L\cos\theta}{\lambda}\sqrt{r} . \tag{39}$$

For MAD $g(\sin\theta)$ is a constant. It equals approximately 0.39 when the array is periodic and M is high. Actually, the first term on the right hand side of Eq. (28) can be recognized as the Cramer-Rao lower bound that holds for the standard deviation of any unbiased estimator[6].

The amplitude and phase comparison curves $g(\sin\theta)$ are plotted in Figure 2 for some values of the parameters δ and d. In all cases both beam comparison estimators are less accurate than the MAD estimator but the loss is quite negligible when θ is near θ_o. Then amplitude comparison is slightly better than phase comparison. When the difference between θ and θ_o becomes large, amplitude comparison accuracy falls and phase comparison becomes more accurate. This effect can be partly avoided by decreasing the angular deviation δ between the look-directions of both beams. But decreasing δ is practically limited by increasing sensitivity of the signal y_1-y_{-1} that occurs in the estimate expression given by Eq. (10) and (11), to accuracy of the beamforming complex weights.

When u=0, i.e. $\theta=\theta_o$, it is seen from Eq. (37) that the optimal value of the distance d between the centers of both array extreme parts used for phase comparison, is 2L/3: then only both extreme thirds of the aperture are used.

NUMERICAL SIMULATION

A Monte-Carlo simulation has been performed to compare actual values taken by $\varepsilon\sqrt{r}L\cos\theta/\lambda$ with its asymptotic value $g(\sin\theta)$ given by the previous equations for amplitude and phase comparison estimation for a linear periodic array. r takes the values 10, 100 and 1000, and $|\sin\theta-\sin\theta_0|$ takes the values 0 and $\lambda/2L$. The results given in Table 1 show that a first-order expansion of ε with respect to $1/\sqrt{r}$ gives a good approximation except for phase comparison at low signal-to-noise ratio, especially when θ is far from θ_0. Then deviations result from the discontinuity of the estimate $f(Z)$ at any vector Z such that $\arg(z_1/z_2)$ equals to $\pm\pi$, see Eq. (14). Such vectors are more frequently observed when r is decreasing and θ is far from θ_0.

TABLE 1

10 log r

$\lvert\sin\theta-\sin\theta_0\rvert$	10 dB	20 dB	30 dB	∞	COMPARISON
0	0,42	0,39	0,39	0,39	amplitude
$\lambda/2L$	0,69	0,72	0,71	0,71	$\delta=\lambda/L$
0	0,40	0,39	0,39	0,39	amplitude
$\lambda/2L$	0,54	0,54	0,54	0,54	$\delta=\lambda/2L$
0	0,48	0,45	0,45	0,45	phase
$\lambda/2L$	0,60	0,50	0,50	0,50	$d=L/2$
0	0,47	0,42	0,41	0,41	phase
$\lambda/2L$	0,94	0,44	0,43	0,43	$d=2L/3$

$\varepsilon\sqrt{r}\,\dfrac{L\cos\theta}{\lambda}$

RESULTS SUMMARY

The maximum likelihood, amplitude comparison and phase comparison bearing estimators have been studied in the case of a single monochromatic plane wave impinging on a linear array of sensors. The maximum likelihood estimator based on sensor signals is the look-direction of the maximal amplitude beam and has been called Maximal Amplitude Direction. The amplitude comparison and phase comparison estimators, whose two observed vector components are beams linearly formed from sensor signals, are less accurate than MAD, except when the signal direction of arrival θ is near the mean look-direction θ_0 of the two beams. Then amplitude comparison is slightly less accurate, even when optimal performance is achieved by using only the sensors located at both extreme thirds of the aperture. But when θ is far from θ_0, the amplitude comparison accuracy decreases faster than the phase comparison one, so that phase comparison is becoming more accurate; amplitude comparison off-axis accuracy can be improved by decreasing the angular deviation between the look-directions of the two beams.

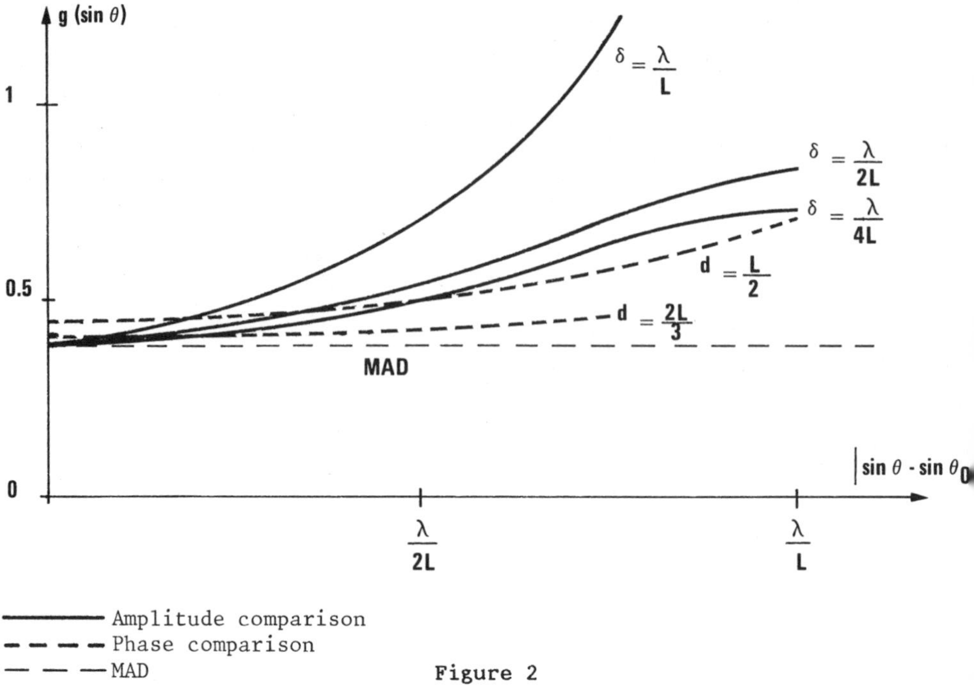

Figure 2

REFERENCES

1. M.H. Carpentier, "Radars: New Concepts", Gordon and Breach, New York (1968).
2. M.I. Skolnik, "Radar Handbook", McGraw Hill (1970).
3. D.R. Rhodes, "Introduction to Monopulse", McGraw Hill (1959).
4. G. Bienvenu and L. Kopp, "Optimality of High Resolution Array Processing Using the Eigensystem Approach", IEEE Trans. ASSP 31:1235 (1983).
5. E. Mosca, "Maximum Likelihood Angle Estimation in Amplitude Comparison Monopulse Systems Operating on a Single Pulse Basis", Alta Frequenza 37:408 (1968).
6. L.E. Brennan, "Angular Accuracy of a Phased Array Radar", IRE Trans. AP 9:268 (1961).
7. S. Sharenson, "Angle Estimation Accuracy with a Monopulse Radar in the Search Mode", IRE Trans. ANE 9:175 (1962).

ITERATIVE METHODS FOR ENHANCEMENT OF MULTICHANNEL DATA

Ole E. Naess

Statoil, Den norske stats oljeselskap a.s
P.O. Box 300, Forus
N-4001 Stavanger, Norway

ABSTRACT

Direct summation of multichannel recordings e.g. in beamforming is not always optimum in terms of detection and signal to noise ratio. An alternative is to use iterative methods. Two such methods, the Iterative Stacking (IS) and the Iterative Median Stacking (IMS) methods will be described and discussed. The first method operates separately on positive and negative amplitudes, while the second method applies an adaptable median operator. Both methods were originally developed for use in the processing of deep seismic data.

INTRODUCTION

Iterative methods for enhancing multichannel data are based on an assumption that the result becomes more correct after an optimized number of iterations. Too many iterations may sometimes give adverse effects. This depends on the criterion involved and its uniqueness relative to the data the sought solution.

In this paper two methods will be considered. These are the Iterative Stacking (IS) and the Iterative Median Stacking (IMS) methods. In the IS method positive and negative amplitudes are treated separately in an iterative manner. As the final step the two polarities are combined. This method represents a powerful way of detecting signals that are in phase, and suppressing signals that are out of phase.

The IMS method is a versatile method which, although based on the median concept, includes the possibility of differential weighting of the input data. This method incorporates the benefits of the median filtering to attenuate sporadic strong noise while at the same time allowing for suppression of different kinds of systematic noise through weighting.

The methods are meant to be applied to a set of recordings. On each recording (in seismic work called "a trace") amplitudes are given as a function of recording time. The basic assumption about the primary signals is that they ideally should be equal in strength and polarity on all the recordings in the set. Furthermore, before applying the iterative methods, it is also assumed that the different recordings have been corrected (or shifted) in such a way that the primary signal occur at the same point on the time axis of each member of the set. When the noise is random and of the same magnitude as the primary, a straight summation will often give sufficient noise attenuation. However, when very strong sporadic noise and/or coherent noise which would be added partly in phase is present, then other ways of enhancing the data should be attempted.

The basic situation for the iterative approach considered here is that we have a set of discrete amplitudes - one from each recording. Each of these amplitudes consists of a constant primary part and a variable noise component.

The object is to do something to the data set which afterwards makes it possible to get an improved estimate of the primary part i.e. the primary signal when the procedure is performed at consecutive points of the time axis. In order to do something meanigful we need quantified estimates of the noise component belonging to each discrete amplitude. Such estimates are found by inference from the data assumptions. Then there is basically two ways of using this information.

a) The data can be filtered by changing of weighting the single amplitudes according to the estimated noise components.

b) The data can be filtered by excluding a certain number of the amplitudes that have the largest estimated noise components, or equivalently extracting those amplitudes that have the lowest noise components for further use.

Based on the data resulting from either of these filter operations, new updated estimates of the primary part and hence the noise components will result. This procedure may be repeated several times in an iterative manner.

The IS method use an updating procedure as in a) while the extraction process in b) is used in the IMS method.

DESCRIPTION OF METHODS

In this paper only a stepwise and non-mathematical description of Iterative Stacking (IS) and Iterative Median Stacking (IMS) will be given. A more theoretical description of IS is found in an article by Naess, O.E. (1979). More information is given in Naess, O.E. (1982) and Naess, O.E. and Bruland, L. (1981).

The IMS method is formally described in Naess, O.E. and Bruland, L. (1985 I) and (1986). Several other methods for enhancing multichannel seismic data are reviewed and discussed in Naess, O.E. and Bruland, L. (1985 II

Both methods can be summarized in five steps. Each input consists of one discrete amplitude from each recording in the data set.

<u>Iterative Stacking (IS)</u>

1. Sort the input data into two sets consisting only of positive and only of negative amplitudes.

2. Calculate sums of only positive and only negative amplitudes and divid both sums by the same optional constant M.

3. If amplitudes are larger (in absolute value) than the sum of the same

polarity divided by M, change these amplitudes by setting them equal to the sum divided by M.

4. Repeat 2. and 3. with the (partly) new amplitudes. (The number of repetitions or iterations is optional).

5. Calculate final discrete amplitude by adding the last positive and negative sum divided by M.

Iterative Median Stacking (IMS)

1. Choose appropriate initial median operator.

2. Reorder initial data input sequence with respect to intended weighting.

3. Run median operator through the reordered data sequence.

4. Change the length of the median operator proportionally to the reduction in the number of amplitudes. Then repeat 2. and 3. on the present set of amplitudes.

5. Stop the process when the number of amplitudes in equal to or less than the number of elements in the median operator. (Alternatively one may choose to stop earlier and average the present amplitudes).

DISCUSSION OF METHODS

In this part each step of the two methods will be further explained and discussed concurrently.

Step 1

The reason for sorting the data according to their polarity in the IS method is due to the fact that both polarities can not be correct. Hence, although we do not know which polarity is representing the primary, they should still be treated differently and separately. In IMS the initial choice of median operator is important. A typical choice in the case of 24 channel data would be an operator with 3 elements spanning 13 input points. The choice of operator determines the amount of filtering and the degree of weighting.

The distribution in amplitude values in IS will tell us about the amount of filtering or adjustment to be performed. In the IMS the data itself does not directly influence the amount of filtering. This will be related to the chosen operator. However, the choice of operator will be based on an evaluation of the data and how strong filtering is needed.

Step 2

In this IS method the two single-polarity sums are divided by a chosen constant M. A usual choice of M would be to set it equal to or slightly less than the number of input channels i.e. input amplitudes at each point in time. Together M and the sums will determine the limiting values for the consecutive filtering. One should especially note that since a constant M is used, the amplitudes with the least represented polarity will have a lower limiting value and hence be subjected to a more severe filtering.

The sorting or reordering of the input data in IMS is necessary in order to exploit the weighting characteristics of the chosen median operator. The weighting is in reality caused by edge effects and is of a purely

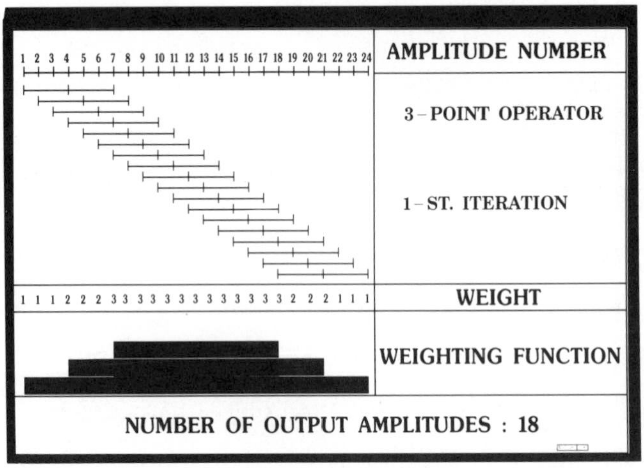

Fig. 1. Graphical example of IMS filtering.

statistical type. In fig. 1 is shown a graphical example of how a 3 element median operator of length 7 points runs through a data set of 24 amplitudes. In the lower part is shown the resulting weights and the weighting function we get in this case. By weights we simply mean the number of times each input amplitude contribute to the operator as it moves across the data. The weighting function shows that it is always the centermost positioned amplitudes that have the highest weights. This can be exploited by placing the amplitudes we want to give more weight in the upweighted part before the filter is run. That is, the sorting in this step of IMS gives us the possibility to give those amplitudes which by their location in the dataset are more reliable, a better chance of surviving the filter operation.

In the IS method we do not assume any model of the noise distribution as a function of position in the dataset. The limiting values (the two sums divided by M) are only influenced by the data and our choice of M. This constant will determine the degree of tolerance, that is, how large deviations in amplitude we will accept without reducing or weighting them down.

Step 3

In this step of the IS method all amplitudes are compared to the sum (divided by M) of the same polarity. Depending on the result a sort of "shaving" operation is then performed whereby all amplitudes above a certain limit is reduced to this limit. This reduction is only performed when the presence of (estimated) noise on a single amplitude leads to a higher total absolute value of the amplitude. If the presence of noise results in a lower amlitude value nothing is done. The reason for this is partly that we are dealing with highly uncertain estimates, and partly that the possible effect of noise in the two situations is different. An amplitude that is to large has in principle no upper limit and may for example be higher in value than the sum of all the other amplitudes in the set. Such an amplitude may then completely change the primary estimate if normal averaging across the set is used. However, when an amplitude is smaller in absolute value than the estimated primary, then its maximum effect is equal to that of an amplitude of primary value but opposite polarity participating in the averaging process.

In the filter operation in IMS we choose not to define an output unless the whole operator is located within the data set. This will cause a reduction in the number of output amplitudes relative to the input data. The convergence in IMS is based upon this property.

An important distinction between the two methods is hereby indicated. While the IMS filter the data set by reducing its number of amplitudes, the IS works by reducing the values of those amplitudes that are above certain limits. The IMS process discards amplitudes that does not adhere to the basic assumption about the consistency of primaries, while the IS forces them to become more consistent.

These changes in amplitude will depend on the relevant limiting value. This limit is highly influenced by the number of amplitudes having the appropriate polarity.

Step 4

A certain adjustment of several amplitudes have now occurred with the IS method. Further optimization is then sought by repeating the procedure. The number of repetitions would depend on the actual data and be subject to testing. In most cases only a limited number of iterations (1-5) would be used. Too many iterations might give adverse effects in terms of distortion of pulses. In most cases both primaries and noise will be subject to a gradual reduction in amplitude value with increasing iterations. However, the noise will undergo a faster reduction which is related to it's representativity in the data set. When choosing the number of iterations it is then a question of balancing between an efficient noise suppression and minimal effect on primaries.

The change in length of the median operator in IMS is fully defined when the necessary initial parameters have been decided. This means that the total number of iterations can be precalculated from the initial parameters. One should note that it is only the length of the operator that is changed. The number of elements remains constant during the iterations.

Step 5

In IS the successive iterations were performed on positive and negative amplitudes separated in two subsets. These amplitudes have been changed in an overall fashion according to the representativity of the polarity, and each amplitude has been subjected to a possible reduction which is a function of its relative size in the subset. The method does not make any choice between the polarities. Hence, the final output is the sum of all the resulting (partly) adjusted amplitudes.

In IMS the iterative process is stopped when the number of output values hav been reduced to a minimum which will depend on the initial parameters. If a 3-element operator is used the final output is directly defined when the number of output values gets below 4. Final output is then set equal to the value itself, the average value or the median value, in the case of 1, 2 and 3 resulting amplitudes. It isalso possible to stop the IMS at an earlier stage and simply average the present amplitudes in order to get the sinal estimate. The version with only one iteration is called Weighted Median Stack (Naess, O.E. and Bruland, L., 1986).

Other aspects

The IMS method automatically gives the possibility of weighting the data in a systematic way according to the positions in the input data set. The IS method as normally used is completely invariant to amplitude positions. However, a sort of weighting with positions can also be achieved with IS if the final output is differently defined.

The change in amplitudes occurring during the iterations is a function of the degree of similarity between the amplitudes. In effect all amplitudes then contribute to the filtering of each other. Instead of using all amplitudes to define the final result in step 5, we could limit ourselves to only a smaller number of selected amplitudes. This selection could be done according to the positions in the data set. These amplitudes would then in effect get a higher weight. A version of IS where only one of the resulting amplitudes was used as final output is described and discussed in Naess, O.E (1982). In such variations of the IS method a distinction is done between amplitudes that are allowed to contribute to the filtering only, and those who are also contributing directly to the fianl output.

DATA EXAMPLES

A synthetic data example of the use of the IS method is shown in fig.3. The input data is displayed in fig. 2. This set of "recordings" have two primary events P_1 and P_2. Each pulse belonging to the same primary is situated at the same recording time, i.e. occur at the same horizontal level in fig. 2. The uppermost event is a simple model of the waterbottom reflecte signal as it appear after necessary corrections of the input data. The other coherent events are multiples of this signal. These are marked as M_1, M_2 and M_3 on fig. 3. A major problem in deep seismic work is to suppress such multiple noise. The scale on the vertical axis on both figures is recording or reflection time in seconds.

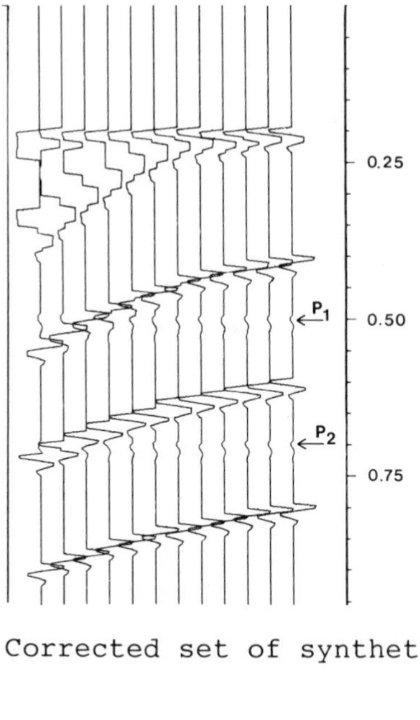

Fig. 2. Corrected set of synthetic data.

The right curve on fig. 3 shows the result after a simple addition of the "recordings" in fig. 2. Considerable noise that partly buries the two primaries, and which is caused by the out of phase summation of multiple events, is present on this curve. The next curves from left to right on fig. 3 shows the result after 2, 3, 4 and 5 iterations of IS. After 2 iterations considerable improvement has occurred and only a slight amount of noise can be seen. Further iterations seems to remove the noise completely and no further improvement is possible after 3 iterations (middle curve) in this example.

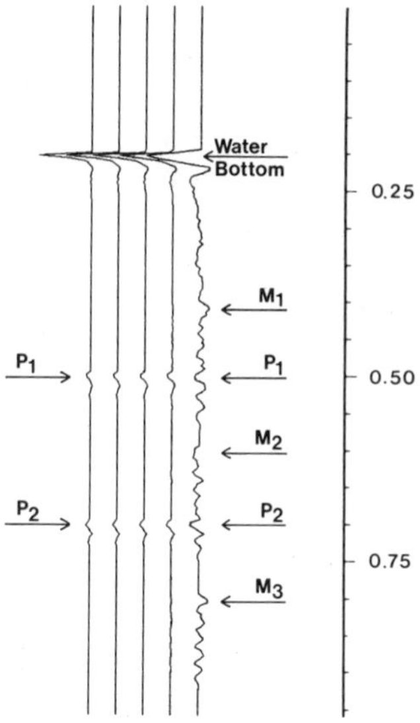

Fig. 3. Effect of using IS on data set in fig. 2.

An example of the application of the IMS method is shown on the right side of fig. 5. The left side of this figure shows the result when a straight summation approach has been used. Fifteen input data sets with different random noise patterns and strong sporadic noise was used as input. The basic input data set is exemplified in fig. 4 except that the strong sporadic noise is not shown. This noise was constructed by changing the polarity of one vertical curve (trace) in the input data sets (fig. 4) and multiplying its amplitudes by 10 before doing the IMS and summation to get the results in fig. 5. In this example the object was to test the methods ability to attenuate both the strong multiples M_1 and M_2 and the sporadic noise. The two multiples are indicated on fig. 4 and 5. The primary signals appear as

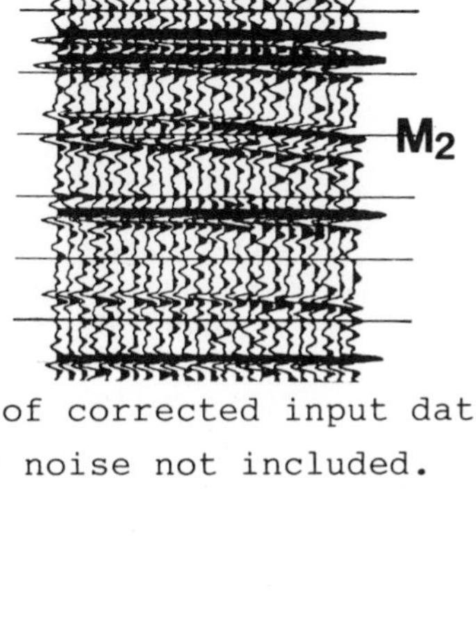

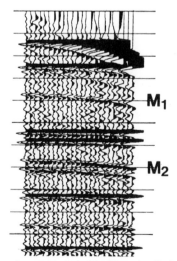

Fig. 4. Example of corrected input data set. Simulated sporadic noise not included.

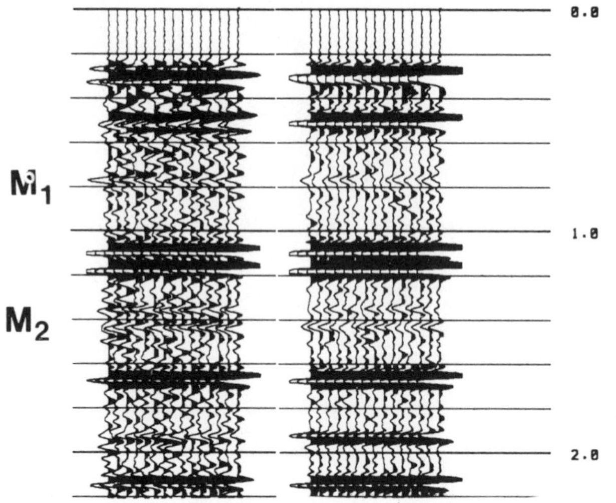

Fig. 5. Comparison between straight summation result (to the left) and IMS (to the right).

horisontal events in fig. 4. In fig. 5 we notice that a marked improvement in the suppression of the multiples result when IMS (to the right) is used instead of direct sum (to the left). In addition the general signal to noise ratio has been considerably increased when IMS was used. This means that the influence of sporadic noise is significantly reduced when the iterative median method is used as a substitute for summation.

Further examples on real data can be found in the references.

CONCLUSION

In this paper two iterative methods for enhancement of multichannel data have been schematically described. Each step of the two methods have been discussed concurrently and their main properties were illustrated by examples. Both methods have a non-linear filter response and they must be used with this in mind. The selection of the appropriate parameters and operators should be partly based on an evaluation of the input data and partly on the objective of the filtering. The Iterative Stacking (IS) method will work toward making all amplitudes more similar by reducing the value of the more extreme amplitudes. With the other method, Iterative Median Stacking (IMS), one is progressively trying to extract the most reliable amplitudes and discard the rest. The final goal is to detect the single amplitude from each input that most correctly represents the true value. The IMS method is hence based upon an implicit assumption that a least one of the original input amplitude must be very similar to the true one. An underlying assumption in the IS method is that the most correct value is the one we get when no further change occurs with more iterations. However, if the constant M is larger than the number of amplitudes of the most represented polarity, then the process will converge towards zero. This means that if to many iterations are used we will not achieve the objective of increasing the signal to noise ratio since effectually all amplitudes will become very small.

The IMS process converges by decreasing the number of amplitudes resulting from each iteration, while IS converges by a reduction in the amplitude values. Both methods can be used in such a way that a higher weight or more emphasis is placed upon a limited group of amplitudes when this group is specified by the positions of its member in the input data set. Such weighting is an inherent property of IMS, while it necessitates a certain adjustment in the IS process.

ACKNOWLEDGEMENTS

The author would like to thank Statoil, Den norske stats oljeselskap a.s., for permission to publish this paper.

REFERENCES

Naess, O.E. (1979), "Superstack - an iterative stacking algorithm", Geophysical Prospecting 27, 16-18

Naess, O.E. and Bruland, L. (1981), "Velocity Analysis using Iterative Stacking", Geophysical Prospecting 29, 1-20

Naess, O.E. (1982), "Single-trace processing using iterative CDP-stacking", Geophysical Prospecting 30, 641-652

Naess, O.E. and Bruland, L. (1985 I), "Iterative Median Stack", Technical program extended abstracts and biographies, SEG (1985), Washington DC

Naess, O.E. and Bruland, L. (1985 II), "Stacking methods other than simple summation", Development in Geophysical Exploration Methods - 6, Elsevier Applied Science Publisher Ltd. Barking, Essex, England

Naess, O.E. and Bruland, L. (1986), "Improvement of multichannel seismic data through application of the median concept," in Proceedings of the Offshore Technology Conference Houston, May 1986.

LINEAR PREDICTION FOR SENSOR RECOVERY IN LINE ARRAY BEAMFORMING

R.S. Walker † and D.N. Swingler*

† Defence Research Establishment Atlantic
P.O. Box 1012, Dartmouth, N.S.
Canada, B2Y 3Z7

* Division of Engineering, Saint Mary's University
Halifax, N.S., Canada, B3H 3C3

ABSTRACT

A method is proposed for recovering signals from faulty sensors in a linear acoustic array, in order to prevent degradation of beamformer sidelobes. A linear-predictive interpolator is developed based on the measurements at the working sensors, and is subsequently used to interpolate the missing data. Both simulations and real data are used to show recovery is consistently achieved for at least ~10% faulty sensors, while the additional computational requirements are small compared to the beamforming load. Finally, the application of the same linear-prediction concept to aperture extrapolation is shown to lead to an increased resolution capability for the array.

INTRODUCTION

The linear array of uniformly spaced sensors is a common receiver for examining the directionality of the underwater acoustic field. The beamforming operation, whereby the field power is estimated as a function of angle, consists of phasing the narrowband sensor outputs to a set of plane-wavefronts whose bearings span the directions of interest. The accuracy of the beampower estimates is controlled in part by the accuracy in setting the sensor phases and gains; if the sensor positions or sensitivities are perturbed, the quality of the directionality map can be degraded. One extreme example of this problem arises when a sensor actually fails electrically or mechanically. As will be shown, simple recovery procedures such as zeroing the sensor input to the beamformer are unsatisfactory when strong directional components are present in the field.

This paper presents an approach to sensor recovery that enables satisfactory beamforming despite a relatively large number of failures. A "high-resolution" algorithm is used to predict the strong directional signals that otherwise would be measured at the faulty sensors. These predicted data can then be incorporated in the beamforming operation. In particular, the procedure relies on a Linear Prediction algorithm through which a linear-prediction model of the field is developed from the working sensors. This model can then be applied directly to interpolate the missing data. We will show that the method is reliable, robust and computationally simple. Finally, while the beamformer and sensor-recovery procedure discussed herein assumes only narrowband processing, extension to broadband processing can be realized by integration over frequency of the narrowband results.

Adaptations of this concept have been proposed for recovering missing or bad data in time-series records, with application to spectral estimation (Nuttall, 1976), radar (Bowling and Lai, 1979), and digital audio (Veldhuis, et.al, 1985). The application to beamforming is believed to be new.

THE PHASE-SHIFT BEAMFORMER

Beamforming for an N-sensor linear array of inter-sensor separation d consists of electrically steering the array in direction θ from its axis; specifically, a linearly-varying phase shift is applied across the (complex) sensor signals, $x_{n,m}(f)$, measured in a narrow band about frequency f. Index n indicates sensor number, and m is the observation period, or snapshot. The phased samples are summed, squared and averaged over M snapshots to obtain the beampower estimate, P (boldface italic type has been used to denote vectors):

$$P = 1/M \sum_{m=1}^{M} \left|\sum_{n=1}^{N} x_{n,m}\, e^{jn\beta}\right|^2 = Z^{\dagger} \left(1/M \sum_{m=1}^{M} \boldsymbol{X}_m \boldsymbol{X}_m^{\dagger}\right) Z = Z^{\dagger} \boldsymbol{R}_M Z \qquad (1)$$

where $\beta = 2\pi df\cos\theta/c$ is the spatial frequency, $-\pi \leq \beta \leq \pi$, and c is the speed of sound. In effect, beam steering is a spatial DFT, where the steering vector $\boldsymbol{Z}$ contains the phasings (and typically amplitude shading for sidelobe reduction). $\boldsymbol{R}_M$ is an estimate of the sensor covariance matrix, $\boldsymbol{R}$, based on the M observations. In practice, $\boldsymbol{X}_m$ consists of coefficients at f of temporal DFT's applied to the mth segment of each sensor time-series, where the length of the segment is the reciprocal of the bandwidth of the narrowband filter at f. Hence, M is equivalent to the *bandwidth-time product* for the measurement. Integration of the narrowband beampowers over contiguous frequency bins can be used to obtain a broadband beam map.

EFFECTS OF SENSOR FAILURE

We will assume that B sensors in the array have failed, with the faulty data values replaced by a zero. Then the rows and columns of $\boldsymbol{R}_M$ corresponding to the faulty sensors will contain zeroes also.

Figure 1 demonstrates through simulation the effect of such zeroing on the beamformer performance. The simulation uses an N=32 array with half-wavelength sensor separation so that $2\pi df/c = \pi$, and the spatial frequency $-\pi \leq \beta \leq \pi$ spans exactly the acoustic space from $-1 \leq \cos\theta \leq 1$. The field is modelled to consist of four plane waves in spatially-uncorrelated noise at beam signal-to-noise ratios (SNRs) of 30, 20, 10 and 0 dB, at spatial frequencies of 0.87π, 0, 0.71π and 0.17π respectively. The beamformer incorporates Hamming shading and uses M=500 snapshots to estimate the directionality map. The temporal variations of the sensor data across snapshots are modelled as narrowband Gaussian processes. The dashed curve, labelled B=0, is the desired map when no sensors have failed, and in this and subsequent figures it has been offset by -10 dB for readability. The four plane waves are identified as the peaks in this map. The remaining 10 curves have been obtained by beamforming with 4 of the 32 sensors zeroed, where for each curve the faulty sensors have been selected randomly. Degraded sidelobes now make the two weaker signals indiscernible.

If Q is the probability of sensor failure, then Nuttall (1980) has shown that the *average* sidelobe level is $\sim Q(1-Q)/N_{eff}$. N_{eff} is the number of *effective* sensors in the array for the chosen shading, which for Hamming shading is 0.73N. In Figure 1, Q=4/32, and $10\log Q(1-Q)/N_{eff} = -23.3$ dB. This level relative to the level of the strongest signal is seen to describe reasonably the average of the degraded beampowers in all directions but for the two strongest signals. Nuttall's result indicates that even one sensor failure can limit the achievable sidelobe level of a 32 sensor array to no better than about -28 dB.

RECOVERY METHODS

Methods of sensor recovery can be grouped into data-domain methods, i.e. those that replace missing data on each snapshot, and covariance-domain methods, or those that replace the missing rows and columns of $\boldsymbol{R}_M$. One covariance-domain approach exploits the fact that for a spatially-stationary field, $\boldsymbol{R}_M$ is Toeplitz as M $\rightarrow\infty$. Therefore, zeroes in $\boldsymbol{R}_M$ can be replaced with the average of the available terms along the same matrix diagonal, provided there is at least one non-zero entry in the diagonal. As M $\rightarrow\infty$, this method achieves the same map as when B=0. However, the stability of the beam estimates for small M can be severely degraded (Walker, 1980), with the possibility of introducing negative beampowers.

A data-domain method has been proposed recently by Stockhausen and Farrell (1984) that uses the DFT of each snapshot to identify dominant plane waves; these are subsequently used to interpolate the missing data. The approach has similar performance to that discussed below, but at a greater computational burden.

LINEAR PREDICTION FOR SENSOR RECOVERY

The linear-prediction approach presented below can be applied to either data- or covariance-domain recovery, although only the data-domain version is treated here.

In the linear-prediction model, the pth-order prediction for the kth sensor signal is a linear sum of the signals at the p adjacent sensors:

$$\hat{x}_k = \sum_{n=1}^{p} a_n x_{k-n}, \quad p+1 \leq k \leq N \tag{2}$$

where we have dropped the signal index, m, appearing in Eqn. (1) to indicate the snapshot. The coefficients, **a**, are selected to minimize the squared error $|\hat{x}_k - x_k|^2$ over the N-p possible values of k. These cofficients may be obtained using the computationally-efficient Burg algorithm (cf. Kay and Marple, 1981). This algorithm also has the property that the coefficients so selected guarantee that Eqn.(2) is stable. This is obviously desirable if the $\hat{x}_k$ is used explicity, for example as discussed below for sensor recovery.

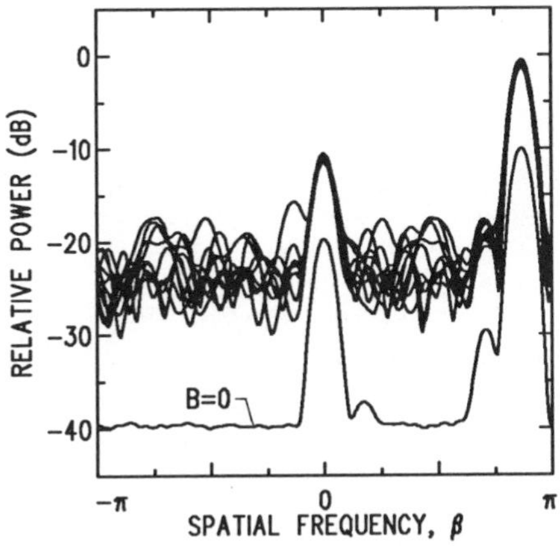

Fig. 1. Beam maps for 4 zeroed sensors, 10 faulty-sensor configurations, 32 sensors, 500 snapshots, d/λ=0.5. Hamming shading.

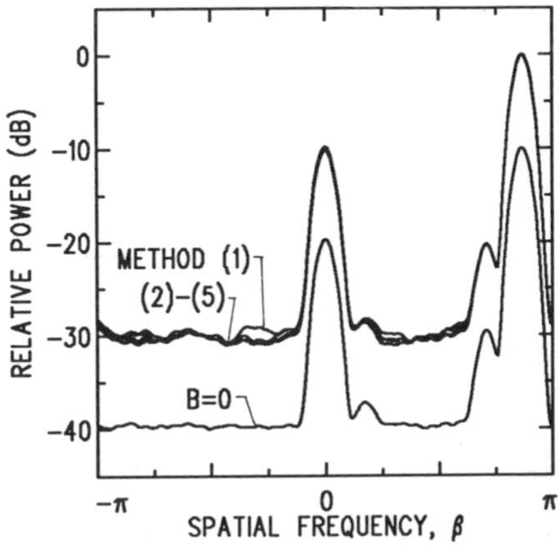

Fig. 2. Comparison of generation methods for **a**. Sensors 10,12,18,25 faulty. 32 sensors,500 snapshots,d/λ=0.5. Hamming shading.

Since the model in Eqn.(2) can be shown to possess an all-pole spectrum in spatial frequency, it is most sensitive to strong plane-wave components in the field. Since it is precisely such components whose sidelobes are responsible for the degraded beam map when sensors fail, the method is intuitively well suited to sensor recovery.

Once **a** is obtained for an assumed model order, p, the missing data for the jth faulty sensor can be predicted with one-step forward- and backward-prediction filters:

$$\hat{x}_j = 1/2\,(\hat{x}_j^{\,f} + \hat{x}_j^{\,b})\,, \qquad \text{where } \hat{x}_j^{\,f} = \sum_{n=1}^{p} a_n\, x_{j-n}\,, \tag{3}$$

and $\hat{x}_j^{\,b}$ is defined similarly in the backward direction. With the B faulty-sensor positions arranged in increasing order, the prediction filters can step along the array in either direction. Then predictive interpolation of adjacent bad sensors can be realized by incorporating $\hat{x}_j^{\,f}$ (or $\hat{x}_j^{\,b}$) in the prediction equation for $\hat{x}_{j+1}^{\,f}$ (or $\hat{x}_{j-1}^{\,b}$). Note that a constraint on the faulty-sensor geometry arises since the prediction filters for each direction cannot be initiated until a faulty-sensor position is encountered with at least p working sensors preceding it. Where this is not achieved, the prediction from the other direction can be used alone. This constraint requires that at least one p-length segment exists with no faulty sensors in order for the interpolation to be applied to each bad sensor. In the worst-case failure geometry this requires $B \leq N/(p+1)$ (for $N=32$, $p=4$, $B \leq 6$), whereas in the best case $B \leq N-p$ (for $N=32$, $p=4$, $B \leq 28$). Therefore, the interpolator can accomodate a relatively large fraction of bad sensors.

Methods for estimating **a**

A remaining issue is the method of dealing with the missing data when solving for **a** via the Burg algorithm. Five possible methods are discussed here. The first three work entirely within the current snapshot. However, the last two use **a** obtained from previous snapshots to recover the missing data for the current observation before estimating the new coefficients. The methods are as follows:

Method (1): proceed with the Burg algorithm, including the zeroed data in the calculation;
Method (2): use **a** from Method (1) to first predict the missing data, then calculate a new **a** incorporating these predictions;
Method (3): include only the working-sensor data in the calculation of **a** (Nuttall, 1976);
Method (4): use a running average of previous **a**'s to predict the missing data for the current snapshot before calculating the new **a**;
Method (5): use **a** from the current snapshot in the predictions for the next several snapshots, incorporating the predicted data in the update of **a**.

In Figure 2 the performance of these five methods is compared for the field model of Figure 1 and for one faulty-sensor configuration (sensors 10, 12, 18, 25 faulty). A model order p=4 has been assumed arbitrarily. Method (4) averages **a** exponentially with a time constant of 10 snapshots, while Method (5) only updates **a** every *5th* snapshot. The figure reveals that the latter four methods lead to essentially identical and complete recovery, while Method (1), not unexpectedly, still shows some minor sidelobe degradation. The apparent robustness of the technique to the selection of **a** is due to the fact that sidelobe degradation is only a problem when the plane-wave SNR is high; fortuitously, this is the condition under which the Burg algorithm is most reliable.

Sensitivity to the number of failed sensors

The effect of the number of failed sensors is presented in Figure 3, once again with the data set of the previous figures but using Method (4), p=4, to estimate **a**. Four faulty-sensor configurations have been selected at random, corresponding to B=0, 4, 6 and 8. The resulting beam maps have been stacked in the figure with an offset of 10 dB. Shown in Figure 3a are the unrecovered maps, while Figure 3b contains the maps after sensor recovery has been applied. It appears that the recovery for even 8 bad sensors is reasonable, suggesting that the method is reliable in terms of sidelobe recovery for some 10-20% failed sensors.

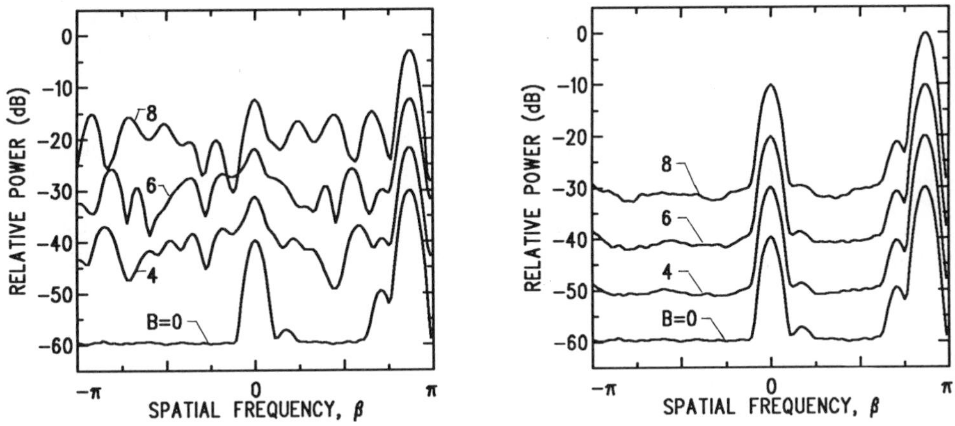

(a) (b)

Fig. 3. Sensitivity to number of faulty sensors, B, for B=0, 4, 6, 8. (a) B sensors zeroed; (b) sensors recovered via Method (4), with **a** averaged over 10 snapshots. 10 dB offset. Faulty sensor assignments are as follows: B=4: 10,12,18,25. B=6: 7,8,12,14,20,25. B=8: 2,8,9,10,15,18,19,28.

Sensitivity to model order

So far we have assumed that the "best" model order is somehow known *a priori*, which is generally not the case. Therefore, the sensitivity to model order is examined in Figure 4. The faulty-sensor configuration of Figure 2 has been used (B=4), with **a** estimated via Method (4). In this case, however, the model order takes on values of p=1, 2, 4 and 6. For p=1, there is still evidence of degraded sidelobes. For $p \geq 2$, the results appear essentially independent of p. In other words, p must be chosen at least equal to the number of *strong* plane waves, that is, those with a beam SNR greater than ~20 dB, which in this case is 2. In practice, this condition can often be ensured with p fixed at a small value (say $p \approx 4$), so that the selection of the order based on the actual measurements is not essential.

Sensitivity to faulty-sensor geometry

Figure 5 presents beam maps for the same set of four-sensor failures considered in Figure 1, but using Method (4), p=4, in the data-recovery algorithm. Consistently good

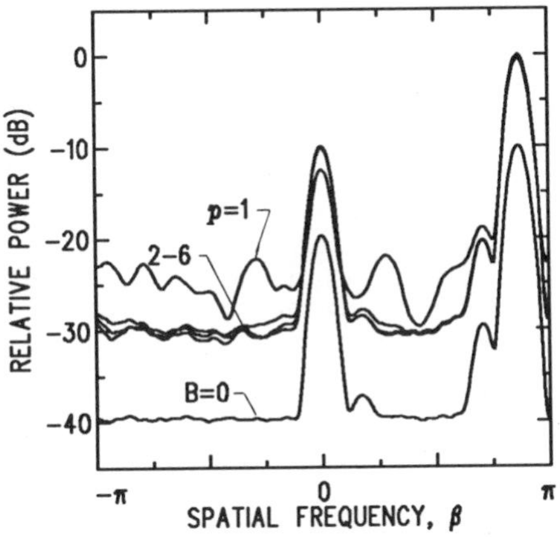

Fig. 4. Sensitivity to model order, p=1, 2, 4 and 6. Sensors 10,12,18,25 zeroed and recovered via Method (4).

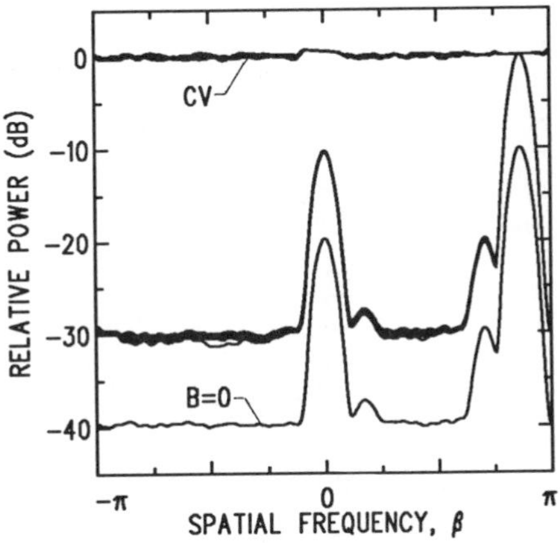

Fig. 5. Sensitivity to faulty-sensor geometry for the 10 configurations of Fig. 1. Recovered using Method (4), p=4.

sidelobe behaviour is evident. Note that power levels in directions other than for strong signals are slightly less than for B=0, since the linear-prediction model does not treat the total field. This discrepancy amounts to an SNR loss for weak plane-wave signals, relative to the SNR for no failures. It is given by $\sim 10\log(N-B)/N$, which for this data set is -0.6 dB.

Also included in Figure 5 are estimates of the Coefficient of Variation, CV, obtained from the 500 snapshots over each of the 10 failure geometries. CV is the ratio of the beampower's standard deviation to mean. For the Gaussian-signal model assumed in these simulations, the beampowers for each snapshot should ideally be Chi^2-distributed (2 degrees of freedom), so that the CV has a value of unity (0 dB) (cf. Stockhausen and Farrell, 1984). Decreased stability in the beamformer estimate due to the recovery procedure would be seen in an increase in CV. No significant increase is evident.

Computational Load

Finally, we note that the main computational load in this recovery approach lies in the generation of **a**. For the Burg algorithm this requires $\sim 5Np$ multiplications, to which we add the $2pB$ multiplications required for the interpolator. We assume N^2 multiplications are required to form N beams. With the approach of Method (5), updates of **a** could be calculated less frequently than every snapshot, at a rate depending on the dynamics of the field. For an update in **a** every sth snapshot, the fractional increase in beamformer load due to sensor recovery is given by $p(2B+5N/s)/N^2$. For example, for p=4, B=4, N=32 and s=5, the fractional increase is only 0.15.

PERFORMANCE WITH REAL DATA

Shown in a waterfall format in Figure 6 are beam maps obtained over about 1 hour of data collected with a DREA research line array. The array consists of 30 sensors, and the analysis has been conducted at a frequency such that the sensor separation d=0.57 wavelengths. Each map is an average of M=16 snapshots, with a total of 39 maps offset progressively up the figure by 5 dB. As in the previous analysis with synthetic data, a Hamming shading has been used in beamforming. All sensors were determined to be functioning normally.

The strong plane-wave arrival in Figure 6 seen tracking across array broadside (β=0) was from a towed sound projector. The propagation was such that the projector sound arrived by more than one path having similar, but not identical, incident angles at the array.

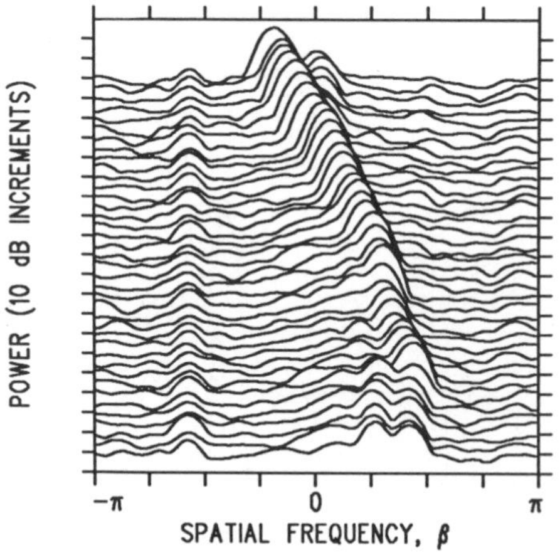

Fig. 6. Beam maps for real data from DREA research array. All sensors working. 30 sensors, d/λ=0.57,16 snapshots per map. 5 dB offset.

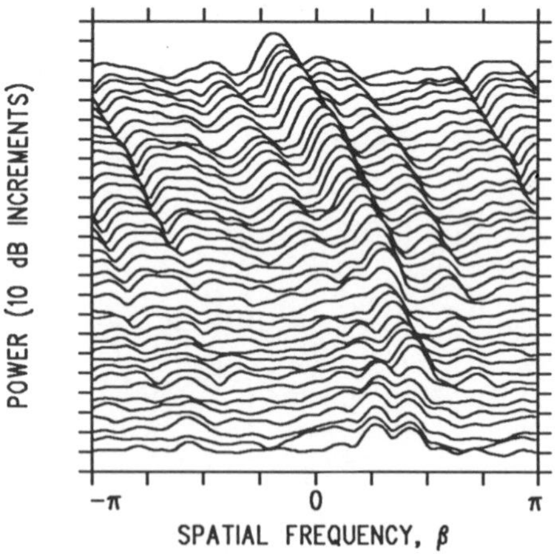

Fig. 7. Same data as Fig. 6, but with four sensors zeroed (10,12,18,25). 5 dB offset.

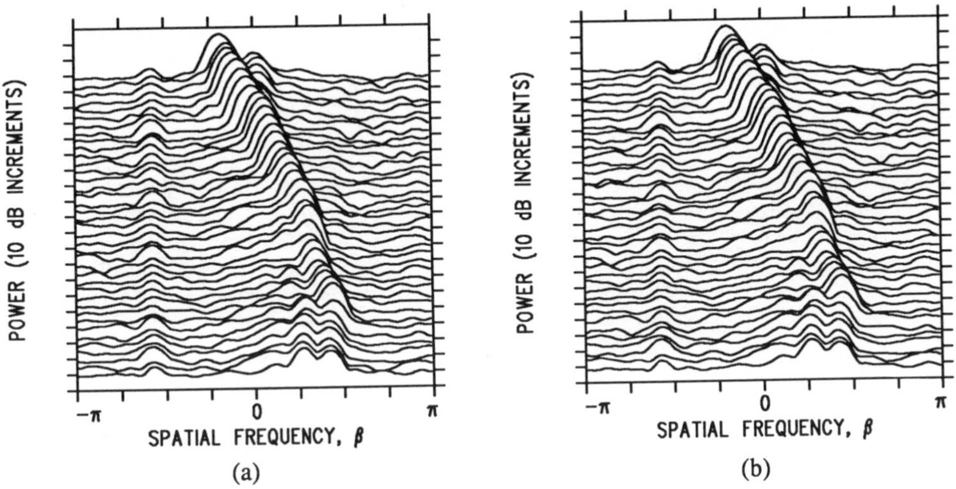

Fig. 8. Recovery from Fig. 7 (sensors 10,12,18,25 zeroed) using the linear-prediction algorithm. (a) Method (4), with **a** averaged over 10 snapshots; (b) Method (5) with **a** recalculated every 5*th* snapshot. p=4. 5 dB offset.

The path with the dominant power has a beam SNR varying from about 15-25 dB over the analysis period. The weaker signal of constant bearing near β=-1.8 is a synthetic plane wave with a beam SNR of about 5 dB. We have added the latter to the real data to permit examination of the effects of sensor failure on the detectability of weak signals.

The significant degradation in the beam maps when four of the 30 sensors have their data values zeroed (sensors 10, 12, 18 and 25) is revealed in Figure 7. The sidelobe effects are particularly evident beyond the middle of the record, where the projector SNR has increased to ~25 dB. However, even near the start of the record (lower map), detectability of the weak synthetic signal is seen to be compromised by the increase in sidelobe level.

Figure 8 shows the results of applying two versions of the linear-prediction algorithm to the four zeroed sensors of the previous figure. In Figure 8a, Method (4) is used to average the prediction coefficients over the previous 10 snapshots. In Figure 8b, a new set of coefficients are calculated every 5*th* snapshot (Method (5)), which is a reasonable update rate for the observed field dynamics. A model order p=4 is used throughout. Both procedures lead to essentially complete recovery of the original beam maps of Figure 6. No significant differences in the map structure or ability to detect the weak signal are evident in comparing these figures. Of course, Method (5) has a decided computational advantage, and hence would appear to be the preferred implementation.

LINEAR PREDICTION FOR SENSOR EXTRAPOLATION

Finally, we remark that by using the prediction coefficients to run the prediction filter off the ends of the array, we have a simple method of extending, or extrapolating the aperture (Swingler, et.al., 1986). Beamforming on the extrapolated array will result in increased gain and resolving power for the *strong* plane-wave signals which the prediction model has identified. Hence weaker signals, otherwise masked by these stronger interferences, may have increased detectablility even though the prediction filter has not accounted for them directly in the extrapolation. As an added benefit, the array shading now applied to the extrapolated array suppresses the sidelobes of the strong plane waves as desired. However, since the shading over the unextrapolated portion of the array is less severe, the higher gain and narrower beamwidth of the nominally unshaded (but unextrapolated) array acts on these weaker signals.

The gains resulting from aperture extrapolation are demonstrated in Figure 9 for the real data analysed previously. Figure 9a again displays the beam maps in the waterfall format of

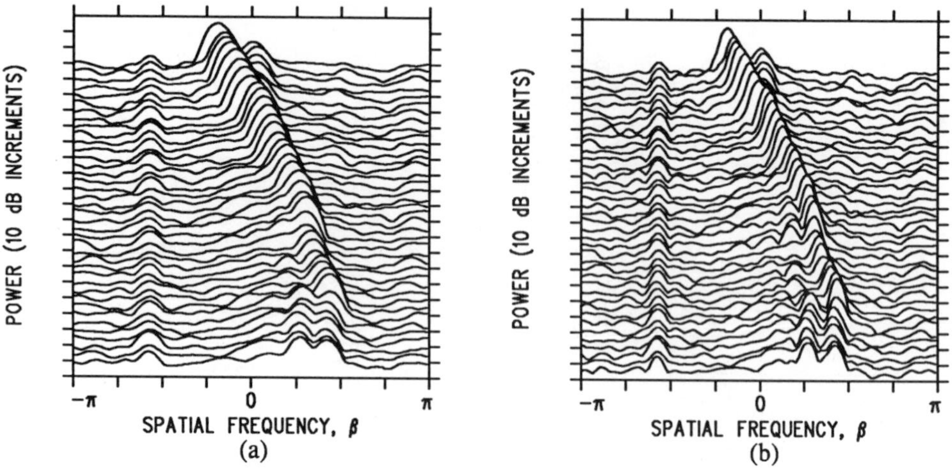

Fig. 9. Aperture extrapolation. (a) Original 30 sensor data of Fig. 6; (b) aperture extrapolated to N=60 sensors. Extrapolation uses Method (5), updating **a** every *5th* snapshot. p=4. 5 dB offset.

Figure 6. On the other hand, Figure 9b uses the same data but with the aperture extrapolated to double its original length (15 sensors added to each end, for a total of 60 sensors). The model order is p=4, and the prediction coefficients are updated every *5th* snapshot via Method (5). There appears to be an increase in both the resolution and SNR of the plane-wave signals, as suggested by the arguments presented above. Moreover, the statistical stability of the beam maps does not appear unduly decreased by the extrapolation.

SUMMARY

We have shown that linear prediction can be used for recovery of data from faulty sensors in a linear array, so that the beamformer's sidelobe degradation is minimal. The method is insensitive to the configuration of faulty sensors, to the selection of model order, and to the method of treating the missing sensors in the generation of the prediction coefficients. The computational overhead added to the the beamforming load can be small. Finally, the technique also offers promise for linear-array aperture extrapolation in order to increase the resolution capability of the array.

REFERENCES

Bowling, S.B., and Lai, S.T., 1979, The use of linear prediction for the interpolation and extrapolation of missing data and data gaps prior to spectral analysis, Lincoln Lab TN1979-46.

Kay, S.M., and Marple, S.L., 1981, Spectrum analysis - a modern perspective, Proc. IEEE, 69:1380.

Nuttall, A.H., 1976, Spectral analysis of univariate process with bad data points via maximum entropy and linear predictive techniques, NUSC TR5303.

Nuttall, A.H., 1980, Effects of random shadings, phasing errors and element failures on the beam patterns of linear and planar arrays, NUSC TR6191.

Stockhausen, J.H., and Farrell, J.B., 1984, FIX - a beamforming technique for line arrays with missing elements, DREA TM 84/O.

Swingler, D.N., Walker, R.S., and G.B. Lewis, 1986, Linear prediction for aperture extrapolation in line array beamforming, Proc. IEEE ICASSP'86, Tokyo, paper 47.8.

Veldhuis, R.N.J., Janssen, A.J.E.M., and Vries, L.B., 1985, Adaptive restoration of unknown samples in certain time-discrete signals,Proc. IEEE ICASSP'85, Tampa, paper 27.1.1.

Walker, R.S., 1980, A study of filled and sparse line array beamformers, DREA TM 80/G.

AN ALGORITHM FOR HIGH-RESOLUTION BROADBAND SPATIAL PROCESSING

J. Newcomb and R.A. Wagstaff

Naval Ocean Research and Development Activity
Code 245, NSTL, MS, U.S.A. 39529

ABSTRACT

Past emphasis in high-resolution spatial signal processing has been on coherent narrowband techniques. In general, these techniques fail to perform optimally with broadband incoherent acoustic energy. An incoherent high-resolution broadband processing algorithm based on the Wagstaff-Berrou-Broadband technique (WB^2) is being developed at NORDA. The new WB^2 code has been applied to both narrowband and broadband simulated data in recent tests of the technique. The results indicate the great potential of the algorithm for broadband applications. These results, as well as results of application to measured broadband data, will be presented.

INTRODUCTION

Past emphasis in high-resolution spatial signal processing has been on coherent narrowband techniques (e.g., maximum entropy and maximum likelihood). Typically, such techniques produce a beamformed output for only a single frequency as does a standard FFT beamformer. Therefore, these techniques, in general, tend to perform well with high signal-to-noise ratio narrowband signals, but by their very nature fail to perform optimally with broadband incoherent acoustic energy.

The need for high-resolution spatial signal processing is more important for broadband signals than it is for narrowband signals since the ability to discriminate by frequency between different sources on the same beam is not possible in broadband processing as it is in narrowband processing. A high-resolution broadband signal processing technique would, therefore, not only improve broadband source localization but also decrease masking by broadband sources along other azimuths. The problem of masking is particularly acute when the competing sources have low frequency content and, thus, dominate a broad range of azimuths by virtue of the wide system response at lower frequencies.

DISCUSSION

Algorithm

A technique that is uniquely suited for processing broadband incoherent acoustic signals is the Wagstaff Iterative Technique (WIT) as implemented in the Wagstaff-Berrou-Broadband (WB^2) high-resolution algorithm. The WB^2 algorithm accepts as input the output from a conventional narrowband or broadband processor (i.e., an FFT or a time delay-and-sum beamformer). This inherent versatility of the WB^2 algorithm seems to indicate that WB^2 might be applied successfully to a limited number of broadband signals with few modifications to the algorithm. The incoherent high-resolution broadband processing algorithm, currently being developed at NORDA, is based on WB^2.

The technique utilized by WB^2 and the new code has its basis in the fact that the true acoustic field at the receiver is observed through some type of receiver system. Since such a system typically does not have a transparent response function, the actual observed outputs inherently reflect not only the characteristics of the true acoustic field but also the characteristics of the system response function. In other words, the observed (or measured) outputs are the true acoustic field convolved with (or filtered by) the system response function. For example, if narrowband processing is to be done and the receiver system is a line array, the observed outputs are the measured beam levels from a conventional narrowband beamformer and the system response function is the beam patterns of the array as a function of frequency and azimuth. Or for broadband processing, the observed outputs are from a conventional broadband beamformer and the system response function is the broadband system response as a function of frequency and azimuth.

WB^2 is an iterative technique that requires only two pieces of information to obtain an estimate of the true field at the array: the measured beam levels of the array and the broadband system response as a function of azimuth. The technique, illustrated in Figure 1, can be briefly outlined as follows:

- An initial estimate of the true field is made,
- The field estimate is convolved with the broadband system response to obtain the estimated beam levels,
- The estimated beam levels are compared with the measured beam levels,
- The differences between the estimated and measured beam levels are used to modify the estimate of the true field, and then
- the previous three steps are repeated until the mean squared differences are minimized.

If the technique is properly implemented in terms of decibel quantities instead of intensities, which includes treating the decibel levels as simple arithmetic quantities manipulated by simple addition and subtraction operations, the use of matrix inversions and other complicated mathematics is avoided. Furthermore, since the inputs can be the measured beam levels from a conventional broadband beamformer, the technique has the potential for being uniquely suited to broadband applications. The only difficulty lies in determining the broadband system response function.

One possible choice for a broadband system response function is the narrowband system response function (i.e., beam patterns) for a given frequency in the frequency band of interest. This choice is not the most accurate or most desirable, but it is easy to implement.

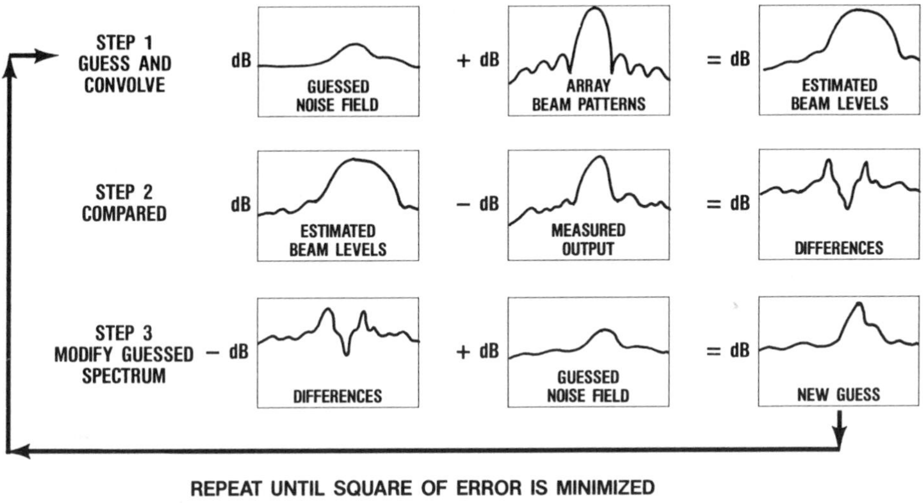

Figure 1: Graphic illustration of WB^2 technique. The iteration is repeated until the square of the system error (differences) is minimized.

Another possible choice is to use a weighted average of the narrowband system responses over the frequency band of interest. The results of this choice are highly dependent upon the number of component frequencies used in the weighted average. The weights for the average, when not unity, would typically be chosen to correspond to the frequency spectrum of the source. This, of course, requires some prior knowledge of the source. This choice corresponds approximately to a matched filter approach where the system response function is matched to the frequency spectrum of the source. Furthermore, this approach is expected to discriminate against those sources that are not of interest because of the mismatch between the system response function and those sources.

Modeled Data

The new algorithm, which is also being called WB^2, has been applied to both narrowband and broadband simulated data in recent tests of the technique. Figure 2 illustrates the results when the system response function used by WB^2 is a single frequency system response function. In this case, the inputs to WB^2 are from a simulated data set where the dots are the time-domain beamformed outputs of the simulator. This data set simulates a broadband (100-500 Hz) source approximately fifty kilometers broadside (i.e., at a relative bearing of ninety degrees) to a thirty-two element horizontal line array. The one hundred twenty-nine beams that are formed and displayed are the result of one hundred time averages. This corresponds to an average over ten seconds of data. Since this version of the simulator did not include any noise background, white noise was added to the averaged beamformed outputs.

The solid line in Figure 2 is the result from the WB^2 broadband algorithm. As can be seen from a comparison of the 3 dB down points of

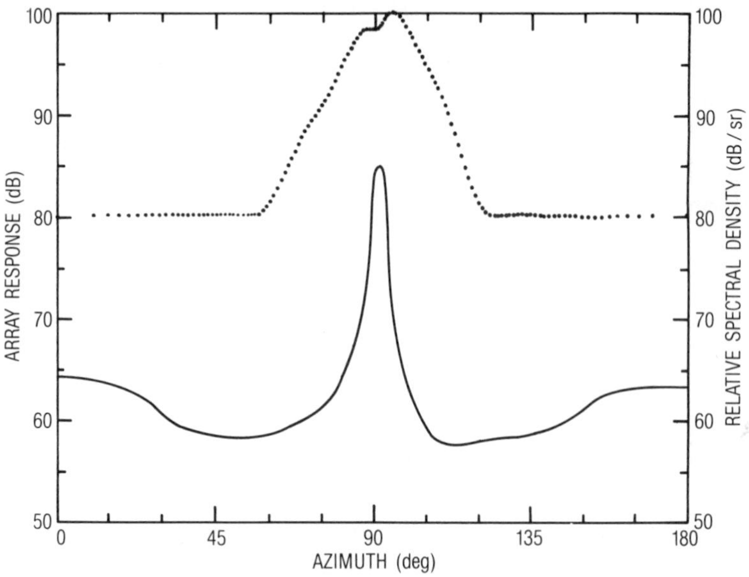

Figure 2: Comparison of simulated time-domain beamformer output with white noise added and WB^2 high-resolution output. The solid line is the WB^2 result.

the spatial spectrum in Figure 2, the WB^2 result does indeed provide better localization of the source than the time-domain beamformer by a factor of about seven. In addition, the source azimuth selected by WB^2 (ninety-one degrees) is closer to the actual source azimuth (ninety degrees) than the source azimuth selected by the time-domain beamformer (ninety-five degrees).

Figure 3 illustrates another example of WB^2 applied to a simulated data set. This data set is the result of only fifty-four time averages (corresponding to an average over four seconds of data). In addition, the data simulator in this case included a random noise background to simulate a more realistic acoustic field. All other parameters are the same as the previous case.

Again by a comparison of the 3 dB down points of the spatial spectrum, it can be seen that WB^2 provides better localization than the time-domain beamformer output by a factor of about five. And, once again, the source azimuth selected by WB^2 (approximately ninety degrees) is closer to the actual source azimuth (ninety degrees) than the source azimuth selected by the time-domain beamformer (approximately ninety-three degrees).

The above results by themselves indicate the great potential of the algorithm for broadband applications. WB^2 is expected to yield even more dramatic results if the matched filter approach for the system response function, as described earlier, is utilized.

<u>Measured data</u>

A field mission was conducted from in May and June 1986 to test the WB^2 algorithm. The main effort was to obtain two sets of beamformed data simultaneously from two co-located arrays each having a different aperture. The beamformed data from the larger aperture array was used to

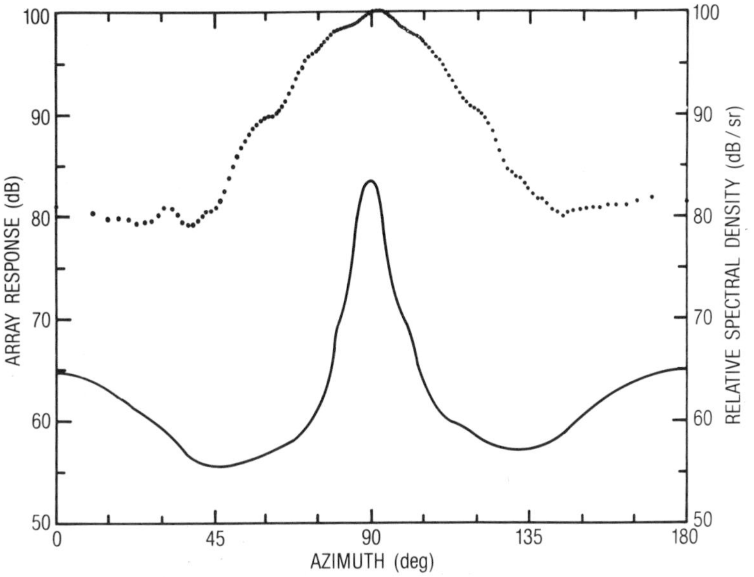

Figure 3: Comparison of simulated time-domain beamformer output with random noise background and WB^2 output. The solid line is the WB^2 result.

compare to the results of the algorithm applied to the beamformed data from the shorter aperture array. The facilities and assets necessary to acquire this data and test the algorithm were available via the NUSC-detachment at the Tudor Hill Laboratory, Bermuda. These assets included the R/V ERLINE for the towing of an acoustic source and a hardware time-domain broadband beamformer.

The acoustic source originally scheduled to be used for this exercise (the HX-90) was unavailable for the exercise. The backup acoustic source (the HX-29R) was approximately 20 dB weaker than the HX-90 as a broadband source. This fact necessitated operating the source in a broadband mode (125 Hz - 250 Hz) only during periods of low ambient noise and simulating the broadband mode (emission of a 240 Hz, a 245 Hz, a 250 Hz, a 255 Hz, and a 260 Hz tonal simultaneously) during periods of medium ambient noise. No source tows were possible during periods of high ambient noise. At least once during each tow date, the source would be operated in a cw mode (emission of a single tonal) to aid in data evaluation and array performance.

The hardware time-domain beamformer utilized for this exercise had the capacity to accept up to sixty hydrophone inputs and provide up to sixty-four beamformed outputs between any two azimuths. For this exercise, the beamformer was set up to accept signals from the middle twenty hydrophones of the array and form forty-one beams and, at the same time, accept signals from the middle ten hydrophones of the array and form twenty-one beams for a total of thirty hydrophone inputs and sixty-two beamformed outputs. The steering angles of the twenty-one beams formed from the middle ten hydrophones corresponded to every other steering angle for the beams formed from the middle twenty hydrophones. Thus, while both subapertures covered the same range of azimuths, the shorter subaperture covered that range with half as many beams having approximately twice the width of the beams of the longer subaperture. In addition, the electronics of the beamformer were such that two ranges of

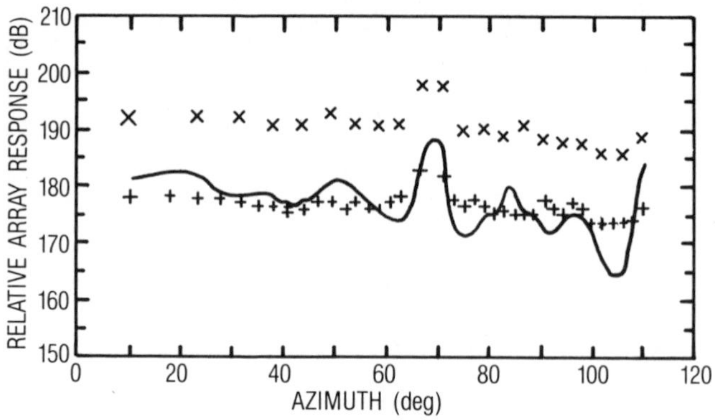

Figure 4: Comparison of actual time-domain beamformer output with WB^2 output. The solid line is the result of the WB^2 algorithm applied to the ten hydrophone beamformed data (x's). The +'s are the twenty hydrophone beamformed data normalized to the WB^2 result. The beamformed data represents a thirty-two second time average.

steering angles, equally spaced in sine space, could be set up. Selection between the two ranges (10°-110° and 10°-71.25°) was accomplished by the flip of a switch. The smaller range of azimuths had approximately twice the beam coverage of the larger range.

Before the data was processed by the WB^2 algorithm, the beamformed data was time averaged. The length of the time average varied depending upon the particular source track being time averaged. The results of one such time average of the beamformed data is illustrated in Figure 4.

In this case the data was averaged for only thirty-two seconds since the source was being towed obliquely to the array. The source was being operated in the broadband mode (125 Hz - 250 Hz) at a range of approximately five nautical miles from the array. There were no large competing sources within the ships radar range at the time. The beamformer range selected was the 10°-110° azimuth range.

The x's in Figure 4 are the time-averaged beamformed outputs for the middle ten hydrophones and the +'s are the time-averaged beamformed outputs for the middle twenty hydrophones. The solid line is the result of the WB^2 algorithm applied to the time-averaged beamformed data for the middle ten hydrophones. The array response values in Figure 4 (and the following figure) are only relative decibel values and are not indicative of the actual intensities of the data. In addition, the time-averaged beamformed data for the middle twenty hydrophones were normalized to the result of the WB^2 algorithm as an aid to easy visual comparison.

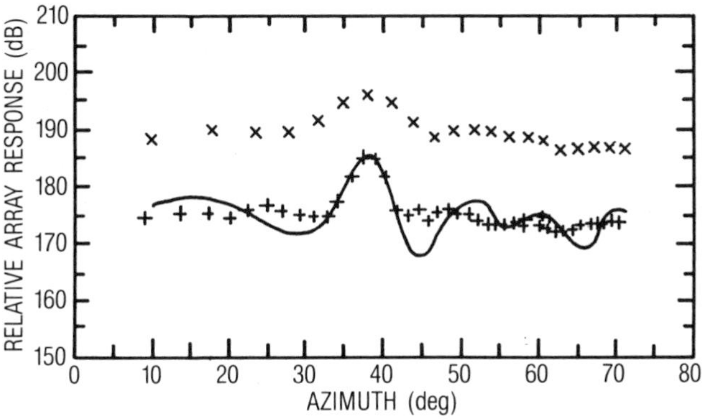

Figure 5: Comparison of actual time-domain beamformer output with WB^2 output. The solid line is the result of the WB^2 algorithm applied to the ten hydrophone beamformed data (x's). The +'s are the twenty hydrophone beamformed data normalized to the WB^2 result. The beamformed data represents an eight minute time average.

As is evident from Figure 4, twenty hydrophones do a much better job of localizing the source than ten hydrophones. The source peak width at the three dB down points is improved by almost a factor of two as expected (approximately a seven degree width versus approximately a four degree width). What might not be expected is that the same results were obtained by applying the WB^2 algorithm to the ten hydrophone data. The spurious peaks in the WB^2 trace result from the imperfect representation of the sidelobes in the broadband system response.

Figure 5 is an example of the data taken with the beamformer operating in the 10° - 71.25° azimuth range. The source was being operated in the broadband mode (125 Hz - 250 Hz) and being towed radially toward the array from a point approximately ten nautical miles from the array. Since sporadic competing broadband sources existed in the form of humpback whales during this tow, the beamformed data was averaged for eight minutes. The peak width of the ten hydrophone data is approximately ten degrees compared to a peak width of approximately four degrees for the twenty hydrophone data as expected. Once again the WB^2 algorithm applied to the ten hydrophone data is in excellent agreement with the twenty hydrophone data.

CONCLUSION

A high-resolution broadband spatial processing algorithm is being developed at NORDA. The algorithm (WB^2) has been tested with both simulated and measured broadband data. Examples of both have been presented. Although continued development and refinement is indicated, the results presented here more than demonstrate the improved localization capabilities and potential of the WB^2 algorithm.

With simulated data, an improvement of approximately a factor of five to seven was achieved. With measured broadband data an improvement of approximately a factor of two was obtained. This improvement has been shown to be equivalent to an increase in the number of hydrophones of a factor of two by a direct comparison.

ABSTRACTS

MULTIPLE SCATTERING AT ROUGH OCEAN BOUNDARIES

by John A. DeSanto

ABSTRACT

This paper reviews several multiple scattering theories which are available to treat the problem of the scattering of rough ocean boundaries. They include the connected diagram expansion method, the smoothing method, the Stochastic Fourier transform approach and the spectral expansion method. The interrelationships and differences of these methods are discussed as well as their applicability to the rough ocean boundary problem. The material is developed along lines appropriate for calculating the coherent (specular for homogeneous surface statistics) and incoherent fields for randomly distributed surfaces, but periodic surface components can also be included.

ACOUSTIC REMOTE SENSING OF THE WAVE HEIGHT DIRECTIONAL SPECTRUM OF SURFACE GRAVITY WAVES

by Steven H. Hill

ABSTRACT

A method for measuring the waveheight directional spectra of surface gravity waves using an underwater acoustic sensor is described. The method, which extracts waveheight spectral information from the second order sea echo Doppler spectrum, is analogous to an HF RADAR technique.

The acoustic receiver consists of a multi-element upward looking array which should allow several antenna patterns to be generated. The instrument is able to provide spectra of high directional precision. Preliminary experimental results are presented.

SOUND SPEED PROFILE INVERSION IN THE OCEAN

by Linda Boden and John A. DeSanto

ABSTRACT

For sound speed profiles which vary only in depth, a method has been developed whereby the scattered field data in frequency (or k-space) can be related to the sound speed profile correction (from an assumed profile guess input) as a quasi-Fourier transform pair. The inversion is straightforward. The method uses a Fourier-Bessel representation of the acoustic field, a Born approximation on the depth dependent part of the Green's function, a WKB representation for the wave functions and asymptotics and linearization to derive the transform pair. Small perturbations about the main profile trend are recovered very accurately if the background slope is assumed known.

ATTENUATION OF LOW-FREQUENCY SOUND IN THE SEA: RECENT RESULTS

by D.G. Browning and R.H. Mellen

ABSTRACT

Accurate estimates of low frequency sound attenuation throughout the oceans of the world can now be made based on a three component relaxation formula (magnesium sulphate, magnesium carbonate, boric acid). Comparison of predicted values and measured data are given for the North Atlantic Ocean, North Pacific Ocean, Indian Ocean, Mediterranean Sea, Red Sea, and the sub-arctic.

PARAMETRIC ACOUSTIC ARRAY APPLICATION USING LIQUIDS WITH LOW SOUND SPEED

by A.W.D. Jongens and P.B. Runciman

ABSTRACT

The measurement of insertion loss in small panel samples (immersed in water) is normally restricted at low frequencies by the diffraction of sound around the sample. The use of a parametric array to generate the incident field has been demonstrated by Humphrey (J. Sound Vib. 98, 67-81, 1985). In this present paper the use of a parametric array in a cylinder containing a low sound speed liquid is shown to extend the low frequency capability of such a system. Experimental results using a freon liquid show that good results were obtained at frequencies as low as 6 KHz. - Ed.

PROGRAM

The following is the program for the Symposium on Underwater Acoustics, with cross references to the corresponding manuscript in this book. The original paper number is given in brackets after the title, e.g. [L-1]; and the page number is given at the right. Note that the titles and authors given below are not necessarily the same as in the final manuscript. Several papers presented in the joint sessions L and M will appear in the corresponding proceedings for the Acoustical Imaging meeting [**Acoustical Imaging** Vol. 15, (H.W. Jones, Ed.), Plenum, New York, 1987]; these are indicated by [In **Acoustical Imaging 15**]. A few authors have not provided a manuscript; these are indicated by [Manuscript not available]. Papers that were not presented at the meeting have been deleted from this program.

WEDNESDAY, 16 July
09:00-12:00

JOINT SESSION

15th SYMPOSIUM ON
ACOUSTICAL IMAGING
and
12 ICA ASSOCIATED SYMPOSIUM ON
UNDERWATER ACOUSTICS

SESSION K: PLENARY SESSION

SESSION L: UNDERWATER ACOUSTIC IMAGING I

Chairmen: H.W. Jones, Canada
H.M. Merklinger, Canada

WEDNESDAY, 16 July
13:30-17:20
Room A

SESSION M: UNDERWATER ACOUSTIC IMAGING II (Joint Session)

Chairmen: L.M. Brekhovskikh, USSR
G. Wade, USA

WEDNESDAY, 16 July
13:40-16:40
Room B

SESSION N: CHARACTERIZING THE OCEAN AND ITS BOTTOM I

Chairmen: T. Akal, Turkey
D. Oldenburg, Canada

THURSDAY, 17 July
08:20-12:10
Room A

SESSION P: SCATTERING BY BIOLOGICAL AND OTHER BODIES

Chairmen: A.D. Pierce, USA
Yu. Yu. Zhitkovskii, USSR

THURSDAY, 17 July
08:20-12:10
Room B

SESSION Q: SOUND PROPAGATION IN THE OCEAN I

Chairmen: W.A. Kuperman, USA
E.C. Shang, China

THURSDAY, 17 July
13:20-17:20
Room A

SESSION R: TRANSDUCERS, RADIATION AND ACOUSTIC INSTRUMENTATION I

Chairmen: L. Bjørnø, Denmark
L. Hargrove, USA

THURSDAY, 17 July
13:30-17:00
Room B

SESSION S: CHARACTERIZING THE OCEAN AND ITS BOTTOM II

Chairmen: T.G. Muir, USA
P. Wille, FRG

THURSDAY, 17 July
19:20-22:00
Room A

SESSION T: SOUND PROPAGATION IN THE OCEAN II

Chairmen: R.P. Chapman, Canada
G. Quentin, France

THURSDAY, 17 July
19:20-22:00
Room B

SESSION U: REVERBERATION AND SURFACE SCATTERING

Chairmen: V.A. Akulichev, USSR
R.R. Goodman, USA

FRIDAY, 18 July
08:20-12:00
Room A

SESSION V: SOUND PROPAGATION IN SHALLOW WATER

Chairmen: F.A.A. Fergusson, Canada
D.H. Guan, China

FRIDAY, 18 July
08:20-12:30
Room B

SESSION W: TRANSDUCERS, RADIATION AND ACOUSTIC INSTRUMENTATION II

Chairman: S. Ueha, Japan

SESSION X: SIGNAL PROCESSING AND BEAMFORMING I

Chairman: R.A. Wagstaff, USA

FRIDAY, 18 July
13:30-16:30
Room A

SESSION Y: SOUND PROPAGATION OVER SLOPING BOTTOMS

Chairmen: S.T. McDaniel, USA
D.E. Weston, UK

ADJOURNMENT

FRIDAY, 18 July
13:40-16:50
Room B

SESSION Z: SIGNAL PROCESSING AND BEAMFORMING II

Chairmen: C. Bright, Canada
B.G. Hurdle, USA

ADJOURNMENT

AUTHOR INDEX

This index will lead you to either a paper by the author (in **bold** type) or to a bibliography in this book which cites a paper by the author.

SUBJECT INDEX